Y0-BQT-551

Abstracts of

MICROBIOLOGICAL METHODS

a volume in the series

TECHNIQUES IN PURE AND APPLIED MICROBIOLOGY

edited by

CARL-GÖRAN HEDÉN

Karolinska Institute, Stockholm

Abstracts of MICROBIOLOGICAL METHODS

EDITED BY

V. B. D. SKERMAN

Department of Microbiology
University of Queensland

WILEY-INTERSCIENCE
a Division of John Wiley & Sons
New York • London • Sydney • Toronto

10 9 8 7 6 5 4 3 2

Library of Congress Catalogue Card Number: 69-16128

SBN 471 79385 X

Printed in the United States of America

Series Preface

In an environment that is becoming increasingly Man-made we are only slowly learning our ecological lessons. Ecologists tell us that Nature has many bottlenecks in the cycles with which it attempts to break down sewage, biocides, and industrial pollutants into products existing in the pre-Man state. But similar restrictions also affect the processes which govern biological productivity by influencing the availability of essential elements like nitrogen and phosphorus. All those fundamentally important processes depend on the activities of mixed microbial populations, and it is sobering – in this age of self-confidence in science – to consider how little we really know about them. Normally we cannot even identify all the members of the populations concerned, much less describe their intricate physiological interdependence. This fragmentary knowledge is strangely contrasted by the enormous advances made in the last couple of decades with regard to the structure, the genetics, and the chemistry of individual representative microorganisms.

The reason for this discrepancy is obvious: we must learn to crawl before we can walk. Preparing ourselves for the upright position we crawl slowly over the ground of knowledge and pick up the information we need with relative ease, since we are close to the ground, a fact which also influences our choice of projects. But when starting to walk we must be able to combine with great speed relevant facts from specialties which are often quite distant from one another. We then find that we need shortcuts through the rapidly growing forest of microbiological periodicals, which now approach 800. As anybody knows who has been flooded with titles after having asked an inadequately defined question to MEDLARS, or other advanced abstract services, the computer does not solve this problem. The only one who can do it seems to be the experienced investigator who takes on himself the seemingly thankless task of sifting through the material for the benefit of colleagues all over the world. Similar often quite anonymous efforts provide the cement in the foundation for international microbiology. This is centered around the Microbiology Division of the International Union of Biological Sciences: the International Association of Microbiological Societies (IAMS). Since this body is geared to the practical needs of the profession, it is most appropriate that the first volume in this series is the work of a sectional chairman, Professor V. Skerman. Further contributions of this nature will be a logical consequence of parallel aims and should also help to make the fruits of the UNESCO Microbiology program generally available, for instance a World Directory of Culture Collections which is now in preparation.

In the microbiological laboratory one might need to find a description of a technique, useful for a taxonomical study, an indication of the availability of relevant type strains, a guide with regard to media and cultivation techniques, an outline of methods for studying the anatomy of the cells or for breaking them up to purify their structural and functional components. This is the type of information that this new series of handbooks will try to provide. It is an effort to supply the practical advice which is needed in most laboratories active in the various fields of applied microbiology and to do it without an overdose of theoretical considerations. This, however, does not imply that the books will be of only limited value to a theoretically oriented laboratory. Consider the extensive use of microorganisms as research tools – now common among biophysicists, molecular biologists, immunologists, bioengineers and many others – and you will appreciate the need for a quick guide to the accepted techniques for handling bacteria and viruses.

Pure and applied microbiology go together; they are opposite sides of the same coin. One is a road over forbiddingly steep hills, and one on which the path is always partly hidden from view. The other is the goal, for after all, as Orville Wyss has emphasized, applied microbiology constitutes the backbone of our science, even if "we have responded to the gibes of the humanists who have always objected to the university leaving the cloister and entering the market place. It has never been demonstrated that the cloister is in any way superior to the market place for training a man to think, or that applied science is in any way inferior to pure science as an intellectual effort". There are many signs that the young student generation is more keenly aware of this than most of their professors, but this should not make the students forget Louis Pasteur's famous statement: "Without theory, practice is but routine born of habit. Theory alone can bring forth and develop the spirit of inventions". If the student keeps this in mind he will find that microbiology offers more challenging opportunities to make inventions that will affect Man's future health and wellbeing than most other subjects which he might choose to study.

Carl-Göran Hedén
Vice-president, IAMS
Chairman, Microbiology Panel of UNESCO
Chairman, International Cell Research Organization

Introduction

The Numerical Taxonomy Subcommittee of the International Committee on Nomenclature of Bacteria held a Conference in Quebec in 1964 to which I was invited to contribute a paper on the standardization of methods.

Most taxonomists are agreed that uniformity in the descriptions of bacteria will follow only when there is uniformity in the methods used to examine them. Unfortunately most attempts to achieve uniformity have remained sterile. Meetings rarely achieve anything, since the necessary evidence is rarely available.

Considering the matter too contentious to hazard any unsupported recommendations, a survey was made of the methods which had been employed by various authors and reported in a number of leading journals and books over the previous 25 years (in some cases longer). It was hoped that a document could be prepared which would provide the necessary evidence upon which some decisions could possibly be reached – decisions based on the frequency of use and stability of various methods.

Subsequent to the presentation of the document at the Quebec Conference it was suggested that the information be published. In this volume the information alone is presented without any attempt to assess it critically. This will be done elsewhere. Initially some attempt was made to edit material to eliminate what appeared to be unessential wording, but as the abstracting progressed the undesirability of this practice became evident and the most recent abstracts have been quoted verbatim.

Every endeavour has been made to avoid misquotation where abstracts have been abbreviated and apologies are offered in advance for any omissions or errors. Since the purpose was to present methods, and *not* the conclusions of the paper, reference has been made to the results of research only where results relate to the suitability of the method itself.

Methods are arranged under headings in chronological order irrespective of the source of the material. Journals abstracted were those in English and French which were immediately available to me. They include the following:

Annales de l'Institut de Pasteur, Paris

Vols. 70 – 111 (1944 – 1967)

Antonie van Leeuwenhoek Journal of Microbiology and Serology
Vols. 20 – 33 (1954 – 1966)
Applied Microbiology
Vols. 1 – 16 (1953 – 1966)
Canadian Journal of Microbiology
Vols. 1 – 12 (1954 – 1966)
Journal of Agricultural Research
Vols. 26 – 78 (1923 – 1949)
Journal of Bacteriology
Vols. 1 – 92 (1917 – 1966)
Journal of Dairy Research
Vols. 1 – 33 (1929 – 1966)
Journal of Experimental Medicine
Vols. 69 – 124 (1939 – 1966)
Journal of General Microbiology
Vols. 1 – 45 (1947 – 1966)
Journal of Infectious Diseases
Vols. 64 – 116 (1929 – 1966)
Journal of Pathology and Bacteriology
Vols. 31 – 92 (1928 – 1966)

References cited in the abstract of each method have been included with the abstract. In the case of abstracts from the Annales de l'Institut de Pasteur, references have been quoted exactly as in the original papers. All others have followed a standardized format based on abbreviation for Journals published in The World List of Scientific Periodicals (1900 – 1960, 4th ed. 3 vols., edited by Peter Brown and George Burder Stratton, London, Butterworths, 1963).

Where papers within these journals made reference to important techniques published in other journals not in the above list, the method was abstracted from the relevant Journal when procurable.

I should like to express my appreciation to the following copyright owners for permission to reproduce abstracts from the journals indicated.

American Journal of Clinical Pathology and *Stain Technology,* by courtesy of Williams and Wilkins (Miss A. Widerman).

American Journal of Public Health, by courtesy of the American Public Health Association Inc. (Dr. B. F. Mattison).

Annales de l'Institute de Pasteur, by courtesy of Dr. Fournier and Masson et Cie, Paris.

Antonie van Leeuwenhoek Journal of Microbiology and Serology, by courtesy of the Nederlandsche Vereeniging voor Microbiologie (Dr. A. L. Houwink).

Applied Microbiology and *Journal of Bacteriology,* by courtesy of the American Society for Microbiology (R. A. Day).

Biochemical Journal, by courtesy of The Biochemical Journal (F. Clark).

Canadian Journal of Microbiology, by courtesy of the National Research Council (J. A. Morrison).

International Bulletin of Bacteriological Nomenclature and Taxonomy, by courtesy of the International Committee on Bacteriological Nomenclature (Dr. R. E. Buchanan).

Journal of Agricultural Research and *U.S.D.A. Agricultural Monograph No. 16,* by courtesy of the United States Government Printing Office (C. W. Burkley).

Journal of Biological Chemistry, by courtesy of The American Society of Biological Chemists, Inc. (R. A. Harte).

Journal of Clinical Pathology, by courtesy of the British Medical Association (T. D. V. Swinscow).

Journal of Dairy Research, by courtesy of The National Institute for Research in Dairying (L. A. Mabbitt).

Journal of Dairy Science, by courtesy of the American Dairy Science Association (E. O. Herreid).

Journal of Experimental Cell Research, by courtesy of Academic Press, Inc. (M. McGrath).

Journal of Experimental Medicine, by courtesy of the Rockefeller University Press (W. A. Bayless).

Journal of General Microbiology and *Journal of Hygiene,* by courtesy of Cambridge University Press (American Branch) (E. F. Barney).

Journal of Infectious Diseases, by courtesy of the University of Chicago Press (I. Dupart).

Journal of Laboratory and Clinical Medicine, by courtesy of The C. V. Mosby Company (Mrs. M. W. Martin).

Journal Pathology and Bacteriology, by courtesy of the Pathological Society of Great Britain and Ireland (Professor C. L. Oakley).

Nature, by courtesy of Macmillan (Journals) Limited (J. Maddox).

National Collection of Type Cultures Catalogue, by courtesy of Dr. S. T. Cowan. (Referred to in the text as 'N.C.T.C. Methods 1954.')

Public Health Laboratory, by courtesy of the Public Health Laboratory (B. Elkan Diamond).

In addition to the Journals, methods from the following publications have been reproduced with the assent of the authors and/or publishers to whom I am deeply indebted.

W. H. Ewing (1962). *Enterobacteriaceae. Biochemical Methods for Group Differentiation.* U.S. Department of Health, Education and Welfare, Public Health Service Publication, No. 734 (revised).

N. R. Smith, R. E. Gordon, and F. E. Clark (1952). *Aerobic spore-forming bacteria.* U.S.D.A. Agr. Monograph No. 16.

P. Hauduroy (1951). *Techniques Bactériologiques.* Masson et Cie, Éditeurs, Paris.

A. A. Miles and G. S. Wilson (1964). Topley and Wilson's *Principles of Bacteriology and Immunity,* 5th ed., Arnold, London.

A. R. Prévot (1966). *Techniques pour le Diagnostic des Bactéries anaérobies.* 2nd ed. Éditions de la Tourelle. St. Mandé (Seine).
V. B. D. Skerman (1967). *A Guide to the Identification of the Genera of Bacteria.* 2nd ed. Williams and Wilkins, Baltimore.
Society of American Bacteriologists *Manual of Microbiological Methods* (1957). McGraw-Hill Book Co. Inc., New York.

I extend my thanks to the numerous authors who have given their consent to requests to reproduce abstracts from their papers.

Because of the very large number of abstracts involved and the relatively minor nature of many, copyright owners of some Journals indicated that approaches to authors would not be necessary. In such cases authors were consulted when the abstracts were unduly large. All other abstracts have been carefully rechecked. In all cases where author consent was requested by copyright owners this was sought and obtained, except in a few cases where letters were returned undelivered and subsequent efforts to locate authors have been unavailing. The tolerance of authors who detected errors in abstracting is much appreciated. An apology is offered for errors which may have been missed.

Some may well ask why it was considered desirable to publish the collection of abstracts, particularly where the information given was fragmentary or merely recapitulated information published earlier by other authors. The reasons are twofold.

As indicated in the opening remarks, attempts to discuss the problem of standardization make little progress when the evidence is not available in a form suitable for consideration. It has not been possible to cover all the microbiological Journals – not even all the important ones – because of time available and financial and linguistic incapacity. For this reason *all* the relevant information was taken for such journals and books as were available and the limits of the search have been clearly stated. Subsequent surveys may be based on this.

All statements, irrespective of the detail given, have been quoted, since they give an indication of the areas in which tests have been applied even when details of the test were inadequate or not given.

It is hoped that a number of international groups may be persuaded to examine the information on the various tests and organize investigations which may lay the foundations for recommendations of tests to be used on an international basis. The increasing number of authors who have accepted the proposals in the Manual of Methods of the American Society for Microbiology or similar "standard" texts indicates that an internationally prescribed Manual of Methods may gain wide acceptance. Such a document would also have a decided influence on the manufacture of media by commercial organizations for diagnostic methods.

In connection with these abstracts attention is drawn to a critical assessment of methods as applied to the *Enterobacteriaceae* prescribed by Dr. Costin in the Zentralblatt für Bakteriologie I. Referat, 1965, *198*/5, 385–463.

The compilation of the abstracts has occupied a considerable amount of time of numerous assistants over a period of three years.

Particular appreciation is expressed to three graduates of the University of Queensland – Lynette Winders, Valerie Skerman, and Lesley Jones for their assistance in locating and checking the abstracts and to my secretarial staff members Barbara Heron, Lynne Clark, Elwyn Blemmings, Wendy Loder, and Ann Shervey for their cheerful acceptance of such a mundane task.

Since I have contributed little to the contents of the document, the object of which is to promote action in the sphere of international microbiology, all nett proceeds from the sale of this document will be directed into the work of various Committees of the I.A.M.S.

V. B. D. Skerman

Herston, Brisbane, Australia

Contents

Abstracts of
MICROBIOLOGICAL METHODS

ACETONE

Acetone (production of)—anaerobes

A. -R. Prévot (1966). Techniques pour le diagnostic des Bactéries Anaérobes. Éditions de la Tourelle, St. Mandé.

Acétone: Dans un tube à essais, verser 1 cm^3 de distillat;* puis ajouter 1 cm^3 de réactif de Denigès. Chauffer à ébullition puis refroidir brusquement. Une réaction positive se traduit par l'apparition d'une coloration blanche puis d'un précipité blanc dans les 24 heures.

*Distillat Alcaline sous "Amines (volatile)—anaerobes"—A. -R. Prévot.

AESCULIN HYDROLYSIS

Aesculin hydrolysis—*Streptococcus faecalis, Streptococcus lactis*

P. M. F. Shattock and A. T. R. Mattick (1943–1944). J. Hyg., Camb. 43, 173–188.

Confirmed by the production of a black precipitate with ferric chloride in Lemco broth containing 0.5% aesculin.

Aesculin hydrolysis—*Corynebacterium*

R. F. Brooks and G. J. Hucker (1944). J. Bact. 48, 295–312.

This was determined in cultures grown in Diernhofer's (1932) medium for 2 weeks at 37°C. Daily observations were made for the disappearance of the bluish sheen from the medium, indicating aesculin hydrolysis. Cultures showing this were incubated further for 2 days; then 10 drops of 0.04% bromthymol blue indicator were added to test for acid production from the glucose released by aesculin hydrolysis.

K. Diernhofer (1932). Milchw. Forsch. 13, 368–374.

Aesculin hydrolysis—*Streptococcus*

K. E. Hite and H. C. Hesseltine (1947). A study of the streptococci isolated from the uterus and the vagina. J. Infect. Dis. 80, 105–112.

The hydrolysis of aesculin was tested in the liquid medium described by Harrison and Vanderleck [1909. Zentbl. Bakt. ParasitKde 22, 547–551] and in a modification of the medium, in which the sodium taurocholate was omitted.

Aesculin hydrolysis—*Streptococcus*

Y. Abd-el-Malek and T. Gibson (1948). J. Dairy Res. 15, 233–248.

Hydrolysis was determined by addition of ferric chloride after incubation for 5–7 days in a medium containing 0.1% aesculin, 1% sodium hippurate, and 0.5% each of peptone, tryptone, meat extract, and yeast extract. The two test substances were also used separately but there appeared to be no advantage in doing so. Strains that failed to grow were retested with 0.25% glucose in the medium.

Aesculin hydrolysis—*Streptococcus*

P. F. Swartling (1951). J. Dairy Res. 18, 256–267.

Shown with $FeCl_3$ solution.

Aesculin hydrolysis—*Streptococcus*

G. Andrieu, L. Enjalbert, and L. Lapchine (1954). Annls Inst. Pasteur, Paris 87, 617–634.

Le milieu à l'esculine est la classique gélose à l'esculine des bactériologistes, fraichement préparée.

Aesculin hydrolysis—*Lactobacillus*

G. H. G. Davis (1955). J. gen. Microbiol. 13, 481–493.

Medium: peptone, 1% (w/v); yeast extract, 0.5% (w/v); sodium acetate, 1% (w/v); aesculin, 0.5% (w/v); ferric ammonium citrate, 0.05% (w/v); Tween 80, 0.1% (v/v); Salts A, 0.5% (v/v); Salts B, 0.5% (v/v); at pH 6.8, dispensed in 4-ml quantities into bijou bottles autoclaved at 10 lb for 20 min to sterilize. The tests were inoculated with 1–2 drops, incubated for 14 days, and observed at 48-hr intervals for loss of fluorescence and blackening of the culture, indicating hydrolysis. For Salts A and Salts B see A. C. Hayward (1957). J. gen. Microbiol. 16, 9–15.

Aesculin hydrolysis—*Chromobacterium*

P. H. A. Sneath (1956). J. gen. Microbiol. 15, 70–98.

Iron was used as the indicator of hydrolysis. The medium consisted of 1% peptone, 0.1% aesculin, and 0.05% ferric citrate in distilled water. Both this liquid medium in tubes, and plates of the medium solidified with 1.5% of agar, were inoculated and observed at 25°C, the tubes for 14 days and the plates for 4 days. The media were sterilized by autoclaving at 115°C for 15 min.

Aesculin hydrolysis—*Lactobacillus*

J. Naylor and M. E. Sharpe (1958). J. Dairy Res. 25, 92–103.

Aesculin-tomato broth containing Evans peptone, 1.5%; NaCl, 0.5%; Yeastrel, 0.3%; aesculin, 0.2%; ferric ammonium citrate, 0.5%; and Tween 80, 0.01% (all w/v); and tomato juice, 10% (v/v), at pH 6.6 was used. The inoculated medium was observed for 1 week for blackening.

Aesculin hydrolysis—*Pseudomonas*

M. E. Rhodes (1959). J. gen. Microbiol. 21, 221–263.

The aesculin medium did not contain acid indicator, and the cultures were tested for the presence of the aglycone aesculetin after incubation for 14 days; on adding 2 drops of 5.0% (w/v) $FeCl_3$ the cultures turned densely black when aesculetin was present. This test was not fully satisfactory because a few isolates grew slightly in the medium and the aesculetin test was doubtfully positive. Therefore Sneath's (1956) recommended methods, involving the use of peptone + aesculin agar or broth, and the microtest of Cowan (1953) were used to confirm aesculin hydrolysis.

S. T. Cowan (1953). J. gen. Microbiol. 8, 391.
P. H. A. Sneath (1956). J. gen. Microbiol. 15, 70.

Aesculin hydrolysis—*Fusobacterium*

A. C. Baird-Parker (1960). J. gen. Microbiol. 22, 458–469.

Medium (%, w/v). proteose-peptone No. 3, 1.0; Lab-Lemco, 0.3; Oxoid yeast extract, 0.1; glucose, 0.1; disodium hydrogen phosphate, 0.5; aesculin, 0.5. Cultures were incubated for 14 days. The presence of phenolic aglycone was detected by adding 0.5 ml of a 1.0% (w/v) solution of ferric ammonium citrate and examining for the development of a dark reddish-brown colour in the medium and also loss of fluorescence in the case of aesculin (Barnett, Ingram, and Swain, 1956).

J. A. Barnett, M. Ingram, and T. Swain (1956). J. gen. Microbiol. 15, 529.

Aesculin hydrolysis—*Pediococcus*

H. L. Günther and H. R. White (1961). J. gen. Microbiol. 26, 185–197.

The method of Davis (1955) was used, except that Tween 80 (which according to Jensen and Seeley, 1954, is not required by pediococci) and salt solutions A and B were omitted from the medium, and sodium chloride (0.2%, w/v), manganese sulphate (0.05%, w/v), and magnesium sulphate (0.05%, w/v) were added. The cultures were examined daily for 7 days.

For maintenance of stock cultures, preparation of inocula, and in all experimental work, Oxoid tomato juice broth or tomato juice agar, adjusted to pH 6.6, were used unless otherwise stated. The following were exceptions to this rule: for strain Tc. 1, sodium chloride (5%, w/v) was added to the medium, and for the aerococci, glucose-Lemco broth (Shattock and Hirsch, 1947) or glucose-yeast extract agar (containing, as %, w/v, peptone, 1.0; Yeastrel, 0.3; glucose, 1.0; NaCl, 0.25; agar, 1.0, at pH 7.4) was used. Cultures were incubated aerobically at 30°C with specified exceptions.

G. H. G. Davis (1955). J. gen. Microbiol. 13, 481.
E. M. Jensen and H. W. Seeley (1954). J. Bact. 67, 484.
P. M. F. Shattock and A. Hirsch (1947). J. Path. Bact. 59, 495.

Aesculin hydrolysis—*Xanthomonas*

A. C. Hayward and W. Hodgkiss (1961). J. gen. Microbiol. 26, 133–140.

The liquid and solid media of Sneath (1960) were used and observed for blackening and loss of fluorescence for a period of 14 days.

P. H. A. Sneath (1960). Iowa St. Coll. J. Sci. 34, 243.

Aesculin hydrolysis—*Pasteurella*

H. H. Mollaret (1961). Annls Inst. Pasteur, Paris 100, 685–690.

Nous avons recherché l'action de *P. pseudotuberculosis* en milieu gélosé contenant 1 p. 1000 d'esculine.

Aesculin hydrolysis—*Lactobacillus*

R. E. Smith and J. D. Cunningham (1962). Can. J. Microbiol. 8, 727–735.

Loss of fluorescence of Naylor and Sharpe's medium (1958) when irradiated with ultraviolet light indicated hydrolysis of aesculin.
J. Naylor and M. E. Sharpe (1958). J. Dairy Res. 25, 92–103.

Aesculin hydrolysis—*Vibrio*

G. H. G. Davis and R. W. A. Park (1962). J. gen. Microbiol. 27, 101–119.

Occurs within 7 days in the basal medium: Koser salt solution (Mackie and McCartney, 1953, p. 429) plus Oxoid yeast extract, 0.3% (w/v). Initial pH adjusted to 7.0. Plus 0.5% (w/v) aesculin and 0.005% (w/v) ferric ammonium citrate.
T. J. Mackie and J. E. McCartney (1953). Handbook of Practical Bacteriology, 9th ed., Livingstone.

Aesculin hydrolysis—*Aeromonas*

I. W. Smith (1963). J. gen. Microbiol. 33, 263–274.

Aesculin hydrolysis was tested on the medium of Sneath (1956) after 4 days.
P. H. A. Sneath (1956). J. gen. Microbiol. 15, 70.

Aesculin hydrolysis—*Lactobacillus*

F. Gasser (1964). Annls Inst. Pasteur, Paris 106, 778–796.

L'hydrolyse de l'esculine est recherchée dans un milieu contenant du citrate de fer ammoniacal, decrit par Davis en 1955.
G. H. G. Davis, K. A. Bisset, and C. M. F. Hale (1955). J. gen. Microbiol. 1955, 13, 481.

Aesculin hydrolysis—*Pediococcus*

E. Coster and H. R. White (1964). J. gen. Microbiol. 37, 15–31.

The authors used the methods of Günther and White (1961). Their tomato juice broth or agar with the addition of Tween 80 (0.1%, v/v), adjusted to pH 6.6, were used in maintenance and preparation of inocula with certain specified exceptions. Transfers for preparing inocula were made every 24 hr except for *Pediococcus halophilus* (48 hr) and some brewing strains (fortnightly). All cultures were incubated aerobically at 30°C except for brewing strains, which were incubated in an atmosphere of 95% (v/v) hydrogen and 5% (v/v) carbon dioxide at 22°C.
H. L. Günther and H. R. White (1961). J. gen. Microbiol. 26, 185–197.

Aesculin hydrolysis—*Pseudomonas aeruginosa*

R. R. Colwell (1964). J. gen. Microbiol. 37, 181–194.

The agar plate technique of Sneath (1956) was used. The standard inoculum for all tests was a single drop (0.05 ml) from a sterile disposable pipette (Fisher Scientific Company) of a 24–48-hr broth culture. The stock culture medium (YE) was a modification of that used by Rhodes (1959): Difco yeast extract, 0.3 g; Difco Bacto-proteose-peptone, 1.0 g; NaCl, 0.5 g; agar, 1.5 g (agar omitted from liquid stock media); distilled water, 1 L; adjusted to pH 7.2–7.4 with NaOH. Tests were carried out at room temperature (25°C) except where otherwise indicated.

M. E. Rhodes (1959). J. gen. Microbiol. 21, 221.
P. H. A. Sneath (1956). J. gen. Microbiol. 15, 70.

Aesculin hydrolysis—*Lactobacillus casei* var. *rhamnosus*

W. Sims (1964). J. Path. Bact. 87, 99–105.

The medium of Swan (1954) was used for the hydrolysis of aesculin.

A. Swan (1954). J. clin. Path. 7, 160.

Aesculin hydrolysis—*Vibrio marinus*

R. R. Colwell and R. Y. Morita (1964). J. Bact. 88, 831–837.

Hydrolysis of aesculin was checked by inoculation of a medium consisting of 0.1% aesculin added to synthetic seawater agar, with 0.05% ferric citrate as indicator for the breakdown of aesculin.

Synthetic seawater containing sodium chloride, 2.4%; potassium chloride, 0.07%; magnesium chloride (hydrated), 0.53%; and magnesium sulfate (hydrated), 0.7%, was used as the base. The inoculum was prepared from a 24 hour artificial seawater broth. Cultures were incubated at 18°C.

Aesculin hydrolysis—*Lactobacillus*

M. Gemmell and W. Hodgkiss (1964). J. gen. Microbiol. 35, 519–526.

To the basal medium containing no citrate were added aesculin (1%, w/v), glucose (0.25%, w/v), ferric citrate (0.05%, w/v), and agar (0.15%, w/v). This medium was melted and cooled to 45°C before inoculation. Hydrolysis of aesculin was detected by the appearance of coral-like white crystals in the medium and a loss of fluorescence under the ultraviolet lamp. The crystals were isolated by melting the medium and pouring through filter paper. They dissolved in alkali to give a deep orange solution, which was not flourescent. The crystals were assumed to be aesculetin, one of the hydrolysis products of aesculin. All cultures that produced crystals in the medium also showed loss of fluorescence, and no loss of fluorescence was noted in cultures where no crystals appeared.

The basal medium, which was used throughout with minor modifications, had the following constituents in 1 L of tap water: meat extract (Lab-Lemco), 5 g; Evans peptone, 5 g; Difco yeast extract, 5 g; Tween 80, 0.5 ml; $MnSO_4 \cdot 4H_2O$, 0.1 g; potassium citrate, 1 g; pH 6.5.

Aesculin hydrolysis—*Pediococcus*

R. Whittenbury (1965). J. gen. Microbiol. 40, 97–106.

The author used the method of M. Gemmell and W. Hodgkiss (1964, J. gen. Microbiol. 35, 519).

Aesculin hydrolysis—*Brevibacterium*

R. Chatelain and L. Second (1966). Annls Inst. Pasteur, Paris 111, 630–644.

Hydrolyse de l'esculine: gélose à l'esculine; lecture après 3 jours. La température d'incubation des cultures est de 30°C.

H. R. Olivier (1963). Traité de biologie appliquée, tome II, Maloine édit., Paris.

Aesculin hydrolysis—*Nocardia madurae*

R. E. Gordon (1966). J. gen. Microbiol. 45, 355–364.

The cultures were inoculated into aesculin broth (aesculin, 1 g; ferric citrate, 0.5 g; peptone, 10 g; NaCl, 5 g; water, 1000 ml: Cowan and Steel, 1965), incubated at 28°C, and observed for growth and blackening of the medium at 2, 4, and 6 weeks. A tube of the same broth without aesculin was also inoculated with each culture and used as a control.

S. T. Cowan and K. J. Steel (1965). Manual for the Identification of Medical Bacteria. Cambridge: Cambridge University Press.

Aesculin hydrolysis—RM bacterium

A. J. Ross, R. R. Rucker, and W. H. Ewing (1966). Can. J. Microbiol. 12, 763–770.

Tests for aesculin hydrolysis were made by the method of Vaughn and Levine (1942), in which ferric citrate was incorporated into the medium.

R. H. Vaughn and M. Levine (1942). J. Bact. 44, 487–505.

AGAR HYDROLYSIS

Agar hydrolysis—unspecified "aerobic bacterium"

S. A. Waksmen and W. Bavendamm (1931). J. Bact. 22, 91–102.

The organisms developed best on the lactate-agar of Drew (1915), this medium having the following composition:

Sea water	1000 g
KNO_3	0.5 g
Na_2HPO_4	0.25 g
Calcium lactate	2.0 g
Agar (Difco Bacto)	18.0 g

The organism was isolated on a medium containing 0.5 g of K_2HPO_4 and 12.5 g of agar in 1000 ml of seawater. It was best cultivated on a

medium containing 0.2 g of peptone and 10 g of agar in 1000 ml of seawater. It grew well on seawater-agar with KNO_3, $(NH_4)_2SO_4$, or asparagine as sources of nitrogen, however.
G. H. Drew (1915). Carnegie Institute, Washington 5, 7–45.

Agar hydrolysis—*Pseudomonas, Achromobacter*
H. E. Goresline (1933). J. Bact. 26, 435–457.
The medium used in the isolation had the following composition:

$(NH_4)_2SO_4$	1 g
K_2HPO_4	1 g
NaCl	2 g
$MgSO_4$	0.5 g
$FeSO_4$	trace
$MnSO_4$	trace
H_2O	1000 ml
agar	15 g
pH 7.2	

Agar plates were poured, using as inoculum 1-ml portions of a suspension of material washed from the cinders of an experimental trickling filter receiving creamery wastes. The plates were incubated at 25°C for 5 days. After 3 days of incubation small pits in the agar surface could be seen.

Agar hydrolysis—*Vibrio, Pseudomonas, Cytophaga*
R. Y. Stanier (1941). J. Bact. 42, 527–559.
Medium:

Aged seawater	100 ml
Bacto-peptone	0.5 g
K_2HPO_4	0.2 g
$CaCO_3$	2.0 g

Cultures were incubated at 22°C.
A number of strains, particularly the rapid agar liquefiers, die off rapidly because of the production of acid if the $CaCO_3$ is omitted.
Some bacteria cause liquefaction; others, only depression. A valuable aid in studying this is the iodine reaction discovered by Gran (1902). If one pours an I-KI solution over a plate on which agar-digesting bacteria have been growing, the areas surrounding the colonies take on at most a light straw color, whereas, the unattacked agar stains a reddish-violet, the intensity of the coloration decreasing with the age of the plate.
H. H. Gran (1902). Bergens Mus. Årb., no. 2, 1–16.

Agar hydrolysis—anaerobic myxobacteria
H. Veldkamp (1961). J. gen. Microbiol. 26, 331–342.
Cytophaga fermentans var. *agarovorans* and *C. salmonicolor* var. *agarovorans* grow on and liquify a medium containing 1% (w/v) agar +

0.1% (w/v) yeast extract. Both growth and liquefaction were improved by the addition of corn steep liquor 0.1% (w/v) and nutrient broth 0.1% (w/v). Agar (2%) tends to suppress spread of the colonies. Glucose and galactose (filtered), 0.1–0.2%, are inhibitory to both species.

ALCOHOL and ALDEHYDE (PRODUCTION of)

Alcohol and aldehyde (production of)—anaerobes

A. -R. Prévot (1966). Techniques pour le diagnostic des Bactéries Anaérobes. Éditions de la Tourelle, St. Mandé.

Alcools: dans un tube à essais, verser 1 cm^3 de distillat* et 1 cm^3 d'une solution de bichromate de potasse nitrique. Une réaction positive se traduit par l'apparition d'une coloration bleue après un séjour de 2 minutes au bain-marie à 100°C. Dans ce cas on recherchera l'alcool éthylique.

Alcool éthylique: dans un tube à essais, verser 2 cm^3 de distillat,* ajouter 4 gouttes de lessive de soude; verser goutte à goutte du reactif iodo- ioduré jusqu'à coloration jaune-blanchâtre qu'on fait disparaître par une goutte de lessive de soude; on laisse refroidir. Une réaction positive se traduit par l'apparition immédiate d'une odeur d'iodoforme et la formation, dans les 24 heures, de cristaux jaunes hexagonaux que l'on peut rechercher au microscope dans le dépôt formé.

Aldéhydes: dans un tube à essais, verser 2 cm^3 de distillat* et ajouter lentement ½ à 1 cm^3 de réactif de Schiff. Si le mélange demeure limpide et incolore, on peut affirmer qu'il n'existe pas d'aldéhyde. Lorsqu'une teinte rouge apparaît, le mélange peut contenir soit un aldéhyde, soit un corps voisin, aussi convient-il de compléter cette réaction de groupe par la réaction suivante spécifique des aldéhydes: à 1 cm^3 de distillat ajouter quelques gouttes de liqueur cupropotassique; une réaction positive se traduit par l'apparition à froid d'un précipité rouge d'oxydule de cuivre.
*Distillat Alcaline sous "Amines (volatile)—anaerobes"—A. -R. Prévot.

ALGINATE HYDROLYSIS

Alginate hydrolysis

S. A. Waksman, C. L. Carey, and M. C. Allen (1934). J. Bact. 28, 213–220.

The following methods were used in the isolation of the bacteria from their respective habitats. One per cent solution of purified alginic acid (Schmidt and Vocke, 1926) prepared from different species of *Fucus,* largely *F. vesiculosus,* was dissolved in NaOH or KOH solution and the reaction adjusted pH 7.0–7.2. Sodium nitrate and salts, in concentrations used in Czapek's solution, were added. For the isolation of the marine bacteria, seawater or 3.5% NaCl solution was used; for soil bacteria, distilled water. For a solid medium agar was added.

The determination of the decomposition of alginic acid by bacteria presents no difficulty in the case of distilled water or salt water media; the residual alginic acid is precipitated by means of a mineral acid or by addition of a solution of a calcium salt. In the seawater medium a large part of the alginic acid is precipitated in the medium by the calcium and magnesium of the water; as a result of the development of specific bacteria, the precipitate gradually disappears, because of the action of the organisms and that of the enzyme alginase, which they produce abundantly.

E. Schmidt and F. Vocke (1926). Ber. dt. chem. Ges. 59B, 1585.

Alginate hydrolysis

V. B. D. Skerman (1967). A Guide to the Identification of the Genera of Bacteria. 2nd ed. Baltimore: Williams & Wilkins.

The composition of the following medium is, with minor modifications, that described by Waksman and his associates. The method of distribution is one found most satisfactory in the author's laboratory.

Dissolve 10 g of sodium alginate in 1 L of distilled water. Adjust the pH to 7.0 with HCl and then add $NaNO_3$, 2 g; KCl, 0.5 g; $MgSO_4 \cdot 7H_2O$, 0.5 g; $FeSO_4 \cdot 6H_2O$, 0.01 g; KH_2PO_4, 1.0 g; NaCl, 20–35 g (required only for marine species); agar, 20 g. Dissolve the salts and the agar, and readjust the pH to 7.0 if necessary. Sterilize at 121°C for 15 minutes.

For the preparation of agar plates, prepare a mineral salts agar of the above composition with the omission of the sodium alginate. Distribute in 10-ml amounts in sterile petri dishes and allow to set. Pour 10 ml of the alginate agar onto the surface of the mineral base and allow to set.

Inoculate the plates and incubate for 14 days; examine at intervals for clarification of the agar around the colonies. Where this occurs the digestion of the alginate results in a pitting of the agar, which should not be confused with agar hydrolysis.

Note: The composition of the medium as described restricts the growth of organisms which do not utilize nitrate. The substitution of ammonium sulfate for the nitrate is recommended, although for comparative studies use of the original medium must be made.

S. A. Waksman, C. L. Carey, and M. C. Allen (1934). J. Bact. 28, 213.

Alginate hydrolysis—*Agarbacterium*

A. K. Williams and R. G. Eagon (1962). Can. J. Microbiol. 8, 649–654.

Preparation of alginate: Sodium alginate used in the assay procedures was prepared by washing alginic acid (Kelacid Kelco Co., Clark, N. J.) first with distilled water until inorganic phosphorus could no longer be detected in the wash water, then once with 95% ethanol, and finally three times with acetone. The alginic acid was next collected by filtration and air-dried at room temperature.

This dried, purified alginic acid was suspended in water and neutralized with 10% NaOH as needed. Enough filter aid (Celite, Super-Cel) was added to the solution to produce a thin slurry, which was then filtered through Whatman No. 1 filter paper; the volume of the filtrate was adjusted with water to yield the desired concentration of sodium alginate.

Preparation of culture medium: *A. alginicum* was grown in 24 L Florence flasks containing 20 L of medium of the following composition: commercially prepared sea salt, 300 g; $(NH_4)_2SO_4$, 10 g; $MgCl_2$, 10 g; NH_4Cl, 10 g; Bacto-Casitone, 150 g; sodium alginate, 50 g; yeast extract, 10 g; and distilled water to 20 L. Additional NaOH was added to pH 7.3. The flask was inoculated with 1 litre of a 12 hour adapted culture and aerated vigorously for 14 hours at 35°C.

Preparation of cell-free extracts: The cells were harvested by centrifugation, washed with 0.052 *M* $MgCl_2$ suspended in 0.067 *M* phosphate buffer (pH 7.0), and disintegrated with a Raytheon 200-watt, 10-kc magnetostrictive oscillator for 15 minutes. Cell debris was removed by centrifugation at 35,000 x *g* for 60 minutes. This extract was next dialyzed for three 12-hour periods against changes of distilled water. The dialyzed extract was then centrifuged at 35,000 x *g* for 60 minutes. Extracts prepared in this manner could be frozen for several weeks with no appreciable loss of activity.

Assay procedure for alginase: Hydrolysis of alginate was determined by the increase in reducing power of reaction mixtures containing extract, alginate, and phosphate buffer. Unless otherwise indicated, reactions were allowed to proceed for 10 minutes at 37°C. Reactions were stopped by the addition of 2 ml of the 3,5-dinitrosalycylic acid reagent of Sumner and Howell as modified for the assay of amylases by Noelting and Bernfeld (1948). Color was developed by placing the tubes in a boiling water bath for 5 minutes, after which they were cooled immediately and read in a Klett-Summerson colorimeter with filter No. 54.

G. Noelting and P. Bernfeld (1948). Helv. chim. Acta 31, 286–290.

Alginate hydrolysis—*Enterobacteriaceae*

B. R. Davis and W. H. Ewing (1964). J. Bact. 88, 16–19.

The media used in tests for utilization and liquefaction of alginate were modifications of Simmons' citrate agar and of an alginate medium mentioned by Skerman (1959), respectively. The former was called alginate agar (synthetic), and was prepared by substituting 0.25% sodium alginate for the citrate in Simmons' citrate-agar medium. This was prepared and inoculated in the manner prescribed for Simmons' medium. The second was labeled nutrient alginate medium, and was prepared as follows: sodium alginate, 10.0 g; peptone, 2.0 g; yeast extract, 1.0 g; sodium nitrate, 2.0 g; potassium chloride, 0.5 g; magnesium sulfate, 0.5 g; ferrous sulfate, 0.01 g; potassium phosphate (monobasic), 1.0 g; agar, 15.0 g; and distilled water, 1000 ml; pH 7.0. The medium was distributed in tubes, sterilized at 121°C for 15 min, and allowed to solidify in a slanted position (long slants). Positive results are indicated by zones of clarification along the lines of inoculation. Cultures were incubated for 21 days at 37°C.

V. B. D. Skerman (1959). A Guide to the Identification of the Genera of Bacteria. Baltimore: Williams & Wilkins.

Alginate hydrolysis—RM bacterium

A. J. Ross, R. R. Rucker, and W. H. Ewing (1966). Can. J. Microbiol. 12, 763–770.

The medium for the detection of alginolytic activity was a modification of that mentioned by Skerman (1959), and that employed in tests for utilization of alginate was prepared by substituting 0.25% sodium alginate for the sodium citrate in a medium similar to Simmons' citrate-agar (Davis and Ewing, 1964).

B. R. Davis and W. H. Ewing (1964). J. Bact. 88, 16–19.
V. B. D. Skerman (1959). A Guide to the Identification of the Genera of Bacteria. Baltimore: Williams & Wilkins.

Alginate hydrolysis—marine bacterium

Mme C. Billy (1966). Annls Inst. Pasteur, Paris 110, 591–602.

Bien que l'algine soit un constituant des algues brunes marines ou Phéophycées, l'alginolyse n'est pas une fonction exclusivement marine. Un gel d'alginate de sodium dans l'eau distillée, laissé à l'air libre, peut être liquéfié par des germes banaux.

Les milieux à l'algine se préparent facilement avec de l'eau distillée, formant des solutions plus ou moins visqueuses, mais parfaitement homogènes. Nous avons utilisé des gels additionnés de sels minéraux monovalents pour rechercher l'alginolyse dans une vase d'eau douce provenant de Pavie (Italie). Comme nous l'avons en effet exposé succinctement dans une première note relative à l'alginolyse [1], une association symbiotique entre un *Clostridium* et un vibrion sulfato-réducteur permet une liquéfaction rapide de gels renfermant 8 à 10 p. 1000 d'alginate de sodium, seule source de carbone du milieu.

De nombreux auteurs, tels Waksman, Carey et Allen [21], Preiss et Ashwell [15], Davis et Ewing [6], Kooiman [12], Franssen et Jeuniaux [9], ont proposé des milieux de culture en eau distillée pour l'étude de l'alginolyse. En 1934, Waksman, Carey et Allen ont recherché les germes alginolytiques du sol dans des milieux préparés à l'eau distillée et ceux des fonds marins dans les mêmes milieux additionnés de 3,5 p. 100 de NaCl.

Récemment, Yaphe [23] a fort judicieusement adapté la méthode de Wieringa, sur milieux à double couche, à la recherche des bactéries alginolytiques marines. Par ce procédé, les fonctions alginolytique et agarolytique se trouvent bien séparées; les cations divalents de l'eau de mer, qui précipitent trop facilement l'algine, diffusent ici lentement d'un milieu gélosé à l'eau de mer vers une solution d'alginate de sodium. Cette technique a permis de lever certains doutes sur le pouvoir alginolytique ou agarolytique de quelques bactéries marines aérobies.

Pour les recherches en anaérobiose, le problème de la précipitation de l'algine par les cations divalents de l'eau de mer demeure, et il est, malgré tout, nécessaire de travailler sur des gels homogènes et limpides. L'idéal serait de disposer d'un alginate moyennement méthylé supportant les fortes concentrations en cations de l'eau de mer, analogue à ce qu'est la pectine "ruban violet"* pour l'étude de la pectinolyse (Kaiser [11]). Il est toutefois possible de pallier l'inconvénient de précipitation par l'addition de tampon phosphaté à un alginate pur quelconque. Avec l'alginate utilisé ici,† la quantité de phosphate dipotassique nécessaire à l'obtention d'un pH neutre et d'un minimum de précipitation dans le milieu a été de 20 p. 1000. Pour importante qu'elle puisse paraître, cette quantité n'est certes pas excessive. Preiss et Ashwell [15] ont ajouté 15 p. 1000 de K_2HPO_4 et 5 p. 1000 de NaH_2PO_4 à un milieu d'alginate de sodium pour cultiver un *Pseudomonas* d'une boue de la Baie du Potomac.

Nous avons déjà donné la composition du milieu utilisé pour l'étude de *Cl. alginolyticum* [3]. Nous y ajouterons quelques détails complémentaires de fabrication:

Eau de mer	950 ml
Bouillon d'algues	50
K_2HPO_4	20 g

Chauffer jusqu'à trouble blanchâtre abondant et filtrer, deux fois de préférence:

Peptone	1 g
Extrait de levure Difco	1
Alginate	5 à 10

La quantité d'alginate à introduire dans le milieu varie selon le produit utilisé et la viscosité désirée; avec le Cecalginate Maton HV/KP les gels ont été préparés à 8 ou 9 p. 1000 pour l'étude de l'association *Cl. alginolyticum-D. desulfuricans,* mais à 5 p. 1000 seulement pour l'étude du

*Pectine "ruban violet", Société Unipectine, Paris.

†Alginate de sodium Cecalginate, aimablement fourni par les établissements Ceca, Alginates Maton, Paris.

Clostridium avec filtrat de la culture du germe associé. L'alginate mis dans un *mixer* est mouillé avec un peu d'alcool absolu; les autres éléments de la préparation sont versés tièdes sur l'alginate. Après mélange, le milieu est maintenu au bain-marie environ une heure à 50°, en vue d'une bonne homogénéisation et d'une répartition facile en tubes.

L'alginate de sodium est à pH neutre, le milieu l'est aussi. Si tel n'était pas le cas, il faudrait ajuster le pH vers 7–7.5 avec de la lessive de soude N/10.

La répartition du milieu se fait à raison de 6 cm^3 en tubes de 14 mm. On stérilise momentanément à 120°C, plutôt qu'une demi-heure à 110°C, pour éviter (comme pour la pectine) une trop forte dépolymérisation de l'algine.

Avant l'emploi, chaque tube est régénéré vingt minutes au bain-marie à 100°C. Après ensemencement les tubes sont étirés, vidés et scellés.

Pour l'isolement des colonies de *Cl. alginolyticum,* on utilise des géloses profondes à l'acide alginique, moins polymérisé que l'alginate et donnant par suite des milieux moins compacts avec l'agar.

Ces milieux ont été préparés avec la "Cecalgine" Maton ou avec d'autres qualités d'acide alginique (Prolabo; Hopkin et Williams).

Voici la composition du milieu qui a donné les meilleurs résultats:

Eau de mer	950
Bouillon d'algues	50
K_2HPO_4	2

Chauffer et filtrer:

Peptone	1
Extrait de levure Difco	1
Acide alginique	7
Bacto-agar	5

Le milieu est agité au *mixer* pendant trois minutes après imprégnation par l'alcool de l'acide alginique et de la gélose. Le pH est ajusté à 7–7.5 par une solution de Na_2CO_3 (de préférence à la lessive de soude N/10). Le milieu est réparti dans des tubes de 8–9 mm sur une hauteur de 8 cm et stérilisé momentanément à 120°C. Il reste toujours dans le fond du tube un précipité qui ne gêne ni pour la lecture ni pour l'isolement des colonies.

Les proportions de phosphate et d'algine peuvent, bien entendu, varier selon l'origine de l'eau de mer, son degré de salinité et sa concentration en cations divalents, et selon le poids moléculaire de l'algine, c'est-à-dire de sa viscosité. Dans le cas présent, l'eau de mer avait même origine que le sédiment (Wimereux, Pas-de-Calais).

[1] Billy (C.). C. R. Acad. Sci., 1963, 257, 3700-3701.
[3] Billy (C.). Ann. Inst. Pasteur, 1965, 109, 147-151.
[6] Davis (B. R.) and Ewing (W. H.). J. Bact., 1964, 88, 16-19.
[9] Franssen (J.) et Jeuniaux (Ch.). Bull. Cl. Sci. Acad. R. Belgique,

1964, 5, 713-724.
[11] Kaiser (P.). Bull. Inst. Océanogr., 1961, 58, nº. 1210.
[12] Kooiman (P.). Bioch. Bioph. Acta, 1954, 13, 338-340.
[15] Preiss (J.) and Ashwell (G.). J. biol. Chem., 1962, 237, 309-316.
[21] Waksman (S. A.), Carey (C. L.) and Allen (M. C.). J. Bact., 1934, 28, 213.
[23] Yaphe (W.). Nature, 1962, 196, 1120.

Alginate hydrolysis—anaerobes

A.-R. Prévot (1966). Techniques pour le diagnostic des Bactéries Anaérobes. Éditions de la Tourelle, St. Mandé.

Depuis les travaux de C. Billy, on sait que certaines espèces anaérobies peuvent hydrolyser les alginates. C'est le cas en particulier de *Clostridium alginolyticum.* La détection d'une alginase anaérobie est beaucoup plus délicate que celle d'une alginase aérobie, surtout quand il s'agit d'un anaérobie marin halophile obligé. Voici la méthode mise au point pour ces derniers par C. Billy. Voici la composition des milieux de culture de cette espèce.

Eau de mer	950 cm³
Bouillon d'algues	50 cm³
K_2HPO_4	20 g

Chauffer jusqu'à l'obtention d'un trouble blanchâtre abondant et filtrer 2 fois sur papier.

Ajouter:

Peptone	1 g
Extrait de levure Difco*	1 g
Alginate de sodium	5 à 10 g

L'alginate mouillé avec un peu d'alcool absolu est traité au "mixer" et le milieu liquide tiède est versé sur lui, on maintient au bain-marie 1 heure à 50° et on répartit en tube de 18/14. On stérilise à 120° quelques minutes. On refroidit à 45°, on ensemence, on étire et on scelle sous vide. La liquéfaction du gel d'alginate indique la présence d'une alginase.

On peut également préparer des géloses profondes à l'algine pour obtenir des colonies isolées:

Eau de mer	950 cm³
Bouillon d'algues	50 cm³
K_2HPO_4	2 g

Chauffer et filtrer comme précédemment.

Ajouter:

Peptone	1 g
Extrait de levure	1 g
Alginate	7 g
Bacto agar	5 g

*Cecalginate Maton HV/KP.

[1] F. Bernheim, M. L. C. Bernheim et D. Webster. J. biol. Chem., 1935, 110, 165.
[9] F. Roland, D. Bourbon, et C. Szturm. Ann. Inst. Pasteur, 1947, 73, 914.
[13] J. Singer et R. E. Volcani. J. Bact., 1955, 69, 303.

Phenylalanine—enterobacteria

R. Buttiaux, J. Moriamez and J. Papavassiliou (1956). Annls Inst. Pasteur, Paris 90, 133–143.

La technique recommandée par Henriksen [1] pour déceler la transformation de la phénylalanine en acide phénylpyruvique (A.P.P.) est relativement lente et compliquée. Nous l'avons ainsi simplifiée [10].

On prépare le milieu:

Peptone bactériologique U.C.L.A.F. (ou autre)	10 g
Phosphate bipotassique	1 g
Chlorure de sodium	5 g
Extrait de levures	3 g
Phénylalanine	2 g
Agar en poudre	12 g
Eau distillée	1000 ml

On dissout par chauffage doux et ajuste à pH 7,2. On répartit environ 5 ml par tube de 160 mm x 16 mm. On autoclave vingt minutes à 120°. On incline les tubes de façon à obtenir une grande surface de milieu. Ce dernier peut être conservé à la température du laboratoire pendant plus de deux mois.

Au moment de l'emploi, on ensemence largement avec le germe étudié. On incube durant dix-huit à vingt-quatre heures à 37°. On arrose les stries microbiennes avec 1 ml, environ, d'eau salée physiologique et on les émulsionne. On ajoute alors une pincée de cristaux de sulfate d'ammoniaque jusqu'à saturation, puis I goutte d'acide sulfurique dilué au dixième; on agite; on additionne le tout, enfin, de V gouttes d'une solution aqueuse d'alun de fer ammoniacal à demi-saturation.

La présence d'acide phénylpyruvique se traduit par l'apparition immédiate d'une coloration verte intense. Il ne faut pas tenir compte des autres teintes observées parfois.

[1] S. D. Henriksen. J. Bact., 1950, 60, 225.
[10] R. Buttiaux, R. Frenoy at J. Moriamez. Ann. Inst. Pasteur Lille, 1953-1954, 6, 62.

Phenylalanine (and **malonate**)—*Enterobacteriaceae*

J. G. Heyl (1957). Antonie van Leeuwenhoek 23, 33–58.

Leifson's malonate medium according to Shaw (combined with Henriksen's phenylalanine test).

Ammonium sulfate	2.0 g
Potassium phosphate	0.6 g
Monopotassium phosphate	0.4 g
Sodium chloride	2.0 g
Sodium malonate	3.0 g
DL-phenylalanine	2.0 g
Yeast extract	1.0 g
Bromthymol blue	0.025 g

Cultures are incubated overnight at 37°C; an alkaline reaction is produced by bacteria using malonate as a source of carbon.

Phenylalanine—*Escherichia aurescens*

H. Leclerc (1962). Annls Inst. Pasteur, Paris 102, 726–741.

Désamination de la phénylalanine en acide phénylpyruvique: Elle est recherchée d'après Buttiaux et coll. [6].

[6] Buttiaux (R.), Moriamez (J.) et Papavassiliou (J.). Annls Inst. Pasteur 1956, 90, 133.

Phenylalanine (and **malonate**)—*Vibrio*

G. H. G. Davis and R. W. A. Park (1962). J. gen. Microbiol. 27, 101–119.

Tested by the method of Shaw and Clarke (1955, J. gen. Microbiol. 13, 155).

Phenylalanine—*Enterobacteraceae*

W. H. Ewing (1962). *Enterobacteriaceae. Biochemical Methods for Group Differentiation.* U.S. Department of Health, Education, and Welfare, Public Health Service Publication No. 734 (revised).

Agar (Ewing *et al.* 1957):

Test for deamination of phenylalanine to phenylpyruvic acid.

Yeast extract	3 g
DL-phenylalanine	2 g
(or L-phenylalanine)	1 g
Disodium phosphate	1 g
Sodium chloride	5 g
Agar	12 g
Distilled water	1000 ml

Tube and sterilize at 121°C for 10 minutes and allow to solidify in a slant position (long slant).

Test reagent: 10 per cent (w/v) solution of ferric chloride.

Inoculation: Inoculate the slant of the PA agar with a fairly heavy inoculum from an agar slant culture.

Incubation: 4 hours or, if desired, 18 to 24 hours at 37°C.

Following incubation, 4 or 5 drops of ferric chloride reagent are allowed to run down over the growth on the slant. If phenylpyruvic acid has been formed a green color develops in the syneresis fluid and in the slant.

W. H. Ewing, B. R. Davis and R. W. Reavis (1957). Pub. Hlth Lab. 15, 153.

Phenylalanine (and **malonate**)—*'Bacterium salmonicida' (Aeromonas)*

I. W. Smith (1963). J. gen. Microbiol. 33, 263–274.

Done by the method of Shaw and Clarke (1955).

C. Shaw and P. H. Clarke (1955). J. gen. Microbiol. 13, 155.

Phenylalanine—*Pseudomonas odorans* (28°C)

I. Málek, M. Radochová and O. Lysenko (1963). J. gen. Microbiol. 33, 349–355.

Modifications described in the Report (1958) were used.

Report of the Enterobacteriaceae Subcommittee of the Nomenclature Committee of the I.A.M.S. (1958). Int. Bull. bact. Nomencl. Taxon. 8, 25.

Phenylalanine—*Malleomyces, Pseudomonas*

Bach-Toan-Vinh (1965). Annls Inst. Pasteur, Paris 109, 460–463.

Recherche d'une Phénylalanine Désaminase – Technique rapide de Ben Hamida et Le Minor [1].

[1] Ben Hamida (F.) et Le Minor (L.). Annls Inst. Pasteur, 1956, 90, 671.

Phenylalanine—*Moraxella*

W. J. Ryan (1964). J. gen. Microbiol. 35, 361–372.

The micromethod of C. Shaw and P. H. Clarke (J. gen. Microbiol. 13, 155, 1955) was used to detect phenylalanine deaminase.

Positive and negative controls were included.

Phenylalanine—*Pseudomonas aeruginosa*

R. R. Colwell (1964). J. gen. Microbiol. 37, 181–194.

Production of phenylpyruvic acid from phenylalanine was tested for by the method of Ewing, Davis and Reavis (1957).

W. H. Ewing, B. R. Davis and R. W. Reavis (1957). Publ. Hlth Lab. 15, 153.

Phenylalanine—*Hyphomicrobium neptunium*

E. Leifson (1964). Antonie van Leeuwenhoek 30, 249–256.

The liquid medium consisted of peptone (Casitone, Difco) 0.2%, yeast extract 0.1%, tris buffer (tris (hydroxymethyl) amino methane) 0.05%, phenylalanine 0.1%, agar 1.5%, artificial seawater half strength, pH 7.5. Ten per cent ferric chloride was used to test for phenylalanine deaminase.

Phenylalanine—*Alcaligenes*

R. G. Mitchell and S. K. R. Clarke (1965). J. gen. Microbiol. 40, 343–348.

The combined medium of Shaw and Clarke (1955), incubation for 6 days was used.
C. Shaw and P. H. Clarke (1955). J. gen. Microbiol. 13, 155.

Phenylalanine—*Mima, Herellea, Flavobacterium*

J. D. Nelson and S. Shelton (1965). Appl. Microbiol. 13, 801–807.

The phenylalanine deaminase test was performed as outlined by Edwards and Ewing (1962).
P. R. Edwards and W. H. Ewing (1962). *Identification of Enterobacteriaceae,* 2nd ed. Burgess Publishing Co., Minneapolis.

Phenylalanine—*Erwinia*

A. von Graevenitz and A. Strouse (1966). Antonie van Leeuwenhoek 32, 429–430.

The authors used the method of W. H. Ewing, B. R. Davis and P. R. Edwards, Publ. Hlth Lab. 18, 77–83.

Phenylalanine—*Mycobacterium*

R. E. Gordon (1966). J. gen. Microbiol. 43, 329–343.

Cultures were streaked on slants of the following composition: yeast extract, 3 g; DL-phenylalanine, 2 g; Na_2HPO_4, 1 g; NaCl, 5 g; agar, 12 g; distilled water, 1000 ml (Ewing, Davis and Reavis, 1957). After 14 days, four or five drops of 10% (w/v) of $FeCl_3$ were pipetted over the growth on the slants. The development of a green colour in the agar demonstrated the formation of phenylpyruvic acid from the phenylalanine.
W. H. Ewing, B. R. Davis and R. W. Reavis (1957). Publ. Hlth Lab. 15, 153.

Serine—*Streptococcus*

R. Whittenbury (1965). J. gen. Microbiol. 38, 279–287.

Liquid medium containing L-serine 0.3% (w/v) was adjusted to pH 6.5. Ammonia production was determined by Nessler's reagent and acetoin (indicating dissimilation of pyruvate, the deamination product) by Barritt's (1936) modification of the Voges-Proskauer test.

The basal liquid medium contained meat extract (Lab-Lemco), 0.5 g; peptone (Evans), 0.5 g; yeast extract (Difco), 0.5 g; Tween 80, 0.05 ml; $MnSO_4 \cdot 4H_2O$, 0.01 g; in 100 ml tapwater, adjusted to pH 6.5 and autoclaved at 121°C for 15 min.
M. M. Barritt (1936). J. Path. Bact. 42, 441.

Tryptophane—enterobacteria

P. Thibault and L. le Minor (1957). Annls Inst. Pasteur, Paris 92, 551–554.

Recherche de la tryptophane-désaminase.—Certaines espèces d'Entérobactéries possèdent un enzyme qui provoque une désamination oxydative d'acides aminés et les transforment en leur acide cétonique correspondant.

Bernheim et coll. [8], Henricksen et Closs [9] ont proposé de rechercher la transformation de la L-phényl-alanine en acide phényl-pyruvique, réaction pour l'étude de laquelle l'un de nous a proposé, avec Ben Hamida, une méthode simplifiée [10]. Singer et Volcani [11] ont étudié la désamination oxydative de 8 acides aminés et en concluent que le tryptophane est le meilleur substrat pour objectiver cette réaction.

Nous avons voulu savoir si le milieu de Ferguson additionné de tryptophane pouvait servir comme méthode simple à la recherche de l'uréase, de l'indole et de la tryptophane-désaminase (TDA) à la fois. Les résultats ont été parfaitement satisfaisants.

Méthode. – Une suspension épaisse de la souche à l'étude dans un petit volume de milieu (IV gouttes) est laissée au moins deux heures à l'étuve à 37°, ou bien agitée pendant quinze minutes dans un appareil du type utilisé pour la réaction de Kahn. Pour mettre en évidence l'acide indol-acétique, il suffit d'ajouter I goutte de la solution officinale de chlorure ferrique (26 p. 100) diluée au tiers dans de l'eau distillée. Des essais comparatifs nous ont montré qu'il était inutile d'acidifier par un tampon ou de l'acide chlorhydrique. Une réaction positive se traduit par une coloration brun rougeâtre, une réaction négative par une couleur jaune.

Il est évidemment possible d'effectuer cette réaction sur une fraction d'un milieu ayant servi à la recherche de l'uréase, dont l'autre fraction servira à la recherche de l'indole au moyen du réactif de Kovacs. C'est cette méthode que nous employons maintenant de manière systématique.

[8] Bernheim (F.), Bernheim (M. L. C.) et Webster (D.). J. biol. Chem., 1935, 110, 165.

[9] Henriksen (S. D.) et Closs (K.). Acta path. microb. scand., 1938, 15, 101.

[10] Ben Hamida (F.) et Le Minor (L.). Ann. Inst. Pasteur, 1956, 90, 671.

[11] Singer (J.) et Volcani (R. E.). J. Bact., 1955, 69, 303.

Tryptophane—*Malleomyces, Pseudomonas*

Bach-Toan-Vinh (1965). Annls Inst. Pasteur, Paris 109, 460–463.

Recherche d'une Tryptophane Désaminase – Technique rapide de Thibault et Le Minor [8].

[8] Thibault (P.) et Le Minor (L.). Ann. Inst. Pasteur, 1957, 92, 551.

AMINO ACID DECARBOXYLASES

Amino acid decarboxylases—*Lactobacillus*

A. W. Rodwell (1953). J. gen. Microbiol. 8, 224–232.

Determined manometrically at 30°C in oxygen-free N_2 and pH controlled with McIlvaine buffer.

L-arginine, L-glutamic acid, L-histidine, L-lysine, L-ornithine and L-tyrosine were tested.

Amino acid decarboxylases—*Proteus-Providencia*

C. Shaw and P. H. Clarke (1955). J. gen. Microbiol. 13, 155–161.

Decarboxylases. Two drops suspension, 2 drops 0.03 M solution amino acid adjusted to pH 5.0, 2 drops 0.0125 M-phthalate buffer, pH 5.0 + bromcresol purple. At this pH value CO_2 is liberated so that the reaction mixture will show an indicator colour change when decarboxylation has taken place. Readings were taken at 2, 4 and 24 hr at 37°C. Toluene was added in later tests. Each set of tests included a suspension control without amino acid; when this suspension was alkaline, buffer was added to all tubes until a yellow colour was given by the control.

Amino acid decarboxylases—*Proteus*

L. Ekladius, H. K. King and C. R. Sutton (1957). J. gen. Microbiol. 17, 602–619.

Leucine, valine, norvaline, isoleucine and α-amino-n-butyric acid are decarboxylated by washed cells at pH 7.0. Manometric methods used.

Amino acid decarboxylases—enterobacteria

P. Thibault and L. le Minor (1957). Annls Inst. Pasteur, Paris 92, 551–554.

Recherche de la lysine-décarboxylase.—L'intérêt de la recherche des décarboxylase à été montré par Moeller [3] et Moeller et Kauffmann [4]. Des méthodes simplifiées ont été proposées par Moeller [5] et Carlquist [6]. Mais elles nécessitent des milieux spéciaux. Nous avons employé la réaction à la ninhydrine comparativement sur le milieu liquide préconisé par Carlquist et sur le milieu gélosé lactose-glucose-SH_2.

Principe.—Gale [7] a montré que les décarboxylases des bacilles Gram-négatifs n'agissent que sur les amino-acides ayant au moins un groupement chimique actif autre que le carboxyle terminal et les groupes α-aminés: la lysine est décarboxylée en cadavérine, l'histidine en histamine, etc. Par chromatographie, Carlquist [6] a montré que la réaction à la ninhydrine pratiquée sur un extrait chloroformique d'un milieu complexe alcalinisé se produit essentiellement avec la cadavérine.

Technique.—Les cultures sur milieux à l'hydrolysat de caséine d'une part, au lactose-glucose-SH_2 d'autre part, sont additionnées, après séjour de dix-huit heures à 37°, de 1 cm^3 de NaOH 4 N (le titre de la soude n'a pas besoin d'être déterminé exactement; on obtient les mêmes résultats avec la lessive de soude diluée de moitié) et de 2 cm^3 de chloroforme. Après agitation modérée, on laisse le chloroforme décanter, ou, éventuellement, on centrifuge. On prélève environ 0,5 cm^3 de chloroforme parfaitement limpide (il est important de ne pas emmener de phase aqueuse) qui est transvasé dans un petit tube. On ajoute un volume égal d'une solution à 0,1 p. 100 de ninhydrine dans le chloroforme. Une réaction positive se traduit par une belle coloration violet-améthyste apparaissant au bout de trois à cinq minutes à la température du laboratoire. La réaction est considérée comme négative si, au bout de dix minutes, aucune coloration n'est apparue. Carlquist conseille de ne pas attendre plus de quatre minutes et de ne pas chauffer le tube, car des réactions faussement positives, dues à la présence dans le chloroforme de composés mineurs réagissant avec la ninhydrine, pourraient apparaître. Nous n'avons pas observé, avec la ninhydrine et le chloroforme dont nous disposons, de réactions faussement positives. Seules, certaines souches donnent une coloration très pâle au bout de quinze minutes, qui ne peut prêter à confusion avec la coloration violette.

[3] Moeller (V.). Acta path. microb. scand., 1954, 35, 259.
[4] Moeller (V.) et Kauffmann (P.). Acta path. microb. scand., 1955, 36, 173.
[5] Moeller (V.). Acta path. microb. scand., 1955, 36, 158.
[6] Carlquist (P. R.). J. Bact., 1956, 71, 339.
[7] Gale (E. F.). The Chemical activities of Bacteria, 3e éd., University Tutorial Press, Londres, 1951.

Amino acid decarboxylases—*Pseudomonas*

M. Véron (1961). Annls Inst. Pasteur, Paris 100, Suppl. 6, 16–42.

La recherche d'une lysine-décarboxylase dans une culture de 48 heures à 30°C sur milieu glucose-lactose-SH_2 est réalisée par la technique classique utilisée pour les enterobactéries et décrite par Thibault et Le Minor.

P. Thibault et L. Le Minor (1957). Ann. Inst. Pasteur (Paris) 92, 551.

Amino acid decarboxylases—*Escherichia aurescens*

H. Leclerc (1962). Annls Inst. Pasteur, Paris 102, 726–741.

Décarboxylases des acides aminés. – Les décarboxylases de la lysine, de l'ornithine et de l'acide glutamique, la dihydrolase de l'arginine sont mises en évidence d'après la méthode de Moeller [33]. Ewing, Davis et Edwards [10] ont montré le grand intérêt qu'elles présentent dans la taxinomie des *Enterobacteriaceae.*

[10] Ewing (W. H.), Davis (R.) and Edwards (P. R.). Publ. Hlth Lab., 1960, 18, 4, 77.
[33] Moeller (V.). Acta path. microbiol. scand., 1955, 36, 158.

Amino acid decarboxylases—*Enterobacteriaceae*

W. H. Ewing (1962). *Enterobacteriaceae. Biochemical Methods for Group Differentiation.* U.S. Department of Health, Education, and Welfare, Public Health Service Publication, No. 734 (revised).

Tests for decarboxylation of arginine and ornithine, as well as lysine, are included at this point rather than in Part II, since the basal medium is the same and since it often is advisable to determine the reactions of cultures in media containing each of these amino acids. Whether lysine medium is employed alone or whether all three media are used, a control tube should always be inoculated with each culture tested. This point cannot be overemphasized. The writer is of the opinion that the Moeller (1955) method should be regarded as the standard or reference method for taxonomic work. Other methods, such as the alternates mentioned herein, are of undoubted value in certain areas of the family and can be recommended for general use in those areas, but they may give equivocal results in other areas *(v. infra)*.

Moeller Method (1955)

Basal medium	
Peptone (Orthana special)*	5 g
Beef extract	5 g
Bromocresol purple (1.6%)	0.625 ml
Cresol red (0.2%)	2.5 ml
Glucose	0.5 g
Pyridoxal	5 mg
Distilled water	1000 ml
Adjust pH to 6.0	

The basal medium is divided into four equal portions, one of which is tubed without the addition of any amino acid. These tubes of basal medium are used for control purposes. To one of the remaining portions of basal medium is added 1 per cent of L-lysine dihydrochloride, to the second 1 per cent of L-arginine monohydrochloride, and to the third portion 1 per cent of L-ornithine dihydrochloride. If DL-amino acids are used they should be incorporated into the media in 2 per cent concentration, since the microorganisms apparently are only active against the L forms. The amino acid media may be tubed in 3 or 4 ml amounts in small (13 x 100 mm) screw-capped tubes and sterilized at 121°C for 10 minutes. A small amount of floccular precipitate may be seen in the ornithine medium. This does not interfere with its use.

Inoculation: Inoculate lightly from a young agar slant culture. After inoculation add a layer (about 4 or 5 mm in thickness) of sterile mineral (paraffin) oil to each tube including the control. A control tube should be inoculated with each culture under investigation.

Incubation: 37°C. Examine daily for 4 days. Positive reactions are indicated by alkalinization of the media and a consequent change in the color of the indicator system from yellow to violet or reddish-violet. The majority of positive reactions with *Enterobacteriaceae* occur within the first day or two of incubation but sufficient delayed reactions occur to warrant a 3- or 4-day incubation and observation period.

*Peptone Special 'Orthana' Meat USP XV from A/S Orthana Kemish Fabrik, Copenhagen, Denmark.

Falkow Method (1958). This method, which may be regarded as a modification of the Moeller method given above, was originally devised as a lysine medium only. However, Falkow (personal communication, 1957) stated that arginine could be substituted successfully for lysine and the writer has employed Falkow's basal medium with lysine, arginine, and ornithine. Comparative studies (author's unpublished data) have shown that the Falkow modification can be used successfully in most areas of the family. Results obtained with cultures that belong to the *Klebsiella* and *Aerobacter* groups may be equivocal, however, and the method cannot be recommended for use with members of these groups.

Basal Medium	
Bacto peptone	5 g
Yeast extract	3 g
Glucose	1 g
Bromocresol purple (1.6% soln)	1 ml
Distilled water	1000 ml

Adjust pH to 6.7 – 6.8 if necessary.

The basal medium is divided into four parts and treated in the same manner as that given above for the Moeller medium except that only 0.5 per cent of L amino acid is added. After the addition of the amino acids to three of the four portions of basal medium, the media are tubed in small (13 x 100 mm screw-capped) tubes and sterilized at 121°C for 10 minutes. The remaining portion of basal medium, without amino acid, serves as a control.

Inoculation: Inoculate lightly from a young agar slant culture. Oil seals are not used with this method.* A control tube should be inoculated with each culture under investigation.

Incubation: 37°C. Examine daily for 4 days. These media first become yellow due to acid production from glucose, later if decarboxylation occurs, the medium becomes alkaline (purple). The control tubes should remain acid (yellow).

The Ninhydrin test† (Carlquist, 1956).

This method is a modification of the extraction method of Moeller (1955).

Trypticase or casitone	15 g
Dibasic potassium phosphate	2 g
Glucose	1 g
Distilled water	1000 ml

The medium is dispensed in 5 ml amounts in screw-capped tubes and sterilized at 121°C, 15 minutes.

**Authors note:* (personal communication) "if 24 hr readings (only) are made – otherwise oil seals should be used."

†W. H. Ewing (personal communication) states that test for Lysine iron agar is to be preferred.

Inoculation: The medium may be inoculated from an agar slant culture.
Incubation: 37°C, 18 to 24 hr.

After incubation 1 ml of 4 N NaOH is added and mixed with each culture. Then 2 ml of chloroform is added to each tube, the tubes are shaken vigorously, and centrifuged lightly for a few minutes or until the chloroform separates into a distinct clear layer. The cadaverine is extracted by the chloroform. One-half ml of the clear chloroform layer is removed with a clean pipette and placed in a 13 x 100 mm or similar small tube. To this is added 0.5 ml of a 0.1% chloroform solution of ninhydrin (1,2,3-triketohydrindene). This mixture is observed for 4 minutes at room temperature. A positive reaction is evidenced by the appearance of a deep purple color. Incubation of the tests for longer than 4 minute periods, or at higher temperatures may result in the production of false positive reactions caused by the presence of small quantities of other reacting compounds.

The ninhydrin test for decarboxylation is of value in the differentiation of members of the *Salmonella* and *Arizona* groups on the one side and *Citrobacter* (including *Bethesda-Ballerup*) strains on the other. The majority of cultures that belong to the *Salmonella* and *Arizona* groups are ninhydrin positive (they possess a lysine decarboxylase) while, with rare exceptions, *Citrobacter* cultures are ninhydrin negative. The test is also of value in the differentiation of shigellae and anaerogenic nonmotile, *E. coli* cultures such as the *Alkalescens-Dispar* biotypes. Shigellae are ninhydrin negative and the majority of *Alkalescens-Dispar* cultures are ninhydrin positive (Edwards, Fife, and Ewing, 1956). Other substances present in the peptone of the medium may be reacted upon in such a way as to give a positive ninhydrin test. For this reason the ninhydrin test cannot be recommended for use in other areas of the family.

P. R. Edwards, M. A. Fife and W. H. Ewing (1956). Am. J. med. Technol. 22, 28.
S. Falkow (1958). Am. J. clin. Path. 29, 598.
V. Moeller (1955). Acta path. microbiol. scand. 36, 158.

Amino acid decarboxylases

G. S. Wilson and A. A. Miles (1964). Topley and Wilson's *Principles of Bacteriology and Immunity,* 5th edition – Arnold.

Tests for the decarboxylation of lysine, arginine, ornithine, and glutamic acid for use in the differentiation of the enterobacteria were described by Moeller (1954a, b, 1955), to whose papers reference should be made (see also p. 819, Chap. 27).

V. Moeller (1954a). Acta path. microbiol. scand. 34, 102.
(1954b). Ibid. 35, 259.
(1955). Ibid. 36, 158.

Amino acid decarboxylases—*Malleomyces, Pseudomonas*

Bach-Toan-Vinh (1965). Annls Inst. Pasteur, Paris 109, 460–463.

Recherche d'une lysine décarboxylase. – Méthode de Thibault et Le Minor [8] et méthode de Moeller [7].

Recherche de l'arginine décarboxylase. – Méthode de Gasser [3] et méthode de Moeller [7].

Recherche de l'ornithine décarboxylase. – Méthode de Moeller [7].

[3] Gasser (F.). Ann. Inst. Pasteur, 1964, 106, 778.
[7] Moeller (V.). Acta path. microbiol. scand., 1955, 36, 158.
[8] Thibault (P.) et Le Minor (L.). Ann. Inst. Pasteur, 1957, 92, 551.

Amino acid decarboxylases—*Bacteroides oralis*

W. J. Loesche, S. S. Socransky and R. J. Gibbons (1964). J. Bact. 88, 1329–1337.

Lysine, arginine, and ornithine decarboxylases were determined by the method of Ewing (1960).

W. H. Ewing (1960). *Enterobacteriaceae. Biochemical Methods for Group Differentiation.* U.S. Public Health Service Publ. 734.

Amino acid decarboxylases—*Alcaligenes*

R. G. Mitchell and S. K. R. Clarke (1965). J. gen. Microbiol. 40, 343–348.

Moeller's (1955) method was used.

V. Moeller (1955). Acta path. microbiol. scand. 36, 158.

Amino acid decarboxylases—*Providencia, Rettgerella*

C. Richard (1966). Annls Inst. Pasteur, Paris 110, 105–114.

Pour des raisons de commodité, les milieux ont été incubés à 37°C.

Recherche des décarboxylases de la lysine (LDC) et de l'ornithine (ODC), de la dihydrolase de l'arginine (ADH).

LDC: technique de Le Minor-Thibault à le ninhydrine [29].

ODC et ADH: par alcalinisation du milieu de Falkow [7] sans peptone (lectures limitées à quatre jours).

[7] Falkow (S.). Am. J. clin. Path., 1958, 29, 598.
[29] Thibault (P.) et Le Minor (L.). Ann. Inst. Pasteur, 1957, 92, 551.

Amino acid decarboxylases—RM bacterium

A. J. Ross, R. R. Rucker and W. H. Ewing (1966). Can. J. Microbiol. 12, 763–770.

Moeller's medium was used for decarboxylase tests and readings were made daily for 4 days.

Arginine decarboxylase—*Aeromonas shigelloides*

H. Leclerc and R. Osteux (avec la collaboration technique de Mme Larivière) (1966). Annls Inst. Pasteur, Paris 110, 737–754.

The authors' experiments show the existence of an arginine decarboxylase in *Aeromonas shigelloides.* Electrophoresis, study of CO_2 release and urea level have demonstrated the mechanism and kinetics of arginine splitting.

Conditions for decarboxylase activity have been defined: optimal pH and temperature, action of co-enzymes, inhibitors, velocity of the reaction according to the substrate concentration.

The best conditions (pH, temperature, culture medium) for the enzyme formation have been studied.

Arginine decarboxylase—*Staphylococcus*

M. Kocur, F. Přecechtěl and T. Martinec (1966). J. Path. Bact. 92, 331–336.

Splitting of arginine was studied by Moeller's (1955) method.

V. Moeller (1955). Acta path. microbiol. scand. 36, 158.

Glutamic acid decarboxylase

H. Proom and A. J. Woiwod (1951). J. gen. Microbiol. 5, 681–686.

The authors described a paper chromatographic method for detecting the γ-aminobutyric acid produced in cultures.

Glutamic acid decarboxylase

E. Rowatt (1955). J. gen. Microbiol. 13, 552–560.

Decarboxylation of glutamate. Washed suspensions (2–6 mg dry wt in 3.5 ml) were incubated in manometer flasks with 0.2 M-acetate buffer containing 0.0056 M-glutamate and 0.09% (w/v) cetavlon. CO_2 output was measured directly in air at 37°C.

Note: The author has asked that it be noted that the above method be credited to Dr. Krebs (see Biochem. J. 43, 1948, 51.)

Histidine decarboxylase—*Lactobacillus*

B. M. Guirard and E. E. Snell (1964). J. Bact. 87, 370–376.

The authors describe manometric methods for estimation of histidine decarboxylase.

Lysine decarboxylase—enterobacteria

P. R. Carlquist (1956). J. Bact. 71, 339–341.

After several trials, a medium consisting of pancreatic digest of casein (U.S.P. XIV)*, 15 g; dipotassium phosphate, 2 g; glucose, 1 g; and distilled water, 1000 ml, was found to be satisfactory for demonstration of lysine decarboxylase activity. Five ml of the medium is dispensed into screw-cap tubes (15 x 125 mm) and sterilized in the autoclave. The medium is inoculated with the organism under study and incubated at 37°C for 18 to 24 hours. After incubation 1 ml of 4 N NaOH is added and mixed. Two ml of chloroform is added and the tube shaken vigorously.

*Several different lots of Trypticase, BBL, and Casitone, Difco, were tested and all yielded uniform results.

The cadaverine is found in the chloroform phase after extraction. The tube is centrifuged to break the emulsion and separate sufficient clear chloroform extract for testing below the heavy layer of denatured protein present at the chloroform-medium interphase. One-half ml of the clear chloroform extract is removed with a pipette and placed in a 13 x 100 mm tube. To this is added 0.5 ml of 0.1 per cent ninhydrin (1,2,3-triketo-hydrindene) in chloroform. The reaction mixture is observed at room temperature for 4 minutes. A positive reaction is evidenced by the occurrence of a deep purple color. Incubation of the reaction mixture at higher temperature or for longer periods of time may result in the production of a false positive result because of the presence of minor ninhydrin-reacting compounds in the chloroform extract that were not identified.

This method was designed to separate the *Arizona* and *Bethesda-Ballerup* groups. This is also referred to as the ninhydrin test. It depends on the production of cadaverine by decarboxylation of the lysine present in the pancreatic digest of casein.

Lysine decarboxylase—*Vibrio*

S. Szturm-Rubinsten, D. Piéchaud and M. Piéchaud (1960). Annls Inst. Pasteur, Paris 99, 309–314.

Moeller [4] et Carlquist [5] ont montré l'intérêt de la recherche de la lysine-décarboxylase pour les entérobactéries. Edwards et coll. [6] recommandent le test à la ninhydrine de Carlquist pour différencier le groupe *Alkalescens-Dispar* des *Shigella.* Thibault et coll. [7] proposent une méthode simple de recherche de la lysine-décarboxylase pour la différenciation rapide des entérobactéries. C'est cette dernière que nous avons utilisée pour rechercher la lysine-décarboxylase de nos souches.

[4] Moeller (V.). Acta path. microb. scand., 1954, 35, 259.
[5] Carlquist (P. R.). J. Bact., 1956, 71, 339.
[6] Edwards (P. R.), Fife (M. A.) et Ewing (W. H.). Amer. J. Med. Techn., décembre 1956, 28.
[7] Thibault (P.) et Le Minor (L.). Ann. Inst. Pasteur, 1957, 92, 551.

Lysine decarboxylase—*Vibrio*

G. H. G. Davis and R. W. A. Park (1962). J. gen. Microbiol. 27, 101–120.

Lysine decarboxylase activity. Method of Carlquist (1956). Medium: Basal medium plus 0.5% (w/v) lysine hydrochloride. Tested after 5 days. P. R. Carlquist (1956). J. Bact. 71, 339–341.

Lysine decarboxylase—*Moraxella*

W. J. Ryan (1964). J. gen. Microbiol. 35, 361–372.

The micromethod of C. Shaw and P. H. Clarke (J. gen. Microbiol. 13, 155, 1955) was used to detect lysine decarboxylase.

Positive and negative controls were included.

Lysine decarboxylase—*Hyphomicrobium neptunium*

E. Leifson (1964). Antonie van Leeuwenhoek 30, 249–256.

The liquid medium consisted of peptone (Casitone, Difco) 0.2%, yeast extract 0.1%, tris buffer (tris(hydroxymethyl) amino methane) 0.05%, artificial seawater half strength, pH 7.5. To the base medium was added 0.1% L-lysine hydrochloride for the lysine decarboxylase test and the final pH compared with that of a control without added lysine.

Lysine and ornithine decarboxylases—*Vibrio comma*

J. C. Feeley (1965). J. Bact. 89, 665–670.

These were determined in Falkow's medium with mineral oil seal (Ewing, Davis and Edwards, 1960).

W. H. Ewing, B. R. Davis and P. R. Edwards (1960). Publ. Hlth Lab. 18, 77–83.

Lysine decarboxylase—*Erwinia*

A. von Graevenitz and A. Strouse (1966). Antonie van Leeuwenhoek 32, 429–430.

The authors used the method of W. H. Ewing, B. R. Davis and P. R. Edwards. Publ. Hlth Lab. 18, 77–83.

Lysine decarboxylase—*Brevibacterium*

R. Chatelain and L. Second (1966). Annls Inst. Pasteur, Paris 111, 630–644.

Lysine-décarboxylase: recherchée après 2 jours de culture selon la technique de Thibault et Le Minor [22].

La température d'incubation des cultures est de 30°C.

[22] Thibault (P.) et Le Minor (L.). Ann. Inst. Pasteur, 1957, 92, 551.

Ornithine decarboxylase—enterobacteria

J. G. Johnson, L. J. Kunz, W. Barron and W. H. Ewing (1966). Appl. Microbiol. 14, 212–217.

Ornithine decarboxylase test medium consisted of Decarboxylase Medium (Difco) with 1% ornithine and 0.3% agar. This semisolid medium was inoculated by a single stab and was incubated no longer than 24 hr; it did not require a paraffin oil seal. Results of decarboxylase tests with this medium compared favorably with those obtained with Moeller's medium.

Ornithine decarboxylase—enterobacteria

C. Richard (1966). Annls Inst. Pasteur, Paris 110, 114–119.

Pour mettre en évidence l'O.D.C. des bactéries, nous avons utilisé un milieu simple, dérivé du milieu classique de Falkow [2] par suppression de la peptone. Selon Taylor [10], en effet, l'emploi de peptone comme source d'azote pourrait, à la faveur d'une ammoniogénèse, alcaliniser les milieux, même si l'amino-acide n'est pas dégradé, et conduire à des résultats faussement positifs.

Sur une centaine de souches appartenant à diverses espèces d'*Enterobacteriaceae,* nous avons vérifié que le milieu adopté et désigné comme "milieu de Falkow à l'ornithine sans peptone" (en abrégé milieu F.O.S.P.) fournissait, en ce qui concerne la recherche de l'O.D.C. les mêmes résultats que les milieux classiques de Moeller et de Falkow [7, 2].

Matériel et méthodes.

1° Souches. – Les 297 souches examinées proviennent toutes de la collection du Service des Entérobactéries à l'Institut Pasteur de Paris, où elles avaient été étudiées (sérotypes, biotypes et lysotypes) [4, 5, 6].

Ces souches appartiennent à 9 sérotypes OB différents, soit 26:B_6 (29 souches); 55:B_5 (67 souches); 86:B_7 (10 souches); 111:B_4 (71 souches); 119:B_{14} (10 souches); 125:B_{15} (34 souches); 126:B_{16} (7 souches); 127:B_8 (53 souches); 128:B_{12} (16 souches), subdivisés sur la base de leur antigène H ou éventuellement de leur biotype.

2° Milieu de Falkow a l'ornithine sans peptone (milieu F.O.S.P.) pour la recherche de l'ornithine décarboxylase. – Ce milieu est préparé comme suit:

L-ornithine (dichlorhydrate) pure pour analyse	5 g
Extrait de levure Difco	3 g
Glucose pur	1 g
Solution de pourpre de bromocrésol à 1.6 g pour 100 ml d'alcool à 90°	1 ml
Eau distillée	1 ml

On ajuste le pH à 6,3 au moyen de soude normale pour corriger la forte acidité due au dichlorhydrate de L-ornithine.

Réparti à raison de 5 ml par tube (tube à vis 160 x 12,5), le milieu est stérilisé à l'autoclave pendant quinze minutes à 120°C. Après refroidissement il présente une teinte violet-pourpre pâle (zone de virage de l'indicateur: jaune à pH = 5,2, violet pourpre à pH = 6,8).

3° Ensemencement du milieu F.O.S.P. et lecture des résultats. – On ensemence avec un faible inoculum prélevé à partir d'une culture de 18 heures sur gélose nutritive; la surface du milieu est ensuite recouverte d'une couche de 5 mm d'huile de paraffine stérile. L'incubation est faite à l'étuve à 37°C. Les tubes sont examinés, chaque jour, pendant quatre jours. Les résultats sont notés de la manière suivante:

O.D.C. –: le milieu présente une teinte jaune après quatre jours.

O.D.C. +: teinte violette après un à deux jours.

O.D.C. (+): teinte violette après trois à quatre jours.

Rappelons brièvement le mécanisme de la décarboxylation d'un amino-acide dans les conditions opératoires indiquées. Dans un premier temps les bactéries font fermenter le glucose et acidifient le milieu (teinte jaune de l'indicateur). Cette acidité favorise la formation des aminoacide-décarboxylases. Si la bactérie possède une O.D.C., il se produit dans un second temps une décarboxylation de l'ornithine suivant la réaction:

$$NH_2{-}(CH_2)_3{-}CH\begin{matrix} \diagup COOH \\ \diagdown NH_2 \end{matrix} \rightarrow CO_2 + NH_2{-}(CH_2)_4{-}NH_2$$

(ornithine) (putrescine)

Le milieu est alcalinisé et l'indicateur vire du jaune au violet.

[2] Falkow (S.). Amer. J. clin. Path., 1958, 29, 598.
[4] Le Minor (S.). Étude bactériologique *d'Escherichia coli* isolés au cours de gastro-entérites infantiles. *Thèse Pharmacie,* Paris, 20 juin 1953.
[5] Le Minor (S.), Le Minor (L.), Nicolle (P.) et Buttiaux (R.). Ann. Inst. Pasteur, 1954, 86, 204.
[6] Le Minor (S.), Le Minor (L.), Nicolle (P.), Drean (D.) et Ackermann (H.). Ann. Inst. Pasteur, 1962, 102, 716.
[7] Moeller (V.). Acta path. microbiol. scand., 1955, 36, 158.
[10] Taylor (W.). Appl. Microbiol., 1961, 9, 487.

Tyrosine decarboxylase—*Streptococcus*

W. D. Bellamy and I. C. Gunsalus (1945). J. Bact. 50, 95–103.

By control of the medium, a tyrosine decarboxylase enzyme free of its coenzyme (pyridoxal phosphate) has been produced during the growth of a culture of *Streptococcus faecalis* R.

Refer to the original for details.

Tyrosine decarboxylase—*Streptococcus*

P. Morelis and L. Colobert (1958). Annls Inst. Pasteur, Paris 95, 568–587.

L'activité tyrosine-décarboxylasique. – Gale [12], en 1940, a montré que certaines souches étaient capables de transformer la tyrosine en tyramine par décarboxylation. Divers auteurs ont pu attribuer à cette amine certaines intoxications alimentaires en rapport avec la présence d'entérocoques dans les semi-conserves de viande [10].

Cette activité décarboxylante peut être suivie grâce aux variations du pH qu'elle suscite; en effet, dans un milieu acidifié par la fermentation de glucides, le départ du gaz carbonique et la formation d'une amine soluble provoquent secondairement une forte élévation du pH [29].

Pour pratiquer cette épreuve nous avons utilisé le milieu suivant: peptone Vaillant 5B, 10 g; mannitol, 1 g; glucose, 2 g; eau, 1000 cm^3; on a ajouté de la tyrosine de façon à obtenir la concentration de 0,5 g p. 100*; le bleu de bromothymol servit d'indicateur coloré. Le milieu fut réparti en tubes et autoclavé à 115°. Les tubes de milieu furent ensemencés à partir d'une culture de 24 heures sur bouillon.

[10] Dack (G. M.) et Niven (C. F.). J. inf. Dis., 1949, 85, 131.
[12] Gale (E. F.). Bioch. J., 1940, 34, 853.
[29] Sharpe (M. E.). Proc. Soc. appl. Bact., 1948, 11, 13.

*Cette concentration est trop élevée pour que la tyrosine se dissolve entièrement.

Tyrosine decarboxylase—*Streptococcus*

C. W. Langston, J. Gutierrez and C. Bouma (1960). J. Bact. 80, 693–695.

Tyrosine decarboxylase activity was determined by a procedure similar to the method of Sharpe (1948).

M. E. Sharpe (1948). Proc. Soc. appl. Bact. 19th Ann. Conf., 13–17.

Tyrosine decarboxylase—*Streptococcus*

R. H. Deibel, D. E. Lake and C. F. Niven, Jr. (1963). J. Bact. 86, 1275–1282.

Decarboxylation of tyrosine was determined by use of modified Eldredge tubes (Williams and Campbell, 1951). A yeast extract-Tryptone basal medium containing 1.0% glucose was employed, to which was added 0.5% L-tyrosine. Tyrosine decarboxylation was indicated by the abundance of precipitated barium carbonate from the tyrosine-containing cultures, but not from the inoculated basal medium. Where there was doubt, back-titrations of the barium hydroxide were performed for both the inoculated basal media and the tyrosine-containing media.

Cultures were incubated at 37°C.

O. B. Williams and L. L. Campbell (1951). Fd Technol., Champaign 5, 306.

TYROSINE BREAKDOWN

Tyrosine decomposition—*Mycobacterium*

R. E. Gordon and J. M. Mihm (1959). J. gen. Microbiol. 21, 726–748.

Decomposition of tyrosine: Each culture was streaked once across a plate containing approximately 20 ml of the following medium: peptone, 5 g; beef extract, 3 g; agar, 15 g; tyrosine, 5 g; distilled water, 1000 ml; pH 7.0. An even distribution of the crystals of tyrosine was ensured by placing 0.5 g of tyrosine in a 250 ml flask, adding 100 ml of the remaining components, autoclaving, cooling to 47°C and carefully mixing the agar before and during pouring. After 14 and 21 days of incubation at 28°C, the plates were observed for the disappearance of the crystals underneath and bordering the growth.

Tyrosine breakdown—*Vibrio*

G. H. G. Davis and R. W. A. Park (1962). J. gen. Microbiol. 27, 101–119.

Tyrosine breakdown, tested by the ability to dissolve tyrosine (0.5%, w/v) suspended in nutrient agar within 7 days.

AMINO-ACID DEHYDROGENASES

Alanine dehydrogenase—*Bacillus*

N. G. McCormick and H. O. Halvorson (1964). J. Bact. 87, 68–74.

The authors give a detailed description of the methods for purification and assay of L-alanine dehydrogenase of *Bacillus cereus.*

Arginine dihydrolase—*Pseudomonas aeruginosa*

R. R. Colwell (1964). J. gen. Microbiol. 37, 181–194.

Arginine dihydrolase was tested for by the method of Thornley (1960). M. J. Thornley (1960). J. appl. Bact. 23, 37.

Threonine dehydrogenase—*Eggerthella, Bacteroides, Fusiformis*

H. Beerens and Mme M. M. Tahon-Castel (1965). Annls Inst. Pasteur, Paris 108, 682–684.

Solution tampon phosphate de pH 7,6. – Mélanger:

Solution de PO_4H_2K à 9,075 g p. 1000 d'eau distillée: 132 ml.

Solution de $PO_4Na_2H,12H_2O$ à 23,9 g p. 1000 d'eau distillée: 868 ml.

Gélose. – Agar (Difco) 20 g, eau distillée 1000 ml. Dissoudre en chauffant. Répartir en tubes. Stériliser à l'autoclave.

Suspension microbienne. – 30 ml de milieu VL (viande, levure) liquide glucosé à 2 p. 1000 répartis en tubes à centrifuger de 45 ml, sont ensemencés avec 2 à 3 ml d'une culture jeune de la souche étudiée en milieu de Rosenow. On recouvre d'une couche d'huile de vaseline stérile. On incube à 37°. Au cours de la phase logarithmique de croissance, dès que la culture est suffisante, on centrifuge et on lave deux fois le culot microbien avec une solution de tampon phosphate de pH 7,6 régénéré; les bactéries sont finalement émulsionnées dans 1 ml de solution tampon; la mise en suspension est facilitée par introduction dans le tube de deux billes de verre. Ces opérations doivent être effectuées rapidement, en moins d'une heure.

Réaction. – Introduire dans un tube de 180 x 9 mm: suspension microbienne, 0,5 ml; solution de bleu de Nil (BDH) à 1/200 en tampon phosphate de pH 7,6 *préparée extemporanément,* 0,5 ml; solution de DL-thréonine à 10 p. 100 en tampon phosphate, 0,5 ml.

Témoin. – Introduire dans un tube de 180 x 9 mm: suspension microbienne, 0,5 ml; solution de bleu de Nil (BDH) à 1/200 en tampon phosphate de pH 7,6, 0,5 ml; solution de tampon phosphate pH 7,6, 0,5 ml.

Couler à la surface de ces suspensions, dans chaque tube, environ 2 ml de gélose. Incuber au bain-marie à 37°. La lecture s'effectue après dix-huit à vingt heures d'incubation.

La thréonine est utilisée lorsque la suspension témoin reste bleue et que la suspension additionnée de thréonine vire au jaune.

Elle ne l'est pas quand la suspension témoin et la suspension + thréonine restent bleues. La réaction est stable et ne subit pas de modification même après un séjour de vingt-quatre heures à 37°.

Arginine dihydrolase—*Pseudomonas*

R. Y. Stanier, N. J. Palleroni and M. Doudoroff (1966). J. gen. Microbiol. 43, 159–271.

Assay of the arginine dihydrolase system. We assayed this dihydrolase system as follows. The organisms, grown on mineral + lactate medium, were suspended in M/30 phosphate buffer (pH 6.8) to an optical density of 200 Klett units (green filter 54). Four ml of bacterial suspension in a test tube were gassed with N_2, and 1 ml of 10^{-3} M-arginine was added. The mixture was gassed again with N_2 and the tube stoppered and incubated at 30°C for 2 hr. At the end of the incubation, the tube was immersed in a boiling water bath for 15 min, and arginine determined in a 1 ml sample. As control, we used an identical suspension which was heated immediately after the addition of arginine, and thereafter subjected to the same treatment. Arginine was determined by the quantitative method of Rosenberg, Ennor and Morrison (1956).

H. Rosenberg, A. H. Ennor and V. F. Morrison (1956). Biochem. J. 63, 153.

AMMONIA FROM ARGININE

Ammonia from arginine—*Streptococcus*

C. F. Niven, Jr., K. L. Smiley and J. M. Sherman (1942). J. Bact. 43, 651–660.

Yeast extract	0.5 %
Tryptone	0.5 %
K_2HPO_4	0.2 %
Glucose	0.05%
d-arginine-HCl	0.3 %
pH 7.0	

The whole medium is autoclaved. A medium with 4.0% tryptone but no arginine serves as a control.

Incubate for 2 days at 37°C.

Test with Nessler's reagent in a spot plate.

Ammonia from arginine—*Streptococcus*

E. J. Hehre and J. M. Neill (1946). J. exp. Med. 83, 147–162.

The medium was 0.5 per cent tryptone peptone, 0.5 per cent bacto-yeast extract, 0.2 per cent K_2HPO_4, 0.05 per cent glucose, plus 0.3 per cent arginine*. The cultures were incubated 5 days and then tested with Nessler's solution. All of the strains reported as positive, reacted when diluted 1:50 or higher, whereas the negative fluids failed even with undiluted material.

*C. F. Niven, K. L. Smiley and J. M. Sherman (1942). J. Bact. 43, 651.

Ammonia from arginine—*Streptococcus*

Y. Abd-el-Malek and T. Gibson (1948). J. Dairy Res. 15, 233–248.

The method suggested by Niven, Smiley and Sherman (1942) was used except that the medium was modified, 0.5% yeast extract being replaced by 0.25% of the yeast autolysate, yeastrel.

In the earlier stages of the work, tests were made for the formation of ammonia in 4% peptone, but after the appearance of the paper by Niven *et al.* (1942) the arginine medium, which gives identical but clearer results, was adopted.

C. F. Niven, K. L. Smiley and J. M. Sherman (1942). J. Bact. 43, 651.

Ammonia from arginine—*Staphylococcus* and *Micrococcus*

Y. Abd-el-Malek and T. Gibson (1948). J. Dairy Res. 15, 249–260.

Niven, Smiley and Sherman's (1942) method as modified previously (1948) was used. Cultures were incubated for 3 days at 37°C or 5 days at 30°C. Staphylococci and micrococci may form small amounts of ammonia from the yeast extract of the medium, and doubtful reactions were controlled by parallel cultures in a medium differing only in containing no added arginine.

Y. Abd-el-Malek and T. Gibson (1948). J. Dairy Res. 15, 233.
C. F. Niven, K. L. Smiley and J. M. Sherman (1942). J. Bact. 43, 651.

Ammonia from arginine—*Staphylococcus*

J. B. Evans and C. F. Niven, Jr. (1950). J. Bact. 59, 545–550.

Arginine hydrolysis was detected by growing the culture for 3 days at 30°C in 1.0 per cent tryptone, 0.5 per cent yeast extract, 0.5 per cent sodium chloride, and 0.3 per cent arginine, and testing qualitatively for ammonia production with Nessler's reagent.

Ammonia from arginine—*Streptococcus*

P. F. Swartling (1951). J. Dairy Res. 18, 256–267.

The method of Niven, Smiley and Sherman (1942) was used with 0.25% yeastrel instead of the yeast extract in the original medium.

C. F. Niven, K. L. Smiley and J. M. Sherman (1942). J. Bact. 43, 651.

Ammonia from arginine

M. Briggs (1953). J. gen. Microbiol. 9, 234–248.

Tomato glucose broth

Tomato juice	10	ml
Neopeptone Difco	1.5	g
Glucose	2.0	g
NaCl	0.5	g
Tween 80*	0.1	ml
Yeastrel	0.6	g
Soluble starch	0.05	g
Distilled H_2O	100	ml
pH 6.8		

15 pounds/sq. in. for 20 minutes.
0.3% (w/v) L-arginine HCl is added before autoclaving.

A control medium was also used without arginine.

Nessler's solution was used after 14 days incubation and read immediately. (A greenish colour develops on standing.)

*polyoxyethylene – sorbitan monooleate

Ammonia from arginine—*Lactobacillus*

G. H. G. Davis (1955). J. gen. Microbiol. 13, 481–493.

Medium:

peptone, 0.5% (w/v); yeast extract, 0.3% (w/v); glucose, 0.3% (w/v); sodium acetate, 1% (w/v); L-arginine hydrochloride, 0.3% (w/v); Tween 80, 0.1% (v/v); Salts A, 0.5% (v/v); Salts B, 0.5% (v/v); pH 7.4, dispensed in 4 ml quantities in 1/4 oz (bijou) bottles and steamed 20 min or 3 consecutive days to sterilize. Tests were inoculated with 1 to 2 drops, incubated for 10 days and tested by adding 1 ml of Nessler's reagent to detect ammonia.

For Salts A and Salts B see A. C. Hayward (1957). J. gen. Microbiol. 16, 9–15.

Ammonia from arginine—*Pseudomonas*

J. C. Sherris and J. G. Shoesmith (1959). J. gen. Microbiol. 21, 389–396.

The qualitative method of Moeller (1955). Medium containing (wt/L): Evans' peptone, 5 g; Lab Lemco, 5 g; pyridoxal, 0.005 g; glucose, 0.5 g; L-arginine monohydrochloride, 10 g; bromcresol purple, 0.01 g; and cresol red, 0.005 g; adjusted to pH 6.0, was distributed in narrow tubes to give columns 2 cm in height. After autoclaving (10 lb/sq. in. for 10 min) the tubes were sealed with sterile liquid paraffin. A tube of this medium and a control without arginine were inoculated with a straight wire from a nutrient agar culture. Organisms which produced acid from glucose under the conditions of the test rapidly turned the control tube yellow, and those which did not attack arginine also gave a yellow colour in the arginine containing medium. With organisms, such as pseudomonads, which did not produce acid from glucose in Moeller's medium, the control tube remained slate grey in colour for the first 1–2 days, and sometimes became faintly mauve on further incubation. Arginine breakdown was indicated by the appearance of a violet colour in the tube containing the test medium.

All tests on members of the *Enterobacteriaceae* were carried out only at 37°C; other organisms were tested at 22°C and also at 37°C. Tubes were examined daily for 5 days. A definite violet colour in the test medium with a yellow slate grey or faintly mauve colour in the control tube was taken as evidence of arginine breakdown. After incubation for 4 days the tubes were tested for ammonia with Nessler's reagent as described by V. Moeller (1955, Acta path. microbiol. scand. 36, 158).

The Manchester qualitative technique. The workers in the University Dept. of Bacteriology developed a qualitative test for arginine breakdown, using a medium modified from that described for urease tests by W. B. Christensen (1946, J. Bact. 52, 461). It consisted of: peptone (Evans, 1.0 g; NaCl, 5.0 g; KH_2PO_4 (anhydrous), 2.0 g; phenol red, 2.5 ml of a 0.4% (w/v) solution in water; distilled water to 1 litre. The pH value was adjusted to 6.8 and glucose (1.0 g) and L-arginine monohydrochloride (10.0 g) were added. The medium was distributed in 5 ml amounts in 14 mm diameter tubes, and autoclaved at 10 lb/sq. in. for 10 minutes. Two drops (*c.* 0.04 ml) of a 24 hr peptone water culture were inoculated into a tube of this medium and into a control without added arginine. The tubes were incubated for 24 hr in most instances at 37°C but tests on organisms with a lower optimum growth temperature were carried out at 30°C or 22°C whichever was appropriate. The colour which developed after incubation was compared visually with that given by buffer solutions of pH 7.0, 7.4, 8.0 and 8.4 containing the same concentration of phenol red as the medium.

Ammonia from arginine—*Streptococcus* and *Lactobacillus*

C. W. Langston and C. Bouma (1960). Appl. Microbiol. 8, 212–222.

The method and medium described by Niven *et al.* (1944) was used to determine ammonia production from arginine. After 3 days of incubation, ammonia was detected by placing 3 or 4 drops of culture into a spot plate and adding 1 drop of Nessler's reagent.

C. F. Niven, Jr., K. L. Smiley and J. M. Sherman (1944). J. Bact. 43, 651–660.

Ammonia from arginine—*Leuconostoc*

E. I. Garvie (1960). J. Dairy Res. 27, 283–292.

The methods described by Abd-el-Malek and Gibson (1948) for the production of ammonia from arginine and for the production of gas from glucose were adopted.

Y. Abd-el-Malek and T. Gibson (1948). J. Dairy Res. 15, 233.

Ammonia from arginine—*Pediococcus*

H. L. Günther and H. R. White (1961). J. gen. Microbiol. 26, 185–197.

The method described by Niven, Smiley and Sherman (1942) was used.

Maintenance of stock cultures and methods of cultivation. For maintenance of stock cultures, preparation of inocula and in all experimental work, 'Oxoid' tomato juice (TJ) broth or tomato juice (TJ) agar, adjusted to pH 6.6, were used unless otherwise stated. The following were exceptions to this rule: for strain Tc. 1 sodium chloride (5%, w/v) was added to the medium; and for the aerococci glucose Lemco broth (Shattock and Hirsch, 1947) or glucose yeast extract (GY) agar (containing, as %, w/v; peptone, 1.0; Yeastrel, 0.3; glucose, 1.0; NaCl, 0.25; agar, 1.0; at pH 7.4) was used.

Cultures were incubated aerobically at 30°C with specified exceptions.

C. F. Niven, Jr., K. L. Smiley and J. M. Sherman (1942). J. Bact. 43, 651.

P. M. F. Shattock and A. Hirsch (1947). J. Path. Bact. 59, 495.

Ammonia from arginine—*Pseudomonas*

M. Véron (1961) Annls Inst. Pasteur, Paris 100, Suppl. 6, 16–42.

L'hydrolyse de l'arginine est recherchée selon la méthode originale de Sherris et Shoesmith; pour les details techniques, on pourra se reporter au rapport suivant (Buttiaux).

Sherris (J. C.) et Shoesmith (J. G.). J. gen. Microbiol., 1959, 21, 389.

Buttiaux (R.). Ann. Inst. Pasteur, 1961, Suppl. au numéro de juin, p. 43.

Ammonia from arginine—*Streptococcus*

W. E. Sandine, P. R. Elliker and H. Hays (1962). Can. J. Microbiol. 8, 161–174.

The method of Niven *et al.* (1942) was used.

C. F. Niven, Jr., K. L. Smiley and J. M. Sherman (1942). J. Bact. 43, 651–660.

Ammonia from arginine—*Vibrio*

G. H. G. Davis and R. W. A. Park (1962). J. gen. Microbiol. 27, 101–119.

Medium (%, w/v): Oxoid tryptone, 0.2; NaCl, 0.5; K_2HPO_4, 0.03; L-arginine-HCl, 0.5; glucose, 0.1; pH 7.1. Tested with Nessler reagent after 5 days.

Ammonia from arginine—*Pseudomonas odorans*

I. Málek, M. Radochová and O. Lysenko (1963). J. gen. Microbiol. 33, 349–355.

According to Moeller over 7 days at 28°C.

Ammonia from arginine—*Lactobacillus*

F. Gasser (1964). Annls Inst. Pasteur, Paris 106, 778–796.

Recherche de la production d'ammoniac à partir de l'arginine. Un milieu M.R.S.M.* ne contenant que 0,5 p. 100 de glucose et auquel on ajoute extemporanément 0,3 p. 100 d'arginine stérilisée par filtration sur bougie L_3 est incubé pendant vingt-quatre heures après inoculation. La culture est centrifugée, le surnageant rejeté, le culot microbien est lavé dans 10 ml de tampon Sörensen $M/20$ pH 6 et centrifugé à nouveau. Le culot lavé est repris par 1 ml du même tampon et distribué en deux parties égales dans deux tubes de Kahn. L'un d'eux reçoit 0,1 ml d'arginine à 10 p. 100. Ces tubes sont portés au bain-marie à 37°C pendant deux heures. Leur contenu est ensuite alcalinisé par 3 gouttes de NaOH 4*N* (ou de lessive de soude diluée de moitié) puis additionné de 4 gouttes de réactif de Nessler. Dans ces conditions, il n'apparaît aucun précipité ni aucune coloration dans les tubes témoins et les tubes négatifs, tandis que le précipité orange spécifique de l'ammoniac caractérise les tubes positifs.

La lecture peut être contrôlée après quelques heures de repos par l'examen du culot de sédimentation: uniquement microbien dans les tubes témoins et négatifs, ou de couleur orangée et abondant dans les tubes positifs.

*See **Basal medium for growth**—*Lactobacillus* p. 112

Ammonia from arginine—*Lactobacillus casei* var. *rhamnosus*

W. Sims (1964). J. Path. Bact. 87, 99–105.

Nessler's reagent was used to detect ammonia after the organisms had been grown for 3 days in Hiss's serum water containing 0.2 per cent arginine hydrochloride.

Ammonia from arginine—*Aerococcus catalasicus*

O. G. Clausen (1964). J. gen. Microbiol. 35, 1–8.

Cultures were incubated aerobically at 37°C unless otherwise stated.

Hydrolysis of arginine in broth accompanied by formation of NH_3 was demonstrated by mixing equal parts of Nessler's reagent and 48 hr arginine-containing broth culture.

Ammonia from arginine—*Lactobacillus*

M. Gemmell and W. Hodgkiss (1964). J. gen. Microbiol. 35, 519–526.

Hydrolysis of arginine. Glucose (0.5%, w/v), arginine (0.3%, w/v) and agar (0.15%, w/v) were added to the basal medium.* Inoculation was done as in the fermentation tests.* One drop of each culture was tested

*See **Fermentation**—*Lactobacillus* p. 314

for ammonia with Nessler's reagent after incubation for 3 days. Although 0.5% glucose inhibited the production of ammonia from arginine by some other strains, the activity of the strains described here was not affected.

The basal medium which was used throughout with minor modifications had the following constituents in 1 L tap water: meat extract (Lab Lemco), 5 g; Evans peptone, 5 g; Difco yeast extract, 5 g; Tween 80, 0.5 ml; $MnSO_4 \cdot 4H_2O$, 0.1 g; potassium citrate, 1 g; pH 6.5.

Ammonia from arginine—*Pediococcus*

E. Coster and H. R. White (1964). J. gen. Microbiol. 37, 15–31.

The authors used the methods of H. L. Günther and H. R. White (J. gen. Microbiol. 26, 185, 1961). Their Tomato Juice (TJ) broth or agar with the addition of Tween 80 (0.1%, v/v), pH adjusted to 6.6 were used in maintenance and preparation of inocula with certain specified exceptions.

Transfers for preparing inocula were made every 24 hours except for *Pediococcus halophilus* (48 hrs) and some brewing strains (fortnightly).

All cultures were incubated aerobically at 30°C except for 'brewing strains' which were incubated in an atmosphere of 95% (v/v) hydrogen and 5% (v/v) carbon dioxide at 22°C.

Brewing strains were incubated for 21 days.

Ammonia from arginine—*Streptococcus*

R. Whittenbury (1965). J. gen. Microbiol. 38, 279–287.

The method of Niven, Smiley and Sherman (1942) was routinely used with the exception that Tween 80 was included in the medium. In some studies liquid medium containing L-arginine monohydrochloride 0.3% (w/v), with or without glucose 0.1 or 1% (w/v), adjusted to pH 6.5, was used. Ammonia production was determined with Nessler's reagent on a spot plate and glucose disappearance with Benedict's reagent. Cultures were held at 37°C in a water bath. Growth was measured nephelometrically at 10–15 min intervals at the end of the lag phase; pH values were measured electrometrically.

The basal liquid medium contained meat extract (Lab-Lemco), 0.5 g; peptone (Evans), 0.5 g; yeast extract (Difco), 0.5 g; Tween 80, 0.05 ml; $MnSO_4 \cdot 4H_2O$, 0.01 g; in 100 ml tapwater, adjusted to pH 6.5 and autoclaved at 121°C for 15 min.

C. F. Niven, K. L. Smiley and J. M. Sherman (1942). J. Bact. 43, 651.

AMMONIA FROM ASPARAGINE

Ammonia from asparagine—*Xanthomonas*

F. A. Wolf (1924). J. agric. Res. 29, 57–68.

Stock agar plus 1 per cent of asparagine was used in these tests. The indicator consisted of a sufficient quantity of 4 per cent solution of rosolic acid, and sufficient NaOH was added to give the medium a decided orange color and a reaction of about pH 6.0. In poured plate cultures incubated for about 10 days, the orange color gave way to a beautiful brilliant red. This change begins with a halo around each colony and comes to involve the entire plate. The change in color is due to the liberation of ammonia in the decomposition of asparagine as a result of the activity of the enzyme amidase.

AMMONIA FROM ORGANIC NITROGEN

Ammonia from organic nitrogen—anaerobes

A.-R. Prévot (1966). Techniques pour le diagnostic des Bactéries Anaérobes. Éditions de la Tourelle, St. Mandé.

Ammoniac: l'ammoniac étant un produit constant du métabolisme des germes anaérobies, sa mise en évidence qualitative n'offre pas d'intérêt; son dosage quantitatif s'effectue par la méthode de Foremann dont la technique sera indiquée plus loin.

Toutes ces réactions sont faites immédiatement après avoir obtenu le distillat mais les résultats définitifs ne sont notés comme positifs, faibles ou négatifs, que le lendemain quand les précipités recherchés ont eu le temps de se déposer.

Dosage de l'ammoniac: Le dosage s'effectue selon la méthode de Foremann, par entraînement à la vapeur d'eau. Verser successivement dans un ballon à fond rond de 250 cm³ et à long col, les produits suivants:

alcool à 96°	80 cm³
filtrat de culture	5 cm³
solution alcoolique de phtaléine du phénol	11 gouttes
solution de soude à 10 p. 100	QS

jusqu'à virage au rouge franc du mélange, puis en ajouter un léger excès. Recueillir le distillat, sortant d'un réfrigérant descendant dans un vase à précipiter de 250 cm³ contenant quelques cm³ d'eau distillée et 3 à 4 gouttes d'une solution d'alizarine sulfonate de soude à 5 p. 1000. Le distillat doit tomber directement dans ce liquide par un tube plongeant. Le mélange à distiller doit rester rouge pendant toute la durée de l'opération. Quand l'alizarine vire au violet (brutalement), continuer encore la distillation pendant 10 minutes. Retirer alors le tube plongeant dans le distillat et y doser l'ammoniac avec de l'acide sulfurique N/10 qu'on laisse tomber goutte à goutte au moyen d'une burette graduée jusqu'au virage au jaune paille de l'alizarine. Soit n le nombre de cm³ d'acide sulfurique N/10 utilisés: la quantité d'ammoniaque, exprimée en g pour

1000 cm^3 de culture, est donnée par la formule suivante:

$$\frac{17 \times n}{10 \times 5} \text{ ou plus simplement } \frac{34 \times n}{100}.$$

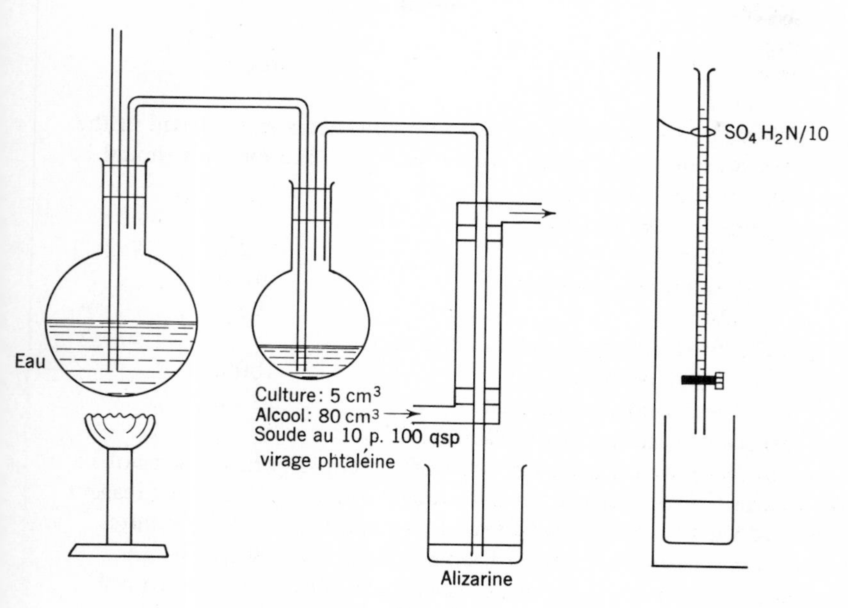

Fig. 1
Dosage de l'ammoniac

AMMONIA FROM PEPTONE

Ammonia from peptone—bacterial spot of tomato

M. W. Gardner and J. B. Kendrick (1921). J. agric. Res. 21, no. 2, 123–156.

Cultures in 5 per cent peptone bouillon after 5 days' incubation at 22°C, tested for ammonia with Nessler's reagent.

Ammonia from various substrates—*Xanthomonas*

F. R. Jones and L. McCulloch (1926). J. agric. Res. 33, 493–521.

Cultures grown in beef bouillon, nitrate bouillon, yeast agar, milk, whey agar, and various other media of various ages, gave negative results in ammonia tests made with Nessler's solution.

Ammonia from peptone—various

P. Arne Hansen (1930). J. Bact. 19, 223–229.

The author recommends the growth of the organism on a slant of agar of the following composition –

Peptone	40.0 g
Glucose	2.0 g
K_2HPO_4	5.0 g
Agar	15.0 g
Water	1000 ml

pH 7.5

The following test, designed for use on liquid media can be used in the same general way for solid media. Growth on the solid medium should be suspended in the first reagent.

The reagents for this test are:

I.	Thymol	2 g
	Sodium hydroxide 2 N	10 ml
	Water	90 ml
II.	Bromine water (saturated at room temperature)	100 ml
	Sodium hydroxide 2 N	35 ml

The test may be carried out as follows:

One ml of reagent I is added to 5 ml of the liquid (which must not be acid) to be tested for ammonia and thoroughly mixed. One ml of reagent II is then added and shaken and allowed to stand for 15 to 20 minutes. If ammonia is present the mixture becomes blue or greenish blue. A small amount of ether is added, shaken well and allowed to stand until the ether rises to the top. It is important to add the two solutions separately and not to add the ether until the color has developed. The blue color is extracted by means of the ether in which it is soluble, resulting in a deep red-violet color.

Neither this test nor the Nessler method will give a reaction in acid media. The acidic media should be made either alkaline or neutral, by the addition of alkali.

Ammonia from peptone—*Fusobacterium*

E. H. Spaulding and L. F. Rettger (1937). J. Bact. 34, 535–548.

The authors used:

Proteose peptone	10 g
Liebig's meat extract	3 g
Potato extract (aq)	100 ml
Distilled water	900 ml

pH 7.6

The Thomas test was used for ammonia but no time was specified.

Ammonia from peptone—*Streptococcus*

C. E. Safford, J. M. Sherman and H. M. Hodge (1937). J. Bact. 33, 263–274.

sont extraites à l'éther privé de peroxyde dans un extracteur continu, pendant vingt-quatre à quarante-huit heures. Dans le ballon à distiller, on a ajouté à l'éther d'extraction 1 ml d'une solution aqueuse d'acide phosphorique à 4 p. 100.

Après le temps d'extraction la solution acide aqueuse est soustraite, diluée à l'eau et amenée à pH 7, puis saturée en mélange PO_4K_3, SO_4Na_2, et les amines sont extraites dans un petit volume de *n*-butanol. Celui-ci est alors séché sur SO_4Na_2, filtré et reçoit un barbotage suffisant de ClH gazeux pour salifier les amines. Le butanol est alors concentré sous vide et amené à un volume tel que 1 ml de milieu de culture correspond à 10 µl d'extrait, qui est chromatographié en phase descendante avec le système: butanol, 40; acide acétique, 10; eau, 50. L'isobutylamine produit de décarboxylation de la valine a un R_F de 0.57; la β-méthylbutylamine provenant de la leucine, un R_F de 0.77.

H. Proom et A. J. Woiwod. J. gen. Microbiol.,1949, 3, 319.

Amines—volatile—anaerobes

A.-R. Prévot (1966). Techniques pour le diagnostic des Bactéries Anaérobes. Éditions de la Tourelle, St. Mandé.

Dès que la souche est obtenue en culture rigoureusement pure, on ensemence une fiole d'Erlenmeyer contenant 750 cm^3 de bouillon VF glucosé à 10 p. 1000. et privé d'air. On laisse à l'étuve de 5 à 8 jours, jusqu'à la fin de la fermentation (dépôt des germes sur le fond du récipient). On décante le liquide clair, on le passe sur papier filtre Laurent avec terre d'infusoire et sur le filtrat on opère ainsi:

Distillation alcaline. prendre 200 cm^3* du filtrat que l'on amène à pH 8.2 avec de la lessive de soude au 1/10e (virage au rouge du rouge de phénol). Les verser dans un ballon de 1000 cm^3 et ajouter 8 à 10 g de talc comme antimousse. Distiller pendant environ soixante minutes jusqu'à obtention de 30 cm^3 de distillat environ et en ne dépassant pas le moment où l'alcalinité du distillat cesse. La distillation doit se faire avec un réfrigérant descendant et le distillat doit être recueilli dans une fiole d'Erlenmeyer entourée de glace (Fig. 2).

Le produit à distiller mousse en général très abondamment, aussi convient-il de régler le chauffage et de refroidir le col du ballon par un courant d'air de telle sorte que la mousse ne l'atteigne pas. Si néanmoins cet incident se produisait, il faudrait arrêter immédiatement la distillation, rincer le réfrigérant et recommencer une nouvelle distillation en chauffant plus modérément. Le distillat contient des produits volatils neutres ou alcalins: amines, alcools, aldéhydes, acétone, phénols, indole, scatole, crésols, ammoniac. On met en évidence ces différents produits de la façon suivante.

*Au cas où l'on prendrait un volume plus grand de filtrat de culture il conviendrait d'en tenir compte dans le calcul de l'acidité volatile rapportée à 1 litre de bouillon de culture.

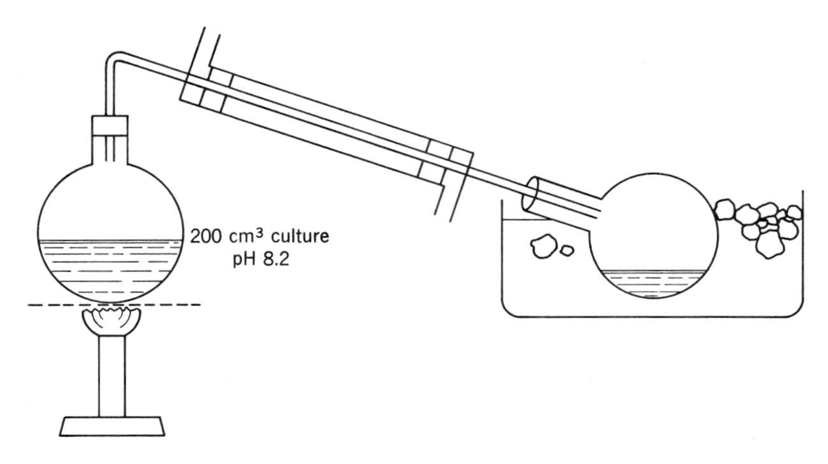

Fig. 2
Distillation alcaline

Amines. a) dans un tube à essais, verser 1 cm^3 de distillat et 1 cm^3 de réactif de Mayer. Chauffer jusqu'à ébullition. Refroidir brusquement. Une réaction positive se traduit par un trouble blanchâtre immédiat et un dépôt de cristaux brunâtres que l'on observe au bout de 24 heures.

b) dans un tube à essais, verser 1 cm^3 de distillat et 1 cm^3 de réactif de Nessler. Une réaction positive se traduit par l'apparition d'une coloration jaune immédiate et d'un sédiment jaune le lendemain.

Aminoacide-décarboxylases.

La multiplication des cantines collectives industrielles et scolaires a fait apparaître une éclosion d'intoxications alimentaires dont les laboratoires d'analyses et les laboratoires de toxicologie ont à s'occuper de plus en plus. Nous disons bien intoxications alimentaires, car les toxi-infections alimentaires dont les 2 principaux agents anaérobies sont *W. perfringens* et *Cl. botulinum* relèvent uniquement de la bactériologie et de l'immunologie (voir plus loin aux chapitres concernant ces espèces). Ce que les anciens toxicologues appelaient "ptomaïnes" et qu'ils devinaient être des produits de métabolisme bactérien d'aliments mal conservés sont maintenant bien connues dans leur constitution chimique, le mécanisme de leur formation et leur pharmacodynamie. Ce sont des amines toxiques résultant de la décarboxylation des acides aminés de la nutrition protéique par les amino-décarboxylases bactériennes. Les anaérobies sont particulièrement bien équipés en amino-decarboxylases et on trouve dans les aliments toxiques des quantités importantes d'histamine, de tyramine, de tryptamine, de β-phénylamine, de putrescine, de cadavérine, de mono-, di- et triméthyl-amines etc.

Pour détecter les amino-acides-décarboxylases d'une espèce anaérobie, il faut la cultiver sur un milieu riche en acides aminés capables d'être décarboxylés en amines toxiques: histidine, tyrosine, tryptophane, phénylalanine, ornithine, lysine etc. Le bouillon VF est particulièrement indiqué pour cela. On cultive la souche dans un Erlenmeyer de 1 litre et on y recherche les amines de décarboxylation par la technique de Blass et Sarraf.

ÉLECTROPHORÈSE

On se sert de l'appareil de Macheboeuf, Rebeyrotte et Dubert et des grands bacs de Lérès. La séparation des bases aminées sera réalisée dans les tampons suivants:

tampon de pH 2.4: acide acétique N;
tampon de pH 4 : phtalate acide de potassium 0.1 N, 100 ml, NaOH 0.1 N, 8 ml; eau q.s. pour compléter à 200 ml;
tampon de pH 4 : acide sulfosalicylique 0.05 M; on amène à pH 4 par addition de soude;
tampon de pH 8.6: véronal acide: 1.84 g véronal sodique, 10.3 g q.s. pour compléter à 1000 ml;
tampon de pH 10.0: ammoniaque 0.065 N (7 ml d'ammoniaque pur 22° dans 1 litre d'eau).

Seul le tampon de pH 10.0 (ammoniaque 0.065 N) avec du papier Whatman 3, pendant 4 heures a permis de séparer à partir du bouillon 4 amines fixes: la tyramine, l'histamine, la putrescine et la cadavérine.

On dépose les gouttes du bouillon sur une ligne distante de 8 cm du point médian, c'est-à-dire à 12 cm de l'extrémité anodique de la feuille. Les amines migrent vers la cathode beaucoup plus loin que les protides et les acides aminés.

Les gouttes à analyser sont déposées avec des micropipettes graduées: 5, 10, 15 mm^3; on sèche ces gouttes sur le papier avec un séchoir électrique après chaque 5 mm^3 pour que le diamètre de ces gouttes ne dépasse pas 5 mm; on installe la feuille sur le chevalet médian de l'appareil tout en laissant ses 2 extrémités tremper suffisamment dans le tampon, puis on l'humecte délicatement et complètement avec ce même tampon à l'aide d'une pipette de 10 ml, en prenant soin particulièrement de laisser couler le tampon goutte à goutte et loin des taches séchées du bouillon. On attend que ces taches soient mouillées par capillarité et on branche l'appareil.

Après écoulement du temps voulu, le papier est séché à l'étuve à 100° puis révélé par pulvérisation d'une solution de ninhydrine (tricétohydrindène) à 0.1 p. 100 dans du butanol saturé d'eau, additionné de 1 p. 100 d'acide acétique puis remis à l'étuve pendant quelques minutes pour obtenir des colorations violettes, bleues ou gris-violet suivant la nature de l'amine.

A côté du révélateur général des amines primaires et secondaires qui est la ninhydrine, nous avons utilisé des révélateurs spécifiques tels que:

1) *Le réactif de Pauly,* formule de Dent qui colore la tache de tyramine sur le papier d'électrophorèse en rouge orangé et l'histamine en rose foncé; on le prépare de la façon suivante:

– on mélange 5 cm^3 d'une solution d'acide sulfanilique à 1 p. 100 dans l'acide chlorhydrique N refroidi à 0° dans la glace avec 0.5 cm^3 d'une solution aqueuse de nitrite de sodium à 5 p. 100 et 40 cm^3 d'une solution aqueuse de carbonate de sodium à 6.5 p. 100. On prépare ce mélange juste avant l'usage et on le pulvérise immédiatement sur le papier d'électrophorèse séché. Les taches colorées apparaissent à l'instant et sont stables pendant longtemps.

2) *Le réactif d'Ehrlich,* formule de Smith donne une coloration pourpre avec la tryptamine; c'est une solution à 1 p. 100 de *p*-diméthylaminobenzaldéhyde dans un mélange de 90 cm^3 d'acétone et de 10 cm^3 d'acide chlorhydrique pur.

On prépare ce réactif et on le pulvérise immédiatement sur le papier d'électrophorèse séché; la tache pourpre de la tryptamine apparaît en 2 minutes à la température du laboratoire.

3) *Le réactif de Sakaguchi,* formule de Roche, donne une coloration rose (fleur de pêcher) avec l'agmatine, on pulvérise successivement sur le papier séché:

a) une solution renfermant 0.2 cm^3 NaOH à 40 p. 100; 0.2 cm^3 de solution alcoolique d'α-naphtol à 1 p. 100; 0.2 cm^3 de solution aqueuse d'urée à 40 p. 100 et 10 cm^3 de *n*-butanol;

b) une solution de BrONa provenant de la dilution au 1/5 de celle obtenue par l'addition de 0.9 cm^3 de brome à 100 cm^3 de NaOH à 10 p. 100.

La coloration des taches n'est stable qu'à condition d'employer des solutions finement pulvérisées et d'opérer la dessiccation rapide et complète à froid après la pulvérisation de chacun de ces 2 réactifs.

4) *Le réactif de Chargaff* pour la révélation des amines tertiaires qui ne se révèlent pas avec la ninhydrine; il donne sur le papier des taches bleu foncé sur un fond bleu clair; on révèle avec ce réactif de la façon suivante:

1. On pulvérise une solution aqueous d'acide phosphomolybdique à 2 p. 100 sur le papier.
2. On lave par immersion dans le butanol pendant 5 minutes.
3. On lave dans l'eau courante pendant 5 minutes.
4. On plonge ensuite dans une solution de chlorure stanneux à 0.4 p. 100 dans l'acide chlorhydrique 3 N.

Les taches bleu foncé apparaissent immédiatement sur un fond clair.

La mobilité d'une amine, c'est-à-dire la distance parcourue par ce corps sur le papier à électrophorèse dépend de la nature de l'amine, du pH du tampon, de la durée de l'électrophorèse; mais il faut noter qu'elle n'est pas la même si cette amine est en solution pure ou en présence d'excès d'autres substances.

Pour rechercher la mobilité des amines dans le bouillon, la technique des témoins latéraux a été utilisée; sur le papier à électrophorèse est mis successivement, le bouillon pur qui ne contient pas initialement d'amines puis le même bouillon additionné des différentes amines témoins.

Pour avoir des résultats valables, il faut que chaque expérience d'électrophorèse, de chromatographie ou de chromatophorèse soit accompagnée et contrôlée par des témoins internes et latéraux.

Chromatographie sur papier.

La séparation des amines fixes se fait par la chromatographie ascendante et descendante dans le solvant butanol acide acétique, eau 125 : 30 : 125, formule de Woiwod. On laisse migrer le solvant jusqu'au bord de la feuille, on la sèche à l'étuve à 100°, ensuite on la développe avec la ninhydrine acétique ou avec un révélateur spécifique approprié comme indiqué précédemment pour l'électrophorèse sur papier.

Voici les R_F de ces amines en solutions pures sous forme de chlorhydrate sur papier Whatman 3:

Putrescine	0,12
Cadavérine	0,15
Histamine	0,15
Agmatine	0,18
Tyramine	0,1
Tryptamine	0,66

Pratiquement si on chromatographie un mélange de ces 6 amines, on obtient 2 grosses taches, la première est formée par la putrescine, la cadavérine, l'histamine et l'agmatine et la deuxième par la tyramine et la tryptamine.

Voici les R_F de quelques amino-acides déposés sur le même chromatogramme:

Lysine	0,13
Arginine	0,17
Alanine	0,32
Tyrosine	0,43
Phénylalanine	0,58

De ce qui précède, on déduit que la chromatographie seule ne permet pas d'identifier les amines dans un bouillon de culture puisqu'il contient des acides aminés et des peptides qui ont des R_F très voisins des R_F des amines; elle ne permet pas non plus, en l'absence d'amino-acides, la résolution complète d'un mélange de bases aminées.

Chromatoionophorèse.

Pour la chromatoionophorèse, une feuille de papier Whatman n° 3 de

57 cm x 46,5 cm est utilisée. On effectue dans la première dimension une chromatographie descendante et dans la deuxième dimension une électrophorèse sur papier, le sens du courant étant perpendiculaire au sens d'écoulement du solvant.

On dispose la solution à analyser à 7 cm de la ligne médiane MN, vers le bord qui sera plongé dans le compartiment anodique et à 11 cm du bord AB. On trace les lignes EF et GH suivant la largeur de l'appareil à électrophorèse qui doit recevoir ces bandes.

Le solvant utilisé pour la chromatographie est le mélange butanol, acide acétique-eau 125 : 30 : 125 phase supérieure; on met la phase inférieure dans un petit bécher pour saturer l'atmosphère du bac de chromatographie.

On laisse migrer le solvant généralement 24 heures jusqu'à ce qu'il commence à s'écouler du papier; on sèche alors le papier dans une étuve à 100° suffisamment pour chasser les dernières traces d'acide acétique.

La bande CDEF découpée est placée dans l'appareil à électrophorèse en utilisant un des tampons cités plus haut, les tampons pH 8,6 ou 10,8 de préférence et en mouillant le papier comme indiqué dans le chapitre de l'électrophorèse.

La révélation peut être faite avec le révélateur général qui est la ninhydrine et les révélateurs cités plus haut.

Amines volatiles.

Les amines volatiles suivantes en solutions aqueuses de leurs chlorhydrates 0,01 M peuvent être séparées:

Amines primaires: méthylamine, éthylamine, propylamine, butylamine, isobutylamine, isoamylamine, phényléthylamine et allylamine.

Amines secondaires: diméthylamine (en solution 0,1 M) et diisoamylamine.

Amines tertiaires: triméthylamine (en solution 0,2 M) et triéthylamine (en solution 0,02 M).

On dépose sur le papier 5 à 10 mm^3 de ces solutions. Le distillat alcalin du bouillon à analyser qui est présumé contenir les amines volatiles est recueilli dans un récipient refroidi à 0° neutralisé par l'acide sulfurique en excès, avec la phénolphtaléine comme indicateur, puis évaporé à sec sur un bain-marie; le résidu blanc qui est le sulfate des diverses amines volatiles présentes dans ce distillat est dissout dans 5 ml d'eau distillée; ainsi ces amines volatiles sont assez concentrées pour permettre leur identification par la chromatographie et l'électrophorèse. On décèle ainsi des quantités allant jusqu'à 2 μg. On révèle les amines primaires et secondaires avec une solution de ninhydrine à 0,1 p. 100 dans le butanol saturé d'eau additionnée de 1 p. 100 d'acide acétique par le même procédé de révélation cité plus haut pour les amines fixes; mais en chauffant plus longtemps, pour révéler les couleurs des amines secondaires, on obtient des colorations violettes, bleues ou rouges suivant la nature de l'amine.

Les amines tertiaires sont révélées par le réactif de Chargaff.
CHROMATOGRAPHIE SUR PAPIER.
Amines primaires et secondaires.

On peut séparer et identifier ces amines avec la chromatographie descendante sur le papier W 1 tamponne avec l'acétate de soude à 8,2 g par litre; avec ce papier tamponné on empêche la formation des taches allongées et difficilement identifiables on obtient ainsi des taches parfaitement rondes et denses; pour tamponner le papier, on le plonge dans la solution tampon et on le sèche immédiatement dans l'étuve à 100°. Le solvant est le mélange butanol – acide acétique – eau 125 : 30 : 125; pour saturer l'atmosphère, on met la phase inférieure (aqueuse) dans un récipient à l'intérieur du bac de chromatographie.

Les solutions des amines obtenues par distillation des milieux de culture comme indiqué plus haut sont déposées sur le papier Whatman n° 1 tamponné en quantités variables entre 5 mm^3 et 20 mm^3; on ne doit pas déposer sur le papier plus de 5 mm^3 à la fois pour ne pas avoir de taches d'un diamètre dépassant 0,5 mm; les dépôts supérieurs forment sur le chromatogramme des taches trop diffuses; il est préférable qu'on sèche le papier avec un séchoir électrique après chaque application de 5 mm^3. On arrête la chromatographie quand le solvant migre jusqu'au bord inférieur du chromatogramme, le papier est ensuite séché à l'étuve à 100°, puis pulvérisé avec la ninhydrine acétique indiquée plus haut et remis à l'étuve de nouveau 5 à 15 minutes pour développer les couleurs.
Amines tertiaires.

Les triméthylamine et triéthylamine sont chromatographiées dans le mêmes conditions que les amines primaires et secondaires sur Whatman n° 1 tamponné avec le solvant butanol-acétique, mais révélées avec le révélateur de Chargaff puisqu'elles ne se révèlent pas avec la ninhydrine; elles ont les R_F suivants: triméthylamine 0,21; triéthylamine 0,68.
ÉLECTROPHORÈSE SUR PAPIER.
Amines primaires et secondaires.

L'électrophorèse est pratiquée comme pour les amines fixes sur papier Whatman n° 1 et révélée par la ninhydrine-acétique.

Dosage des amines. Divers procédés peuvent être appliqués pour l'évaluation quantitative des amines dans le bouillon de culture.

1) *Evaluation directe.* Elle consiste à comparer visuellement l'intensité et l'ordre de grandeur des taches obtenues par électrophorèse et chromatographie avec des quantités croissantes de témoins.

Avec cette méthode, on peut déterminer des quantités d'amines allant jusqu'à ± 2 μg ce qui donne une évaluation finale dans un litre de bouillon de ± 20 p. 100 dans des conditions moyennes.

En faisant varier les quantités déposées du distillat et des solutions témoins, on peut arriver à avoir des taches de la même intensité et du même ordre de grandeur jusqu'à ± 1 μg et on calcule la quantité d'amine dans 1 litre du bouillon avec une précision de ± 20 p. 100.

2) Dosage spectrophotométrique après élution. On fait éluer la partie du papier d'électrophorèse ou de chromatographie présumée contenir l'amine; sur cet éluat, on fait agir un réactif approprié et on mesure sa densité optique au spectrophotomètre; parallèlement, on prépare une courbe de concentration de cette même amine rigoureusement dans les mêmes conditions et en reportant la densité optique de l'échantillon étudié sur cette courbe, on a quantitativement cette amine dans l'échantillon et on calcule la quantité dans 1 litre de bouillon.

Elution. On fait une électrophorèse ou une chromatographie après avoir déposé plusieurs gouttes l'une à côté de l'autre de la solution étudiée, c'est-à-dire du bouillon ou du distillat, puis on révèle 2 bandes étroites aux 2 extrémités, supérieure et inférieure de ce papier pour localiser l'emplacement de l'amine ou des amines qu'on veut doser; ensuite on découpe la bande de papier qui contient des amines.

Ces bandes découpées peuvent être éluées par immersion dans l'eau distillée avec agitation mécanique; ou mieux, on élue par capillarité en opérant de la manière suivante: dans un vase clos de chromatographie, on plonge dans l'eau distillée l'extrémité supérieure de cette bande découpée et on recueille de l'extrémité inférieure, l'eau qui tombe goutte à goutte par capillarité; on laisse l'opération continuer pendant 1-4 jours et avant de l'arrêter, on s'assure que l'eau qui coule ne contient plus d'amines, avec le réactif de Pauly, pour l'histamine et la tyramine et la ninhydrine pour les autres amines.

On peut doser la base aminée qui se trouve dans cet éluat par différents procédés:

1° *Dosage de l'histamine et de la tyramine par le réactif de Pauly.* C'est la méthode citée par Fraenkel-Conrat et Singer pour l'histidine et la tyrosine. La solution à analyser est diluée à 2 ml, traitée par 0,2 ml de l'acide sulfanilique à 1 p. 100 dans l'acide chlorhydrique N et par 0,2 ml du nitrite de sodium à 5 p. 100; après 30 minutes à la température du laboratoire, on ajoute 5 ml d'une solution de carbonate de soude, 6,5 p. 100 pour le dosage de l'histamine, 2,5 ml de cette même solution pour celui de la tyramine; la lecture est faite au spectrophotomètre à 5100 A après 5 minutes. Parallèlement, on prépare avec des quantités de 2 à 30 de ces 2 amines, une courbe de concentration (qui est droite) en diluant ces quantités dans 2 ml d'eau distillée et avec les mêmes réactifs dans les mêmes conditions.

Cette courbe de concentration est effectuée en mettant en abscisses la densité optique indiquée par le spectrophotomètre et en ordonnées, la concentration de l'amine dosée en μg; on reporte sur cette courbe la densité optique de l'échantillon dosé, qui doit contenir de 2 à 15 μg et on lit directement sa concentration.

Il convient de faire cette courbe de concentration avec chaque série de dosage en utilisant les mêmes lots de réactifs et d'eau distillée.

La précision est de ± 1 μg ce qui donne dans 1 litre de bouillon de culture microbienne une précision finale de ± 5 p. 100.

2° *Dosage des amines par la ninhydrine-hydrindantine. Méthode de Moore et Stein pour la détermination photométrique des acides aminés et leurs dérivés.* La solution du réactif est préparée avec les produits suivants:

a) Ninhydrine: pure.

b) Hydrindantine: on mélange en agitant une solution de 8 g de ninhydrine dissous dans 200 ml d'eau distillée à 90° avec une solution de 8 g d'acide ascorbique dissous dans 40 ml d'eau à 40°; la cristallisation de l'hydrindantine commence immédiatement sans chauffage supplémentaire; 1/2 heure plus tard, le produit obtenu est filtré et lavé à l'eau puis séché jusqu'à poids constant dans un dessiccateur sous vide, protégé de la lumière, sur SO_4H_2 et KOH en pastilles, puis conservé dans une bouteille opaque.

c) Tampon pH 5,6: acétate de soude 4 N: on dissout 272 g d'acétate de soude dans 200 ml d'eau distillée en chauffant sur un bain-marie; après refroidissement, on ajoute 50 ml d'acide acétique glacial et on complète jusqu'à 500 ml, la solution doit avoir un pH 5,5.

d) Méthylcellosolve: redistillé pur.

Solution du réactif extemporanément préparé en dissolvant 0,1 g de ninhydrine et 0,02 d'hydrindantine dans 8,5 ml de méthylcellosolve et 3 ml de tampon pH 5,5; la solution rougeâtre obtenue est faite dans une bouteille opaque et utilisée tout de suite.

Mode opératoire. – La solution à analyser doit contenir de 5 à 15 μg d'une amine qui donne une coloration avec la ninhydrine est diluée à 2 ml, puis traitée par 1 ml de la solution du réactif; le tube est bouché, capuchonné, agité brièvement et chauffé dans un bain-marie bouillant couvert pendant 15 minutes; après refroidissement rapide dans l'eau froide, la lecture est faite au spectrophotomètre sans ou après dilution avec une solution d'alcool éthylique-eau distillée 50 p. 100 suivant l'intensité de la coloration obtenue; la lecture est faite à 5700 A = 570 mμ.

Comme dans le dosage avec le réactif de Pauly, on fait une courbe de concentration avec des quantités croissant régulièrement d'amines de 5 à 30 μg en portant en abscisses la densité optique et en ordonnées la concentration en μg.

Les dosages avec cette méthode n'ont de valeur que si on fait simultanément et rigoureusement dans les mêmes conditions, une courbe de concentration avec chaque série de détermination, puisque des traces infimes d'ammoniaque ou de ses composés dans l'atmosphère ou dans l'eau changent considérablement la coloration obtenue.

La putrescine, la cadavérine, la méthylamine, l'éthylamine, la butylamine, l'isoamylamine et la β-phényléthylamine ont toutes 2 pics à 410 et à 570 mμ.

AMIDES

Amides—*Pseudomonas aeruginosa*

R. R. Colwell (1964). J. gen. Microbiol. 37, 181–194.

De-amidation of acetamide was tested for by the method outlined by Bühlmann, Vischer and Bruhin (1961).

X. Bühlmann, W. A. Vischer and H. Bruhin (1961). J. Bact. 82, 787.

Amides—*Mycobacterium buruli* n.sp.

J. K. Clancey (1964). J. Path. Bact. 88, 175–187.

Basically the method of Bönicke was employed to demonstrate the enzymatic breakdown of acetamide, allantoin, benzamide, malonamide, nicotinamide, pyrazinamide, salicylamide and succinamide, but Nessler's reagent was used to detect the presence of ammonia. A small amount of the washed bacterial deposit was also tested to ensure the absence of ammonia before the addition of the amide.

R. Bönicke (1962). Bull. Un. int. Tuberc. 32, 32.

Amides—*Mycobacterium*

A. Tacquet, F. Tison and B. Devulder (1965). Annls Inst. Pasteur, Paris 108, 514–525.

The authors used the methods of Bönicke (Zentbl. Bakt. ParasitKde Abt I, Orig. 1960, 178, 186–194 and Bull. Un. int. Tuberc., 1962, 32, 13–76.

Amides—*Mycobacterium*

P. Hauduroy, F. Bel and A. Hovanessian (1966). Annls Inst. Pasteur, Paris 111, 84–86.

Recherche des amidases [3] (série des 10 amides de Bönicke; lecture des résultats après douze, quinze et parfois vingt heures).

[3] Bönicke (R.). Bull. Un. int. Tuberc., 1962, 32, 13.

Amides—*Mycobacterium*

M. Tsukamura (1966). J. gen. Microbiol. 45, 253–273.

Bönicke's amidase pattern was tested according to the description of Bönicke (1962). Incubation time was 16 hr. The following amides were used: acetamide; benzamide; urea; isonicotinamide; nicotinamide; pyrazinamide; salicylamide; allantoin; succinamide; malonamide.

R. Bönicke (1962). Bull. Un. int. Tuberc. 32, 13.

ΓIBIOTICS AND MISCELLANEOUS INHIBITORY SUBSTANCES

Antibiotics—*Serratia*

B. Bizio (1823). Translated by: C. P. Merlino (1924). J. Bact. 9, 527–543.

Therefore, before doing anything else, I subjected the piece of colored polenta to a very strong camphor vapor and then repeated the experiment with this bit of polenta; but in spite of this the red coloring appeared as heretofore. I obtained like results even after subjecting the colored polenta to the strongest odors of turpentine and of tobacco, and only after prolonged subjection to sulphur fumes was the power to produce the phenomenon taken from the colored bit of polenta. Still I could not draw a sure deduction from this, since the strong acid produced in this case might well injure the germs of microscopic plants, in case these latter were the cause of the phenomenon.

Antibiotics—*Streptococcus*

G. H. Chapman (1936). J. Bact. 32, 41–46.

Tolerate for 1 hour
1/500,000 merthiolate
1/2,500 phenol
1/5,000 basic fuchsin
1/50,000 hexylresorcinol
1.0% sodium carbonate

Antibiotics—*Leptospira*

J. A. H. Wylie and E. Vincent (1947). J. Path. Bact. 59, 247–254.

Method: The organisms were grown in a medium of 12 per cent rabbit serum in water double distilled from glass. All the serum required for the investigation was taken from rabbits whose sera had previously been shown to support the growth of the leptospirae under investigation – a necessary precaution, because some rabbit sera do not support the growth of leptospirae and a few contain antibodies which agglutinate *L. icterohaemorrhagiae.* One batch of the serum-water medium was prepared and divided into five equal volumes, one of which was left as a control. To the remaining four either penicillin or streptomycin was added according to the following scheme:

(1) 1 unit penicillin per ml
(2) 10 units penicillin per ml
(3) 5 units streptomycin per ml
(4) 50 units streptomycin per ml

The four separate solutions and the control were then put into tubes measuring 25 x 2 cm, each tube receiving 5 ml of the medium. The species of *Leptospira* to be tested were separately inoculated into one tube of each of the above groups, the inoculum being 1 ml of the stock culture. Since the stock cultures were of approximately equal density the number of leptospirae delivered in each case was about the same. All the tubes were incubated at 28°C and the cultures examined macroscopically and microscopically after 2 and 3 days.

Whereas streptomycin is stable, penicillin deteriorates on incubation; therefore, to check the loss of activity over the period of investigation, all tubes to which penicillin had been added were tested for potency at the end of the experiment by the cylinder plate method (Heatley, 1944) against the Oxford H strain of *Staphylococcus aureus.* The samples from the tubes to which 10 units had been added were diluted 10 times before assay. Upon each plate a standard of 1.0, 0.75 or 0.5 unit per ml was placed in addition to 6 or 7 samples. A graph constructed from the size of the standard rings showed that after 3 days' incubation approximately 1/4–1/3 of the penicillin had been lost, whether the leptospirae had grown or not.

N. G. Heatley (1944). Biochem. J. 38, 61.

Antibiotics—*Streptococcus*

K. E. Hite and H. C. Hesseltine (1947). A study of aerobic streptococci isolated from the uterus and vagina. J. infect. Dis. 80, 105–112.

Penicillin sensitivity: The sensitivity of strains to penicillin was tested by the tube method to Kolmer, (1945) using Penicillin-sodium (Abbott) and dextrose-veal infusion broth. In all tests a comparison was made with the H strain of *Staphylococcus aureus.* Tubes were observed after 1, 2 and 3 days incubation because some strains of streptococci grew slowly. The 72 hour incubation period undoubtedly increased the apparent resistance of strains to some extent since the staphylococcus was observed to grow in slightly higher concentrations after 48 and 72 hours than after 24 hours. However, results of the test have comparative value.

J. A. Kolmer (1945). Penicillin Therapy, Appleton-Century, New York.

Antibiotics—*Mycobacterium*

P. Hauduroy (1951). Techniques Bactériologiques. Masson et Cie, Éditeurs, Paris.

SENSIBILITE DU BACILLE TUBERCULEUX A LA STREPTOMYCINE

Une partie du produit pathologique – traité ou non, suivant sa pollution – est mélangée soigneusement par agitation à volume égal:

1° Avec une solution de streptomycine à 10 mg par cm^3, soit 10,000 γ (concentration finale; 5000 γ par cm^3).

2° Avec une solution de streptomycine à 1 mg par cm^3, soit 10000 (concentration finale; 500 γ par cm^3).

On ensemence directement ces mélanges sur un ou plusieurs tubes d'un milieu solide (Löwenstein) en ayant soin de faire un ou plusieurs tubes témoins.

En définitive, on a trois séries de tubes:

1° Produit pathologique sans streptomycine (témoins).

2° Produit pathologique avec streptomycine (5000 γ par cm^3 du mélange).

3° Produit pathologique avec streptomycine (500 γ par cm^3 du mélange).

On porte à l'étuve et on lit les résultats après une quinzaine de jours environ. Les conclusions que l'on peut tirer de l'expérience sont les suivantes:

a) Témoin = +;
produit + 5000 γ/cm^3 de streptomycine = 0;
produit + 500 γ/cm^3 de streptomycine = 0;
Souche sensible à la streptomycine.

b) Témoin = +;
produit + 5000 γ/cm^3 de streptomycine = +;
produit + 500 γ/cm^3 de streptomycine = +;
Souche resistante à la streptomycine.

c) Témoin = +;
produit + 5000 γ/cm^3 de streptomycine = 0;
produit + 500 γ/cm^3 de streptomycine = +;
Souche que l'on doit considérer comme à la limite inférieure de la résistance.

Cette technique est assez sommaire. Il est certain qu'une grande partie de l'antibiotique est perdue dans le milieu de culture, que le contact entre le germe et la streptomycine n'est pas intime, et ce procédé ne permet pas de mesurer exactement la sensibilité. Il s'agit beaucoup plus d'une analyse qualitative que d'une analyse quantitative.

On contrôle les résultats ainsi obtenus par un titrage pratiqué avec la technique des dilutions en milieu liquide. *(C. C. T. M.).*

(Méthode de titration in vivo). – *a)* Méthode de test splénique chez la souris. – Cinq souris sont inoculées par voie intrapéritonéale avec approximativement 0,1 mg de bacilles (poids sec) contenu dans 0,4 cm^3 d'une culture homogène de bacilles provenant de la souche H 37 Rv sensible. Deux souris servent de témoins; les trois autres reçoivent tous les jours respectivement 1500, 3000 et 4500 γ de streptomycine en deux injections. Dans les mêmes conditions, une souche moyennement résistante (souche F 189 titrant 40) est inoculée à un deuxième lot de souris du même poids. Après 12 jours de traitement, les animaux sont sacrifiés. L'autopsie montre des rats de dimensions variables. Pour la souche H 37 Rv, les rates des souris infectées et non traitées sont 10 à 12 fois plus grosses que celles des souris témoins non infectées et les rates des souris infectées et traitées sont à peine augmentées de volume. Inversement, chez les souris infectées par la souche résistante, les rates des animaux, traités ou non, sont de dimensions comparables.

b) Méthode du test intradermique chez le cobaye. – Quatre cobayes sont infectés par voie intradermique avec 0,2 cm^3 de culture homogène d'une souche H 37 Rv sensible, ou de souches résistantes. Un cobaye témoin n'est pas traité par la streptomycine. Les autres sont traités respectivement par 5000, 10000 et 15000 γ par jour en deux injections. On compare les lésions locales consécutives à des infections intradermiques. Les souches résistantes provoquent chez les témoins et chez les animaux traités des abcès caséeux au point d'inoculation. Ces abcès se fistulisent vers le 10e jour. La souche sensible ne produit pas d'abcès chez les animaux traités. A l'autopsie on constate, malgré le traitement, des adénopathies correspondant au foyer cutané chez les animaux inoculés avec les souches résistantes. Chez les animaux inoculés avec la souche sensible et traités, il n'y a pas d'adénopathie satellite. Entre les mains de E. Bernard et B. Kreis, cette méthode s'est montrée d'interprétation assez délicate et peu recommandable pour la pratique. *(I. P. L.).*

Antibiotics—*Staphylococcus*

G. I. C. Ingram (1951). J. gen. Microbiol. 5, 30–38.

Following the techniques of Foster and Woodruff (1943, J. Bact. 46, 187) and Frisk (1945, Nord. med. (Hygiea). 28, 2249), a series of penicillin dilutions were made in molten agar, and poured in Petri dishes. When set, surface light streaks were made with the modified Heatley pen. After inoculation the results were recorded by noting the highest concentration of penicillin that permitted growth, and the lowest that inhibited growth, in the series of plates that was used, for example, '0.01–0.02'. This value was taken as the cell sensitivity.

Antibiotics

S. D. Henriksen (1952). J. gen. Microbiol. 6, 318–328.

For the study of sensitivity to antibiotics, three different methods were used.

(1) A filter-paper disk method (Jensen and Kiaer, 1948). Antibiotic solutions used to moisten the disks were: penicillin 175 units/ml, streptomycin 2800 μg/ml, sulphathiazole 6.6 mg/ml, chloromycetin and aureomycin 1000 μg/ml each. Incubation is at 28° for 48 hr.

(2) An antibiotic tablet method (Lund, Funder-Schmidt, Christensen and Dupont, 1951). The tablets (Roskilde Medical Co. Ltd., Roskilde, Denmark), which contain the above antibiotics and also terramycin, are deposited on peptone-free 10% horse blood agar plates, which have been inoculated with a suspension of the organism to be tested. Inhibition zones are measured after incubation for 24 hr at 37°C in a closed jar

containing some water.

(3) Titration in 10% (v/v) serum broth containing twofold dilutions of the antibiotics. Incubation at 28°C for 48 hr. Readings were made after 24 and 48 hr.

K. Jensen and I. Kiaer (1948). Acta path. microbiol. scand. 25, 168.
E. Lund, H. Funder-Schmidt, H. Christensen and A. Dupont (1951). Acta path. microbiol. scand. 29, 221.

Antibiotics—*Pseudomonas* and *Vibrio*

J. M. Shewan, W. Hodgkiss and J. Liston (1954). Nature, Lond. 173, 208–209.

Differentiation of *Pseudomonas* and *Vibrio* by use of antibiotics.

This method employs the normal sensitivity test for antibiotics, using antibiotic tablets, or in the case of the vibriostatic agent, a saturated aqueous solution in an assay cup. All pigment-producing *Pseudomonas* spp so far studied from a variety of sources, even although they may have lost the power of pigment production on subsequent culturing have been found to be insensitive to antibiotics and the vibriostatic agent at the concentrations listed in the table (below). On the other hand, all the vibrios so far tested react to both terramycin and the vibriostatic agent. The remaining group of *Pseudomonas* spp which grow somewhat slowly and never form pigment under any conditions so far tested can also be distinguished by their sensitivity to terramycin.

Type of organism	Inhibiting Agent		
	Penicillin (2.5 IU)	Terramycin (10 γ)	Vibriostatic Compound 0/129* (saturated aqueous solution)
Pigment-producing *Pseudomonas*	–	–	–
Non-pigment-producing *Pseudomonas*	–	+	–
Vibrios	–	+	+

*2,4-diamino-6,7-diisopropylpteridine. –, insensitive; +, sensitive.

Antibiotics—luminous bacteria

R. Spencer (1955). J. gen. Microbiol. 13, 111–118.

The solid media used consisted of tap-water Lemco agar (Lemco, 10 g; peptone (Oxoid), 10 g; NaCl, 5 g; tap water, 1000 ml; pH 7.6) and sea water Lemco agar (Lemco, 10 g; peptone, 10 g; tap water, 250 ml; aged sea water, (ZoBell, 1946) 750 ml; pH 7.6). Tap water peptone water (Peptone, 10 g; NaCl, 0.5 g; tap water, 1000 ml; pH 7.6) or sea water peptone water (peptone, 10 g; tap water, 250 ml; aged sea water, 750 ml; pH 7.6) were used for liquid media. For the biochemical tests tap water media alone were used as it was found that these gave growth in all cases.

For the antibiotic sensitivty tests, tap water Lemco agar was used as sea water inhibited the action of terramycin to some extent. Tablets containing the indicated amounts of antibiotic were used for penicillin, chloramphenicol, streptomycin and terramycin, and the cup assay method with a saturated aqueous solution for the vibriostatic substance 0/129 (2-4-diamino, 6-7-di-isopropyl-pteridine; Shewan *et al.* 1954). The agar plates were surface-poured with a young culture and dried at room temperature for some hours before the antibiotics were applied.

C. E. ZoBell (1946). *Marine Microbiology*, Waltham, Mass.; Chronica Botanica Co.
J. M. Shewan, W. Hodgkiss and J. Liston (1954). Nature, Lond. 173, 208.

Antibiotics—*Chromobacterium*

P. H. A. Sneath (1956). J. gen. Microbiol. 15, 70–98.

Nutrient agar plates containing 100 units sodium benzylpenicillin/ml. Incubated for 2 days 25°C and 1 day 37°C.

Antibiotics—*Bacillus*

K. L. Burdon (1956). J. Bact. 71, 25–42.

The sensitivity to penicillin was determined by inoculating slants of tryptose agar, or trypticase soy agar, to which a solution of penicillin G had been added in the amount of 10 units per ml of agar. The slants were inoculated by a single streak and observed for growth during incubation for 48 to 72 hr.

Antibiotics—*Bacillus*

E. R. Brown, M. D. Moody, E. L. Treece and C. W. Smith (1958). J. Bact. 75, 499–509.

Growth on penicillin agar. Plates of heart infusion agar containing 6 units of penicillin G (Wyeth) per ml were prepared, inoculated with a loopful of growth from agar slant cultures (18 hr) of the organisms tested, and the plates were incubated at 37°C for 18 hr before examination. One disadvantage of this technique is the fact that secondary growth on the part of *B. anthracis* has led to erroneous results in the diagnosis of the anthrax organism.

Antibiotics—*Pseudomonas*

M. E. Rhodes (1959). J. gen. Microbiol. 21, 221–263.

Resistance to the pteridine derivative 0/129.

Shewan, Hodgkiss and Liston (1954) recommended the use of the vibriostatic compound 0/129 (2:4-diamino-6:7-di-isopropyl-pteridine) to differentiate *Vibrio* from *Pseudomonas,* particularly the non-chromogenic pseudomonads. The compound (kindly supplied by Dr. H. O. J. Collier) was tested by placing filter-paper strips soaked in a saturated aqueous solution of compound 0/129 (sterilized by filtration through a sintered-glass filter) across streak inocula of the individual isolates on yeast extract agar plates. Five *Vibrio* spp cultures (from NCTC) and one *V. icthyodermis* kindly given by Mr. Hodgkiss were used as controls.

J. M. Shewan, W. Hodgkiss and J. Liston (1954). Nature, Lond. 173, 208.

Antibiotics—*Mycobacterium* (atypical)

A. Beck (1959). J. Path. Bact. 77, 615–624.

Sensitivity to isoniazid (I.N.H.), streptomycin and para-aminosalicylic acid (P.A.S.).

This was tested on Lowenstein media containing the following drug concentrations per ml of medium: I.N.H. 0.2, 1, 5, 10, and 50 μg; streptomycin 1, 3, 10, and 30 μg; P.A.S. 4, 16, and 64 μg.

Antibiotics—*Pasteurella septica*

J. M. Talbot and P. H. A. Sneath (1960). J. gen. Microbiol. 22, 303–311.

Antibiotic sensitivities were determined with "Sentest" tablets (Evans and Co. Ltd. Liverpool, England) on a blood agar plate flooded with an 18 hr culture. Sulphonamide sensitivity was investigated on the medium of Jewell and Pearmain (1954, J. clin. Path. 7, 308) with sulphadimidine tablets. Results were recorded as "sensitive" or as "resistant".

Antibiotics—*Staphylococcus*

H. Williams Smith and W. E. Crabb (1960). J. Path. Bact. 79, 243–249.

Each strain of *Staph. aureus* was inoculated over a portion of the surface of an agar plate and discs containing either 40 μg of oxytetracycline hydrochloride or 2 units of sodium penicillin G placed upon them. The plates were incubated at 37°C for 24 hr. Cultures with zones of inhibition of growth significantly less than those exhibited by the Oxford staphylococcus were recorded as resistant. Such cultures were retained and the minimum inhibitory concentration (M.I.C.) of the drug to which they were resistant determined. For this agar plates containing two-fold falling concentrations of drug were inoculated lightly with a number of strains under test and incubated at 37°C for 24 hr. The lowest concentration of drug that prevented growth was taken as the M.I.C.

Antibiotics—*Pasteurella*

G. R. Smith (1961). J. Path. Bact. 81, 431–440.

A preliminary test of the sensitivity of strains to various antibiotics was made by the absorbent paper disk technique (Morley, 1945) using "Multodisks" (Oxoid Division, Oxo Ltd.) which contained 10 μg chloramphenicol; 10 μg erythromycin; 5 μg oleandomycin; 10 μg streptomycin; 10 μg tetracycline; or 1.5 units penicillin. Five per cent sheep blood agar plates were flooded with 1 in 10 dilutions of 5 hr infusion broth cultures. All *P. haemolytica* broth cultures of this age were known to contain roughly similar numbers of viable organisms. The excess moisture was removed from the plates and they were subsequently dried in the incubator. A Multodisk paper was then placed on the surface of each culture and tests were read after 18 hours' incubation.

Sensitivity to penicillin was further investigated by a broth dilution method. Two-fold dilutions of the sodium salt of crystalline penicillin G were prepared in infusion broth and 1.0 ml volumes of these were inoculated with 1.0 ml volumes of 6 hr broth cultures diluted 1 in 10^6 in fresh broth, so that each strain was tested in the presence of 1.25, 0.63, 0.31, 0.16 and 0.08 units per ml. After 18 hours incubation the tubes were examined visually for the presence or absence of growth.

D. C. Morley (1945). J. Path. Bact. 57, 379.

Antibiotics—rhizosphere bacteria

P. G. Brisbane and A. D. Rovira (1961). J. gen. Microbiol. 26, 379–392.

Sensitivity to antibiotics was tested by surface seeding basal agar plates with cultures and then placing one Oxoid 'Multodisk' (11–14D)/plate. Each disk contains eight peripheral disks – one for each of the following: chloramphenicol (50 μg), erythromycin (50 μg), sulphafurazole (500 μg), novobiocin (30 μg), oleandomycin (10 μg), penicillin (5 units), streptomycin (25 μg), and tetracycline (50 μg). The zones of inhibition were recorded after incubation for 7 days at 26°C.

Antibiotics—*Vibrio*

G. H. G. Davis and R. W. A. Park (1962). J. gen. Microbiol. 21, 101–119.

Resistance to antibacterial agents was tested using Evans "Sentest" or Mast disks on surface seeded nutrient agar. Antibiotics: penicillin, 1 and 2.5 i.u.; streptomycin, 20 and 80 μg; chloramphenicol, 40 and 100 μg; Aureomycin, 10 and 100 μg; Terramycin, 10, 25 and 100 μg; erythromycin, 1 and 10 μg; tetracycline, 100 μg; neomycin, 100 μg; bacitracin, 5.5 units; novobiocin, 5 and 10 μg; oleandomycin, 5 and 20 μg. Also tested 0/129 using an aqueous suspension (c. 50 mg in 20 ml) and agar well technique. (NOTE: a saturated aqueous solution of 0/129 proved non-inhibitory against *Vibrio comma* in preliminary tests).

Antibiotics—*Actinobacillus actinomycetemcomitans* and *Haemophilis aphrophilus*

E. O. King and H. W. Tatum (1962). *Actinobacillus actinomycetemcomitans* and *Haemophilus aphrophilus.* J. infect. Dis. 111, 85–94.

Antibiotic sensitivity tests were performed using Difco 3-dilution sensitivity disks placed on blood agar plates which had been inoculated with a swab dipped in a broth culture. The tests were read after 24 and 48 hours incubation.

Antibiotics—*Pseudomonas odorans*

I. Málek, M. Radochová and O. Lysenko (1963). J. gen. Microbiol. 33, 349–355.

The sensitivity to antibiotics was determined by the tablets method routinely used in diagnostic laboratories on 2.0% (v/v) blood agar, the only exception being that the drop method was used for testing polymyxin. 28°C.

Antibiotics—*Aeromonas salmonicida*

I. W. Smith (1963). J. gen. Microbiol. 33, 263–274.

Sensitivity to antibiotics. Evans Sentests were applied to the seeded surface of nutrient agar plates and the results read after incubation for 1 day. The vibriostatic agent, 0/129 (Shewan, Hodgkiss and Liston, 1954) was incorporated in filter-paper discs and applied to the agar and read as above.

J. M. Shewan, W. Hodgkiss and J. Liston (1954). Nature, Lond. 173, 208.

Antibiotics—*Staphylococcus aureus*

R. L. Brown and J. B. Evans (1963). J. Bact. 85, 1409–1412.

Antibiotic sensitivity spectra. Antibiotic sensitivity spectra were determined by spreading 0.1 ml of culture on the surface of TSB agar* dropping Multidiscs (Colab Laboratories, Inc., Chicago Heights, Ill.) onto the surface of the agar, inverting the petri dishes, and incubating for 3 days.

*Trypticase Soy Broth plus 1.5% agar.

Antibiotics—*Listeria monocytogenes*

M. Füzi (1963). J. Path. Bact. 85, 524–525.

The organisms from 24-hr broth cultures were streaked on the surface of blood agar plates (5 per cent human blood, 2 per cent agar fibre, 1 per cent Richter peptone, pH = 7.4) that had been dried for 30 min in the incubator. Paper discs containing 50 μg neomycin sulphate ("Biotest" sensitivity discs, Human Oltóanyagtermelö és Kutató Intézet, Budapest, Hungary) were then placed upon the surface of the inoculated medium. After standing 30 min at room temperature to allow diffusion of the antibiotic into the medium, the plates were incubated at 37°C for 20 hr.

Antibiotics—*Staphylococcus aureus*

W. E. Allen and I. McVeigh (1963). Can. J. Microbiol. 9, 179–186.

Prior to use, clonal stock cultures of each strain were prepared. These were maintained on agar slants of Antibiotic Assay medium* with the following composition: peptone (BBL gelysate), 5.0 g; yeast extract, 1.5 g; beef extract, 1.5 g; NaCl, 3.5 g; K_2HPO_4, 3.68 g; KH_2PO_4, 1.32 g; glucose, 1.0 g; agar, 16.0 g; and distilled H_2O, 1000.0 ml. The Antibiotic Assay medium, either in the liquid state (hereafter called broth medium) or in the solid state by the addition of 1.6% agar (hereafter identified as agar medium), was used for the experimental work.

The maximum amount of penicillin tolerated by each strain was determined by the twofold serial dilution method. The inoculum used in the test consisted of 0.2 ml of a maximum stationary phase culture per 50 ml of broth medium. Sterile buffered crystalline potassium penicillin G was dissolved and diluted to the desired concentration in sterile phosphate buffer solution (K_2HPO_4, 2.0 g; KH_2PO_4, 8.0 g; and distilled H_2O, 1000.0 ml; pH 6.0). The highest concentration of the antibiotic in which growth was evident (as manifested by visible turbidity) after 48 hours of incubation at 37°C was considered to be the maximum amount tolerated. (*Baltimore Biological Laboratory, Inc., Baltimore 18, Maryland).

Antibiotics—*Staphylococcus*

A. T. Willis and G. C. Turner (1963). J. Path. Bact. 85, 395–405.

Antibiotic sensitivity tests were performed on meat infusion agar using Multodiscs (Oxoid) at the following concentrations: penicillin, 1.5 units; streptomycin, tetracycline, chloramphenicol and erythromycin, 10 μg; novobiocin, 5 μg. All strains were tested for sensitivity to methicillin (Celbenin, Beecham Research Laboratories) with 10 μg discs.

Antibiotics—*Mycobacterium buruli* n.sp.

J. K. Clancey (1964). J. Path. Bact. 88, 175–187.

The authors describe in detail, or cite references to, methods used to study the sensitivity of the new species and strains of *M. tuberculosis, M. ulcerans, M. balnei, M. fortuitum* to streptomycin, p-aminosalicylic acid, isoniazid, viomycin, ethionamide, thiosemicarbazone, phenazine, 2-furoic-acid-hydrazide, sodium salicylate, nicotinamide and tetrazolium salts.

Antibiotics—*Staphylococcus*

S. Krynski, W. Kedzia and M. Kaminska (1964). Some differences between staphylococci isolated from pus and from healthy carriers. J. infect. Dis. 114, 193–202.

Antibiotic sensitivity was determined by the paper disc method.

Antibiotics—*Lactobacillus*

M. Gemmell and W. Hodgkiss (1964). J. gen. Microbiol. 35, 519–526.

Two drops of a 24-hr broth culture were spread on plates of basal agar medium containing glucose (0.5%, w/v). Oxoid Multodiscs 11-15 F or 11-14 D were placed on the plates, which were incubated in the inverted position for 24-48 hr at 30°C. Sensitivity was measured by the size of the zone of inhibition around the disc.

The basal medium which was used throughout with minor modifications had the following constituents in 1 L tap water: meat extract (Lab Lemco), 5 g; Evans peptone, 5 g; Difco yeast extract, 5 g; Tween 80, 0.5 ml; $MnSO_4 \cdot 4H_2O$, 0.1 g; potassium citrate, 1 g; pH 6.5.

Antibiotics—*Moraxella*

W. J. Ryan (1964). J. gen. Microbiol. 35, 361–372.

With the exception of polymyxin, all tests were done by the disc method on blood agar plates. Commercially available discs were used. Results were recorded after 2 days of incubation.

Antibiotics—*Desulfovibrio, Clostridium*

A. M. Saleh (1964). J. gen. Microbiol. 37, 113–121.

Test of inhibitors. A bacteriostatic screening method was used. Tests were carried out in Baars's medium with yeast extract and cysteine in conformity with the practice at the National Chemical Laboratory over a number of years. Four substances known to have strong bacteriostatic or bactericidal effects on sulphate-reducers (Saleh *et al.* 1964) were tested against the 45 strains. These were bis-(p-chloro-phenyl-diguanido-) hexane diacetate (Hibitane, Imperial Chemical Industries Ltd.), cetyltrimethylammonium bromide (CTAB), 5,5′-dichloro-2,2′-dihydroxy-diphenyl methane (Panacide; British Drug Houses Ltd.), and 2,4-dinitrophenol. Solutions as obtained were assumed to be sterile; CTAB and 2,4-dinitrophenol were made into concentrated solutions, allowed to stand for a few hours and assumed to be sterile. Serial dilutions were made in sterile distilled water, 2 ml of the appropriate dilution added to 18 ml sterile Baars's medium, and the medium adjusted when necessary to pH 7.0–7.2 (bromothymol blue) with sterile NaOH or HCl. Each experimental medium was dispensed into three test tubes and inoculated with enough stock culture of the appropriate organism to give an initial concentration of between 10^5 and 5×10^5 bacteria/ml. After anaerobic incubation the tubes were examined for blackening (formation of FeS); cultures which showed doubtful growth were examined microscopically.

A. M. Saleh, R. Macpherson and J. D. A. Miller (1964). J. appl. Bact. 87, 1073.

Antibiotics—*Pseudomonas aeruginosa*

R. R. Colwell (1964). J. gen. Microbiol. 37, 181–194.

Sensitivity to antibiotics and a 'vibriostatic compound'. The strains were tested for sensitivity to penicillin (high concentration), dihydrostreptomycin (10 μg), chloramphenicol (30 μg), chlortetracycline (30 μg), oxytetrocycline (30 μg), polymyxin B (300 units), tetracycline (30 μg; Bauer, Roberts & Kirby, 1959–60) and the vibriostatic compound (2,4-diamino-6,7-diisopropylpteridine) described by Shewan, Hodgkiss & Liston (1954). YE agar plates were used, with sensitivity discs (Baltimore Biological Laboratory, Inc.) for the antibiotics listed.

The standard inoculum for all tests was a single drop (1/20 ml) from a sterile disposable pipette (Fisher Scientific Company) of a 24-48 hr broth culture. The stock culture medium (YE) was a modification of that used by Rhodes (1959): Difco yeast extract, 0.3 g; Difco Bacto-Proteose Peptone, 1.0 g; NaCl, 0.5 g; agar, 1.5 g (agar omitted from liquid stock media); distilled water 1 L; adjusted to pH 7.2 – 7.4 with NaOH. Tests were carried out at room temperature (25°C) except where otherwise indicated.

A. W. Bauer, C. E. Roberts, Jr. and W. M. M. Kirby (1959–60). Antibiotics, 1959–1960, p. 574.
M. E. Rhodes (1959). J. gen. Microbiol. 21, 221.
J. M. Shewan, W. Hodgkiss and J. Liston (1954). Nature, Lond. 173, 208.

Antibiotics—*Vibrio marinus*

R. R. Colwell and R. Y. Morita (1964). J. Bact. 88, 831–837.

The strains were tested for sensitivity to penicillin (10 units), dihydrostreptomycin (10 μg), chloramphenicol (30 μg), oxytetracycline (30 μg), chlortetracycline (30 μg), erythromycin (15 μg), novobiocin (30 μg), polymyxin B (300 units), kanamycin (30 μg), tetracycline (30 μg), and the vibriostatic compound, 2,4-diamino-6,7-diisopropylpteridine (Shewan, Hodgkiss, and Liston, 1954). Sensi-Discs of the antibiotics (BBL) were used; the pteridine compound was sprinkled, in crystal form, onto moist, freshly streaked plates of the cultures tested.

J. M. Shewan, W. Hodgkiss and J. Liston (1954). Nature, Lond. 173, 208.

Antibiotics—*Staphylococcus*

R. Sutherland and G. N. Rolinson (1964). J. Bact. 87, 887–899.

Sensitivity disc tests. Sensitivity disc tests were carried out in the usual fashion by flooding agar plates with suitable dilutions of overnight broth cultures of the test organisms and placing freshly prepared filter-paper discs (diameter, 6 mm) containing 5 μg of cloxacillin or 10 μg of methicillin on the surface of the plates. The diameters of zones of inhibition were measured after overnight incubation at 37°C.

Minimal inhibitory concentrations (MIC). MIC levels of antibiotics were determined by serial dilution of graded concentrations of the drugs in 5 ml volumes of broth or 18 ml volumes of agar. Inocula comprised 1 drop (approximately 0.03 ml) of an overnight broth culture of the test organism diluted as specified, and the cultures were incubated at 37°C for 24 and 48 hr.

Antibiotics—*Staphylococcus*

J. Donaldson, A. J. Moriarity, N. Joshi and D. G. Dale (1964). Can. J. Microbiol. 10, 163–168.

Antibiotics were dried *in vacuo* and dissolved in phosphate buffer, pH 6.0 (0.2 M KH_2PO_4 and 0.2 M K_2HPO_4) to a concentration of 1000 μg/ml. Two milliliter (2 ml) quantities of each were dispensed in parafilm-covered tubes, stored at −20°C, used once, and discarded. Dilution in broth was then made to achieve the desired final concentrations. Dilutions of penicillin G ranged from 0.01 to 200 μg/ml. Methicillin, oxacillin, and cephalosporin C dilutions ranged from 0.1 to 100 μg/ml. Minimum inhibitory concentration (MIC) tests were carried out using a twofold tube dilution technique with brain-heart-infusion or nutrient broth as the diluent.

Following inoculation, the suspensions were incubated at 37°C for 24 hours. The concentration of drug in the first tube showing no turbidity was read as the MIC. In some experiments, MIC was determined using whole and skim milk as the diluent. Inhibition end points were detected in such media by the use of methylene-blue stained smears.

The use of whole milk, skim milk, and nutrient broth gave results for MIC's which did not differ significantly from broth alone.

Antibiotics—*Staphylococcus*

E. F. Harrison and C. B. Cropp (1965). Appl. Microbiol. 13, 212–215.

Minimal inhibitory concentration (MIC) test. The antistaphylococcal MIC for each antibiotic was determined by conventional tube-dilution methods, with Trypticase Soy broth (BBL). Each tube was inoculated with a diluted (10^5 organisms per milliliter) 24-hr broth culture of the isolate. Tubes were incubated at 37°C, and the MIC end point was determined after 18 hr according to the criterion of English and Gelwicks (1951).

Antibiotics tested were penicillin G, methicillin, ristocetin, vancomycin, erythromycin glycoheptonate, ampicillin and lysostaphin.

A. R. English and P. C. Gelwicks (1951). Antibiotics Chemother. 1, 118–124.

Antibiotics—general

R. A. Beargie, E. C. Bracken and H. D. Riley, Jr. (1965). Appl. Microbiol. 13, 279–280.

The authors describe the use of 9-well spot plates in place of tubes for antibiotic tests.

Antibiotics—methods of testing

A. Branch, D. H. Starkey and E. E. Power (1965). Appl. Microbiol. 13, 469–472.

The tube dilution method of performing antibiotic sensitivity tests is commonly employed as an accurate method for defining the minimal inhibitory concentration in relation to pathogenic organisms. It is also used as a reference for comparing minimal inhibitory concentrations with the size of the zone of inhibition in the agar diffusion test. Although surveys have shown that there is no standardized method and technique of performing the tube dilution test, it is generally assumed that all of the diversified methods will yield the same results and interpretations. With the assistance of five experts, seven different tube dilution methods were compared; 16 antibiotics and three organisms for each antibiotic were used. The conclusions drawn are that, although the accuracy of a single method within its own confines is acknowledged, the minimal inhibitory concentrations and interpretations cannot be interpolated from one laboratory to another where a different technique is employed. The results are frequently discrepant. It is suggested that a uniform method be developed and promulgated for general use.

Antibiotics—*Streptomyces, Waksmania, Streptosporangium, Micromonospora*

S. T. Williams and F. L. Davies (1965). J. gen. Microbiol. 38, 251–261.

Selection of antibiotics. The antibiotics to be used in detailed tests were selected by a preliminary experiment in which they were tested against a range of actinomycetes. The antibiotics used were Nystatin (Squibb); Actidione (Light); Polymyxin B sulphate (Burroughs Wellcome); sodium benzyl penicillin (Glaxo); Streptomycin sulphate (Glaxo); Chloramphenicol (Allen and Hanburys); Chlortetracycline hydrochloride (Cyanamid).

The test organisms were 45 *Streptomyces* species, *Micromonospora melanosporea, Waksmania rosea* and a *Streptosporangium* sp. These tests and subsequent ones in this work were done with a starch + casein medium adjusted to pH 7.0 (Küster and Williams, 1964). Solutions of the sterile antibiotics were prepared in sterile water; for the insoluble nystatin a suspension was made. Antibiotic solutions were added to molten medium at 45°C to give concentrations of 0, 1.0, 5.0, 10.0, 30.0, 50.0, and 100.0 μg/ml. One-ml samples of spore suspensions of test organisms were mixed with 9 ml samples of media and plates poured. Spore suspensions were prepared by incorporating one loopful of a mature sporing culture into 10 ml of sterile water. Plates were incubated at 25°C for 7 days; development of micro-organisms on antibiotic plates were compared with that on control plates and any inhibition of growth noted.

E. Küster and S. T. Williams (1964). Nature, Lond. 202, 928.

Antibiotics and other inhibitory substances—*Bacillus*

G. R. F. Hilson (1965). J. gen. Microbiol. 39, 407–421.

Isoniazid sensitivity:

Dubos-Davis fluid medium* containing isonicotinyl hydrazide 500 μg/ml was used. Screw-capped ½ oz bottles containing 10 ml of this medium were inoculated and incubated for 2 days. *Other drugs.* Sensitivity to penicillin, streptomycin, tetracycline, chloramphenicol, erythromycin and sulphadimidine was tested by the disc diffusion plate technique of Fairbrother and Martyn (1951) on the medium of Jewell and Pearmain (1954). The amount of each drug incorporated in the disc was such as to produce inhibition zones of 15-20 mm diameter after overnight incubation on plates inoculated with the standard sensitive Oxford strain of *Staphylococcus aureus.*

R. W. Fairbrother and G. Martyn (1951). J. clin. Path. 7, 37.

P. Jewell and G. E. G. Pearmain (1954). J. clin. Path. 7, 308.

**Mackie and McCartney's Handbook of Bacteriology* (1960), 10 ed. Ed. by R. Cruickshank, pp. 211, 212, 214. Edinburgh and London: Livingstone.

Antibiotics—*Alcaligenes*

R. G. Mitchell and S. K. R. Clarke (1965). J. gen. Microbiol. 40, 343–348.

Antibiotic sensitivities were determined by the agar gel diffusion method, with 'Mast' antibiotic discs.

Antibiotics—*Thiobacillus*

M. Hutchinson, K. I. Johnstone and D. White (1965). J. gen. Microbiol. 41, 357–366.

Sensitivity to ampicillin, bacitracin, celbenin, chloramphenicol, novobiocin and streptomycin was tested using "Sentest" tablets (Evans Medical Ltd.) on the S6 medium.

The basal mineral salts medium, S0, was (g/L distilled water): Na_2HPO_4, 1.2; KH_2PO_4, 1.8; $MgSO_4$, 0.1; $(NH_4)_2SO_4$, 0.1; $CaCl_2$, 0.03; $FeCl_3 \cdot 6H_2O$, 0.02; $MnSO_4 \cdot 4H_2O$, 0.02. The S6 medium was prepared by adding $Na_2S_2O_3.5H_2O$, 10 g, to this basal medium.

(No indication is given of the method used to solidify the medium for the S6 "plates").

Cultures were incubated for 28 days at 28°C.

Other inhibition tests were conducted with the liquid S6 medium using mixed phosphate 4%; NaCl, 5%; KNO_3, 1.0%; NH_4Cl, 2.5%; NH_4 thiocyanate, 200 ppm; and phenol, 25 and 100 ppm.

Test for effectiveness was made by comparison with controls.

Antibiotics—*Enterobacteriaceae*

R. J. Roantree and J. P. Steward (1965). J. Bact. 89, 630–639.

Testing of resistance to penicillin. Escherichia coli strains A27 and A48 and their penicillin-resistant mutants were tested by inoculating 10^6 organisms from an overnight broth culture into 2-ml quantities of broth containing varying concentration of benzylpenicillin. The greatest concentration of drug permitting visible growth at 48 hr was taken as the value for degree of resistance.

Antibiotics—marine bacteria

R. M. Pfister and P. R. Burkholder (1965). J. Bact. 89, 863–872.

Sensitivity of the organisms to six antibiotics – bacitracin, novobiocin, oleandomycin, penicillin, tetracycline, and viomycin – was determined by use of standard Difco sensitivity discs impregnated with 'medium' levels of concentration for each antibiotic. Cultures used to form the seed layer for testing the sensitivity discs were grown in seawater broth on a rotary shaker at 25°C overnight, and then layered on the surface of poured agar plates. The sizes of zones of inhibition were not taken into account, but inhibition was recorded as either present or absent. At the same time, a solution of pteridine 0/129 (Shewan, Hodgkiss, and Liston, 1954) was tested in moistened paper discs placed on the seeded plates, to aid in the differentiation of *Vibrio* and *Pseudomonas.*

J. M. Shewan, W. Hodgkiss and J. Liston (1954). Nature, Lond. 173,208–209.

Antibiotics—*Pseudomonas piscicida*

A. J. Hansen, O. B. Weeks and R. R. Colwell (1965). J. Bact. 89, 752–761.

Ability to grow in the presence of the vibriostatic compound, pteridine 0/129 (2,4-diamino-6,7-diisopropylpteridine) (Shewan, Hodgkiss, and Liston, 1954), was established by sprinkling the compound lightly upon seawater media previously inoculated with bacteria being tested.

J. M. Shewan, W. Hodgkiss and J. Liston (1954). Nature, Lond. 173, 208–209.

Antibiotics—*Staphylococcus*

S. Mitsuhashi, H. Oshima, U. Kawaharada, and H. Hashimoto (1965). J. Bact. 89, 967–976.

Determination of drug resistance. All bacterial suspensions were prepared in nutrient broth and grown in stationary culture for 18 hr at 37°C. Each culture was spotted onto HI agar plates, containing serial twofold dilutions of each drug. Drug resistance was expressed as the maximal concentration of each drug which allowed the growth of bacteria.

Antibiotics—*Bacillus anthracis*

R. F. Knisely (1965). J. Bact. 90, 1778–1783.

Tryptose Agar (Difco) slants were prepared to contain 10 units of penicillin G (The Upjohn Co., Kalamazoo, Mich.) per ml. The slants were inoculated by a single streak, incubated at 37°C, and observed for growth during a 24 to 48-hr period.

Antibiotics—*Escherichia, Enterobacter, Proteus, Pseudomonas* and *Staphylococcus*

D. E. Mahony and P. Chadwick (1965). Can. J. Microbiol. 11, 829–836.

This paper describes a method of antibiotic sensitivity testing in which sensitivity is measured by inhibition of microcolony formation on agar plates containing antibiotic.

Antibiotics. Antibiotics were obtained in powder form from Dr. L. Greenberg, Laboratory of Hygiene, Tunney's Pasture, Ottawa. Antibiotic powder was added to 20 ml amounts of sterile, distilled water to make a stock solution of 1000 μg/ml. These stock solutions were then stored at 4°C for a period not exceeding a month.

Preparation of Antibiotic Plates. Antibiotic plates were prepared by the addition of antibiotic stock solutions to molten heart infusion agar at 45°C. For most antibiotics two separate concentrations were used. These were as follows (in μg or units per ml) – penicillin 0.1, 1.0; erythromycin 0.8, 3.0; novobiocin 2, 36; tetracycline 5, 16; streptomycin 5, 15; chloramphenicol 8, 16; kanamycin 6, 16; neomycin 6, 16; colimycin 3, 10; polymyxin B 20, solframycin 20. Before use, the plates were allowed to dry with lids raised for ½ hour at 37°C. The backs of the plates were divided into a number of sectors by means of a rubber ink stamp, since up to 16 organisms were to be inoculated on one plate.

Inoculation and Reading of Plates. Portions of a few bacterial colonies of each strain to be tested were suspended in 1 – 2 ml of sterile broth. One milliliter of the broth was transferred to a sterile plastic screw-cap 15 mm in diameter, 16 of these caps being held in a wooden block. A multiple inoculator consisting of 16 glass rods held in another wooden block could then be used to transfer droplets of broth from the caps to the agar surface. After the inoculum had dried into the agar the plates were incubated at 37°C for 4 hours, after which time they were examined for microcolonial growth using the 16 mm objective lens and 10 x eyepiece of a binocular microscope, giving a magnification of 100. A control plate containing no antibiotic was included in each set of tests.

The authors subsequently discuss the use of the method in direct examination of pathological material (1966. Can. J. Microbiol. 12, 683–690) and on mixed cultures (1966. Can. J. Microbiol. 12, 699–702).

Antibiotics—*Staphylococcus aureus*

A. A. Richtarik, K. Lindemulder and B. De Boer (1965). Can. J. Microbiol. 11, 637–640.

The culture medium was Trypticase Soy Broth (Baltimore Biological Laboratory). Fifteen chemical agents (heavy metals, antimalarials, and diuretics) were tested for antimicrobial action. Six which were effective at concentrations of 0.125 μg/ml or lower were studied in combination with potassium phenoxymethyl penicillin. The agents were dissolved in sterile distilled water. The concentration of the inhibiting agent in the first tube of the dilution series (using the serial broth dilution method of Grove and Randall, 1955) was 0.125 μg/ml for a single agent or 0.0625 μg/ml for the agent and 0.0625 μg/ml for penicillin, giving a combined concentration of 0.125 μg/ml in the first tube. The final volume in each tube was 2 ml (0.5 ml broth + 1.5 ml of 1:1000 broth dilution of an 18 hour culture). The tubes were incubated for 18 hours, after which the least concentration which prevented or inhibited growth was recorded as the minimal inhibitory concentration. The procedure was carried out in duplicate.

D. C. Grove and W. A. Randall (1955). *Assay methods of antibiotics: a laboratory manual.* Medical Encyclopedia, Inc., New York. pp. 190–196.

Antibiotics—*Mycobacterium*

R. M. McCune, F. M. Feldmann and W. McDermott (1966). J. exp. Med. 123, 469–486.

Colonies of tubercle bacilli isolated from animal tissues onto solid oleic acid-albumin agar were subcultured in Tween-albumin medium. After 7 days' incubation, the culture was diluted five and one half times in Tween basal medium (without albumin or glucose) at pH 5.55 for pyrazinamide and pH 7.2 for isoniazid. An inoculum of 0.1 ml of this suspension was used for each tube in the test series. (This is the equivalent of from 2 x 10^5 to 1 x 10^6 culturable units per tube.)

An autoclaved stock solution of isoniazid in distilled water, 10 mg per ml, was used. Serial twofold dilutions were made in Tween-albumin medium (pH 7.2) to give a final volume of 5 ml per tube. If the culture to be tested was believed to be in the susceptible range, the concentrations of isoniazid used ranged from 0.008 to 2.0 μg per ml, whereas if the test culture was believed to be resistant, isoniazid concentrations that ranged from 0.8 to 200 μg per ml were employed. One tube of Tween-albumin medium containing no drug was included in each test series.

The stock solution of pyrazinamide was 5 mg per ml of albumin-glucose (4.5% bovine serum albumin fraction V plus 5% glucose in physiologic saline) sterilized by Seitz filtration. This solution was diluted in albumin-glucose, and 0.5 ml of each dilution was added to 4.5 ml of Tween basal medium which had been adjusted to pH 5.55 with hydrochloric acid. The final concentrations of pyrazinamide were 500, 100, 20 and 10 μg per ml. One tube of Tween-albumin medium, adjusted to 5.55, was included in each series. A 7 day culture of *Mycobacterium tuberculosis* H37Rv was included as a control for each drug.

All cultures were examined after 7 and 14 days' incubation at 37°C. The drug susceptibility of the culture was defined as the lowest concentration of drug in which no macroscopic growth could be detected after 14 days' incubation. If this concentration was below 0.064 μg for isoniazid or 100 μg for pyrazinamide, the culture was considered to be susceptible to the corresponding drug.

Antibiotics—*Azotobacter*

D. A. van Schreven (1966). Antonie van Leeuwenhoek 32, 67–93.

The following media were used:

I. Yeast-mannitol-agar (YMA): nutrient agar, 15.0 g; mannitol, 15.0 g; $CaCO_3$, 2.0 g; K_2HPO_4, 0.5 g; $MgSO_4 \cdot 7H_2O$, 0.2 g; NaCl, 0.2 g; yeast extract, 100 ml; water, 900 ml.

II. YMA + 0.25% peptone.

III. Mannitol-glucose-agar (MGA): nutrient agar, 15.0 g; mannitol, 2.0 g; glucose, 0.2 g; $CaCO_3$, 0.5 g; K_2HPO_4, 1.0 g; $MgSO_4 \cdot 7H_2O$, 1.0 g; soil extract, 200 ml; water, 800 ml.

IV. MGA + 0.25% peptone.

V. Potato flour-agar (PFA): nutrient agar, 20.0 g; potato flour, 20.0 g; water, 1000 ml.

VI. PFA + 0.25% peptone.

VII. Medium C of Morales, 1957: nutrient agar, 15.0 g; 10% mannitol, 20 ml; 10% glucose, 20 ml; ethyl alcohol, 8 ml; soil extract, 100 ml; and 852 ml of a solution containing the following amounts of salt per liter of distilled water: K_2HPO_4, 0.5 g; $MgSO_4 \cdot 7H_2O$, 0.2 g; NaCl, 0.5 g; $MnSO_4$, trace; $FeCl_3$, trace.

VIII. Medium VII + 0.25% peptone.

IX. YMA + 20% soil extract +0.5% glucose + 0.25% peptone + 10% horse serum (1.3% agar).

X. Medium IX + 2.5% NaCl.

XI. Nutrient agar, 13 g; glucose, 5.0 g; $CaCO_3$, 2.0 g; K_2HPO_4, 0.5 g; $MgSO_4 \cdot 7H_2O$, 0.2 g; NaCl, 25 g; peptone, 2.5 g; soil extract, 200 ml; water, 800 ml.

On a number of plates a drop of sterile water containing penicillin was deposited on the surface of the medium at a marked point near the edge of the agar plate. After 4 days a loopful of a suspension of *Azotobacter* was streaked from the centre of the plate outward to this point. This method will be referred to as the streak-test.

On other plates a drop of sterile water containing penicillin was deposited in the centre of the medium three days before or immediately after the inoculation of the whole surface of the plate with *Azotobacter*.

In other cases penicillin was included at the time of preparation of the medium. It was added after sterilization of the agar media as a sterile aqueous solution. The amounts used are mentioned in the text.

All media were sterilized for 30 min at 120°C. Glucose and horse serum were sterilized separately by filtration. Petri dishes were used with an inner diameter of 13.5 cm, containing 40 ml agar medium.

In most cases *Azotobacter* strain C 41 that had been isolated from garden soil was used. In many cases also strains C 76 and C 77 were used, which were kindly supplied by the Microbiological Laboratory of the Institute for Soil Fertility in Groningen. Ten strains were tested on Medium IX. Before using the strains their purity was tested in peptone-water and on milk-agar according to the method of Gray (1953).

The cultures were incubated at 29°C and studied by means of a Wild phase-contrast microscope. A number of micrographs were made with a Philips electron microscope using gold-palladium for shadow casting.

E. A. Gray (1953). Nature, Lond. 171, 1163.
J. Morales (1957). Nuevo estudio biologico del genero *Azotobacter* (descubrimiento de sus formas L). Thesis. Madrid.

Antibiotics—*Pseudomonas*

R. R. Gillies and J. R. W. Govan (1966). J. Path. Bact. 91, 339–345.

Pyocines are naturally occurring antibiotic substances produced by many strains of *Pseudomonas pyocyanea (Ps. aeruginosa)* and active mainly against other strains of this species.

Culture media. Tryptone soya agar (Oxoid) was prepared according to the maker's instructions and 5 per cent of horse blood was incorporated. We now use this tryptone soya blood agar (TSBA) as a routine for growing producer strains. Infusion broth (Cruickshank *et al.*, p. 190) was used to grow the indicator strains.

Pyocine typing technique.

The strain to be typed (a potential pyocine producer strain) is streaked diametrically across a plate of TSBA in such a way that the width of the inoculum is *c.* 1 cm. The plate is then incubated at 32°C for 14 hr. The macroscopic growth is removed with a glass slide, 3 – 5 ml $CHCl_3$ is placed in the lid of the petri dish and the dish with the medium is replaced on the lid for 15 min so that microscopic remnants of the culture are killed.

The plate is then opened and the residual $CHCl_3$ is decanted and retrieved for future use by filtering through filter paper. Traces of $CHCl_3$ vapour are eliminated from the culture plate by exposing it to air for a few minutes.

Cultures of the 8 indicator strains grow in infusion broth for 4–6 hr at 37°C are streaked on to the $CHCl_3$-treated medium at right-angles to the line of the original inoculum; 5 strains are applied on the left side of the line and the other 3 strains on the right side. The plate is then incubated at 37°C for 8–18 hr.

Any pyocines produced by the original inoculum diffuse into the medium during the first period of incubation and then exert their inhibitory action on the indicator strains during the second. The pyocine types of the strains under test are recognised from the patterns of inhibition they produce on the indicator strains.

A table showing the activity of producer strains against indicator strains is given.

R. Cruickshank *et al.* (1960). *Mackie and McCartney's Handbook of Bacteriology*, 10th ed., Edinburgh and London, pp. 190 and 195.

Antibiotics—*Staphylococcus*

M. Kocur, F. Přecechtěl and T. Martinec (1966). J. Path. Bact. 92, 331–336.

Sensitivity to antibiotics was investigated in diffusion tests on nutrient agar; the quantity of antibiotic per tablet was: penicillin, 10 units; streptomycin, 20μg; tetracycline, 50 μg; chloramphenicol, 20 μg; erythromycin, 20 μg. Spontin, 30 μg per tablet, was obtained from Abbott Laboratories, Chicago, U.S.A.

Antibiotics—*Staphylococcus*

A. T. Willis, J. A. Smith and J. J. O'Connor (1966). J. Path. Bact. 92, 345–358.

Antibiotic sensitivity tests. Antibiotic sensitivity tests were performed on sensitivity test agar (Oxoid) with "Multodisks" (Oxoid) containing the following amounts of antibiotics: penicillin, 1.5 units; streptomycin, tetracycline, chloramphenicol and erythromycin, 10 μg; novobiocin and oleandomycin, 5 μg. All strains were also tested for sensitivity to neomycin with low potency "Sentests" tablets (Evans).

Antibiotics—*Mycobacterium*

H. Boisvert (1966). Annls Inst. Pasteur, Paris 111, 180–192.

Sensibilité aux antibiotiques mesurée par la méthode des proportions [12] Pour l'identification, on doit éprouver la sensibilité de la souche à l'INH, à l'hydrazide de l'acide tiophene 2-carboxylique (TCH), au pyrazinamide (PZA) pour les souches l'NH resistantes, et à la cyclosezine en ce qui concerne le BCG.

[12] Canetti (G.), Rist (N.) et Grosset (J.). Rev. Tuberc., 1963, 27, 263.

Antibiotics—*Mycobacterium*

R. E. Gordon (1966). J. gen. Microbiol. 43, 329–343.

Sensitivity to penicillin. A tube containing 20 ml of 'yeast dextrose agar' was inoculated with one or two loopsful, depending on the turbidity, of a 14- to 28-day-old culture in glucose broth. The agar was carefully mixed and poured into a sterile Petri dish. After the agar solidified, a penicillin disc (Bacto-sensitivity disc, 10 units) was placed on the agar with sterile forceps. Because the agar was sometimes thin at the centre of the plate, the disc was placed approximately 2 cm away from the centre. Plates were incubated and observed at 3, 5, and 7 days for a zone of inhibition around the disc. Some cultures were so sensitive to penicillin they did not grow on the plate. In such instances the cultures used as inoculum were tested for viability.

Antibiotics—*Mycobacterium*

M. Tsukamura (1966). J. gen. Microbiol. 45, 253–273.

The niacin test was done according to Konno (1963).

K. Konno (1963). Kyubu-Shikkan (in Japanese), 7, 525.

Antibiotics—*Staphylococcus aureus*

W. A. Zygmunt, H. P. Browder and P. A. Tavormina (1966). Can. J. Microbiol. 12, 341–345.

Antibiotic Sensitivity Tests.

Minimal inhibitory concentrations (m.i.c.) for each antibiotic were determined after a 20-hr incubation at 37°C by conventional twofold serial dilution methods in a liquid medium (Antibiotic Medium No. 3, Difco). An inoculum level approximating 1.0×10^6 viable cells/ml of medium was used.

Twelve antibiotics were studied and are listed:—

Pencillin G, Ampicillin, Propicillin, Methicillin, Ancillin, Cloxacillin, Nafcillin, Oxacillin, Cephalothin, Ceporin, Fusidic acid, Lysostaphin.

The lysostaphin preparation used had a microbiological potency of 250 units/mg and assayed about 95% protein by the method of Lowry *et al.* (1951) when lysozyme was used as a standard. By disc electrophoresis (Reisfeld *et al.* 1962) the major protein component was shown to be the antibiotic.

The procedures used for the preparation of all antibiotic solutions were similar to those previously described (Zygmunt, Harrison and Browder 1965).

O. H. Lowry, N. J. Rosebrough, A. L. Farr and R. J. Randall (1951). J. biol. Chem. 193, 265–275.

R. A. Reisfeld, U. J. Lewis and D. E. Williams (1962). Nature, Lond. 195, 281–283.

W. A. Zygmunt, E. F. Harrison and H. P. Browder (1965). Appl. Microbiol. 13, 491–493.

Antibiotics—various

P. Chadwick and D. E. Mahony (1966). Can. J. Microbiol. 12, 683–690.

Recently (1) we reported a rapid method for testing the antibiotic sensitivity of bacteria in which the criterion of sensitivity was the inhibition of microcolony formation on agar plates containing known concentrations of antibiotic. The results of the tests were available 4 hours after inoculation of the plates. The tests were performed on pure cultures so that preliminary isolation of the organisms was necessary.

Performance of the microcolony test for antibiotic sensitivity directly on pathological material would constitute a significant advance. A report on the antibiotic sensitivity of an organism would be available 4 hours after receipt of a specimen in the laboratory.

In this paper we report a pilot study which was carried out to assess the magnitude of some of the problems involved in devising a direct, rapid, antibiotic sensitivity test based on microcolony formation.

Materials and Methods

Antibiotics

These were obtained in powder form from Dr. L. Greenberg, Biologics Control Laboratory, Laboratory of Hygiene, Ottawa, Ontario, with the exception of lincomycin, which was obtained from the Upjohn Company, Kalamazoo, Michigan, U.S.A., and ampicillin, obtained as "Penbritin" from Ayerst, McKenna and Harrison Ltd., Montreal, Que. Solutions of antibiotics were made at a strength of 1 mg per ml in sterile distilled water. Antibiotic plates were made by adding the appropriate volume of fresh solution to molten agar at 50°C and pouring plates immediately. The concentrations of antibiotics in the agar were penicillin, 0.1 and 1.0 unit/ml; erythromycin, 0.4 and 0.8 μg/ml; novobiocin, 2 and 36 μg/ml; tetracycline, 5 and 16 μg/ml; streptomycin, 5 and 15 μg/ml; chloromycetin, 8 and 16 μg/ml; kanamycin, 6 and 16 μg/ml; neomycin, 6 and 16 μg/ml; colymycin, 3 and 10 μg/ml; polymyxin B, 20 units/ml; framycetin, 20 μg/ml; lincomycin, 1.6 μg/ml; oxacillin, 1 μg/ml; methicillin, 5 μg/ml; and ampicillin, 6 and 20 μg/ml.

Source of Organisms

Organisms were obtained directly from pathological specimens, which consisted of infected urines, blood from patients with bacteremia or septicemia, or swabs from infected lesions.

Direct Antibiotic Sensitivity Tests

Swabs were rinsed in 1 ml of 1/4 strength Ringer solution. Urines were diluted 10^{-1} and 10^{-2} in 1/4 strength Ringer solution. Blood specimens were centrifuged at about 500 *g* and the supernatants were diluted 10^{-1} and 10^{-2} in 1/4 strength Ringer solution. The swab rinsings and the dilutions of urines and blood were transferred to antibiotic and control plates by means of a multiple inoculator designed in our laboratory

(1). The plates were marked with a grid so that the inoculation sites could be found quickly during subsequent microscopy. Sixteen inocula (representing from 8 to 16 specimens depending on the type of specimen) were placed on one plate. All specimens were inoculated onto infusion agar and onto a set of antibiotic plates. Some of the early specimens in the series were tested against only a few antibiotics; later, most or all of the antibiotics in the above list were used. Plates were incubated for 4 hours at 37°C. The agar surface was then examined with the 16-mm objective of an ordinary light microscope. The infusion agar control plate was examined first and if heavy or moderate growth of microcolonies was visible on this plate the antibiotic plates were examined similarly. The growth patterns on the antibiotic plates were classified in one of five categories.

1. Microcolonies equal or nearly equal in numbers and size to those on the control infusion agar plate. This appearance was interpreted as resistance to the antibiotic.
2. Individual bacterial cells scattered evenly over the inoculated agar surface with no formation of microcolonies. This appearance was interpreted as sensitivity to the antibiotic.
3. Individual bacterial cells as in (2) above but with, in addition, a few microcolonies usually smaller than, but in some cases equal in size to, those on the control plate.
4. Microcolonies equal or nearly equal in numbers to those on the control plate but very much smaller and of virtually uniform size.
5. Microcolonies somewhat smaller than those on the control plates, almost equal in numbers but showing a distorted morphology. This distortion occurred only with Gram-negative bacilli and consisted in a different general shape of the microcolony (usually elongated), with longer and often more refractile individual bacilli. It was seen exclusively on the novobiocin and ampicillin plates.

To avoid, as far as possible, subjectivity in the interpretation of microcolonial appearances, definite opinions on antibiotic sensitivity were expressed only when the appearances were unequivocal, i.e. showed the pattern in 1 or 2 above. Tests in which intermediate or distorted growth occurred were regarded as invalid.

After the plates had been read, they were returned to the 37°C-incubator, and reexamined after a total incubation period of 20 hours, for the presence or absence of growth visible by naked eye. At this stage, growth patterns on the antibiotic plates were classified in one of four categories.

1. Heavy growth equal to or nearly equal to that on the control plate. This was interpreted as "resistance".
2. No growth, interpreted as "sensitivity".
3. A few well-developed colonies, the control plate showing confluent growth. This was interpreted as a sensitive population containing a few

resistant organisms (S (R) growth).

4. A slight, confluent growth much less than that on the control plate (± growth).

Indirect Antibiotic Sensitivity Tests

Specimens examined by the direct test were plated on appropriate media for isolation of their bacterial flora. Representative colonies from this flora were tested for antibiotic sensitivity by the microcolony method (1). An infusion agar plate was again included as a control. Plates were incubated at 37°C, examined microscopically at 4 hours and by naked eye at 20 hours.

(1) D. E. Mahony and P. Chadwick (1965). Can. J. Microbiol. 11, 829–836.

Antibiotics—clinical specimens

J. V. Bennett, J. L. Brodie, E. J. Benner, and W. M. M. Kirby (1966). Appl. Microbiol. 14, 170–177.

The authors describe apparatus for the performance of large numbers of tests for antibiotic concentration in human serum simultaneously.

Antibiotics—*Staphylococcus*

M. Bals and M. Bratu (1966). Appl. Microbiol. 14, 582–583.

The action of lincomycin was determined by dilutions in nutrient agar distributed in petri dishes. The following concentrations were used: 0.01, 0.1, 0.5, 1, 10, 25, 50, and 100 μg/ml. The inoculum for the study of the bacteriostatic and bactericidal action was, as a rule, of the order 10^8 to 10^{10} organisms per milliliter. The inhibition was considered slight (or partial) when separate colonies were obtained, marked when fewer than 10 colonies grew, and total when no colonies developed.

To appraise the bactericidal effect upon staphylococci in the logarithmic phase (3-hr culture) and in the stationary phase (24-hr culture), inoculations were made on media containing increasing concentrations of lincomycin. In all cases, the inocula were approximately 10^3 organisms per milliliter. At various intervals, transplants with cellophane (Chabbert and Patte, 1960) were made to fresh, antibiotic-free media. When fewer than 10 colonies per milliliter developed on the transplants, the bactericidal action was appraised as being marked; when the transplant remained sterile, the bactericidal action was considered to be total.

Y. A. Chabbert and J. C. Patte (1960). Appl. Microbiol. 8, 193–199.

Antibiotics—anaerobes

A.-R. Prévot (1966). Techniques pour le diagnostic des Bactéries Anaérobes. Éditions de la Tourelle, St. Mandé.

Antibiogramme anaérobie.

La recherche des antibiotiques inhibant de la façon la plus active les anaérobies est devenue obligatoire depuis que l'on sait:

1° que pour beaucoup de ces infections très graves et à marche rapide, l'antibiotique préférentiel est le seul moyen de sauver le malade et qu'il faul l'administrer précocement et à fortes doses.

2° que pour une même espèce anaérobie, certaines souches repondent mieux à tel antibiotique qu'à tel autre.

Diverses techniques ont été proposées. Elles ont toutes été étudiées comparativement dans notre service. C'est celle qui nous donne quotidiennement les résultats les plus rapides et les plus valables que nous décrirons. Nous pratiquons des antibiogrammes étendus avec les 25 antibiotiques classiques en utilisant les disques de Chabbert. On peut disposer 4 disques par boîte de Petri. C'est donc 6 boîtes de Petri qu'il faut manipuler. (On peut également opérer avec 4 ou 5 disques par boîtes, mais cela rend certaines lectures douteuses).

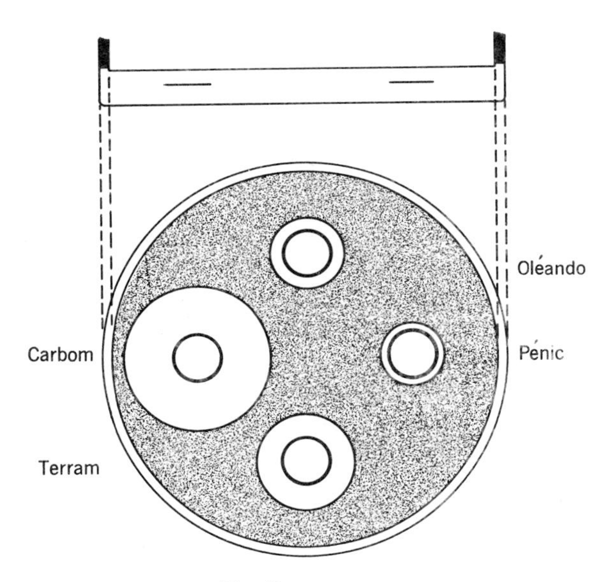

Fig. 3.
Antibiogramme anaérobie

Dans le couvercle on dépose 4 gouttes de gélose dure, fondue, aux sommets d'un carré tel que la distance d'un des sommets au bord du couvercle soit environ la 1/2 de la longueur d'un des côtés du carré.

On colle une pastille sur chacune de ces gouttes et on laisse la gélose se coaguler. Puis délicatement on coule sur l'ensemble une gélose VF glucosée anaérobie encore fluide ensemencée extemporanément avec une culture encore jeune de la souche (autant que possible en phase exponentielle). Il faut que la hauteur de la couche soit environ 1/2 cm. Très délicatement on dépose à la surface de la gélose encore fluide la boîte inversée, qu'on maintient quelques secondes jusqu'au moment où la gélose est recoagulée.

Il faut faire très attention afin que aucune bulle d'air ne soit emprisonnée entre gélose et verre. On scelle à la paraffine l'intervalle entre boîte et couvercle en vue d'empêcher la dessication et on porte à l'étuve. La lecture se fait en général dans les 12 heures pour les espèces à croissance rapide, dans les 24 heures pour les espèces à croissance moyenne, dans les 48 heures pour les espèces à croissance lente. Pour les espèces très gazogènes comme *W. perfringens* il est recommandé d'employer une gélose VF additionnée de 3 à 4 ‰ de nitrate de sodium; celui-ci servant d'accepteur d'hydrogène, il n'y a pas de dégagement gazeux pour gêner la lecture.

Acetylacetic acid esters—*Mycobacterium*

J. Soleil, J.-G. Marchal and V. Loppinet (1966). Annls Inst. Pasteur, Paris 111, 373–375.

The authors report the activity of 32 esters of acetyl-acetic acid on two strains of *Mycobacterium.* They used the method of dilution in broth and diffusion in agar. No details of methods are given.

Bile and brilliant green—enterobacteria

V. R. Miller and G. J. Banwart (1965). Appl. Microbiol. 13, 77–80.

The paper reports a study of the combined action of bile and brilliant green, each in varying concentrations, on the growth of enterobacteria.

The following basal medium and components were used.

The basal medium contained in a volume of 1 liter: 5 g of Proteose Peptone No. 3 (Difco), 3 g of D-mannitol (BBL), and 0.02 g of bromocresol purple (MCB) ND180.

Brilliant green (1 g; 98% dye content; Difco) was dissolved in 100 ml of distilled water and further dilutions were made as required to add to the basal medium.

Bile salts (Difco) were weighed into those media containing 2 or 4% levels, whereas a 1% solution was dispensed with the basal medium for lower concentrations.

When the higher concentrations of brilliant green and bile salts were added without prior dilution, a precipitate formed during preparation. This difficulty was overcome by adding the brilliant green to the basal medium and diluting to approximately 1 liter with distilled water, then adding the required amount of bile salts, and diluting to 1 liter.

The various media were tubed in 9-ml amounts and autoclaved at 15 psi for 15 min. The media were freshly prepared each week and stored in a dark 15°C walk-in cooler until tested.

The cultures were grown in nutrient broth for 24 hr at 37°C. After incubation, 1 ml of each of these cultures was transferred, respectively, to 100 ml of 5% Proteose Peptone No. 3 (Difco). The inoculated peptone was incubated at 37°C for 24 hr. A three-tube most probable number (MPN) technique (Hoskins, 1934) was used to determine the ability of each of the test media to support the growth of the organisms. The organisms were diluted in sterile 0.1% peptone water, and 1 ml of inoculum was added to each of three tubes of media. These inoculated tubes were incubated at 37°C for 24 hr. Color change of the broth was recorded as positive growth. Brain Heart Infusion (BHI) broth (Difco) was used as a control. A plate count was also performed, on Plate Count Agar (Difco), to use as a comparison of the number of microorganisms determined with BHI and the basal medium.

Aerobacter strains behaved like *Salmonella* strains in all but three combinations of bile and brilliant green. These three contained respectively:—

(i) brilliant green, 10 mg/L; bile salts, 0.25 g/L, (ii) brilliant green, 20 mg/L; bile salts, 0.5 g/L and (iii) brilliant green, 20 mg/L; bile salts, 1 g/L.

These media gave complete inhibitation of *Escherichia* and considerable inhibitation of *Aerobacter*. *Salmonella* species were not, or only slightly, inhibited.

J. K. Hoskins (1934). Publ. Hlth Rep. Wash. 49, 393–405.

Cadmium sulfate—*Pseudomonas aeruginosa*

A. H. Wahba and J. H. Darrell (1965). J. gen. Microbiol. 38, 329–342.

The organisms were tested for growth on 0.2% (w/v) cadmium sulfate in nutrient agar No. 1.

Nutrient agar No. 1 (Oxoid No. 1) contained Lab Lemco beef extract, 1 g; yeast extract (Oxoid L 20), 2 g; peptone (Oxoid L 37), 5 g; sodium chloride, 5 g; agar, 15 g; distilled water to 1 L; pH 7.4. The test was modified from one containing cadmium chloride by I. L. Shkeir, F. L. Losse and A. N. Bahn (1963). Bact. Proc. p. 71.

Author's note: More recent work by J. H. Darrell and in other centres would indicate that 0.2% cadmium sulphate is a little too high. A concentration of 0.1% in nutrient agar provides a more sensitive test with only slight reduction in specificity.

Carbamate test—*Brucella*

M. J. Pickett, E. L. Nelson and J. D. Liberman (1953). J. Bact. 66, 210–219.

Several representative strains of brucellae were examined according to the procedure of Renoux (1952) with satisfactory results. For convenience, however, subsequent tests were made with the reagent tableted, as with the dyes, and with the sodium diethyldithiocarbamate at a concentration (weight: weight) of 1:200. For the test, one-quarter of a petri plate was inoculated uniformly with one 2 by 10 mm loopful of a stock suspension of brucellae. A carbamate tablet then was placed on this inoculated quadrant, and the plate was incubated (without prior overnight refrigeration as with the dyes) for two days. The results obtained from these tests are expressed as the width, in mm, of the outer zone of inhibited growth.*

*In sterile control quadrants there are four rings or zones surrounding each tablet. Inhibition occurs with nearly all strains of *B. melitensis* and *B. suis* in the first wide zone; inhibition is variable in the second and third zones; most strains of brucellae show enhanced growth in the fourth zone; and only strains of *B. melitensis* are strongly inhibited beyond the fourth zone.

G. Renoux (1952). Annls Inst. Pasteur, Paris 82, 556–562.

Carbonyl and thiocarbamyl disulfides—*Aerobacter*

J. D. Buckman, B. S. Johnson and L. Field (1966). Can. J. Microbiol. 12, 1263–1267.

Thirteen unsymmetrical disulfides of the structure –C (=X)SS –, where X is O or S, were evaluated for bactericidal and fungistatic activity under conditions simulating an industrial environment. Some carbonyl disulfides, RC(O)SSR′, were active fungistatic agents and promising bactericides. Some of the thiocarbamyl disulfides, $R_2NC(S)SSR'$, were found to be promising fungistatic agents and also showed significant bactericidal activity. Correlations of biological activity with chemical properties are discussed.

Chloral hydrate—*Bacillus anthracis*

R. F. Knisely (1965). J. Bact. 90, 1778–1783.

A 10% solution of chloral hydrate (USP) was freshly prepared, seitz-filtered, and added to Heart infusion agar (Difco) in a final concentration of 0.25%.

Ethanol tolerance

R. Steel and T. K. Walker (1957). J. gen. Microbiol. 17, 445–452.

Glucose, 3% (w/v); Difco yeast extract, 0.5% (w/v); dist. H_2O; + 2–14% (v/v) ethanol (95%) in increments of 2%, added asceptically to sterile medium.

Ethyl hydrocuprine—*Streptococcus*

G. Andrieu, L. Enjalbert and L. Lapchine (1954) Annls Inst. Pasteur, Paris, 87, 617–634.

Le milieu à l'éthyl-hydrocuprine (E.H.C.) est préparé au moment de l'emploi. La solution mère, en eau distillée stérile à 1/5000, se garde bien en glacière. A partir de cette dilution on prépare une gélose-sérum contenant 1/50000 de H.E.C. et l'on coule en boîtes de Petri.

Fatty acids—*Staphylococcus*

S. J. Edwards and G. W. Jones (1966). J. Dairy Res. 33, 261–270.

The susceptibility of coagulase-negative and coagulase-positive staphylococci to various fatty acids, found by Adams and Rickard (1963) to be present in the keratin-like material of the bovine teat duct, was investigated. Sodium salts of saturated and unsaturated fatty acids (C_6–C_{20}) were prepared and incorporated at concentrations of 1, 10, 50 μg/ml in a liquid medium composed of 20 g Bacto Casamino acids, 1.29 g KH_2PO_4, 7.06 g K_2HPO_4, 500 ml yeast dialysate, and water to final volume of 1 L. The yeast dialysate was prepared by dialysing a 4% (w/v) solution against 4 times its volume of distilled water at 4°C for 7 days.

E. W. Adams and C. G. Rickard (1963). Am. J. vet. Res. 24, 122.

Glycine tolerance—*Vibrio sputorum*

W. J. Loesche, R. J. Gibbons and S. S. Socransky (1965). J. Bact. 89, 1109–1116.

Thioglycollate medium without added dextrose (BBL) supplemented with 0.2% yeast extract (Difco) and 0.1% KNO_3 was selected as a basal medium because of the characteristic growth of the oral vibrios in the upper third of the tube. To this medium 1% glycine was added.

All cultures were incubated at 35 to 37°C in Brewer jars containing an atmosphere of 95% H_2 and 5% CO_2.

Hydrogen peroxide—*Serratia marcescens*

See p. 464 of these abstracts.

Hydrazide of thiophene-2-carboxylic acid—*Mycobacterium*

A. Tacquet, F. Tison and B. Devulder (1965). Annls Inst. Pasteur, Paris 108, 514–525.

The authors studied the sensitivity by culture on a medium of Löwenstein-Jensen containing 10 mg of thiophene-2-carboxylic acid hydrazide (T.C.H.) per ml. [see Bönicke (R.). Bull. Un. int. Tuberc., 1962, 32, 13–76.]

Hydroxylamine—*Mycobacterium*

M. Tsukamura (1966). J. gen. Microbiol. 45, 253–273.

Hydroxylamine resistance was defined as an ability of slow-growing mycobacteria to grow on the Löwenstein-Jensen medium or the Ogawa egg medium containing 0.025% (w/v) hydroxylamine hydrochloride after incubation for 3 weeks or an ability of rapid-growing mycobacteria to grow on the media containing 0.05% hydroxylamine after incubation for 1 week (Tsukamura, 1965).

M. Tsukamura (1965). J. Bact. 90, 556.

Hypocholesteraemic compounds—various

S. Aaronson (1965). J. gen. Microbiol. 39, 367–371.

Micro-organisms from the collection of Haskins Laboratories and the Biology Department, Queens College, New York, were used. For disc assays, filter-paper discs (12.7 mm diam) were soaked in solutions of the test compound, dried and autoclaved. Various agar media appropriate for the growth of the micro-organisms, e.g. nutrient agar for the bacteria, and Sabouraud's glucose agar for the fungi, were seeded with test micro-organisms. Cultures were incubated at 25°C for 5–10 days. In some experiments chemically defined liquid media were used appropriate to the protozoan or alga. Benzmalecene (BM; *N*-[1-methyl-2,3-di-(*p*-chlorophenyl)-propyl] maleamic acid) was supplied by Dr. D. Hendlin (Merck, Sharp and Dohme Laboratories, Rahway, N. J.). Triparanol (TR; 1-[*p*-(β-diethyl-aminoethoxy)-phenyl]-1-(*p*-tolyl)-2-*p*-chlorophenyl ethanol) was supplied by Dr. F. J. Murray (W. S. Merrell, Cincinnati, Ohio). Benzmalecene was dissolved in dilute alkali, and triparanol in 95% (v/v) ethanol in water. Oleic acid (99% pure by gas-liquid chromatography) was bought from the Hormel Institute, Austin, Minnesota. The other chemicals were of a commercial grade.

Mercury sensitivity test—*Staphylococcus*

A. T. Willis and G. C. Turner (1963). J. Path. Bact. 85, 395–405.

Moore recommended spot-inoculation of overnight broth cultures on to freshly prepared peptone agar containing 1 in 27,500 mercuric chloride. Staphylococci that grew in 12–18 hr were regarded as mercury-resistant. We have found this method unsatisfactory because, as Moore showed, mercuric chloride is rapidly inactivated in the medium so that plates cannot be stored for more than 3 days. We have developed a disc method which has the advantage that ordinary nutrient agar plates can be used.

Preparation of mercuric chloride discs. Discs 6 mm in diameter were punched from sheets of Whatman No. 1 filter paper and were dry-sterilised in freeze-drying tubes in batches of 500. A 1 in 1000 solution of mercuric chloride was prepared by dissolving a tablet of mercuric chloride, BPC, in 600 ml distilled water. The solution was added in 5 ml volumes to batches of 500 discs. The whole volume of fluid was absorbed (see Gould and Bowie, 1952) so that each disc contained about 10 μg mercuric chloride. The discs were then freeze-dried by the single-stage method described by Flewett *et al.* (1955), transferred to sterile screw-capped bottles and stored at room temperature. Similar control discs, impregnated with 0.9 per cent sodium chloride, were prepared.

T. H. Flewett, K. S. Zinnemann, M. W. C. Oldfield, H. S. Shucksmith and F. Dexter (1955). Lancet 1, 888.

J. C. Gould and J. H. Bowie (1952). Edinb. med. J. 59, 178.

B. Moore (1960). Lancet 2, 453.

Nalidixic acid—*Streptococcus, Listeria, Erysipelothrix*

H. Beerens and Mme M. M. Tahon-Castel (1966). Annls Inst. Pasteur, Paris 111, 90–93.

Milieu de culture. – Il s'agit d'une gélose nutritive au sang de cheval additionnée de 40 μg d'acide nalidixique (1) par millilitre. On la prépare ainsi:

Extraite de viande	3 g
Tryptose (Difco)	10 g
Chlorure de sodium	5 g
Acide nalidixique	0.04 g
Agar en poudre	15 g
Eau distillée	1000 ml

Dissoudre par chauffage doux. Ajuster le pH à 7.2. Répartir à raison de 20 ml par tube de 200 x 20. Autoclaver vingt minutes à 120°C. Au moment de l'emploi, faire fondre au bain-marie bouillant, laisser refroidir à 50° environ, ajouter stérilement 1,5 ml de sang de cheval. Agiter pour obtenir un mélange homogène. Couler en boîtes de Petri.

L'acide nalidixique supporte la stérilisation; on peut cependant l'ajouter extemporanément, en même temps que le sang, en solution dans l'eau distillée stérile.

Technique d'étude. – Le comportement de 300 souches de bactéries aérobies appartenant aux genres les plus fréquemment recontrés en bactériologie médicale a été envisagé. Toutes avaient été isolées sur milieux ne contenant pas d'acide nalidixique. Chaque souche a été ensemencée à la fois sur le milieu sélectif et sur milieu témoin sans antibiotique à partir d'une culture en bouillon de dix-huit heures. L'incubation a été faite à 37°C.

L'absence de pouvoir inhibiteur de l'acide nalidixique, utilisé à la concentration de 40 μg/ml, vis-à-vis des streptococci, enterococci, *D. pneumoniae, Listeria, Erysipelothrix,* a été mise en évidence en ensemençant 0,1 ml d'une dilution d'une culture en bouillon à la surface, d'une part du milieu décrit, d'autre part d'un milieu témoin sans acide nalidixique. L'inoculum a été réparti à l'étaleur. Le dénombrement des colonies a été effectué après dix-huit heures d'incubation à 37°C.

(1) L'acide nalidixique nous a été fourni par les Laboratoires Winthrop.

Nitrite—*Brucella*

M. J. Pickett and E. L. Nelson (1954). J. Bact. 68, 63–66.

Nitrite tablets were prepared and employed according to the procedure already described for carbamate tablets (Pickett *et al.* 1953). Initial trials with tablets containing 10, 5, 4, 2, 1 and 0.5 per cent of sodium nitrite led to the selection of 2 per cent nitrite as the concentration which gave most satisfactory results.

M. J. Pickett, E. L. Nelson and J. D. Liberman (1953). J. Bact. 66, 210–219. [See **Carbamate test**—*Brucella* p. 88.]

Phenethyl alcohol—*Bacillus anthracis*

R. F. Knisely (1965). J. Bact. 90, 1778–1783.

PEA medium. PEA (2-Phenylethanol, Matheson, Coleman and Bell, Cincinnati, Ohio) was added to Heart infusion agar (Difco) in a final concentration of 0.3%.

Phenethyl alcohol—*Bacillus cereus*

C. C. Remsen, D. G. Lundgren and R. A. Slepecky (1966). J. Bact. 91, 324–331.

The substance inhibits the development of the spore septum and membranes of *Bacillus cereus.*

Phenol—*Chromobacterium*

P. H. A. Sneath (1956). J. gen. Microbiol. 15, 70–98.

A sample (0.8 ml) of a nutrient broth culture grown at 25°C for 2 days and then well shaken was mixed with 0.2 ml of 5% (w/v) aqueous phenol. It was incubated at 20°C and after 5 and 10 min 0.05 ml samples were transferred to 5 ml lots of nutrient broth, of which a loopful was plated on to blood agar. The plates and broths were incubated at 25°C for 4 days, the broths being then plated for sterility.

Phenol—*Lactobacillus*

R. E. Smith and J. D. Cunningham (1962). Can. J. Microbiol. 8, 727–735.

Phenol was added to (Difco) APT broth to give final concentrations of 0.2, 0.3, 0.4, 0.5, and 0.6% (w/v). Inoculated tubes were incubated for 2 weeks and examined for growth, using appropriate controls for comparison. The authors used 24 hr lactic broth cultures (Elliker, Anderson and Hannesson, (1956). J. Dairy Sci. 39, 1611–1612) as inocula.

Picric acid—*Mycobacterium*

M. Tsukamura (1966). J. gen. Microbiol. 45, 253–273.

Picric acid (0.1%, w/v) tolerance was defined as ability to grow on the Sauton agar medium containing 0.1% (w/v) picric acid after incubation for 3 weeks (Tsukamura, 1965).

Picric acid (0.2%, w/v) tolerance. Ability to grow on Sauton agar containing 0.2% picric acid after 3 weeks.

M. Tsukamura (1965). Am. Rev. resp. Dis. 92, 491.

Pteridine derivative 0/129—*Pseudomonas odorans*

I. Málek, M. Radochová and O. Lysenko (1963). J. gen. Microbiol. 33, 349–355.

The method of Rhodes (1959) was used for determining resistance toward the pteridine derivative 0/129 (2,4-diamino-6,7-di-isopropylpteridine).

M. E. Rhodes (1959). J. gen. Microbiol. 21, 221.

Sodium azide—*Moraxella*

W. J. Ryan (1964). J. gen. Microbiol. 35, 361–372.

Resistance to sodium azide was examined by incorporating it in blood agar. A drop of bacterial suspension sufficient to give semiconfluent growth on control plates was used as inoculum. Results were read after 3–4 days of incubation.

Steroids (nitrogen containing)—various

R. F. Smith, D. E. Shay and N. J. Doorenbos (1963). J. Bact. 85, 1295–1299.

Bacteria were cultured on nutrient agar and Eugonagar (BBL, 01-265), depending upon their growth requirements. Yeasts and molds were grown on Sabouraud agar, and the clostridia were tested on anaerobic agar using the Brewer anaerobic petri dish. Heat-sensitive steroids were sterilized by filtration and added to the culture media. Aqueous stock solutions of the steroids were prepared by using 95% ethanol or dilute HCl.

Because of the limited quantity of each compound that was tested (100 mg), the gradient plate technique described by Szybalski and Bryson (1952) was employed. The use and applicability of the gradient plate technique for screening purposes were described by Braude, Banisster, and Wright (1955). All the tests were repeated two to six times.

The nomenclature and structure of each steroid are given. The steroids were prepared in the Pharmaceutical Chemistry Laboratory under the direction of Norman J. Doorenbos.

Inhibitory properties were found to be specific and potent in four compounds, with inhibitory concentrations as low as 0.37 μg/ml. Three of the active steroids are 4-aza cholestanes and one is a 4-nor-3,5-secocholestane amide. Sensitivity to the compounds was greatest in the gram-positive bacteria, followed by the yeasts and molds. The gram-negative bacteria were not inhibited. All 16 steroids interfered to some extent with pigmentation in *Serratia marcescens* but not with pigment production in *Pseudomonas aeruginosa.* In a few instances, some of the molds were stimulated by the steroids at a concentration of 250 μg/ml.

A. I. Bruade, J. Banisster and N. Wright (1955). Antibiotics A. 1133–1140.

W. Syzbalski and V. Bryson (1952). Science, N.Y. 116, 45–51.

Sulfur dioxide—*Zymomonas*

N. F. Millis (1956). J. gen. Microbiol. 15, 521–528.

Growth was delayed 8 days at 500 μg/ml; concentrations over 750 μg/ml inhibited growth.

Tellurite—*Streptococcus*

N.C.T.C. Methods 1954.

Inoculate McLeod medium containing potassium tellurite in a concen-

tration of 1:2500. Incubate at 37°C for 20 hours. The tellurite resistant organisms show a heavy growth of fat, jet-black colonies. The tellurite sensitive organisms fail to grow, or produce only a very light growth.
K. Skadhauge (1950). *Studies on Enterococci with Special Reference to the Serological Properties,* p. 64. Copenhagen: Einar Munksgaards.

Tellurite—*Streptococcus*

P. Morelis and L. Colobert (1958). Annls Inst. Pasteur, Paris 95, 569–587.

La résistance au tellurite de potassium a été éprouvée, d'une part, en milieu liquide, d'autre part, en milieu solide. La composition du premier est la suivante: peptone Vaillant 5 B, 10 g; lactose, 5 g; phosphate disodique, 5 g; eau, 1000 cm^3. La solution de tellurite fut préparée à part par dissolution dans l'eau stérile sans autoclavage ultérieur (le tellurite se décompose au-dessus de 50°) et ajoutée de façon à obtenir exactement une concentration finale de 1 p. 2500.

Les souches qui se développent dans ce milieu donnent un abondant dépôt noir et un noircissement diffus du milieu surnageant. Mais comme même les germes sensibles au tellurite provoquent la formation d'un dépôt plus ou moins abondant au fond du tube, il est, dans la pratique, difficile de différencier les souches résistantes des souches sensibles. En revanche, cette distinction est beaucoup plus aisée si l'on emploie un milieu solide qui n'est autre que le milieu liquide gélifié par l'adjonction de 16 g d'agar-agar par litre. C'est au moment de couler que le tellurite de potassium est ajouté au milieu encore fluide, mais refroidi au-dessour de 50°. Dans ces conditions, les souches résistantes forment des colonies noires brillantes, alors que les souches sensibles ne se développent pas ou donnent des colonies très maigres, de couleur blanchâtre. On a observé que 35 des 67 souches retirées des excréments humains se sont développées sur le milieu solide, tandis que seulement 18 des 62 souches d'origine animale l'ont fait. De même, 20 des 22 souches isolées à partir de produits pathologiques se sont montrées capables de pousser vigoureusement sur ce milieu.

Tellurite—*Streptococcus*

F. M. Ramadan and M. S. Sabir (1963). Can. J. Microbiol. 9, 443–450.

The authors tested growth on 1:2500 potassium tellurite blood-agar following the method of K. Skadhauge (1950). *Studies on Enterococci with Special Reference to the Serological Properties.* Einar Munksgaards, Copenhagen.

Tellurite—*Staphylococcus aureus*

R. L. Brown and J. B. Evans (1963). J. Bact. 85, 1409–1412.

Counts on tellurite glycine agar were determined by surface-plating 0.1-ml amounts of 10^{-5}, 10^{-6}, and 10^{-7} dilutions. All black colonies were

counted after 24 hr of incubation. These counts were compared with those from duplicate samples on TSB agar* after 48 hr of incubation.
*Trypticase Soy Broth plus 1.5% agar

Tellurite—*Aerococcus catalasicus*

O. G. Clausen (1964). J. gen. Microbiol. 35, 1–8.

Cultures were incubated aerobically at 37°C unless otherwise stated.

Tellurite resistance was investigated by inoculating the strains on to McLeod's blood agar containing 0.04% potassium tellurite. Growth or lack of growth was recorded after 20 hr incubation.

Tellurite—*Moraxella*

W. J. Ryan (1964). J. gen. Microbiol. 35, 361–372.

Resistance to tellurite was examined by incorporating it in blood agar. A drop of bacterial suspension sufficient to give semiconfluent growth on control plates was used as inoculum. Results were read after 3–4 days of incubation.

Tellurite—*Streptococcus*

R. Whittenbury (1965). J. gen. Microbiol. 38, 279–287.

Basal medium agar containing glucose, 0.5% (w/v), tellurite, 0.04% (w/v) (the concentration recommended by Skadhauge, 1950), adjusted to pH 6.5, was used. The results, erratic at first, became constant when the concentration of the meat extract in the medium was doubled.

The basal liquid medium contained meat extract (Lab-Lemco), 0.5 g; peptone (Evans), 0.5 g; yeast extract (Difco), 0.5 g; Tween 80, 0.05 ml; $MnSO_4 \cdot 4H_2O$, 0.01 g; in 100 ml tapwater, adjusted to pH 6.5 and autoclaved at 121°C for 15 min. Cultures were incubated at 30°C.

K. Skadhauge (1950). *Studies on Enterococci with Special Reference to the Serological Properties.* Copenhagen: Einar Munksgaards.

Thallous acetate—*Moraxella*

W. J. Ryan (1964). J. gen. Microbiol. 35, 361–372.

Resistance to thallous acetate was examined by incorporating it in blood agar. A drop of bacterial suspension sufficient to give semiconfluent growth on control plates was used as inoculum. Results were read after 3–4 days of incubation.

Trialkylgermanium acetates—fungi and bacteria

A. K. Sijpesteijn, F. Rijkens and G. J. M. van der Kerk (1964). Antonie van Leeuwenhoek 30, 113–120.

A continued study of the antimicrobial activity of a series of trialkylgermanium acetates has confirmed earlier findings that the most active compounds against fungi are the ethyl and propyl derivatives, whereas for optimum antibacterial activity the chain length has to be somewhat greater.

All fungi but only a few of the bacteria tested appear to be sensitive.
Incorporation of blood in the medium abolishes antibacterial activity but leaves antifungal activity unaltered.

Zinc sulfate—*Moraxella*
W. J. Ryan (1964). J. gen. Microbiol. 35, 361–372.
Resistance to zinc sulfate was examined by incorporating it in blood agar. A drop of bacterial suspension sufficient to give semiconfluent growth on control plates was used as inoculum. Results were read after 3–4 days of incubation.

ARBUTIN BREAKDOWN

Arbutin breakdown—*Fusobacterium*
A. C. Baird-Parker (1960). J. gen. Microbiol. 22, 458–469.
Medium (%, w/v): proteose peptone No. 3, 1.0; Lab. Lemco, 0.3; Oxoid yeast extract, 0.1; glucose, 0.1; disodium hydrogen phosphate, 0.5; arbutin, 0.5. Cultures were incubated for 14 days. The presence of a phenolic aglycone was detected by adding 0.5 ml of a 1.0% (w/v) solution of ferric ammonium citrate and examining for the development of a dark reddish brown colour in the medium.

AROMATIC RING CLEAVAGE

Aromatic ring cleavage—*Pseudomonas*
R. Y. Stanier, N. J. Palleroni and M. Doudoroff (1966). J. gen. Microbiol. 43, 159–271. [Revised by M. Doudoroff (personal communication, October 3, 1967).]
In recent years it has become evident that aerobic pseudomonads can dearomatize two of the central intermediates in the metabolism of aromatic compounds, (catechol, protocatechuic acid), by two different mechanisms, *ortho* and *meta* cleavage (Dagley, Evans and Ribbons, 1960). These two modes of cleavage channel the degradation of a given aromatic compound after ring fission into two completely different metabolic pathways.
The authors used a method which they attribute to Dr. Keiichi Hosokawa.
Strains are grown on chemically defined media containing an aromatic substrate as sole source of carbon and energy. The authors most commonly use *p*-hydroxybenzoate as a growth substrate when the cleavage of protocatechuate is to be tested and benzoate when the cleavage of cate-

chol is to be tested. Bacteria are either centrifuged from liquid medium or scraped off a plate and resuspended in 0.2 *M* Tris buffer (pH 8.0). The suspension is toluenized and 2 ml samples are furnished with 20 μmoles of either catechol or sodium protocatechuate. When *meta* cleavage occurs, the mixture turns bright yellow within a few minutes. If no yellow color develops within 5 minutes, the tube is shaken for 1 hr at 30°C, and then tested for the presence of β-ketoadipate (indicative of an *ortho* cleavage) by the Rothera reaction (Kilby, 1948). A deep purple Rothera reaction is given by this β-keto acid. The development of this color is strong presumptive evidence of *ortho* cleavage. A parallel control experiment, using patches of organism grown on yeast agar plates, was always done to determine whether the observed activity had been specifically induced by growth on the aromatic substrate.

S. Dagley, W. C. Evans and D. W. Ribbons (1960). Nature, Lond. 188, 560.
B. A. Kilby (1948). Biochem. J. 43, v.

ARYLSULFATASE

Arylsulfatase—*Mycobacterium*

J. E. M. Whitehead, P. Wildy and H. C. Engbaek (1953). J. Path. Bact. 65, 451–460.

Examination for production of arylsulfatase.

Preparation of medium: The original Tween-albumin medium described by Dubos and Davis (1946) was distributed in 4.5 ml amounts in 15 ml screw-capped bottles, sterilized by autoclaving at 15 lb/sq in. for 20 min and stored at 5°C until required. Two days before use, 0.2 ml of a sterile 10 per cent solution of bovine albumin fraction V (Armour Laboratories) which had been heated to 56°C for 30 min was added to each bottle. Where, in the test medium, substrate was incorporated, 0.5 ml of a solution of either tripotassium phenolphthalein disulfate or potassium 1-naphthylsulfate sterilized by Seitz filtration was added to each bottle with sterile precautions so as to produce the required final concentration. The bottles were then incubated at 37°C for 2 days to detect possible contamination.

Preparation of inoculating cultures. An inoculating culture of each strain was prepared by emulsifying in the above medium, without substrate, a colony from a Löwenstein-Jensen slope. After incubation at 37°C for 10–14 days most strains were inoculated into the test medium. Murine strains, however, grew more slowly and were not ready for testing before 28 days. The test medium was inoculated with 0.2 ml of the inoculating culture, using a dropping pipette. Except with 14 strains of saprophytic mycobacteria, whose growth was granular, the opacities of the inoculating cultures were compared with a series of barium-sulfate standards (Lehmann, 1950), and it was calculated that the bacterial content of the inocula lay between 0.04 and 0.15 mg dry weight of mycobacteria. A sample of each inoculum was plated on blood agar to detect whether contamination had occurred.

Detection of arylsulfatase activity

(a) *In medium containing phenolphthalein disulfate.* After incubation at 37°C in the test medium containing phenolphthalein disulfate, a sample of each culture was plated out on blood agar to determine whether contamination had occurred. Cultures were then tested for arylsulfatase activity by adding alkali (2 N NaOH) and observing whether a red color developed due to phenolphthalein liberated by enzymic hydrolysis. Because decolorisation of phenolphthalein occurs in strongly alkaline solutions, the alkali was added carefully drop by drop, with shaking, until the maximum color had developed. The intensity of the color was assessed visually and recorded according to the scale: ++++ = deep red; +++ = red; ++ = pink; + = pale pink; ± = faint tinge of pink.

(b) *In medium containing naphthyl sulfate.* Each strain to be tested in the medium containing naphthyl sulfate was also inoculated into the medium without substrate as a control, because it was found that faintly positive diazo reactions occurred after growth, even in the absence of substrate. Each culture was plated out after incubation to detect possible contamination.

The naphthol liberated by hydrolysis of the naphthyl sulfate was detected by means of a diazo reaction. To each test and control culture was added 1 ml of 2 N NaOH and 0.5 ml of freshly prepared diazo reagent (1 ml of 0.8 per cent sodium nitrite and 9 ml of 0.25 per cent sulfanilic acid dissolved in N HCl). The cultures were shaken and the difference in depth of color between the test and control cultures was assessed visually and recorded according to the scale: ++++, +++, ++, + and ± = degrees of color difference between control and medium containing substrate on addition of alkali and diazo reagent.

R. J. Dubos and B. D. Davis (1946). J. exp. Med. 83, 409.
J. Lehmann (1950). XIe. Conférence de l'Union internationale contre la Tuberculose, Copenhagen, 1950.

Arylsulfatase—*Chromobacterium*

P. H. A. Sneath (1956). J. gen. Microbiol. 15, 70–98.

Tested in peptone water containing 0.001 M-potassium phenolphthalein disulfate after growth for 7 days at 25°C by adding N-NaOH drop by drop.

J. E. M. Whitehead, A. R. Morrison and L. Young (1952). Biochem. J. 51, 585.

Arysulfatase—*Mycobacterium* (atypical)

A. Beck (1959). J. Path. Bact. 77, 615–624.

Arylsulfatase was tested in 2-week-old Dubos Tween-albumin cultures to which 0.005 M phenolphthalein disulfate was added*, the color changes from white to faint pink, pink or red, observed after addition of N NaOH were recorded by +, ++ and +++ signs.

*J. E. M. Whitehead, P. Wildy and H. C. Engbaek (1953). J. Path. Bact. 65, 451.

Arylsulfatase—*Mycobacterium*

R. J. Jones and D. E. Jenkins (1965). Can. J. Microbiol. 11, 127–133.

The substrate, tripotassium phenolphthalein disulfate in 0.01 M aqueous solution, was sterilized by passage through an ultrafine fritted-disc buchner funnel. The substrate was aseptically added to Tween-albumin broth in a final concentration of 0.001 M. Medium was dispensed in 5-ml amounts into screw-capped culture tubes. Culture medium was inoculated with 0.1 ml of the inoculum suspension. After incubation at 37°C for 4 days, 0.4 ml of 2 N Na_2CO_3 was added to the culture tube to test for arylsulfatase activity. A pale pink to a bright red color indicated a positive reaction, and no color change indicated a negative reaction. The brighter the red color the greater the enzyme activity.

If not otherwise indicated all reagents and media were sterilized by autoclaving at 121°C for 15 minutes.

Arylsulfatase—*Mycobacterium parafortuitum* n.sp.

M. Tsukamura (1966). J. gen. Microbiol. 42, 7–12.

Three-day arylsulfatase test was carried out according to the method of Wayne (1961), and 3-week arylsulfatase test according to the method of Kubica and Vestal (1961).

G. P. Kubica and A. L. Vestal (1961). Am. Rev. resp. Dis. 83, 728.

L. G. Wayne (1961). Am. J. clin. Path. 36, 185.

Arylsulfatase—*Mycobacterium*

M. Tsukamura (1966). J. gen. Microbiol. 45, 253–273.

Three-day arylsulfatase test was done according to the method of Wayne (1961).

Two-week arylsulfatase test was done according to the method of Kubica and Vestal (1961) with a final concentration of 0.001 M-tripotassium phenolphthalein disulfate.

G. P. Kubica and A. L. Vestal (1961). Am. Rev. resp. Dis. 83, 728.

L. G. Wayne (1961). Am. J. clin. Path. 36, 185.

Arylsulfatase—*Proteus*

F. H. Milazzo and J. W. Fitzgerald (1966). Can. J. Microbiol. 12, 735–744.

The authors describe quantitative methods for estimating the enzyme.

BASAL MEDIA FOR DETERMINATION OF SUITABILITY OF CARBON AND NITROGEN COMPOUNDS FOR GROWTH

Basal medium for carbon compounds—organic acids

S. A. Koser (1923). J. Bact. 8, 493–520.

$(NH_4)H_2PO_4$	1.0 g
K_2HPO_4	1.0 g
$MgSO_4 \cdot 7H_2O$	0.2 g
NaCl	5.0 g
Distilled water	1000 ml

This gives a colorless clear solution having a pH of 6.7–6.8. To this solution the various organic acids were added and the reaction brought back to pH 6.8 by the addition of *N* NaOH.

Basal medium for carbon compounds—organic acids

H. E. Goresline (1933). J. Bact. 26, 435–457.

	Bactopeptone	10 g
or	$(NH_4)_2SO_4$	2 g
	K_2HPO_4	5 g
	H_2O	1000 ml

For utilisation of organic acids 0.2% was added and the final pH adjusted to 6.6. For carbohydrate utilisation 0.2–0.6% carbohydrate was added and the pH adjusted to 7.2. Andrade's indicator (concentration not specified) was added to one series to detect acid production and a quinhydrone electrode used to follow pH changes in another.

Basal medium for carbon and nitrogen compounds

H. E. Sagen, A. J. Riker and I. L. Baldwin (1934). J. Bact. 28, 571–595.

$MgSO_4 \cdot 7H_2O$	0.2 g
NaCl	0.2 g
$CaCl_2$	0.1 g
K_2HPO_4	0.2 g
Nitrogen source	5 g
Carbon source	5 g
Distilled water	1000 ml

The nitrogen source, and the salt-carbohydrate were prepared separately in double strength, neutralized, sterilized, mixed and tubed aseptically.

Basal medium for carbon compounds—*Moraxella*

A. Lwoff and A. Aurdureau (1941). Annls Inst. Pasteur, Paris 66, 417–424.

Milieu. Techniques. – Le milieu de base est le suivant:

Sulfate d'ammoniaque	0.75 g
Phosphate monopotassique	4.5 g
Chlorure de potassium	0.5 g
Sulfate de magnésium	0.05 g
Eau bidistillée	1000 cm^3
Soude pour pH 7.4.	

Après stérilisation, on ajoute pour 4 cent. cubes de milieu I goutte de la solution suivante, stérilisée à part:

Citrate ferrique	1 g
Chlorure de calcium	1 g
Eau bidistillée	1000 cm^3

Au milieu ainsi constitué, on ajoute l'aliment carboné en proportions convenables, soit par exemple I à V gouttes d'éthanol à 100° pour 4 cent. cubes. L'éthanol à concentration élevée se montre plus ou moins toxique. C'est ainsi qu'avec X gouttes d'éthanol pour 4 cent. cubes de milieu, le développement de la culture se trouve légèrement retardé. Il n'est d'ailleurs pas nécessaire de fournir à la bactérie une surabondance d'aliment carboné. On obtient une très bonne culture en ajoutant II gouttes d'éthanol dilué au 1/10 pour 4 cent. cubes de milieu et un développement encore appréciable avec II gouttes d'éthanol dilué au 1/100 (ce qui donne une concentration finale de 1 p. 5000 environ). Pour l'entretien de la souche destinée aux expériences, nous avons utilisé un milieu additionné de I à II gouttes d'éthanol au 1/10 pour 4 cent. cubes. Dans ces conditions, l'ensemencement dans un milieu non additionné d'aliment carboné n'est suivi que d'un développement négligeable.

En présence de l'aliment carboné favorable, on obtient en général, en vingt-quatre heures, une culture abondante. Certains composés, comme par exemple l'acide propionique, se sont révélés toxiques à concentration élevée. L'absence de développement en présence d'une substance déterminée pouvant être due à la toxicité de cette substance, nous avons, pour tous les corps "négatifs", effectué une épreuve de nontoxicité: un tube témoin contenant de l'éthanol et la substance à étudier doit montrer un développement normal. Toutes les expériences ont été répétées plusieurs fois. Les tubes dans lesquels ne se produisait aucun développement ont été suivis pendant deux mois avant que l'on conclue à la non-assimilation d'une substance.

Les corps à étudier sont stérilisés, soit à l'autoclave vingt minutes à 115° pour les corps thermostables, soit par filtration sur bougie Chamberland L3 pour les corps thermolabiles. Nous avons utilisé, soit des produits Merck pour analyses, soit des produits Fraenkel et Landau, soit des produits du Département de Biochimie de l'Université d'Illinois. Certains corps ont été purifiés à nouveau.

Basal medium for nitrogen compounds—*Micrococcus*

B. P. Cardon and H. A. Barker (1946). J. Bact. 52, 629–634.

Isolation medium:

Glycine	0.5	g
Yeast autolysate	3	vols
With the following salts		
pH 7.1, M/1 phosphate buffer	0.5	vol
$CaSO_4$ (saturated solution)	0.25	vol
Cysteine HCl	0.02	g
$MgSO_4 \cdot 7H_2O$	0.005	g
$FeSO_4 \cdot 7H_2O$	0.001	g
In distilled water	100	ml

Growth medium:

Glycine	0.3	g
Bacto yeast extract	0.5	g
Bacto peptone	0.5	g
With salts (as above)		
In glass distilled water	100	ml

The optimum pH for growth in this medium is about 7.2, the range being 6.0 to 8.5. Incubated 48 – 72 hours at 30 or 37°C.

The organism obtains energy by a fermentation of glycine according to the equation

$$4\ CH_2NH_2COOH + 2\ H_2O = 4\ NH_3 + 3\ CH_3COOH + 2\ CO_2.$$

Basal medium for nitrogen compounds—*Clostridium propionicum*

B. P. Cardon and H. A. Barker (1946). J. Bact. 52, 629–634.

Isolation medium:

Alanine	1.0	g
Yeast autolysate	1	vol
pH 7.0, M/1 phosphate buffer	2	vol
Tap water	100	ml

Growth medium:

Alanine	0.3	g
Bacto peptone	0.3	g
Bacto yeast extract	0.4	g
With salts (as in preceding abstract)		
Distilled water	100	ml

Growth will occur between pH 5.8 and 8.6, the optimum being between 7.0 and 7.4.

The equation for the decomposition of alanine is

$$3\ CH_3CHNH_2COOH + 2\ H_2O = 3\ NH_3 + 2\ CH_3CH_2COOH + CO_2 + CH_3COOH.$$

Basal medium for the utilisation of thiamine

R. P. Williams and S. A. Koser (1947). Bacterial destruction of thiamine. J. infect. Dis. 81, 130–134.

$(NH_4)_2HPO_4$	4.0 g
KH_2PO_4	1.5 g
NaCl	5.0 g
$MgSO_4$	0.1 g
$FeSO_4 \cdot 7H_2O$	0.1 mg
$MnCl_2 \cdot 4H_2O$	0.1 mg
Redistilled H_2O	1000 ml

The pH was approximately at the neutral point after sterilization in the autoclave. To this basal medium a filtered solution of thiamine hydrochloride was added (usually 0.1%). When quantities up to 0.3% of the vitamin were used it was unnecessary to adjust the pH.

Basal medium for nitrogen compounds

P. Hauduroy (1951). Techniques Bactériologiques. Masson et Cie, Éditeurs, Paris.

MILIEU POUR L'UTILISATION DE L'AZOTE ORGANIQUE OU INORGANIQUE:

$NH_4H_2PO_4$	1.0 g
KCl	0.2 g
$MgSO_4$	0.2 g
Gélose	13 g
Eau distillée	1000 ml

Ajuster à *p*H 7. Ajouter 12 ml d'une solution à 0.04 de pourpre de bromocrésol comme indicateur. Après répartition et stérilisation à l'autoclave du milieu précédent, on ajoute à ce milieu les hydrates de carbone que l'on désire utiliser et que l'on a stérilisé par filtration à travers une bougie Chamberland L_2. En ajouter suffisamment pour avoir une concentration de 0.5 p. 100 (1). Placer les tubes ainsi préparés à 37° pendant une nuit, vérifier s'ils sont restés stériles. Ensemencer. Observer la croissance, la production d'acide et de gaz, aux 3e, 7e, 14e et 21e jours.

Si une couche ne produit pas d'hydrolyse du glucose dans ce milieu contenant de l'azote inorganique, elle est ensemencée alors sur un tube de gélose inclinée contenant 0.5 p. 100 de glucose avec comme indicateur du pourpre de bromocrésol, et sur un tube de gélose inclinée ne contenant que l'indicateur, ceci à titre de contrôle. S'il se forme de l'acide à partir du glucose, on doit interpréter ce résultat en disant que la souche ne peut pas utiliser l'azote ammoniacal et un milieu contenant de l'azote organique doit être utilisé. (*A. T. C. C.*)

(1) Modification du milieu de Ayers, Rupp, and Johnson (1919. U.S. Dept. Agr. Bull. 82, 39.)

Basal medium for carbon compounds—*Sporovibrio*

J. Pochon and M. A. Chalvignac (1952). Annls Inst. Pasteur, Paris 82, 399–403.

Physiologie. – Pour préciser les caractères de cette souche et surtout pour tenter de déterminer sa place dans la systématique, nous avons essayé tous les donateurs d'hydrogène indiqués par Baars, puisque telle est la base même de la classification. Les cultures ont été réalisées en milieu liquide, les différentes substances ajoutées à la concentration M/30; trois passages successifs ont été réalisés sur chaque milieu afin d'éliminer toute trace de lactate apportée par le premier ensemencement. Au troisième passage, seuls étaient utilisables le méthanol, l'éthanol, l'acide pyruvique, l'acide tartrique.

Baars. Thèse de Delft, 1930.

Basal medium for nitrogen compounds—ammonia

N.C.T.C. Methods 1954.

All inoculations to be made with a straight wire from suspension used for citrate test. Incubate at 30°C and 37°C and read daily for 5 days.

Method. Medium:

Washed agar	1.0%
Ammonium phosphate	0.1%
Glucose	1.0%
Potassium chloride	0.02%
Magnesium sulphate	0.02%

Add bromcresol purple as indicator.

G. J. Hucker (1948). In *Bergey's Manual of Determinative Bacteriology*, 6th ed., p. 236.

Basal medium for carbon and nitrogen compounds—*Pseudomonas (Malleomyces) pseudomallei*

L. Chambon and P. de Lajudie (1954). Annls Inst. Pasteur, Paris 86, 759–764.

a) Milieux synthétiques utilisés: – Pour l'étude du métabolisme carboné et azoté, des sources de carbone et d'azote sont ajoutées au milieu salin de base de Lwoff, milieu dont la composition est la suivante:

PO_4KH_2	1.2 g
$PO_4Na_2H \cdot 2\ H_2O$	8.6 g
$SO_4Mg \cdot 7\ H_2O$	0.2 g
Cl_2Ca	0.01 g
$SO_4Fe \cdot 7\ H_2O$	0.0005 g
Eau bidistillée	Q.S.P. 1000 ml

Les milieux synthétiques ainsi constitués sont répartis dans des tubes à essai à raison de 5 ml par tube, puis stérilisés à 110° ou par trois chauffages à 100° pour ceux dont la formule comporte un glucide. Le pH est ajusté à 7.5 par addition de soude.

L'hydrolyse de l'urée, l'utilisation du carbone, du citrate de soude et des alcools éthylique et méthylique, la production d'hydrogène sulfuré à partir de la cystéine ont été étudiées comparativement en milieux synthétiques et en milieux semisynthétiques gélosés dont nous donnons plus loin la composition.

b) Ensemencement. – *Milieux synthétiques.* – Un premier tube de milieu synthétique est ensemencé avec un inoculum très faible d'une culture de vingt-quatre heures sur gélose ordinaire; deux repiquages successifs effectués à l'öse dès apparition d'une culture macroscopiquement visible (vingt-quatre heures en général) permettent d'éliminer les métabolites qui auraient pu être entraînés avec l'inoculum initial. La lecture définitive est faite sur le 3e tube.

Milieux semi-synthétiques géloses. – L'ensemencement est effectué à partir d'une culture sur gélosé de vingt-quatre heures.

Basal medium for carbon compounds—organic acids

R. E. Gordon and J. M. Mihm (1957). J. Bact. 73, 15–27.

NaCl	1 g
$MgSO_4$	0.2 g
$(NH_4)_2HPO_4$	1 g
KH_2PO_4	0.5 g
Agar	15 g
Distilled water	1000 ml

2 g of the sodium salt of the organic acid to be used were added to the above, pH adjusted to 6.8, then 20 ml of a 0.04% solution of phenol red (indicator solution) added. Use of an organic acid as a source of carbon by a culture was demonstrated by the alkaline color of the indicator after 4 weeks' incubation at 28°C.

Basal medium for nitrogen compounds—*Leptotrichia buccalis*

R. D. Hamilton and S. A. Zahler (1957). J. Bact. 73, 386–393.

Medium:

(Difco) Peptone	20 g
(Difco) Yeast extract	2 g
Glucose	5 g
NaCl	5 g
K_2HPO_4	10 g
Sodium thioglycolate	1 g
KNO_2	0.002 g
Distilled water	1 litre

pH adjusted to 7.2 before autoclaving.

Nitrite tested for after 72 hr at 37°C.

For method of detection of nitrite disappearance see Committee on Bacteriological Technic of the Society of American Bacteriologists (1953).

Manual of Methods for Pure Culture Study of Bacteria. Williams and Wilkins Co., Baltimore, Maryland.

Basal medium for carbon compounds—*Butyrivibrio* and *Bacteroides*

J. Gutierrez, R. E. Davis and I. L. Lindahl (1959). Appl. Microbiol. 7, 304–308.

Basal medium (%) in tap water; NH_4Cl, 0.05; NaCl, 0.1; $MgSO_4$, 0.005; $CaCl_2$, 0.005; resazurin, 0.0001; peptone, 0.5; Difco yeast extract, 0.25. Cysteine HCl, 0.04%; $NaHCO_3$, 0.5%; and alfalfa-saponins, 0.5% were autoclaved separately and added to the liquid medium at the time of inoculation. Incubated under CO_2, anaerobic techniques used were described by Hungate (1950).

R. E. Hungate (1950). Bact. Rev. 14, 1–49.

Basal medium for carbon compounds—*Escherichia*

S. Schäfler and S. Benes (1959). Annls Inst. Pasteur, Paris 96, 231–237.

L'utilisation des substances organiques comme unique source de carbone (0,2–0,5 p. 100) a été testée dans le milieu salin de Koser dépourvu de citrate avec 0,001 p. 100 de bleu de bromothymol.

Basal medium for growth factor—*Pediococcus*

H. L. Günther and H. R. White (1961). J. gen. Microbiol. 26, 185–197.

The method used was a modification of that outlined in the *Difco Manual* (1953). Tests were carried out in triplicate in 4 ml amounts in 12 x 80 mm EEL colorimeter tubes. Growth after 18 hr of incubation was measured turbidimetrically with an EEL colorimeter. The folinic acid used in these experiments was supplied as 'leucovorin' by Lederle Laboratory Division Ltd. and was used in concentrations of 0, 0.15, 0.3 and 0.6 mμg/ml. Twenty isolates were examined for this requirement.

Maintenance of stock cultures and methods of cultivation. For maintenance of stock cultures, preparation of inocula and in all experimental work, 'Oxoid' tomato juice (TJ) broth or tomato juice (TJ) agar, adjusted to pH 6.6, were used unless otherwise stated. The following were exceptions to this rule: for strain Tc. 1 sodium chloride (5%, w/v) was added to the medium; and for the aerococci glucose Lemco broth (Shattock and Hirsch, 1947) or glucose yeast extract (GY) agar (containing, as %, w/v; peptone, 1.0; Yeastrel, 0.3; glucose, 1.0; NaCl, 0.25; agar, 1.0; at pH 7.4) was used.

Cultures were incubated aerobically at 30°C with specified exceptions.

Difco Manual (1953). 9th ed., Difco Lab. Inc., Detroit 1, Mich., U.S.A.

P. M. F. Shattock and A. Hirsch (1947). J. Path. Bact. 59, 495.

Basal medium for nitrogen compounds—*Pediococcus*

H. L. Günther and H. R. White (1961). J. gen. Microbiol. 26, 185–197.

The medium and method described by Hucker (1924) were used. Incubation continued for 14 days.

Maintenance of stock cultures and methods of cultivation. For maintenance of stock cultures, preparation of inocula and in all experimental work, 'Oxoid' tomato juice (TJ) broth or tomato juice (TJ) agar, adjusted to pH 6.6, were used unless otherwise stated. The following were exceptions to this rule: for strain Tc. 1 sodium chloride (5%, w/v) was added to the medium; and for the aerococci glucose Lemco broth (Shattock and Hirsch, 1947) or glucose yeast extract (GY) agar (containing, as %, w/v; peptone, 1.0; Yeastrel, 0.3; glucose, 1.0; NaCl, 0.25; agar, 1.0; at pH 7.4) was used.

Cultures were incubated aerobically at 30°C with specified exceptions.

G. J. Hucker (1924). Bull. N.Y. St. agric. Exp. Stn. 101, 36.
P. M. F. Shattock and A. Hirsch (1947). J. Path. Bact. 59, 495.

Basal medium for carbon compounds—*Arthrobacter*

E. G. Mulder and J. Antheunisse (1963). Annls Inst. Pasteur, Paris 105, 46–74.

Pour déterminer l'utilisation de différents corps carbonés on a fait usage du "replica test". On transfère du matériel bactérien d'une plaque, ayant une composition polyvalente et portant des colonies de différentes souches d'*Arthrobacter* rangées régulièrement, à une série d'autres plaques contenant l'extrait de levure (0,1 p. 100) et un des corps carbonés (0,5 p. 100) qu'on veut tester. Pour ce transfert on emploie un disque d'une taille sensiblement identique à celle de la boîte de Petri et portant un certain nombre de pointes dont la position correspond au groupement des colonies. Avec ce disque, qui peut être stérilisé simplement par la chaleur, on peut transférer en même temps un grand nombre de souches. Pour déterminer l'utilisation d'un corps carboné on compare la croissance sur des plaques contenant seulement l'extrait de levure.

Milieu minéral A

HPO_4K_2	1.0 g
$SO_4Mg \cdot 7\ H_2O$	0.3 g
$Cl_2Ca \cdot 2\ H_2O$	0.05 g
$Cl_3Fe \cdot 6\ H_2O$	0.01 g
$SO_4Cu \cdot 5\ H_2O$	0.1 mg
$SO_4Zn \cdot 7\ H_2O$	0.1 mg
$SO_4Mn \cdot 7\ H_2O$	1.0 mg
MoO_4Na_2	0.01 mg
BO_3H_3	0.01 mg
Cl_2Co	0.01 mg
Eau distillée en verre	1 litre
pH	7.0

Basal medium for carbon and nitrogen compounds—*Staphylococcus epidermidis*

D. Jones, R. H. Deibel and C. F. Niven (1963). J. Bact. 85, 62–67.

In experiments concerned with the anaerobic utilization of the various substances as energy sources, the basal medium consisted of Tryptone (Difco), 10 g; yeast extract (Difco), 5 g; K_2HPO_4, 5 g; NaCl, 5 g; energy source, 10 g; distilled water, 1 liter; pH 7.0 to 7.2. For anaerobiosis, broth cultures were placed in a 6 liter desiccator and flushed twice with hydrogen. The final gas phase consisted of a mixture of 90% hydrogen and 10% carbon dioxide. Growth was estimated by determining the optical density of cultures in a Spectronic 20 spectrophotometer at a wavelength setting of 600 mμ.

Compounds tested were – pyruvate, L-serine, arginine, malate, citrate, gluconate, ascorbate and glycerol.

Basal medium for carbon compounds—*Clostridium glycolicum* n.sp.

L. W. Gaston and E. R. Stadtman (1963). J. Bact. 85, 356–362.

The minimal medium finally adopted contained (per 100 ml of medium): 1.0 ml of ethylene glycol; 0.06 M potassium phosphate buffer (pH 7.4); 5.0 ml of Parenamine; the eleven vitamins, thiamine, pyridoxine, pyridoxamine, pyridoxal, pantothenate, riboflavine, nicotinic acid, nicotinamide, *p*-aminobenzoic acid, biotin, and pteroylglutamate, in the concentrations used by Kline and Barker (1950); 0.02% $MgSO_4 \cdot 7H_2O$, 0.002% $CaCl_2 \cdot 2H_2O$, 0.002% $FeSO_4 \cdot 7H_2O$, 0.0006% $MnSO_4 \cdot 4H_2O$, 0.0006% $Na_2MoO_4 \cdot 2H_2O$ and 0.00024% $CoCl_2 \cdot 6H_2O$ and 0.01% $Na_2S \cdot 9H_2O$. K_2CO_3, 10% and 5 N HCl were autoclaved separately, and 1.5 ml of K_2CO_3 and 2 drops of HCl were added (per 100 ml of medium) just prior to inoculation.

L. Kline and H. A. Barker (1950). J. Bact. 60, 349–363.

Basal medium for physiological tests—*Butyrivibrio, Borrelia, Bacteroides, Selenomonas, Succinivibrio*

R. S. Fulghum and W. E. C. Moore (1963). J. Bact. 85, 808–815.

The salts-rumen fluid-protein (SRP) medium of Fulghum (1958) was similar to the rumen fluid-glucose-cellobiose agar medium of Bryant and Burkey (1953). A stock salts solution for the SRP medium was prepared by dissolving 1.0 g of K_2HPO_4, 1.0 g of KH_2PO_4, 2.0 g of NaCl, and 10.0 g of $NaHCO_3$ in 1 liter of boiled, distilled water equilibrated with oxygen-free CO_2. Then, 0.25 g of $CaCl_2 \cdot 2H_2O$, 0.20 g of $MgSO_4 . 7H_2O$, and 2 ml of a 0.1% (w/v) resazurin solution were added separately, complete solution being effected between additions.

The SRP-basal medium used for determining cultural and physiological characteristics contained 50% stock salts solution, 20% clarified rumen fluid, and 30% oxygen-free, CO_2-equilibrated, distilled water. Variations of this basal medium are described with the results obtained. These media were designed to duplicate the conditions provided by similar media described by Bryant and Doetsch (1954) and Bryant and Small (1956).

Clarified rumen fluid was prepared from rumen fluid expressed through two layers of cheesecloth. The liquor was heated at 121°C for 20 min under an oxygen-free CO_2 atmosphere in a sealed container. Particulate debris was then removed by centrifugation at 22,000 x *g* for 20 min.

The anaerobic methods of Hungate (Bact. Rev. 14, 1–49, 1950) were used in handling and incubation of the cultures. The atmosphere in the tubes was oxygen-free nitrogen. Incubation was at 39°C with observations at 24, 48 and 72 hours. The chemical tests used were those suggested in the Society of American Bacteriologists, Committee on Bacteriological Technic *Manual of Microbiological Methods.* McGraw-Hill. 1957.

M. P. Bryant and L. A. Burkey (1953). J. Dairy Sci. 36, 205–217.
M. P. Bryant and R. N. Doetsch (1954). J. Dairy Sci. 37, 1176–1183.
M. P. Bryant and N. Small (1956). J. Bact. 72, 16–21.
R. S. Fulghum (1958). *The development of differential media for the isolation of proteolytic bacteria from the rumen.* M. S. thesis, Virginia Polytechnic Institute, Blacksburg.

Basal medium for nitrogen compounds—*Pseudomonas*

E. J. Behrman and E. J. Stella (1963). J. Bact. 85, 946–947.

The enrichment medium consisted of 0.03% yeast extract, 0.02 M potassium phosphate buffer (pH 7), 0.02% $MgSO_4 \cdot 7H_2O$, and 0.1% trace element solution (E. J. Behrman and R. Y. Stanier, J. biol. Chem. 228, 923, 1957). This medium was supplemented with 0.3% tryptophan or neutralized kynurenic acid, inoculated, and incubated under stationary aerobic conditions.

D-, L-, and DL-tryptophan were used, together with the Kynurenic acid in order to select organisms which could utilise the three known pathways for dissimilation of tryptophan by strains of *Pseudomonas* (E. J. Behrman, Nature, Lond. 196, 150, 1962).

Three groups of organisms were recognized: aromatic pathway organisms, which metabolize L-tryptophan via anthranilic acid; racemase-aromatic organisms, which metabolize both D- and L-tryptophan via anthranilic acid; and quinoline pathway organisms, which metabolize both D- and L-tryptophan via kynurenic acid.

Basal medium for carbon compounds—*Mycobacterium*

H. B. Lukins and J. W. Foster (1963). J. Bact. 85, 1074–1087.

The mineral salts medium contained, per liter of deionized water: $(NH_4)_2SO_4$, 1.0 g; Na_2CO_3, 0.1 g; KH_2PO_4, 0.5 g; $MgSO_4 \cdot 7H_2O$, 0.2 g; $CaCl_2$, 10 mg; $FeSO_4 \cdot 7H_2O$, 5 mg; $MnSO_4$, 2 mg; Cu, 50 μg (as $CuSO_4 \cdot 5H_2O$); B, 10 μg (as H_3BO_3); Zn, 70 μg (as $ZnSO_4 \cdot 7H_2O$); Mo, 10 μg (as MoO_3); pH 7.0. Water-washed agar was added when solid media were desired. Ordinarily, the liquid growth cultures were prepared in "prescription" bottles or suction flasks. Procedures for cultivation in gaseous substrates are given elsewhere (Lukins and Foster, *in press;* Kester and Foster, 1963). Washed cellular suspensions were prepared from liquid cultures, using 0.067 M phosphate buffer (pH 7) as the suspension fluid. Propane or butane was used in a 50% gas–50% air mixture, and liquid alkanes were added at 0.2% (v/v). The hydrocarbons were at least 99 moles % pure.

Substances metabolised include: n-propane, n-butane, n-pentane, n-hexane, n-undecane, acetone, 2-butanone, 2-pentanone, 2-tridecanone and 2-octadecanone.

The organisms could oxidise a wider range of the substances than they were able to use for growth.

A. S. Kester and J. W. Foster (1963). J. Bact. 85, 859–869.

Basal medium for nitrogen compounds—*Vibrio*

L. Ringen and F. W. Frank (1963). J. Bact. 86, 344–345.

Growth with glycine was determined by adding 1% glycine to (i) Brucella agar (Albimi Laboratories, Inc. Flushing, N.Y.) containing 0.1% $FeSO_4 \cdot 7H_2O$.

(ii) Albimi Brucella broth containing 0.15% agar.

Basal media for carbon compounds—*Pediococcus pentosaceus*

W. J. Dobrogosz and R. D. DeMoss (1963). J. Bact. 85, 1356–1364.

The following media were used throughout this investigation. Medium A contained (per liter): NZ Case (Sheffield), 10 g; yeast extract, 10 g; K_2HPO_4, 1 g; mineral solution, 20 ml; pH adjusted to 7.0. The mineral solution had the following composition (per liter): NaCl, 200 g; $MgSO_4 \cdot 7H_2O$, 8 g; $FeSO_4 \cdot 7H_2O$, 0.4 g; $MnCl_2 \cdot 4H_2O$, 0.15 g. Medium A-1 had an identical composition, except that 2% potassium phosphate was added with the pH at 6.0. The semisynthetic medium B contained the following (per liter): NZ Case (Sheffield), 10 g, Na acetate, 2 g; K_2HPO_4, 4 g; adenine, guanine, uracil, thymine, xanthine, and niacin, 5 mg each; Ca panthothenate, pyridoxine, pyridoxal, riboflavine, and thiamine, 1 mg each; biotin, 100 μg; leucovorin, 750 μg; vitamin B_{12}, 2 μg; mineral solution, 20 ml; pH adjusted to 7.0. All growth substrates were prepared at a 5 to 10% solution and autoclaved separately. Sufficient amounts were added to the basal media to give the desired concentration just before inoculation.

Growth determinations were made by use of an Evelyn colorimeter with a 660-mμ filter. Dry weight measurements were obtained with the use of a previously prepared standard curve relating dry weight to optical density.

Basal medium for growth factors—*Staphylococcus aureus*

R. L. Brown and J. B. Evans (1963). J. Bact. 85, 1409–1412.

Nutritional studies. The methods and basal medium of Gretler *et al.* (1955) were used, unless otherwise indicated. The basal medium contained acid-hydrolyzed casein, tryptophan, cysteine, glucose, citrate, phosphate, adenine, guanine, xanthine, uracil, and mineral salts. Tubes of test media were inoculated with 0.05 ml of a 1:100 dilution of a 24-hr broth culture. Growth was measured as optical density at 600 mμ in a Bausch and Lomb Spectronic 20 colorimeter. Measurements were made after 24 and 48 hr of incubation. Serial transfers were made after 48 hr from tubes showing growth, by transferring 0.05 ml to fresh tubes of the same media.

A. C. Gretler, P. Mucciolo, J. B. Evans and C. F. Niven, Jr. (1955). J. Bact. 70, 44–49.

Basal medium for carbon compounds—*Streptococcus*

R. H. Deibel, D. E. Lake and C. F. Niven, Jr. (1963). J. Bact. 86, 1275–1282.

Tests for the utilization of specific substances as sources of energy for growth were conducted in the same basal medium* employed for the detection of proteolysis, but with the omission of glucose. The test compounds were added at the 1% level prior to sterilization of the media. In these experiments, the test strains were diluted 1:100, and one drop of the diluted cultures was used as the inoculum. After incubation for 24 hr, growth response was estimated by optical density determinations in a Bausch and Lomb Spectronic-20 colorimeter, at a wavelength setting of 600 mμ.

Cultures were incubated at 37°C.

*See **Protein digestion**—*Streptococcus* p. 759

Basal medium for carbon compounds—*Micrococcus roseus*

R. C. Eisenberg and J. B. Evans (1963). Can. J. Microbiol. 9, 633–642.

The utilisation of glucose and other substrates was determined using basal medium consisting of 0.2% Trypticase (BBL) and 0.1% yeast extract (BBL). Substrates were sterilized separately by autoclaving and were added aseptically to the growth flasks to give a 1% concentration.

Basal medium for growth—*Lactobacillus*

F. Gasser (1964). Annls Inst. Pasteur, Paris 106, 778–796.

1° Le *milieu de culture* utilisé est un milieu M.R.S. [14] modifié (M.R.S.M.) dont la composition est la suivante:

Peptone trypsique de caséine	15	g
Macération de viande	500	ml
Extrait de levure	5	g
Acétate de sodium	5	g

Citrate d'ammonium		2	g
Phosphate bipotassique		2.4	g
Tween 80		1	ml
Sulfate de manganèse		50	mg
Sulfate de magnésium		200	mg
Glucose		20	g
Eau distillée	q.s.p.	1000	ml

Le pH final est 6.3–6.5. Le milieu est réparti en tubes de 17 x 170 et stérilisé à 115°C pendant trente minutes. La macération de viande est préalablement ajustée à pH 6.5, précipitée par autoclavage à 120°C pendant trente minutes, puis filtrée avant d'entrer dans la composition du milieu.

· Ce milieu est utilisable sous forme solide par addition de 12 p. 1000 d'un agar en poudre résistant bien à l'hydrolyse en milieu acide (agar Biomar). L'incubation des milieux solides est effectuée dans un dessiccateur de chimie contenant 90 p. 100 d'azote et 10 p. 100 de CO_2.

[14] De Man (J. C.), Rogosa (M.) and Sharpe (M. E.). J. appl. Bact. 1960, 23, 130.

Basal medium for carbon and nitrogen compounds—*Chromobacterium violaceum*

W. A. Corpe (1964). J. Bact. 88, 1433–1441.

The salt mixture used in defined media contained (in grams per liter): KH_2PO_4, 1.36; Na_2HPO_4, 2.13; $MgSO_4 \cdot 7H_2O$, 0.2; $FeSO_4 \cdot 7H_2O$, 0.0005; and $CaCl_2$, 0.005. The salts were dissolved in distilled water, adjusted to pH 7.0, and autoclaved in tubes or flasks. Ammonium sulfate or other nitrogen source was added as sterile solution to give a final concentration of 0.05 to 0.1% (w/v). Glucose served as the carbon source, and was also added aseptically to the sterile salts solution to give a final concentration of 1% (w/v).

Flasks containing 50 ml of complete medium were inoculated and incubated at 30°C on a rotary shaker. The inoculum was grown for 24 hr in synthetic medium, centrifuged, washed once in sterile phosphate buffer (pH 7.0), and resuspended in buffer at the same concentration as was the original culture. Absorbance of the culture was about 0.300 at 600 mμ. Portions (0.5 ml) were used to inoculate flasks.

Growth of cultures was measured turbidimetrically at 600 mμ.

Basal medium for carbon and nitrogen compounds—*Vibrio marinus*

R. R. Colwell and R. Y. Morita (1964). J. Bact. 88, 831–837.

Nutritional requirements of the organisms for growth were tested on a 0.4% vitamin-free Casamino Acids and synthetic seawater medium and on media containing synthetic seawater, 1.5% agar (pH 7.2), and each of the following amino acids at 1.0% concentration: DL-alanine, β-alanine,

L-arginine, L-lysine, L-proline, L-phenylalanine, and tyrosine. Ammonium phosphate and sodium formate were also tested.

Synthetic seawater containing sodium chloride, 2.4%; potassium chloride, 0.07%; magnesium chloride (hydrated) 0.53% and magnesium sulfate (hydrated) 0.7% was used as the base. The inoculum was prepared from a 24 hour artificial seawater broth. Cultures were incubated at 18°C.

Basal medium for nitrogen compounds—*Bacteroides oralis*

W. J. Loesche, S. S. Socransky and R. J. Gibbons (1964). J. Bact. 88, 1329–1337.

L-Threonine utilization (Béerens, Guillaume, and Petit, 1959) was determined in the basal medium to which 0.5% L-threonine was added.

Thioglycollate Medium without Dextrose (BBL), supplemented with 0.2% yeast extract (Difco) and hemin, was used as a basal medium. All cultures were incubated at 35°C in Brewer jars containing an atmosphere of 95% H_2 and 5% CO_2.

H. Béerens, J. Guillaume and H. Petit (1959). Annls Inst. Pasteur, Paris 96, 211–216.

Basal medium for carbon compounds—*Pseudomonas aeruginosa*

K. Morihara (1964). J. Bact. 88, 745–757.

The medium used was composed of 1% NH_4Cl, 1% $Na_2HPO_4 \cdot 12H_2O$, 0.2% KH_2PO_4, 0.05% $CaCO_3$, 0.05% $MgSO_4 \cdot 7H_2O$, and 1% of the carbon compound to be tested. The medium was adjusted to pH 7.2. The culture was observed for 7 days at 30°C.

Basal medium—*Veillonella*

M. Rogosa (1964). J. Bact. 87, 162–170.

The V17 medium was simply a broth medium with agar and basic fuchsin omitted, but otherwise having the same composition as the V15 medium (cited under **Oxidase**—*Veillonella* p. 698).

Basal medium for carbon and nitrogen compounds—*Vibrio extorquens*

P. K. Stocks and C. S. McCleskey (1964). J. Bact. 88, 1065–1070.

Determination of carbon and nitrogen sources suitable for growth were made in the mineral salts medium of Brown (1958): KNO_3, 1.0 g; $MgSO_4 \cdot 7H_2O$, 0.2 g; $K_2HPO_4 \cdot 3H_2O$, 0.5 g; $FeCl_3 \cdot 6H_2O$, 0.05 g; and NaCl, 0.2 g per liter of distilled water. Mineral salts-agar was prepared by adding 15 g of agar (Difco) per liter. Tryptone Glucose Extract (TGE) agar (Difco) and other common bacteriological media were employed in tests for purity of cultures and in some of the biochemical tests. For purposes of comparison, the mineral salts media of Mevius (1953) and Jayasuriya (1955) were employed in some tests. Carbon sources were routinely added to give concentrations of 0.1 and 0.5%; in some cases, where inhibition was feared, concentrations of 0.01% were also used.

Inocula for testing carbon and nitrogen sources consisted of washed cells which had been grown in mineral salts-methanol (0.5%) medium on a rotary shaker at 28°C for 48 hr. Tests for the utilization of substrates were also incubated at 28°C, and observations were made at intervals for 21 days. Cultures showing no growth after 21 days were recorded as negative.

L. R. Brown (1958). *Isolation, characterization and metabolism of methane oxidizing bacteria.* Ph.D. Thesis, Louisiana State University.
G. C. N. Jayasuriya (1955). J. gen. Microbiol. 12, 419–428.
W. Mevius, Jr. (1953). Arch. Mikrobiol. 19, 1–29.

Basal medium for carbon and nitrogen compounds—*Leptospira*

O. H. V. Stalheim and J. B. Wilson (1964). J. Bact. 88, 48–54.

The authors used a slight modification of the synthetic medium 198E of Vogel-Hunter (1961) to study nutrition of *Leptospira.*

H. Vogel and S. H. Hunter (1961). J. gen. Microbiol. 26, 223–230.

Basal medium for carbon compounds—*Pseudomonas, Flavobacterium, Achromobacter, Xanthomonas*

H. H. Tabak, C. W. Chambers and P. W. Kabler (1964). J. Bact. 87, 910–919.

The material containing organisms subjected to preliminary enrichment was eventually inoculated in baffled Erlenmeyer flasks containing 50 ml of Gray and Thornton's (1928) mineral salts medium to which 0.025 μg of vitamin B_{12} had been added. The specific phenolic compound served as the sole source of carbon. The medium was prepared by aseptically adding concentrates to sterile distilled water. Final pH was 7.0 to 7.2. All cultures were incubated at room temperature on an orbital shaker.

Organisms grown on phenol at 100–500 ppm were tested manometrically for their ability to oxidise other related compounds.

Dihydric phenols were generally oxidized; trihydric phenols were not. Cresols and dimethylphenols were oxidized; adding a chloro group increased resistance. Benzoic and hydroxybenzoic acids were oxidized; sulfonated, methoxylated, nitro, and chlorobenzoic acids were not; *m*-toluic acid was utilized but not the *o*- and *p*-isomers. Benzaldehyde and *p*-hydroxybenzaldehyde were oxidized. In general, nitro- and chloro-substituted compounds and the benzenes were difficult to oxidize.

P. H. H. Gray and H. G. Thornton (1928). Zentbl. Bakt. ParasitKde Abt. II, 73, 74.

Basal medium for carbon compounds—*Pseudomonas, Achromobacter*

D. Claus and N. Walker (1964). J. gen. Microbiol. 36, 107–122.

The following basal inorganic media were used, with the addition of the appropriate organic compound and, when required, of agar (2%). Medium A was a slight modification of Tausson's (1929) medium and compounded from solution *a:* $(NH_4)_2SO_4$, 1.2 g; $CaCl_2 \cdot 2H_2O$, 0.1 g; $MgSO_4 \cdot 7H_2O$, 0.1 g; Fe (as ferric citrate), 0.002 g; distilled water, 1 L and solution *b:* K_2HPO_4, 0.2 g; KH_2PO_4, 0.1 g; distilled water, 200 ml, sterilized separately and mixed aseptically in the proportion of 5 vol *a* to 1 vol *b.* Medium B contained: K_2HPO_4, 0.8 g; KH_2PO_4, 0.2 g; $CaSO_4 \cdot 2H_2O$, 0.05 g; $MgSO_4 \cdot 7H_2O$, 0.5 g; $FeSO_4 \cdot 7H_2O$, 0.01 g; $(NH_4)_2SO_4$, 1.0 g; distilled water, 1 L. Because of marked changes in pH value which occurred with different substrates, 1 g $CaCO_3$/L was added to this medium for the cultures to be used in oxygen-uptake experiments.

Cultural conditions for growing bacteria on toluene or other liquid aromatic hydrocarbon made use of Tausson's 'Methode des Diffusionszuflusses'. Agar plate cultures were incubated in a closed desiccator, over a saturated aqueous solution of toluene kept saturated by contact with liquid toluene contained in an inverted glass dish. For shaken-flask cultures, a small test tube with a hole in its wall and containing a little toluene was suspended inside the flask; or an open glass tube (5–6 mm bore) was inserted through the cotton wool plug, dipped into the medium in the flask and a few drops of toluene were introduced into this tube. Toluene thus diffused as vapour or by solution into the medium without liquid toluene itself coming into direct contact with the bacteria. Toluene was supplied to larger cultures in flasks (5 or 10 L) or in the continuous culture apparatus (described by Skinner and Walker, 1961) by aerating them with air previously saturated with water and toluene vapours.

F. A. Skinner and N. Walker (1961). Arch. Mikrobiol. 38, 339.
W. O. Tausson (1929). Planta 7, 735.

Basal medium for carbon and nitrogen compounds—*Pseudomonas aeruginosa*

R. R. Colwell (1964). J. gen. Microbiol. 37, 181–194.

The standard inoculation for all the tests was carried out by spotting agar plates containing basal salts medium (NaCl, 5.0 g; $MgSO_4 \cdot 7H_2O$, 0.2 g; K_2HPO_4, 1.0 g; distilled water 1L; adjusted pH 6.8) with added organic acids and ammonium-N or amino acids. A control plate with only the basal salts medium was tested at the same time. Vitamin-free Casamino acids, proline, DL-alanine, β-alanine, arginine, lysine, phenylalanine (California Biochemical Corp.) and sodium formate were tested for ability to support growth.

Basal medium for carbon compounds—*Agrobacterium*

V. Sundman (1964). J. gen. Microbiol. 36, 171–183.

Isolation of the bacteria. As previously described (Sundman, 1962), the α-conidendrin-decomposing bacteria were isolated from soil samples by enrichment through several subcultures in a nitrate + mineral salts + vitamin basal medium to which 0.5% (w/v) of α-conidendrin crystals were added as the carbon + energy source. The base solution (K-solution) contained: $NaNO_3$, 2.5 g; KH_2PO_4, 1.0 g; $CaCl_2$, 0.1 g; $MgSO_4 \cdot 7H_2O$, 0.3 g; NaCl, 0.1 g; $FeCl_3$, 0.01 g; vitamin solution (thiamine 10 mg, Ca-panthothenate 10 mg, biotin 20 μg, B_{12} vitamin 40 μg, pyridoxin 10 mg, nicotinic acid 10 mg, *p*-aminobenzoic acid 10 mg, folic acid 2 mg, dissolved in 200 ml of water) 5 ml; water 1000 ml; final pH 6.5. For isolation, the enrichment cultures were plated on KYE agar (K-solution + Bacto yeast extract 0.1% (w/v) + agar 2% (w/v)). The isolates were replated on the same medium, and the ability to decompose α-conidendrin was noted as a darkening upon growth in the liquid enrichment medium with α-conidendrin as sole source of carbon + energy. About 10 α-conidendrin-decomposing isolates from each soil sample were kept as stock cultures for further work, with monthly transfers on: (a) semi-solid KYE agar (KYE agar with agar content decreased to 0.3% w/v); (b) the liquid α-conidendrin-containing medium.

Manometric studies have been carried out on α-conidendrin and related substances in this and the succeeding paper (V. Sundman (1964). J. gen. Microbiol. 36, 185–201.)

Note: Dr. Sundman (personal communication) advises that the mineral part of the base solution in the above abstract was adopted from Konetzka *et al.* (1952). J. Bact. 63, 771–778. She has since reduced the $CaCl_2$ content to 0.01 g/liter. She further states that the main part of the work referred to in the abstract was published in: V. Sundman (1965). Acta Polyt. Scand. Ch 40, 116pp "A study on lignanolytic soil bacteria with special reference to α-conidendrin decomposition".
V. Sundman (1962). Meddn finska KemSamf, 71, 26.

Basal medium for carbon compounds—*Rhizobium*

P. H. Graham (1964). Antonie van Leeuwenhoek 30, 68–72.

Media. All strains were maintained on slopes of a yeast extract mannitol medium containing: agar, 20 g; mannitol, 10 g; K_2HPO_4, 0.5 g; $CaCl_2$, 0.2 g; $MgSO_4 \cdot 7H_2O$, 0.1 g; NaCl, 0.2 g; $FeCl_3 \cdot 6H_2O$, 0.01 g; boiled yeast infusion, 100 ml; distilled water, 900 ml. The medium used in testing for carbohydrate utilisation was similar to the yeast extract medium but contained no carbohydrate and only 25 ml/liter yeast infusion.

The compounds tested were: L-arabinose, dextrin, dulcitol, D-fructose, galactose, D-glucose, lactose, maltose, D-mannitol, D-mannose, raffinose, L-rhamnose, sodium citrate, sodium fumarate, sodium malate, sodium pyruvate, sodium succinate, sucrose, trehalose, and D-xylose. Of each compound a 10% solution was prepared or, where the solubility was less than 10%, a saturated solution. Each was sterilised by Seitz filtration or by steaming (dextrin). Tubes of the yeast extract test medium were liquefied by autoclaving and allowed to cool to 45°C. They were then inoculated, poured into petri dishes, and allowed to set. A size 5 cork borer, sterilised by flaming, was used to cut holes in the agar medium. Three drops of a solution of one of the carbon compounds under test were placed in each hole, and the plates incubated at 28°C for seven days. Zones of growth stimulation were observed by holding the plates over a viewing box in which a circular filament provided indirect lighting.

Basal medium for carbon compounds—*Pseudomonas*

D. S. Robinson (1964). Antonie van Leeuwenhoek 30, 303–316.

Chemicals of A.R. grade were used when these were available. Owing to the prohibitive cost of highly purified hydrocarbons, those used were of the Laboratory Reagent Grade of British Drug Houses Ltd. Nonane however (95% pure), 1-octanol and 2-octanone were purchased from L. Light and Co. Before addition to growth media they were heated to 100°C for 5 min to ensure sterility. The paraffin (5 ml/100 ml medium) was added to a sterile basal culture medium of the following composition: 1 liter of distilled water; NH_4Cl, 2 g; KH_2PO_4, 4 g; $NaH_2PO_4 \cdot 2H_2O$, 6 g; $MgSO_4 \cdot 7H_2O$, 0.2 g; $FeSO_4 \cdot 7H_2O$, 1 mg; $ZnSO_4 \cdot 7H_2O$, 0.1 mg; $CuSO_4 \cdot 5H_2O$, 0.1 mg; H_3BO_3, 40 μg; $MnSO_4 \cdot 7H_2O$, 40 μg; $MoO_3 \cdot H_2O$, 20 μg. The pH value was adjusted to 7.0 with NaOH by means of a glass electrode. Glucose was used at 2% (w/v). Octoic acid, acetic acid and the decarboxylic acids used were dissolved at a concentration of 0.1% (w/v) prior to sterilisation.

Growth of the organism. The standard procedure was to transfer cells from the stock slopes to 10 ml of the sterile basal culture medium supplemented with either glucose, octane or octoic acid contained in a conical flask (100 ml capacity). The cultures were incubated at 30°C on a shaker revolving at 110 rev/min. Three serial transfers were made at 24 hr intervals using the loop of an inoculating needle. 1 ml samples of the resultant cultures were used to inoculate the main bulk of the medium (100 ml) containing the respective substrates. The cultures were contained in widenecked conical flasks (capacity 500 ml) and incubated on the rotary shaker. The growth of the organism was followed by measuring the optical densities of the cultures with a Hilger Absorptiometer using H508 filters.

Basal medium for carbon and nitrogen compounds—*Methanomonas methanooxidans*

L. R. Brown, R. J. Strawinski and C. S. McCleskey (1964). Can. J. Microbiol. 10, 791–799.

A mineral salts broth, consisting of KNO_3, 1.0 g; $K_2HPO_4 \cdot 3H_2O$, 0.5 g; $MgSO_4 \cdot 7H_2O$, 0.2 g; $FeCl_3 \cdot 6H_2O$, 0.05 g/L distilled water was used for the enrichment and propagation of the methane oxidizer.

Methane and other gases employed (Matheson Company, East Rutherford, New Jersey) were of high purity: methane, c.p. grade, 99.0% pure; ethane, 95.0%; *n*-propane and *n*-butane, instrument grade, 99.9%; oxygen, extra dry grade, 99.6%; carbon dioxide, bone dry grade, 99.9%. The methane supplied to the organism as the carbon and energy source was routinely mixed with oxygen and carbon dioxide in the proportion of 65:30:5 on a volume basis.

The only carbon source other than methane which allowed growth was methanol.

Various nitrogenous compounds (in concentration of 25 mg atoms per L) were screened to determine their ability to serve as sole nitrogen source for the methane oxidizer. KNO_3, NH_4Cl, peptone, L(+)-arginine, L(−)-cystine, L(−)-leucine, and L-glutamate but not DL-aspartate supported growth. More methane was consumed when NH_4Cl or glutamate was used as nitrogen source but more carbon dioxide was formed when peptone was the nitrogen source. These results suggest that peptone may be used as a carbon source although the organism is unable to use it as sole carbon source.

Basal medium for carbon compounds—*Klebsiella aerogenes*

S. C. Agarwal (1964). J. Path. Bact. 87, 186–190.

Dissimilation of citrate. The method described by Dagley and Dawes (1953) was used. D475 cells were grown aerobically or anaerobically in a medium in which the concentration of citrate was limiting for growth. The medium was: Na citrate dihydrate, 7 g; KH_2PO_4, 5.4 g; $(NH_4)_2SO_4$, 1.2 g; $MgSO_4 \cdot 7H_2O$, 0.4 g; and glass-distilled water, 1000 ml; the pH was adjusted to 7.1. $MgSO_4 \cdot 7H_2O$ was added sterilely as a 10 per cent solution.

S. Dagley and E. A. Dawes (1953). J. Bact. 66, 259.

Basal medium for carbon compounds—*Haemophilus pleuro-pneumoniae, Haemophilus parainfluenzae.*

D. C. White, G. Leidy, J. D. Jamieson and R. E. Shope (1964). J. exp. Med. 120, 1–12.

The defined medium is the simplified medium of Herbst and Snell (1949) to which histidine was added (Leidy, Jaffe and Alexander, 1962). Sodium bicarbonate was incorporated in this medium to a concentration of 0.02 M where subsequently indicated.

Fermentation reactions are difficult to use as a taxonomic criterion with *Haemophilus* species since these bacteria grow in such complex media that the effects of added carbohydrate give variable results. However, for those species which grow in the defined medium, growth is dependent upon the carbohydrate added.

Lactate, D-glucose, D-xylose, maltose, sucrose, mannitol, and galactose at 0.01 M concentration with 0.02 M HCO_3^- were utilised by one or more strains.

E. J. Herbst and E. E. Snell (1949). J. Bact. 58, 379.
G. Leidy, I. Jaffe and H. E. Alexander (1962). Proc. Soc. exp. Biol. Med. 111, 725.

Basal medium for carbon compounds—*Escherichia coli*

H. Nakamura (1965). J. Bact. 90, 8–14.

A minimal medium, S_1, was composed of 3.5 g of Na_2HPO_4, 1.5 g of KH_2PO_4, 1 g of NH_4Cl, 1 ml of 1 M $MgSO_4$, and 1 ml of 10^{-3} M $FeCl_3$ per liter. It was made up by mixing stock solutions which had been sterilized separately. Another minimal medium, S_2, contained less phosphate to avoid precipitation of Mg; it consisted of 2 g of K_2HPO_4, 1 g of $(NH_4)_2SO_4$, 1 ml of 1 M $MgSO_4$, and 1 ml of 10^{-3} M $FeCl_3$ per liter. The media were supplemented with either glucose or another carbohydrate (at concentrations of 0.4 and 0.5%, respectively), previously sterilized separately from the minerals. The initial pH was 7.2 in S_1 and 7.8 in S_2.

For testing sugar utilization, S_1 agar in which the test sugar was the sole carbon source was used. Eosin-methylene blue (EM) agar without succinate (as modified by Hirota, 1960), plus a given sugar, was sometimes used in diagnosing sugar utilization for basic dye-resistant strains; the growth of sensitive cells was inhibited by this medium, although such strains could utilize the sugar in it.

Cultures were incubated at 37°C.

Y. Hirota (1960). Proc. natn. Acad. Sci. U.S.A. 46, 57–64.

Basal medium for nitrogen compounds—marine bacteria

R. M. Pfister and P. R. Burkholder (1965). J. Bact. 89, 863–872.

Amino acids were tested as a carbon source by inoculating a medium of synthetic seawater containing, per liter, KH_2PO_4, 0.025 g, and N-Z-Case, 4.0 g.

Ammonium nitrate was tested as a nitrogen source by use of a medium of seawater with additions of the following (per liter): KH_2PO_4, 0.05 g; NH_4NO_3, 5.0 g; glucose, 4.0 g; fructose, 4.0 g; and sodium citrate, 4.0 g.

Basal medium for carbon and nitrogen compounds—*Alcaligenes viscolactis*

J. D. Punch, J. C. Olson, Jr., and J. V. Scaletti (1965). J. Bact. 89, 1521–1525.

Basal salts. The basal salts solution contained (per liter) K_2HPO_4, 5 g; KH_2PO_4, 3 g; Mg^{++}, 3 x 10^{-3} M as $MgSO_4 \cdot 7H_2O$, 0.74 g; Mn^{++}, 2 x 10^{-4} M as $MnSO_4 \cdot H_2O$, 0.34 g; Fe^{++}, 2 x 10^{-4} M as $FeSO_4 \cdot 7H_2O$, 0.056 g; and NaCl, 2 x 10^{-4} M, 0.012 g. When media were prepared containing basal salts solution, the $MgSO_4 \cdot 7H_2O$ was sterilized separately and added aseptically after sterilization to the rest of the particular medium.

Supplements to basal salts. Lactose, glucose, peptone (Difco), yeast extract (Difco), Vitamin Free Casamino Acids (Difco), and various inorganic and organic nitrogenous compounds were used to supplement the basal salts solution. Concentrations and combinations of these additives are given in the original paper. All nitrogenous compounds were sterilized separately and added aseptically.

Basal medium for nitrogen compounds—various

T. Tosa and I. Chibata (1965). J. Bact. 89, 919–920.

To determine the occurrence of "cyclic amide hydrolase," we screened a number of microorganisms utilizing the cyclic amides (γ-butyrolactam, δ-valerolactam, ϵ-caprolactam, and pyroglutamic acid) for the production of ω-amino acids. The media contained 0.1% glucose, 0.05% sodium citrate, 0.7% K_2HPO_4, 0.2% KH_2PO_4, 0.01% $MgSO_4 \cdot 7H_2O$, and the cyclic amide as the sole source of nitrogen. The utilization of cyclic amides was measured by growth of the microorganism, and the formation of ω-amino acids was detected by paper chromatography of the broth.

Basal medium for nitrogen compounds—*Mycobacterium*

M. Tsukamura and J. Tsukamura (1965). J. Bact. 89, 1442.

The medium was used as follows: $NaNO_2$, 1.38 g (0.02 M); KH_2PO_4, 0.5 g; $MgSO_4 \cdot 7H_2O$, 0.5 g; sodium citrate, 2.0 g; purified agar (Wako Pure Chemical Co. Osaka, Japan), 30 g; glycerol, 50 ml; and distilled water, 950 ml (pH was adjusted to 7.2 by addition of 10% NaOH). The medium was sterilized prior to addition of $NaNO_2$ by autoclaving at 115°C for 30 min. $NaNO_2$ was dissolved in distilled water at a concentration of 13.8%, sterilized by Seitz filtration (the same data were obtained by heating the $NaNO_2$ at 100°C for 5 min), and added to the medium aseptically. Slants were prepared, and the medium was inoculated with one loopful of the stock cultures and incubated at 37°C. Growth was observed every week. A control medium without $NaNO_2$ was employed, and growth was always decided by comparison with growth on the control medium. To avoid mistakes in reading, it was important to use a purified agar.

Basal medium for carbon compounds—*Bacillus*

G. R. F. Hilson (1965). J. gen. Microbiol. 39, 407–421.

The ammonia basal medium of Knight and Proom (1950) was used: this medium contains ammonium phosphate as the only source of nitrogen, and whichever carbohydrate is added constitutes the only source of carbon. It was found convenient to adopt the following modifications: the medium was adjusted to pH 7.4, and phenol red 0.01% (w/v) and agar 1.5% (w/v) were added. Stock solutions of the carbohydrates, 10% (w/v) in distilled water, were sterilized by Seitz filtration and were added to the melted basal medium to a final concentration of 1% (w/v). The medium was dispensed as slopes in ½ oz screw-capped bottles. The bottles were incubated for 3 days; surface growth indicated utilization of the carbohydrate tested, and yellow coloration of the medium indicated the development of an acid reaction.

B. C. J. G. Knight and H. Proom (1950). J. gen. Microbiol. 4, 508.

Basal medium for carbon compounds—*Pediococcus*

R. Whittenbury (1965). J. gen. Microbiol. 40, 97–106.

Acetic acid + acetate agar was a modification of the medium proposed by Keddie (1951) as being selective for lactobacilli and contained meat extract (Lab Lemco), 0.5 g; peptone (Evans), 0.5 g; yeast extract (Difco),

0.5 g; Tween 80, 0.05 ml; glucose, 1 g; agar (Davis), 1.5 g; tap water to 90 ml; adjusted to pH 5.4 and autoclaved at 121° for 15 min. Before plating the medium 10 ml of 2 M-acetic acid + sodium acetate buffer (pH 5.4) were added.
R. M. Keddie (1951). Proc. Soc. appl. Bact. 14, 157.

Basal medium for carbon compounds—soil bacteria (unnamed)
J. Ooyama and J. W. Foster (1965). Antonie van Leeuwenhoek 31, 45–65.

Organisms were routinely cultured in a mineral salts solution (Leadbetter and Foster, 1958) with a single hydrocarbon added as the sole source of carbon and energy. Liquid cultures, usually in rubber-stoppered vessels, were incubated on a continuous shaker at 30° or 37°C. Gaseous substrates were furnished as 50–50 (v/v) mixtures in desiccators or sealed suction flasks. Solid media were obtained by addition of 2% water-washed agar to the salts solution.

Hydrocarbons tested: The following were purchased from Phillips Petroleum Co., Bartlesville, Oklahoma and were 99 mol per cent minimum purity: C_1 through C_{13} aliphatic n-alkanes, trimethylmethane, 2,2-dimethylpropane, 2-methylbutane, 2,2-dimethylbutane, 2,3-dimethylbutane, 2-methylpentane, 3-methylpentane, 2,4-dimethylpentane, 2,2,5-trimethylhexane, cyclopentane, methylcyclopentane, cyclohexane, methylcyclohexane. From the same source, only 95 mol per cent purity: 2,3-dimethylpentane and 3-methylhexane. From Humprey-Wilkinson, Inc., North Haven, Connecticut: n-dodecane, n-tetradecane, n-hexadecane, n-octadecane, n-docosane and n-eicosane. From the Matheson Co., Cincinnati, Ohio: cyclopropane and cyclohexanol. From Distillation Products, Inc., Rochester, N.Y.: cyclopentanone and cyclohexanone. From Aldrich Chemical Co., Milwaukee, Wisconsin: cyclopentene, cyclohexene, cyclohexene oxide, cycloheptane, cyclooctane, D-3-methylcyclopentanone, trans-1,2-cyclohexanediol, 2-methylcyclohexanone, and cyclooctene.
E. R. Leadbetter and J. W. Foster (1958). Arch. Mikrobiol. 30, 91–118.

Basal medium for carbon compounds—*Mycobacterium, Nocardia*
R. W. Traxler, P. R. Proteau, and R. N. Traxler (1965). Appl. Microbiol. 13, 838–841.

The paper describes studies on the metabolism of asphalt.

Materials and Methods

Organisms. The two organisms used in most of this investigation were *Mycobacterium ranae* and *Nocardia coeliaca.* All organisms were carried as stock cultures on Trypticase Soy Agar (TSA). Inocula were prepared from TSA slants grown at 30°C, washed in saline three times, and suspended to a turbidity of 300 Klett units.

Thin-layer and asphalt-emulsion techniques. The early experiments used a thin layer of asphalt suspended on the mineral salts medium, as described previously (Phillips and Traxler, 1963). This method provided a considerable surface area for microbial action but did not subject much of the asphalt to microbial action unless an extremely thin (100 to 200 μ) film was used. With such thin films, the quantity of asphalt which can be exposed to microbial action is so low as to limit physical testing.

To avoid this difficulty, an emulsion of asphalt in an aqueous dispersion of bentonite (clay) particles was used to provide the greatest possible surface-to-volume ratio of asphalt for microbial action. A 5% (w/v) suspension of bentonite in warm (80°C) water was prepared and placed in a Waring Blendor. The asphalt to be tested was dissolved in the minimal amount of benzene which allowed the asphalt to be poured, and this asphalt solution was added slowly to the warm bentonite suspension with slow agitation. After the total volume of asphalt had been added, the mixture was agitated at high speed for approximately 30 min, or until a smooth, homogeneous emulsion was obtained. The resulting bentonite-asphalt emulsion was autoclaved at 121°C for 30 min, which served to remove all traces of benzene. The product was stable and did not separate upon standing or autoclaving.

For each asphalt studied, three 1,500-ml flasks, each containing 100 ml of bentonite-asphalt emulsion, were sterilized, and to each flask were added 300 ml of sterile mineral salts medium. One flask of each series was inoculated with 3.0 ml of the washed suspension of *M. ranae,* a second with *N. coeliaca,* and the third flask served as a sterile control. All flasks were then incubated under stationary conditions at 30°C for 4 months.

Asphalt recovery from the emulsion. After incubation, any clear liquid was carefully decanted from the culture. The bentonite-asphalt slurry was transferred to a Waring Blendor and extracted with three to five 150 to 200-ml portions of a 3:1 mixture of benzene and ethyl alcohol. Slow agitation was used to obtain the extraction, since high-speed agitation would cause further emulsification. The benzene-alcohol extractions from each flask were continued until all traces of asphalt had been recovered. With some samples, the bentonite was carried over in the extracts and had to be removed by centrifugation at 2,000 $\times$ g for 30 min. The benzene-alcohol was stripped from the extract by vacuum distillation to a volume of 200 to 300 ml. The final solution was then cleared of solvent in a flash evaporator. All samples were then checked for occluded water and dried if necessary.

Testing the asphalts. Viscosity measurements were made at 25°C with a Hallikainen microfilm viscometer (Labout and van Oort, 1956). One advantage of this method is that a small sample (0.2 to 0.5 g) of asphalt is required for the determination. A limitation to our study is the quantity of asphalt that can be exposed to microbial action; therefore, physical tests which require a small quantity of asphalt are necessary.

J. W. A. Labout and W. P. van Oort (1956). Analyt. Chem. 28, 1147–1151.

U. A. Phillips and R. W. Traxler (1963). Appl. Microbiol. 11, 235–238.

Basal medium for carbon and nitrogen compounds—*Spirillum gracile* n.sp.

E. Canale-Parola, S. L. Rosenthal and D. G. Kupfer (1966). Antonie van Leeuwenhoek 32, 113–124.

A basal medium used in some of the nutritional studies contained the following components (g/100 ml distilled water): K_2HPO_4, $MgCl_2 \cdot 6H_2O$, 10^{-2} each; $MnSO_4 \cdot H_2O$, 5 x 10^{-3}; $FeSO_4 \cdot 7H_2O$, 2 x 10^{-4}; thiamine HCl, nicotinic acid, calcium pantothenate, 10^{-5} each; pyridoxine HCl, riboflavin, vitamin B_{12}, 10^{-6} each; biotin, folic acid, 5 x 10^{-7} each. Various carbon and nitrogen sources were added to this medium as described below, and the pH was adjusted to 7.2–7.3.

In all nutrition experiments the organisms were grown at 30°C in 16 x 150 mm test tubes containing 10 ml of liquid medium. Each culture tube was inoculated with approximately 10^6 cells.

The only substances tested which supported growth in this carbon and nitrogen free base were L-glutamic acid, L-aspartic acid or L-cysteine.

With 0.1% NH_4Cl, the following substances (0.05%) acted as both carbon and energy sources:– glucose, D-xylose, L-arabinose, α-keto-glutaric acid, sodium succinate, potassium pyruvate and sodium lactate.

Glycerol and acetate (0.05%) were also used in a growth limiting medium consisting of peptone, 15 mg; yeast extract, 2 mg; K_2HPO_4, 10 mg; water, 100 ml.

Basal medium for carbon and nitrogen compounds—*Thiobacillus thiooxidans*

R. G. Butler and W. W. Umbreit (1966). J. Bact. 91, 661–666.

The authors used labelled radioactive compounds to study assimilation in a basal medium.

The growth medium of Vogler and Umbreit (1941) was used, with 1% (w/v) elemental sulfur as the energy source. The basal inorganic medium was dispensed in 100 ml quantities into 250 ml Erlenmeyer flasks and was then autoclaved. Sulfur, previously sterilized by intermittent steaming, was added aseptically at 1 g per flask. Radioactive compounds were sterilized by passage through a sterile Millipore filter and were added to the flask at the time of inoculation.

The flasks were inoculated with 1 ml of a 7 day-old culture of *T. thiooxidans* and were allowed to stand for 3 days to permit contact between the organisms and the sulfur. The flasks were then put on a rotary shaker at 200 rev/min and were shaken for an interval of 4 to 7 days. It has been shown that little growth results if the flasks are shaken immediately (Cook, 1964). At the termination of experiments, samples from each flask were plated on nutrient agar to detect heterotrophic contaminants, but none was found in the experiments reported.

T. M. Cook (1964). J. Bact. 88, 620–623.
K. G. Vogler and W. W. Umbreit (1941). Soil Sci. 51, 331–337.

Basal medium for carbon compounds—*Methylococcus capsulatus*

J. W. Foster and R. H. Davis (1966). J. Bact. 91, 1924–1931.

The following mineral salts solution was used in the media: $NaNO_3$, 2.0 g; $MgSO_4 \cdot 7H_2O$, 0.2 g; KCl, 0.04 g; $CaCl_2$, 0.015 g; Na_2HPO_4, 0.21 g; NaH_2PO_4, 0.09 g; $FeSO_4 \cdot 7H_2O$, 1.0 mg; $CuSO_4 \cdot 5H_2O$, 5 μg; H_3BO_4, 10 μg; $MnSO_4 \cdot 5H_2O$, 10 μg; $ZnSO_4 \cdot 7H_2O$, 70 μg; MoO_3, 10 μg; deionized water, 1 liter. For a solid medium, 2% water-washed agar (Difco) was added. Gas atmospheres of the appropriate compositions were provided in closed desiccators, or in flasks closed with rubber stoppers fitted with glass tubing through which gassing could be done. The desiccator was partially evacuated, and the air was replaced with the desired gas to a final pressure of 1 atm. Petri plates and cotton-plugged flasks or tubes were incubated in the desiccators at appropriate temperatures.

Cultures were grown under 50% methane: 50% air at an optimal temperature of 37°C. Methanol could replace methane for growth.

Basal medium for carbon compounds—*Desulfovibrio africanus* n.sp.

L. L. Campbell, M. A. Kasprzycki and J. R. Postgate (1966). J. Bact. 92, 1122–1127.

Carbon source utilization studies were carried out in the medium C of Butlin, Adams, and Thomas with lactate replaced by the appropriate carbon source at a concentration of 0.2% (w/v). Filter-sterilized (Millipore) Na_2S (1 mM) was added to poise the oxidation-reduction potential except in the case of *D. gigas,* for which sodium ascorbate (0.5 mM) was used. For strains of marine origin, all media contained 2.5% NaCl.

Anaerobiosis was obtained with pyrogallol plugs made alkaline with a solution containing K_2CO_3 (15%, w/v) and NaOH (10%, w/v).

K. R. Butlin, M. E. Adams and M. Thomas (1949). J. gen. Microbiol. 3, 46–59.

Basal medium for organic compounds for energy and growth—*Pseudomonas*

R. Y. Stanier, N. J. Palleroni and M. Doudoroff (1966). J. gen. Microbiol. 43, 159–271.

Every strain was tested for ability to grow at the expense of 146 different organic compounds. These tests were performed by replica plating (Lederberg and Lederberg, 1952). The test media were prepared by adding each organic compound at the appropriate concentration to the standard mineral base, solidified by the addition of 1% (w/v) of Ionagar. In order to avoid browning of the agar, it was sterilized separately from the mineral base (both at double strength), and mixed with it after removal from the autoclave. Each test series included a control plate without added organic compound. The use of Ionagar in place of less highly purified agars almost completely eliminated background growth on the control plates, which greatly facilitated the reading of the results.

The standard mineral base contained, per liter: 40 ml of Na_2HPO_4 + KH_2PO_4 buffer (M; pH 6.8); 20 ml of Hutner's vitamin-free mineral base (Cohen-Bazire, Sistrom and Stanier, 1957); and 1.0 g of $(NH_4)_2SO_4$. This basal medium is easy to prepare, and provides all necessary minerals, including trace elements. It is heavily chelated with nitrilotriacetic acid and EDTA, and forms a copious precipitate upon autoclaving. The precipitate redissolves as the medium cools, to form a water-clear solution.

Chemically defined media were prepared by adding an appropriate organic carbon and energy source to the standard mineral base; the concentration was generally 1–3 g/L.

The utilization of hydrocarbons as sole sources of carbon and energy cannot easily be determined on solid media because of the difficulty of obtaining a properly emulsified preparation of these water-insoluble compounds. This property was therefore determined in tubes containing 4–5 ml liquid mineral medium to which a few drops of the compound had been added under aseptic conditions from a filter-sterilized sample. In our tests we have only used highly purified n-dodecane and n-hexadecane. The tubes were incubated at 30°C on a rotary shaker and the results recorded daily for a period of 7 days.

The reader should consult the original paper for the lengthy discourse on preparation of such media for the pseudomonads.

G. Cohen-Bazire, W. R. Sistrom and R. Y. Stanier (1957). J. cell. comp. Physiol. 49, 25.

J. Lederberg and E. M. Lederberg (1952). J. Bact. 63, 399.

Basal medium for nitrogen compounds—*Mycobacterium parafortuitum* n.sp.

M. Tsukamura (1966). J. gen. Microbiol. 42, 7–12.

Utilization of nitrogen compounds as sole nitrogen source (including sodium nitrite) was tested on the following medium: glycerol, 50.0 ml; KH_2PO_4, 0.5 g; $MgSO_4 \cdot 7H_2O$, 0.5 g; sodium citrate, 2.0 g; purified agar, 30.0 g; distilled water, 950 ml (pH 7.0). Nitrogen compounds were sterilized by heating at 100°C for 5 min or Seitz filtration (sodium nitrite) and added to medium aseptically to a final concentration of 0.02 M. Growth was observed after 2 weeks' incubation at 37°C.

Basal medium for carbon and nitrogen compounds—*Mycobacterium*

M. Tsukamura (1966). J. gen. Microbiol. 45, 253–273.

Utilization for growth of nitrogen compounds as sole nitrogen and carbon sources. This was shown recently to be very useful for classification of rapidly growing mycobacteria (Tsukamura, 1965). The tests were carried out on the following medium: KH_2PO_4, 0.5 g; $MgSO_4 \cdot 7H_2O$, 0.5 g; purified agar, 20.0 g; distilled water, 1000 ml. The medium was supplemented with a nitrogen compound to a final concentration of 0.02 M, and the medium was adjusted to pH 7.0 and sterilized by autoclaving at 115°C for 30 min. After sloping, the medium was inoculated with one loopful of the stock cultures and incubated. Growth was read after incubation for 2 weeks (rapid-growing mycobacteria) or after incubation for 4 weeks (slow-growing mycobacteria). The results were insensitive to inoculum size and the readings were clear. The following compounds were used: sodium L-glutamate; L-serine; glucosamine-HCl; acetamide; benzamide; nicotinamide; monoethanolamine (0.1 M); trimethylene diamine (0.1 M).

Incubation was at 37°C unless otherwise noted. As the source of inoculation for tests 3- to 4-week-old cultures of slow-growing mycobacteria or 1-week-old cultures of rapid-growing mycobacteria grown on Löwenstein-Jensen medium were used.

M. Tsukamura (1965). Rep. Conf. Taxon. Bact. Brno. p. 369.

Basal medium for nitrogen compounds—*Mycobacterium*

M. Tsukamura (1966). J. gen. Microbiol. 45, 253–273.

Utilization for growth of nitrogen compounds as sole nitrogen sources (a selected series from Tsukamura, 1965), where this method was shown to be useful. The tests were carried out in the following medium: glycerol, 30 ml; KH_2PO_4, 0.5 g; $MgSO_4 \cdot 7H_2O$, 0.5 g; sodium citrate, 1.0 g; purified agar, 20.0 g; distilled water, 970 ml. Nitrogen compounds were added to the above medium to a final concentration of 0.02 M, and the medium was adjusted to pH 7.0. The medium was sterilized by autoclaving at 115°C for 30 min and sloped. Only nitrite medium was sterilized by heating at 100°C for 5 min or else nitrite solution was sterilized by Seitz-filtration and added to medium aseptically. The medium, together with control medium without nitrogen source, was inoculated with one loopful of the stock cultures and incubated. Growth was scored in comparison with growth on the control medium after incubation for 2 weeks (rapid-growing mycobacteria) or after incubation for 4 weeks (slow-growing mycobacteria). The following were used for the tests: sodium L-glutamate, L-serine; L-methionine; acetamide; benzamide; urea; pyrazinamide; isonicotinamide; nicotinamide; succinamide; $NaNO_3$; $NaNO_2$ (Tsukamura and Tsukamura, 1965).

M. Tsukamura (1965). Rep. Conf. Taxon. Bact. Brno. p. 369.
M. Tsukamura and J. Tsukamura (1965). J. Bact. 89, 1442.
M. Tsukamura and J. Tsukamura (1966). Am. Rev. resp. Dis. 94, 104.

Basal medium for carbon compounds—*Mycobacterium*

M. Tsukamura (1966). J. gen. Microbiol. 45, 253–273.

Utilization of carbohydrates (sugars, alcohols, glycols) as sole carbon sources. This was observed by using the medium: $(NH_4)_2SO_4$, 2.64 g; KH_2PO_4, 0.5 g; $MgSO_4 \cdot 7H_2O$, 0.5 g; purified agar, 20 g; distilled water, 1000 ml. Sugars were added to a final concentration of 0.5% (w/v), and alcohols and glycols to a final concentration of 0.1 M. Growth was observed after incubation for 2 weeks (rapid-growing mycobacteria) or after incubation for 4 weeks (slow-growing mycobacteria). The following were used: glycerol; glucose; fructose; sucrose; mannose; galactose; arabinose; xylose; rhamnose; trehalose; raffinose; inositol; mannitol; sorbitol; ethanol; propanol; propyleneglycol; 1,3-butyleneglycol; 1,4-butyleneglycol; 2,3-butyleneglycol.

Basal medium for carbon compounds—*Escherichia, Alcaligenes*

B. E. Gustafsson, T. Midtvedt and A. Norman (1966). J. exp. Med. 123, 413–432.

The metabolism of 24 bile acids was studied using radioactive substrates in Oxoid Todd-Hewitt broth.

Basal medium for carbon compounds—*Providencia, Rettgerella*

C. Richard (1966). Annls Inst. Pasteur, Paris 110, 105–114.

Pour des raisons de commodité, les milieux ont été incubés à 37°C.

Métabolisme des glucides. Détermination de la voie métabolique d'attaque des glucides à l'aide du milieu Mevag [32].

[32] Veron (M.) et Chatelain (R.). Ann. Inst. Pasteur, 1960, 99, 253.

Basal medium for carbon compounds—*Actinobacillus mallei*

D. H. Evans (1966). Can. J. Microbiol. 12, 625–629.

The authors describe 2 complex basal media (one liquid and one solid) and two chemically defined media together with methods of use.

Basal medium for carbon compounds—*Micrococcus sp.*

N. P. Jayasankar and J. V. Bhat (1966). Can. J. Microbiol. 12, 1031–1039.

The mineral basal solution described by Betrabet and Bhat (1958), to which 0.1% phenol was added as the sole source of carbon (except for the addition of 0.05% yeast extract), was used for the isolation of organisms capable of utilizing phenol.

S. M. Betrabet and J. V. Bhat (1958). Appl. Microbiol. 6, 89–93.

BILE AND SAPONIN SOLUBILITY

Bile solubility—*Diplococcus*

J. M. Alston (1928). J. Bact. 16, 397–407.

The question of bile solubility was decided by the addition of 0.1 ml of a sterile 10 per cent solution of sodium taurocholate to 1 ml of the broth culture twenty-four to forty-eight hours old, and examination was made after an hour in an incubator at 37°C. In cases of doubt, when the broth culture was not very turbid, a stained film preparation was used to make sure whether or not lysis of the organisms had occurred. Control tests of the solubility of pneumococci were carried out.

Bile and saponin solubility—*Diplococcus*

S. J. Klein and F. M. Stone (1931). J. Bact. 22, 387–401.

From the one day plain broth cultures, 0.5 ml was pipetted into each of three test tubes (100 x 13 mm). To each was added either 0.5 ml of a 10 percent solution of saponin in saline (saponin Merck pure), 0.5 ml saline or 0.5 ml of 10 percent bile in saline. The saponin solution was usually freshly prepared, and was heated for fifteen minutes in boiling water before use. It was found that such heating does not affect the bacteriolytic activity. The tests were observed for two to three hours at room temperature, were then placed in the refrigerator and examined again the following day.

The development of complete transparency in the tube containing the saponin-treated bacteria was regarded as evidence of complete lysis. Lysis was often observed in a few minutes and was usually complete within thirty minutes.

To test for dissolution of the bacterial bodies, a loopful of the saponin-treated suspension was emulsified in a loopful of Loeffler's methylene blue on a cover slip, and examined microscopically in the hanging drop.

Bile solubility—*Diplococcus*

P. Hauduroy (1951). Techniques Bactériologiques. Masson et Cie, Éditeurs, Paris.

BILE.

(Solubilité dans la). – Prendre une culture en bouillon de 6 heures Ajuster à *p*H 7, si le *p*H est plus acide. Ajouter 10 p. 100 d'une solution à 10 p. 100 de taurocholate de soude dans de l'eau physiologique (0.9 p. 100 de chlorure de sodium). Les cultures de pneumocoques doivent être lysées en 30 minutes environ. *(W. I. C.)*

Bile solubility—*Streptococcus*

N.C.T.C. Methods 1954.

For differentiation of the streptococci from the pneumococci.

Grow organisms in serum or blood broth for 24 hours. Centrifuge and discard supernatant. Resuspend organisms in 0.5 ml isotonic saline adjusted to pH 7.0 (a pH range of 6.5 to 8.0, however, is satisfactory). Add 0.5 ml of a 10% solution of sodium deoxycholate.

Incubate at 37°C for 30 minutes.

Under these conditions pneumococci cause rapid clearing of the suspension. If clearing has not taken place in 30 minutes the organisms are definitely streptococci, not pneumococci.

Bile solubility—*Diplococcus*

V. B. D. Skerman (1967). *A Guide to the Identification of the Genera of Bacteria.* 2nd ed. Baltimore: Williams & Wilkens.

This test has been employed to differentiate the pneumococcus, *Diplococcus pneumoniae,* from the other α-hemolytic streptococci. Living cultures of this organism are soluble in bile salts at neutral pH.

Centrifuge the growth from a 5 ml culture of the organism in glucose-tryptic broth, and resuspend the growth in 0.5 M phosphate buffer, pH 7 to 7.6, containing 2 per cent sodium chloride and 0.05 per cent sodium deoxycholate. Incubate at 37°C for 60 minutes and examine. Cultures of pneumococci lyse under these conditions.

A. B. Anderson and P. D. A. Hart (1934). Lancet 2, 359.

Bile solubility—*Diplococcus pneumoniae*

C. V. Z. Hawn and E. Beebe (1965). J. Bact. 90, 549.

The medium was conventional blood-agar in a petri dish. The fresh colony suspected of being *D. pneumoniae* was examined under a hand lens. On this colony was placed a loopful of 2% sodium deoxycholate (pH 7.0). The plate was incubated at 37°C for 30 min. On re-examination, the suspect colony, if *D. pneumoniae,* was found to have disappeared, leaving only the partially hemolyzed medium to help identify its previous location.

ALKALI SOLUBILITY

Alkali solubility—*Neisseria gonorrhoeae*

A. Cantor, H. A. Shelanski and C. Y. Willard (1942). J. Bact. 44, 236–240.

The microscopic alkali-solubility of "oxidase positive" and other colonies on the primary plate was determined as follows. Growth from one colony was emulsified as thoroughly as possible in about 0.02 ml (or 3 loopfuls) of water in the center of a slide. The suspension was so prepared as to show not more than a just perceptible turbidity. A loopful of this suspension was then mixed with two loopfuls of water previously placed on the same slide but to the left of the original suspension. Another loopful of the original suspension was then mixed with a loopful of N/10 or N/5 sodium hydroxide previously placed on the same slide but to the right of the original suspension. After 30 seconds to 1 minute a loopful of hydrochloric acid of corresponding strength was mixed with the alkali treated suspension on the right. The three preparations on the slide were then air dried, heat fixed, and stained by Gram's method in the usual manner. The smears were then examined microscopically. The center smear was examined first, to establish the presence of gram-negative diplococci. The smear of the water dilution of the original suspension, that is, the control smear toward the left end of the slide, was examined to determine the normal appearance, degree of autolysis, and the approximate number of organisms before alkali treatment. The smear of the alkali-treated suspension on the right was then examined to determine whether or not the cells were dissolved by the alkali. It was found that when testing soluble neisserias there was a maximum number of cells which could be dissolved by a given volume of the alkali. Therefore, in the preparation of smears, no more than the cells from one small colony or from part of a large colony were emulsified on the slide.

BILE TOLERANCE

Bile tolerance—*Streptococcus*

A. C. Evans (1936). J. Bact. 31, 423–437.

All cultures were tested for ability to grow on infusion agar containing 5 per cent rabbit blood and 10 per cent bile. The bile was prepared with dehydrated "bactooxgall" (Digestive Ferments Company) made up to the original density. Those cultures which grew on 10 per cent bile were tested again for ability to grow on blood-agar containing 40 per cent bile. Control plates of blood-agar without the addition of bile were always inoculated at the same time to insure the viability of the inoculum. Readings were made after 48 hours incubation.

Bile tolerance—anaerobic streptococcus

M. L. Stone (1940). J. Bact. 39, 559–582.

Blood-bile plates were prepared by the additions of 5 per cent horse blood and 10 or 40 per cent ox bile respectively, to 3 per cent beef-heart agar. These plates were inoculated and incubated anaerobically for 48 hours at the end of which time they were inspected for growth. Control blood plates without bile were included in order to verify the viability of the organisms inoculated.

Bile tolerance—*Streptococcus*

K. E. Hite and H. C. Hesseltine (1947). A study of aerobic streptococci isolated from the uterus and vagina. J. infect. Dis. 80, 105–112.

Bile tolerance was tested by observing surface growth on 10% sheep blood agar plates containing 10, 30 and 40% ox bile. The bile was prepared from the dehydrated product (Difco).

Bile tolerance—*Streptococcus*

Y. Abd-el-Malek and T. Gibson (1948). J. Dairy Res. 15, 233–248.

Inhibition by 40% bile. The medium consisted of 4 parts of ox bile and 6 parts of a broth containing 0.5% each of lactose, peptone, tryptone and meat extract.

Bile tolerance—*Streptococcus*

G. Andrieu, L. Enjalbert and L. Lapchine (1954). Annls Inst. Pasteur, Paris 87, 617–634.

Le milieu bilié est préparé avec de la bile desséchée (bile extract Difco), réhydratée et ajoutée à la gélose fondue pour atteindre une concentration de 40 p. 100 de bile.

Bile tolerance—*Lactobacillus*

D. M. Wheater (1955). J. gen. Microbiol. 12, 123–132.

Tomato glucose broth of the following composition – tomato juice 10%, Neopeptone 1.5%, glucose 2%, sodium chloride 0.5%, Tween 80 0.1%, Yeastrel 0.6%, and soluble starch 0.05% – was prepared, and to one-half of this basal medium 2% sodium tauroglycocholate was added, the other half serving as a control. Both the control and the bile medium were filled in 5 ml amounts into ¼ oz McCartney bottles and sterilized for 20 mins at 15 psi. The inoculated bottles were incubated at 37° and read after 24 hr.

Bile tolerance—*Streptococcus*

P. Morelis and L. Colobert (1958). Annls Inst. Pasteur, Paris 95, 568–587.

Le milieu à 40 p. 100 de bile est constitué par de l'eau peptonée à laquelle on ajoute de la bile de boeuf [39]. Après vingt-quatre heures d'incubation à 37°, toutes les souches s'y développèrent.

[39] Wahl (R.) et Mayer (P.). Ann. Inst. Pasteur, 1956, 91, 1, 147, 279.

Bile tolerance—*Pseudomonas fluorescens*

M. Rhodes (1959). J. gen. Microbiol. 21, 221–263.

The effect of 0.5% (w/v) sodium taurocholate (B.P.C. Evans) on the growth of isolates in 5 ml yeast extract broth was recorded.

Bile tolerance—*Mycobacterium*

R. E. Gordon and J. M. Mihm (1959). J. gen. Microbiol. 21, 736–748.

MacConkey agar: 7- to 14-day-old cultures in glucose broth were streaked on a slant of MacConkey agar (Difco) and a control slant of glycerol agar with a loop, 2.5 mm in outside diameter. Growth and colour change of the neutral red indicator were noted after 7 and 28 days at 28°.

Bile tolerance—*Pasteurella, Ristella, Capsularis, Spherophorus, Fusiformis*

H. Béerens and M. M. Castel (1960). Annls Inst. Pasteur, Paris 99, 454–456.

Les bactéries étudiées sont préalablement cultivées sur base semisolide glucosée décrite antérieurement [5] et dont nous rappelons la formule: peptone trypsique, 10 g; extrait de viande, 3 g; extrait de levures, 5 g; glucose, 10 g; cystéine (Chte), 0.30 g; chlorure de sodium, 5 g; agar (Difco), 0.40 g; eau distillée, 1000 ml; pH 7.2–7.4. Elle est répartie en tubes de 160 ×16 à raison de 12 ml approximativement et stérilisée par autoclavage à 115° pendant trente minutes. Elle est régénérée avant l'emploi.

Le milieu bilié se prépare ainsi: protéose peptone n° 3 (Difco), 10 g; extrait de levures, 0.50 g; glucose, 5 g; chlorure de sodium, 5 g; agar (Difco), 5 g; eau distillée, 1000 ml; pH, 7.2–7.4. Il est réparti en tubes de 180 × 9 à raison de 4.5 ml par tube et stérilisé par autoclavage à 115° pendant trente minutes. Au moment de l'emploi, après fusion au bain-marie bouillant, on ajoute à chacun d'eux 0.5 ml de bile de boeuf stérile pour obtenir une concentration finale de 10 p. 100. Un milieu témoin est également préparé; il ne contient pas de bile.

En partant d'une culture en base glucosée de vingt-quatre à quarante-huit heures, on inocule 2 tubes de chacun de ces 2 milieux (bile et témoin), au moyen d'une effilure de pipette Pasteur fermée (méthode de l'épuisement). On les incube à 37°, et on observe durant trois ou quatre jours.

[5] Rosenfeld (W. D.). J. Bact. 1947, 54, 664.

Bile tolerance—*Vibrio*

G. H. G. Davis and R. W. A. Park (1962). J. gen. Microbiol. 21, 101–119.

Growth on nutrient agar in 7 days on 5 or 10% (w/v) sodium taurocholate.

Bile tolerance—*Aerococcus catalasicus*

O. G. Clausen (1964). J. gen. Microbiol. 35, 1–8.

Cultures were incubated aerobically at 37°C unless otherwise stated.

Ox bile resistance was studied by inoculation on to 40% ox bile blood agar (40% autoclaved ox bile, 8% defibrinated horse blood, and 52% agar medium with 2.6% agar, quality Kobe I) and incubation for 20 hr.

Bile tolerance—*Moraxella*

W. J. Ryan (1964). J. gen. Microbiol. 35, 361–372.

The effect of bile salts was determined by incubation for 4 days on MacConkey agar plates prepared as described in the Ministry of Health

Report on the Bacteriological Examination of Water Supplies (1940). Report (1940). Rep. Pub. Hlth Med. Subj., Min. Hlth London, no. 71. Revised ed., p. 52.

Bile tolerance—*Bacteroides oralis*

W. J. Loesche, S. S. Socransky and R. J. Gibbons (1964). J. Bact. 88, 1329–1337.

Tolerance of 10% bile (Oxgall, Difco) was determined in the basal medium supplemented with 0.5% glucose.

Thioglycolate Medium without Dextrose (BBL), supplemented with 0.2% yeast extract (Difco) and hemin, was used as a basal medium. All cultures were incubated at 35°C in Brewer jars containing an atmosphere of 95% H_2 and 5% CO_2.

Bile tolerance—*Lactobacillus casei* var. *rhamnosus*

W. Sims (1964). J. Path. Bact. 87, 99–105.

The medium of Swan (1954) was used for testing for growth in the presence of 40% bile.

A. Swan (1954). J. clin. Path. 7, 160.

Bile tolerance—*Vibrio sputorum*

W. J. Loesche, R. J. Gibbons and S. S. Socransky (1965). J. Bact. 89, 1109–1116.

Thioglycolate medium without added dextrose (BBL) supplemented with 0.2% yeast extract (Difco) and 0.1% KNO_3 was selected as a basal medium because of the characteristic growth of the oral vibrios in the upper third of the tube. 10% bile was added to this medium.

All cultures were incubated at 35 to 37°C in Brewer jars containing an atmosphere of 95% H_2 and 5% CO_2.

2,3-BUTYLENE GLYCOL (2,3-BUTANEDIOL)

2,3-Butylene glycol—*Aerobacter, Aerobacillus, Aeromonas*

R. Y. Stanier and S. Fratkin (1944). J. Bact. 47, 412, G5.

The oxidation of 2,3-butylene glycol in the genera *Aerobacter, Aerobacillus* and *Aeromonas.*

Representatives of the three genera characterized by a butylene glycol fermentation of carbohydrates all oxidize 2,3-butylene glycol under aerobic conditions.

Aerobacillus polymyxa and *Aeromonas hydrophila* cause an oxidation only to acetoin. *Aerobacillus polymyxa,* which produces pure levo-butylene glycol, oxidizes this isomer about 12 times as rapidly as the meso form. *Aeromonas hydrophila,* which produces mainly meso-butylene glycol, cannot attack the levo isomer at all.

Aerobacter aerogenes oxidizes meso but not levo-butylene glycol, carrying the oxidation to CO_2 and water.

2,3-Butylene glycol

F. D. di Accadia (1947). Annls Inst. Pasteur, Paris 73, 1114–1116.

See **Acetoin and 2,3-butylene glycol estimation** p. 569.

2,3-Butylene glycol—*Cytophaga*

B. J. Bachmann (1955). J. gen. Microbiol. 13, 541–551.

The 2,3-butylene glycol occasionally found among the fermentation products was detected by the method of Kniphorst and Kruisheer (1937).

L. C. E. Kniphorst and C. I. Kruisheer (1937). Untersuch. Lebensmitt, 73, 1.

2,3-Butylene glycol—*Aeromonas*

J. P. Stevenson (1959). J. gen. Microbiol. 21, 366–370.

The organisms were cultured for 48 hr at 37°C in a tryptic meat digest with the addition of 1.0% (w/v) glucose. Butanediol was estimated by Neish's (1952) method and by the method recommended by Pirt (1957).

A. C. Neish (1955). Report National Research Council of Canada No. 46-8-3. 2nd Revision N.R.C. 2952.

S. J. Pirt (1957). J. gen. Microbiol. 16, 59.

2,3-Butylene glycol—*Aeromonas*

I. W. Smith (1963). J. gen. Microbiol. 33, 263–274.

Production of 2,3-butanediol was determined after 1, 3 and 7 days' growth in glucose medium by Bullock's method (1961, Progr. Fish. Cult. 23, 147).

2,3-Butylene glycol—*Bacillus polymyxa*

S. K. Long and R. Patrick (1965). Appl. Microbiol. 13, 973–976.

The product was estimated by the method of Desnuell and Naudet (1945).

P. Desnuelle and M. Naudet (1945). *Analytical methods for bacterial fermentations,* 2nd rev., p. 37–38. Natl. Res. Council Can., No. 2952.

2,3-Butylene glycol—RM bacterium

A. J. Ross, R. R. Rucker and W. H. Ewing (1966). Can. J. Microbiol. 12, 763–770.

The procedure given by Neish (1952) was used for the detection of 2,3-butanediol, in addition to the usual Voges-Proskauer test using O'Meara's reagent.

A. C. Neish (1952). Natn. Res. Coun. A. Rep. No. 46-8-3 (2nd rev.)

CAPSULES

Capsules—*Streptococcus*

J. E. Morison (1940). J. Path. Bact. 51, 401–412.

Examination of isolated organisms: Organisms were deposited from fluid media in an angle centrifuge. The presence or absence of capsules was demonstrated by the India ink method of Butt *et al.* (1936). The essential feature is the use of 6 per cent glucose solution. A loopful each of this and of a very dense suspension of organisms were mixed with a loopful of India ink and spread like a blood film. After drying without heat most stains could be used, but alkaline methylene blue was satisfactory. Too opaque an ink necessitates an undesirably thin film if the capsules are to be outlined fully. Organisms grown on solid media may be suspended in glucose. The capsule-staining methods of Hiss, as used by Hobby and Dawson (1937) and of Pradhan (1937), as well as the prolonged application of Leishman's stain, were very uncertain. Churchman and Emelianoff (1933), in an extensive study, commented on the uncertainty of results with the modified Wright's stain they employed. They were extremely reluctant to conclude that because capsules had not been stained they were not present.

E. M. Butt, C. W. Bonynge and R. L. Joyce (1936). J. infect. Dis. 4, 426.
J. W. Churchman and N. V. Emelianoff (1933). J. exp. Med. 57, 485.
G. L. Hobby and M. H. Dawson (1937). Br. J. exp. Path. 18, 212.
M. G. Pradhan (1937). Br. J. exp. Path. 18, 90.

Capsules—negative staining with nigrosin

A. Fleming (1941). J. Path. Bact. 53, 293–296.

Nigrosin (water-soluble Gurr's), 10 g; Water, 90 ml; Formalin (preservative only), 10 ml.

Capsules—*Pasteurella*

A. M. Jasmin (1945). J. Bact. 50, 361–363.

Suspend a small amount of culture picked up by a fine platinum wire in a loopful of physiological saline containing 0.5 to 1.0 per cent phenol and 10 per cent blood serum, and spread as a thin film on a slide. Dip rapidly in methyl alcohol, drain and flame to burn off excess alcohol. Stain for 30 seconds to 1 minute with any common stain, and wash with water.

Capsules—various

P. B. White (1947). J. Path. Bact. 59, 334–335.

Capsules

1. Free the slide from grease by heating it in the Bunsen flame, after

which it may be quickly cooled by laying it on a smooth, glazed tile.

2. Place on the slide a suitable drop of an almost saturated aqueous solution of congo red containing about 10 per cent of serum and mix in a particle of the agar-grown culture to be examined.

3. Spread the mixture in a film of varied thickness, dry with gentle warmth and then fix thoroughly in the flame.

4. Flood the film when cooled with a 0.5 per cent aqueous solution of HCl; drain, blot gently, and drive off excess acid with gentle warmth.

5. Stain the film for 15–20 seconds with a 1 per cent aqueous solution of methylene blue, which may be acidulated with acetic acid – 1 small drop of "glacial" acetic to 20 ml of stain.

6. Drain off the stain – do not wash – and blot the film gently but thoroughly; dry and examine, mounting first if there is any question of preserving the preparation.

To these instructions it need hardly be added that the stains and serum employed should be clean, bacteriologically and otherwise; that when blood or other serous material is examined no other addition of serum to the congo red is required; and that in the examination of broth cultures the organisms must first be separated from peptone, salts, and so on, the water-solubility of which would threaten the integrity of the film. The process takes very much the same amount of time and labour as does Gram staining. The film, at first red, then rendered blue by the acid, assumes a deep purple tint through interaction of the two dyes, the color later maturing to red-purple on intervention of alkali from the glass.

Capsules—*Bacillus*

J.-P. Aubért and Mlle J. Millet (1950). Annls Inst. Pasteur, Paris 79, 468–469.

Pour les colorations, nous avons employé la méthode de Churmann et Emelianoff modifiée par Mme Lenormand [2] : on étale la suspension microbienne sur la lame et on laisse sécher à l'air. On plonge la lame dans une solution de Giemsa R pendant une à deux minutes, on lave au tampon de Clark-Lubs (pH 6.4) et on laisse sécher à l'air. Les corps bactériens apparaissent en bleu et la capsule en rose pâle.

[2] Mme Lenormand. *Ann. Parasitol.* 1948, 23, 55.

Capsules—general

P. Hauduroy (1951). Techniques Bactériologiques. Masson et Cie, Éditeurs, Paris.

COLORATION DES CAPSULES.

Les capsules des bactéries sont assez difficiles à colorer. Les méthodes les plus simples et les plus sûres sont les suivantes.

Méthode à l'encre de Chine (Méthode de Burri). – Déposer sur une lame une petite goutte d'encre de Chine (encre spéciale pour la bactériologie).

Mélanger à cette goutte une goutte de volume égal de la culture à étudier (culture liquide, sérosité, culture sur milieu solide diluée en eau physiologique). Étaler le mélange sur la lame à l'aide du bord d'une lamelle. Laisser sécher. Examiner à l'immersion.

Les capsules apparaissent incolores sur le fond noir de la préparation (capsules dites négatives).

Cette méthode peut servir pour la coloration des spirilles et des spirochètes.

Capsules—*Acetobacter*

C. R. Marshall and V. T. Walkley (1952). J. gen. Microbiol. 6, 377–381.

To a 5 per cent (w/v) solution of gelatin in water at 60°C add sufficient Cotton Blue powder to give an intense dark blue color. Then add 1–2 per cent powdered eosin and mix thoroughly.

A little of the gelatin stain is placed on a wet film of the organism on a warm stage. Cover with a warm cover glass. Cells are pink and capsules colourless against a blue background.

The culture may be diluted with very dilute NaOH before use, to help separate capsulated cells.

Capsules—*Bacillus, Klebsiella, Diplococcus, Cryptococcus*

J. Tomcsik and S. Guex-Holzer (1954). J. gen. Microbiol. 10, 97–109.

A study of non-specific pH-dependent capsular reactions of several bacteria in the presence of proteins was made. Several proteins were tested over a range of pH values. Three types of capsules were recognised.

Good results with *K. pneumoniae* Types A and C, *Diplococcus pneumoniae*, and some *Bacillus* spp. were obtained by the following method.

Bacteria were suspended in 0.85 per cent NaCl from 24-hour cultures and diluted to *ca.* 5×10^6/ml. Equal volumes of suspension and 1/20 bovine serum were mixed; then equal volumes of this mixture and Teorell and Stenhagen buffer, pH 3.6, were mixed and loopfuls sealed under coverglasses and observed immediately or up to 1 hour or more afterward by phase contrast.

Buffers of pH 2.4–2.8 should be used for *Cryptococcus* and *Klebsiella* type B. The final pH of the mixture is determined by the buffering action of the serum.

Capsules—*Pseudomonas*

C. K. Williamson (1956). J. Bact. 71, 617–622.

In mucoid cultures of *P. aeruginosa*, capsules were demonstrated in relief using india ink for the background and crystal violet to stain the cell proper. The crystal violet was prepared by adding 0.1 g of dye (C.I. no. 681) to 100 ml of distilled water containing 0.25 ml glacial acetic acid. The bacteria were suspended in india ink, and thin smears were prepared on clean slides so that the ink dried in a few moments. The slides were then flooded with the stain and permitted to stand for 5 minutes. The stain was drained from the slide, and the smear was rinsed with a 20 per cent aqueous copper sulfate solution.

Capsules—*Bacillus*

K. L. Burdon (1956). J. Bact. 71, 25–42.

The capacity to form smooth, mucoid colonies containing capsulated bacilli on bicarbonate-containing agar was determined by use of the medium of Thorne *et al.* (1952). These authors, and Chu (1952), have shown that *virulent* anthrax bacilli, but not avirulent strains, will form their characteristic polypeptide capsule *in vitro* in the presence of excess CO_2 in the medium or in the atmosphere. Plates were inoculated rather heavily, then tightly sealed with tape before incubation. The medium consisted of: dehydrated nutrient broth (Difco), 0.8 per cent; yeast extract (Difco), 0.3 per cent; glucose, 0.5 per cent; and agar 2.5 per cent. A 7 per cent solution of sodium bicarbonate, sterilized by filtration, was added to the sterile agar medium in a final concentration of 0.7 per cent. The final pH was 8.5.

The colony forms most distinctive to the naked eye were the giant colonies obtained when Frazier's gelatin-agar plates (Frazier, 1926) were inoculated in a center spot only. Agar colonies on streak plates were routinely examined under a binocular dissecting microscope.

The characteristics of the growth in broth, in bromcresol purple milk, on potato slants and on coagulated (Löffler's) serum slants were observed and recorded in detail.

H. P. Chu (1952). J. Hyg., Camb. 50, 433–444.
W. C. Frazier (1926). J. infect. Dis. 39, 302–309.
C. B. Thorne, C. A. Gomez, R. D. Housewright (1952). J. Bact. 63, 363–368.

Capsules—*Streptococcus*

A. P. MacLennan (1956). J. gen. Microbiol. 14, 134–142.

A loopful of culture and a loopful of cobalt blue ink (Uno water-proof ink) were mixed on a grease-free slide, smeared, and then dried without heating. After staining with 1/6 Löffler's methylene blue for 3 minutes and washing in water, the preparation was dried without heating. Dr. Nuala Crowley (personal communication) suggested the use of 'Uno' ink, which gives a more uniform background than other inks.

Capsules—staining methods

Society of American Bacteriologists Manual of Microbiological Methods – McGraw-Hill Book Co. Inc., New York 1957.

(i) *Leifson's Stain (Leifson 1930)*

$KAl(SO_4)_2 \cdot 12H_2O$, or $NH_4Al(SO_4)_2 \cdot 12H_2O$ (sat aqu solution) –	20 ml
Tannic acid (20 per cent aqu solution)	10 ml
Distilled water	10 ml
Ethyl alcohol, 95 per cent	15 ml
Basic fuchsin (sat solution in 95 per cent ethyl alcohol)	3 ml

Mix ingredients in order named. Keep in tightly stoppered bottle, and the stain may be good for a week.

Staining schedule:

1. Prepare smears and dry them in air.
2. Flood slides with the above solution, and allow to stand 10 min at room temperature in warm weather or in an incubator in cold weather.
3. Wash with tap water. (If a counterstain is desired, borax methylene blue may be applied, without heat, followed by another washing.)
4. Stain 5-10 min, without heating, in borax methylene blue (methylene blue, 90 per cent dye content, 0.1 g; borax, 1 g; distilled water, 100 ml).
5. Wash in tap water.
6. Dry, and examine.

Results: capsules, red; cells, blue.

(ii) *Anthony's Method*

with Tyler's Modification – Anthony (1931).

Original formula

Crystal violet (85 per cent dye content)	1 g
Distilled water	100 ml

*Tyler's modification**

Crystal violet (85 per cent dye content)	0.1 g
Glacial acetic acid	0.25 ml
Distilled water	100 ml

Staining schedule:

1. Prepare smears and dry them in the air.
2. Stain 2 min in the above aqueous crystal violet or, according to Tyler, 4–7 min in the above acetic crystal violet.
3. Wash with 20 per cent aqueous $CuSO_4 \cdot 5H_2O$.
4. Blot dry, and examine.

Results: capsules, blue violet; cells, dark blue.

*See W. H. Park and A. W. Williams (1933). *Pathogenic Microorganisms,* 10 ed., Lea and Febiger, Philadelphia.

(iii) *Hiss's Method.* Hiss (1905)

Original statement of formula

Sat alc basic fuchsin or gentian violet	5–10 ml
Water to make	100 ml

Emended formula

Basic fuchsin (90% dye content)	0.15–0.3 g
Distilled water	100 ml

or

Crystal violet (85 per cent dye content)	0.05–0.1 g
Distilled water	100 ml

Staining schedule:

1. Grow organisms in ascitic fluid or serum medium, or mix with drop of serum and prepare smears from this mixture.
2. Dry smears in the air, and fix with heat.
3. Stain with one of the above solutions a few seconds by gently heating until steam rises.
4. Wash off with 20 per cent aqueous $CuSO_4 \cdot 5H_2O$.
5. Blot dry, and examine.

Results: Capsules, faint blue; cells, dark purple.

E. E. Anthony (1931). Science, N.Y. 73, 319.
P. J. Hiss, Jr. (1905). J. exp. Med. 6, 317–345.
E. Leifson (1930). J. Bact. 20, 203–211.

Capsules—*Escherichia aurescens*

H. Leclerc (1962). Annls Inst. Pasteur, Paris 102, 726–741.

Capsules. – Leur présence est recherchée par la technique à l'encre de Chine, après culture de 24 heures sur le milieu saccharosé de Worfel Ferguson décrit par Ewing [9].

[9] Ewing (W. H.). *Enterobacteriaceae, biochemical methods for group differentiation.* Public Health Service Publication, n° 734, U. S. Govt Printing Office, 1960.

Capsules—general

N. C. Dondero (1963). J. Bact. 85, 1171–1173.

A convenient and simple substitute for nigrosin in dark-phase contrast microscopy is skim milk. A few drops allowed to flow beneath the cover slip quickly provide a turbid background which delineates the capsule. A 10% suspension of dried milk powder, as prepared and sterilized for milk culture tubes, is satisfactory. Bromcresol purple milk, methylene blue milk, and litmus milk serve equally well. In contrast to India ink, the sterilized milk does not introduce extraneous encapsulated bacteria into the field.

Ten per cent concentration of dried milk at pH 6.8–7.4 is optimal.

Capsules—*Lactobacillus casei* var. *rhamnosus*

W. Sims (1964). J. Path. Bact. 87, 99–105.

Capsules were demonstrated by the eosin-serum method of Howie and Kirkpatrick (1934) and the wet film India ink method (Duguid, 1951). Staining with alcian blue for 2 min also revealed capsules, the cell body remaining unstained.

J. P. Duguid (1951). J. Path. Bact. 63, 673.
J. W. Howie and J. Kirkpatrick (1934). J. Path. Bact. 39, 165.

Capsules—*Staphylococcus aureus*

S. Mudd and S. J. DeCourcy, Jr. (1965). J. Bact. 89, 874–879.

India ink negative stain technique. A small loopful of the appropriately grown suspension of organisms was thoroughly mixed with an equal volume of black Pelikan Waterproof Drawing Ink (Günther, Wagner) on a cover slip, overlaid with a microscope slide, and the excess mixture was squeezed out by gentle pressure on the cover slip and blotted up. Because of the very fine particle size of solids in this ink, the thickness of the wet-mount film proved to be critical. Optimal contrast was obtained when a thin film of undiluted ink was used rather than a thicker film of diluted ink. After teasing out all the bubbles, the edges of the cover slip were sealed with colorless nail polish (Cutex).

CARRAGEENIN HYDROLYSIS

Carrageenin hydrolysis

W. Yaphe and B. Baxter (1955). Appl. Microbiol. 3, 380–383.

The enzyme was prepared by growing the organism in a culture medium containing carrageenin (Sea Kem No. 6), 0.5%; NaCl, 3.0%; $MgCl_2$, 0.05%; NaH_2PO_4, 0.05%; $NaNO_3$, 0.05%; and $FeCl_3$, 0.002%; dissolved in distilled water.

Cultures were incubated in Erlenmeyer flasks, at 20 to 25°C on a rotary shaker for 3 to 4 days. The material was then passed through Sharples supercentrifuge, and the effluent was filtered through a Selas porcelain filter No. 03 to obtain a cell-free filtrate. This and the Sharples effluent were both tested for enzymic activity.

The rate of hydrolysis was followed by measuring the increase in reducing power (Somogyi, 1952) with galactose as standard. One volume of the enzyme solution was added to 2 volumes of 0.6% carrageenin in M/20 sodium phosphate buffer. Aliquots of 5 ml were removed at intervals for analysis. Activity was expressed as the amount of galactose formed after hydrolysis of carrageenin for 2 min at 25°C and pH 7.5.

N. Somogyi (1952). J. biol. Chem. 195, 19–23.

CATALASE

Catalase—*Pseudomonas*

M. Levine and D. Q. Anderson (1932). J. Bact. 23, 337–347.

The presence of catalase was determined by adding 1 or 2 drops of 3 per cent hydrogen peroxide to a twenty-four hour old culture growing in a Petri dish on agar. Effervescence indicated the presence of catalase.

Catalase—*Propionibacterium*—semi-quantitative method

E. R. Hitchner (1934). J. Bact. 28, 473–479

Tubes containing 10 ml of sodium-lactate broth were inoculated with a 6-mm loop of three day old broth cultures of the respective strains, which – to insure vigorous growth – had been carried through several successive transfers in the same media. After 10 days' incubation the tubes were thoroughly shaken; 5 ml were transferred to a 50 ml volumetric flask and diluted to the mark with M/150 phosphate solution, pH 6.5. An aliquot, the amount depending on the catalase content as determined by a preliminary test, was transferred to a 100 ml Erlenmeyer flask; 5 ml of approximately 0.1 N solution of neutral hydrogen peroxide were added, and the mixture was placed immediately in an ice-bath. After a definite time interval, again depending on the activity of the culture, the flask was removed, the contents were acidified with dilute sulfuric acid, and the residual hydrogen peroxide was estimated by titration with 0.025 N potassium permanganate. To compensate for any decomposition of the hydrogen peroxide by the organic matter in the solution, a similar test with a heated culture was run as a check.

Catalase—*Lactobacillus, Bacteroides*

K. H. Lewis and L. F. Rettger (1940). J. Bact. 40, 287–307.

Catalase production was tested by flooding actively growing glucose-cysteine agar plate cultures with a 3 per cent solution of hydrogen peroxide.

Catalase—anaerobic gram-negative diplococci

G. C. Langford, Jr., J. E. Faber, Jr., and M. J. Pelczar, Jr. (1950). J. Bact. 59, 349–356.

Evidence of catalase activity was determined by the following method: Slants of trypticase soy agar were inoculated with each of the cultures and after 48 hours' incubation in anaerobic jars the cells were washed off with physiological saline. Equal amounts of 3 per cent hydrogen peroxide (freshly prepared from 30 per cent solution) were added to the cell suspensions. Evolution of bubbles indicated a positive catalase reaction. Six representative cultures that gave a negative catalase reaction were examined by the Warburg manometric method, which confirmed the original indication that the organisms were catalase-negative. The catalase activity of those organisms that showed a positive reaction could be inhibited by the addition of a small amount (0.2 to 0.5 ml) of 2 per cent potassium cyanide to 2 ml of cell suspension. Suitable positive and negative controls were included with each biochemical determination.

Catalase—*Staphylococcus*

C. Shaw, J. M. Stitt and S. T. Cowan (1951). J. gen. Microbiol. 5, 1010–1023.

'Ten vol.' H_2O_2 was run down the surface of a culture grown on nutrient agar for 24 hr. Gas production, however slight, was recorded as positive.

Catalase—*Staphylococcus*

C. R. Marshall and V. T. Walkley (1952). J. gen. Microbiol. 6, 377–381.

Used method of T. K. Walker and J. Tošić (1943, Biochem. J. 37, 10) using sterile malt extract buffered with $CaCO_3$ and solidified with 1.5% agar.

Catalase—microtests

P. H. Clarke and S. T. Cowan (1952). J. gen. Microbiol. 6, 187–197.

Ten volumes peroxide and suspension were drawn into a capillary tube. Gas evolved immediately. If it did not do so in 10 seconds, the tube was sealed and observed for longer periods.

Catalase—*Actinomyces, Corynebacterium*

H. Beerens (1953). Annls Inst. Pasteur, Paris 84, 1026–1032.

Action sur l'eau oxygénée. – La catalase est mise en évidence par la formation de mousse observée après addition de X gouttes d'H_2O_2 à 10 volumes à 2 ou 3 ml de culture en milieu V. F. (viande et foie) liquide en tube scellé sous vide. Cette catalase a été décelée soit après le premier repiquage des cultures provenant des collections étrangères, soit, pour *A. israeli* var. *liquefaciens,* immédiatement après son isolement.

Catalase—*Pseudomonas (Malleomyces) pseudomallei*

P. de Lajudie and E. R. Brygoo (1953). Annls Inst. Pasteur, Paris 84, 509–515.

Catalase:– Elle fut recherchée par l'action de 1 ml d'eau oxygénée à 10 vol. sur une culture en gélose inclinée de quarante-huit heures.

Catalase—*Microbacterium*

R. A. McLean and W. L. Sulzbacher (1953). J. Bact. 65, 428–433.

Formation of catalase was confirmed by visible reaction with hydrogen peroxide, by reactions on blood agar with benzidine, and by Penfold's (1922) method. However, because of the importance of catalase formation in differentiating between the genera *Lactobacillus* and *Microbacterium,* a need for more quantitative evidence seemed to be indicated. Washed cells from 18 hour veal infusion agar cultures were suspended in 0.06 M phosphate buffer and their ability to decompose hydrogen peroxide was measured in a Warburg manometer by observing the rate of oxygen liberation. The density of the bacterial suspension was determined by direct microscopic counts, and catalase capability was calculated by the graphical method of Van Schouwenburg (1940).

W. J. Penfold (1922). Med. J. Aust. 2, 120.
K. L. Van Schouwenburg (1940). Enzymologia 8, 344–352.

Catalase—*Pediococcus*

E. A. Felton, J. B. Evans, and C. F. Niven, Jr. (1953). J. Bact. 65, 481–482.

Production of catalase was tested by streaking these cultures on APT agar plates (Evans and Niven, 1951) containing yeast extract, tryptone, citrate, mineral salts, "tween 80", and 1.0 per cent glucose and incubating the plates at 30°C for varying lengths of time. After incubation the plates were flooded with 5 ml of a 3 per cent hydrogen peroxide solution and observed for the appearance of gas bubbles for a period of 10 minutes. All six cultures grew luxuriantly on this agar medium, but five gave no evidence of producing catalase. However, one culture appeared to be weakly positive.

The six cultures also were streaked on a medium containing 1.0 per cent tryptone, 0.5 per cent yeast extract, 0.5 per cent sodium chloride, 0.2 per cent K_2HPO_4, 1.5 per cent agar, and only 0.05 per cent glucose (YTG agar) and tested for catalase after 24 and 48 hours. On this medium the cultures grew poorly, but all exhibited a weak but definitely positive catalase test.

J. B. Evans and C. F. Niven, Jr. (1951). J. Bact. 62, 599–603.

Catalase

N.C.T.C. Methods 1954.

Method 1. Grow organism on agar (or serum agar) slope; run down 1 ml 10 vol H_2O_2 and let stand; read after 5 and 30 minutes.

(Control medium if any other than agar is used.)

Method 2. Grow organism in broth; add 1 ml 10 vols H_2O_2, and read after 5 and 30 minutes.

Variables	T°	Time
	30	1
	37	1
	30	3
	37	3
	30	5
	37	5

Controls:	Positive	=	86
	Negative	=	370

Catalase—*Microbacterium*

V. Bolcato (1957). Antonie van Leeuwenhoek 23, 351–356.

Formation of catalase was observed by visible reaction with hydrogen peroxide, i.e., by flooding with 10 ml of a 3 per cent hydrogen peroxide solution saccharose agar slant colonies.

Catalase

Society of American Bacteriologists *Manual of Microbiological Methods* – McGraw-Hill Book Co. Inc., New York 1957.

A plate culture of the organism in question is flooded with a 10 per cent solution of H_2O_2. The evolution of gas bubbles from the colonies denotes the presence of catalase.

Catalase—*Actinomyces bovis*

S. King and E. Meyer (1957). J. Bact. 74, 234–238.

The presence of the constitutive enzyme catalase was determined as follows: A loopful of culture from each organism grown anaerobically on a brain heart infusion agar slant was emulsified in a drop of saline on a glass slide, after which a drop of hydrogen peroxide was added. Liberation of oxygen was shown by bubbling, indicating the presence of catalase. Organisms too hard to emulsify were either broken up by shaking with glass beads or ground in a mortar.

In order to verify the validity of this test for the presence of catalase, a more sensitive standard titration method described by Sumner and Somers (1947) was used for measuring catalase activity as follows: The organisms were weighed (wet wt), triturated in a mortar, suspended in a phosphate buffer of pH 6.8, and stored at –20 C until ready for use. Then the suspension was thawed and centrifuged, the supernatant was decanted, and the cells were resuspended in 1 ml of the same buffer. This was added to 25 ml of 0.01 N H_2O_2 held in an ice bath. At 2 min intervals, from 0 to 10 min, 5 ml of this solution were transferred into 5 ml of 2 N H_2SO_4 to stop the reaction. At the end of 10 min, the H_2O_2 in each flask was titrated with 0.005 N $KMnO_4$. If catalase were present, the quantity of H_2O_2 decreased with time as indicated by subsequent titrations with $KMnO_4$.

J. B. Sumner and G. F. Somers (1947). *Chemistry and Methods of Enzymes,* pp. 24–25. Academic Press, New York, N. Y.

Catalase—*Bacillus*

E. R. Brown, M. D. Moody, E. L. Treece and C. W. Smith (1958). J. Bact. 75, 499–509.

Estimated by the addition of approximately 1 ml of hydrogen peroxide (3 per cent) to the surface of a slant of tryptone yeast-extract-glucose medium which had been inoculated and incubated at 37°C for 24 hr.

Catalase—*Acetobacter*

J. de Ley (1958). Antonie van Leeuwenhoek 24, 281–297.

The presence of catalase was checked by adding one drop of a 3% H_2O_2 solution directly on the colony or on about 40 mg living bacteria from a centrifuge tube. The H_2O_2 solution was checked with a drop of blood or with a colony of a catalase-positive *Acetobacter.*

Catalase—*Pseudomonas fluorescens*

M. E. Rhodes (1959). J. gen. Microbiol. 21, 221–263.

A loopful of solid growth taken from a 24 hr yeast extract agar slope was removed into a drop of '10 volumes' hydrogen peroxide solution and examined for the production of gas bubbles.

Catalase—atypical mycobacteria

A. Beck (1959). J. Path. Bact. 77, 615–624.

Examined by Middlebrook's (1954) method. The varying amounts of oxygen liberated were marked by + to +++ signs.

G. Middlebrook (1954). Am. Rev. Tuberc. pulm. Dis. 69, 471.

Catalase—*Bacillus megaterium*

C. Weibull, H. Beckman and L. Bergstrom (1959). J. gen. Microbiol. 20, 519–531.

Determined by the permanganate titration method of – R. K. Bonnichsen, B. Chance and H. Theorell (1947). Acta chem. scand. 1, 685.

Catalase

V. B. D. Skerman, *A Guide to the Identification of the Genera of Bacteria.* 2nd ed. Williams and Wilkins Book Co., Baltimore, 1967.

Catalase is an enzyme containing a hematin as the prosthetic group and is capable of decomposing hydrogen peroxide to gaseous oxygen and water. It is widely distributed in nature and is present in most aerobic cells. The function of catalase is to remove the toxic product H_2O_2 resulting from coupled oxidation-reduction processes involving oxygen.

$2\ H_2O_2 \rightarrow 2\ H_2O + O_2$

Pour 1 ml of H_2O_2 (10 volumes per cent) over the surface of a 24-hour agar slope culture. If catalase is present, bubbles of oxygen will be released from the surface of the growth.

Catalase—*Mycobacterium*

L. Lugosi (1959). Annls Inst. Pasteur, Paris 97, 597–606.

Détermination de l'activité catalasique. – Andrejew, Tacquet et Gernez-Rieux ont constaté que, pour étudier l'activité catalasique des mycobactéries, il fallait prendre en considération les règles essentielles de la cinétique enzymatique [11]. Nous avons déterminé l'activité catalasique des souches employées dans nos recherches selon les méthodes dont se sont servis ces auteurs et selon les principes concernant les bases théoriques de la détermination de Sumner et Somers [12]. Un poids donné de la culture âgée de 10 jours est placé dans 25 ml de solution tampon phosphate à pH 6.8. On y ajoute 45 ml d'eau distillée stérile. Après addition de 5 ml de H_2O_2 à 1 p. 100 on détermine l'activité catalasique sur des fractions aliquotes de 5 ml prélevées à des intervalles définis après une acidification par 5 ml 2 N de H_2SO_4 par un titrage de 0.05 $KMnO_4$. Nous déterminons la valeur de la constante de la vitesse de réaction au point de la section extrapolée au temps zéro de la courbe du système de coordonnées log k-t. Nous obtenons la valeur de l'activité catalasique des souches examinées en cherchant le quotient de la constante de vitesse de la réaction par le poids de bactéries en grammes.

[11] Andrejew (A.), Tacquet (A.) et Gernez-Rieux (Ch.). Ann. Inst. Pasteur, 1956, 91, 767.
[12] Sumner (J. B.) et Somers (G. F.). *Chemistry and Methods of Enzymes.* Acad. Press, Inc., New York, 1953.

Catalase—*Leptospira*

S. Faine (1960). J. gen. Microbiol. 22, 1–9.

Qualitative Catalase Test:

Approx. 2 cm length of fluid or semi-solid culture was taken up in a capillary tube, followed by a similar length of 10% H_2O_2.

Cultures were grown in 20–200 ml volumes of modified Korthof medium containing hemoglobin from laked red cells and pooled rabbit or sheep serum. Catalase activity due to the laked red cells was destroyed by heating the medium at 56° for 60 min after Seitz filtration.

A quantitative method is described.

Catalase—*Fusobacterium*

A. C. Baird-Parker (1960). J. gen. Microbiol. 22, 458–469.

Tested by flooding 3-day agar cultures with a '20 volumes' solution of hydrogen peroxide and examining for the evolution of gas bubbles from the colonies (Society of American Bacteriologists, 1957. *Manual of Microbiological Methods,* New York: McGraw-Hill).

Catalase—*Gaffkya, Aerococcus,* tetrad-forming cocci, *Pediococcus*

R. H. Deibel and C. F. Niven, Jr. (1960). J. Bact. 79, 175–180.

Catalase production was determined in APT broth (Evans and Niven, 1951) and on an agar medium containing only 0.05% glucose (Felton *et al.* 1953).

J. B. Evans and C. F. Niven, Jr. (1951). J. Bact. 62, 599–603.
E. A. Felton, J. B. Evans and C. F. Niven, Jr. (1953). J. Bact. 65, 481–482.

Catalase—*Streptococcus* and *Lactobacillus*

C. W. Langston and C. Bouma (1960). Appl. Microbiol. 8, 212–222.

Catalase production was determined on broth cultures and on agar streak plates containing low carbohydrate (0.05%). To determine catalase from cultures growing in broth, a few drops of the broth were transferred to a spot plate, and 3 per cent hydrogen peroxide was added. The evolution of gas constituted a positive test. Precautions should be taken in reading tests, especially when the cultures are weakly catalase-positive. Flaming of the pipette before transferring the culture to the spot plate should be avoided, and the test should be observed for at least 5 min before discarding. Streak plates were flooded with 3 per cent hydrogen peroxide and observed for gas evolution. Strains that produced catalase always showed greater activity on plates than in broth.

Catalase—*Bacillus*

J. Szulmajster and P. Kaiser (1960). Annls Inst. Pasteur, Paris 98, 774–777.

Activité catalasique. – Une propriété importante de ce microorganisme (anaérobie strict) est son activité catalasique. Cette activité, basée sur la décomposition de l'eau oxygénée, a été mesurée avec un extrait bactérien dialysé par la méthode spectrophotométrique de Patrick et Wagner [6] modifiée par Dolin [7]. Cette activité catalasique est inhibée par le cyanure de potassium et l'azoture de sodium à la concentration de 10^{-3} M. Il faut toutefois souligner que l'activité catalasique spécifique (μmoles H_2O_2 décomposées/mg protéines/unité temps) de cette bactérie est environ 2000 fois plus petite que celle d'un extrait de *B. subtilis,* par exemple, mesurée dans les mêmes conditions.

[6] Patrick (W. H.) et Wagner (H. B.) Anal. Chem. 1949, 21 1279.
[7] Dolin (M. I.). J. biol. Chem. 1957, 225, 357.

Catalase—*Shigella*

S. Szturm-Rubinstein (1960). Annls Inst. Pasteur, Paris 99, 305–309.

Nous avons recherché la catalase par plusieurs techniques.

1. A la surface d'une culture sur gélose inclinée de vingt-quatre heures on met 1 ml d'eau oxygénée à 3 p. 100. La gélose est gardée en position presque horizontale. La réaction est positive quand des bulles de gaz se dégagent en trente à soixante secondes. Elle est négative quand il n'y a pas de production de gaz.

2. La deuxième technique, plus simple et qui fournit des résultats aussi nets, consiste à mettre deux gouttes d'eau oxygénée à 3 p. 100 sur

une lame pour agglutination. Une anse de culture sur gélose de vingt-quatre heures y est suspendue. Dans les cas où la catalase est positive un très fort dégagement de gaz se produit immédiatement; on peut même apprécier l'importance du dégagement.

3. La troisième technique permet de mesurer quantitativement la production de catalase. Nous avons utilisé la variante de la technique de Blom et coll. [3] préconisée pour *Vibrio foetus* [2].

La densité des cultures en bouillon de vingt-quatre heures a été mesurée avec un électrocolorimètre, afin d'être sûr qu'elle ne varie pas beaucoup d'une souche à l'autre. Après avoir noté la quantité de culture, on ajoute une quantité égale d'eau oxygénée à 3 p. 100. On remplace le bouchon de coton par un bouchon en caoutchouc perforé muni d'une pipette effilée. Le tube est immédiatement renversé et on marque avec un crayon gras le niveau du liquide. On mélange énergiquement le contenu du tube en le renversant plusieurs fois et on le place en le retournant dans un support au-dessus de l'évier du laboratoire. Au bout de dix, trente, soixante minutes on mesure la hauteur de liquide déplacé par le gaz et celui-ci est pris comme indice de catalase. Nous avons fait des mesures à des périodes plus tardives (jusqu'à six heures après le début de l'épreuve), mais à partir de trente minutes les différences entre les résultats sont insignifiantes.

[2] Organisation des Nations Unies pour l'Alimentation et l'Agriculture Rome, 1956. (Edité sous la direction de J. H. Lang, consultant auprès de la F. A. O.)
[3] Blom (E.) et Christensen (N. O.). Skand. vet. Tidskr. 1947, 37, 1.

Catalase—*Pediococcus*

H. L. Günther and H. R. White (1961). J. gen. Microbiol. 26, 185–197.

Felton, Evans and Niven (1953) found that a medium of low carbohydrate content (YTG) gave a greater number of positive reactions than a medium of high carbohydrate content (APT) used by Evans and Niven (1951). Gutekunst, Delwiche and Seeley (1957) recommended that cultures to be used for catalase tests should be neutralized after incubation. In the present work, preliminary tests were carried out with 12 isolates of pediococci to compare nutrient broth (containing, as %, w/v: Yeastrel, 0.3; peptone, 1.0; NaCl, 0.5; at pH 7.0) with TJ broth and GY broth as media for catalase tests. No qualitative differences were found but the reactions in nutrient broth were sometimes stronger. In view of this and of the recommendations of the above workers, nutrient broth was retained as the experimental medium. Two loopfuls (about 4 mm diameter) of a vigorously growing culture were used as inoculum for 5 ml medium and incubation was carried out for 24 hr, or 72 hr when necessary.

Two ml of freshly prepared 3% (10 vol.) hydrogen peroxide were added and the cultures examined up to 30 min for visible gas bubbles.

Maintenance of stock cultures and methods of cultivation.

For maintenance of stock cultures, preparation of inocula and in all experimental work, 'Oxoid' tomato juice (TJ) broth or tomato juice (TJ) agar, adjusted to pH 6.6, were used unless otherwise stated. The following were exceptions to this rule: for strain Tc. 1 sodium chloride (5%, w/v) was added to the medium; and for the aerococci glucose Lemco broth (Shattock and Hirsch, 1947) or glucose yeast extract (GY) agar (containing, as %, w/v: peptone, 1.0; Yeastrel, 0.3; glucose, 1.0; NaCl, 0.25; agar, 1.0; at pH 7.4) was used.

Cultures were incubated aerobically at 30°C with specified exceptions.

J. B. Evans and C. F. Niven, Jr. (1951) J. Bact. 62, 599.

E. A. Felton, J. B. Evans and C. F. Niven, Jr. (1953). J. Bact. 65, 481.

R. R. Gutekunst, E. A. Delwiche, and H. W. Seeley (1957). J. Bact. 74, 693.

P. M. F. Shattock and A. Hirsch (1947). J. Path. Bact. 59, 495.

Catalase—*Xanthomonas*

A. C. Hayward and W. Hodgkiss (1961). J. gen. Microbiol. 26, 133–140.

Loopfuls of agar growth were emulsified in 10 volumes H_2O_2 on a slide and examined microscopically for evolution of oxygen.

Catalase—*Escherichia aurescens*

H. Leclerc (1962). Annls Inst. Pasteur, Paris 102, 726–741.

e) Catalase. – On la met en évidence après addition de quelques gouttes d'eau oxygénée à 10 volumes sur une suspension bactérienne en eau physiologique stérile, d'une culture de 24 heures.

Catalase—*Actinobacillus actinomycetemcomitans* and *Haemophilus aphrophilus*

E. O. King and H. W. Tatum (1962). *Actinobacillus actinomycetemcomitans* and *Haemophilus aphrophilus.* J. infect. Dis. 111, 85–94.

The catalase test was carried out by pouring 3% hydrogen peroxide over the growth upon a 24 to 48-hour heart infusion agar slant.

Catalase—*Vibrio*

G. H. G. Davis and R. W. A. Park (1962). J. gen. Microbiol. 27, 101–119.

Tested with 10 volumes H_2O_2 upon 24 hr nutrient agar growth.

Catalase—*Caryophanon*

P. J. Provost and R. N. Doetsch (1962). J. gen. Microbiol. 28, 547–557.

Medium

Hy – Case SF	10.0 g
Bacto yeast extract	5.0 g
Na Acetate (anh.)	0.5 g
Agar	15.0 g

pH 7.8 – 7.9

This test was done as outlined in the *Society of American Bacteriologists' Manual of Microbiological Methods* (1957) New York: McGraw-Hill.

Catalase—*Lactobacillus*

R. E. Smith and J. D. Cunningham (1962). Can. J. Microbiol. 8, 727–735.

This characteristic was investigated with Hayward's fermentation medium containing dextrose, 0.1% (w/v); agar, 2.0% (w/v); peptone (Oxoid), 0.5% (w/v); yeast extract (Difco), 0.3% (w/v); salts A, 0.5% (v/v); salts B, 0.5% (v/v); Tween 80, 0.1% (v/v); sodium acetate (hydrated), 0.5% (w/v). The constituents were diluted in glass-distilled water and the pH value of the medium adjusted to 6.8–7.0. Solution A contained 10 g KH_2PO_4 and 10 g K_2HPO_4 in 100 ml of distilled water; solution B contained 11.5 g $MgSO_4 \cdot 7H_2O$, 2.4 g $MnSO_4 \cdot 2H_2O$ and 0.68 g $FeSO_4 \cdot 7H_2O$ in 100 ml of distilled water. Plates were poured, streaked, and incubated for 3 days, then tested with hydrogen peroxide for catalase production.
A. C. Hayward (1957). J. gen. Microbiol. 16, 9–15.

Catalase—*"Bacterium salmonicida" (Aeromonas)*

I. W. Smith (1963). J. gen. Microbiol. 33, 263–274.

Catalase activity was tested by emulsifying a loopful of 2-day nutrient agar culture in 20 volumes hydrogen peroxide (Sneath, 1956).
P. H. A. Sneath (1956). J. gen. Microbiol. 15, 70.

Catalase—*Pseudomonas odorans*

I. Málek, M. Radochová and O. Lysenko (1963). J. gen. Microbiol. 33, 349–355.

Formation of catalase was detected by adding 2 drops of strong hydrogen peroxide solution to 1-day broth cultures – 28°C.

Catalase—*Vibrio*

L. Ringen and F. W. Frank (1963). J. Bact. 86, 344–345.

Catalase production was determined by adding 1 to 2 ml of 3% H_2O_2 to 5 day cultures in Albimi Brucella broth containing 0.15% agar.

Catalase—*Mycobacterium*

D. A. Mitchison, J. B. Selkon and J. Lloyd (1963). J. Path. Bact. 86, 377–386.

The catalase activity of the cultures was measured in a Warburg apparatus. The volume of a 9-day-old culture in 7H-10 Tween-albumin liquid medium (Cohn *et al.*, 1959) that contained 1.43 mg (dry weight) of tubercle bacilli, as estimated by nephelometry, was filtered through a Millipore membrane filter of 0.45 μ average pore diameter and the bacilli were washed twice with sterile distilled water. The washed bacilli were then resuspended in 1 ml sterile distilled water and shaken with glass beads. In the test, 0.7 ml of this suspension, containing 1 mg dry weight of bacilli, was added to the side-arm of a Warburg flask and 0.3 ml of 0.3 per cent hydrogen peroxide and 3.0 ml of M/15 Sørensen phosphate buffer at pH 6.8, containing 0.025 per cent Tween 80, were added to the main chamber of the flask. After their contents had been mixed, the manometers were read at intervals for 2 hr. The reaction was carried out at 10°C. A thermobar and a manometer without hydrogen peroxide were included in each experiment.

M. L. Cohn, G. Middlebrook and W. F. Russell, Jr. (1959). J. clin. Invest. 38, 1349.

Catalase—*Spirillum*

W. A. Pretorius (1963). J. gen. Microbiol. 32, 403–408.

Growth from a 24 hr nutrient agar slope was placed into a drop of '10 volumes' H_2O_2 solution and examined for production of gas bubbles.

Catalase—*Lactobacillus, Streptococcus, Pediococcus, Leuconostoc*

E. A. Delwiche (1963). J. gen. Microbiol. 31, vii.

Catalase activity is demonstrable in certain cultures of *Pediococcus, Streptococcus, Lactobacillus* and *Leuconostoc.* It is suppressed in sugar media where acid production is pronounced.

It is less sensitive to substrate inactivation than the haem-iron catalase of *E. coli.*

Some properties are given.

Catalase

G. S. Wilson and A. A. Miles (1964). Topley and Wilson's *Principles of Bacteriology and Immunity,* 5th edition – Arnold.

Tested on a 24-hours' agar slope culture at 37°C. One ml of H_2O_2 (10 volumes) is poured over the growth, and the tube is set in an inclined position.

Gas bubbles produced = positive

No gas produced = negative.

Catalase—*Lactobacillus*

F. Gasser (1964). Annls Inst. Pasteur, Paris 106, 778–796.

L'absence de catalase est vérifiée sur lame en émulsionnant une colonie de quarante-huit heures dans une goutte d'eau oxygénée à 10 volumes. Un témoin positif (entérobactérie) permet de contrôler la validité de l'épreuve.

Catalase—*Mycobacterium*

A. Andrejew, S. Kwiek, and Ch. Gernez-Rieux (1964). Annls Inst. Pasteur, Paris 107, 503–519.

Mésure de l'activité catalasique (procédé chimique). – L'activité catalasique est déterminée d'après la méthode de Jolles, modifiée par Sumner et Dounce [9], dans les conditions suivantes : on mélange 2.5 ml de H_2O_2 0,1 N (concentration finale = 0,01 N) et 1,25 ml de tampon d'un pH donné 0,2 M (concentration finale = 0,01 M) avec 20,75 ml d'eau bidistillée. Après quinze minutes d'équilibration thermique ce mélange reçoit 0,5 ml de surnageant plus ou moins dilué suivant son activité. On prélève aussitôt 5 ml du mélange et on les verse dans 2 ml de SO_4H_2 4N (temps 0). On agite le mélange réactionnel à la température choisie et, après un temps donné (généralement quatre minutes), on arrête la réaction en introduisant à nouveau 5 ml du mélange réactionnel dans 2 ml de SO_4H_2 4N.

Chaque échantillon reçoit alors 0,5 ml d'une solution d'iodure de K à 10 p. 100 et 0,1 ml de molybdate d'Am à 1 p. 100 dans SO_4H_2 0,05 N. On agite et après trois minutes on titre avec l'hyposulfite de Na 0,005 N, en ajoutant, vers la fin, quelques gouttes d'amidon à 1 p. 100 dans NaCl saturé, qui rend le virage plus perceptible. Dans certains cas, on utilise la microméthode correspondante.

On calcule la constante de vitesse de réaction d'après l'équation bien connue, employée par von Euler et Josephson [4] :

$$k = \frac{1}{T} \log_{10} \frac{X_0}{X}$$

où T = durée de la réaction en minutes; X_0 = ml d'hyposulfite versés dans l'échantillon correspondant au temps 0, et X = ml d'hyposulfite versés après T minutes.

[4] Euler (H. von) und Josephson (K.). Ann. Chem., 1927, 452, 158.

[9] Sumner (J. B.) and Dounce (A. L.) dans Colowick (S. P.) et Kaplan (N. O.). *Methods in Enzymology,* Academic Press, New York, 1955, vol. 2, p. 780.

Catalase—*Pseudomonas aeruginosa*

K. Morihara (1964). J. Bact. 88, 745–757.

Young cells grown on bouillon-agar slants for 18 to 24 hr were dipped in a 3% aqueous solution of hydrogen peroxide, and examined for production of bubbles. Cultures were incubated at 30°C.

Catalase—*Corynebacterium pseudotuberculosis*

C. H. Pierce-Chase, R. M. Fauve and R. Dubos (1964). J. exp. Med. 120, 267–281.

The presence of catalase was determined by suspending sediment growth from Pfanstiehl peptone (PF) broth or a loopful of growth from PF agar into a well containing 0.1 ml of 30 per cent hydrogen peroxide.

Pfanstiehl peptone agar—beef heart infusion containing 1 per cent Pfanstiehl peptone, 0.5 per cent NaCl, and 1.5 per cent agar.

Catalase—*Lactobacillus*

M. Gemmell and W. Hodgkiss (1964). J. gen. Microbiol. 35, 519–526.

Production of catalase: (1) Slopes of basal medium to which had been added glucose (0.05%, w/v) and agar (1.5%, w/v) were inoculated and incubated for 2 days. Peroxide-splitting activity was judged visually by effervescence on the addition of hydrogen peroxide 10 volumes to heaped growth. (2) Agar (1.5%. w/v) and glucose (0.5%, w/v) were added to the basal medium. To 90 ml of this medium (melted) were added 10 ml of a 1:1 (by vol) mixture of defibrinated ox-blood and tap water. The medium was heated at 100°C for 15 min to denature the blood and destroy the blood catalase. Cultures were streaked on plates of this medium and after incubation were tested as above.

The basal medium which was used throughout with minor modifications had the following constituents in 1 L tap water: meat extract (Lab-Lemco), 5 g; Evans peptone, 5 g; Difco yeast extract, 5 g; Tween 80, 0.5 ml; $MnSO_4 \cdot 4H_2O$, 0.1 g; potassium citrate, 1 g; pH 6.5.

Catalase—*Aerococcus catalasicus*

O. G. Clausen (1964). J. gen. Microbiol. 35, 1–8.

Cultures were incubated aerobically at 37°C unless otherwise stated.

Ten per cent H_2O_2 was added to 24 hr agar slant cultures and observed for up to 5 min to determine whether gas was formed. Control tests were performed on non-inoculated agar slant media after 24 hr incubation.

Catalase—*Pediococcus*

E. Coster and H. R. White (1964). J. gen. Microbiol. 37, 15–31.

The authors used the methods of H. L. Günther and H. R. White (J. gen. Microbiol. 26, 185, 1961). Their tomato Juice (TJ) broth or agar with the addition of Tween 80 (0.1%, v/v), pH adjusted to 6.6 were used in maintenance and preparation of inocula with certain specified exceptions.

Transfers for preparing inocula were made every 24 hours except for *Pediococcus halophilus* (48 hr) and some brewing strains (fortnightly).

All cultures were incubated aerobically at 30°C except for 'brewing strains' which were incubated in an atmosphere of 95% (v/v) hydrogen and 5% (v/v) carbon dioxide at 22°C. As some strains grew poorly in the nutrient broth media used in the test all cultures were also tested after subculturing for 24 hr (7 days for brewing strains) on YTG agar (Felton, Evans and Niven, 1953) modified by the addition of Tween 80 (0.1%, v/v), $FeSO_4$ (0.04%, w/v), $MgSO_4$ (0.08%, w/v) and $MnCl_2$ (0.014%, w/v).

E. A. Felton, J. B. Evans and C. F. Niven, Jr. (1953). J. Bact. 65, 481.
H. L. Günther and H. R. White (1961). J. gen. Microbiol. 261, 185.

Catalase—*Pseudomonas aeruginosa*

R. R. Colwell (1964). J. gen. Microbiol. 37, 181–194.

Catalase was tested for by adding a drop of ('10 vol') hydrogen peroxide to a smear of a 24–48-hr yeast extract (YE) agar colony on a glass slide.

The standard inoculum for all tests was a single drop (1/20 ml) from a sterile disposable pipette (Fisher Scientific Company) of a 24-48 hr broth culture. The stock culture medium (YE) was a modification of that used by Rhodes (1959): Difco yeast extract, 0.3 g; Difco Bacto-Proteose Peptone, 1.0 g; NaCl, 0.5 g; agar, 1.5 g (agar omitted from liquid stock media); distilled water, 1 L; adjusted to pH 7.2–7.4 with NaOH. Tests were carried out at room temperature (25°C) except where otherwise indicated.

M. E. Rhodes (1959). J. gen. Microbiol. 21, 221.

Catalase and Pseudocatalase—*Lactobacillus, Pediococcus, Aerococcus, Streptococcus* and *Leuconostoc*

R. Whittenbury (1964). J. gen. Microbiol. 35, 13–26.

Detection of hydrogen peroxide-splitting activity. Four media were used: HB agar, hematin agar (for heme-requiring catalase) and two basal medium agars (ph 6.8–7), one containing 0.05% (w/v) glucose (for pseudocatalase) and the other 1% (w/v) glucose (all lactic acid bacteria are negative on this medium). The first two media were prepared as plates, the last two as slopes. All were inoculated by streaking with a capillary pipette containing an 18 hour old culture. Activity was recognized visually by effervescence on the addition of hydrogen peroxide ('10 vol') to heaped growth. Negative results were checked by placing heaped growth into H_2O_2. The media themselves showed no effervescence on adding H_2O_2.

Heated blood (HB) agar was basal medium agar (95 ml; pH 6.8–7) containing 1% (w/v) glucose autoclaved at 121°C for 15 min. After adding to the molten medium 5 ml of a 1 + 1 mixture of defibrinated ox blood + water, the complete medium was heated at 100°C for 15 min.

Cultures were incubated at 30°C.

Note: This catalase was observed in certain strains of homo- and heterofermentative lactobacilli, *Pediococcus* and *Leuconostoc* only when grown on a heme-containing medium. Unlike most catalase producing bacteria these organisms cannot synthesize the heme-component of the catalase.

A pseudocatalase which has no heme prosthetic group and is often acid sensitive could be found in strains of *Leuconostoc, Pediococcus, L. plantarum* (one strain only).

Catalase—*Moraxella*

W. J. Ryan (1964). J. gen. Microbiol. 35, 361–372.

A small amount of growth from nutrient agar or from Löffler's serum was suspended in one drop of '20 vol' hydrogen peroxide on a glass slide and the production on effervescence observed.

Catalase—*Actinomyces*

L. K. Georg, G. W. Robertstad and S. A. Brinkman (1964). J. Bact. 88, 477–490.

Catalase tests were performed according to standard procedures.

Catalase—motile marine bacteria and *Hyphomicrobium neptunium**

E. Leifson, B. J. Cosenza, R. Murchelano and R. C. Cleverdon (1964). J. Bact. 87, 652–666.

Catalase was determined by addition to agar-slant cultures of commercial 40% hydrogen peroxide diluted 1:10 (v/v) with distilled water.

Culture media. The culture broth and the plating agar were prepared with seawater taken from Noank Harbor and filtered through a 0.45-μ Millipore filter. To the water were added 0.2% Casitone (Difco) and 0.1% yeast extract; the pH was adjusted to 7.5; the mixture was then boiled, filtered through paper, and sterilized by autoclaving. This was the broth used for primary culture and for flagellar staining. For slants and plates, 1.5% agar was added to the broth. In studies of the isolated bacteria, artificial seawater was used in all media. Comparative studies showed that all the isolates grew equally well on media prepared with artificial seawater diluted with an equal quantity of distilled water, compared with media made with undiluted artificial seawater or with undiluted natural seawater. Since the acid-base buffer content of these media is low, it was advantageous to add additional buffer. Tris (hydroxymethyl)aminomethane (tris) buffer in 0.05% concentration proved to be satisfactory.

Note: The composition of the artificial seawater is not given or cited.

*E. Leifson (1964). Antonie van Leeuwenhoek 30, 249–256.

Catalase—*Veillonella*

M. Rogosa (1964). J. Bact. 87, 162–170.

'Catalase' tests were done repeatedly on washed cells from V17 broth, from growth on anaerobic streak plates, and from colonies in pour-plates; 5% H_2O_2 freshly diluted from refrigerated 30% H_2O_2 was used, and special care was taken to avoid mixing peroxide and cells with any metal objects. These tests were done immediately after removing cultures or cells from an anaerobic environment, and also after at least 1-hr exposure in the air.

For V17 medium see **Basal medium**—*Veillonella* p. 114.

Catalase—*Mycobacterium buruli* n. sp.

J. K. Clancey (1964). J. Path. Bact. 88, 175–187.

The qualitative test of Middlebrook (1954) was used to determine the presence of catalase. The tests were carried out in duplicate on strains grown on plain Löwenstein-Jensen medium and the same medium containing 5 μg per ml isoniazid.

G. Middlebrook (1954). Am. Rev. Tuberc. pulm. Dis. 69, 471.

Catalase—*Bacteroides oralis*

W. J. Loesche, S. S. Socransky and R. J. Gibbons (1964). J. Bact. 88, 1329–1337.

The authors used the methods in the *Manual of Microbiological Methods* (1957).

Society of American Bacteriologists *Manual of Microbiological Methods* – McGraw-Hill Book Co. Inc., New York 1957.

Catalase—*Dermatophilus*

M. A. Gordon (1964). J. Bact. 88, 509–522.

The catalase test was performed by touching one-half loopful of growth on BHI Agar to a drop of Superoxol (30% hydrogen peroxide) on a glass slide and observing for evolution of bubbles.

The inoculum was taken from Brain Heart Infusion agar slants.

Catalase—*Vibrio marinus*

R. R. Colwell and R. Y. Morita (1964). J. Bact. 88, 831–837.

The authors used the method of the Society of American Bacteriologists. 1957. *Manual of Microbiological Methods.* McGraw-Hill Book Co., Inc., New York – modified by the addition to the media of the following salts to produce a synthetic seawater: – sodium chloride, 2.4%; potassium chloride, 0.07%; magnesium chloride (hydrated) 0.53% and magnesium sulfate (hydrated) 0.7%.

A standard inoculum was 1 drop from a pasteur pipette (*c.* 0.05 ml) of a 24 hour artificial seawater broth.

Cultures were incubated at 18°C.

Catalase—*Leptospira*

P. J. Rao, A. D. Larson and C. D. Cox (1964). J. Bact. 88, 1045–1048.

The paper describes an assay procedure for catalase in *Leptospira pomona.*

Catalase—*Alcaligenes*

R. G. Mitchell and S. K. R. Clarke (1965). J. gen. Microbiol. 40, 343–348.

A loopful of culture from nutrient agar was held in a drop of '10 vol' H_2O_2 on a slide and examined macroscopically for effervescence.

Catalase and Pseudocatalase—*Streptococcus*

R. Whittenbury (1965). J. gen. Microbiol. 38, 279–287.

The method of R. Whittenbury (1964). J. gen. Microbiol. 35, 13, was used (see p. 156).

Catalase—*Mycobacterium*

A. Tacquet, F. Tison and B. Devulder (1965). Annls Inst. Pasteur, Paris 108, 514–525.

Used the method of A. Andrejew, Ch. Gernez-Rieux and A. Tacquet (Ann. Inst. Pasteur, 1956, 91, 586–589).

Catalase—*Lactobacillus, Streptococcus, Pediococcus*

M. A. Johnston and E. A. Delwiche (1965). J. Bact. 90, 347–351.

Certain strains of lactobacilli and pediococci incorporated hematin during growth, with the concomitant formation of cyanide- and azide-sensitive catalase. Three of five strains of lactobacilli and five of twenty-five strains of pediococci were capable of this biosynthesis. The pediococci required the heme component of blood, whereas the lactobacilli could incorporate the heme component in the form of purified and solubilized hemin or from blood. In all cases where inhibitor-sensitive enzyme was produced, it was accompanied by the production of inhibitor-insensitive enzyme. In the absence of hematin, only insensitive enzyme was obtained. Two catalase-positive strains of *Streptococcus faecalis* were found incapable of the synthesis of a heme-type enzyme, as was one member of the genus *Leuconostoc.* Iron and manganese in the growth medium stimulated the production of the insensitive catalase, but significant quantities of these metals could not be found in purified enzyme preparation obtained from *Lactobacillus planatarum.* Aeration had little or no effect on growth, but it consistently doubled the amount of cyanide- and azide-resistant catalase. By means of conventional enzyme fractionation techniques, it was possible to separate the two different enzymes present in the cell-free extract of a strain of *Pediococcus homari* which had been grown in the presence of blood.

See also – R. H. Deibel and J. B. Evans (1960). J. Bact. 79, 356–360.

E. A. Delwiche (1960). Bact. Proc. p. 168.

M. A. Johnston and E. A. Delwiche (1965). J. Bact. 90, 352–356.

D. Jones, R. H. Deibel and C. F. Niven, Jr. (1964). J. Bact. 88, 602–610.

R. Whittenbury (1960). Nature, Lond. 187, 433–434.

R. Whittenbury (1964). J. gen. Microbiol. 35, 13–26.

Catalase—*Mycobacterium*

F. G. Winder (1966). J. Bact. 92, 413–417.

The paper describes the extraction and assay of catalase.

A qualitative test for catalase (Gałasinski, Wołosowicz and Tysarowski, 1962) was used on all fractions.

Quantitative assay was by the method Winder and O'Hara.

W. Gałasinski, N. Wołosowicz and W. Tysarowski (1962). Acta biochim. pol. 9, 199–204.
F. G. Winder and C. O'Hara (1964). Biochem. J. 90, 122–126.

Catalase—*Bacteroides, Veillonella, Neisseria*

J. van Houte and R. J. Gibbons (1966). Antonie van Leeuwenhoek 32, 212–222.

Catalase was determined by addding one drop of 3% hydrogen peroxide to a colony, and examining for bubbles with a dissecting microscope.

The basal medium was BBL thioglycolate medium, without added dextrose, supplemented with 5% horse serum and hemin (5 μg/ml) menadione (0.5 μg/ml).

Catalase—*Spirillum gracilis*

E. Canale-Parola, S. L. Rosenthal and D. G. Kupfer (1966). Antonie van Leeuwenhoek 32, 113–124.

Catalase tests were performed by microscopic observation of O_2 evolution from colonies or sections of colonies flooded with a 3% solution of H_2O_2.

Catalase—*Mycobacterium*

R. M. McCune, F. M. Feldmann and W. McDermott (1966). J. exp. Med. 123, 469–486.

A mixture of equal proportions of 32% hydrogen peroxide and 10% Tween 80 was used as reagent. One drop was applied to a colony growing on an oleic acid albumin agar plate. The reaction was recorded as positive if the colony and the reagent reacted by bubbling.

Catalase—*Mycobacterium*

M. Tsukamura (1966). J. gen. Microbiol. 45, 253–273.

Catalase activity was tested by immersing one loopful of the organism into a 30% aqueous solution of H_2O_2 and observing the occurrence of bubbling.

Catalase—*Brevibacterium*

R. Chatelain and L. Second (1966). Annls Inst. Pasteur, Paris 111, 630–644.

Catalase: suspension d'une öse de culture dans une goutte d'eau oxygénée à 10 volumes (environ 3 p. 100).

La température d'incubation des cultures est de 30°C.

Catalase and peroxidase—*Mycobacterium*

H. Boisvert (1966). Annls Inst. Pasteur, Paris 111, 180–192.

Catalase et peroxydase avant et après chauffage à 70° [25]. 10 mg environ de culture jeune sont introduits dans deux tubes à hémolyse contenant 0,1 ml d'eau distillée. L'un de ces tubes est porté au bain-marie à 70°, 15 min et refroidi aussitôt. On introduit dans les deux tubes 1 ml de la solution de Bogen [2]: pyrochatéchine 0,1 g, perhydrol 110 vol 0,5 ml, eau distillée 100 ml, à laquelle on ajoute 1,25 ml de Tween 80.

Au bout d'une heure on note la hauteur de mousse et la coloration du culot, brun rouge si la réaction peroxydasique est positive. On examine une deuxième fois le culot à la dix-huitième heure.

[2] Bogen (E.). Amer. Rev. Tuberc., 1957, 76, 1110.

[25] Kubica (G. P.) et Gleason (L. P.) Amer. Rev. respir. Dis., 1960, 81, 837.

Catalase—anaerobes

A.-R. Prévot (1966). Techniques pour le diagnostic des Bactéries Anaérobes. Éditions de la Tourelle, St. Mandé.

Recherche des catalases.

La recherche des catalases se fait très simplement en versant de l'eau oxygénée du Codex à 10 volumes, soit dans la culture liquide, soit sur un culot de centrifugation, soit sur des colonies de surface en gélose inclinée incubée dans le vide. La décomposition est rapide et donne un dégagement de l'O_2.

CATALASE EFFECT

Catalase effect—*Chromobacterium*

P. H. A. Sneath (1956). J. gen. Microbiol. 15, 70–98.

Nutrient agar ditch plates were prepared the ditch containing 1 μg pure horse liver catalase/ml. Drops of tenfold dilutions (from 1/10 to $1/10^6$) of 24 hr broth cultures were run across the plate and ditch, and the plates were incubated for 2 days at 25°C. Organisms showing the catalase effect give, with the more dilute inocula, colonies only upon the ditch.

Chromobacterium strains, like *Pasteurella pestis,* are very sensitive to traces of hydrogen peroxide, as shown by the inability of single organisms to grow aerobically on nutrient agar, and the small number of separated colonies obtained when strains are plated out on this medium. The inhibition is overcome by blood, hematin or catalase (Sneath, 1955). The strains reported on by Sneath (1955) were all mesophils; but six psychrophils have since been examined and all show the effect.

P. H. A. Sneath (1955). J. gen. Microbiol. 13, 1.

CELL WALL STAIN

Cell wall stain

K. A. Bisset and C. M. F. Hale (1953). Expl Cell Res. 5, 449–454.

Smears, usually of cultures less than 24 hours old, were made on cover slips, rather thickly, to allow for the loss of a proportion of the material by washing, and were immersed in 1 per cent phosphomolybdic acid at room temperature (20°C approximately) for three to five minutes, and stained with 1 per cent methyl green for a similar period. The cell walls stained dark green or purple, and the cytoplasm was unstained.

Cell wall stain—"Bifid bacteria"

V. Sundman, K. af Björksten and H. G. Gyllenberg (1959). J. gen. Microbiol. 21, 371–384.

Several cell-wall staining methods were tried. The ordinary tannic acid-crystal violet method (Robinow, 1946) gave diffuse and unsatisfactory preparations of most strains studied. The stronger crystal violet employed in the method of Webb (1954) gave better results, but the mordanting (1 hour 5%, w/v, tannic acid) was obviously insufficient for the actinomycetes and some of the bifid bacteria. The phosphomolybdic acid + methyl green method (Hale, 1953) the crystal violet-Congo red method (Chance, 1953) and the direct staining method with Victoria blue (Robinow and Murray, 1953) did not give satisfactory results. After some experiments the following two modifications of the methods mentioned above were found most suitable.

1. Phosphomolybdic acid + crystal violet method. The unfixed print or smear was mordanted in 1% (w/v) phosphomolybdic acid for 5-10 min, washed in tap water, stained with 0.5% (w/v) crystal violet for 15 sec, washed in tap water; mounted in water.

2. Tannic acid + crystal violet method. The unfixed print or smear is mordanted in 5% (w/v) tannic acid for one day, washed in tap water, stained with 0.5% (w/v) crystal violet for 3 min washed in tap water, studied unmounted. The latter method has some disadvantages; it is time consuming and the prolonged mordanting tends to give a background on the slide. It was used when method No. 1 failed to give satisfactory preparations.

H. L. Chance (1953). Stain Technol. 28, 205.
C. M. F. Hale (1953). Lab. Pract. 2, 115.
C. F. Robinow (1946). In *The Bacterial Cell,* ed. by R. J. Dubos, *Addendum,* p. 372. Cambridge, Mass.: Harvard University Press.
C. F. Robinow and R. G. E. Murray (1953). Expl Cell Res. 4, 390.
R. B. Webb (1954). J. Bact. 67, 252.

CELLULOSE HYDROLYSIS

Cellulose hydrolysis

J. R. Sanborn (1926). J. Bact. 12, 343–353.

China Blue-Rosolic acid – Cellulose medium

K_2HPO_4	1 g
$MgSO_4$	1 g
Na_2CO_3	1 g
$(NH_4)_2SO_4$	2 g
H_2O	1000 ml
Raw cotton	30 g

The salts are dissolved in 500 ml of distilled water and 0.5% agar prepared with the other 500 ml water. Before sterilization the two are mixed and the cotton added cut into small fragments. With constant stirring 10 ml of the "CR"* indicator is added. The medium is then distributed in petri dishes and autoclaved. The pH is 8.4 and the medium is a brilliant red.

With production of acid the medium turns a deep blue near pH 7.0.

*"CR" indicator is obtained by mixing equal parts of 0.5 per cent aqueous solution of China blue with 1 per cent solution of rosolic acid in 95 per cent alcohol. On the alkaline side China blue is water clear. Rosolic acid gives on the acid side different shades of pale yellow, which is masked by the deep blue of the China blue and gives sharp color values in media. In alkaline environment, China blue being colorless, the rosolic acid gives a pure red.

Cellulose hydrolysis—various

L. A. Bradley and L. F. Rettger (1927). J. Bact. 13, 321–345.

For maintenance of pure cultures,

Tryptic Casein-digest	100 ml
Tap water	900 ml
Beef extract	1 g
pH 7.4.	

Dispensed in tubes and a strip of partially immersed filter paper used to supply cellulose. The organisms which hydrolyzed cellulose in this medium could also grow without it.

The medium is not satisfactory for growing enrichment cultures. A mineral medium was more satisfactory.

K_2HPO_4	1 g
$MgSO_4 \cdot 7H_2O$	1 g
NaCl	1 g
$CaCO_3$	2 g
KNO_3	2 g

Cellulose	strip of filter paper
Water	1000 ml

Cellulose hydrolysis—*Sporocytophaga* and *"Bacterium fimi"*

R. J. Dubos (1928). J. Bact. 15, 223–234.

1. It is shown that a rapid and abundant growth of aerobic cellulose decomposing bacteria is obtained by the use of the following medium which is very simple and very specific: $NaNO_3$, 0.05 g; K_2HPO_4, 1.0 g; $MgSO_4 \cdot 7H_2O$, 0.50 g; KCl, 0.50 g; $FeSO_4 \cdot 7H_2O$, 0.01 g; distilled water, 1000 g. Five milliliter portions of this medium are introduced into test tubes containing a strip of filter paper partly immersed.

2. The slightly alkaline reaction of the medium (pH 7.5) favours the growth of bacteria while it retards the growth of fungi.

3. The low concentration of nitrogen salts shortens the incubation period of bacteria, since, in almost all cases, cellulose decomposition can be recorded after thirty-six to seventy-two hours at 28°C.

4. Growth is obtained even when only one or very few cells are used for inoculation.

Cellulose hydrolysis—thermophilic cellulose decomposing organisms

P. A. Tetrault (1930). J. Bact. 19, 15.

Tetrault's modification of the Morse-Kopeloff anaerobic chamber was employed. Viljoen, Fred and Peterson's medium, to which was added agar and cellulose in the form of ground filter paper or filter paper rounds, was used for most experiments. Incubation was at 60° to 65°C.

With the streak method of inoculation the following technique was followed. A layer of clear agar was poured in the bottom of the chamber and allowed to solidify. A thin layer of agar containing finely ground paper was then added and also allowed to solidify. A round of sterile filter paper may be substituted for this layer. Streak inoculations were made with a straight wire. A third layer of agar was then poured over the inoculated surface. If a filter paper round was used, this third layer was clear agar. After solidification of the agar the chamber was inverted, sealed and incubated.

Cellulose hydrolysis—*Trichoderma, Azotobacter, "Spirochaeta cytophaga"*

C. E. Skinner (1930). J. Bact. 19, 149–159.

The basic medium used by Sanborn (1926a, b, 1927) was as follows:

K_2HPO_4	1 g
$MgSO_4$ (anhydrous)	1 g
Na_2CO_3 (anhydrous)	1 g
$(NH_4)_2SO_4$	2 g
H_2O (distilled)	1000 ml

This is a modification of the well known solution of McBeth and Scales (1913) who added an excess of $CaCO_3$. It was found that some of the organisms did not grow in either of the media due to the extremely high pH. Therefore, McBeth and Scales' solution was modified by omitting the $CaCO_3$ and substituting NaCl for Na_2CO_3. In certain cases the reaction was changed further by the addition of NaOH or HCl. All four of my organisms grew well in this solution with filter paper as the source of energy. In some instances, to be explained later, an equivalent amount of KNO_3 was substituted for the $(NH_4)_2SO_4$. In every case, the sterilized nitrogen compound was added aseptically after sterilization of the medium.

Eaton and Dikeman No. 615 filter paper was used as a source of cellulose.

I. G. McBeth and F. M. Scales (1913). U. S. D. A. Bur. Plant Indus. Bul. 266.
J. R. Sanborn (1926a). J. Bact. 12, 1–12.
J. R. Sanborn (1926b). Ibid. 12, 343–353.
J. R. Sanborn (1927). Ibid. 13, 113–121.

Cellulose hydrolysis—*Clostridium cellulosolvens*

P. B. Cowles and L. F. Rettger (1931). J. Bact. 21, 167–182.

This paper deals with the isolation, growth upon solid medium, and study of an anaerobic organism which is morphologically similar to those described by W. Omelianski (1902). Zentbl. Bakt. ParasitKde II, 8, 193; 225; 257; 289; 321; 353; 385; 605.

Media:

1. Omelianski's medium

Ammonium sulphate or peptone	1.0 g
Di-potassium phosphate	1.0 g
Magnesium sulphate	0.5 g
Sodium chloride	trace
Calcium carbonate	excess
Water	1000 ml

2. Fecal extract medium

Peptone	1.0 g
Di-potassium phosphate	1.0 g
Sodium chloride	1.0 g
Calcium carbonate	excess
Fecal extract	250 ml
Water	750 ml

The fecal extract was prepared by extracting horse feces with ten parts of water, filtering through paper, autoclaving, and passing through a Berkefeld candle.

3. Beef infusion broth

Beef infusion	1000 ml

Peptone	5.0 g
Di-potassium phosphate	1.0 g
pH	7.0

In all of the above media cellulose was added in the form of a strip of filter paper placed in each tube.

Cellulose hydrolysis—various

M. Aschner (1937). J. Bact. 33, 249–252.

M. Aschner describes a method for the use of the cellulose membrane synthesized by *Acetobacter xylinum* in a mineral salts base or as a direct substitute for an agar plate.

Cellulose hydrolysis—thermophilic bacteria

H. C. Murray (1944). J. Bact. 47, 117–122.

Enrichment medium:

1 g NaCl, 0.5 g $MgSO_4 \cdot 7H_2O$, 2.5 g KH_2PO_4, 15–20 g ground filter paper, 1000 ml distilled water. In liquid media $CaCO_3$ was frequently added as a neutralizing agent, or the pH was adjusted to 7.4.

When a solid medium was required, filter paper was comminuted in a ball mill and agar (0.8%)* added without carbonate. The nitrogen source was generally 0.5% peptone. Incubated at 60°C.

*The 0.8% agar which permits more extensive digestion than higher agar concentrations is poured over a solidified layer of 2% agar in distilled water in a petri dish. The resulting preparation is sufficiently rigid to be inverted and incubated.

Cellulose hydrolysis—*Clostridium*

R. E. Hungate (1944). J. Bact. 48, 499–513.

Isolation: Evidence of the organism was first observed when rumen contents were inoculated into a dilution series of shake tubes containing agar and finely divided cellulose in a solution of inorganic salts. The cellulose was prepared by treating absorbent cotton with strong hydrochloric acid until it disintegrated into small particles, which were then washed free of chlorides. According to Farr and Eckerson (1934), these particles give an X-ray diffraction pattern similar to that of untreated cellulose. They were ground with water in a pebble mill to give a finely divided suspension. The salt solution was composed of the following: NaCl, 0.6 g; $(NH_4)_2SO_4$, 0.1 g; KH_2PO_4, 0.05 g; K_2HPO_4, 0.05 g; $MgSO_4$, 0.01 g; $CaCl_2$, 0.01 g; and tap water, 100 ml. Sterile oxygen-free nitrogen containing 5 per cent carbon dioxide was bubbled through the medium after sterilization, and sufficient sterile sodium carbonate solution to give a pH of 7.4 was added. The tubes were stoppered without admitting air and incubated at 38°C for three weeks.

W. K. Farr and S. H. Eckerson (1934). Contr. Boyce Thompson Inst. Pl. Res. 6, 309–313.

Cellulose hydrolysis—preparation of cellulose

J. Pochon, Y. T. Tchan, T. L. Wang and J. Augier (1950). Annls Inst. Pasteur, Paris 79, 376–380.

Si l'on veut obtenir de la cellulose précipitée, incorporable à la gélose, n'ayant pas été modifiée dans ses propriétés chimiques, il faut utiliser des moyens mécaniques jusqu'à obtention de particules du même ordre de grandeur que les bactéries.

Dans ce but nous avons broyé du papier filtre dans de l'eau, avec des billes de verre, pendant vingt-quatre heures. Comme le pH a tendance à s'élever, par libération des ions alcalins du verre, nous opérons en présence d'un excès de CO_2 (les tampons ne peuvent être employés car ils ont un effet agglutinant sur la suspension). Dans ces conditions nous obtenons une suspension de particules dont la dimension est de l'ordre de quelques μ. Cette suspension s'incorpore facilement à la gélose et peut également servir pour imprégner des plaques de silico-gel.

Cellulose hydrolysis—aerobes and anaerobes

J. Pochon and Mme. Baÿ (1951). Annls Inst. Pasteur, Paris 81, 179–186.

La gamme des techniques utilisées est très étendue.

I. Cellulolyse en aérobiose. – 1. Plaques de silico-gel imprégnées avec la solution saline standard de Winogradsky additionnée d'azote nitrique et recouvertes d'une feuille de papier filtre (Durieux III), puis ensemencées avec des grains de terre ou des particules de contenu de panse.

2. Plaques de silico-gel imprégnées de la même façon et recouvertes d'une suspension de cellulose précipitée. Celle-ci est ellemême préparée selon deux procédés:

a) 200 cm^3 d'acide sulfurique concentré sont additionnés de 120 cm^3 d'eau distillée.. Le mélange, porté à 57°, est versé sur 10 g de coton. Contact dix secondes, puis on verse 1 litre d'eau froide. On laisse déposer; on lave la cellulose qui est ensuite mise à sécher et remise en suspension.

b) On étend à 300 cm^3, avec de l'eau, 270 cm^3 d'acide chlorhydrique concentré. On ajoute du coton en quantité suffisante pour absorber complètement l'acide. Contact de vingt-quatre heures à froid. Laver. Sécher. Suspension aqueuse broyée pendant trente-six heures dans un agitateur avec billes de verre.

3. Papier enfoui dans la terre, soit avec la technique de la colonne de terre, soit papier enfoui directement dans un pot de terre, verticalement (le papier est plié en quatre épaisseurs, les prélèvements sont faits sur les plis intérieurs, gagnés par la cellulolyse). Dans les deux cas, l'attaque en aérobiose correspond à la partie supérieure, proche de la surface du papier. De la même façon nous avons enfoui du papier dans du contenu de panse.

II. Cellulolyse en anaérobiose. – 1. Papier enfoui, comme il vient d'être dit. Les prélèvements sont faits à la partie inférieure du papier, loin de la surface, correspondant à la zone d'anaérobiose.

2. Milieu liquide de Hungate additionné de papier filtre.

3. Milieu de Hungate gélosé à la cellulose précipitée.

4. Milieu de Hungate gélosé avec feuille de papier filtre contre la paroi du tube.

5. Milieu K. S. G. A. de Sijpesteijn (1948) avec différentes concentrations de filtrat de panse et d'eau de levure.

Les cultures ont été faites à 23° pour le sol et à 37° pour la panse. On notait soigneusement les modifications (consistance, couleur) du papier, l'aspect des colonies et leur auréole claire sur milieu solide à la cellulose précipitée. Les prélèvements étaient effectués régulièrement et les examens microscopiques faits à l'état frais et après coloration par l'érythrosine phéniquée ou par la méthode de Gram, en lumière ordinaire et en lumiere polarisée.

A. K. Sijpesteijn (1948). Cellulose-decomposing bacteria from the rumen of cattle. Thesis. Leiden University.

Cellulose hydrolysis—*Ruminococcus*

A. K. Sijpesteijn (1951). J. gen. Microbiol. 5, 869–879.

The author describes the isolation of *Ruminococcus flavefaciens* using the following medium.

Conical flasks (100 ml) contained 25 ml of a solution of the following composition (in %, w/v): Davis's New Zealand agar (1.5), K_2HPO_4 (0.06), KH_2PO_4 (0.04), $(NH_4)_2SO_4$(0.04), NaCl (0.12), $MgSO_4 \cdot 7H_2O$ (0.02), $CaCl_2 \cdot 6H_2O$ (0.02). To each flask 19 ml tap water was added. The medium was boiled out and a reducing agent added, either 25 mg Na thioacetate per flask in the case of strain S, or 50 mg L-cysteine HCl adjusted to pH 7, for strain D. After autoclaving, 3 ml yeast autolysate and 3.5 ml 7% $NaHCO_3$ were added per flask to give a final concentration of 6% (v/v) and 0.5% (w/v) respectively in the final liquid volume of *c.* 50 ml. The bicarbonate solution was sterilized by filtration. The filled flasks were put in a waterbath at 38-40°C and gassed for about 5 min with CO_2 freed from traces of oxygen by passage over reduced copper turnings heated to *c.* 400°C. The medium was subsequently distributed between six sterile tubes each containing a strip of Whatman no. 1 filter paper.

After inoculation each tube was gassed with CO_2 for about 4 min. This proved sufficient to ensure saturation of the medium with CO_2 (producing a final pH *c.* 6.6), to remove traces of oxygen from the medium, and to replace the air above it by CO_2. Methylene blue became decolorized almost immediately when added to this medium. During the final gassing procedure the plug was pushed into the tube, the Pasteur pipette withdrawn, and a rubber stopper quickly fitted to the tube. The tubes were then put into cold water to solidify the agar and after sealing the stoppers with paraffin-wax, incubated at 38°C.

Cellulose hydrolysis

E. R. Hall (1952). J. gen. Microbiol. 7, 350–357.

Cellulose utilization in domestic rabbits. Methods employed are similar to those used by Hungate (1950).

The basic culture medium contained the following final percentages (w/v) of ingredients: NaCl, 0.025; $(NH_4)_2SO_4$, 0.01; K_2HPO_4, 0.02; KH_2PO_4, 0.01; $CaCl_2$, 0.003; $MgSO_4 \cdot 7H_2O$, 0.003; $NaHCO_3$, 0.5; sodium thioacetate, 0.02 or L-cysteine monohydrochloride, 0.01. Either cotton or filter paper cellulose in the form of a fine suspension was added in concentrations of 0.1–0.2% (w/v) in liquid and 0.6% (w/v) in agar media. Rumen fluid or cecal extract in a concentration of 20–30% (v/v) was included in all culture media.

The medium minus $NaHCO_3$ was prepared and sterilized in 200 ml Florence flasks. After sterilization, the $NaHCO_3$ solution which had been sterilized by filtration, was added and the medium transferred to sterile rubber-stoppered tubes. The final pH value of the medium was 6.8–7.1. Anaerobiosis was maintained during preparation and transfer of the medium by passing oxygen-free CO_2 or N_2 through it and the receiving tubes. R. E. Hungate (1950). Bact. Rev. 14, 1.

Cellulose hydrolysis—anaerobes

L. R. Maki (1954). Antonie van Leeuwenhoek 20, 185–200.

Best results were obtained when sludge was serially diluted into a medium prepared as follows: 40 ml tapwater containing 0.4 g finely divided cellulose (Whatman's No. 1 filter paper ground wet in a pebble mill) were added to 30 ml of supernatant liquid obtained by letting the solids settle out of the effluent from the primary digester of the Pullman sewage plant. Thirty ml of a solution of inorganic salts were added so that the final concentrations were 0.1% NaCl, 0.1% $(NH_4)_2SO_4$, 0.05% K_2HPO_4, 0.01% $MgSO_4$, and 0.01% $CaCl_2$. Yeast extract 0.2%, agar 1.5% and resazurin 0.0001%, were the other ingredients of the medium. Sterile sodium carbonate (0.25% final concentration) was added after sterilization, and a reducing agent was introduced just after inoculation. Cysteine, sodium sulfide, and sodium formaldehyde sulfoxylate have been used with equal effectiveness as reducing agents. The inoculum was obtained from the sewage plant and cultured within a few hours of collection. Cultures were incubated at 38°C. Cellulose-decomposing colonies were detected by the clearing of the surrounding cellulose.

Cellulose hydrolysis—rumen microorganisms

G. Halliwell (1957). J. gen. Microbiol. 17, 153–165.

Several different forms of cellulose were used:

Native cotton fibres (Peruvian Tanguis, kindly supplied by the British Cotton Industry Research Association) were used untreated, and also de-waxed as follows. Fibres were purified by Soxhlet extraction with redistilled ethanol for 8 hr and with ether for 6 hr, followed by refluxing under nitrogen in 1% (w/v) sodium hydroxide for 8 hr. After washing with boiled water until neutral, washing was continued successively with cold water, 1% (w/v) acetic acid, and water. The product was finally dried in air. Cellulose powder, referred to in the text as cellulose powder (Whatman), was standard grade ashless powder for chromatography and was used untreated. *Swollen cellulose* was prepared from absorbent cotton and also from cellulose powder (Whatman) by soaking in phosphoric acid (A. R., 90%) at 1°C for 2 hr, followed by washing until acid and phosphate free (Walseth, 1952). *Hydrocellulose* was prepared from absorbent (de-waxed) cotton by immersion in 75 vol of 11 N-HCl (A. R.) for 48 hr at 18°C. After rejecting the supernatant fluid the fine powder was poured into water and then filtered off on a Büchner funnel, washed acid- and chloride-free and finally dried in vacuum at room temperature. A product similar to this has been widely used in work on cellulolytic microorganisms (Hungate, 1942, 1950a,b; Sugden, 1953), after further degradative treatment by grinding.

Determination of the cellulolytic activity of rumen microorganisms: The standard assay for cellulolytic activity. Fifty mg of insoluble cellulose (or 30 mg with some forms of cellulose) was added to a CO_2-saturated, $NaHCO_3$ + salts buffer with or without addition of rumen microorganisms (1 ml). The assay tubes, 150 mm × 15 mm, were closed with a Bunsen valve. The total volume was made up to 8 ml (initial pH value 6.8) and the tubes were incubated at 37°C for about 40 hr. Unless stated to the contrary, $NaHCO_3$ buffer or medium had the following composition: 0.2 M-$NaHCO_3$, 100 ml; 0.154 M-KCl, 4 ml; 0.154 M-KH_2PO_4, 1 ml; 0.154 M-$MgSO_4$, 1 ml; 0.154 M-$(NH_4)_2HPO_4$, 5 ml; 0.11 M-$CaCl_2$, 3 ml (Elsden 1945).

Method of analysis of cellulose breakdown. Breakdown of insoluble cellulose was followed by gravimetric determination of the residual cellulose.

S. R. Elsden (1945). J. exp. Biol. 22, 51.
R. E. Hungate (1942). Biol. Bull., mar. biol. Lab. Woods Hole 83, 303.
R. E. Hungate (1950a). Bact. Rev. 14, 1.
R. E. Hungate (1950b). A. Rev. Microbiol. 4, 53.
B. Sugden (1953). J. gen. Microbiol. 9, 44.
C. S. Walseth (1952). Tech. Pap., Pulp. Pap. Ind. N. Y. 35, 228.

Cellulose hydrolysis

V. B. D. Skerman, *A Guide to the Identification of the Genera of Bacteria.* 2nd ed. The Williams and Wilkins Book Co., Baltimore, 1967.

Cellulose Strip-Peptone Medium:

Peptone	5.0 g
NaCl	5.0 g
Tap water	1000 ml

Steam until dissolved. Adjust the pH to 7.4 and filter if necessary. Dispense in 5 ml amounts in 150- by 13-mm tubes.

In each tube place a 70- by 10-mm strip of Whatman No. 1 filter paper. Sterilize at 121°C for 15 minutes.

Preparation of precipitated cellulose:

Carefully add 100 ml of concentrated sulfuric acid to 60 ml of water in a flask cooled by running water, with care to direct the neck of the flask away from the body. Cool to 70°C and to a 5-ml sample add 0.3 g of Whatman No. 1 filter paper. If the paper dissolves rapidly and subsequently chars, proceed as follows (if the paper does not dissolve prepare a slightly more concentrated acid).

Place a series of ten 150- by 18-mm tubes in a rack in a water bath at 70°C. Into these tubes pipette 0.2, 0.4, 0.6, 0.8, 1.0, 1.2, 1.4, 1.6, 1.8 and 2.0 ml of water, respectively. To each add 10 ml of the diluted sulfuric acid and mix. To each tube add 0.3 g of filter paper (torn into small pieces). In the higher acid concentrations the paper usually dissolves and chars. In the lower concentrations it will disintegrate but may not char. When dispersed in water this disintegrated paper yields a mass of fine fibers which scintillate in reflected light and settle out rapidly on standing. These fibers are too coarse for a good cellulose medium. At an intermediate acid concentration the paper is reduced to a gelatinous mass which slowly turns brown if allowed to stand. If this is rapidly diluted a mass of finely divided particles is produced which remains suspended for some period and does not scintillate in reflected light. This material is suitable for the preparation of precipitated cellulose agar.

Note the degree of dilution of the acid and dilute the bulk accordingly. Hold at 70°C and add 3 g of Whatman No. 1 filter paper (torn in small pieces) to each 100 ml of acid. Rotate the flask until slight yellowing (charring) is evident and then empty the contents rapidly into a 5-L flask containing 2000 ml of tap water. The resulting opalescent suspension should remain dispersed for some time and show no evidence of scintillation in reflected light. Allow this to stand until the cellulose settles and siphon off the supernatant. Centrifuge the deposit and resuspend in water to wash it free of acid. Repeat the process several times until 10 ml of the suspension will no longer produce an acid reaction after the addition of 1 ml of 0.1 N NaOH.

Titrate a sample with 0.1 N NaOH and then neutralize the bulk.

Centrifuge the cellulose and then resuspend in 20 ml of water. Progressively dilute a 1-ml sample until 5 ml of the diluted sample dispensed in a 10-cm petri dish still gives a perceptible opalescence. Note the degree of dilution and dilute the bulk of the cellulose by *half* this amount. (It is further diluted in the preparation of the medium.)

Sterilize at 121°C for 20 minutes.

Cellulose hydrolysis—anaerobes

F. A. Skinner (1960). J. gen. Microbiol. 22, 539–554.

Preparation of finely divided cellulose. Fifteen g Whatman ashless cellulose powder (Standard grade) and 80 ml distilled water were treated in a macerator (MSE) for 20 min. The resulting slurry was divided into several portions and each shaken with 250 ml distilled water in a measuring cylinder. After settling, the turbid supernatant fluids were collected by siphoning and each residue treated as before with another 250 ml water. This process was continued until supernatant fluids were only slightly turbid. Bulked fluid was then made acid with a few drops of N-HCl and allowed to stand overnight. The flocculated cellulose (*c.* 300 ml) was concentrated and washed by centrifugation with several changes of distilled water until the deflocculation point, as indicated by the appearance of slightly turbid centrifugate was reached. The cellulose was then re-suspended in distilled water, made up to 2% (w/v), and stored in the refrigerator until required. Three batches made at different times gave yields of fine cellulose of 23.3, 22.7 and 23.2% of the original 15 g used. Higher yields could not be obtained because the fine material seems to derive from the broken ends of the original fibres. When these ends become polished the macerator has almost no further action. The yield of fine cellulose could be increased up to *c.* 50% by applying the above method to short fibres prepared by soaking absorbent cotton wool in near-concentrated HCl (Hungate, 1950).

Almost all the cellulose-containing media used were made up with cellulose prepared as described. Only large liquid cultures (for analysis of fermentation products) were made with cellulose prepared by ballmilling (Hungate, 1950) because a suitable mill only became available at a late stage in the work.

Buffered mineral salt solution for preparation of cellulose media. This contained $(NH_4)_2SO_4$, 1.0 g; K_2HPO_4, 13.0 g; KH_2PO_4, 7.0 g; $CaCl_2$, 0.1 g; $MgSO_4 \cdot 7H_2O$, 0.1 g; NaCl, 2.0 g; distilled water, 1.0 L; pH 7.0.

Liquid cellulose medium A. Buffered mineral salt solution, 50.0 ml; resazurin solution (0.1%, w/v), 0.1 ml; soil extract (Fred and Waksman, 1928), 15 ml; yeast extract (Difco), 0.01 g; cellulose suspension (2%, w/v), 20 ml; cysteine hydrochloride, 0.01 g; distilled water to 100 ml.

Liquid cellulose medium B. Buffered mineral salt solution, 50.0 ml; resazurin solution (0.1%, w/v), 0.1 ml; yeast extract (Difco), 0.1 g; cellulose suspension (2%, w/v), 20 ml; cysteine hydrochloride, 0.05 g; distilled water to 100 ml.

Addition of sodium carboxymethylcellulose (CMC) to cellulose media. CMC (0.5 g) was added slowly to 50 ml of mechanically stirred buffered mineral salt solution. Complete solution was effected by stirring the mixture for 15 min after all the CMC had been added. A suitable volume was added to cellulose agar medium to give the required CMC concentration.
Enrichment culture medium. This contained: $(NH_4)_2SO_4$, 1.0 g; K_2HPO_4, 1.0 g; $MgSO_4 \cdot 7H_2O$, 0.5 g; $CaCO_3$, 2.0 g; NaCl, trace; distilled water, 1.0 L (Omelianski, 1902).
Nutrient agar. Peptone, 10 g; Lab-Lemco, 3.0 g; yeast extract, 1.0 g; NaCl, 5.0 g; Bacto-agar, 15.0 g; distilled water, 1.0 L; pH adjusted to 7.0–7.2.
Yeast-peptone agar. Yeast extract (Difco), 15.0 g; peptone, 5.0 g; Bacto-agar, 15.0 g; distilled water, 1.0 L; pH adjusted to 7.0–7.2.
Incubation. All cultures were incubated at 35°C.

E. B. Fred and S. A. Waksman (1928). *Laboratory Manual of General Microbiology.* McGraw-Hill Book Co. Inc. New York.
R. E. Hungate (1950). Bact. Rev. 14, 1.
W. Omelianski (1902). Zentbl. Bakt. ParasitKde (II. Abt. Orig.). 8, 225.

Cellulose hydrolysis—soil bacteria

M. Charpentier (1960). Annls Inst. Pasteur, Paris 99, 153–155.

Différentes méthodes ont jusqu'ici été employées pour mettre en évidence cette flore: méthode utilisant un substrat solide: le silico-gel, préconisé par Winogradsky [1], par Imchenetski et Solntseva [2] ; ou la gélose dont la manipulation est plus aisée: Omelianski [3], Pouchkinskaya [4], Stanier [5]. Des milieux liquides ont également été utilisés [5, 6] mais ne permettent pas de distinguer les différents organismes cellulolytiques.

Une méthode utilisant comme support solide le silico-gel nous semble donner la meilleure sélection, à condition de modifier le mode d'ensemencement habituel. Le silico-gel préparé selon la technique originale de Winogradsky [1] est réparti dans une boîte de Pétri (30 ml), puis rigoureusement dialysé et imprégné par 2 ml de la solution minérale suivante:

PO_4HK_2		10 g
NO_3NH_4		3 g
SO_4Mg		5 g
ClNa		1 g
Cl_2Ca		1 g
Cl_3Fe		1 ml
OH_2	Q. S. P.	1000 ml

et placé à l'étuve à 56° jusqu'à évaporation complète de la solution minérale en excès. On ensemence alors chaque boîte de II gouttes de suspension-dilution de terre au 1/10, 1/50 et 1/100 qu'on répartit uniformément sur la surface du gel. On recouvre, donc après ensemencement, d'une feuille de papier filtre stérile (papier Durieux) qui doit adhérer parfaitement à la surface. Les boîtes sont conservées à l'étuve à 28°, en atmosphère humide, pendant deux à trois semaines.

[1] Winogradsky (S.). Ann. Inst. Pasteur, 1929, 43, 549.
[2] Imchenetsky (A. A.) et Solntseva (L. I.). C. R. Acad. Sci. U.R.S.S., série Biol. 1936, nº 6, 1115.
[3] Omelianski (V. L.). *Manuel pratique de microbiologie.* Moscou-Leningrad, 1940.
[4] Pouchkinskaya (O. I.). Mikrobiologuia, U.R.S.S., 1954, 23, 34.
[5] Stanier (R. Y.). Bact. Rev. 1942, 6, 143.
[6] Barjac (H. de). Pédologie (Belgique) 1957, 7, 138.

Cellulose hydrolysis

Y. Henis, P. Keller and A. Keynan (1961). Can. J. Microbiol. 7, 857–863.

Basic medium for testing cellulose-degradation. The medium used contained 0.1% $NaNO_3$, 0.1% K_2HPO_4, 0.05% KCl, 0.05% $MgCl_2 \cdot 6H_2O$, 0.01% Bacto yeast extract (Difco), in 1 liter of distilled water, pH 7.0. For a solid medium, 1.5% Bacto agar (Difco) was added. The carbon sources added to this medium were: Whatman filter paper No. 1; cellulose film, Grade P.T. 300;* carboxymethylcellulose (ICI), specific viscosity of 5.6 at a concentration of 1%, or glucose in various concentrations. The cellophane was treated overnight with 0.1 M versene at pH 8.0, and then washed thoroughly with distilled water and boiled for 30 minutes.
Estimation of cellulose-decomposition on agar media – Strips of filter paper or cellophane, 1 x 8 cm, were placed on agar plates and two soil crumbs (average diameter 2 mm) were placed at equal distances on each strip. Cellulose decomposition was estimated by lifting the strips carefully by means of sterile forceps and determining the time required for decomposition to proceed to such an extent that the two portions of the strip on either side of the soil crumb came apart. When pure cultures were tested, suspensions of the organisms were placed on one location in the middle of the strip.

*Kindly supplied by the British Cellophane Co., Ltd., Bridgewater, Somerset, England.

Cellulose hydrolysis—soil organisms

J. Rivière (1961). Annls Inst. Pasteur, Paris 101, 253–258.

a) *Silico-gel.* – On prépare du silico-gel selon la technique de Winogradsky [14]. Après imprégnation, les plaques sont stérilisées en les soumettant cinq minutes à l'action d'une lampe à rayons ultra-violets. L'ensemencement se fait de la façon suivante: 5 g de la culture d'enrichissement sont soigneusement broyés dans un mortier avec 45 ml d'eau ajoutés en plusieurs fois, puis recueillis dans une fiole Erlenmeyer de 250 ml agitée cinq minutes dans un Microid Flask Shaker avant de faire les dilutions. II gouttes des diverses dilutions sont réparties à la surface des plaques; si celles-ci sont bien sèches, les gouttes sont absorbées au bout de quelques heures (on utilise un compte-gouttes donnant XX gouttes au millilitre). L'incubation se fait à 28°C pendant huit jours. Ce milieu convient aux genres suivants: *Sporocytophaga, Cytophaga et Cellvibrio* dont la croissance est difficile en présence de gélose [3].

b) *Gélose à la cellulose.* – Sa composition est la suivante:

Cellulose en poudre*	5	g
$NaNO_3$	1	g
PO_4HNa_2: 2 H_2O (Tampon M/150 pH: 6.8)	1.18	g
PO_4H_2K (Tampon M/150 pH: 6.8)	0.9	g
$SO_4Mg \cdot 7\ H_2O$	0.5	g
KCl	0.5	g
Extrait de levure (Difco)	0.5	g
Hydrolysat de caséine (Difco)	0.5	g
Gélose Noble (Difco)	10	g
Eau q. s. p.	1	litre

Après stérilisation (vingt minutes à 120°C), ce milieu est coulé en plaques après avoir été bien agité pour que la poudre de cellulose soit répartie de façon homogène. Quand elles sont bien sèches, on y ensemence II gouttes de chaque dilution que l'on étale soigneusement sur toute la surface. L'incubation se fait à 28°C pendant huit jours. Ce milieu convient à toutes les autres bactéries cellulolytiques aérobies qui donnent des colonies généralement entourées d'une très petite zone de cellullolyse.

Milieux de purification. – La purification des bactéries cellulolytiques aérobies se fait aussi avec deux types de milieux différents suivant la façon dont elles ont été isolées.

a) *Gélose inclinée.* – Il s'agit d'une gélose à base de glucose stérilisée par filtration (0.1 p. 100).

Sa composition est voisine de la précédente, mais sans cellulose, et contient des peptones (0.1 p. 100 bacto-tryptone Difco). Après plusieurs passages on obtient des colonies dont la pureté est vérifiée par examen microscopique et étalement sur plaques.

b) *Milieu liquide additionné d'antibiotiques.* – Dans le cas des souches isolées sur silico-gel *(Cytophaga* et *Sporocytophaga)*, la purification sur gélose inclinée (Stanier [12]) est insuffisante. On obtient de bons résultats par passage dans des tubes de milieu liquide contenant une bande de papier-filtre de 1 x 6 cm (Durieux n° 111), et des concentrations croissantes de sulfate de kanamycine (kanamycine Bristol): 2.5, 5, 10 mg par litre.

*La poudre de cellulose (Whatman) est mise à tremper dans de l'acide chlorhydrique N pendant douze heures à froid, puis lavée sur Buchner pour éliminer l'acide avant d'être utilisée dans la préparation du milieu.

La composition du milieu est la suivante:

Bacto-tryptone (Difco)	1	g
NO_3Na	1	g
PO_4H_2K (Tampon M/150 pH: 6.8)	0.9	g
$PO_4HNa_2 \cdot 2\ H_2O$ (Tampon M/150 pH: 6.8)	1.18	g
$SO_4Mg \cdot 7\ H_2O$	0.5	g
$SO_4Fe \cdot 7\ H_2O$	0.5	g
$SO_4Mn \cdot H_2O$	trace	
Extrait de levure (Difco)	0.5	g
Eau q. s. p.	1	litre

Après cinq passages (chaque repiquage se fait après huit jours d'incubation à 28°C), la pureté des souches est vérifiée par examen microscopique, étalement sur plaques et culture dans du bouillon nutritif. En effet, les *Cytophaga* et les *Sporocytophaga* ne présentent aucune croissance dans ce dernier milieu.

Test d'activité cellulolytique. – Selon Siu [10], le test le plus simple consiste à inoculer des bandes de papier filtre partiellement immergées dans une solution minérale. C'est ce que nous avons fait en utilisant le milieu précédent sans antibiotique: après huit jours d'incubation à 28°C, une légère agitation doit amener une désintégration du papier pour permettre de conclure qu'il y a un pouvoir cellulolytique notable.

[3] Charpentier (M.). Ann. Inst. Pasteur, 1960, 99, 163.
[10] Siu (R. G. H.). *Microbial Decomposition of Cellulose,* Reinhold Publishing Corporation, New York, 1951, 106.
[12] Stanier (R. Y.). Bact. Rev., 1942, 6, 143.
[14] Winogradsky (S.). Ann. Inst. Pasteur, 1929, 43, 549.

Cellulose hydrolysis—*'Bacterium salmonicida' (Aeromonas)*

I. W. Smith (1963). J. gen. Microbiol. 33, 263–274.

Cellulose activity was estimated by the digestion of filter-paper strips in half-strength peptone water after 7 days (Skerman, 1957, see p. 1019).

V. B. D. Skerman (1957). In *Bergey's Manual of Determinative Bacteriology,* 7th ed., London, Bailliere, Tindall and Cox Ltd.

Cellulose hydrolysis—*Bacteroides* and *Ruminococcus*

G. Halliwell and M. P. Bryant (1963). J. gen. Microbiol. 32, 441–448.

Culture media: Cellulose broth was used to grow the organisms and also as a basal medium to examine the cell-free enzymic breakdown of the different forms of cellulose. Broth (100 ml) containing: 30 ml whole rumen fluid (prepared by expressing rumen fluid through surgical gauze, allowing to stand overnight at 1°C and separating off the middle phase from the lower sediment and upper floating material); 3.75 ml mineral solution 1 (0.6%, w/v, K_2HPO_4); 3.75 ml mineral solution 2 (w/v: 0.6% KH_2PO_4; 1.2, $(NH_4)_2SO_4$; 1.2, NaCl; 0.25, $MgSO_4 \cdot 7H_2O$; 0.12, $CaCl_2$);

0.1 ml 0.1%, w/v, resazurin solution; 5 ml 8%, w/v, Na_2CO_3; 2 ml 2.5%, w/v, cysteine hydrochloride; water to 100 ml. The broth was equilibrated with, and maintained under, O_2-free CO_2. Ground cellulose powder, cellulose powder (Whatman) or dewaxed Texas cotton fibres were incorporated in the medium at a final concentration of about 0.2%, w/v.

Cellulose hydrolysis—fecal organisms

J. E. Emerson and O. L. Weiser (1963). J. Bact. 86, 891–892.

An agar medium was prepared containing the following: distilled water, 1000 ml; K_2HPO_4, 1 g; $MgSO_4$, 1 g; Na_2CO_3, 1 g; $(NH_4)_2SO_4$, 2 g; and agar, 15 g. (Except for the agar, this is McBeth's cellulose ammonium sulfate medium.) After autoclaving, 12- to 15-ml amounts were poured into sterile petri plates. Portions (0.1 ml) of dilutions of homogenized feces from each subject were streaked over the surface. The plates were incubated at 36°C for 1 hr to dry the surface; then they were overlaid with melted 1.5% agar containing 5% peptone and 1.5% carboxymethylcellulose, incubated at 36°C, and examined daily. Growth was evident in 24 to 48 hr as small crescents or breaks in the overlay agar. By 4 to 5 days, these areas were pitted and easily countable.

Cellulose hydrolysis—*Vibrio marinus*

R. R. Colwell and R. Y. Morita (1964). J. Bact. 88, 831–837.

The authors used the method of the Society of American Bacteriologists 1957. *Manual of Microbiological Methods.* McGraw-Hill Book Co., Inc., New York – modified by the addition of the media of the following salts to produce a synthetic seawater:– sodium chloride, 2.4%; potassium chloride, 0.07%; magnesium chloride (hydrated) 0.53% and magnesium sulfate (hydrated) 0.7%.

A standard inoculum was 1 drop from a pasteur pipette (*c.* 0.05 ml) of a 24 hour artificial seawater broth.

Cultures were incubated at 18°C.

Cellulose hydrolysis—*Hyphomicrobium neptunium*

E. Leifson (1964). Antonie van Leeuwenhoek 30, 249–256.

The liquid medium consisted of peptone (Casitone, Difco) 0.2%; yeast extract 0.1%; tris buffer (tris(hydroxymethyl) amino methane) 0.05%; artificial seawater half strength; pH 7.5. To the liquid culture medium was added a strip of filter paper to determine cellulose decomposition.

Cellulose hydrolysis—*Cellvibrio gilvus*

M. L. Schafer and K. W. King (1965). J. Bact. 89, 113–116.

The culture and media used were those described by Hulcher and King (1958), except that the vitamins were omitted and cellobiose was included at 0.2% (w/v).

F. H. Hulcher and K. W. King (1958). J. Bact. 76, 565–570.

Cellulose hydrolysis—*Streptomyces antibioticus* strain C2A

M. D. Enger and B. P. Sleeper (1965). J. Bact. 89, 23–27.

For production of cellulase, the organism was grown on the following medium: KH_2PO_4, 2 g; NaCl, 3 g; KNO_3, 3 g; $MgSO_4 \cdot 7H_2O$, 0.5 g; $CaCl_2 \cdot 2H_2O$, 0.4 g; $FeSO_4$, 20 mg; $MnSO_4 \cdot H_2O$, 10 mg; water, 1,000 ml; and cellulose paste prepared as described by Hungate (1950), 20 g; pH 7.2. Erlenmeyer flasks (1 liter) containing 350 ml of the above medium were inoculated with 10 ml of cellulose-grown culture and incubated on a rotary shaker at 30°C.

Appreciable cellulase activity occurred in 2 to 7 days. Five electrophoretically distinct active cellulolytic components were isolated from the crude extracellular cellulase system.

R. E. Hungate (1950). Bact. Rev. 14, 1–49.

Cellulose hydrolysis—*Brevibacterium*

R. Chatelain and L. Second (1966). Annls Inst. Pasteur, Paris 111, 630–644.

Cellulolyse: culture en eau peptonée à 0,5 p. 100 contenant une bande de papier filtre (I) [19]. Lecture après 10 et 30 jours.

La température d'incubation des cultures est de 30°C.

[19] Skerman (V. B. D.). *A guide to the identification of the genera of bacteria.* Williams et Wilkins édit., Baltimore, 1959.

Cellulose hydrolysis—fungi

G. S. Rautela and E. B. Cowling (1966). Appl. Microbiol. 14, 892–898.

The cellulose substrate used in the clearing test was prepared according to the procedure of Walseth. Whatman powdered cellulose, swollen in 85% *o*-phosphoric acid for 2 hr at 4°C, was regenerated and washed in the cold by repeated suspending and decanting in distilled water, 1% (w/v) Na_2CO_3, and distilled water again until neutral. A 5-g amount (dry weight) of the resulting suspension of cellulose particles was combined with 2.0 g of $NH_4H_2PO_4$, 0.6 g of KH_2PO_4, 0.4 g of K_2HPO_4, 0.89 g of $MgSO_4 \cdot 7H_2O$, 100 μg of thiamine HCl, 0.5 g of yeast extract, 4.0 mg of adenine, 8.0 mg of adenosine, 17 g of agar, and distilled water to make 1 liter. Sterile, uniformly opaque, vertical columns (3 to 6 cm in height) of this medium were prepared in 18 mm test tubes and inoculated with discs of mycelium and agar, 12 mm in diameter, from petri-plate cultures of the organisms to be tested. Depth of clearing (DC) in the tubes was measured every 7 days for 35 days with a millimeter scale after incubation at 26 ± 1°C and 66 ± 2% relative humidity.

Whatman powdered cellulose and Avicel (a microcrystalline cellulose prepared by treating a very pure α-cellulose with 2.5 N HCl for 15 min at 105°C; American Viscose Corp., Marcus Hook, Pa.) failed in various concentrations to give sufficiently distinct clearing, probably because of their high degree of crystallinity. Walseth cellulose, prepared as described, was used successfully in our tests. Larger values of DC were obtained with concentrations of 0.125% and 0.25% than with 0.50% or 1.0% (w/v) Walseth cellulose. DC values for the same fungi tested in replicate tubes but inoculated and incubated at different times were only slightly more variable than those obtained in tubes inoculated and incubated at the same time. Tests with batches of Walseth cellulose prepared at different times gave different values of DC with the same fungus and incubation time. Thus, comparisons of relative cellulolytic activity should be made for values ues obtained on the same batch of substrate.

An incubation time of 21 days gave optimal sharpness of clearing, but correlation analyses showed that a period of 35 days gave values of DC more closely correlated with weight loss of native cellulose. A substrate concentration of 0.125% or 0.25% was more advantageous than 0.50% or 1.0% (w/v) for fungi of low cellulolytic activity.

C. S. Walseth (1952). TAPPI 35, 228–233.

Cellulose hydrolysis—soil organisms

J. C. Went and F. de Jong (1966). Antonie van Leeuwenhoek 32, 39–56.

Rate of decomposition of cellulose in the soil. The process of cellulose decomposition was estimated by inspecting at intervals cellophane sheets sticking to slides which were buried in the soil. The cellophane (number PT 300) from AKU Ltd., Arnhem, was boiled and washed before it was dried on to the slides. The cellophane-covered slides were buried in a vertical position with only the upper 1 cm sticking out of the soil (Tribe, 1957). In every experiment we buried a row of 10–30 slides in one plane, to place them against an undisturbed soil surface. At first, slides were taken every week from the soil and examined in the laboratory, later at intervals of 5–10 weeks. The results were expressed in numbers, 0 meaning that no cellophane had disappeared, 9, that decomposition was complete.

Microscopic observation of cellophane decomposition in the soil. Cellophane-covered slides could be examined under the microscope when taken from the soil within the first three to four weeks. After 4 weeks too many fungi were growing over the cellophane to study it under the microscope.

Soil-extract agar. In order to study the effect of soil extracts on cellophane decomposition by fungi in pure culture three media with agar were prepared.

Soil extracts were either heat-sterilized or filter-sterilized. The heat-sterilized soil extract was prepared by heating 1 kg of clay soil with 1 liter water at 120°C for 30 min, then adding 0.5 g $CaSO_4$ before filtering, and heat-sterilizing the filtrate. The filter-sterilized soil extract was prepared by extraction at room temperature of 1 kg of clay soil with 1 liter of water overnight, followed by filtering through a bacteriological G5 filter.

We used:

1) Agar with inorganic salts and 5% of heat-sterilized soil extract.
2) Agar with inorganic salts and 50% of heat-sterilized soil extract.
3) Agar with inorganic salts and 50% of filter-sterilized soil extract.

Pieces of sterile cellophane were placed on these agars and inoculated with the fungus under examination.

H. T. Tribe (1957). *Microbial Ecology*, 7th Symposium Soc. Gen. Microbiol. University Press, Cambridge. p. 287–298.

Cellulose hydrolysis—anaerobes

A.-R. Prévot (1966). Techniques pour le diagnostic des Bactéries Anaérobes. Éditions de la Tourelle, St. Mandé.

Milieux de détection et de culture des Cellulolytiques.

Pour les cellulolytiques chimiotrophes, un substrat salin du type Oméliansky est utilisé, auquel on ajoute de la cellulose sous forme de papier-filtre ou sous forme précipitée. Il est bon d'ajouter un peu d'azote organique sous forme d'autolysat de levure. Enfin plusieurs d'entre eux nécessitent un facteur de croissance tel que: liquide de panse pour ceux qu'on veut isoler de la panse des ruminants; extrait de matière fécale humaine pour *Caduceus cellulose dissolvens;* extrait de broyat de doryphore pour *Caduceus leptinotarsae,* etc.

L'essai industriel des substances antiseptiques destinées à protéger les produits cellulosiques finis, se fait de préférence avec les cellulolytiques thermophiles dont le délai de destruction est très court (5 à 10 jours). La cellulose des milieux d'essai est alors une bandelette de tissu de coton écru, stérilisée dans un substratum minéral du type Oméliansky additionné d'un peu d'autolysat de levure. L'incubation se fait à 55°.

CELLULOSE PRODUCTION

Cellulose production—*Acetobacter xylinum*

M. Schramm and S. Hestrin (1954). J. gen. Microbiol. 11, 123–129.

Medium for culture:

Glucose, 2.0% (w/v); Bacto peptone (Difco), 0.5% (w/v); yeast extract (Difco), 0.5% (w/v); agar, 2.5%; pH 6.0 initially.

CO_2 is liberated at the rate of 0.44 μ-mole/mg dry wt of cells/hour.

Cellulose is synthesised by non-proliferating cells from glucose.

Max. production under 100% O_2.

For method of cellulose determination see M. Schramm and S. Hestrin (1954). Biochem. J. 56, 163.

Cellulose production—*Acetobacter xylinum*

B. Millman and J. R. Colvin (1961). Can. J. Microbiol. 7, 383–387. The Formation of Cellulose microfibrils by *Acetobacter xylinum* in agar surfaces.

The formation of extracellular cellulose microfibrils by *Acetobacter xylinum* in agar surfaces is remote from the cell membrane and does not involve an intermediate, amorphous high polymer, in agreement with conclusions from studies of liquid suspensions. Growth of individual microfibrils is at the tip(s) only and the rate of extension (0.2 μ per bacterial cell per minute at 34°C) is comparable with that in liquid medium. The rate of nucleation of new microfibrils is about 40 per 10^3 bacteria per minute at 34°C. Both rates are constant after an induction period of about 30 seconds. Newly nucleated microfibrils could be identified unequivocally down to a length of 0.5 μ. A characteristic feature of growth of cellulose on agar surfaces is the formation of bundles of microfibrils with their axes roughly parallel. The results suggest that the rate-limiting step in the formation of these microfibrils has an activation energy of about 15 kc.

CHITIN HYDROLYSIS

Chitin hydrolysis

A. G. Benton (1935). J. Bact. 29, 449–465.

Chitin was prepared as follows:– Crab shells were scrubbed as free as possible of flesh and dirt and decalcified in cold 1% HCl, which was changed 3 times within a period of about 1 week. The limp leathery shells were washed, cut into pieces of suitable size and shape, and soaked in 2% KOH for 10 days, during which period they were stirred several times and on 3 occasions brought to a temperature just below boiling and then allowed to cool. This should remove protein and other organic matter except chitin; most of the pigment dissolves out in alkali. As traces of pigment would not render the product unsuitable for this work, while any chemical alteration of the chitin itself should be avoided for theoretical reasons, the often recommended complete bleaching by permanganate and bisulphite was omitted. The material was washed free of alkali, extracted 3 times in boiling ethyl alcohol, and dried, after which treatment it appears colorless. In enrichment and cultivation media it is always advisable, and in the case of some strains actually necessary, that the chitin protrude above the liquid, as the most vigorous growth and the first visible disintegration of chitin takes place usually at surface. A strip about 5–7 mm wide and 4–5 cm long placed in a test tube containing about 5 ml of liquid meets this requirement admirably and comparisons of amount of turbidity and presence of a pellicle or sediment are much easier when liquid is not filled with a mass of small irregular bits. Moreover the beginning of chitin destruction is much more easily observed on a piece with a straight entire edge. Crab shells can be bought in the market and are convenient to handle; a number of strips can be cut from each one, and the remaining portions ground fine and used in the plating medium.

The saline fluid used as a basis in all cultures was distilled water containing 0.03% each of K_2HPO_4, $MgSO_4 \cdot 7H_2O$ and NaCl. The tubes used for enrichment and transfer of mixed cultures and for isolation of colonies fished from plates, each contained a strip of chitin nearly covered with the fluid. After autoclave sterilization the pH was 7.6–7.8. Obviously any organism which can grow in pure culture in such a medium must derive its energy and carbon from the chitin and also its nitrogen unless it is able to fix this element from the air. A useful variation is the addition of a trace of peptone. In the case of most of the organisms studied, this hastens the growth of pure cultures. It is not advisable to use it for isolations from plates, however, as it tends to favour the persistence of organisms devoid of chitinase. Apparently these almost always accompany the true chitin destroyers, and mixed colonies are the rule rather than the exception on the first platings.

For obvious reasons the ordinary nutrient agar and gelatin plates used in Rammelberg's and Benecke's method of strewing sterilized ground chitin over the surface of plates, are far from ideal. A plating medium containing finely ground chitin evenly dispersed through a 3% solution of agar in the basic saline solution has proved entirely satisfactory. A dissecting microscope is almost a necessity for fishing colonies, as they are often too small to be seen well with the naked eye, particularly when obscured by chitin particles.

In general, the material to be investigated was inoculated into the simple enrichment medium, transferred to a second tube as soon as growth was evident (usually within 10 days if at all), then to a third tube and plated out. Further transfers before plating were apparently of no advantage. Representative colonies were fished, incubated and replated until the plates appeared to contain but one type of colony. Transplants from such homogeneous plates were regarded as pure cultures unless more than one morphological type appeared in stained smears.

W. Benecke (1905). Bot. Zbl. 63, 227–261.
G. Rammelberg (1931). Bot. Arch. 32, 1–37.

Chitin hydrolysis

C. E. ZoBell and S. C. Rittenberg (1938). J. Bact. 35, 275–287.

Chitinoclastic bacteria were detected by inoculating chitin medium with samples of raw sea water, bottom sediments and other marine materials. The chitin medium was prepared by partly covering 1 x 5 cm strips of purified chitin in test tubes with sea water and sterilizing at 124°C for 20 minutes. The chitin strips were prepared from lobster shells by methods similar to those used by Benton (1935). After successive prolonged treatments with 1 per cent hydrochloric acid, 2 per cent potassium hydroxide, several changes of boiling alcohol and finally water, the chitin was colorless and reacted positively to the tests for chitin listed by Buchanan and Fulmer (1930). Simple nitrogen and carbon compounds were added to some of the media as sources of readily available nutrients. All media were prepared with sea water.

The inoculated media were incubated at 21°C and examined periodically for evidence of chitinoclastic activity.

A. G. Benton (1935). J. Bact. 29, 449–463.
R. E. Buchanan and E. I. Fulmer (1930). *Physiology and Biochemistry of Bacteria.* Williams and Wilkins Co., 1, 98.

Chitin hydrolysis

P. H. Clarke and M. V. Tracey (1956). J. gen. Microbiol. 14, 188–196.

Organisms were grown at Rothamsted on nutrient broth (10 g peptone; 3 g Lemco; 1 g Marmite; 5 g NaCl; made to 1 L with distilled water; pH 7.2) or on nutrient agar. Organisms grown by P. H. C. working at the National Collection of Type Cultures (NCTC) were grown in nutrient broth (10 g peptone; 10 g Lemco; 5 g NaCl; water to 1 L; pH 7.4) with the exception of the clostridia which were grown in meat broth containing particles of meat. Cultures were incubated with toluene overnight before use.

Chitin. The shells of *Sepia officinalis* were used as a source (they are commercially available as cuttlefish "bones," imported usually from Portugal or India where they are gathered in a dry bleached condition from the beaches). After soaking in dilute HCl each shell disintegrates into one thick chitinous layer (the pro-ostracum) and a large number of chitinous sheets (about 7 μ thick) with irregular surfaces. The yield from the shells is about 3%, and about two-thirds of the dry weight of the sheets is chitin. This material was dispersed in cold conc. HCl and, after centrifugation or filtration through glass wool, poured into 10 vol of water, with vigorous stirring. The precipitated finely divided chitin was washed and concentrated by decantation and a thick suspension finally dialysed against running distilled water. A suspension containing 100 μg chitin N/ml was used as the stock substrate and preserved with toluene.

Chitinase. After incubation overnight with toluene cultures were centrifuged and 2 ml of the supernatant mixed with 2 ml of pH 5 acetate buffer (0.2 M) and 1 ml of chitin suspension in centrifuge tubes. Controls of the same total volume containing culture fluid without chitin, and with chitin and buffer alone were also prepared. To all tubes a few drops of toluene were added, the tubes stoppered and the contents mixed. After incubation at 37°C for 1, 3 and 8 days the tubes were centrifuged and the acetylglucosamine content of the supernatant fluid determined. When the organisms produced indole a determination before incubation with chitinase was made. The presence of acetylglucosamine was detected by a modification of the Morgan and Elson (1934) method (Tracey, 1955). In this method a color is formed by the reaction of p-dimethylaminobenzaldehyde (Ehrlich reagent) with the product formed by heating acetylglucosamine in alkaline solution. A color is also produced in this reaction by acetylgalactosamine and by blood group mucoids (Aminoff, Morgan and Watkins, 1952) but not by the disaccharide (N,N′-diacetylchitobiose) which is the simplest product, other than acetylglucosamine, of the incomplete enzymic hydrolysis of chitin (Reynolds, 1954; Kuhn, Gauhe and Baer, 1954). Under the conditions used indole gives a very similar color (indole color absorbs maximally at 565 mμ; acetylglucosamine color at 580 mμ) with about 2.5 times the intensity of that given by acetylglucosamine of equimolar concentration. Thus the appearance of a pink color after the reaction has been carried out indicates the presence of free acetylglucosamine, indole or possibly breakdown products of chitin other than N,N′-diacetylchitobiose. In centrifuged fluids from toluene-killed cultures there was no further production of indole on incubation while in the presence of chitinase the production of material which reacted with Ehrlich reagent after alkaline heating was continued for many days. Only in the presence of indole was any color produced in the controls. Consequently, chitinase may be detected in culture fluids from indole-producing organisms since in its presence there is a steady increase in its color produced in the presence of chitin but no increase in its absence. It seems probable that the material responsible for increased color production as a result of incubation with chitin is acetylglucosamine alone, though it is possible that large fragments resulting from the breakdown of chitin may react in a manner similar to the blood group mucoids. Should this be so, color production is still indicative of chitin breakdown and hence of the presence of chitinase, for this term may apply to a number of enzymes of different properties with different end products of hydrolysis.

In the absence of indole the presence of 0.25 μg acetylglucosamine -N can be detected in 1 ml of fluid after incubation with chitin corresponding to the breakdown of a total of 18 μg chitin (1.25 μg chitin-N). In the presence of indole a rather greater breakdown must have occurred for a positive result to be recognized.

D. Aminoff, W. T. J. Morgan and W. M. Watkins (1952). Biochem. J. 51, 379.
R. Kuhn, A. Gauhe and H. H. Baer (1954). Ber. dt. chem. Ges. 87, 1138.
W. T. J. Morgan and L. A. Elson (1934). Biochem. J. 28, 988.
D. M. Reynolds (1954). J. gen. Microbiol. 11, 150.
M. V. Tracey (1955). *Modern Methods of Plant Analysis* ed. K. Paech and M. V. Tracey, 2, 264.

Chitin hydrolysis

V. B. D. Skerman *A Guide to the Identification of the Genera of Bacteria.* 2nd ed. Williams and Wilkins Book Co., Baltimore, 1967.

Preparation of chitin:

Scrub and clean crab or lobster shells and soak them for 1 week in 1 per cent HCl; change the acid three times during this period. The acid decalcifies the shells which should become limp and leathery. Wash the shells well with water and then cut them into strips .5 to 1 inch wide.

Soak the strips in 2 per cent KOH for 10 days. During this period heat five times to a temperature just below the boiling point and then cool. This treatment extracts some pigment and removes protein and other organic matter except chitin.

Wash the strips free of alkali and extract with several changes of ethyl alcohol until all color is removed. Dry.

The strips are suitable for use in a liquid mineral salts medium for the growth of crude cultures and maintenance of chitinoclastic organisms.

For use in agar media the chitin must be finely dispersed. Take some of the strips and place them in a 3.0-L flask. Hold 1500 ml of cold water in readiness in a second flask. Add 100 ml of 50 per cent H_2SO_4 to the chitin strips.

As soon as the chitin disappears, dilute the acid rapidly with the 1500 ml of water. Allow the reprecipitated chitin to settle overnight, decant the supernatant, and wash the chitin by repeated centrifugation until neutral to litmus. Alternatively, dialyse the chitin free of acid by suspending it in a sheet of nonwater-proofed cellophane in a stream of running water.

Resuspend the chitin in water to a density such that after diluting 1:10, 5 ml dispensed in a 10-cm petri dish, will give a slight but distinct opalescence.

Distribute in 10-ml quantities and sterilize at 121°C for 15 minutes.

Preparation of Media:

1. Chitin strip-mineral salts liquid medium

K_2HPO_4	1.0 g
$MgSO_4 \cdot 7H_2O$	0.5 g
NaCl – fresh, 0.5 g; marine,	30.0 g
$CaCl_2 \cdot 2H_2O$	0.1 g

$FePO_4$	0.001	g
NH_4Cl	1.0	g
H_2O	1000	ml

Dissolve the mineral salts, adjust the pH to 7.0, and distribute in bottles or test tubes. Add a strip of chitin to each container and sterilize at 121°C for 15 minutes.

2. *Chitin agar*

(a) Prepare a mineral salts agar base by the addition of 1.5 to 2.0 per cent agar to the mineral salts medium in 1 above. Dissolve the agar by autoclaving, adjust the pH to 7.0, and distribute in 5- and 10 ml amounts in suitable containers. Sterilize at 121°C for 15 minutes.

(b) To prepare chitin agar plates, melt one 10-ml and one 5-ml mineral salts agar deep and cool to 45°C in a water bath. Pour the 10-ml quantity into a sterile petri dish and allow to set.

To the 5-ml quantity add 0.5 ml of the warmed chitin suspension, mix well, and then pour as a layer on the surface of the mineral salts agar base.

This practice minimizes the time taken for chitinoclastic organisms to bring about a visible change in the suspended chitin in the medium.

Isolation of chitinoclastic bacteria

To obtain enrichment cultures, inoculate soils or marine muds into the mineral medium containing the chitin strips. Incubate at the desired temperature until obvious decomposition has occurred. Subculture to another tube of the same medium and reincubate. When decomposition becomes rapid, inoculate the surface of a chitin agar plate, and pick off the colonies of chitin-hydrolysing organisms. Purify in the usual manner and maintain in the liquid medium; subculture when the strips of chitin have nearly disappeared.

A. G. Benton (1935). J. Bact. 29, 449.
L. L. Campbell, Jr. and O. B. Williams (1951). J. gen. Microbiol. 5, 894.

Chitin hydrolysis—*Pseudomonas* and marine organisms

J. Brisou, C. Tysset, Y. de Rautlin de la Roy, R. Curcier and R. Moreau (1964). Annls Inst. Pasteur, Paris 106, 469–478.

La recherche de chitinolyse sur des fragments de chitine purifiée introduits dans des eaux peptonées nécessite une très longue observation, la désintégration est lente et souvent difficile à lire.

En revanche, la préparation de chitine colloïdale est à la portée de tout laboratoire. Les résultats peuvent être lus dès les premiers jours.

Il y a quelques années on préparait la chitine brute à partir de carapaces de crustacés selon une technique mise au point par Benton. Actuellement, il y a avantage à faire usage des chitines blanches livrées par le commerce. Cette matière première est dissoute soit dans l'acide chlorhydrique concentré, soit dans l'acide sulfurique à 50 p. 100. Une fois la chitine solubilisée, on la précipite dans de l'eau distillée froide (refroidie à + 4°). On lave ensuite le précipité toujours à froid, à grande eau dans un cristallisoir, puis sur un filtre mouillé pour éliminer l'excès d'acide. On neutralise à la soude et on rince de nouveau à l'eau, soit sur un filtre, soit par centrifugation.

La pâte de chitine obtenue est remise en suspension dans de l'eau neutre et stérilisée par trois chauffages successifs à 80°. On répartit dans des petits flacons de 20 ml. Cette chitine colloïdale est émulsionnée dans l'eau et on la conserve au frigorifique à + 4°. Cette émulsion titrée à environ 5 à 6 mg de chitine par ml sera incorporée à la gélose nutritive au moment de l'emploi selon la technique suivante.

On dispose d'une gélose nutritive ainsi composée:

Peptone	0.2 a 0.5	g
$K_2H\ PO_4$	0.005	g
Extrait de levure	0.1	g
Agar	2.5	g
Eau douce ou eau de mer	100	ml

pH = 7.3.

Stériliser à 120° pendant quinze minutes.

On verse quelques gouttes de chitine colloïdale dans une boîte de Petri de 10 cm de diamètre. On coule immédiatement la gélose fondue et on égalise le mélange aussi rapidement que possible. On laisse solidifier, et l'on fait évaporer à l'étuve l'excès d'humidité. Les boîtes sont alors prêtes à l'emploi.

Chitin hydrolysis—*Vibrio marinus*

R. R. Colwell and R. Y. Morita (1964). J. Bact. 88, 831–837.

The authors used the method of A. G. Benton (1935). J. Bact. 29, 449–465; M. S. Pohja (1960). Acta agralia Fennica 96, 1–80 – modified by the addition to the media of the following salts to produce a synthetic seawater:– sodium chloride, 2.4%; potassium chloride, 0.07%; magnesium chloride (hydrated) 0.53% and magnesium sulfate (hydrated) 0.7%.

A standard inoculum was 1 drop from a pasteur pipette (*c.* 0.05 ml) of a 24 hour artificial seawater broth.

Cultures were incubated at 18°C.

Chitin hydrolysis—RM bacterium

A. J. Ross, R. R. Rucker and W. H. Ewing (1966). Can. J. Microbiol. 12, 763–770.

The method used for determination of chitinase also was mentioned by Skerman (1959).

V. B. D. Skerman (1959). *A guide to the identification of the genera of bacteria.* Williams & Wilkins Book Co., Baltimore, Md.

CHOLERA-RED REACTION

Cholera-red reaction—*Vibrio*

J. Gallut (1948). Annls Inst. Pasteur, Paris 74, 27–39.

The author has made a detailed study of the factors affecting the Cholera-Red test, and gives the following Résumé:

Le dosage séparé des deux facteurs (nitrites et indol) de la réaction du C.-R. nous a permis de préciser l'horaire optimum de la recherche de cette réaction en différentes conditions de culture du vibrion cholérique. Le cas des milieux glucosés a été examiné particulièrement: le mécanisme du C.-R. y est simple et ne requiert pas d'autre explication que celle de la pré-utilisation de glucose qui doit être totale, ceci n'étant possible que dans certaines limites.

Compte tenu de cette restriction, il est toujours possible de mettre en évidence le C.-R. dans une culture de *V. cholérique* en activité à un instant déterminé qui varie suivant le mode (aérobie, anaérobie ou aéré) de cette culture.

Cholera-red reaction—*Vibrio*

P. Hauduroy (1951). Techniques Bactériologiques. Masson et Cie, Éditeurs, Paris.

EAU PEPTONEE POUR LA RECHERCHE DE LA REACTION INDOL-NITREUSE

Peptone de Witte	10 g
Nitrate de potassium	1 g
NaCl	3 g
Eau distillée	1 litre

Cholera-red reaction—*Vibrio*

G. H. G. Davis and R. W. A. Park (1962). J. gen. Microbiol. 27, 101–119.

Cholera-red reaction, tested by adding concentrated H_2SO_4 (0.5 ml) to a 5 day old culture in peptone water containing 0.001% (w/v) KNO_3 (Beam, 1959).

W. E. Beam (1959). J. Bact. 77, 328.

CHOLINESTERASE

Chlorinesterase—*Pseudomonas fluorescens*

D. B. Goldstein and A. Goldstein (1953). J. gen. Microbiol. 8, 8–17.

Grown in mineral medium consisting of KH_2PO_4, 0.1%; NH_4Cl, 0.1%; $MgSO_4$, 0.05%; pH 6; 0.1% acetylcholine, (isolated from fermenting cucumber).

Esterase determination was done manometrically at 27°C in atm. of 95% N_2 and 5% CO_2. Intact cells or acetone dried powder were suspended in bicarbonate Ringer's solution (NaCl, 9.0 g; KCl, 0.3 g; $CaCl_2$, 0.25 g; $NaHCO_3$, 2.1 g/L) 0.2 ml acetylcholine or other ester in the side arm.

The enzyme was not the same as animal cholinesterases.

It was sensitive to prostigmine, not inhibited by excess substrate and had a unique substrate specificity.

The esterase hydrolyses propionyl, acetyl and acetyl-β-methyl esters of choline in decreasing order of activity. It does not hydrolyse butyryl and benzoyl esters.

It hydrolyses triacetin and tributyrin.

COAGULASE

Coagulase—*Staphylococcus*

G. H. Chapman, C. Berens, A. Peters and L. Curcio (1934). J. Bact. 28, 343–363.

One loopful of the 24 hour culture on solid medium was mixed with 0.5 ml of citrated human plasma and incubated 3 hours at 37°C. The tubes were inspected at ½, 1, 2 and 3 hours.

Coagulase—*Staphylococcus*

E. Neter (1937). J. Bact. 34, 243–254.

Serial dilutions of the supernatant of plain broth cultures (18 hours at 37°C) in the volume of 0.5 ml were mixed with 0.25 ml of plasma (10 ml human, rabbit or guinea pig blood and 1 ml of 2% potassium oxalate solution shaken thoroughly and centrifuged) and incubated at 37°C and read after 1 hour.

Coagulase—*Staphylococcus*

R. Thompson and D. Khorazo (1937). J. Bact. 34, 69–79.

Oxalated plasma was obtained by placing 10 ml of fresh human blood into a centrifuge tube containing 0.02 g of potassium oxalate, thoroughly mixing, and then centrifuging out the cells. Then 0.3 ml of the plasma was mixed with 0.2 ml of an 18–24 hour broth culture of the organism to be tested. Partial or complete clotting after 2 hours' incubation at 37°C indicated the production of coagulase.

Coagulase—*Staphylococcus*

R. W. Fairbrother (1940). J. Path. Bact. 50, 83–88.

There is no standard method for carrying out the coagulase test; citrated human plasma undiluted or diluted from 1:2 to 1:10 with saline or broth has been used by different workers. After various trials, a fixed dilution of one part of citrated plasma (0.5–1.5 per cent citrate) and two parts of physiological saline was used in this work. This was inoculated with either a loopful of a 24-hour growth on agar or blood agar, or with several drops of a 24-hour culture in glucose broth and incubated at 37°C; readings were made at hourly intervals for 4–6 hours and finally after 18–24 hours.

Coagulase—*Staphylococcus*

G. H. Chapman, C. Berens and M. H. Stiles (1941). J. Bact. 41, 431–440.

The following conditions should give more constant results. Fresh rabbit plasma diluted 1:4, or a suitable dilution of human plasma, with a minimum amount of anticoagulant, should be used. The culture should be grown on a suitable medium, such as 1 per cent lactose agar made with proteose peptone, only long enough to get sufficient growth for the test (3 to 6 hours, depending on the heaviness of the inoculum). The mixture of plasma and culture should be incubated at 37°C and inspected hourly for 4 hours and then kept on the laboratory table and inspected again the following day. Plastridge (personal communication) found that some cultures did not clot plasma until 18 to 24 hr under these conditions. Experiments showed that the plasma dried on blotting paper loses little or none of its clotting power after many months in the refrigerator.

The papers are prepared as follows: Pipette 0.3 ml of sterile, fresh, oxalated rabbit plasma into a series of sterile 100 x 12 mm test tubes. Drop a piece of sterile blotting paper 5 x 25 mm into each tube. Put the tubes upright in a large culture dish and put it in a desiccator containing the desiccant (Dessigel S, Central Scientific Company). Exhaust until the tubes froth. Seal, bleed, shut off the pump, disconnect the desiccator, and set it aside until the following day. Then open the desiccator and take out the tubes. Tap them gently on a hard surface to loosen the papers from the sides of the tubes. Slide the papers into sterile vials, putting several into each vial. Stopper the vials and store them in a refrigerator. Do not keep a cold vial open too long, or moisture will condense in it and the strips will not remain dry.

The following method has given constant results in testing the plasma-coagulating power of staphylococci.

Fresh plasma: centrifuge fresh human or rabbit blood, with a minimum of anticoagulant, under aseptic precautions. Transfer the plasma to small sterile bottles and keep them in the refrigerator. When ready for use, put 0.2 ml into a small test tube and add 0.6 ml of 0.4 per cent NaCl.

Dried plasma: put one of the papers, torn in two pieces, into a 100 x 12 mm test tube, add 0.5 ml of water and let stand 30 minutes in the incubator to redissolve the plasma. Remove the paper from the tubes.

The resulting solution is then ready for use.

Pour 1 per cent lactose agar made with proteose peptone, Difco, into petri dishes to a depth of 5 mm. Inoculate the culture on the surface of the medium, spread with a glass spreader, and incubate until a faint growth is visible (3 to 6 hours).

Put a loopful of the growth into a tube of diluted, fresh or dried, plasma and disperse the bacterial mass as much as possible by shaking vigorously. Put the mixture in the incubator at 37°C, keeping the door open as little as possible. Remove the tubes from the incubator in 1 hr and tilt them to see whether any of them have clotted. In the majority of instances of clotting the tubes can be inverted without loss of the mixture. In a few instances the clot is not so firm and appears as a jelly-like mass which floats when the tube is tilted. Tubes which show no evidence of clotting are returned to the incubator quickly and reexamined in 2, 3, or 4 hours after they were first put in. If the results are still negative the tubes may be left on the laboratory table and reexamined up to 24 hours later. If no clot forms by that time the result is negative.

Coagulase—*Staphylococcus*

J. B. Penfold (1944). J. Path. Bact. 56, 247–250.

The author describes a solid medium consisting of a 2.5% agar base with 25 per cent human plasma incorporated. Thin plates are poured and can be stored in a refrigerator for at least 14 days. Opaque rings round the colonies develop in 96 per cent of coagulase positive and no coagulase negative strains. Opacity underneath the colonies is not significant.

Coagulase—*Staphylococcus* and *Micrococcus*

Y. Abd-el-Malek and T. Gibson (1948). J. Dairy Res. 15, 249–260.

Rabbit plasma was used in the method recommended by Chapman, Berens and Stiles (1941). The majority of the strains were also examined by a rapid slide test (Berger), using fibrinogen prepared from rabbit plasma. The two methods gave the same result in every instance in which they were compared. The use of whole plasma in a slide test was also tried, but was given up because of the agglutination of occasional cultures which were negative in the tube and the fibrinogen tests.

F. M. Berger (1943). J. Path. Bact. 55, 435.
G. H. Chapman, C. Berens and M. H. Stiles (1941). J. Bact. 41, 431.

Coagulase—Gram-negative rods

E. M. Harper and N. S. Conway (1948). J. Path. Bact. 60, 247–251.

0.5 ml of human citrated plasma diluted 1/1 to 1/16 with saline was inoculated with two drops of 24-hour broth cultures of various Gram-negative organisms and incubated at 37°C. The results were read at intervals: within 18 hours soft clots had formed in dilutions to 1 in 8.

Accordingly, all subsequent experiments were carried out with plasma diluted 1 in 2.

The following three experiments were carried out.

(i) Concentrations of citrate ranging from 0.25 to 5 per cent were prepared in plasma diluted 1 in 2 with saline and inoculated with a strain previously shown to clot the routine citrated plasma. After 18 hours' incubation clots formed in tubes up to and including 1% citrate, but never above this concentration, although there was good growth.

(ii) The organisms were grown in oxalated plasma; no clot formed.

(iii) Use was then made of plasma passed through a Seitz pad (such a plasma does not clot on recalcification but coagulates on addition of thrombin, showing that fibrinogen is present) and plasma in which the thrombin mechanism was inactivated by heparin (2.5 Toronto units per ml). In both cases the plasma failed to clot when inoculated with the organism under investigation.

These findings indicate that clotting of plasma by bacteria may be due either to the production of a coagulating enzyme or, as in the present instance, to the breakdown of the anti-coagulant – citrate. Accordingly coagulase tests, which aim at revealing the presence of *Staphylococcus aureus* by demonstrating the action of its clotting enzyme, should always be performed with heparinized plasma. If, as is usual, only citrated plasma is available, heparin should be added.

Coagulase—*Staphylococcus*

J. B. Evans and C. F. Niven, Jr. (1950). J. Bact. 59, 545–550.

Only young, vigorous cultures were used. Stock cultures were transferred daily in broth for at least 3 days before testing. This procedure, however, was omitted for those strains that had recently been isolated from foods. For the test 2 drops of a 24-hour culture were mixed with 0.5 ml of citrated rabbit plasma in a 9 by 75 mm vial and incubated for 3 hours at 30°C. Lyophilized plasma was reconstituted with distilled water for the test. When unknown cultures were tested, a known coagulase-positive strain was included for control purposes.

Coagulase—*Staphylococcus*

C. Shaw, J. M. Stitt and S. T. Cowan (1951). J. gen. Microbiol. 5, 1010–1023.

Equal volumes of 24 hr broth culture were added to 1/10 dilutions of oxalated plasma. Readings were made after incubation at 37°C for 4 hr and after standing overnight at room temperature.

Coagulase—*Staphylococcus*

P. Hauduroy (1951). Techniques Bactériologiques. Masson et Cie, Éditeurs, Paris.

ACTION COAGULANTE SUR LE PLASMA

(Staphylocoque). – Prélever du sang humain (10 cm^3) en présence de 0.02 g d'oxalate de potassium. Centrifuger 20 minutes à 6000 tours/minute. Recueillir le plasma dans un tube à hémolyse, mettre 0.3 cm^3 de plasma et 0.2 cm^3 de filtrat ou de culture à éprouver. Porter à l'étuve à 37°. Observer après 30 minutes, 1 heure, 2, 4, 24 heures et noter le moment où la coagulation apparaît (technique de Kourilsky et Mercier). Le plasma peut être conservé pendant 5 à 6 semaines. Le plasma de lapin peut être utilisé.

(C. C. T. M.)

Coagulase—*Staphylococcus*

J. Marks (1952). J. Path. Bact. 64, 175–186.

In the tests, mixtures of one volume of an overnight broth culture with five volumes of a 1 in 10 saline dilution of human plasma were incubated in a 37°C water bath for 6 hours. No more positive reactions appeared after keeping tubes over night at room temperature. Quantitative coagulase tests were performed in the same way using dilutions of broth culture with a constant concentration of plasma, except that readings were made after 4 hours' incubation and complete coagulation only was recorded as a positive result. A single batch of plasma was used for all the quantitative tests.

Coagulase—*Pseudomonas (Malleomyces) pseudomallei*

P. de Lajudie and E. R. Brygoo (1953). Annls Inst. Pasteur, Paris 84, 509–515.

Coagulase:– La technique utilisée fut celle recommandée par Pochon (Dumas, 1951) pour la recherche de la staphylocoagulase.

J. Dumas (1951). *Bactériologie médicale,* Flammarion, Paris.

Coagulase—gram negative intestinal bacilli

R. Mushin and V. J. Kerr (1954). J. gen. Microbiol. 10, 445–451.

False positive tube tests can arise through metabolism of citrate of citrated plasma, and liberation of Ca^{++} and not through a coagulase enzyme. Slide tests were uniformly negative.

Methods: Rabbit and human plasmas were tested. Rabbit blood was collected aseptically by cardiac puncture and mixed in the proportion 9 ml blood with 1 ml of 10% (w/v) sodium citrate (final concentration 1%). Human plasma was obtained from a blood bank and contained 4 vol blood to 1 vol of 2% (w/v) sodium citrate (final concentration 0.4%). In tests with heparinized plasma 4 drops of the anticoagulant were added to 20 ml citrated human plasma. The citrate media used included Koser's (1924) ammonium citrate solution and citrate in Bacto-peptone basal medium (Kauffmann, 1951). Slide coagulase tests were performed with undiluted citrated rabbit plasma. For tube coagulation experiments rabbit plasma was diluted 1/10 or 1/5 with normal saline; the human plasma was diluted 1/4 or 1/2 or used undiluted. In tube tests 0.5 ml of appropriate plasma was added to 0.5 ml of an 18 hr culture of the required organism in Hartley's tryptic broth. The results were recorded after 24 hr incubation at 37°C; negative tubes were kept under observation for 3 days. Each batch of plasma was tested for clotting capacity by adding 1 drop of 5% calcium chloride to 0.5 ml plasma. Control slide and tube coagulase tests were included in which known coagulase-positive and coagulase-negative strains of staphylococci were used.

F. Kauffmann (1951). *Enterobacteriaceae.* Copenhagen: Ejnar Munksgaard.
S. A. Koser (1924). J. Bact. 9, 59.

Coagulase—*Staphylococcus*

E. S. Duthie (1954). J. gen. Microbiol. 10, 427–436.

Free Coagulase. The free coagulase was a lyophil dried preparation (16,000 M.C.D./mg) from the "Newman" strain of *Staphylococcus aureus* (Duthie and Lorenz, 1952) and was normally used at a concentration of 0.5 or 1.0 mg/ml. Tests were made by adding 0.2 ml to 0.2 ml of a suitable coagulable material in tubes 5.0 x 0.8 cm. The tubes were held against a black background in a warm illuminated box, and tapped until a clot appeared, the time being measured. For times up to 3 min the reciprocal of the coagulase concentration plotted against the clotting time gave a straight line (Duthie, 1954).

Bound Coagulase. Unless otherwise stated, the strain used was "Newman," previously described as a good producer of free coagulase (Duthie and Lorenz, 1952). The cells from shaken digest broth cultures were washed in saline and resuspended to a standard density (4 x 10^9 cells/ml); 0.2 ml of this suspension was added to 0.2 ml of diluted plasma or to 0.2% bovine fibrinogen in 5.0 x 0.8 cm tubes and shaken in a special rack in a Kahn shaker. Against a black background, using direct lighting and a lens, the degree of clumping could be graded quantitatively and was recorded as –, tr., ±, +, + ± or ++.

E. S. Duthie and L. Lorenz (1952). J. gen. Microbiol. 6, 95.
E. S. Duthie (1954). J. gen. Microbiol. 10, 437.

Coagulase—*Staphylococcus aureus*

H. J. Rogers (1954). J. gen. Microbiol. 10, 209–220.

The technique which is commonly employed for estimating the potency of coagulase by finding the dilution of the supernatant which just fails to coagulate either suitable plasma, or a combination of fibrinogen and suitable "activator," was not sufficiently precise for the purposes of the present investigation. Therefore, coagulase was measured by the time taken by dilutions of the culture supernatants to clot fresh, whole, pooled human plasma under standard conditions. Clotting time was estimated by dropping ballotini glass balls, no. 12 (Chance Bros., Ltd., Birmingham), into a tube containing the plasma-culture supernatant mixture. The tubes were incubated at 37 ± 0.5°C in an ordinary glass-sided flocculation bath with a strong light on the side opposite the operator and focused on the tubes. When clotting occurred, the brilliantly refractile balls could be seen to stop abruptly in their downward course. A plot of the logarithm of the time taken to clot the plasma against the logarithm of the volume of supernatant used, gave a satisfactory straight line, provided the clotting time was between 1 and 10 min. This double logarithmic relationship was first observed for thrombin clotting by Fischer (1935). An arbitrary unit, the "clot," was defined as the volume of supernatant which, when adjusted to 0.5 ml with normal physiological saline, would clot 1.0 ml of human plasma in 240 sec at 37° in a tube 0.5 cm in diameter, agitated by constant tapping.

A. Fischer (1935). Biochem. Z. 278, 320.

Coagulase—*Staphylococcus*

C. H. Lack and D. G. Wailling (1954). J. Path. Bact. 68, 431–443.

Strains were originally selected by the slide method of Cadness-Graves *et al.* (1943), but all were subsequently tested on plasma-agar plates. Human plasma from blood rejected by the blood bank was mixed with nutrient agar in a concentration of 12–15 per cent (v/v) of plasma at 46°C. The mixture was poured into Petri dishes and spot-inoculated. Coagulase production was indicated by the development of opacity around spots after over night incubation. All slide-positive strains were also plate-positive. One hundred of these strains were also tested by Fisk's tube method (Fisk, 1940) and found to be tube-positive.

B. Cadness-Graves, R. Williams, G. J. Harper and A. A. Miles (1943). Lancet 1, 736.
A. Fisk (1940). Br. J. exp. Path. 21, 311.

Coagulase—*Staphylococcus*

F. S. Thatcher and W. Simon (1956). Can. J. Microbiol. 2, 703–714.

The criterion for the production of coagulase was the formation within six hours at 37°C of a firm gel in 0.5 ml of rabbit plasma (Difco) after addition of the cells scraped from a "slant" culture on nutrient agar incubated at 37°C for 24 hr.

Coagulase—*Streptococcus*

M. Wood (1959). J. gen. Microbiol. 21, 385–388.

Some strains of Group D streptococci are capable of coagulating citrated rabbit plasma. This reaction is not due to citrate utilization, but to a specific agent elaborated by the organisms; this agent is probably enzymic in nature. The addition of 5 units heparin/ml to the citrated rabbit plasma inhibits coagulation by streptococci and makes the reaction more specific for staphylococci.

Citrated rabbit plasma was prepared by mixing equal volumes of a 4% (w/v) sodium citrate solution and whole blood; thus an approximately ½ dilution of the plasma was obtained, in which the citrate concentration was between 2 and 3%. The inoculum used was 4 drops of a suspension washed from an overnight culture on a blood agar slope by nutrient broth and this was added to 0.5 ml of 1/2, 1/6 and 1/12 dilutions of the plasma in saline. The tubes were observed for production of a solid clot after fixed time intervals (½, 2, and 5 hr) of incubation in a water bath at 37°C and then after 24 hr at room temperature.

Coagulase-positive, late positive and coagulase-negative staphylococcal strains were used as controls and a plasma control with nutrient broth was incubated with each test. About half of the 29 streptococcal strains tested caused plasma to clot within 24 hours and occasionally within 5 hours.

Coagulase—*Staphylococcus aureus*

W. J. Fahlberg and J. Marston (1960). Coagulase production of *Staphylococcus aureus.* I. Factors influencing coagulase production. J. infect. Dis. 106, 111–115.

Titration of Coagulase. Human plasma from outdated citrated blood-bank blood was withdrawn and stored in individual bottles at room temperature. Aliquots of the plasma were diluted 1:5 with M/100 phosphate-buffered saline at pH 7.4 just prior to addition to assay tubes.

Serial dilutions of the supernatants of broth cultures in 0.5 ml amounts were mixed with an equal volume of the diluted plasma, and the greatest dilution of coagulase which would clot the plasma in 6 hours at 37°C was recorded.

Culture mediums and methods. A solution of trace elements was prepared according to Tager and Hales (1947). Bacto Brain Heart Infusion Difco (BHI) was prepared according to the manufacturer's directions and trace elements added in the amount of 1 ml per L of infusion.

The above medium and several variations of it were tested.

M. Tager and H. B. Hales (1947). Yale J. Biol. Med. 20, 41–49.

Coagulase—*Staphylococcus*

H. A. Tamimi (1961). Can. J. Microbiol. 7, 674–675.

The test was performed by pipetting 0.5 ml of the material tested into a 75 x 13 mm (Wassermann) tube. Each tube received 0.5 ml of normal rabbit plasma diluted 1:4 with saline, and was incubated at 37°C for 24 hours.

Coagulase—*Staphylococcus*

R. D. Comtois (1962). Can. J. Microbiol. 8, 201–211.

Three drops of an overnight culture in Trypticase Soy Broth (BBL) were mixed into 0.5 ml of a freshly prepared dilution of dried coagulase (supplied by Hyland Laboratories, Los Angeles, Calif.) and the mixture incubated in a 37°C water bath and examined for clotting after 1 to 2 hours. Reactions negative after 4 hours were re-examined after 24 hours.

Coagulase—*Staphylococcus* and *Micrococcus*

D. A. A. Mossel (1962). J. Bact. 84, 1140–1147.

Lyophilized rabbit plasma (Difco) was used for all tests, which were carried out following the directions of the manufacturer. In all tests a coagulase-positive and coagulase-negative reference strain were used as controls. In over 80 series of tests, not a single erratic result was obtained.

Coagulase—*Staphylococcus*

S. Fletcher (1962). J. Path. Bact. 84, 327–335.

Serial doubling dilutions of the solution under test were made in 0.5 ml volumes of a diluent consisting of 70 parts saline solution, 25 parts broth and 5 parts 5 per cent "Merthiolate" solution. To these dilutions were added amounts of a 0.1 per cent solution of bovine fibrinogen (Armour) in the above diluent with 1 per cent of fresh human plasma as activator. The coagulase titers were determined after 12 hrs incubation and were expressed as minimum clotting doses (MCD) per ml, as described by Lominski and Roberts (1946). When the solution to be titrated contained serum, 1000 units of heparin (Roche) were added to each 100 ml of fibrinogen solution.

I. Lominski and G. B. S. Roberts (1946). J. Path. Bact. 58, 187.

Coagulase—*Staphylococcus*

K. J. Harrison (1963). J. Path. Bact. 85, 341–348.

The authors described a method for the bulk production and assay of coagulase.

Coagulase—*Staphylococcus aureus*

R. L. Brown and J. B. Evans (1963). J. Bact. 85, 1409–1412.

The coagulase test was conducted by mixing 0.1 ml of a 24-hr culture with 0.5 ml of reconstituted Difco Coagulase Plasma in 10-mm test tubes and incubating for 3 hr.

Coagulase—*Staphylococcus*

M. C. Drummond and M. Tager (1963). J. Bact. 85, 628–635.

The mechanism of fibrinogen clotting by staphylocoagulase and the coagulase-reacting factor (CRF) was studied from the standpoint of the products of the reaction, and compared with thrombin clotting of fibrinogen. Both clotting principles effect clotting with the release of non-protein nitrogenous material, which can be resolved into three peptides by Dowex-50 chromatography. On the basis of the sedimentation constants, electrophoretic mobilities, and end-group analyses of these peptides, it appears that the products of fibrinogen clotting by coagulase-CRF and by thrombin are similar.

Coagulase—*Staphylococcus*

J. W. McLeod (1963). J. Path. Bact. 86, 35–53.

Coagulase tests were done with human plasma obtained from blood to which heparin and bovine fibrinogen, or EDTA and dextran, had been added. Mixtures of equal parts of 20 per cent plasma in saline and 24-hr broth cultures of the staphylococci were incubated and observed up till 6 hr. Except in a few instances these results coincided with observations on coagulase production obtained by Dr. J. C. Gould using rabbit plasma in tests by the method suggested by Cruickshank *et al.* (1960).

R. Cruickshank *et al.* (1960). Mackie and McCartney's *Handbook of bacteriology,* 10 ed., Edinburgh and London, p. 469.

Coagulase—*Staphylococcus*

L. M. Carantonis and M. S. Spink (1963). J. Path. Bact. 86, 217–220.

The slide plasma clumping test was carried out by thickly emulsifying the strain in fresh, citrated human plasma diluted 1 in 10 in distilled water. The tube coagulase test was set up at the same time; some of the culture was inoculated into a tube containing 10 drops of nutrient broth; 2 drops of the same 1 in 10 dilution of plasma were added and the tubes were incubated for 3 hr at 37°C and then left at 22°C overnight. All strains that produced a coagulum, at either 37° or 22°C, were recorded as coagulase-positive (CP).

Coagulase—*Staphylococcus*

D. D. Smith (1963). J. Path. Bact. 86, 231–236.

All strains were tested by the tube method of Fisk against human and mouse plasma (Gorrill). The inoculated tubes were incubated for 3 hr at 37°C and then observed after they had stood overnight at room temperature; note was taken of the quality of clot formed, whether firm and solid ('strong') or wispy and easily disrupted ('weak').

Thirty-six of the strains were also tested in cow and sheep plasma.

Cell-free coagulase was prepared from broth cultures of rough variants from human, cow and sheep strains (Smith *et al.*, 1952). Quantitative measurements were made by determining the highest dilution, in two-fold series, that clotted a 0.1 per cent w/v solution of bovine fibrinogen to which was added 1 per cent v/v human plasma. Coagulase dilutions and fibrinogen solution were in 25 per cent broth in saline with 1 in 2000 thiomersalate.

A. Fisk (1940). Br. J. exp. Path. 21, 311.
R. H. Gorrill (1951). Br. J. exp. Path. 32, 151.
D. D. Smith, R. B. Morrison and I. Lominski (1952). J. Path. Bact. 64, 567.

Coagulase—*Staphylococcus*

D. C. Smith, V. D. Foltz and T. H. Lord (1964). J. Bact. 87, 188–195.

The human plasma used for the coagulase test was obtained from outdated bank blood. The coagulase test was performed by the method of Todd, Sanford, and Wells (1954), with observations after 2, 4, and 24 hr. The appearance of a firm clot within 24 hr is considered positive.

J. C. Todd, A. H. Sanford and B. B. Wells (1954). *Clinical diagnosis by laboratory methods,* p. 704, 12th ed. W. B. Saunders Co., Philadelphia.

Coagulase—*Aerococcus catalasicus*

O. G. Clausen (1964). J. gen. Microbiol. 35, 1–8.

Cultures were incubated aerobically at 37°C unless otherwise stated.

Tests for this enzyme were performed by mixing equal parts of 24 hr broth culture and a 1/10 dilution of citrated human plasma in sterile saline solution, followed by incubation in a water bath at 37°C for 4 hr and storage overnight at room temperature.

Coagulase—*Staphylococcus*

S. Krynski, W. Kedzia and M. Kaminska (1964). Some differences between staphylococci isolated from pus and from healthy carriers. J. infect. Dis. 114, 193–202.

Coagulase was determined by the tube method with citrated human plasma and the readings were made after 3 and 24 hours.

Coagulase—*Staphylococcus*

S. I. Jacobs, A. T. Willis and G. M. Goodburn (1964). J. Path. Bact. 87, 151–156.

Strains were tested for coagulase production by the tube and slide methods (Cowan, 1938; Cadness-Graves *et al.,* 1943).

B. Cadness-Graves, R. Williams, G. J. Harper and A. A. Miles (1943). Lancet 1, 736.
S. T. Cowan (1938). J. Path. Bact. 46, 31.

Coagulase—*Staphylococcus*

R. Zemelman and L. Longeri (1965). Appl. Microbiol. 13, 167–170.

One loopful of culture was emulsified in 0.3 ml of citrated rabbit plasma at a dilution of 1:4 in sterile saline. Tubes were then incubated at 37°C, and final results were observed after 3 hr.

Coagulase—*Staphylococcus*

W. R. Chesbro, F. P. Heydrick, R. Martineau and G. N. Perkins (1965). J. Bact. 89, 378–389.

Coagulase was determined by observing the action of serial dilutions of crude and purified preparations in 0.2-ml amounts upon 0.2 ml of citrated rabbit blood after a 2-hr incubation period at 37°C.

Coagulase—*Streptococcus, Serratia, Escherichia*

B. G. Bayliss and E. R. Hall (1965). J. Bact. 89, 101–105.

Cultures were grown on BBL Infusion Broth (+0.3% agar for fastidious pathogens) or on Infusion agar slants at 35°C to 37°C for 18 to 21 hr. A positive control of *Staphylococcus aureus* was used throughout.

Plasma. Acid-citrate-dextrose (ACD) plasma was obtained from outdated blook-bank blood. Citrated rabbit plasma was prepared by placing 35 to 40 ml of blood obtained by cardiac puncture into a sterile flask containing the dried residue from 1 ml to a 20% solution of sodium citrate. EDTA rabbit plasma contained 0.01 ml of a 10% solution of EDTA (disodium) per milliliter of whole blood. Plasma treated with balanced oxalate, potassium oxalate, or sodium oxalate was prepared in the conventional manner. Heparinized plasma was obtained from blood mixed with a commercial preparation of heparin (Anticlot) according to the directions prescribed by the manufacturer (Clinton Laboratories, Los Angeles, Calif.). The plasma from at least two different animals was pooled, dispensed in samples (2 to 5 ml) in small screw-capped bottles, and stored in a conventional deep freeze at approximately −20°C. Rabbit plasma was diluted with three parts of 0.9% saline, and human plasma was diluted with one part of 0.9% saline before being used for coagulation tests.

Coagulase testing. The method used was as follows. Into a sterile, aluminum-capped test tube was placed 0.5 ml of an 18 to 21 hr broth culture of the organism. To this was added 0.5 ml of diluted plasma. Organisms taken from the agar slant were added directly to 0.5 ml of diluted plasma. The contents of the tubes were mixed and placed in a water bath (37°C). Readings were made in 30 min, and at hourly intervals thereafter, unless the exact time of clotting was desired. If no clotting was observed at the end of 8 hr, the rack of tubes was refrigerated for an additional 16 hr before the final readings were made. Controls for each test included 0.5 ml of the uninoculated broth plus the diluted plasma, 0.5 ml of the diluted plasma alone, and a known coagulase-positive strain of *S. aureus* treated in the same manner as were the organisms being tested.

Coagulase—activity in serum

W. W. Yotis (1965). Antonie van Leeuwenhoek 31, 84–94.

Two to 4 mg/ml of an antibacterial agent occurring in the serum of humans and animals caused the clotting of citrated rabbit, human, calf, ox, sheep, goat, guinea pig, rat, mouse, horse, chicken, pigeon and swine plasma. Heparinized plasmas of the same species were found resistant to the clotting action of the antibacterial agent. Citrated plasma previously heated at 56°C for 30 min, treated with 640 units/ml tyrosinase for 60 min, or absorbed with 0.2 M $Ca_3(PO_4)_2$ and 0.2 M $Mg(OH)_2$ was found resistant to the clotting action of the antibacterial agent. The clotting action of the antibacterial agent was not affected by heating at 66°C for 60 min, or by multiple passage through Seitz filters. The coagulation of citrated plasma proceeded most rapidly with 0.2 M Tris(Hydroxymethyl)-amino-methane buffer, pH 7 to 9, or distilled water as the antibacterial agent solvent; 0.2 M phosphate buffer, pH 6 to 8, reduced the clotting action of the antibacterial agent, while 0.2 M citrate-phosphate buffer, pH 4.0, or 0.2 M carbonate-bi-carbonate buffer, pH 10.0 entirely inhibited the clotting activity of this agent.

Coagulase—*Staphylococcus*

A. K. Highsmith and E. B. Shotts (1965). Appl. Microbiol. 13, 34–36.

Coagulase tests. Three methods were used for the determination of coagulase activity: the conventional tube test, described by Smith and Hale (1944); the slide clumping factor, or slide test (Cadness-Graves *et al.*, 1943); and the modified clumping factor, or the plate test.

Tube test. The test was performed by the addition of 0.1 ml of a 12 to 18-hr-old broth culture of staphylococci to 0.5 ml of rabbit plasma (Difco) in a sterile tube (13 x 100 mm). The tube was then incubated at 37°C for 3 hr. A positive reaction was indicated by any degree of clotting in the mixture.

Slide test. The slide test was performed as described by Cadness-Graves *et al.* (1943). A small portion of a staphylococcal colony was mixed in approximately 0.05 ml of water on a slide; a loopful of rabbit plasma was added and the components were mixed thoroughly. A suspect colony was considered to have coagulase activity if clumping of the bacterial cells occurred within 5 to 20 sec. If clumping is delayed or appears doubtful, the reaction must be confirmed by the tube test.

Plate test. This test utilizes the principle of the slide test and was developed as a result of the need for a rapid technique to process large numbers of staphylococcal cultures in conjunction with bacteriophage typing. At the time the "phage" plate was inspected for the presence of lysis, it was also tested for coagulase activity in the following manner. In an area of the plate where no phage was placed, a loopful (approximately 0.05 ml) of undiluted rabbit plasma was placed on the bacterial "lawn," and the plasma was agitated with a circular motion to create a suspension of cells. A positive reaction was indicated by a distinct clumping of the plasma-bacterial mixture; this should take place within 10 to 20 sec for a suspect culture to be considered coagulase-positive.

The plate test, a modification of the slide test described by Cadness-Graves was developed for the rapid identification of coagulase-positive staphylococci in conjunction with bacteriophage typing. An evaluation of 1,145 cultures by three coagulase-determination methods, the slide, tube, and plate tests, indicates that the plate test is as accurate as the slide tests, and the plate test agrees 97.7% with the tube test.

B. Cadness-Graves, R. Williams, C. J. Harper, and A. A. Miles (1943). Lancet 2, 736–738.
W. Smith and J. H. Hale (1944). Br. J. exp. Path. 25, 101–110.

Coagulase—*Staphylococcus*

F. D. Crisley, J. T. Peeler and R. Angelotti (1965). Appl. Microbiol. 13, 140–156.

Confirmation on a limited scale was accomplished by fishing characteristic colonies to TSA* slants from plates at both levels of contamination after all incubation periods, incubating the slants for 24 hr at 35°C, and testing for coagulase by the tube method, with the use of fresh frozen human plasma and a 4-hr incubation period at 37°C. Since the coagulase-negative cocci and *P. vulgaris* strains used in the mixed culture study were those bacteria previously noted to resemble coagulase-positive staphylococci on the tellurite-containing media, it was felt that the method constituted a fairly stringent test.

The authors used five selective media for the isolation of the staphylococci. These were:

(1) Staphylococcus medium SM110 (Difco);

(2) Tellurite glycine agar (TGA Difco) of Zebovitz, Evans and Niven (1955);

(3) TEA medium of Innes (1960);

(4) Egg-tellurite-glycine-pyruvate agar (ETGPA) of Baird-Parker (1962); and

(5) Tellurite-polymyxin-egg-yolk agar (TPEY) of Crisley *et al.* (1964).

A. C. Baird-Parker (1962). J. appl. Bact. 25, 12–19.
F. D. Crisley, R. Angelotti and M. J. Foter (1964). Publ. Hlth Rep. Wash. 79, 369–376.
A. G. Innes (1960). J. appl. Bact. 23, 108–113.
E. Zebovitz, J. G. Evans and C. F. Niven (1955). J. Bact. 70, 686–690.
*Trypticase Soy Agar (BBL)

Coagulase—*Staphylococcus*

M. E. McDivitt and N. W. Jerome (1965). Appl. Microbiol. 13, 157–159.

The authors describe a fibrinogen-polymyxin medium for isolation of coagulase positive staphylococci.

The base for the fibrinogen-polymyxin medium (FPM) contained the following ingredients (per liter): 25.0 g of Heart Infusion Broth (Difco), 17.5 g of agar (Difco), 0.4 g of cycloheximide (The Upjohn Co., Kalamazoo, Mich.), and 10.0 mg of polymyxin B sulfate (Burroughs-Wellcome and Co., Tuckahoe, N.Y.). A 0.5-ml portion of fibrinogen preparation was added per plate. The fibrinogen preparation was made by sterilizing 3% bovine fibrinogen (Armour Pharmaceutical Co., Kankakee, Ill.) in physiological saline by Seitz filtration; 3% (by volume) of sterile rabbit plasma (Colorado Serum Co., Denver, Colo.) was added after filtration.

Samples of mixed raw milk were obtained over a period of 3 months from the processing plant of the Department of Dairy and Food Industries at the University of Wisconsin. Samples were diluted with a 0.1% peptone dispersion (Difco) and plated in duplicate on FPM and on SM110. Plates were incubated at 37°C for 48 hr. Plate Count Agar (PCA) was used in the determination of total numbers of bacteria present.

Morphology of organisms isolated. The ratio of rods to cocci in raw milk was determined by the examination of stained smears of all colonies on randomly selected portions of PCA plates. Smears of all colonies removed from the selective media were also prepared and stained. Examination revealed the ability of FPM and SM110 to select cocci from a mixed culture.

Presumptive evidence of the presence of coagulase-positive staphylococci on FPM. FPM plates were examined for halo formation 24 hr after incubation and replaced in the incubator to permit full colony development for other investigations. Preliminary studies had shown that, under the conditions of this study, incubation for 24 hr was optimum for determining the number of halos formed.

Confirmatory tests for the presence of coagulase-positive staphylococci on FPM and SM110 (Difco). All visible, removable colonies were picked from randomly selected one-half portions of FPM and SM110 plates containing samples of comparable sample dilution. These were inoculated into Brain Heart Infusion (Difco) and incubated at 37°C for 18 hr. After incubation, tube coagulase tests were performed by adding 0.2 ml of the broth culture to 0.5 ml of fibrinogen-plasma preparation (1.5% [w/v] bovine fibrinogen and 1.5% rabbit plasma in physiological saline). At the time of transfer, the colonies which had shown halo development were identified.

Coagulase—*Staphylococcus*

M. Kocur, F. Přecechtěl and T. Martinec (1966). J. Path. Bact. 92, 331–336.

The free coagulase test was performed according to the recommendations of the Subcommittee on the Taxonomy of Staphylococci and Micrococci (1965). The bound coagulase test was performed by slide agglutination (Cadness-Graves *et al.*, 1943). Both tests were made with human and rabbit plasma.

B. Cadness-Graves, R. Williams, C. J. Harper and A. A. Miles (1943). Lancet 1, 736.
Subcommittee on Taxonomy of Staphylococci and Micrococci (1965). Int. Bull. bact. Nomencl. Taxon. 15, 109.

Coagulase—*Staphylococcus aureus*

R. A. Altenbern (1966). On the nature of albumin-promoted coagulase release by *Staphylococcus aureus.* J. infect. Dis. 116, 593–600.

Coagulase assay. Cultures were centrifuged in an angle centrifuge (1640 rcf) for 15 minutes, and the supernatant medium containing coagulase was decanted and stored at 4°C until titrated. The supernatant fluid was diluted serially in 1.5-fold steps in 0.85% saline containing 50 μg chloramphenicol per ml so that 0.25 ml of dilution remained in each tube. All tubes then received 0.5 ml of coagulase substrate consisting of 1.5% bovine fibrinogen (Armour) and 1.5% human plasma (Warner-Chilcott Laboratories) dissolved in the chloramphenicol-saline just described. Following agitation the titration series was incubated at 37°C for 3 hours. The highest dilution at which there was visible coagulation was designated as the end point, and the reciprocal of this dilution represented the number of units of coagulase per ml of the original sample. Chloramphenicol was included in the diluting fluid and coagulase substrate because experience had shown that false low titers were often obtained as a result of incomplete sedimentation of cells during the centrifugation step.

Coagulase—*Staphylococcus aureus*

F. J. Stutzenberger, C. L. San Clemente and D. V. Vadehra (1966). J. Bact. 92, 1005–1009.

The paper describes a nephelometric method for the determination of catalase.

Coagulase—*Staphylococcus*

H. Raj (1966). Can. J. Microbiol. 12, 191–198.

The authors describe methods for selective isolation of staphylococci from seafoods and testing for coagulase.

The composition of the medium (MSSA) thus developed is yeast extract 0.5%, tryptone 1.5%, sorbic acid 0.15%, mannitol 1%, NaCl 7.5%, thioglycolic acid 0.03%, and cystine 0.005% in distilled water. After the pH is adjusted to 6.7–7.0, the medium is distributed in 15-ml amounts in screw-capped tubes (16 by 150), sterilized by autoclaving for 15 minutes at 15 psi, cooled quickly to 25°C, and capped tightly until used.

The MSSA medium was tested as a primary isolation and enrichment medium for coagulase-positive staphylococci in a two-step procedure in which suitable dilutions of the sample to be tested are inoculated into triplicate tubes of the medium for the MPN count. Tubes showing growth after 48 hours incubation at 37°C are streaked onto S110-Ey plates (Carter, 1960). Initially these plates were incubated for 48 hours at 37°C. As a result of test runs of the specificity of S110-Ey at 37°C and 45°C incubation temperatures, the higher temperature was used in all later experiments.

H. C. Carter (1960). J. Bact. 79, 753–754.

COLLAGENASE

Collagenase (K-toxin)—*Cl. perfringens*

C. L. Oakley, G. H. Warrack and W. E. van Heyningen (1946). J. Path. Bact. 58, 229–235.

Abstractor's note:

The specificity of the reaction in the following test is determined by neutralization of the collagenase with specific antiserum. Mere demonstration of the presence of the collagenase does not require the use of antiserum. Attack on the dyed collagen (Azocoll) released the dye into the fluid.

Preparation of azocoll: Sieve 1 lb hide powder (Baird and Tatlock) to 60 mesh; about 120 g sievings should be obtained. To a solution of 0.575 g benzidine in 100 ml water containing 3 ml conc. HCl cooled in an ice bath, add slowly a solution of 0.45 g sodium nitrite in 10 ml water. Allow to stand for 10 minutes, then pour the tetrazotized benzidine into a chilled solution of 6.25 g sodium acetate ($CH_3COONa{\cdot}3H_2O$) in 500 ml water. To the mixture add slowly, with constant stirring, a solution of 1.1 g R-salt (sodium salt of 2-naphthol-3,6-disulfonic acid) in 100 ml water, followed by 20 ml N K_2CO_3. A brick red dye is formed. In the meantime wash 80 g of 60 mesh hide powder by repeated suspension in and filtration from 10 litres of water and finally resuspended in 500 ml water containing 30 ml 2N K_2CO_3. Add the dye to the suspension in 6 equal lots at intervals of 10 minutes. The color changes from brick-red to purple-red. After all the dye has been added, add 25 ml 2N K_2CO_3 and allow the mixture to stand for ten minutes. Centrifuge or filter off the dyed hide powder and wash by resuspending five times in 5-litre quantities of water. The wash liquors are colored pink. Then resuspend in about 300 ml water, stir constantly, and slowly add 1.7 litres of acetone. Repeat washing with acetone until acetone washings are no longer pink, filter off azocoll and remove residual acetone at 37°C. Sieve the azocoll to 60 mesh, dry over P_2O_5 under reduced pressure and keep in 5–10 g amounts in rubber-capped bottles under nitrogen. Azocoll required for indicator should be suspended as needed in 1 per cent Manucol IV* (300 mg to 100 ml). The Manucol is used to obtain an even suspension of the azocoll; in addition it may help to prevent the dye from diffusing too far into the liquid when azocoll disintegrates.

Titration of sera: Make mixtures in Lambeth tubes of the test dose of filtrate at the level chosen (usually 2 units), amounts of serum differing by 10 per cent, and saline to make up to constant volume. Allow them to stand for half an hour, then add 1 ml azocoll suspension to each. Incubate overnight in a water-bath at 37°C (to prevent convection currents the water in the bath should reach the highest liquid level in the tubes). The end point is taken as the tube which contains the maximum amount of antitoxin showing the development of a well marked red coloring above the azocoll. It has proved easy to use this test as a routine.

*Manucol IV is a polymer of d-mannuronic acid (Allbright and Wilson, Birmingham.)

Collagenase—*Clostridium*

M. Delaunay, M. Guillaumie and A. Delaunay (1949). Annls Inst. Pasteur, Paris 77, 220–227.

Nous avons renoncé à conserver–avant emploi–nos pastilles de collagène A dans l'éther, car ce liquide tend à les dénaturer. En effet, après deux mois de séjour dans l'éther, elles sont lysées plus lentement qu'aussitôt après leur préparation: leur destruction totale, par exemple, se produit en sept jours au lieu de trois; cette destruction, d'autre part, peut rester partielle, alors qu'elle serait complète avec une pastille fraîchement préparée. Désormais, nous nous bornons à conserver ces pastilles à l'état sec et à l'abri de la poussière.

Collagenase—general

M. Delaunay, M. Guillaumie and A. Delaunay (1949). Annls Inst. Pasteur, Paris 76, 16–23.

1. Méthode personnelle de titrage. – Nous avons choisi, comme substrat, le collagène purifié selon la technique de J. Nageotte et L. Guyon [12], que ces derniers désignent sous le nom de collagène A.

On retire les tendons de la queue de rats et on les place dans une solution d'acide acétique au 1/10.000. Ils sont laissés dans ce milieu, à la glacière, pendant vingt-quatre heures en présence de toluène. A ce moment, la préparation est constituée par l'eau acétifiée qui renferme, en suspension, des débris de tendon gonflés et, en solution, du collagène. On élimine les débris par filtration et on ajoute au filtrat un volume égal d'une solution de ClNa à 2.5 p. 100. Le collagène dissous commence à précipiter aussitôt. Cette précipitation sera presque complète au bout de quelques heures. On fait alors une seconde filtration, cette fois sur une membrane de collodion. Le collagène précipité se dépose sur la membrane de manière assez homogène, en donnant une pellicule mince. Celle-ci, après dessiccation convenable, est détachée du collodion; elle ressemble alors à une feuille de papier à cigarettes. Dans cette pellicule, sont régulièrement découpées, avec emporte-pièce, de petites pastilles formées de collagène solidifié, semi-transparent, pur. Ces pastilles sont conservées et stérilisées dans l'éther. Elles serviront, au fur et à mesure des besoins, après avoir été lavées dans de l'eau physiologique, pour rechercher la présence de collagénases dans les différents milieux qu'on désire examiner. En présence d'une collagénase, les pastilles de collagène A subissent un processus de lyse progressive qui peut amener leur disparition complète. Si le milieu est riche en ferment, cette destruction est particulièrement rapide.

Nous avons employé regulièrement, pendant plusieurs mois, cette méthode de titrage. Nous pouvons lui reconnaître les avantages suivants.

a) Elle est *fidèle*. Etant donné qu'on met en oeuvre un collagène très pur, la rapidité et l'intensité de la destruction des pastilles sont seulement fonction de la richesse du milieu en diastase. Le phénomène observé est bien spécifique.

b) Elle donne des *résultats très commodes à lire,* compte tenu de la régularité et du faible volume des pastilles.

c) Enfin, elle n'offre, dans son emploi, *aucune difficulté,* la préparation du collagène A ne réclamant au'un léger entraînement.

[12] Nageotte (J.) et Guyon (L.). Arch. Biol. 1930, 41, 1.

Collagenase—*Clostridium*

J. Brisou (1951). Annls Inst. Pasteur, Paris 81, 117–118.

Le réactif que nous avons utilisé jusqu'ici était préparé à partir de tendons de veau ou de peau de jeune boeuf. Il faut souligner que le collagène est ce qu'il reste de cette matière première traitée successivement par des lavages répétés à l'eau salée, un séjour dans une solution active de trypsine ou de pepsine, une déshydratation par l'alcool, l'alcool-acétone et l'éther. Nous sommes donc loin du précipité salin d'un produit acidosoluble résultant de la macération des tendons dans une eau faiblement acétifiée. L'analyse chimique montre du reste que ce précipité ne contient que des traces d'azote.

Dans une nouvelle série d'expériences, la matière première choisie fut successivement: des tendons de queues de rats, des membranes foetales, et du cordon ombilical humains. Il est possible, en partant de ces organes, de préparer un collagène très purifié titrant 90 à 93 p. 100 d'azote (1). Les fragments de collagène ainsi préparés ont été soumis à l'action de cultures totales et de filtrats de culture des germes anaérobies suivants (2):

(1) Faux établis en multipliant par le coefficient moyen d'azote total des acides aminés (6, 3), le chiffre d'azote total trouvé à l'analyse.

(2) Souches provenant des collections de l'Institut Pasteur de Paris.

Welch. perfringens	3 souches
Cl. histolyticum	2 souches
Cl. sporogenes	2 souches
Cl. oedematiens	2 souches
Cl. bifermentans	1 souches
Cl. aerofoetidum	1 souches

Les bactéries étaient cultivées en milieu Vf papaïnique [2] qui, nous l'avons montré, respecte leur caractère pathogène. Dans ces conditions, seul *Cl. histolyticum* montre une activité collagénolytique indiscutable.

[2] Brisou (J.). Ann. Inst. Pasteur, 1950, 79, 331.

Collagenase—*Bacillus*

D. G. Evans and A. C. Wardlaw (1953). J. gen. Microbiol. 8, 481–487.

Preparation of collagen substrate. Limb-bone shafts from adult rabbits were freed from marrow and connective tissue and dried over P_2O_5. After preliminary fragmentation the bone was ground to a fine powder in a low-temperature ball mill as used by Evans and Prophet (1950) for the preparation of powdered dentine. The powder was decalcified by suspension in a large volume of 0.2 N–HCl for 1 hr and then centrifuged. The deposit was washed free from acid by resuspending in water and centrifuging several times; the final suspension was dialyzed against running water and dried from the frozen state to yield a very finely divided preparation of decalcified bone.

Test for collagenase production. The decalcified bone powder was sterilized by exposing thin layers to ultra-violet radiation. A saline suspension of the powder was mixed with melted 3% agar in saline which had been cooled below 45°C and the mixture was poured into petri dishes. For each plate 10 ml of agar and 20 mg of decalcified bone powder were used. Disks 6 mm in diameter were cut from the agar plate with a sterile cork-borer and transferred to tubes containing 10 ml sterile 1% Evans peptone in saline at pH 7.6. Two disks were placed in each tube. The tubes of medium after the addition of the disks were incubated for 1 week to check sterility; they were then inoculated and incubated at 37°C. With aseptic precautions one disk was removed from each culture after 3 days, the other disk after 5 days and each examined microscopically at 15 x magnification. The degree of disintegration of the collagen particles was determined by comparison with a control disk from a tube of uninoculated medium and was estimated as none, partial or complete.

Collagenase—*Vibrio*

G. H. G. Davis and R. W. A. Park (1962). J. gen. Microbiol. 27, 101–119.

Hide-powder lysis. Method of Evans (1947) in double layer plates of basal medium with hide-powder (Baird and Tatlock) incorporated in top layer. Examined daily.

D. G. Evans (1949). J. gen. Microbiol. 1, 378.

Collagenase—*Pseudomonas aeruginosa*

A. H. Wahba and J. H. Darrell (1965). J. gen. Microbiol. 38, 329–342.

Production of collagenase. This was tested by a modification of the method devised by Oakley, Warrack & van Heyningen (1946). Layered nutrient agar No. 1 plates were prepared, the upper layer containing 0.5% azocoll (hide powder coupled to a red azo dye). Sixteen strains were inoculated on each plate, incubated for 3 days, and observed daily. The test was later performed by spotting melted azocoll-containing nutrient agar No. 1 on a basal layer of 2% agar in water and inoculating the solidified islets of medium with the strains.

Nutrient agar No. 1 (Oxoid No. 1) contained Lab-Lemco beef extract, 1 g; yeast extract (Oxoid L 20), 2 g; peptone (Oxoid L 37), 5 g; sodium chloride, 5 g; agar, 15 g; distilled water to 1 L; pH 7.4.

C. L. Oakley, G. H. Warrack and W. E. van Heyningen (1946). J. Path. Bact. 48, 229.

Collagenase—anaerobes

A. -R. Prévot (1966). Techniques pour le diagnostic des Bactéries Anaérobes. Éditions de la Tourelle, St. Mandé.

Recherche des collagénases et procollagénases.

La collagénase vraie n'existe que dans l'espèce *Cl. histolyticum.* Pour la mettre en évidence on découpe des petits cubes de tendon d'Achille de cheval ou de boeuf qu'on immerge dans du bouillon VF. On stérilise 30 minutes à 110°. On ensemence avec une culture de 24 heures en bouillon VF. On scelle sous vide et on met à 37°. La dissolution du tendon est parfois lente, parfois rapide, 5, 8, 15 jours suivant les souches et se suit à vue d'oeil : le cube devient transparent, puis diminue de taille et bientôt il ne reste qu'un résidu informe, grisâtre, parfois noirâtre.

La procollagénase de *W. perfringens* se recherche sur le procollagène (collagène A. de Nageotte) qu'on extrait du tendon de queue de rat par la méthode de Delaunay. On place les pastilles de procollagène en eau peptonée ou en bouillon VF et on scelle dans le vide après ensemencement avec une souche jeune. La pastille devient rapidement transparente et disparaît peu à peu.

DEOXYRIBONUCLEASE

Deoxyribonuclease—*Staphylococcus*

S. I. Jacobs, A. T. Willis and G. M. Goodburn (1964). J. Path. Bact. 87, 151–156.

Deoxyribonucleic acid (DNA) agar. An 8 per cent solution of the sodium salt of thymus gland deoxyribonucleic acid (B. D. H.) was prepared in sterile distilled water; 10 ml of this solution was added to the cooled nutrient agar base to give a final concentration of DNA of 0.2 per cent, and after thorough mixing plates were poured immediately. Since we have not been troubled by contaminating organisms in this medium, no special precautions were taken to sterilize the DNA itself. Thus, the DNA agar used in this investigation differed from that employed in previous studies in that it consisted of a richer basal medium and contained unautoclaved DNA.

After incubation at 37°C for 18 hr, deoxyribonuclease activity of cultures was determined by flooding the plates with 1.5 N-HCl. This precipitates the undegraded DNA so that enzymatic activity is shown by wide zones of clearing about areas of growth.

The nutrient agar base was a heart infusion broth (Difco) solidified with 1.5 per cent Davis New Zealand agar. The medium was sterilized at 121°C for 15 min, and reagents were mixed with the sterilized base medium when it had cooled to 50–55°C.

The medium was inoculated from a fresh 18 hr broth culture wide a sterile throat swab.

DEXTRAN AND DEXTRIN

Dextran—*Leuconostoc*

E. I. Garvie (1960). J. Dairy Res. 27, 283–292.

Examined on agar plates incubated at 22°C for 5 days. The basal medium was 1% tryptone, 0.5% Yeastrel, 0.5% K_2HPO_4, 0.5% ammonium citrate and 5% sucrose.

Dextrin—*Bacillus*

E. B. Tilden and C. S. Hudson (1942). J. Bact. 43, 527–544.

Full details are given of methods for characterizing amylases converting starch to non-reducing crystalline dextrins.

Dextrin—general

P. Hauduroy (1951). Techniques Bactériologiques. Masson et Cie, Éditeurs, Paris.

FORMATION DES CRISTAUX DE DEXTRINE A PARTIR DE L'AMIDON

Ensemencer les souches dans le milieu suivant (15 ml de milieu par tube):

Farine d'avoine	50 g
$CaCO_3$	20 g
Eau distillée	1.000 ml

Mettre à l'étuve à 37°C pendant 2 semaines.

Prendre 0,5 ml de la partie claire surnageante dans la culture en farine

d'avoine. Placer à 40°C avec 1 ml d'une solution à 3 p. 100 d'amidon (Takamine ou White-rose). Toutes les 15 minutes, ajouter à 3 gouttes du mélange précédent une goutte d'une solution N d'iode ou une goutte de liqueur de Gram. Une öse de ce dernier mélange est desséchée sur une lame et examinée au microscope.
("Iodine test" de Tilden et Hudson. J. Bact., 1942, 43, 527).

(A. T. C. C.)

Dextrin—*Bacillus*

N. R. Smith, R. E. Gordon and F. E. Clark (1952). U.S.D.A. Agr. Monograph No. 16.

The formation of crystalline dextrins from starch was detected by an adaption of the iodine test of Tilden and Hudson. The cultures were incubated at 37°C in rolled wheat and wheat-grain mashes, because they proved more satisfactory than oatmeal mash. About 0.5 g of rolled wheat, 0.2 g of $CaCO_3$ and 15 ml of distilled water were placed in large test tubes and sterilized. The other medium consisted of several grains of wheat, about 0.5 g of $Ca_3(PO_4)_2$ and 15 ml of distilled water. After 3,6 and 10 days of incubation a drop of the clear supernatant liquid of the culture and a drop of Lugol's iodine were mixed on a clean microscope slide and allowed to dry. If crystalline dextrins were present in small amounts, brown to blue hexagonal crystals were demonstrable (magnification, x 100) at the edge of the film. If these dextrins were present in larger amounts, long crystalline needles in fanlike arrangements also appeared at the edge and extended into the film.

In some cases what seemed to be long filaments were observed around the edge of the film. These were not crystalline dextrins, but cleavage lines caused by the presence of noncrystalline products. This method was found to be more sensitive and simpler than the technique used by Tilden and Hudson.

E. B. Tilden and C. S. Hudson (1942). J. Bact. 43, 527–544.

DYE ABSORPTION

Dye absorption—*Agrobacterium tumefaciens*

A. J. Riker, W. M. Banfield, W. H. Wright, G. W. Keitt, H. E. Sagen (1930). J. agric. Res. 41, 507–540.

The absorption of Congo red was observed by Kellerman (1913) and later confirmed by Smith (1917).

Yeast-infusion mannitol agar, to which was added an aqueous solution of Congo red to make a final concentration of 1 g to 40,000 ml, was used for the absorption tests. The various strains were studied as ordinary colonies in loop dilution plates and as giant colonies in plate cultures. The dilution plates and giant colonies were made from suspensions of the organisms which had grown for three days on yeast-infusion mannitol-agar slants. The plates were incubated, under favourable moisture conditions, for 14 days at 21°C.

Brom-thymol blue, crystal violet, basic fuchsin, and methylene blue were not absorbed as selectively as was Congo red.

K. F. Kellerman (1913). U.S. Dept. Agr., Bur. Plant Indus. Circ. 130, 15–17.
E. F. Smith (1917). J. agric. Res. 8, 165–188.

Dye absorption—*Agrobacterium tumefaciens*

A. A. Hendrickson, I. L. Baldwin and A. J. Riker (1934). J. Bact. 28, 597–618.

Since various dyes have been useful (Kellerman, 1913; Riker *et al.*, 1930) in differential studies, fifteen dyes were used in preliminary trials at concentrations of 1 to 10000 and 1 to 20000 with yeast-water mannitol agar. The various dyes used were aniline blue, acid fuchsin, basic fuchsin, Bismark brown, crystal violet, dahlia eosin, fluorescein, orcein, methyl orange, night blue, nigrosin, rosaniline, safranin, and thionine. The only dyes that showed differential absorption or bacteriostatic effects were dahlia, thionine, Bismark brown and aniline blue. Of these only aniline blue was absorbed. In preliminary work it was observed that the crown gall organism absorbed the aniline blue dyes while the hairy root and radiobacter cultures did not.

Water suspensions of these organisms were streaked on yeast-water mannitol aniline-blue agar plates and then incubated at room temperature. The medium had the following composition: mannitol, 5.0 g; magnesium sulfate, 0.2 g; dipotassium phosphate, 0.2 g; sodium chloride, 0.2 g; calcium sulfate, 0.1 g; aniline blue, 0.1 g; agar, 15.0 g; 100 ml of 10 per cent yeast water; and 900 ml of distilled water; adjusted to pH 7.0, from which the crown gall organism absorbed the dye within the limits of 1 to 1000, 1 to 10000 and 1 to 20000 employed – within the pH range 4.4 – 8.0.

No absorption of dye occurred from potato glucose agar containing 1/10000 aniline blue but slow absorption occurred from Patel's (1926) medium.

K. F. Kellerman (1913). U.S. Dept. Agr., Bur. Plant Indus. Circ. 130, 15–17.
M. K. Patel (1926). Phytopathology 16, 577.
A. J. Riker, W. M. Banfield, W. H. Wright, G. W. Keitt, and H. E. Sagen (1930). J. agric. Res. 41, 507–540.

Dye absorption—*Saccharomyces*

R. M. Lycette and L. R. Hedrick (1963). J. Bact. 85, 1–6.

The paper describes the adsorption and fluorescence of fat-soluble fluorescent dyes on class I and class III *Saccharomyces cerevisiae.*

Quantitative studies of the uptake of the fat-soluble fluorescent dye- a 2-(Stilbyl-4′)-4,-5-arylo-1,2,3-trazole compound by *Saccharomyces cerevisiae* is reported. It was previously reported by the same authors in Science 134, 1415, 1961.

Dye absorption—*Micrococcus lysodeikticus*

R. F. Beers, Jr. (1964). J. Bact. 88, 1249–1256.

This paper describes the adsorption isotherms obtained with the bacterium, *Micrococcus lysodeikticus* (Fleming, 1922), and acridine orange, and correlates these findings with the staining characteristics of the dye as observed by fluorescence microscopy.

A. Fleming (1922). Proc. R. Soc. Series B 93, 306–317.

Dye absorption—*Mycobacterium buruli* n. sp.

J. K. Clancey (1964). J. Path. Bact. 88, 175–187.

The cytochemical neutral red reaction was tested by the method of Hughes *et al.* (1954).

D. E. Hughes, E. S. Moss, M. Hodd and M. Hendon (1954). Am. J. clin. Path. 24, 621.

Dye absorption—yeasts

S. Nagai (1965). J. Bact. 89, 897–898.

The author recommends the use of the following red and blue dye combinations in the method described by the author (1963) for the detection of respiratory deficiency in yeast cells.

The method is as follows:

Single dyes in various concentrations and mixtures of two dyes (red and blue) in various combinations were added to nutrient agar medium to make color plates. Basal medium and dye solutions were sterilized separately by steaming at 100°C for 75 and 50 min, respectively, and were mixed together after cooling to about 55°C. In routine practice, 20-ml portions of appropriately diluted (to 20-fold of desired final concentration) single-dye solutions and water or two-dye solutions in 16-mm tubes were added to 360-ml of basal agar medium in 500-ml Erlenmeyer flasks to make 400 ml of color medium, which was, after thorough mixing, divided into 12 petri dishes (9-cm). Final nutrient composition was (w/v): glucose, 2.0%; peptone, 0.15%; dehydrated yeast extract, 0.15%; potassium dihydrogen phosphate, 0.15%; ammonium sulfate, 0.15%; magnesium sulfate, 0.1%; plus agar, 1.2% (to adequate hardness). *Saccharomyces cervisiae* IFO 0044, *S. chevalieri* 0210, and *S. microellipsodes* 1016 were used for general tester organisms. Several other yeasts were later used for comparative survey. Mixtures of normal and respiration-deficient (RD) mutant cells were spread on the color plates so as to produce about 100 to 150 colonies per plate, and were incubated at 30°C for 3 to 7 days.

All dyes were obtained from the National Aniline Division, Allied Chemical Corp., New York, N. Y.

1) Erythrosin B, 8–10 mg/L: Aniline blue, water soluble, 8–12 mg/L.
2) Phloxine B, 8–10 mg/L: Evans blue, 8–12 mg/L.
3) Rose bengal, 8–15 mg/L: Niagara sky blue, 8–12 mg/L.
4) Amaranth, 15–20 mg/L.
5) Eosin, 8–10 mg/L: Trypan blue, 10–15 mg/L.

Dye absorption—*Rhizobium*

N. J. Hahn (1966). Can. J. Microbiol. 12, 725–733.

The basal medium employed, yeast extract-mannitol (YEM) agar and broth was a variation of the yeast water-mannitol medium of Fred and Waksman (1928), and contained Difco yeast extract (3 g), mannitol (10 g), K_2HPO_4 (0.5 g), $MgSO_4 \cdot 7H_2O$ (0.2 g), NaCl (0.1 g), and Congo red (0.025 g) per liter.

The effects of temperature of incubation, variation in the composition of the basal medium, and storage of the prepared medium before inoculation were evaluated: growth on nitrogen deficient sucrose agar (Bryan, 1938), Giltner (1921), Löhnis (1913) and on this medium supplemented with nitrate was investigated; and the reactions of the dye with aqueous solutions of mono-, di-, and tri-valent cations under acid and alkaline conditions were determined.

The medium must be used within 6 hours of the addition of the dye.

C. S. Bryan (1938). Soil Sci. 45, 185–187.
E. B. Fred and S. A. Waksman (1928). *Laboratory manual of general microbiology.* 1st ed. McGraw-Hill, Comac Press, Inc., Brooklyn, N. Y. p. 33.
W. Giltner (1921). *Laboratory manual in general microbiology.* 2nd ed. John Wiley and Sons, New York. p. 263, p. 377.
F. Löhnis (1913). *Laboratory methods in agricultural bacteriology.* Translated by W. Stevenson and J. H. Smith. Chas. Griffin & Co., London, p. 112.

DYE SENSITIVITY

Dye sensitivity—coliforms

C. N. Stark and C. W. England (1933). J. Bact. 25, 439–445.

Used:	Peptone	0.5%
	K_2HPO_4	0.5%
	KH_2PO_4	0.1%

Lactose 0.5%

Crystal violet 1:700,000

No pH adjustment.

Tubes were inoculated, incubated at 37°C for 24 and 48 hr and examined for gas production.

Dye sensitivity—*Streptococcus*

L. A. Rantz (1942). The serological and biological classification of hemolytic and nonhemolytic streptococci from human sources. J. infect. Dis. 71, 60–68.

Growth in the presence of 0.1% methylene blue – One-tenth g of methylene blue was added to 100 ml of skimmed milk, which was autoclaved. Tubes of this medium were inoculated and incubated at 37°C for 48 hours. If growth occurred, the methylene blue was reduced to the leucobase, and the milk appeared white.

Dye sensitivity—*Rhizobium*

H. Z. Gaw (1944). J. Bact. 48, 483–489.

Buchanan's peptone-sucrose solution at pH 7.0 containing concentrations of crystal violet ranging from 1:5000 to 1:100,000 was used. Cultures were incubated 2 weeks or longer at 28°C and then transferred to yeast-water glucose agar to observe growth after 20 days.

All showed resistance to crystal violet, as is common among gram negative organisms.

Dye sensitivity—*Brucella*

J. C. Cruickshank (1948). J. Path. Bact. 60, 328–329.

The ability of *Br. melitensis, Br. abortus* and *Br. suis* to grow in the presence of certain aniline dyes forms one of the most important criteria for distinguishing these species. The dye-sensitivity test, first described by Huddleson and Abell (1928), was further developed by Huddleson (1929, 1931), and, although subjected to some criticism by certain workers, has been favourably reported upon by the majority of those who have used it (Topley and Wilson, 1946).

The technique usually employed for determining the bacteriostatic action of the dyes is that recommended by Huddleson. Dyes obtained from the National Aniline Chemical Company of New York, or standardized against dyes obtained from that source, should be used. Stock solutions in 50 per cent alcohol of thionine, basic fuchsin and methyl violet (1.0 per cent) and of pyronine (0.5 per cent) are prepared. For use, appropriate amounts of these solutions are added to liver-infusion agar, pH 6.6, to make dye plates containing 1 in 30,000 and 1 in 60,000 of thionine, 1 in 25,000 and 1 in 50,000 of basic fuchsin, 1 in 50,000 and 1 in 100,000 of methyl violet and 1 in 100,000 and 1 in 200,000 of pyronine. Heavy inocula of the organisms to be tested are spread over areas 1 cm in diameter on these plates and on a control plate without dye. The amount of growth is observed after 3 days' incubation.

For this technique, nine plates are required if the four dyes are used. Furthermore the preparation of the plates is troublesome, since the agar and the dye solutions have to be kept near boiling point before mixing, the glassware has to be handled hot and the plates have to be rocked gently from time to time during the setting process to prevent precipitation of the dye.

Simplified method

The method now described was devised as a rapid and economical method of testing strains without elaborate preparation. The filter-paper strips which are used can be prepared ahead and stored until required.

Strips of filter paper (Postlip 633) measuring 6 × 0.5 cm are placed in Petri dishes and sterilized in the autoclave. Each strip is then picked up with sterile forceps and one end dipped in an aqueous solution of the dye, which rapidly saturates it. The papers are replaced in the dish and dried in the 37°C incubator overnight. The dye strips, which can be kept in the Petri dishes or in screw-capped bottles, remain effective for an indefinite period. The concentrations of the dyes which have been found satisfactory for impregnation of the paper are as follows: thionine 1 in 800, basic fuchsin 1 in 200, methyl violet 1 in 400 and pyronine 1 in 800.

For use, strips impregnated with each of the dyes are laid in parallel, equally spaced, on the surface of a plate of liver agar, and a tube of the same medium (12 ml), melted and cooled to 50°C, is poured on top. When the agar has set the plate is dried in the 37°C incubator.

Milky suspensions of the *Brucella* strains to be tested, including known strains of *Br. melitensis, Br. abortus* and *Br. suis*, are prepared by adding about one ml of broth to liver-agar slope cultures grown for 2 days in screw-capped bottles and emulsifying some of the growth with a loop. The dye plate, previously marked into sections with a grease pencil, is tilted slightly and inoculated with the *Brucella* suspensions in strips at right angles to the filter papers by means of a flat loop spreader 8 mm in width. When dry the plates are incubated at 37°C in 10 per cent CO_2 and inspected after 2 and 3 days. The results are usually distinct after 2 days' incubation. Up to six strains may be tested on one 4-in. (10 cm) plate.

I. F. Huddleson and E. Abell (1928). J. infect. Dis. 43, 81.

I. F. Huddleson (1929). Mich. State College agric. Exp. Sta. Tec. Bull. No. 100.

I. F. Huddleson (1931). Am. J. publ. Hlth 21, 491.

W. W. C. Topley and G. S. Wilson (1946). *Principles of Bacteriology and Immunity,* 3rd ed., revised by G. S. Wilson and A. A. Miles, London, vol. i, pp. 820 and 821.

Dye sensitivity—*Brucella*

H. B. Levine and J. B. Wilson (1949). The identification of *Brucella abortus* strain 19 by dye bacteriostasis. J. infect. Dis. 84, 10–14.

All reagent bottles and pipettes were chemically clean and dry. Detergents were used to clean petri dishes. The same control lots of Bacto-Tryptose, Bacto-Tryptose agar, and thionine-blue were used throughout these experiments.

(a) *Stock culture medium* – All cultures were carried on Bacto-Tryptose agar. Cultures received from other laboratories were transferred before study three times at 24 hour intervals on this medium. Strain 19 was similarly transferred to serve as a control.

(b) *Inoculum medium* – Cultures to be assayed for dye tolerance were grown for 24 hours at 37°C on Bacto-Tryptose agar slants modified to contain 2% agar.

(c) *Inoculum* – A small amount of the 24 hour growth was suspended in 0.1% Tryptose, 0.5% sodium chloride solution and turbidimetrically standardized and diluted so that a standard 0.3 cm diameter transfer loop carried sufficient inoculum to give rise to 80 to 130 colonies. Using the Evelyn photoelectric colorimeter adapted with a 660 mμ filter, it was found that a 91% light transmittance suspension diluted 1:10^4 accomplished this for all strains.

(d) *Stock dye solution* – Thionine-blue was prepared in a 1:10,000 dilution with sterile, distilled water and shaken mechanically in a tightly stoppered bottle for 48 hours at 37°C in the dark. The dye solution was never steamed or autoclaved because this tends to concentrate it to an extent manifested in an altered level of bacterial inhibition. The above precedure apparently accomplished sterilization, or at least eliminated those organisms which would normally grow on the medium used under the conditions of the test. When not in use the dye solution was kept at 5°C. It remained bacteriostatic in the same concentration for over 6 months.

(e) *Preparation of dye plates* – The different strains of *Br. abortus* were tested on a solid medium containing graded amounts of the dye. The medium was always prepared in a standard volume of 50 ml so that the amount of evaporation due to autoclaving would be constant.

Table 1 – The preparation of Tryptose-agar mediums containing varying quantities of thionine-blue.

	mg dye per liter medium			
	0.0	0.2	0.4	0.6
Weight of medium in grams	2.05	2.05	2.05	2.05
Ml 1:10,000 stock dye solution added	0.0	0.1	0.2	0.3
Ml distilled water added	48.9	48.8	48.7	48.6
Ml of water displaced by dehydrated medium	1.1	1.1	1.1	1.1
Total volume of dye-medium in ml	50.0	50.0	50.0	50.0

Bacto-Tryptose agar was prepared to formula, but deficient by that volume of water which was added after autoclaving in the form of dye solution. A 50 ml preparation requires 2.05 g of the dehydrated medium which displaces a volume of water equal to 1.1 ml. The amount of water added was determined by the volume displaced by the dehydrated medium and the desired volume of dye solution. Thus the total volume of the dye-medium was 50 ml and the only variable in its composition was the weight of the dye. This is exemplified by table 1.

The appropriate volumes of water were added by burette to 6-ounce prescription bottles containing 2.05 g of the dehydrated medium. The mixtures were then shaken well and steamed in the Arnold sterilizer with intermittent shaking for 20 minutes. They were then autoclaved at 15 pounds of steam pressure for 20 minutes.

Immediately upon removal of the medium-containing bottles from the autoclave, while the temperature was still above 90°C, the desired volumes of dye solution were added aseptically to the appropriate bottles using a standard pipette. The stock dye solution was previously warmed to 55°C in a water bath. The dye-medium bottles were shaken vigorously to assure homogeneous distribution of the dye and were placed in a water bath at 45°C until temperature equilibration. This was sufficient time for the foam incurred by shaking to subside. Three petri plates were poured from each 50 ml preparation and allowed to stand for three hours at 37°C.

(f) *Inoculation of the dye-medium* – One loopful of the prepared inoculum was spread over an area of agar approximately 4 square cm. The plates were then allowed to stand for 4 hours at room temperature before inversion for incubation. This was ample time for the inoculum diluent either to be absorbed by the agar or to evaporate and thus prevent the mingling of different strains tested on the same plate.

Dye sensitivity—*Brucella*

G. Renoux and H. Quatrefages (1951). Annls Inst. Pasteur, Paris 80, 182–188.

1. *Action bactériostatique des colorants :* C'est la technique devenue classique d'Huddleson ; conformément aux données déjà établies par le C. R. F. O. (Taylor, Lisbonne, Vidal et Hazemann), nous n'avons recherché que l'inhibition par la thionine et la fuchsine (National Aniline).

Dye sensitivity—*Brucella*

G. Renoux and L. Carrère (1952). Annls Inst. Pasteur, Paris 82, 277–288.

Le milieu de culture est le Bacto-tryptose-agar (Difco) d'un lot vérifié valable (*) a pH 6,9, coulé en boîtes de Petri; les colorants, toujours les mêmes, proviennent de la "National Aniline Co."

*L. Carrère, G. Renoux et H. Quatrefages (1951) Ann. Inst. Pasteur (Paris) 80, 321.

Dye sensitivity—*Brucella*

M. J. Pickett, E. L. Nelson and J. D. Liberman (1953). J. Bact. 66, 210–219.

Differential dye tests. The general procedure was that already described for thionine, basic fuchsin, crystal violet, and pyronine (Pickett *et al.* 1952). Additional dyes examined were eosin Y, safranin O, rose bengal, and azure A.

M. J. Pickett, E. L. Nelson, R. E. Hoyt and B. E. Eisenstein (1952). J. Lab. clin. Med. 40, 200–205.

Dye sensitivity—*Brucella*

A. Chodkowski and J. Parnas (1957). Annls. Inst. Pasteur, Paris 93, 266–269.

Technique. – La technique employée a été décrite par Chodkowski et Parnas [1955] (1). Nous décrirons ici seulement quelques modifications que nous avons apportées à la méthode de typage par l'emploi d'un milieu bactériostatique constitué par des bandes de papier imprégnées de thionine (1/800) et de fuchsine basique (1/300). A la surface du milieu, verticalement par rapport aux bandes de papier, on étale une anse de souches standards de *Br. melitensis, Br. suis* et *Br. abortus,* en suspension dans une solution physiologique, à raison de 100 millions par millilitre; cet étalement est répété quatre fois. Puis la souche type de forme smooth, cultivée quarante-huit heures, en suspension en eau physiologique (1 milliard/ml) est étalée à la surface du milieu; la même souche est alors mise en suspension en eau physiologique à raison de 100 millions de germes par millilitre, puis étalée quatre fois avec une anse; avec la même anse, on fait ensuite deux nouveaux étalements à l'extrémité libre du milieu. La culture est incubée à 37°C, avec 10 p. 100 de CO_2, pendant quatre à cinq jours, et on lit le résultat.

(1) A. Chodkowski et J. Parnas (1955). Ann. Univ. Mariae-Curie Sklodowska 1955, 10, 1.

Dye sensitivity—*Streptococcus*

P. Morelis and L. Colobert (1958). Annls Inst. Pasteur, Paris 95, 568–587.

Le test du développement dans du lait contenant 0.1 p. 100 de bleu de méthylène a été réalisé de la façon suivante: dans des tubes contenant 9 cm^3 de lait frais, écrémé et neutralisé, on a ajouté au moment de l'emploi 1 cm^3 d'une solution stérile de bleu de méthylène à 1 p. 100. Les tubes furent incubés à 37°C après ensemencement. La réduction, lorsqu'elle se produit, débute vers la quatrième heure et se manifeste d'abord au fond du tube, de sorte que finalement les souches fortement réductrices ne laissent subsister qu'une collerette bleue en surface. La coagulation éventuelle se manifeste plus tardivement, quoique de façon variable, en général vers la vingt quatrième ou la quarante huitième heure. Elle s'accompagne d'une réduction intense, qu'il convient de bien distinguer de celle se produisant avant la coagulation, sous peine de risquer de considérer comme réducteur un germe qui ne l'est pas ou très faiblement [36]. Dans notre étude, il s'avère que les souches qui n'ont pas réduit le lait au bleu de méthylène se sont rencontrées principalement parmi les entérocoques isolés de fèces animales.

(36) Sladhauge (K.). Studies on Enterococci. Einar Munskgaard, edit., Copenhague, 1950.

Dye sensitivity—*Mycobacterium*

R. E. Gordon and J. M. Mihm (1959). J. gen. Microbiol. 21, 736–748.

Sensitivity to dyes. Malachite green oxalate, methyl violet, and pyronine B agars were made by adding 0.01% (w/v) of each dye to glycerol agar. Slants of each of the three agars and one of glycerol agar as a control were inoculated from a 7- to 14-day old culture in glucose broth with a loop, 2.5 mm in outside diameter. After incubation at 28°C for 28 days, growth on the agars was recorded.

Dye sensitivity—*Azotobacter*

V. Callao and E. Montoya (1960). J. gen. Microbiol. 22, 657–661.

	1/25,000 Pyronine	1/50,000 Diamond Fuchsin
A. chroococcum	+	−
A. vinelandii	+	+
A. agile	−	+
A. beyerinckii	−	−

+ = growth − = no growth

Other dyes and reactions are given.

Brilliant green	1/50,000
Malachite green	1/25,000
Safranin	1/100,000

Dyes were incorporated in the solid medium 77 of Allen (1951) with the mannitol replaced with 0.5% (w/v) glucose and 0.5% (w/v) maltose.
O. N. Allen (1951). *Experiments in Soil Bacteriology.* Burgess Publishing Co., Minnesota, U. S. A.

Dye sensitivity—thionine blue for *Brucella*

W. J. B. Morgan (1961). J. gen. Microbiol.. 25, 135–139.

1/500,000 to 1/1,000,000 in serum glucose agar (Res. Vet Sci. 1, 51, 1960) differentiated between FAO/WHO reference strains of *Brucella* strain 19 (sensitive) and *Brucella abortus* 544, *B. melitensis* 16M, and *Brucella suis* 1330 (all insensitive).

Testing of 171 isolates of *Brucella* revealed significant variation in the response of *Brucella abortus* to the dye, the majority of strains being insensitive to it.

Dye sensitivity—*Brucella*

B. E. Huntley, R. N. Philip and J. E. Maynard (1963). Survey of Brucellosis in Alaska. J. infect. Dis. 112, 100–106.

The primary culture medium used in the study of material from caribou was tryptose agar (Difco) plus crystal violet in a 1:700,000 final concentration to inhibit Gram-positive contaminants. Plain tryptose agar and liver infusion agar were employed in the study of pure cultures. Aniline dye sensitivity studies were based on the methods recommended by Huddleson (1943). Actual dye content was used as the basis for calculating the final concentrations of basic fuchsin, thionine, and crystal violet (Pickett and Nelson, 1951).

I. F. Huddleson (1943). *Brucellosis in Man and Animals.* New York, The Commonwealth Fund.

M. J. Pickett and E. L. Nelson (1951). J. Bact. 61, 229–237.

Dye sensitivity—*Brucella suis*

M. Moreira-Jacob (1963). J. Bact. 86, 599–600.

By a series of preliminary experiments, it was inferred that the limits of selective concentration of safranine (not corrected for impurities) were 1:5,000 to 1:20,000, with Albimi-Brucella agar as the basic medium. A 1% solution of the dye (National Aniline Div., Allied Chemical Corp., New York, N.Y.; 87% total content) was therefore used. After incubation for 72 hr at 37°C, this solution (in sterile distilled water) can be added without further sterilization to the basic agar melted and kept at about 95°C. Petri dishes, divided to take six strains and well dried in an incubator, were inoculated with 48-hr cultures in slants emulsified in 1.5 ml of saline. Every emulsion was streaked in plates of agar-safranine at 1:10,000, and most of them in additional plates with safranine concentrations of 1:5,000 and 1:20,000.

In this experiment, a collection of 362 stock strains was tried. It was found that safranine O very distinctly differentiates *B. suis* from the other species or varieties of the genus.

Dye sensitivity—*Brucella*

G. S. Wilson and A. A. Miles (1964). Topley and Wilson's *Principles of Bacteriology and Immunity* 5th edition. Arnold.

To Huddleson and Abell (1928) and Huddleson (1929, 1931) we owe a valuable method of distinguishing between the *melitensis, abortus,* and *suis* types, depending on their ability to grow in the presence of certain aniline dyes. Without entering into the detailed technique of the method, we may say that the general procedure is to prepare plates of liver agar, pH 6.6, containing 1/30,000 and 1/60,000 thionine, 1/25,000 and 1/50,-000 and 1/100,000 methyl violet, and 1/100,000 and 1/200,000 pyronine. The dyes used must be obtained from the National Aniline Chemical Company of New York, or standardized against these dyes. The organisms are inoculated rather heavily onto the plates, which are then incubated for 3 days aerobically, or in 10 per cent CO_2, according to the probable nature of the strains under examination.

Alternatively, and for some purposes more conveniently, strips of filter paper impregnated with stronger solutions of these dyes can be incorporated in liver agar plates, which are then streaked transversely with suspensions of the strains to be tested; the degree of sensitivity of the strain is indicated by the amount of growth over and on either side of the underlying strips of paper (Cruickshank, 1948).

This method is recommended by Wundt (1958), who regards it as preferable to the incorporation of dyes in the medium itself.

J. C. Cruickshank (1948). J. Path. Bact. 60, 328.
I. F. Huddleson and E. Abell (1928). J. infect. Dis. 43, 81.
I. F. Huddleson (1929). Mich. State College agric. Exp. Sta. Tec. Bull. No. 100.
I. F. Huddleson (1931). Am. J. publ. Hlth 21, 491.
W. Wundt (1958). Zentbl. Bakt. ParasitKde 171, 166.

Dye sensitivity—*Brucella*

L. M. Jones (1964). J. Bact. 88, 1527.

Safranin O (National Aniline Div., Allied Chemical Corp., New York, N. Y.) was used at a final concentration of 1:5,000 in Trypticase Soy Agar (BBL). Known numbers of organisms from 10 to 10^{10} were inoculated on media with and without the dye.

Dye sensitivity—*Moraxella*

W. J. Ryan (1964). J. gen. Microbiol. 35, 361–372.

Resistance to crystal violet was examined by incorporating it in blood agar. A drop of bacterial suspension sufficient to give semiconfluent growth on control plates was used as inoculum. Results were read after 3–4 days of incubation.

Dye sensitivity—*Bacillus*

G. R. F. Hilson (1965). J. gen. Microbiol. 39, 407–421.

Inhibition by malachite green in Löwenstein-Jensen medium. Several techniques were used to test the capacity of form 2 organisms and *Bacillus* strains to grow on Löwenstein-Jensen medium with and without malachite green; these will be described elsewhere. In general, assessments were made of the power of vegetative forms, unheated spores and spores heated at 80°/30 min to initiate growth from inocula varying in size from about 10 up to 10^6 viable organisms on Löwenstein-Jensen medium, with and without malachite green, dispensed in bottles or in plates.

Dye sensitivity—*Brucella*

L. M. Jones, V. Montgomery and J. B. Wilson (1965). Characteristics of carbon dioxide-independent cultures of *Brucella abortus* isolated from cattle vaccinated with strain 19. J. infect. Dis. 115, 312–320.

Sensitivity tests. – Trypticase-soy agar (Baltimore Biological Laboratory, Baltimore, Maryland) plates were prepared on the day they were to be inoculated. Following autoclaving the trypticase-soy agar was cooled to 47°C, and the appropriate amount of prewarmed thionin blue, penicillin, safranin O, or erythritol was added, mixed, and the plates poured. All substances were tested at several concentrations before decisions were made as follows: thionin blue (obtained from British Drug Houses but no longer available) was added in a final concentration of 1:500,000; penicillin (penicillin G potassium, E. R. Squibb and Sons, New York, N. Y.) was added to give a final concentration of 5 units per ml; safranin O (National Aniline Division, Allied Chemical Corporation, New York, N. Y.) was used in a final concentration of 1:5000; a 10% aqueous solution of erythritol (*i*-erythritol, Nutritional Biochemicals Corporation, Cleveland, Ohio) was filtered through a membrane filter, frozen in small amounts for storage, and added to agar to give a final concentration of 1 mg per ml.

Cultures were grown on trypticase-soy agar slants at 37°C for 24 hours and suspended in saline. A sterile cotton swab was soaked in the suspension for a few seconds, and a single streak of the swab was made on each of the plates. This method did not always differentiate between tolerance to the agent and overgrowth by resistant mutants. Most cultures were retested by placing 0.02-ml amounts of tenfold dilutions of a standardized suspension on each of the plates. Plates were examined on several occasions during 2 to 5 days of incubation at 37°C in air or in 10% added carbon dioxide, depending upon the requirements of the culture. The effect of the substances on the growth of brucella was estimated from the size of the colonies on the test plates in comparison with the control plate of trypticase-soy agar.

EIJKMAN TEST

Eijkman test—*Escherichia coli*

C. A. Perry and A. A. Hajna (1933). J. Bact. 26, 419–429.

The authors modified the Eijkman test by adding a buffer and reducing the amount of glucose.

Peptone (Difco)	15 g
Glucose	3 g
K_2HPO_4	4 g
NaCl	5 g
Distilled water	1000 ml

Cultures were incubated for 24 hours at 46°C.

Eijkman test—enterobacteria

C. E. Skinner and J. W. Brown (1934). J. Bact. 27, 191–200.

Eijkman-glucose, 1.25 per cent, Difco peptone 1.25 per cent, NaCl, 0.625 per cent; incubated at 46°C.

Bulír-mannitol, 0.66 per cent, Witte peptone, 0.55 per cent, NaCl, 0.33 per cent; added to beef infusion inoculated with *Bacterium aerogenes* for twelve hours to destroy muscle sugar, brought to pH 7.0, and neutral red added. Incubated at 46°C.

Lactose broth-lactose, 1.0 per cent, Difco peptone, 1.0 per cent, Leibig's beef extract, 0.3 per cent, NaCl, 0.5 per cent, adjusted to pH 7.0 and incubated at 37.5°C.

These media were inoculated, as described in our previous work (1930), in Durham fermentation tubes.

J. W. Brown and C. E. Skinner (1930). J. Bact. 20, 139–150.

Eijkman test—enterobacteria

C. A. Stuart, A. Zimmerman, M. Baker and R. Rustigian (1942). J. Bact. 43, 557–572.

Difco dehydrated Eijkman broth with brom-thymol blue indicator. Cultures were rapidly transplanted to the Eijkman broth, and brought to 45.5°C in approximately 3½ minutes. Incubated at 45.5°C ± 0.1°C for 24 hours.

Eijkman test—*Proteus*

C. A. Stuart, E. van Stratum and R. Rustigian (1945). J. Bact. 49, 437–444.

In view of the unexpected reactions of *P. morganii* in urea medium at

45°C, it was decided to reinvestigate the Eijkman reactions of *Proteus,* including *P. rettgeri,* which was not tested in the previous work (Stuart, Zimmerman, *et al.,* 1942). Buffered glucose broth with bromthymol blue as indicator was used. Inoculations were made from agar slants (straight needle) and from broth culture (2-mm loop). Cultures used for inoculations were 24 hours old. A total of 327 cultures, some of which were used in the previous work, were inoculated in groups of six and immediately placed in a 45°C water bath. (In the previous work a temperature of 45.5°C was used.) Recordings were made at 24 and 48 hours as no growth, growth, weak acid, and strong acid. As in the previous work, gas production was not recorded because many of the *P. mirabilis* strains produced no gas after 48 hours at 45°C. After 48 hours' incubation at 45°C, all tubes except those of *P. mirabilis* were incubated at room temperature and examined for 9 days for viability as shown by acid or acid and gas production.

C. A. Stuart *et al.* (1942). J. Bact. 43, 557–572.

Eijkman test—*Proteus*

G. T. Cook (1948). J. Path. Bact. 60, 171–181.

Glucose MacConkey broth. The method described in the Ministry of Health Report (1939) for the preparation of single-strength lactose MacConkey broth was followed in making up this medium with the substitution of glucose for lactose. The tubes were incubated for 48 hours at 44°C in a water-bath and the presence of growth, with or without gas production, was then recorded.

Ministry of Health Reports (1939). The bacteriological examination of water supplies, Rep. publ. Hlth med. Subj., Lond., No. 71.

FAT HYDROLYSIS

Fat hydrolysis

M. A. Collins and B. W. Hammer (1934). J. Bact. 27, 473–485.

The medium ordinarily used for studying the action of Nile blue sulfate on simple tri-glycerides, fatty acids and natural and hydrogenated fats, and also for investigating the action of bacteria on simple tri-glycerides and fats, was beef-infusion agar adjusted to a pH of 6.8 to 7.0. A 0.1 per cent aqueous solution of Nile-blue sulfate was added to the agar, in the proportion of 10 to 100 ml of medium, and the agar then put into tubes or flasks and sterilized. Lower concentrations of Nile blue sulfate were also used with satisfactory results, but the colors were less intense.

The emulsions of natural and hydrogenated fats were prepared as follows: The fat to be used was filtered with a hot water funnel and added to a melted 0.5 per cent agar solution in the proportion of 10 ml of fat to 90 ml of the solution. The mixture was sterilized at 15 pounds for twenty-five minutes, allowed to cool until it was solidified and then vigorously shaken to secure an emulsion of the fat. The fat emulsion was stored in this condition and just before use was heated to a temperature that would give a soft jelly-like mass which could be easily transferred with a pipette.

When plates were to be poured, agar containing Nile-blue sulfate was melted and the fat emulsion added to the hot agar in the proportion of 1 ml of the emulsion to 20 ml of the agar. After the dye had been added to the agar, the medium was allowed to remain hot for a few minutes before it was poured.

The liquid tri-glycerides and fatty acids were dispersed in agar containing Nile-blue sulfate in the same general manner as the natural and hydrogenated fats, except that smaller quantities were used because of the cost. The tri-glycerides and fatty acids which are solid at 21°C were dispersed as follows: Agar containing Nile-blue sulfate was added to the plates and kept hot over a low Bunsen flame, while a small amount of the solid tri-glyceride or fatty acid was added and vigorously stirred into it. The agitation was continued until the fat or fatty acid had solidified in small globules or masses.

In the study of the effect of various bacteria on the simple tri-glycerides and natural and hydrogenated fats, the plates poured with the materials were left at room temperature until the surface of the medium was dry (at least twelve hours) to prevent an abnormal spreading of the bacterial colonies. Several organisms were inoculated on each plate, using a small loopful of a forty-eight-hour litmus milk culture of each organism to be studied. The plates were inverted, incubated at 21°C and examined frequently for evidence of lipolysis.

All the examinations for color changes and for disappearance of the globules were made with a hand lens or a wide-field binocular.

Fat hydrolysis

G. Knaysi (1941). J. Bact. 42, 587–589.

In demonstrating lipolysis by microorganisms, it has been the custom to shake an agar medium containing a small concentration of Nile blue sulfate or chloride with a certain amount of the neutral fat to be tested. The microorganisms are then either plated in this medium or grown on poured plates in the form of a streak or a giant colony. Hydrolysis is indicated by the colour change of the fat droplets from red to blue or purple.

The principal disadvantage of this method is the toxicity of the salts of Nile blue, which limits their usable concentrations. Long (1936) found that even a concentration of 10^{-4} of the dye inhibits the growth of certain bacteria.

In order to impart to fat droplets sufficient color without the danger of making the medium toxic to microorganisms, we have been using the following technique for nearly a decade. It consists in shaking the neutral fat to be tested with a saturated aqueous solution of the sulfate or the chloride of Nile blue, in the proportion of 1 ml of the dye solution to 10 ml of the neutral fat. The deep red fat is then separated from the blue solution with a separatory funnel, washed several times with distilled water, and sterilized in the autoclave. One milliliter of this red fat can then be shaken with 100 ml of the agar medium and used for plating. The stock of sterile red fat can be stored in an ice box for several weeks.

It is evident that this technique does away with toxicity, for the red substances of Nile blue are insoluble in water and are present only in the fat phase. Stark and Scheib (1936) used their technique, described above, in an extensive investigation of lipolytic bacteria and found no evidence of toxicity.

Another method uses almost exclusively the oxazine base of Nile blue for detection of fat hydrolysis. This method consists in liberating the base from a solution of its salts by means of sodium hydroxide, dissolving the washed precipitate in the fat, and using the fat as above. If this technique is used, it can be found that detection of lipolysis is not an exclusive property of Nile blue and a few related oxazine dyes, but that a number of other common basic dyes can be used for the same purpose. We have thus been able to differentiate neutral fat from free fatty acids by means of methylene blue, neutral red and malachite green. Those and other basic dyes have in common the property that their free bases are insoluble in water, but soluble in alcohol, fat solvents and fat. It can therefore be said that in general, when the free base of a basic dye is of a different color from its salts, the dye can be used to differentiate neutral fats from free fatty acids.

Neutral fats assume the color of the base, and free fatty acids combine with the base to form soaps having the color of the salts of that dye. In the above cases neutral red, methylene blue and malachite green color neutral fats, respectively, orange-yellow, red, and olive green and free fatty acids red, blue, and blue-green. In all cases the base is liberated with normal NaOH, added drop by drop, until complete precipitation. The precipitate is then filtered out and washed with distilled water adjusted to a pH of about 7.5. The precipitate can then be dried and stored or dissolved in the fat, sterilized and used at one's convenience. It is desirable that the fat be saturated with the base, or nearly so, in order to get a maximum of contrast. By using this technique it may be found that some dye not hitherto used for detecting fat hydrolysis may have a distinct advantage over Nile blue for a given purpose. For instance, some microorganisms that reduce Nile blue may not be able to reduce neutral red. In all instances, the medium used should be neutral or slightly alkaline, highly buffered, and should not contain fermentable sugars, for if acids are present in the aqueous phase, the base can be extracted from the fat and a salt of the dye will be formed in the medium, leaving the fat droplets colorless.

H. F. Long (1936). Iowa St. Coll. J. Sci. 11, 78–80.

C. N. Stark and B. J. Scheib (1936). J. Dairy Sci. 19, 191–213.

Fat hydrolysis—*Corynebacterium*

R. F. Brooks and G. J. Hucker (1944). J. Bact. 48, 295–312.

Determined by organoleptic observations of cultures in raw whole milk after 7 and 14 days' incubation at 37°C.

The cotton plugs were removed momentarily from the tubes, and the presence or absence of butyric acid odor noted.

Fat hydrolysis—*Staphylococcus, Micrococcus*

Y. Abd-el-Malek and T. Gibson (1947–1948). J. Dairy Res. 15, 249–260.

Plates of nutrient agar containing 5% emulsified fat (butterfat, cottonseed oil or olive oil) were inoculated on the surface, and after 7 days at 30°C they were tested with $CuSO_4$ (Berry, 1933).

J. A. Berry (1933). J. Bact. 25, 433.

Fat hydrolysis (including Tweens)—*Staphylococcus*

J. Marks (1952). J. Path. Bact. 64, 175–186.

Medium for demonstrating lipolysis – 0.5 ml of 1 per cent "Tween 80", 0.25 ml tributyrin and 4 ml of isotonic saline were mixed and sterilized by heating in a boiling water-bath or by autoclaving at 10 lb. The tributyrin was then emulsified by vigorous shaking by hand for one minute. The emulsion was mixed with 20 ml of melted nutrient agar and a plate poured. Lipolysis was demonstrated by clearing of the emulsion around stab-inoculated colonies.

Wires used to inoculate plates were charged from cultures on solid media. To show lipolysis, plates were incubated for 48 hours in air plus CO_2.

Fat hydrolysis—*Staphylococcus*

M. E. Davies (1954). J. gen. Microbiol. 11, 37–44.

Fat emulsion. Horse fat was used because, in fatty-acid composition, it resembles human fat more closely than any other mammalian fat available (Hilditch, 1947). The emulsion was prepared from the following: horse fat, 4.0 ml; 6.0% (w/v) aqueous agar, 3.0 ml; distilled water, 93.0 ml. The melted fat and aqueous agar were mixed and boiling water added gradually, the mixture being passed repeatedly through a hand-driven domestic mechanical emulsifier ("Cream Maker, Empire Series". British Emulsifiers Ltd., Greek Street, London, W.1.)

Nutrient fat agar. (N.F.A.) was prepared by adding an emulsion of horse fat to beef infusion agar at pH 7.2–7.4 to give a final fat concentration of 0.3% (w/v). 7.5 ml emulsion were added to 92.5 ml melted agar. Twenty-five ml of medium were poured into Petri dishes (7.0 cm diam) which had been dried for 1 hr before use. On this medium lipolytic activity was shown by a zone of clearing around the colony.

For human strains of staphylococci the author prefers the medium without the Nile base.

T. P. Hilditch (1947). *Chemical Constitution of Natural Fats*, London: Chapman and Hall.

Fat hydrolysis—various

C.J.E.A. Bulder (1955). Antonie van Leeuwenhoek. 21, 433–445.

The author describes a method whereby small fatty oil droplets are sprayed with a nebulizer onto the surface of a previously inoculated plate.

In the vicinity of colonies of fat hydrolyzers the oil droplets become irregular in shape, granular and opaque, a change which can easily be established with a low-magnification microscope. Plates should not need to be incubated more than 5 days; in most cases the change in the droplets became visible in 24 hours.

Fat hydrolysis—*Micrococcus cerolyticus*

H. Friedmann and J. Kern (1956). Can. J. Microbiol. 2, 515–517.

Two methods were employed to demonstrate the lipolytic activity of the organism. When the Nile blue – triglyceride – infusion agar medium described by Turner (1929) was employed, deep blue zones were found to surround the growth of the organism after 18–20 hr at a temperature of 37°C. This is contrasted to the pale blue color of the uninoculated medium. A recent method described by Bulder (1955) was employed to confirm this biochemical activity. Droplets of sterile olive oil are sprayed on the surface of a nutrient agar plate immediately following the streaking of the test organism; aseptic precautions are taken. After a suitable incubation period, lipolysis is manifested by a change in the appearance of the oil droplets when the plate is examined with the low power microscope. It is found that after two days at a temperature of 37°C, the droplets in contact with or in the vicinity of the growth became granular, irregular in shape, and opaque, indicative of fat hydrolysis. The droplets on the uninoculated control plates are unchanged in appearance.

C.J.E.A. Bulder (1955). Antonie van Leeuwenhoek 21, 433–445.
R. H. Turner (1929). J. infect. Dis. 44, 126–133.

Fat hydrolysis (Tweens)—various

G. Sierra (1957). Antonie van Leeuwenhoek 23, 15–22.

A peptone agar medium was prepared with Tween 40, Tween 60 and Tween 80. The basal medium is composed as follows: Difco Bacto-peptone, 10.0 g; NaCl, 5.0 g; $CaCl_2 \cdot H_2O$, 0.1 g; distilled water, 1000 ml. The pH was adjusted to 7.4.

500 ml of the medium was sterilized in the flasks and cooled to 40–50°C. The Tweens were sterilized separately by autoclaving twenty minutes at 120°C. The five ml of each Tween were added to the peptone agar flasks (about 1% final concentration of Tween) and shaken until complete solution. From this clear and transparent medium (pH 7.0–7.4) plates were poured, on which a suspension of the microorganism to be tested was streaked. The cultures were maintained by frequent transfers on peptone agar slants, a fresh slant being prepared eighteen to twenty hours before the inoculation of the plates.

The plate cultures were incubated at 30°C and 37°C according to the optimal growth temperature for each microorganism. It was found that if the microorganism tested had a lipolytic activity, an opaque halo could be easily observed around the colonies. When studied microscopically these opaque halos turned out to consist of crystals of the formed calcium soaps. The various Tweens can be distinguished from each other by their characteristic crystals formed in the halos.

Fat hydrolysis—various

V. B. D. Skerman *A Guide to the Identification of the Genera of Bacteria* – Williams and Wilkins Book Co., Baltimore 1959.

Media for the Detection of Lipolytic Organisms

Basal medium

Peptone	15 g
Yeast extract	5 g
Sodium chloride	5 g
Agar	15 g
Water	1000 ml

Dissolve the ingredients at 121°C for 20 minutes. Adjust the pH to 7.5 and filter. Add the required amount of fat, mix well, and sterilize at 121°C for 20 minutes.

Preparation of nile blue sulfate

Prepare a saturated aqueous solution of Nile blue sulfate. Precipitate the oxazine base from it by the dropwise addition of N NaOH until precipitation is complete. Filter and wash the precipitate with distilled water with the pH adjusted to 7.5. Dry and store for use.

Fats vary in the intensity with which they stain. Tripropionine, tributyrin, tricaproine, tricapyrlin, triolein, beef tallow, butterfat, coconut oil, corn oil, cottonseed oil, lard, linseed oil, and olive oil all stain bright red.

Tricaprin, trilaurin, trimyristin, tripalmitin, and tristearin decrease in the intensity of staining with increase in molecular weight.

Preparation of the dyed fat

Prepare a saturated solution of the Nile blue sulfate oxazine base and mix 1 ml with 10 ml of the fat. If necessary, work in a heated water bath to liquefy the fat and maintain it in the liquid state throughout the washing process.

With tributyrin and similar triglycerides which are liquid at room temperature, add double the quantity of ether to the fat-dye mixture in a separatory funnel; separate the red ether-soluble fat layer from the water layer and wash several times with water. Finally separate the ether-fat layer and evaporate off the ether. (Keep away from flames!) Separate the fat from the residual water and sterilize at 121°C. Store in a refrigerator. For use add 1 ml of the dyed fat to 10 ml of the melted basal medium, mix well, and pour.

With fats which solidify at room temperature wash the dyed fat with several changes of heated water and then disperse the fat in a melted 0.5 per cent neutral agar solution in the proportion of 10 ml of fat to 90 ml of agar.

Sterilize the mixture at 121°C for 20 minutes, cool until solidified, and then shake vigorously until the fat is emulsified.

For use melt the fat emulsion and the culture medium, and add 1 ml of the fat emulsion to 20 ml of the basal medium. Mix well and allow to set.

Note: With Nile blue sulfate the fat stains red and fatty acids stain blue. Hydrolysis of the fat will result in a blue halo around the colonies. In some cases the change in cloudiness of the medium may be more dramatic than the color change.

Similar methods can be employed with the bases of neutral red, methylene blue, and malachite green prepared as for Nile blue sulfate. In the respective cases the neutral fats are orange-yellow, red, and olive green, and the fatty acids are red, blue, and blue-green. One of these may prove suitable if Nile blue is reduced.

M. A. Collins and B. W. Hammer (1934). J. Bact. 27, 473.
G. Knaysi (1941). J. Bact. 42, 587.

Fat hydrolysis (including **Tweens**)—*Pseudomonas*

M. E. Rhodes (1959). J. gen. Microbiol. 21, 221–263.

Hydrolysis of tributyrin. A stable emulsion of 1.0% (v/v) tributyrin in yeast extract agar was prepared by blending in a hand emulsifier; streak cultures were made on poured plates of this medium. Clear zones of hydrolysis were measured after different incubation periods.

Hydrolysis of olive oil. Because the triglycerides of olive oil are composed of higher molecular-weight fatty acids than is tributyrin, it was thought that this might affect the ease with which they were hydrolyzed by pseudomonad enzymes. To encourage potential ability to hydrolyze olive oil, this was incorporated as the major added carbon source in the following medium (%, w/v); $NH_4H_2PO_4$, 0.1; KCl, 0.02; $MgSO_4 \cdot 7H_2O$, 0.02; yeast extract (Oxoid), 0.3; agar, 2.0; olive oil, 5.0; adjusted to pH 7.8 with NaOH. The olive oil contained the base of Night Blue to a final concentration in the medium of 1/15,000. This method was adapted from that of Jones and Richards (1952, Proc. Soc. appl. Bact. 15, 82). Streak plate cultures were examined for growth and the production of a blue salt (formed by the reaction of the liberated fatty acids with the weak base of the dye). Controls of all isolates were grown on plates of similar medium but without olive oil.
Hydrolysis of margarine. The test method was similar to that of the previous one and again depended upon staining the margarine with the base of Night Blue and then incorporating 5.0% (v/v) of it into yeast extract agar.
Hydrolysis of Tween 80. The use of the water soluble Tweens for easy detection of lipolytic activity by microorganisms, including Pseudomonas aeruginosa, was described by Sierra (1957, Leeuwenhoek ned. Tijdschr. 23, 15). His technique, using 1.0% (v/v) Tween 80 (polyoxyethylene sorbitan mono-oleate) in calcium chloride + peptone agar was adopted; parallel cultures on tributyrin + yeast extract agar were simultaneously inoculated. After incubation, the Tween cultures were examined for the presence of opaque halos, due to the formation of conspicuous precipitates of calcium oleate around the zones of growth.

Fat hydrolysis—*Clostridium*

M. Sebald and A. -R. Prévot (1960). Annls. Inst. Pasteur, Paris 99, 386–400.

Technique ; Nous avons utilisé le milieu de Sierra légèrement modifié : peptone, 10 g ; NaCl, 5 g ; CaCl, 0.1 g ; eau distillée, 1000 ml; adjuster à pH : 7.4, ajouter gélose 12 g, précipiter à l'autoclave à 120°C, filtrer ; ajouter le tween à la concentration finale de 0.5 p. 100. Répartir à raison de 4.5 ml en tubes de 8–9 mm ; stériliser 30 min à 110°C. Après régénération, les tubes sont ramenés à 45°C, ensemencés à pipette fermée et refroidis pour réaliser un isolement correct. Incuber à 37°C, pendant un à vingt jours (en raison de l'observation prolongée parfois nécessaire, nous préférons l'utilisation de tubes de 8–9 mm à celle également possible de milieux coulés entre couvercle renversé et fond de boîte de Petri, selon la technique des antibiogrammes). Chaque souche est ensemencée sur trois milieux contenant tweens 40, 60 et 80. Les tubes sont examinés régulièrement et l'on note la date d'apparition du halo autour des colonies lipidolytiques ainsi que l'intensité de la réaction. Il est possible de distinguer 4 types de réactions :

+ + = réaction très intense généralement précoce (un à quatre jours [type *Cl. botulinum* A].

+ = réaction positive mais plus discrète, généralement plus tardive (type *Cl. cauteretsensis*).

± = réaction douteuse macroscopiquement, nécessitant le contrôle microscopique *(Cl. caproicum)*.

0 = réaction négative.

Conformément à la nomenclature proposée par Hoffman, les enzymes attaquant respectivement les tweens 40, 60 et 80 seront nommés : sorbitol-palmitase, sorbitol-stéarase et sorbitol-oléase.

Fat hydrolysis—*Mycobacterium*

A. Andrejew, Ch. Gernez-Rieux and A. Tacquet (1960). Annls Inst. Pasteur, Paris 99, 56–68.

Méthode simplifiée. – Pour les estimations approximatives ou qualitatives de l'activité lipasique, nous avons mis au point et employé la méthode chimique suivante : on dissout 2.42 g de tris(hydroxyméthyl) aminométhane dans 50 ml d'eau ; on ajoute 26.8 ml de HCl 0.2 *M,* on complète à 1 000 ml avec de l'eau et on ajuste le pH à 8 avec HCl ou NaOH. On obtient ainsi le tampon Tris 0.02 *M,* pH 8.

On mélange 1 partie de la suspension bacillaire assez dense, faite dans le tampon Tris 0.02 *M,* pH 8, avec 5 parties de ce même tampon contenant 10 p. 100 de Tween 20, neutralisé au pH 8 et additionné de phénol-sulfone-phtaléine à 0.001 p. 100. On place le mélange à 38°, en présence d'un témoin dépourvu de bactéries. On agite.

L'activité lipasique est décelée lorsque le virage de l'indicateur de l'essai est nettement plus rapide que celui du témoin. On note le temps et on le rapporte au poids sec des bactéries employées.

Le Tween 20 peut être remplacé par la tributyrine, à la concentration finale de 50 μM/ml, pour la mise en évidence de l'activité des estérases simples chez les Mycobactéries.

Fat hydrolysis—*Xanthomonas*

A. C. Hayward and W. Hodgkiss (1961). J. gen. Microbiol. 26, 133–140.

The method of Sierra (1957) was used; the cultures were discarded after incubation for 6 days.

G. Sierra (1957). Antonie van Leeuwenhoek J. Microbiol. Serol. 23, 15.

Fat hydrolysis (Tweens)—*Lactobacillus*

R. E. Smith and J. D. Cunningham (1962). Can. J. Microbiol. 8, 727–735.

To test for lipolytic activity, Sierra's medium (1957) was modified to support the growth of lactobacilli as follows: neopeptone, 5.0 g; trypticase (BBL), 5.0 g; yeast extract, 3.0 g; sodium acetate, 3.0 g; sodium chloride, 5.0 g; calcium chloride, 0.1 g; pyridoxine, 0.1 g; choline chloride, 0.3 g; ascorbic acid, 0.3 g; thiamine HCl, 0.1 g; salts A* 5.0 ml; salts B* 5.0 ml; agar, 20 g; and distilled water to 1000 ml. The medium was autoclaved at 121°C for 10 minutes, cooled to 50°C, and 1.0% (v/v) sterile Tween 40 or Tween 60 added. Streaked plates were incubated for 2 weeks. At the end of this period, they were inverted, and the medium beneath the colonies examined for the presence of salts of fatty acids. Trials with anaerobic incubation gave results similar to those using aerobic conditions.
G. Sierra (1957). Antonie van Leeuwenhoek 23, 15–22.
*See Starch hydrolysis—*Lactobacillus* p. 812.

Fat hydrolysis—*Vibrio*

G. H. G. Davis and R. W. A. Park (1962). J. gen. Microbiol. 27, 101–119.

Tributyrin hydrolysis was tested by clearing within 7 days of Oxoid glycero-tributyrate (1%, w/v) agar.

Fat hydrolysis—*Vibrio*

G. H. G. Davis and R. W. A. Park (1962). J. gen. Microbiol. 27, 101–119.

Tributyrin hydrolysis was detected by clearing of 0.2% (v/v) tributyrin in nutrient agar after 24 hr and 7 days.

Fat hydrolysis

J. G. Franklin and M. E. Sharpe (1963). J. Dairy Res. 30, 87–99.
Lipolytic bacteria. Numbers of lipolytic bacteria were determined by plating in a tributyrin medium (TA) of the following composition (w/v): Evans' peptone, 1.0%; Lab-Lemco, 0.3%; Yeastrel, 0.3%; sodium chloride, 0.5%; Bacto-tryptone, 1.0%; tributyrin, 1.0%; agar, 2.0%; final pH, 6.0 ± 0.1. The plates were incubated aerobically for 5 days at 30°C, when the colonies exhibiting clear zones of lipolysis were counted. Five of these colonies were picked into yeast dextrose broth (YDB), purified and tested for their ability to hydrolyze butterfat by streaking on the butterfat agar (BFA) of Jones and Richards (1952).
A. Jones and T. Richards (1952). Proc. Soc. appl. Bact. 15, 82.

Fat hydrolysis—*Staphylococcus*

D. B. Shah and J. B. Wilson (1963). J. Bact. 85, 516–521.

Some pathogenic staphylococci produce an opacity reaction when grown in media containing egg yolk. The egg yolk factor is a lipase with a requirement for a fatty acid acceptor, rather than any of the three lecithinases. Highest amounts of egg yolk factor were obtained in Heart Infusion broth rather than nutrient broth or Casman's synthetic medium. The lowest-density lipoprotein (lipovitellenin) isolated by ultracentrifugation of egg yolk has been identified as the opacity-producing substrate. The lipase acts on the lipid moiety, resulting in alterations in the solubility of lipovitellenin. Lipolysis is optimal at pH 8. The optima pH for the opacity reaction are 5.5 and 8. The protein (vitellenin) in lipovitellenin has an isoelectric point at pH 5.5. It is proposed that slight alterations in the stabilizing lipid due to lipolysis and the very low solubility of vitellenin may drastically alter the solubility of lipovitellenin at pH 5.5.

Fat hydrolysis—*Staphylococcus epidermidis*

D. Jones, R. H. Deibel and C. F. Niven (1963). J. Bact. 85, 62–67.

The egg yolk opacity test was detected on Colbeck EY agar plates (Difco), and deoxyribonucleic acid hydrolysis determined by method of Di Salvo (1958).

J. W. Di Salvo (1958). Med. Techns Bull. 9, 191–196.

Fat hydrolysis—*Mycobacterium*

J. Viallier, Mme. C. Maret, A. Sédallian and Mlle. J. Augagneur (1964). Annls Inst. Pasteur, Paris 107, 286–288.

Nakayama et Takeya [4] ont insisté sur l'existence d'une estérase thermostable permettant l'identification de *Mycobacterium kansasii.* Le test proposé serait fortement positif avec les Mycobactéries appartenant à cette espèce et négatif avec les autres.

La technique d'etude préconisée par ces deux auteurs est la suivante : on prélève 5–10 mg d'une culture de mycobactéries que l'on émulsionne dans 1 ml d'eau distillée et que l'on place au bain-marie bouillant pendant, dix à quinze minutes. Après refroidissement par un courant d'eau froide, on ajoute à la culture 1.5 ml d'une solution A de phénol phtaléine dibutyrate que l'on prépare de la façon suivante : 10 mg de phénol phtaléine dibutyrate sont dissous dans 3 cm^3 d'acétone et la solution est portée à 20 cm^3 par adjonction d'eau distillée. Après un temps d'incubation à l'étuve à 37°C, variable entre trois et six heures, on ajoute au mélange suspension bactérienne et solution A, 1 goutte d'une solution de carbonate de sodium anhydre ($N = 1$). La réaction est positive et signe la présence d'une estérase thermostable si on obtient une coloration rouge.

(4) Nakayama (Y.) et Takeya (K.) Nature, 1963, 198, 1113–1114.

Fat hydrolysis—*Moraxella*

W. J. Ryan (1964). J. gen. Microbiol. 35, 361–372.

Egg yolk opacity. Organisms were grown on serum agar plates containing 10% (v/v) egg yolk saline. The egg yolk saline was prepared by emulsifying the yolk of one egg in 250 ml physiological saline and sterilized by Seitz filtration. Readings were taken after 48 hr incubation.

Fat hydrolysis—*Pediococcus*

E. Coster and H. R. White (1964). J. gen. Microbiol. 37, 15–31.

The method of Sierra (1957) was employed to test for the presence of these enzymes in some pediococcus cultures using TJ agar as the basal medium.

The authors used the Tomato Juice (TJ) broth or agar of H.L. Günther and H. R. White (J. gen. Microbiol. 26, 185, 1961), with the addition of Tween 80 (0.1%, v/v), and the pH was adjusted to 6.6 for maintenance and preparation of inocula, with certain specified exceptions.

Transfers for preparing inocula were made every 24 hours.

All cultures were incubated aerobically at 30°C.

Note: The author advises that *Pediococcus halophilus* and the brewing strains were not tested for fat hydrolysis.

G. Sierra (1957). Leeuwenhoek ned. Tijdschr. 23, 15.

Fat hydrolysis—*Pseudomonas aeruginosa*

R. R. Colwell (1964). J. gen. Microbiol. 37, 181–194.

Hydrolysis of Tween 20, 40, 60 and 80. The use of the water-soluble Tweens for easy detection of lipolytic activity was described by Sierra (1957). His technique, using 1.0% Tween 20, 40, 60 or 80 in calcium chloride peptone agar, was followed. Cultures were kept for 14 days, being examined routinely for the appearance of conspicuous precipitates around the colonies.

The standard inoculum for all tests was a single drop (1/20 ml) from a sterile disposable pipette (Fisher Scientific Company) of a 24–48 hr broth culture. The stock culture medium (YE) was a modification of that used by Rhodes (1959): Difco yeast extract, 0.3 g; Difco Bacto-Proteose Peptone, 1.0 g; NaCl, 0.5 g; agar, 1.5 g (agar omitted from liquid stock media); distilled water, 1L; adjusted to pH 7.2–7.4 with NaOH. Tests were carried out at room temperature (25°C) except where otherwise indicated.

M. E. Rhodes (1959). J. gen. Microbiol. 21, 221.

G. Sierra (1957). Antonie van Leeuwenhoek 23, 15.

Fat hydrolysis—*Mycoplasma*

S. Rottem and S. Razin (1964). J. gen. Microbiol. 37, 123–134.

Qualitative assays. Edward medium plates containing 0.4% (w/v) tributyrin were prepared. Tributyrin dissolved in a small volume of acetone was added to basal Edward medium containing Bacto-heart infusion broth, Bacto-peptone, NaCl and Bacto-agar (Razin & Oliver, 1961; Razin, 1963). The medium was autoclaved at 121°C for 20 min, cooled to 50°C, and all the other ingredients (Razin, 1963) added. The medium, still liquid, was transferred to a sterile 50 ml beaker and treated in the ultrasonic disintegrator at 1.5 amp. for 3 min. Twelve ml volumes of this medium were then poured into sterile Petri dishes and allowed to solidify. In some experiments the tributyrin plates were covered with a thin film of polyvinyl formal ('Formvar', Shawinigan Ltd., London, E.C. 3). A solution of 0.5% (w/v) Formvar in chloroform was quickly poured on the surface of the agar, and allowed to drain off by inclining the plates. After evaporation of the chloroform a fine film of Formvar was produced on the agar surface. This film was freely permeable to medium nutrients, but prevented direct contact of Mycoplasma organisms placed on top of it, with the medium underneath. The organisms to be tested were grown in 5 ml volumes of liquid Edward medium for 24–48 hr and washed as described above. The washed organisms obtained from a 5 ml culture were suspended in 0.3 ml of 0.02 M-phosphate buffer (pH 7.5) and 0.02 ml drops of these suspensions were placed on the tributyrin plates. The plates were incubated at 37°C for 24 hr; clear zones that appeared in and around the drop areas indicated lipolytic activity.

A simple semi-quantitative method for estimating lipase activity in chromatographic fractions of Mycoplasma proteins was devised. Tributyrin, dissolved in a small volume of acetone, was added in a final concentration of 0.4% (w/v) to 0.02 M-phosphate buffer (pH 7.5) containing 0.9% (w/v) Bacto-agar. The medium was autoclaved at 121°C for 20 min, cooled to 50°C and treated in the ultrasonic disintegrator at 1.5 amp for 3 min. Twelve ml volumes of the resulting suspensions were poured into sterile Petri dishes and allowed to solidify. Drops (0.02 ml) of each of the chromatographic fractions were placed on the plates, which were then incubated at 37°C for 24 hr. The presence and degree of lipolytic activity in these fractions could be then estimated by the size of the clear zones which appeared in the drop areas.

S. Razin (1963). J. gen. Microbiol. 33, 471.
S. Razin and O. Oliver (1961). J. gen. Microbiol. 24, 225.

Fat hydrolysis—*Bacillus*

G. Sierra (1964). Can. J. Microbiol. 10, 926–928.

The author draws attention to the fact that subtilisin, a *proteolytic* enzyme of *Bacillus subtilis,* catalyzes the hydrolysis of some triglycerides, including tributyrin, and therefore it and similar enzymes may give a false impression of the ability of organisms to hydrolyze triglycerides when the test is based on tributyrin.

Fat hydrolysis—*Vibrio marinus*

R. R. Colwell and R. Y. Morita (1964). J. Bact. 88, 831–837.

The authors used the method of G. Sierra (1957). Antonie van Leeuwenhoek 23, 15–22 – modified by the addition to the media of the following salts to produce a synthetic seawater:– sodium chloride, 2.4%; potassium chloride, 0.07%; magnesium chloride (hydrated), 0.53%; and magnesium sulfate (hydrated), 0.7%.

A standard inoculum was 1 drop from a pasteur pipette (*c.* 0.05 ml) of a 24 hour artificial seawater broth.

Cultures were incubated at 18°C.

Fat hydrolysis—*Bacteroides oralis*

W. J. Loesche, S. S. Socransky and R. J. Gibbons (1964). J. Bact. 88, 1329–1337.

Lipolysis was determined in Spirit Blue Agar medium (BBL) supplemented with Wesson oil (Starr, 1941).

M. P. Starr (1941). Science, N.Y. 93, 333–334.

Fat hydrolysis—*Enterobacteriaceae*

B. R. Davis and W. H. Ewing (1964). J. Bact. 88, 16–19.

The Victoria blue medium mentioned by Hugo and Beveridge (1962) gave clear-cut results; hence it was selected for use. The medium was prepared as follows: peptone, 10 g; yeast extract, 3 g; sodium chloride, 5 g; agar, 20 g; Victoria blue (aqueous solution, 1:1,500), 100 ml; and distilled water, 900 ml; at a final pH of 7.8. Corn oil, or other appropriate substrate, was added in 5.0% (v/v) concentration and was mixed thoroughly into the medium. This was accomplished by means of a magnetic stirplate or a blender, or by brief treatment with a sonifier. The medium was distributed in tubes, sterilized at 115°C (10 lb) for 30 min, and allowed to solidify in a slanted position (long slants). The completed medium was light in color, and positive reactions were indicated by the development of a dark-blue color in the medium, in the growth, or in both. As mentioned by Hugo and Beveridge (1962), the acid salt of Victoria blue is water-soluble and blue, whereas the free base is red and soluble in fats. When Victoria blue and a fat are mixed in a melted medium at about pH 8, the dye is extracted by the fat as the red base. With gelation of the agar, a light pinkish-gray medium is obtained. When the fat is hydrolyzed, fatty acid is formed, which converts the dye to the acid (blue) salt and liberates it. The success of the method depends upon the use of substrates that are insoluble in water.

Media containing tributyrin and triacetin also were prepared and used in the manner outlined by Hugo and Beveridge (1962): tributyrin, 10 ml; peptone, 5 g; yeast extract, 3 g; agar, 20 g; and distilled water, 1,000 ml; at a final pH of 7.8. The tributyrin was dispersed in the medium by an appropriate method of homogenization (see above, corn oil medium), and the medium was tubed and sterilized at 80°C for 30 min on 3 successive days. Finally, the medium was allowed to solidify in a slanted position (long slant). Zones of clearing along the lines of inoculation were interpreted as lipolysis.

Triacetin medium was prepared by adding 1.0% (v/v) of the substrate and 0.01% (w/v) of bromthymol blue to nutrient broth (pH 6.8). The medium was sterilized by filtration. Positive reactions were indicated by pH values less than 6.2.

Cultures were incubated for 7 days at 37°C.

W. B. Hugo and E. G. Beveridge (1962). J. appl. Bact. 25, 72–82.

Fat hydrolysis—*Pseudomonas*

G. L. Bullock, S. F. Snieszko and C. E. Dunbar (1965). J. gen. Microbiol. 38, 1–7.

Lipase was demonstrated by: (a) clearing of 1% (v/v) tributyrin in nutrient agar; (b) change in color of pH indicator in Rhodes' (1959) medium, substituting peanut oil for olive oil.

Cultures were incubated at 20–22°C for 1 week.

M. E. Rhodes (1959). J. gen. Microbiol. 21, 221.

Fat hydrolysis—*Alcaligenes*

R. G. Mitchell and S. K. R. Clarke (1965). J. gen. Microbiol. 40, 343–348.

Sierra's (1957) medium, incubated at 28°C for 6 days, was used.

G. Sierra (1957). Antonie van Leeuwenhoek 23, 15.

Fat hydrolysis—*Staphylococcus*

D. V. Vadehra and L. G. Harmon (1965). Appl. Microbiol. 13, 335–339.

Quantitative methods for studying the action of lipases of *Staphylococcus aureus* on milk fat are given.

Fat hydrolysis—*Streptomyces albus*

T. F. Fryer and M. E. Sharpe (1965). J. Dairy Res. 32, 27–34.

Lipolysis was tested on a tributyrin agar medium (Franklin and Sharpe, 1963) and a butterfat agar (Jones and Richards, 1952).

J. G. Franklin and M. E. Sharpe (1963). J. Dairy Res. 30, 87.

A. Jones and T. Richards (1952). Proc. Soc. appl. Bact. 15, 82.

Fat hydrolysis—*Staphylococcus*

W. R. Chesbro, F. P. Heydrick, R. Martineau and G. N. Perkins (1965). J. Bact. 89, 378–389.

Lipase activity was determined by the egg yolk agar plate method described by Richou, Pantaleon, and Quinchon (1960), except that the agar suspension was made in saline-phosphate-magnesium buffer with the use of Colbeck EY agar (Difco).

R. Richou, J. Pantaleon and C. Quinchon (1960). C.r. hebd. Séanc. Acad. Sci., Paris 250, 1131–1133.

Fat hydrolysis—marine bacteria

R. M. Pfister and P. R. Burkholder (1965). J. Bact. 89, 863–872.

Hydrolysis of triolein and tributyrin was detected by dispersing the fats in a melted 0.5% neutral agar solution (10% fat) and autoclaving. In preparation for use, the fat agar emulsion was melted and added aseptically to the melted basal seawater agar (1 ml of fat emulsion per 20 ml of basal medium), and then shaken and dispersed to plates. The organisms were streaked on the surface, incubated up to 2 weeks and examined for a clearing zone (Tendler and Burkholder, 1960), after staining with Sudan B solution to help intensify any visible zones.

The seawater medium contained (per liter): Trypticase (BBL) or N-Z-Case, 2.0 g; Soy-tone (Difco), 2.0 g; yeast extract, 1.0 g; vitamin B_{12}, 1.0 μg; aged seawater; agar (if desired), 16 g. The pH was adjusted to 7.0 prior to autoclaving.

M. D. Tendler and P. R. Burkholder (1960). Appl. Microbiol. 9, 394–399.

Fat hydrolysis—*Staphylococcus*

D. B. Shah and J. B. Wilson (1965). J. Bact. 89, 949–953.

The authors describe a potentiometric method for determination of lipase activity.

Fat hydrolysis—*Cladosporium*

G. Tuynenburg Muys and R. Willemse (1965). Antonie van Leeuwenhoek, 31, 103–112.

A comparison was made between the Eykman layered plate technique and the Victoria Blue Agar method of Jones and Richards (1952) used for the detection and enumeration of lipolytic microorganisms in edible emulsions. Victoria Blue Agar does not permit the growth of important microorganisms such as *Cladosporium suaveolens* and *Cladosporium butyri* because of the absence of sugar required by the method.

The Eykman-plate method is superior in every respect, because the nutrient agar composition can be varied. A complete description of a modified Eykman technique is given.

Preparation of the butter or margarine extract

It is necessary to extract the microorganisms in butter or margarine by shaking with a diluting liquid. To this end 20 g of margarine or butter (boring or scraping sample) are dispersed in 37 ml of diluting liquid in 150 ml wide-mouthed stoppered bottles or screw cap bottles at 39 ± 1°C. The diluant contains: 0.1% peptone + 0.8% NaCl which is sterilized for 20 min at 115°C. After the margarine has melted, an emulsion is formed by shaking vigorously (1 min, if necessary on a shaker). After standing at room temperature for 10 min the fat has separated off to such an extent that the aqueous extract, 1 ml of which contains the microorganisms from 0.5 g margarine or butter, can be transferred to various culture media with a sterile pipette.

The method of Mossel and Zwart (1958) for the preparation of the extract is also applicable. However, the above method is less time-consuming, and moreover, in the case of strongly salted butter or margarine it provides the necessary dilution of the margarine serum at slightly increased temperature.

Eykman-method (modified)

Petri dishes (diameter 9–10 cm) are cleaned by boiling in a 0.4% solution of a synthetic detergent. The plates are rinsed in distilled water and air-sterilized (2–2.5 hr at 170°C). Plastic petri dishes can be conveniently used directly.

Composition of fat to be used for fat layer (1). 200 g of a partial glyceryl ester of higher fatty acids, called mono/diglyceride (mp 58°C) is dissolved in 2000 g hardened, refined palm oil (mp 45–47°C) and this mixture is filtered at 95°C over a NaCl-layer (10 g NaCl) previously dried for some hours at 180°C. The filtrate (free from salt crystals) is collected in air-sterilized 1 liter flasks after which the fat mixture is heated on a heating plate for 1 hr at 150–155°C (thermometer in the fat), cooled, and transferred into 150 ml sterile flasks.

Agar composition to be used for the detection and enumeration of lipolytic bacteria:

Water – Agar – Lab-Lemco – Peptone (WLP): 5 g Peptone, 3 g of Lab-Lemco Beef Extract, 12.5 g $Na_2HPO_4 \cdot 12\ H_2O$, 15 g agar are dissolved in 965 ml water while heating.

The liquid is boiled, dispensed in culture tubes (8 ml per tube) and sterilized for 20 min at 115°C.

Agar composition to be used for the detection and enumeration of lipolytic yeasts and moulds:

Whey – Yeast autolysate–Agar (WY): 36 g Agar, 18 g NaCl, 3.6 g yeast extract; 180 ml acid whey are dissolved in 1800 ml water while heating. The pH is adjusted to 5.5, the mixture dispensed in culture tubes (8 ml per tube) and sterilized for 20 min at 110°C.

Acid whey: 1 liter soured skim milk (pH 4.3) is heated on a steam bath until the whey separates off. It is then filtered through paper and sterilized for 20 min at 110°C.

Applying the thin fat layer in the petri dishes. Petri dishes are heated on a flat copper water bath to 63 ± 2°C and the fat mixture is likewise brought to this temperature.

Ca. 10 ml fat is poured into a petri dish, so that the bottom is covered completely. The excess is poured over into a second petri dish. The first petri dish is closed and placed in a rack tilted to an angle of ca. 60 degrees in order to cool. The superfluous fat collects in the rim of the dish. The entire bottom must remain covered with a thin, even layer of fat. Sometimes the fat recedes so that parts of the bottom of the petri dish are left uncovered. This may occur when the petri dishes are too hot, or moist, or when there are salt crystals in the fat.

Pouring the plates. For preparing the pour-plates, first 1 ml of diluent is brought onto the fat layer and then 1 ml of the margarine or butter extract. Subsequently 6–7 ml WLP or WY agar is added and thoroughly mixed with the extract by swirling a number of times. In this way spreading of colonies on the fat layer is avoided.

Incubation. The plates with nutrient media WLP and WY are incubated for 5 days at 25°C. To prevent the formation of condensation water the plates are incubated upside down. A piece of white cardboard (4 × 4 cm), moistened with 1 drop of glycerol, is put in the lid and removed after 20–24 hr. The cardboard need not be sterilized.

Victoria Blue Method

The instructions given by Jones and Richards (1952) were closely observed. Only fresh butterfat was used. According to these authors this is essential. The basic medium, i.e. peptone 1%, yeast extract 0.3%, sodium chloride 0.5% and agar 2% (pH 7.8) contains no sugar, because, according to the authors, the presence of fermentable sugar gives rise to completely incorrect results with lactic acid bacteria.

The fat emulsion plate of von Sommaruga

This plate was prepared in accordance with the instructions of von Sommaruga (1894).

1) We understand that the complete fat mixture will be shortly available from "Oxoid".

A. Jones and T. Richards (1952). Proc. Soc. appl. Bact. 15, 82–93.
D. A. A. Mossel and H. Zwart (1958). Ned. Melk– en Zuiveltijdschr. 12, 218–224.
E. von Sommaruga (1894). Z. Hyg. InfektKrankh. 18, 441–456.

Fat hydrolysis (Tween 80)—*Mycobacterium*

A. Tacquet, F. Tison and B. Devulder (1965). Annls. Inst. Pasteur, Paris 108, 514–525.

The authors used the method of A. Andrejew, Ch. Gernez-Rieux, and A. Tacquet (Annls Inst. Pasteur, 1960, 99, 56–58).

Fat hydrolysis (Tween 20)—*Mycobacterium*

A. Andrejew and A. Tacquet (1965). Annls Inst. Pasteur, Paris 108, 652–661.

Méthode Chimique. – On prépare, d'une part, le tampon tris (hydroxyméthyl) aminométhane 0.05 M, additionné de tween 20 à raison de 20 p. 100 et ajusté à pH 8. On dissout, d'autre part, 25 mg de rouge de phénol dans 5 ml d'eau distillée et 1.6 ml de NaOH 0.05 N, et on ajoute ensuite

125 ml d'eau distillée (on obtient ainsi une solution de rouge de phénol a 0.05 p. 100). On ajoute à la solution tamponnée de tween la solution de I'indicateur jusqu'à la concentration finale de 0.003 p. 100 en R. P. Ce mélange (tampon + tween + R. P.) doit être nettement rouge carmin au moment de l'emploi. S'il est jaunâtre, il est nécessaire d'y ajouter un minimum de NaOH pour obtenir cette teinte. On doit renouveler le mélange au moins une fois par semaine.

Test qualitatif. Dans un tube à hémolyse de 3 ml on introduit 1 ml du mélange (tampon + tween + R. P.) et la suspension bactérienne (correspondant environ à 3 ou 4 mg de bactéries, en poids sec). On complète à 3 ml avec de l'eau distillée. On bouche le tube en y laissant le moins d'air possible. On agite et on place à 38°, en présence de témoins (l'un dépourvu de bactéries, l'autre de tween). Le virage de l'indicateur est d'autant plus rapide que l'activité lipasique est plus forte.

Procédé quantitatif. On fait passer un léger courant d'azote à travers le mélange contenant les bactéries, le tween et l'indicateur (R. P.) et on dose l'acide libéré à l'aide de NaOH 0.02 M, après une heure de contact à 38°, en présence de témoins.

Fat hydrolysis—*Streptomyces*

T. F. Fryer and M. E. Sharpe (1965). J. Dairy Res. 32, 27–34.

Lipolysis was tested on a tributyrin agar medium (Franklin and Sharpe) and a butter fat agar (Jones and Richards, 1952).

0.05 ml of an 18–24 hr culture of the organism grown in 5 ml Yeastrel nutrient broth, centrifuged and resuspended in 5 ml distilled water was used as inoculum.

J. G. Franklin and M. E. Sharpe (1963). J. Dairy Res. 30, 87.
A. Jones and T. Richards (1952). Proc. Soc. appl. Bact. 15, 82.

Fat hydrolysis—*Mycobacterium*

P. Hauduroy, F. Bel and A. Hovanessian (1966). Annls Inst. Pasteur, Paris 111, 84–86.

Recherche des lipases préconisée par Andrejew, Tacquet, Gernez-Rieux (1, 2). Nous avons employé la technique simplifée de Wayne (12).

(1) Andrejew (A.), Gernez-Rieux (Ch.) et Tacquet (A.). Ann. Inst. Pasteur, 1960, 99, 56.
(2) Andrejew (A.) et Tacquet (A.). Ann. Inst. Pasteur, 1965, 108, 652–661.
(12) Wayne (L. G.). Amer. Rev. resp. Dis., 1964, 86, 579–581.

Fat hydrolysis—*Vibrio, Pseudomonas, Phytobacterium, Achromobacter, Acinetobacter, Flavobacterium, Empedobacter, Erwinia,* and *Serratia*

C. Tysset, J. Brisou and A. Cudennec (1966). Annls Inst. Pasteur, Paris 111, 363–368.

La lipase a été mise en évidence à l'aide du Tween 80 (ester du sorbitol et de l'acide oléique), par la méthode de Sierra [8]. Le milieu de base indiqué par Sierra a été modifié pour l'adapter aux bactéries marines. Il est composé comme suit :

Peptone	10	g
$CaCl_2$	0.1	g
Gélose	20	g
Eau de mer âgée	1000	ml

Le pH est ajusté à 7.6. Ce milieu chauffé est filtré sur papier et additionné de Tween 80 pour une concentration finale de 1 p. 100. Il est alors réparti à raison de 20 ml par tube de 20 × 22 mm. Les tubes bouchés au coton cardé sont stérilisés vingt minutes à 120°C et stockés à + 4°C.

Pour l'emploi, le milieu en culot est fondu au bain-marie bouillant, refroidi à 45°-50°C, et coulé en boîtes de Petri de 100 mm. Lorsque la surface de la gélose est bien sèche, l'ensemencement est pratiqué en points à partir d'une colonie en culture pure sur gélose de 18 à 24 heures. L'incubation a lieu à 25°C pendant trois jours.

En vingt-quatre heures, et rarement au bout de deux ou trois jours, il se développe une couronne opaque de 2 à 3 mm de rayon autour des colonies lipolytiques. Cette couronne est constituée de cristaux d'oléate de calcium bien décrits par Sebald [7]. Les réactions positives sont caractérisées par la présence des cristaux d'oléate de calcium observés à grossissement 100 dans la couronne opaque ou dans la colonie microbienne (Pour des raisons exposées ailleurs, l'observation macroscopique d'une couronne opacifiée n'est pas toujours suffisante pour établir la présence d'une lipase).

(7) Sebald (M.). C. R. Acad. Sci., 1959, 248, 3363–3365.
(8) Sierra (G.). Antonie van Leeuwenhoek, 1957, 23, 15.

Fat hydrolysis—*Mycobacterium*

F. Tison, A. Tacquet, P. Roos and B. Devulder (avec la collaboration technique de Mlle. M.-M. Delaporte) (1966). Annls Inst. Pasteur, Paris 110, 784–787.

Nos premiers essais consistaient à apprécier le halo de précipitation d'un sel calcique par l'action enzymatique sur les différents tweens (Sierra). Nous avons renoncé à cette technique trop sensible, la cristallisation des oléate, palmitate, stéarate de calcium se produisant dès la végétation de nombreuses espèces de Mycobactéries.

C'est pourquoi nous avons eu recours à une méthode d'appréciation de la diffusion en gélose d'un colorant des graisses, le bleu Victoria B [5, 7].

Matériel et Méthodes

Pour nos essais préliminaires, nous avons préparé un milieu de Dubos gélosé sans tween additionné de tributyrine (1) colorée par le bleu Victoria B. Ce milieu a été réparti en boîtes de Petri et en tubes de Felix.

La végétation des Mycobactéries y donne naissance, autour de la colonie (sur boîte de Petri) ou sous la couche de culture (en tube de Felix) à un halo bleu intense, se développant rapidement.

La très grande sensibilité de cette méthode et sa forte positivité pour les différentes espèces nous ont amenés à lui apporter certaines modifications après étude infructueuse de certains inhibiteurs tels que le flourure de sodium.

Le test a été effectué, non plus sur des Mycobactéries en phase de croissance, mais sur des suspensions de germes intacts.

Technique.

a) Préparation du milieu de diffusion. – Pour éviter l'apport d'éléments nutritifs, il est fait usage d'une ⟨⟨ gélose blanche ⟩⟩ contenant :

Agar purifié	20 g
Eau distillée	1000 ml

stérilisée à l'autoclave, ajustée a pH 7.3.

b) Préparation du substrat coloré. – Ajouter, dans la tributyrine, 0.5 p. 100 en poids de bleu Victoria B ; bien mélanger au bain-marie bouillant, pendant 30 minutes. 1 p. 100 de ce substrat coloré est mélange au milieu à 55°C. Après obtention d'une émulsion homogène, on répartit, à chaud, en tubes de Felix qui sont plongés dans la glace fondante. Le milieu terminé doit être opalescent, mauve rosé. Il ne se conserve pas et doit être utilisé rapidement. La souche à étudier doit être en plein développement sur milieu solide.

Les colonies prélevées sont mises en émulsion dans l'eau distillée stérile à raison de 0.1 mg par ml. 0.1 de cette suspension est déposé sur la surface du milieu à la tributyrine et les tubes sont portés pendant vingt-quatre heures à 37°C.

Les réactions positives se traduisent par l'apparition d'une bande bleue à la partie supérieure du milieu ; les réactions négatives, par l'absence de coloration bleue.

(1) Les essais simultanés avec trioléine et la tricaproïne se sont révélés moins démonstratifs qu'avec la tributyrine.

(5) Jones (A.) and Richards (T.). *Proc. Soc. app. Bact.*, 1952, 15, 82–93.

(7) Tuynenburg Muys (G.) et Willems (R.). *Ant. van Leeuwenhoek*, 1965, 31, 103–112.

Fat hydrolysis—*Pseudomonas fragi, Staphylococcus aureus* and *Geotrichum candidum*

J. L. Smith and J. A. Alford (1966). Appl. Microbiol. 14, 699–705.

Addition of lard or sodium oleate to the medium used for lipase production by *Pseudomonas fragi* resulted in a decreased accumulation of lipase in the culture supernatant fluid without affecting cell growth. The production and activity of lipase were inhibited by lard, sodium oleate, and the salts of other unsaturated fatty acids. Some divalent cations, Tweens, lecithin, and bovine serum prevented oleate inhibition but did not reverse it. Similar inhibitory actions were observed with *Geotrichum candidum* lipase, but not with a staphylococcal lipase or pancreatic lipase. A protective effect by protein in crude enzyme preparations is indicated. The ability of oleate to lower surface tension does not appear to be related to its ability to inhibit lipase.

Production and assay of P. fragi lipase. The strains of *P. fragi* employed, the conditions of lipase production, and the methods of assay were as previously reported (Alford and Pierce, 1963), with the following exceptions.

(i) In some media, lard emulsions to give a final concentration of 0.5% lard were added to the culture medium. The supernatant fluid from these cultures contained FFA released from the fat as the lipase was produced. This titratable acidity was subtracted from the value obtained in the assay to give the net activity shown in the data. (ii) Assay samples were incubated for 1 hr. Unless noted otherwise, the lipase employed was the supernatant fluid from a 4- or 5-day culture of *P. fragi* grown in 1% Case peptone medium from Case Labs., Chicago, Ill.

Production and assay of G. candidum lipase. Maximal production of lipase by *G. candidum* was obtained in a high protein medium as previously described (Alford and Smith, 1965). The following synthetic medium gave good yield of lipase when incubated at 20°C for 4 days: 0.1% NH_4Cl, 0.15% KH_2PO_4, 0.012% $MgSO_4 \cdot 7H_2O$, 0.234% monosodium L-glutamate, 0.16% L-arginine HCl, 0.07% L-lysine HCl, and 0.001% each of $FeSO_4 \cdot 7H_2O$, $ZnSO_4 \cdot 7H_2O$, and $MnSO_4 \cdot H_2O$. After the medium had been autoclaved, sterile glucose was added to give a final concentration of 0.25%. Assay conditions were the same as for *P. fragi.*

Production and assay of S. aureus lipase. The lipase of *S. aureus* was produced in phosphate-buffered Trypticase (BBL) broth incubated on a shaker at 30°C for 24 hr, as previously described (Alford and Smith, 1965).

J. A. Alford and D. A. Pierce (1963). J. Bact. 86, 24–29.
J. A. Alford and J. L. Smith (1965). J. Am. Oil Chem. Soc. 42, 1038–1040.

Fat hydrolysis—*Pseudomonas*

R. Y. Stanier, N. J. Palleroni and M. Doudoroff (1966). J. gen. Microbiol. 43, 159–271.

The production of extracellular lipases was tested by patching strains on the Tween 80 medium recommended by Sierra (1957). Hydrolysis of the detergent is shown by formation in the medium of a precipitate of calcium oleate. The time of appearance of a turbid halo surrounding the patch of growth is very variable; we arbitrarily scored as positive those strains which showed such a halo after not more than 6 days of incubation. A weak reaction occurring after a longer period may simply reflect autolysis of the bacteria and liberation of intracellular esterases.

G. Sierra (1957). Antonie van Leeuwenhoek 23, 15.

Fat hydrolysis—*Staphylococcus*

M. Kocur, F. Přecechtěl and T. Martinec (1966). J. Path. Bact. 92, 331–336.

The egg yolk reaction was investigated on an egg yolk agar medium by the method of Shah, Russell and Wilson (1963). Lipase production was studied in cultures grown on nutrient agar with olive oil for 5–7 days (Bulder, 1955).

C. J. E. A. Bulder (1955). Antonie van Leeuwenhoek 21, 433.
D. B. Shah, K. E. Russell and J. B. Wilson (1963). J. Bact. 85, 1181.

Fat hydrolysis—*Staphylococcus*

A. T. Willis, J. A. Smith and J. J. O'Connor (1966). J. Path. Bact. 92, 345–358.

Culture media. The nutrient base used in all the following media was heart infusion broth (Difco), solidified as required with 1.5 per cent of Davis New Zealand agar. Sterilization was at 121°C for 15 min.

Egg yolk agar was prepared as described by Willis (1960).

Egg yolk broth was prepared as described by Gillespie and Alder (1952), except that heart infusion broth (Difco) was used as the nutrient base.

Tween 80 agar was used as described by Gonzalez and Sierra (1961).

Inoculation of media. The agar media were inoculated from fresh 18-hr broth cultures with sterile throat swabs. This method ensured even inoculation, and rapid multiple subcultures could be made without recharging the swab. A 9 cm petri dish conveniently accommodated 9–12 cultures. Egg yolk broth was inoculated from the 18-hr broth culture with a charged loop.

Egg yolk agar proved more suitable than egg yolk broth. There was absolute correlation between the egg yolk agar and the Tween 80 agar for lipolysis with yellow but not with orange and white strains.

W. A. Gillespie and V. G. Alder (1952). J. Path. Bact. 64, 187.
C. Gonzalez and G. Sierra (1961). Nature, Lond. 189, 601.
A. T. Willis (1960). J. Path. Bact. 80, 379.

Fat hydrolysis—RM bacterium

A. J. Ross, R. R. Rucker and W. H. Ewing (1966). Can. J. Microbiol. 12, 763–770.

Lipolysis was studied by the methods employed by Hugh and Beveridge (1962) and Davis and Ewing (1964).

B. R. Davis and W. H. Ewing (1964). J. Bact. 88, 16–19.
W. B. Hugh and E. G. Beveridge (1962). Appl. Bact. 25, 72–82.

Fat hydrolysis—anaerobes

A. -R. Prévot (1966). Techniques pour le diagnostic des Bactéries Anaérobes. Éditions de la Tourelle, St. Mandé.

Recherche des lipases.

De nombreuses méthodes avaient été proposées pour la mise en évidence des lipases bactériennes (méthodes d'Eijkmann ; de Knaysi ; de Fleming et Neil ; méthode stalagmométrique de German ; méthode de Sierra). En 1960, Sébald et Prévot ont adapté cette dernière aux anaérobies.

Le milieu de Sierra est modifié comme suit :

Peptone	10	g
NaCl	5	g
$CaCl_2$	0.1	g
Eau distillée	1000	cm^3

Ajuster à pH 7.4.

Ajouter gélose 12 g.

Précipiter à l'autoclave à 120° filtrer.

Ajouter l'un des 3 tween (40, 60, 80) à la concentration finale de 0.5%. Répartir à raison de 4.5 cm^3 en tubes de 8–9 mm. Stériliser 30 min à 110°. Après régénération, les tubes sont ramenés à 45°, ensemencés à pipette fermée et refroidis. On met à 37° pendant 1 à 20 jours. Chaque souche est ensemencée sur les 3 milieux contenant chacun un des tweens 40, 60 et 80. La détection d'une lipase correspondante : sorbitol – palmitase, sorbitol-stéarase et sorbitol – oléase est faite par l'apparition d'un halo autour des colonies lipidolytiques, dû à l'hydrolyse du lipide correspondant. On note également la rapidité ou la lenteur de la lipidolyse, par la date d'apparition du halo, et son intensité par la taille du halo.

On peut également couler le même milieu en boîte de Pétri en couche épaisse dans le couvercle de la boîte comme pour l'antibiogramme anaérobie et couvrir l'ensemble après ensemencement par la boîte renversée.

On peut également rechercher les exolipases dans le filtrat des cultures anaérobies d'une souche en déposant ce filtrat dans la cavité découpée dans la gélose au tween à la manière de la méthode d'Oxford pour les antibiotiques. Le halo d'hydrolyse diffuse autour de la rondelle découpée.

Par ces techniques Sébald et Prévot ont montré que les exolipases anaérobies peuvent donner lieu à 2 types de réactions :

1) Réactions précoces, généralement intenses, dues à des enzymes constitutives *(Cl. botulinum* A et B ; *Cl. sporogenes ; Cl. hemolyticum ; Cl. cauteretsensis ; Cl. subterminale)* et décelables dans les filtrats de cultures sur milieux ordinaires dépourvus de tween.

2) Réactions tardives et discrètes dues à des enzymes adaptatives ne se formant qu'en présence de tween *(Cl. oedematiens ; Cl. bifermentans ; W. perfringens).*

FAT STAINS

Fat stain

T. Rettie (1931). J. Path. Bact. 34, 595–596.

A new method of applying nile blue as a fat stain.

Thorpe (1907) prepared "nile pink" by dissolving 3 grams nile blue A in 500 ml water, adding 3 ml sulphuric acid and 100 ml xylol. The mixture is heated on a water bath with continuous shaking. The nile pink dissolves in the xylol and is recovered by distilling off the xylol. Nile pink forms salts with acids but on dilution these are hydrolyzed and the base separates out. In nile blue solutions this hydrolysis is prevented and a considerable amount of the pink is held in solution and a satisfactory fat stain results. When used on tissues containing very fine fat globules the blue tends to mask the pink, differentiation with dilute acid takes out some pink along with the blue and the smaller globules are apt to become invisible. To overcome this difficulty and to convert nile blue with a progressive instead of a regressive stain, the following method has been devised:—

A saturated solution of nile pink in spirit is prepared; this keeps indefinitely and contains roughly about 0.075% of the stain. This is a stock solution: it will not stain. The staining solution is prepared by mixing 2 ml stock stain solution with 1 ml 20% H_2SO_4, 7 ml 5% dextrin solution and 0.2 ml of 0.5% nile blue. This gives a solution containing 0.015% nile pink in 0.01% nile blue, 20% alcohol and 2.0% H_2SO_4. If water is used for dilution the stain very soon crystallizes out (1 hour), but with dextrin the crystallization is delayed and the solution may be kept for several days. Large quantities should not be made as there is a continuous, slow deposition of stain. Frozen sections stained in this for from ½ – 1 hour, washed in water, and mounted in glycerol or farrant solution show all fat bright pink against a grey blue background.

To demonstrate very fine granules, as in fatty degeneration of the heart, stain may be used without addition of nile blue; then by raising the condenser the finest pink granules can be easily observed.

For demonstrating fat in fresh or fixed thin layer tissues such as diaphragm or omentum, a convenient solution may be prepared by taking 1 ml stock stain solution, 2 ml 20% H_2SO_4 and diluting with water to 40 ml. The tissue is floated in this solution and fat alone is stained pink.

Fat stain—*Bacillus, Azotobacter, Rhizobium, Mycobacterium, Oöspora, Spirillum*

T. L. Hartman (1940). Stain Technol. 15, 23–28.

The use of sudan black B as a bacterial fat stain.

The method of staining the fats of bacteria is to suspend the organisms in the dye and prepare flat wet mounts. For this procedure it is of value to have a stain which will readily suspend the bacteria in question and not cause precipitates or plasmolysis of the cells due to evaporation of the dye solvent. The solvents found to meet these requirements are 70% alcohol, 50% acetone and water, 50% dioxan and water, and ethylene glycol.

MATERIALS AND METHODS

The sudan black B used during this study was furnished by the National Aniline and Chemical Company, New York City. The staining solution was prepared by dissolving 0.25 g of the dry sudan black B powder in 100 ml of 70% alcohol at room temperature. Solution of the dye takes place almost immediately. The staining solution, as described by Leach (1938), was prepared by adding excess dye to equal amounts of diacetin and distilled water, followed by two days' incubation at 55°C and filtering before use.

Saturated solutions of sudan black B in 50% acetone and distilled water, 50% dioxane and distilled water, and ethylene glycol (should have a boiling point of 195–7°C) were prepared. For best results it is advisable to let the solutions stand for several days, since the acetone and dioxane require considerable time for solution to take place.

The method used in staining the bacteria was to suspend a loopful of the cells in a drop of the stain solution and prepare flat wet mounts.

Sudan black B dissolved in 70% alcohol or ethylene glycol stains the fat bodies of bacteria a deep blue-black color and is recommended as superior to the other sudans.

The alcoholic solution retained its staining ability for over six months without signs of deterioration.

Incubation of cultures varied from 24 hours to 14 days with different organisms.

E. H. Leach (1938). J. Path. Bact. 47, 635.

Fat stain

G. Knaysi (1942). J. Bact. 43, 365–385.

Stain for lipids—Sudan III.

To test for the presence of fat we used a saturated solution in 70 per cent of alcohol by volume. This solution can be used in two ways: (a) a small loopful of a heavy suspension of the organisms in water may be mixed with a droplet of the Sudan III solution on a slide and observed under a cover-glass, or (b) a smear of the organisms is prepared, allowed to dry in the air and immersed in the Sudan III solution for 20 to 30 minutes, after which it is rapidly rinsed with running water, blotted, dried and examined in a droplet of distilled water under a cover-glass.

Fat stain

K. L. Burdon, J. C. Stokes and C. E. Kimbrough (1942). J. Bact. 43, 717–724.

Sudan black B—safranin:

Hartman (1940) used wet preparations only, suspending the bacteria in a solution of Sudan black B in 70% alcohol or in ethylene glycol – the latter being preferred.

The fat globules appear as blue-black bodies in a colorless cytoplasm.

The authors propose the following method:

Technique—Stock Sudan black B solution.

A saturated solution is made by adding excess (*c.* 0.3 g) of the dry stain (obtained from National Aniline and Chemical Company, N. Y.) to 100 ml of 70 per cent alcohol in a screw-top bottle. The solution is ready for use after standing at room temperature, with occasional shaking, for 24 hours. It keeps without loss of staining power for at least one month.

Cultures: To test the aerobic spore-bearers for capacity to store fat, the organisms are grown upon slants of glycerol infusion agar or glucose infusion agar. If fat is accumulated at all by the cells, it appears in approximately equal amounts on either medium within 24 hours and usually is at a maximum after about 48-hours' incubation. The fat granules in smears from sugar agar cultures usually stain somewhat less intensely than those in preparations from glycerol agar slants.

Procedure: Just before use, the stock Sudan black B solution is filtered into small test tubes in about 0.5 ml amounts. The bacteria from the slant cultures under examination are emulsified directly in the staining solution, and the emulsions are allowed to stand at room temperature for 15–20 minutes. During this time the precipitate present largely settles out. A loopful of the emulsion is then removed from the top of the fluid and smeared with a circular motion on a clean slide so that it dries quickly, leaving precipitated particles at the periphery of the drop. The slide is not heated. A 1 per cent aqueous solution of safranin is now applied briefly, and the smear is then washed with water and dried in the usual manner.

The cytoplasm of the bacterial cells is thus stained pink, while the

fatty material contained in it takes on a bluish-gray or bluish-black color. The stained smears, without coverglasses, when stored in an ordinary slide box, remain unchanged for at least six months.
T. L. Hartman (1940). Stain Technol. 15, 23–28.

Fat stain—*Malleomyces mallei*

G. Worley Jr. and G. Young (1945). J. Bact. 49, 97–100.

Hartman's method (1940) showed that lipids are quite common in cells.
T. Hartman (1940). Stain Technol. 15, 23–28.

Fat stain—general

Society of american Bacteriologists *Manual of Microbiological Methods* – McGraw-Hill Book Co. Inc., New York 1957.

Stain for fat droplets – Burdon's Method (1946). Staining solution: 0.3 g of Sudan black B (commission certified) in 100 ml of 70 per cent ethyl alcohol. After the bulk is dissolved, shake at intervals and allow to stand overnight.

Staining schedule:

1. Prepare smears as usual from 18–24 hr cultures, and fix by heat.
2. Flood the entire slide with the above staining solution and allow it to stand undisturbed at room temperature for 5–15 min. (Exact time is unimportant, as good results are often obtained after only 1 or 2 min; on the other hand, no harm results if the slides stain until the solution is completely dry.)
3. Drain, and blot slide completely dry.
4. Cover with xylene by pouring from a dropping bottle or dipping several times in a staining jar. Blot until dry.
5. Counterstain 5–10 sec with 0.5 per cent aqueous safranin, taking care not to overstain.

Note: For acid-fast organisms, Ziehl's carbon fuchsin diluted 1:10 with distilled water may be applied for 1–3 min, instead of safranin.

6. Wash in tap water, blot, and dry.

Results: Fat droplets stain blue-black or blue-gray; the rest of the cell stains pink.
K. L. Burdon (1946). J. Bact. 52, 665–678.

Fat stain

K. L. Burdon (1946). J. Bact. 52, 665–678.

Sudan black B for fixed preparations

1. Prepare the film, let it dry thoroughly in the air, and fix it by heat in the usual way. (Chemical fixation has no special advantages and may result in some loss of demonstrable lipid.)
2. Flood the entire slide with sudan black solution (0.3 g of the powder-

ed stain in 100 ml of 70 per cent ethyl alcohol), and allow the slide to remain undisturbed at room temperature for 5 to 15 minutes. (A staining period of less than 5 minutes will often suffice, but the intracellular lipid is colored somewhat more intensely when the staining is continued for 5 minutes or longer. No further staining apparently occurs after the solution precipitates and turns a greenish or brownish color, but no harm is done if the stain is allowed to dry completely over the film.)

3. Drain off excess stain and blot the slide thoroughly dry.
4. Clear the slide with cp xylol by dipping it in and out of the solvent in a Coplin jar or by adding xylol from a dropping bottle. Blot the cleared slide dry.
5. Counterstain with safranine (0.5 per cent aqueous solution) for 5 to 10 seconds (for ordinary bacteria or fungi), or with dilute carbol fuchsin (Ziehl's carbol fuchsin diluted 1:10 with distilled water) for 1 to 3 minutes (for acid-fast organisms). (Overstaining with the counterstain must be avoided.)
6. Wash in water, blot, and dry the slide.

Comment: After the bulk of the dye has been dissolved, the sudan black B solution should be thoroughly shaken at intervals, then allowed to stand overnight before use. It remains good for several months at room temperature if kept in a well-stoppered, chemically clean container.

The entire slide is flooded with the staining solution to prevent the too rapid evaporation that otherwise occurs. (For reasons not entirely understood, the staining is generally unsatisfactory when slides are immersed in the sudan black B solution in a Coplin jar.) Since the cellular lipid in most organisms takes up the characteristic blue-black color almost at once, it is possible to complete the whole fat-staining procedure within a minute or two if desired. On the other hand, if the technician is occupied with other tasks, the stain may simply be allowed to dry on the slide; then the clearing with xylol and counterstaining may be carried out later at a more convenient time. If the sudan black B solution is allowed to stand on the slide for about 15 minutes and is then set afire by applying the Bunsen flame to the fluid, followed by blotting and xylol-clearing as usual, the intracellular fat in some organisms (e.g., gonococci) is brought out more clearly. Ordinarily this step is unnecessary.

Examination of the cleared preparation without counterstaining is sometimes interesting and revealing. Care should be taken to avoid obscuring very tiny fat droplets by too strong a counterstain. Films must be examined with the oil immersion lens under critical illumination. To discern the smallest lipoid particles, the observer must have a good sense for the color distinction between the blue-black or blue-gray of the fat droplets and the pink of the counterstain.

Fat stain—*Mycobacterium*

H. L. Sheehan and F. Whitwell (1949). J. Path. Bact. 61, 269–271.

The staining of tubercle bacilli with sudan black B.

1. The sputum or other material is smeared not too thickly on a slide and is fixed, while still wet, with Carnoy's fluid for half a minute. Heat fixation after drying the smear is a satisfactory but tedious method. Most other fixatives of wet smears are also quite good if not allowed to act for too long. The sudanophilic lipids can be removed by treatment with chloroform or carbon tetrachloride in about half-an-hour, with xylol in about 3 hours and with alcohol in about 8 hours, but they are unaffected by treatment with acetone for over 24 hours. Antiformin concentration methods do not interfere with the staining. Paraffin sections of tissues are always negative, as would be expected.
2. The fixative is drained off, and the slide is half covered with a saturated solution of sudan black B in 70 per cent alcohol. The stain is lighted and allowed to burn out, and the slide is then swilled with water. The stain is made by adding 3 parts of water to 7 parts of a thoroughly saturated solution of sudan black B in absolute alcohol and filtering the mixture. The diluted stain deteriorates slowly in the course of a few weeks. The stain is efficient over a pH range of about 5 to 9, and with various strengths of alcohol between 60 and 80 per cent. Other solvents, such as dilutions of acetone, chloroform, dioxane, or ethylene glycol, are less satisfactory. Prolonged exposure of films to heated sudan solutions tends to stain other organisms, which may then be partially acetone-fast.
3. The gross deposit of stain on both sides of the slide is washed off with acetone, using a drop bottle, and the smear is then differentiated by immersion in a jar of acetone for 2 minutes. Differentiation may be continued for 6 hours without affecting the staining of tubercle bacilli. Immersion for half an hour in acid alochol, xylol, or chloroform does not cause any appreciable loss of stain from these organisms.
4. The smear is rinsed with water and counterstained with 1 per cent aqueous solution of pyronine for 10 to 60 seconds. It is then washed quickly with water and dried with blotting paper.

The counterstain should only be strong enough to allow easy focusing. Ziehl-Neelsen cannot be used as a superimposed stain as it removes the Sudan. Furthermore, if it is applied previously, it inhibits the Sudan staining.

Fat stain—*Bacillus*

K. L. Burdon (1956). J. Bact. 71, 25–42.

Stainable fat was observed by preparing smears stained with the improved form of the sudan black B-safranin fat stain (Burdon, 1946) from cultures 6 to 9, 18 to 24, and 36 to 48 hr old. The slides were flooded

with sudan black B solution (0.3 g of the stain in 100 ml of 70 per cent alcohol) and left undisturbed for 20 to 30 min, or until the stain dried on the smear. The smears were then blotted dry, cleared by repeated dipping in xylol in a Coplin jar, again blotted dry, and then counterstained lightly with 0.5 per cent aqueous safranin.
K. L. Burdon (1946). J. Bact. 52, 665–678.

Fat stain—*Nocardia*

J. N. Adams and N. M. McClung (1962). J. gen. Microbiol. 28, 231–241.

Heat fixed impression smears of *Nocardia rubra* were stained according to Burdon's (1946) sudan black B method; this technique was modified by omitting the counterstain. The lipid-staining method of Clark and Aldridge (1960) was also used.

The effects of lipase digestion of heat fixed impression preparations were also studied. These experiments were conducted by subjecting *Nocardia rubra* to digestion with a 0.01% (w/v) solution of lipase in 0.1 M $Na_2HPO_4 + KH_2PO_4$ (pH 7.0) at room temperatures for 15 and 30 minute periods. The preparations were then stained by the modified Burdon technique and examined for the presence of lipid inclusions.

Successive staining:

Two successive staining techniques similar to those of Tronnier (1953) and Knaysi (1955, 1959) were used in the present work. The first method may be more readily described as counterstaining rather than successive staining; this method for the differentiation of lipid inclusions from metachromatic granules was carried out as follows: a heat fixed impression smear of *Nocardia rubra* was stained by the modified Burdon technique. A counterstain was then used by treatment with azure A to reveal metachromatic granules. After the smear had been treated with both staining reagents, the slide was washed in distilled water, blotted and examined. This method allowed simultaneous demonstration of lipid inclusions and metachromatic granules.

A second, more complex method for successive staining was done as follows: an osmium-fixed impression was prepared in the usual way and stained with one of the previously described stains (e.g., azure A for metachromatic granules). The slide was observed under oil, representative fields located, and mechanical stage settings recorded; sketches of the field under observation also aided a return to the same field. Such fields were then photographed. The immersion oil was then removed with xylene, which also partially decolorized the preparation. Further decolorization was carried out with 95% (v/v) ethanol in water until no more colour was removed. The slide was then washed in copious quantities of water and, after drying, restained by one of the methods described above (e.g., Burdon's modified lipid-staining technique); the fields which had previously been photographed were again located. After re-examination and rephotographing, the preparation could again be decolorized, after which staining for a third or fourth time was possible. Although the sequence of staining techniques applied to a single preparation was theoretically unlimited, practical considerations led to definite sequences which had to be used. Acid hydrolysis, as used in the chromatin-staining procedures, partially destroyed cellular integrity, and consequently was limited to the last place in a successively stained sequence. Likewise, mordanting in 10% (w/v) tannic acid for 30 min, as used in cell wall staining techniques, affected the staining affinity of the organism; hence cell wall staining was usually relegated to the last or next to last position of a sequence. Sometimes, however, cell wall staining had to be done first in a sequence.

K. L. Burdon (1946). J. Bact. 52, 665.
J. B. Clark and C. Aldridge (1960). J. Bact. 79, 756.
G. Knaysi (1955). J. Bact. 69, 117 and 130.
G. Knaysi (1959). J. Bact. 77, 532.
E. A. Tronnier (1953). Zentbl. Bakt ParasitKde (Abt I) 159, 213.

"FERMENTATION" OF CARBOHYDRATES

Fermentation

C. S. Mudge (1917). J. Bact. 2, 403–415.

An attempt has been made to determine the extent of hydrolysis on sterilization of certain disaccharides commonly used in bacteriological investigations, with the following results.

By means of Barfoed's method maltose and lactose were found to be hydrolyzed to a considerable extent in water solution. Sucrose and raffinose did not break down.

Heating in streaming steam for three successive days in the Arnold steamer seems to hydrolyze lactose and maltose more than heating in the autoclave at 15 pounds for fifteen minutes.

By the use of bacterial cultures the hydrolysis of these sugars was studied in culture media, with similar results. Peptone broths containing the sugars mentioned were subjected to sterilization in the Arnold steamer in streaming steam and to various heatings in the autoclave at 15 pounds pressure. They were then inoculated with strains of bacteria which would not ferment the sugar introduced.

Lactose and maltose broth, so inoculated, were found to be fermented by the organisms introduced. The amount of fermentation increased as the sterilization increased. It is assumed that this fermentation is due to the presence of monosaccharides in the broth. By filtering sugar media this hydrolysis of the sugar can be avoided.

The sugars found above to be easily hydrolyzable, namely, maltose and lactose, formed acid when heated in culture media, or with alanine or asparagine in water solution, whereas raffinose, which is resistant to hydrolysis, formed no acid under like circumstances. The acidity produced in sterilization seems to be due to a reaction between the sugar and the amino groups.

Fermentation—*Streptococcus*

J. M. Sherman and W. R. Albus (1918). J. Bact. 3, 153–174.

The medium used for the fermentation tests had the following composition:

	per cent
Beef extract	0.3
Peptone	1.0
Dibasic potassium phosphate	0.5
Test substance	1.0

In the tests made with glucose, galactose, levulose, and maltose the dibasic potassium phosphate was omitted.

In making the tests, the cultures were incubated at 33°C and then titrated against N/20 NaOH, with phenolphthalein as indicator, and the results expressed as percentage of normal acid. An increase above the control tube of 1 per cent normal acid was regarded as a positive test for fermentation. The sugars were incubated seven days.

Fermentation—coliforms

L. A. Rogers, W. M. Clark and H. A. Lubs (1918). J. Bact. 3, 231–252.

The procedure used was the following: The medium was composed of 1 per cent Witte peptone, 0.5 per cent K_2HPO_4, and 1 per cent of the carbohydrate or alcohol. This medium was distributed (in 10 ml portions) in test tubes sterilized by the intermittent method, inoculated from agar slopes and incubated for five days at 30°C. The pH values before and after fermentation were determined by the colorimetric procedure described by Clark and Lubs (1917). A sufficient number of electrometric measurements were made to show that in the particular medium used the colorimetric pH values were consistent and essentially correct.

W. M. Clark and H. A. Lubs (1917). J. Bact. 2, 1–34, 109–136, 191–236.

Fermentation—*Xanthomonas cannae*

M. K. Bryan (1921). J. agric. Res. 21, no. 3, 143–152.

In fermentation tubes containing 1 per cent peptone plus 1 per cent saccharose, dextrose, lactose, maltose, glycerol or mannitol.

Fermentation—*Pseudomonas apii*

I. C. Jagger (1921). J. agric. Res. 21, No. 3, 185–188.

Fermentation tubes containing tap water to which have been added 2% peptone and 0.5% sodium chloride plus 1%, respectively, of lactose, dextrose, saccharose, and glycerol, show no growth in the closed arm and no gas formation. Litmus paper is used to test for acid production.

Fermentation—*Pseudomonas*

L. McCulloch (1924). J. agric. Res. 29, 159–177.

Litmus agar with 1 per cent peptone and 1 per cent of sugar was used. Fermentation tubes with 2 per cent Difco peptone and 2 per cent sugar were also used.

Fermentation—*Pseudomonas*

C. Elliott (1924). J. agric. Res. 29, 483–490.

Tests were made in media containing 1 per cent peptone and 2 per cent carbohydrate or 0.5 per cent peptone and 1 per cent carbohydrate and pH determinations were made colorimetrically after 1, 2, 3, 5, 6, 10, and 16 days.

Fermentation—*Xanthomonas*

F. A. Wolf (1924). J. agric. Res. 29, 57–68.

Bouillon to which was added sufficient of the carbon compounds to make a 1 per cent solution was employed to follow the progressive changes in hydrion concentration as an index to fermentative activity. The sugar was added to cooled, sterile bouillon flasked in convenient quantities and was tubed with aseptic precautions in previously sterilized test tubes.

Fermentation—*Clostridium*

G. F. Reddish and L. F. Rettger (1924). J. Bact. 9, 13–57.

Fermentation tests. Acid and gas production were determined in deep tubes of agar (1 per cent agar, pH 7.0), 0.5 per cent of the test substances being used. The tubes were sterilized at 15 pounds for ten minutes, quickly cooled to 45°C, and inoculated as soon as possible. Inoculations were made from diluted (1:10) twenty-four-hour plain broth cultures, 0.5 ml being added. They were then incubated aerobically at 37°C.

Plain agar tubes containing no test substance were inoculated and used as controls. Observations for gas formation were made on the third day, and again on the tenth day in the case of *C. putrificum.* At the end of this time tests were also made for acid production by adding bromthymol blue to 5 ml of the melted agar cultures.

Fermentation—*Xanthomonas*

F. R. Jones and L. McCulloch (1926). J. agric. Res. 33, 493–521.

Bromcresol purple was used instead of litmus as the indicator in beef agar, plus 1 per cent of the "sugar."

Fermentation—cellulolytic organisms

L. A. Bradley and L. F. Rettger (1927). J. Bact. 13, 321–345.

Casein digest broth plus 0.25% or 1% filtered (Berkefeld) carbohydrate. Slightly soluble materials were prepared aseptically and subjected to a shortened sterilization period. (See Kulp and Rettger, J. Bact. 9, 1924, 357–394).

Fermentation—*Pseudomonas*

M. K. Bryan (1928). J. agric. Res. 36, 225–235.

Fermentation was tested on beef extract agar and synthetic (ammonium phosphate) agar containing bromcresol purple with "sugar," and in fermentation tubes containing 1 per cent peptone water with the same sugars. Agar cultures were held for 2 weeks.

Fermentation—*Pseudomonas*

A. J. Riker, W. M. Banfield, W. H. Wright, G. W. Keitt, and H. E. Sagen (1930). J. agric. Res. 41, 507–540.

The basal medium chosen for the studies on the fermentation of different carbohydrates was peptone-salt medium: magnesium sulphate ($MgSO_4 \cdot 7H_2O$), 0.20 g; sodium chloride (NaCl), 0.2 g; calcium sulphate ($CaSO_4$), 0.1 g; dipotassium phosphate (K_2HPO_4), 0.2 g; peptone, 5.0 g; and distilled water, 1000 ml. Five liters of this basal medium were made up to double strength. This was measured in 225 ml portions, placed in flasks, and autoclaved at a pressure of 15 pounds for 1 hour. The various carbohydrates employed were made up to 1 per cent strength in 225 ml portions of distilled water and autoclaved in the same way. A flask containing 1 per cent carbohydrate solution was flamed and poured into a flask of the sterile double-strength basal medium. This mixture was then transferred aseptically, in 10 ml portions, into sterilized test tubes. The cultures were incubated at 21°C for 21 days before the carbohydrate fermentation was measured.

The fermentation of the carbohydrates was measured (1) by change in titratable acidity (2) by change in pH, and (3) by quantitative determinations of the sugar in the cultures.

Fermentation—*Streptococcus*

P. R. Edwards (1932). J. Bact. 23, 259–266.

Meat infusion	900 ml
Casein digest	100 ml
Andrade's indicator	1%
pH	7.6

The "sugars" were sterilized by filtration and added to the previously sterilized broth. Inoculated tubes were incubated at 37°C for 10 days.

Fermentation—*Pseudomonas*

M. Levine and D. Q. Anderson (1932). J. Bact. 23, 337–347.

Fermentation studies were carried out on "sugars" in media composed of peptone, 5 g; dipotassium phosphate, 1 g; the respective sugars, 0.3 per cent; 1.0 per cent Andrade indicator per liter. The reaction was adjusted to pH 7.0 to 7.1. Inoculations were made from agar slants and incubated at 23–25°C for 5 days. Acid production was indicated by a red coloration of the indicator.

Fermentation—*Propionibacterium*

C. H. Werkman and R. W. Brown (1933). J. Bact. 26, 393–417.

The cultures were grown in a medium containing:

Difco yeast extract	10 g
K_2HPO_4	1 g
Agar Difco	15 g
Dist. water	1000 ml

Bromthymol blue was used as an indicator, and the pH was adjusted to 7.0–7.2. Aqueous solutions of carbohydrates were prepared and sterilized separately and added to give a final concentration of 0.3%. Readings were taken after 7 days at 30°C. Acid production was checked by titration.

Fermentation—*Fusobacterium*

L. W. Slanetz and L. F. Rettger (1933). J. Bact. 26, 599–621.

Meat extract cysteine broth was used as the basic medium for fermentation studies. It contained 1 per cent proteose peptone, 0.3 per cent Liebig's meat extract and 0.15 per cent cysteine hydrochloride. The test substances were added in 1 per cent amounts. The reaction was adjusted to pH 7.2, and bromthymol blue indicator added. The medium was tubed in Durham fermentation tubes and sterilized in the autoclave (ten minutes at 10 pounds extra pressure). The tubes were inoculated immediately after cooling and placed in anaerobic jars. The liquid within the inverted tubes was not displaced by the exhaustion of the jars, and gas production could be detected in the usual manner. Freshly heated medium must be used, however, to obtain satisfactory results.

Fermentation—*Pseudomonas, Achromobacter*

H. E. Goresline (1933). J. Bact. 26, 435–457.

In order to study the carbohydrate requirements of these organisms the following medium was used:

Bacto peptone	10 g
K_2HPO_4	5 g
H_2O	1000 ml
Carbohydrates	0.2–0.6 per cent
pH	7.2

Andrade's indicator. Reactions of the indicator were periodically checked by determinations of pH and utilization of carbohydrate. Inoculations were from suspensions prepared from plain agar.

Fermentation—*Lactobacillus*

H. R. Curran, L. A. Rogers and E. O. Whittier (1933). J. Bact. 25, 595–621.

The basic medium used in our work consisted of casein digest broth prepared from c. p. casein by tryptic digestion. Ten per cent solutions of the test substance were made up in distilled water. Levulose solution was sterilized by filtration through a Berkefeld filter candle, while all the other sugar solutions were sterilized by steam under pressure. These were added aseptically to the sterile basic medium in sufficient quantity to yield a final concentration of 1 per cent. In order to furnish more favorable growth conditions, each 10 ml of the medium also received 3 drops of sterile horse serum. This serum was heated to 60°C for one-half hour prior to its use in the maltose solutions. The final reactions of all the media were between pH 6.8 and 7.1.

The inoculum was prepared by growing the culture at 37°C in casein digest broth containing 0.1 per cent glucose. At the end of twenty-four hours the organisms were sedimented by centrifugation, the supernatant fluid was poured off, and the cells were resuspended in sterile distilled water. This process was repeated several times, after which the washed cells were diluted with sterile water to an arbitrary density comparable with that of a standard suspension of barium carbonate. One-tenth milliliter of this suspension was used to seed 10-ml portions of the test medium. After ten days' incubation at 37°C the hydrogen ion concentration was determined colorimetrically by the method described by Brown (1924). A distinct change in pH from that of the inoculated sugar-free control was accepted as evidence of carbohydrate utilization.

Fermentation—*Streptococcus*

R. A. McKinney (1934). J. Bact. 27, 373–401.

Tested in sugar-free veal infusion broth, pH 7.8, containing 1% of filtered "sugar" and bromcresol purple.

Fermentation—*Lactobacillus*

J. E. Weiss and L. F. Rettger (1934). J. Bact. 28, 501–521.

"Sugars" were sterilized in 10% solution and added aseptically to yeast extract broth. Galactose, inulin, xylose, arabinose, levulose, and rhamnose underwent slight breakdown on steam sterilization; hence solutions of these sugars were sterilized by filtration. Bromthymol blue was used as an indicator. Readings were taken at 1, 2, 3, and 7 days.

Fermentation—*Clostridium*

S. E. Hartsell and L. F. Rettger (1934). J. Bact. 27, 497–515.

The basic medium employed in the fermentation studies was nutrient infusion broth to which had been added 0.1 to 0.2 per cent cysteine hydrochloride. Valley (1929) has shown that this medium is quite satisfactory for fermentation studies on anaerobic species of bacteria.
G. Valley (1929). J. Bact. 17, 12–13.

Fermentation—*Clostridium*

I. C. Hall and M. L. Snyder (1934). J. Bact. 28, 181–198.

Fermentation reactions were tested by two methods, as follows:
(a) Hiss serum water media containing 1 per cent of the various carbohydrates were prepared by filling slender tubes (150 mm by 10 mm) half full. Litmus was used as an indicator. The tubes were boiled for a few minutes to expel air, cooled, and inoculated with a few drops of fresh glucose-broth cultures of the various strains. Reaction was shown by coagulation of the serum and by reddening the reduction of the litmus.
(b) The tests were repeated in sugar-free broth in constricted tubes with marble seals to which, after boiling to expel oxygen, sufficient 10 per cent sterile carbohydrate solutions were added to yield 1 per cent concentrations. The same sugars were fermented as before, as indicated by abundant gas production and acid reaction to bromthymol blue on removal to a spot plate.

Fermentation—*Pasteurella tularensis*

C. M. Downs and G. C. Bond (1935). J. Bact. 30, 485–490.

The basic medium was meat extract broth, 2 per cent peptone, and 1.5 per cent agar to which cystine hydrochloride was added to make 0.01 per cent. The reaction was adjusted to pH 7.4, and enough phenol red indicator was added to give a distinct salmon pink shade. The carbohydrates were added to make a final concentration of 1 per cent. The medium was then sterilized in the autoclave and before tubing 5 per cent of horse serum was added, and the tubes slanted. The media were inoculated with the 21 strains of *Pasteurella tularense,* incubated at 37°C and observed daily.

Fermentation—*Bacteroides*

A. H. Eggerth (1935). J. Bact. 30, 277–299.

Medium for carbohydrate fermentation tests. This medium had the following composition: 1 per cent of Parke Davis peptone; 0.5 per cent of NaCl; 0.2 per cent of agar; and bromcresol purple as acid indicator. The medium was sterilized in small flasks in the autoclave. The carbohydrates (enough to make 1 per cent of the medium) were autoclaved separately in distilled water and added to the flasks; finally, 4 per cent of sterile serum was added. The completed medium was then distributed in 100 by 11 mm tubes.

It is important to inoculate the fermentation tubes very heavily; otherwise growth may not take place. For this purpose, a nichrome wire with a spiral at the end was used. Young cultures in brain medium served as the source of the inocula. Incubation of the fermentation tubes was continued for forty days, the jars being opened every five to eight days.

Fermentation—*Streptococcus*

A. C. Evans (1936). J. Bact. 31, 423–437.

The cultures were grown for 4 days in sugar-free beef infusion broth of pH 7.4 containing 1 per cent of the test substance. Tests for fermentation were made by adding 0.5 ml of culture to 8.5 ml of distilled water, with bromthymol blue as indicator, except for the determination of the final pH in glucose broth, for which methyl red was used. If the reaction as indicated by bromthymol blue was more acid than pH 6.6 it was recorded with a positive sign; if it was pH 6.6 or higher the reading was recorded.

Fermentation—*Clostridium*

R. S. Spray (1936). J. Bact. 32, 135–155.

Sugar-free fermentation base

To make 1 L. take

10 g	Difco Neopeptone
10 g	Difco Tryptone
2.5 g	agar flakes
1 L	distilled water

Boil to dissolve and add water to make 1 L. Adjust the reaction to pH 7.3 to 7.4. Andrade or other indicators may be added or omitted. These are usually reduced, although many organisms show typical fermentative reactions.

Divide into lots, and add 1 per cent of the desired sugars. Reserve one lot to be tubed without sugar as a control. Tube about 8 cm deep in 200 × 15 mm, or better, 200 × 13 mm tubes. Autoclave for 15 to 20 minutes at 15 lb pressure.

We utilize only the end reaction as evidence of fermentation. Presence and amount of gas cannot be relied on, as emphasized by Hall, Heller and others. We do not use vaseline caps or attempt any measurement of gas. Of course, great excess of gas and growth, as contrasted with the sugar-free control, is indicative. It may often serve as a fairly reliable index, but nothing more.

In general, 36 hours incubation appears to be the optimum time for testing acidity. Some species require 7 days.

We have tried adding various indicators to the medium. However, as observed by others, these are commonly reduced to an inactive form. By reducing the peptone content some indicators remain reasonably stable and active, but as the peptones are reduced in amount many species, such as *Cl. chauvoei* and *Cl. novyi,* may fail to grow. Hence, we have discontinued their use in the medium, and follow Hall's procedure using the spot plate, or remove aseptically, by pipette, some 5 drops of culture and test this in tubes of properly diluted indicator. If carefully done the culture may be reincubated for subsequent retest. A definite acid reaction, as contrasted with the sugar-free control, is sufficient indication of fermentation.

Hall and Heller – personal acknowledgement.

Fermentation—*Staphylococcus*

R. Thompson and D. Khorazo (1937). J. Bact. 34, 69–79.

Cultures were inoculated into tubes of 1 per cent Difco mannitol in beef infusion broth. After 3 days' incubation the presence of acid was tested for by adding 2 drops of bromcresol purple.

Fermentation—*Streptococcus*

C. E. Safford, J. M. Sherman and H. M. Hodge (1937). J. Bact. 33, 263–274.

Substances for the fermentation tests were sterilized separately in 10 per cent solutions and added to sterile yeast-extract peptone broth so as to give a 1 per cent concentration of the test substance. Maltose, inulin and salicin were sterilized by filtration, while the other test substances were autoclaved.

Fermentation—*Fusobacterium*

E. H. Spaulding and L. F. Rettger (1937). J. Bact. 34, 535–548.

Medium:

Proteose peptone	10 g
Liebig's meat extract	3 g
Cysteine HCl	1 g
Distilled water	1000 ml
pH	7.6

"Sugars" are sterilized by filtration and added to give 1% final concentration.

The medium must be used immediately after sterilization.

Production of acid was determined by the spot plate method after 1, 3 and 7 days at 37°C. A biconvex loopful of culture was added to approximately 0.5 ml distilled water containing one drop of 0.16% indicator in a white porcelain spot plate. The inoculum consisted of 0.1 ml of a 24 hr potato broth culture.

A range of indicators was used to determine final pH.

Fermentation—*Mycobacterium*

R. E. Gordon (1937). J. Bact. 34, 617–630.

Carbohydrates. The ability of these cultures to utilize certain carbohydrates as the sole source of carbon was determined by the method described by Merrill (1931). The medium has the following composition:

	per cent
NaCl	0.5
$(NH_4)_2SO_4$	0.5
$Mg \cdot SO_4$	0.0005
KH_2PO_4	0.04
K_2HPO_4	0.16
Agar	2.0
Phenol red (in alkaline solution)	0.0001

Five milliliters of the above medium were placed in tubes and sterilized by autoclaving. One-half per cent of the carbohydrate, which had been sterilized by Berkefeld filtration, was added to each tube.

Utilization of each carbohydrate was established by growth of the culture on the above medium and by the accompanying acid color of the indicator.

M. Merrill (1931). J. Bact. 21, 361–375.

Fermentation—*Sphaerophorus necorphorus*

G. M. Dack, L. R. Dragstedt, R. Johnson and N. B. McCullough (1938). Comparison of *Bacterium necrophorum* from ulcerative colitis in man with strains isolated from animals. J. infect. Dis. 62, 169–180.

Many difficulties have been encountered in determining the biochemical properties of this group of organisms. Some strains fail to grow in a basic medium or grow with great irregularity. More uniform growth occurs when a fermentable carbohydrate is present. The addition of 10% sheep serum, 0.05% cystine or 0.1% cysteine has been found to alter the basic medium so that it will support the growth of most strains. Some strains produce acid or gas or both in the basic medium without the addition of a test substance. Sometimes the bromcresol purple indicator used in these tests was changed in color, especially when serum was added to the basic medium, so that an accurate indication of the pH change was not obtained.

Tests have been made in both broth and agar. A glass electrode potentiometer was used to determine pH changes for a large number of strains. The tests were made in Wassermann tubes containing 3 ml of medium with 1% of the test substance added. To insure good growth, a large amount of inoculum was necessary. Accordingly in most tests 3 drops of a 24–48 hour culture which was growing in Rosenow's* brain medium was used for each tube. The tests were always made in duplicate. After inoculation, the tubes were packed in a pyrex desiccator and incubated under anaerobic conditions, as previously described.

Under the conditions mentioned, growth usually occurred in 24 hours. The test period was 48 hours in one series, but in many tests the incubation period extended to 2 weeks and longer. The longer incubation period did not appear to possess any advantages.

*Strain 114 failed to grow in this medium. Inoculum was from anaerobic blood agar slant cultures.

Fermentation—gonococci

P. J. Almaden (1938). The mucoid phase in dissociation of the gonococcus. J. infect. Dis. 62, 36–39.

Medium

By combining some of the advantages of the media of Hitchens, Raven, and Price, a substrate was prepared which, after being autoclaved, was capable of supporting a luxuriant growth of gonococci. The base of the medium was Difco veal infusion prepared after the manner of Hitchens. To each liter of infusion, there was added 2 g of potassium nitrate, 2 g of dextrose, and 300 ml of hydrolyzed egg albumin. The quantities of reagents employed for hydrolysis of the egg white were those suggested by Price. However, instead of following his more involved method to hasten cleavage, the mixture was heated on a boiling water bath until the hydrolysis appeared to be complete. After hydrolysis, the albumin was adjusted to pH 7.4 and then autoclaved at 250°F for at least 20 minutes. The albumin was stored at room temperature until it was needed. Small quantities of the egg albumin veal infusion agar were prepared at one time, since it was imperative to use fresh medium to obtain the best results. The necessary amount of egg albumin and 1.5% agar were added to the infusion and, after being heated to melt the agar, the medium was dispensed into test tubes measuring 1 x 8 inches and autoclaved. It was found that the use of large test tubes assured the presence of sufficient moisture on the surface of the slant – a factor of importance.

Tests

Semisolid egg albumin agar containing 0.002% bromthymol blue indicator and 0.5% of the desired sugar was inoculated with the various strains of gonococci and incubated at 37°C for 48 hours. The change in color, if present, was compared to the color of uninoculated tubes of medium which had been similarly incubated. A second method which was employed was the inoculation of slants of medium that did not contain any indicator. After 48 to 72 hours in the incubator, the tubes were tested by running a few drops of phenol red indicator over the surface of the slants. This latter method, suggested by Bayne-Jones, was the more delicate test.

Bayne-Jones (1936). Am. J. Syph. Gonorrhea vener. Dis. 20, 9.
Hitchens (1921). J. infect. Dis. 29, 390.
Price (1935). J. Path. Bact. 40, 345.
Raven (1934). J. infect. Dis. 55, 328.

Fermentation—*Pasteurella multocida*

C. T. Rosenbusch and I. A. Merchant (1939). J. Bact. 37, 69–89.

Difco Proteose peptone	5.0 g
NaCl	5.0 g
K_2HPO_4	0.2 g
$MgSO_4 \cdot 7H_2O$	0.1 g
H_2O	1000.0 ml
pH	7.2–7.4
"Sugar"	10.0 g

The medium is sterilized by filtration through Berkefeld W filters. Readings were taken daily for 10 days and then at 20 and 30 days.

Fermentation—*Streptococcus*

E. V. Keogh and R. T. Simmons (1940). J. Path. Bact. 50, 137–144.

Media

Serum peptone agar. Dissolve 0.25 g shredded agar, 1.0 g neopeptone, 0.85 g NaCl in 100 ml of distilled water, adjust to pH 7.6, filter through paper and sterilize in an autoclave in flask A. The consistency of the contents of this flask, when cold, should be that of porridge and not of a solid gel and the agar content may have to be altered to attain the desired consistency. Into flask B measure 25 ml of tryptic broth, 50 ml of 0.85 per cent, NaCl solution and 25 ml of normal horse serum (not more than 8 weeks old) which has been heated at 56°C for 30 minutes. Melt the contents of flask A, cool to 50–55°C and add to flask B, warmed to the same temperature. Mix and tube in 3.0 ml amounts in sterile 3 x ½ inch test tubes. Store in refrigerator and in any case cool in refrigerator before use.

Serum broth. Mix three parts of horse serum with one part of tryptic digest broth, distribute with a sterile pipette in 3.0 ml volumes in tubes and store in the refrigerator.

Mannitol and cellobiose. These substances are used in 0.5 per cent concentration in a basic medium (pH 7.1) containing proteose peptone (Difco), 10 g; NaCl, 5 g; and water to 1 L, with bromcresol purple as indicator (0.1 per cent of 1.6 per cent solution in 95 per cent ethyl alcohol), and sterilized by Seitz filtration. In the case of cellobiose, 10 per cent ascitic fluid was added. This may be unnecessary, but we have not tested sufficient strains without the addition of ascitic fluid to be positive on this point.

Starch is used in a medium containing proteose peptone (Difco), 10 g; NaCl, 5 g; soluble starch (B.D.H.), 1 g; Andrade's indicator, 10 ml; and water to 1 L warmed to dissolve, tubed in 3 ml quantities, and autoclaved for 15 minutes at 115°C.

Technique

The preparation and inoculation of the serum peptone agar tubes are the only technical processes which present any difficulty. It is essential to obtain a medium of the proper consistency and to distribute the inoculum so that 5–20 isolated colonies are evenly distributed throughout the medium. Using a loop, 1 mm external diameter, of 28-gauge platinum wire, one loopful of an 18-hour broth culture is inoculated into the serum broth tube (3 ml). The loop is removed and flamed and the contents of the serum broth tube thoroughly mixed. One loopful of this dilution is transferred to the serum peptone agar tube. The loop is plunged rapidly to the bottom of the tube and twirled in the medium. The contents of the tube are then thoroughly mixed by rotation and lateral tilting, taking care not to introduce air bubbles. With a little practice and adjustment of the size of the loop, satisfactory results are readily obtained.

The "feathery" colonies are usually spherical, sometimes elongated, semi-transparent masses 3–4 mm in diameter. The compact colonies are tiny dense masses about 1 mm in diameter. These appearances are fully described and figured and the possible fallacies of the test discussed by Ward and Rudd (1938, 1939).

H. K. Ward and G. V. Rudd (1938). Aust. J. exp. Biol. med. Sci. 16, 181.
H. K. Ward and G. V. Rudd (1939). Ibid. 17, 77.

Fermentation—anaerobic streptococcus

M. L. Stone (1940). J. Bact. 39, 559–582.

The ability of the various strains of streptococci to ferment certain sugars was tested by growing the organisms in cooked meat medium to which 1 per cent of the sugar being tested has been added. Cultures were incubated for four days under vaseline seal, and the presence of acid was determined by the use of an indicator consisting of equal parts of 1.6 per cent alcoholic solution of bromcresol purple and a 1 per cent alcoholic solution of cresol red, added directly to the culture. The carbohydrates used were glucose, lactose, maltose, inulin, mannitol and salicin. Controls of inoculated medium without carbohydrate were included.

Fermentation—anaerobic streptococcus

M. L. Stone (1940). J. Bact. 39, 559–582.

The medium used was a casein digest broth consisting of 90 ml beef infusion broth and 10 ml of casein digest. To 7 ml of this basic medium in inverted fermentation tubes 0.78 ml of a 10 per cent solution of *trehalose* or *sorbitol* was added. The sugar previously had been dissolved in distilled water and sterilized by filtration through a Berkefeld "V" filter. The basic medium was adjusted to pH 7.6 and sterilized in the autoclave before the addition of the sugar solution. In addition, during the early work on this test, 1 per cent of Andrade's indicator was added to the medium, but when it was determined that the dye frequently inhibited growth, it was omitted in later studies.

The medium described above was inoculated with a standard loopful of the organism and incubated in the Brown jar for a period of ten days at 37°C. At the conclusion of this period the color reactions were noted (in those cases in which indicator had been added) and the pH of the medium was taken by means of a Beckman pH meter. Appropriate positive and negative controls were included in each jar, the positive control being Group E[4*] hemolytic streptococcus and the negative control being Group C[5*] hemolytic streptococcus. Uninoculated medium incubated with the cultures was used to control possible pH drift in the medium. The final determinations of fermentation were repeated at least twice in each case.

4* Lancefield strain K129.
5* Lancefield strain K104.

Fermentation—*Enterobacteriaceae, Streptococcus, Lactobacillus, Propionibacterium* and *Corynebacterium.*

G. B. Robbins and K. H. Lewis (1940). J. Bact. 39, 399–404.

The medium employed for all tests was prepared as follows:

Bacto peptone	1%
Bacto beef extract	1%
Cysteine hydrochloride	0.05%
Test carbohydrate	0.5%
Distilled water	q.s.

Warm gently to dissolve ingredients, adjust to pH 7.2–7.4, add 1 ml per liter of 1.6 per cent alcoholic solution of bromcresol purple, dispense in approximately 6 ml amounts to small Durham tubes, and sterilize in the autoclave at 15 lb pressure for 15 minutes.

The crystalline aldoses and alcohols were added directly to the medium, while the acids were first converted to the sodium salts. This was done by warming the acid or its lactone with slightly more than an equivalent amount of sodium hydroxide or sodium bicarbonate solution.

Heat sterilization was considered satisfactory since sugar acid salts are highly resistant to decomposition by heat, even in moderately acid or alkaline solutions.

Each medium was inoculated with one loopful of a 24-hr-old broth culture and was incubated at 37°C. All cultures of *Lactobacillus* and *Propionibacterium* species were incubated anaerobically in the presence of 10 per cent carbon dioxide according to the method of Weiss and Spaulding (J. Lab. clin. Med. 22, 1937, 726–728).

Uninoculated tubes of each carbohydrate medium and inoculated tubes of carbohydrate-free medium served as controls.

Observations were made daily for the first week, at three-day intervals during the second week, and a final reading was taken after three weeks. The criteria of fermentation were the occurrence of gas in the inverted vials and change of the bromcresol purple indicator to the yellow color characteristic of acid reactions.

The following substrates were fermented: d-gluconic acid, d-saccharic acid, d-mannonic acid, d-galactonic acid, mucic acid, and 1-arabonic acid.

None of these organisms fermented: d-talonic acid, 1-gluconic acid, 1-mannonic acid, 1-rhamnonic acid, 5-keto-d-gluconic acid or d-arabonic acid.

Fermentation—*Lactobacillus, Bacteroides*

K. H. Lewis and L. F. Rettger (1940). J. Bact. 40, 287–307.

In glucose-cysteine* semisolid agar medium, minus the glucose and disodium phosphate, as the basal medium. The test substances were sterilized separately in 10 per cent aqueous solutions and used in a final concentration of 0.5 per cent. Observations were made after 4 and 16 days incubation at 37°C. Acid production was determined by mixing one loopful of culture with one drop of Brom-thymol-blue solution and five drops of distilled water on a spot plate. Gas formation was shown by the appearance of bubbles or torn areas in the agar. Anaerobic broth cultures in Durham fermentation tubes showed comparable acid production, but gas formation by either method was variable.

In the above-mentioned test one 4 mm loopful of 4 to 6 day old egg-meat cultures served as transfer inoculum, with few exceptions. The cultures were, as a rule, incubated at 37°C for from 4 to 6 days under anaerobic conditions.

*M. Bedell and K. H. Lewis (1938). J. Bact. 36, 567–568 (abst.) The medium referred to in this abstract was later published in full by Lewis, Bedell and Rettger (J. Bact. 40, 309–320, 1940). It has the following composition:

Bacto tryptone	20 g
Bacto Beef Extract	10 g
Glucose	5–10 g
Cysteine HCl	0.5 g
Na_2HPO_4 hydrate	4.0 g
Distilled water	1000 ml

Adjust the pH to 7.4. (For a solid medium add agar.) Sterilize at 120°C for 20 minutes.

Fermentation—*Streptococcus*

J. B. Gunnison, M. P. Luxen, J. R. Cummings and M. S. Marshall (1940). J. Bact. 39, 689–708.

The medium used to test fermentation of carbohydrates was Difco tryptone broth base containing phenol red to which 0.2% agar was added to make it semisolid. 10-20% solutions of test carbohydrates were sterilized by Seitz-filtration and added to the base aseptically so as to give 1% concentrations. The medium was dispensed in small tubes in 2 ml amounts. Inoculations were made from cultures in beef heart infusion broth containing particles of meat, and the tests were placed at 37°C and observed daily for one week.

Fermentation—*Mycobacterium*

A. G. Karlson and W. H. Feldman (1940). J. Bact. 39, 461–472.

NaCl	5.0 g
$(NH_4)_2SO_4$	5.0 g
$MgSO_4$	0.05 g
KH_2PO_4	0.4 g
K_2HPO_4	1.6 g
Carbon Source*	5.0 g
0.1% Phenol red	10.0 ml
Agar	15.0 g

*Sterile concentrated solutions of carbohydrates or polyhydric alcohols were added to the cooled sterile agar before it was placed in tubes. The utilization of various compounds as sources of carbon is indicated by the formation of acid or the development of colonies.

Fermentation—*Streptobacillus moniliformis*

F. R. Heilman (1941). A study of *Asterococcus muris (Streptobacillus moniliformis).* II. Cultivation and biochemical activities. J. infect. Dis. 69, 45–51.

Study of the fermentative activities of the ten strains of *A. muris* was carried out on a liquid medium which contained 1% of peptone, 0.5% of sodium chloride and 10% of horse serum. On this medium the organisms grew slowly but rather well. Carbohydrates were sterilized by filtration through Seitz filters and added to the fermentation tubes in amounts sufficient to make a 1% solution. After the inoculated tubes had been incubated 4 days, bromthymol blue was added and the results were recorded.

Fermentation—*Streptococcus*

F. P. Hadley, P. Hadley and W. W. Leathen (1941).
Variation in peroxide production by beta-hemolytic streptococci. J. infect. Dis. 68, 264–277.

Fermentation tests were carried out in 3% chlorphenol-red beef infusion broth to which 1% of the various carbohydrates were added in sterile 10% aqueous solutions, prepared according to the recommendations of Coffey (1936) for the identification of hemolytic streptococci. All carbohydrates were sterilized by filtration except mannitol and sorbitol which were autoclaved.

Coffey (1936). Am. J. publ. Hlth, supp., 26, 159.

Fermentation of starch—coliforms

R. H. Vaughn and M. Levine (1942). J. Bact. 44, 487–505.

Basal Medium:

Bacto-peptone	5 g
K_2HPO_4	1 g
Andrade's Indicator	10 ml
H_2O	1000 ml

Since it is important that starch be free of reducing sugars Kingsford and Argo corn starch known not to reduce Fehling's solution were used. A concentration of 1% starch was prepared by making a thin paste in a small portion of cold basal medium and adding this paste to the desired quantity of boiling basal medium. With care a satisfactory suspension of starch is obtained. (Method of sterilization is not given.) Soluble starches are not satisfactory for differential purposes as most coliform bacteria readily ferment them since they contain products of starch hydrolysis.

Fermentation of aesculin—coliforms

R. H. Vaughn and M. Levine (1942). J. Bact. 44, 487–505.

Basal medium:

Bacto-peptone	5 g
K_2HPO_4	1 g
Andrade's indicator	10 ml
H_2O	1000 ml
Aesculin	3 g
Ferric citrate	0.5 g

Decomposition of aesculin is denoted by blackening of the medium and the accumulation of gas in the Durham tubes. (No sterilization details.)

Fermentation—*Bacillus*

K. L. Burdon, J. C. Stokes and C. E. Kimbrough (1942). J. Bact. 44, 163–168.

Agar slants with deep butts, inoculated into the butt as well as over the slant, were found superior to broth (Durham fermentation tubes) or soft agar stabs for fermentation tests with the common species of aerobic spore-forming bacilli. Sugar-free infusion broth is solidified with 1.5% agar. Twenty per cent stock solutions of pure sugars in distilled water, sterilized by autoclaving at 15 lb pressure for 15 minutes, were added aseptically just before tubing in the proportion of 1 per cent. At the same time 1 ml per liter of a 1.6 per cent alcoholic solution of brom-cresol-purple was added, and the medium was adjusted to an initial pH of 7.

Cultures were incubated at 37°C for 48 hours (Bacillus) and then at room temperature for one week.

Fermentation—*Streptococcus faecalis* and *Strep. lactis.*

P. M. F. Shattock and A. T. R. Mattick (1943–1944). J. Hyg. Camb. 43, 173.

These were carried out in Lemco broth containing 0.5% of test sugars with litmus as indicator. Tubes were incubated at 37°C for *Str. faecalis* and group D strains and at 28–30°C for *Str. lactis.* The cultures were examined after 24 hours, 48 hours, and 10 days.

Fermentation—*Corynebacterium*

R. F. Brooks and G. J. Hucker (1944). J. Bact. 48, 295–312.

Action in glucose nutrient broth was determined by daily observations for change of reaction in nutrient broth containing 1 per cent glucose and bromthymol blue indicator. Observations were made over a period of 2 weeks' incubation at 37°C, at the end of which time half of each culture was removed and titrated with 0.1 *N* NaOH or H_2SO_4 to determine the amount of acid or alkali formed. The remaining half of each culture was analyzed for utilization of glucose.

Fermentation—*Clostridium cellobioparus*

R. E. Hungate (1944). J. Bact. 48, 499–513.

Tested in a mineral salts medium plus biotin + 0.2% of carbohydrate. pH 6.8–7.0

NaCl	0.6 g
$(NH_4)_2SO_4$	0.1 g
KH_2PO_4	0.05 g
K_2HPO_4	0.05 g
$MgSO_4$	0.01 g
$CaCl_2$	0.01 g
Tap water	100 ml

The quantity of biotin was not stated. Sterile, oxygen-free nitrogen containing 5 per cent of carbon dioxide was bubbled through the medium after sterilization. The tubes were stoppered without admitting air and incubated at 38°C.

Fermentation—*Streptococcus*

E. J. Hehre and J. M. Neill (1946). J. exp. Med. 83, 147–162.

Acid from Carbohydrates – The basic medium was 1 per cent tryptose peptone, 0.5 per cent bacto yeast extract, 0.5 per cent NaCl, 0.1 per cent Na_2HPO_4, 0.0025 per cent bromcresol purple; 1 per cent of the test carbohydrate was added. The cultures were observed at 2 day intervals for 10 days.

Fermentation—*Corynebacterium*

Maurice Welsch, G. Demelenne-Jaminon and J. Thibault (1946). Annls Inst. Pasteur, Paris 72, 203–215.

Techniques de recherche des fermentations. – *Milieu de base :* sérum de Hiss. On prépare de l'eau peptonée comme suit :

Peptone de Diest	0.5 g
Phosphate bi-sodique	0.1 g
Eau distillée pour faire	100 c. c.

Dissoudre à chaud, porter quinze minutes à l'ébullition ; filtrer sur Chardin ; ajuster à pH 7.6 ; stériliser à l'autoclave.

A 100 c. c. de cette eau peptonée stérile, ajouter aseptiquement 25 c. c. de sérum de boeuf, stérilisé par passage à travers filtre Seitz. Chauffer ce mélange pendant trente minutes à 100°, trois jours consécutifs. Eprouver la stérilité par séjour de quarante-huit heures à 37°C.

b) Solutions d'hydrates de carbone : Solutions aqueuses à 20 p. 100 des produits Merck suivants : glucose, galactose, saccharose, maltose, lactose, mannite, stérilisées par filtration sur bougie Chamberland L III. Solutions aqueuses à 10 p. 100 des produits British Drug Houses suivants ; dextrine, amidon, glycogène, stérilisées à l'autoclave.

c) Indicateur : teinture de tournesol stérilisée par filtration sur bougie Chamberland L III.

Pour rechercher les fermentations, on ajoute aseptiquement au sérum de Hiss *(a)* une quantité suffisante de solution hydrocarbonée *(b)* pour obtenir une concentration finale de 1 p. 100 avec les mono- et di-saccharides, de 0.5 p. 100 avec les polysaccharides, et *quod satis* de la solution d'indicateur *(c)*. On répartit en tubes à essai stériles et on éprouve la stérilité par séjour de quarante-huit heures à 37°.

Après ensemencement, on observe les cultures journellement pendant huit à dix jours, exceptionnellement même plus longtemps, et on note le virage de l'indicateur et la coagulation éventuels. Des tubes témoins, non ensemencés, doivent rester inchangés.

Fermentation—*Streptococcus*

K. E. Hite and H. C. Hesseltine (1947). A study of aerobic streptococci isolated from the uterus and vagina. J. infect. Dis. 80, 105–112.

Fermentation reactions – Fermentation tests were carried out in veal infusion broth containing 1.0% of the test carbohydrate with bromcresol purple as the indicator. All the tubes were incubated at 37°C and were observed for 21 days. Determinations of the final pH in dextrose broth were made at the end of this period by means of the Cameron pH electrometer.

Fermentation—*Staphylococcus* and *Micrococcus*

Y. Abd-el-Malek and T. Gibson (1947–1948). J. Dairy Res. 15, 249–260.

Cultures in 1% peptone, 0.5% test substance, with bromcresol purple as indicator, were incubated up to 7 days at the optimum temperature. Pentoses were sterilized in aqueous solution (10%) at 15 lb for 15 min.

Fermentation—*Proteus*

G. T. Cook (1948). J. Path. Bact. 60, 171–181.

Sugar media. A solution consisting of 1 per cent sugar, 1 per cent peptone and 0.5 per cent sodium chloride was distributed in 5 ml quantities in test tubes containing Durham tubes and autoclaved at 10 lb for 15 minutes. The presence of acid and gas was recorded after 1, 2, 10 and 21 days' incubation.

Fermentation – micromethods—*Enterobacteriaceae*

Data taken from J. Hannan and R. H. Weaver (1948). J. Lab. clin. Med. 33, 1338–1341. The C. V. Mosby Co. St. Louis, Missouri.

Basal medium.

Difco beef heart infusion	7.5%	
Proteose peptone 3	1.0%	
KH_2PO_4	0.1%	
NaCl	0.5%	
Distilled water		pH 7.0

Indicator

5 ml of a 1.6% alcoholic solution of bromcresol purple and 5 ml of a 1.6% alcoholic solution of cresol red per litre of medium.

It was necessary to remove the fermentable substance from the beef heart infusion fermentation with *Aerobacter aerogenes.* The beef heart infusion is dissolved in the distilled water. This solution is inoculated with 5 ml of a 24 hr nutrient broth culture of *A. aerogenes.* It is then incubated at 37°C until bubbles cease to appear, approximately 40 hr. The organisms are removed by passage through a filter pad (Seitz serum No. 1). The remaining ingredients are then added to this infusion, and the resultant medium is adjusted to pH 7.0. This basal medium is then sterilized by autoclaving.

To this is added sufficient quantity of sterile (filtered) 20% solution of the fermentable substances to be tested to produce a 5% solution in the final medium. Dispense in small tubes with capillary pipettes. Tests proved to be the most satisfactory with 0.15 ml quantities of medium in 5 × 50 mm tubes.

Inoculum

The inoculum should be large. In this study most of the inoculations were made with suspensions prepared by emulsifying growth from an agar slant culture or from a colony in a small amount of sterile saline.

Method

The inoculated medium is capped with a 3 mm layer of 1% agar in distilled water, to which indicator has been added in the proportion of 1 ml each of 1.6% alcoholic solution of bromcresol purple and a 1.6% alcoholic solution of cresol red per liter. The indicator added to the agar solution detects any change in pH due to the absorption of acid substances from the air during storage.

Incubate at 37°C in a water bath. Acid production may first be noted by the production of a yellow colour below the agar cap. Readings must always be made by comparison with inoculated controls of the basal medium that do not contain the test carbohydrate.

Gas production is evinced by the collection of bubbles below the agar cap.

Incubation periods of 10 minutes to 12 hours, depending on the size of the inoculum have been found to be necessary.

Fermentation—*Pseudomonas alboprecipitans*

A. G. Johnson, A. L. Robert and L. Cash (1949). J. agric. Res. 78, 719–732.

The slightly modified formula of Ayers, Rupp and Johnson for synthetic carbohydrate media* was used. The alcohols and sugars were added at a concentration of 1 per cent and the media were sterilized by steaming 1 hour in an Arnold sterilizer on each of three successive days. Before sterilization, bromcresol purple was added as an indicator. The color change from purple to yellow was used as the index of fermentation. An accurate final change in the pH was recorded by the use of a Beckman pH meter. Durham fermentation tubes were used so that the production of gas could be observed at the same time.

*Soc. Amer. Bact., Tech. (1944). Pure Cult. Study Bact. 12, No. 2, 24 pp. (Manual of Methods: Leaflet II, ed. 9)

Fermentation—*Staphylococcus*

J. B. Evans and C. F. Niven, Jr. (1950). J. Bact. 59, 545–550.

The basal medium contained 1.0 per cent tryptone, 0.5 per cent yeast extract, 0.5 per cent sodium chloride, and 0.004 per cent bromcresol purple. The sugars and alcohols were sterilized separately by autoclaving 5 per cent aqueous solutions, and were added aseptically to the basal medium to a final concentration of 0.5 per cent.

Fermentation—Gram-negative diplococci

G. C. Langford, Jr., J. E. Faber, Jr. and M. J. Pelczar, Jr. (1950). J. Bact. 59, 349–356.

Ten per cent solutions of substrates were made in distilled water and sterilized by Seitz filtration except in the case of inulin, which was autoclaved, Just before use, 1 ml of the 10 per cent solutions was added to a tube containing 9 ml of cystine trypticase agar medium that had been heated in the Arnold sterilizer. These media were inoculated, incubated at 38°C, and reactions were read at 48 hours and also at 5 days, since cultures producing acid in sodium lactate did so within 48 hours and quickly became decolorized afterwards.

Fermentation—*Azotobacter*

P. Hauduroy (1951). Techniques bactériologiques. Masson et Cie, Éditeurs, Paris.

Milieu pour l'étude des fermentations des hydrates de carbone pour *Azotobacter agile* et pour *Azotobacter chrococcum.* – On utilise le milieu n° 1 (voir plus haut) sans mannite et sans dextrine. Douze cm^3 d'une solution à 0.04 p. 100 de pourpre de bromocrésol est ajoutée comme indicateur. Après stérilisation à l'autoclave du milieu sans sucre on ajoute la solution d'hydrate de carbone stérilisée séparément, de façon à avoir une concentration finale de 1 p. 100.

Pour les souches d'*Azotobacter indicum* on utilise le milieu n° 2 préparé sans glucose auquel on ajoute 12 cm^3 d'une solution à 0.04 p. 100 de pourpre de bromocrésol et des hydrates de carbone comme il est indiqué précédemment.

Milieu pour *Azotobacter agile* et *Azotobacter chrococcum* :

N° 1 :

K_2HPO_4	1.0 g
$MgSO_4$	0.2 g
Acétate de calcium	0.2 g
Citrate de soude	0.2 g
$FeSO_4$	traces
Molybdate de sodium	assez pour obtenir une concentration de 1 pour un million de molybdène dans le milieu,
Gélose	15 g
Eau de robinet	1 l

La gélose est fondue et on ajoute 5 g de mannite et 5 g de dextrine. Dans le cas où les *Azotobacter* ne peuvent pas utiliser la mannite ou la dextrine, on ajoute 5 g de glucose et 5 g de saccharose.

Le *p*H final est ajusté à *p*H 7.6.

Milieu pour *Azotobacter indicum.*

N° 2 :

KH_2PO_4	1.0 g
$MgSO_4$	0.2 g
$FeSO_4$	0.2 g
Molybdate de soude	de telle sorte qu'on ait une concentration de 1 pour 1 million de molybdène,
Gélose	15 g
Eau de robinet	1 000 cm^3

Faire dissoudre la gélose. Ajouter 10 g de glucose.

Le *p*H final est ajusté à 6.0.

Fermentation—General

P. Hauduroy (1951). Techniques Bactériologiques. Masson et Cie, Éditeurs, Paris.

Fermentation aerobie des sucres

De l'eau peptonée (produit de digestion de muscles de cheval par la papaïne diluée) est additionnée de 0.5 p. 100 du sucre approprié et de 0.0012 p. 100 de pourpre de bromocrésol comme indicateur. Chauffer pendant 10 minutes à 100°C après distribution dans des tubes contenant eux-mêmes un tube de Durham. Pour les bactéries nécessitant la présence de sérum ajouter 5.0 p. 100 de sérum de lapin stérile. *(W. I. C.)*

Fermentation des sucres, des alcools, des glucosides

Substances utilisées pour les essais. – Une assez grande quantité d'-alcoöls et d'hydrates de carbone purs peut être utilisée pour les tests de fermentation. Dans la pratique courante le choix est souvent limité aux substances les plus ordinaires et les moins coûteuses. Dans les recherches particulières la question d'économie est moins importante. Les trois sucres – glucose, saccharose, lactose – et les alcoöls – glycérine et mannite sont largement employées parce qu'ils sont les plus courants.

Ces substances donneront des résultats qui dépendent, bien entendu, du groupe d'organismes étudiés. Si ce groupe (par exemple : groupe des colibacilles) est capable de faire fermenter presque toutes ces substances, celles-ci n'ont de ce fait que peu de valeur pour séparer les espèces les unes des autres. Il faudra alors utiliser une ou plusieurs des substances plus rares.

Voici une liste des différentes substances qui peuvent être utilisées.

Monosaccharides.	Pentoses : *l*-arabinose, xylose, rhamnose.
	Hexoses : glucoses, fructose, mannose, galactose.
Disaccharides.	Saccharose, maltose, lactose, tréhalose, cellobiose, mélibiose.
Trisaccharides.	Raffinose, mélézitose.
Polysaccharides.	Amidon, innuline, dextrine, glycogène.
Alcoöls.	Trois fonctions OH : glycérine.
	Quatre fonctione OH : érythrite.
	Cinq fonctions OH : adonite, arabite.
	Six fonctions OH : mannite, dulcite, sorbite.
Glucosides.	Salicine, coniférine, esculine.

Nombre de ces substances sont hydrolysées et décomposées à la température nécessaire pour la stérilisation. Pour un travail soigneux la stérilisation doit être faite séparément par filtration ou par autoclavage de solutions aqueuses concentrées (habituellement à 20 p. 100 sauf si la viscosité est trop grande), légèrement acides (*p*H 6.8).

On ajoute ces solutions aseptiquement au milieu de base.

Dans d'autres cas, on peut autoclaver pendant 15 minutes à 120° et plonger aussitôt dans l'eau froide. Les sucres sont particulièrement sensibles à des modifications chimiques en présence de solutions phosphatées ou alcalines.

Habituellement on utilise des concentrations à 1 p. 100 des substances mais dans certains cas on peut, pour des raisons d'economie, utiliser des concentrations plus faibles.

Milieu de base. – Les substances à étudier doivent être ajoutées à un milieu de base permettant la culture des micro-organismes du groupe étudié. Pour le travail courant, il est bon d'employer deux milieux de base: le bouillon à l'extrait de viande et la gélose peptonée avec extrait de viande. On peut ainsi mieux étudier les fermentations, certains germes poussant mieux sur les milieux solides, d'autres dans les milieux liquides. Voici la formule de ces milieux.

Bouillon a l'extrait de viande.

Extrait de viande	3 g
Peptone	5 g
Eau distillée	1 000 cm^3

Gélose peptonée à l'extrait de viande. – A 1 000 cm^3 du milieu précédent ajouter 12 g de gélose desséchée (ou 15 g de gélose commerciale). Dissoudre en chauffant jusqu'à 100°C. Filtrer jusqu'à ce qu'il n'y ait plus de sédiment.

Un autre milieu de base important a été indique par Ayers, Rupp, et Johnson. Voici sa formule.

Milieu de Ayers, Rupp et Johnson.

$NH_4H_2PO_4$	1 g
KCl	0.2 g
$MgSO_4 + 7H_2O$	0.2 g
Eau	1 000 cm^3
Sucre	10 g

ajuster a *p*H, 7 par addition de NaOH. En général, 6 cm^3 de NaOH normale sont nécessaires.

Ce milieu peut être utilisé seulement par les germes qui utilisent les sels d'ammonium comme source d'azote. Mais, il peut également servir pour les germes qui provoquent de nombreux changements dans les réactions de protéines ou qui produisent une quantité d'acide si petite qu'elle ne peut être dépistée dans les milieux hautement tamponnés.

On doit noter particulièrement s'il y a une culture abondante ou pas de culture dans de tels milieux après adjonction de la substance sur laquelle porte la recherche. Si la culture est très maigre ou s'il n'y a pas de culture en bouillon ou sur gélose, il est nécessaire de suivre les recommandations suivantes: souvent la pauvreté des cultures est due à l'absence de sels inorganiques nécessaires pour le développement des germes ou à l'absence de quelque facteur inconnu dans la peptone. La meilleure façon de suppléer à cette déficience des milieux indiqués précédemment est l'emploi du milieu à la levure dont la formule est indiquée ci-dessous.

Milieu à la levure. – Dans les formules de bouillon et de gélose indiquées plus haut on remplace l'extrait de viande par 2.5 g d'extrait de levure par litre. Il est nécessaire d'employer un extrait sous forme de poudre comme par exemple celui qui est fabriqué par Difco.

Si l'étude de la souche se fait en milieu liquide, celui-ci doit être stérilisé dans des tubes à fermentation; s'il s'agit de milieux solides, ceux-ci doivent être distribués sous forme de gélose inclinée. La gélose inclinée peut être ensemencée seulement à la surface ou en partie sur la surface, en partie en enfonçant le fil d'ensemencement à la base du milieu. Il a été montré en effet que dans la pratique une grande quantité de gaz peut se former au pied de la gélose inclinée même s'il y a une culture abondante à la surface du milieu. Si l'on a quelque raison de suspecter une production de gaz que l'on n'a pas vue, ajouter une culture que l'on a agitée au tube contenant la gélose inclinée.

Le milieu à l'extrait de levure est un milieu très satisfaisant pour l'étude des bactéries propioniques, des streptocoques, ou des lactobacilles.

Dans le cas de quelques microbes microaérophiles, un autre milieu de base peut être employé. Il s'agit de la gélose semi-solide, milieu qui ne contient que 0.2 a 0.5 p. 100 de gélose.

Quelques bactéries peuvent ne pas pousser dans le bouillon standard indiqué plus haut ou sur la gélose par suite de la présence d'une trop grande quantité de matière organique. Pour ces germes, le milieu au phosphate d'ammonium (ou la gélose) donnent de bons résultats.

Il est souvent nécessaire de prévenir une augmentation considérable des ions H. On y arrive en général en ajoutant un excès de $CaCO_3$ stérilisé à chaque culture ou en tamponnant suffisamment le milieu. Si on ajoute du carbonate de calcium, il est nécessaire de l'utiliser en poudre fine pour avoir une grande surface d'adsorption des acides formés. Le carbonate doit être mélangé au milieu par agitation, faute de quoi il n'est pas un agent neutralisant effectif.

Études des résultats obtenus. – Dans l'étude des fermentations les déterminations suivantes sont habituellement pratiquées: concentration finale en ions H, sucres résiduels, variétés et quantités d'acides organiques, d'acétone, d'alcoöls éthylique, butyl et isopropyl, de carbone dioxyde.

Pour le travail courant, dans le cas des microorganismes sur lesquels on ne possède pas de très grands renseignements, l'utilisation des indicateurs est spécialement recommandée pour savoir s'il y a ou s'il n'y a pas eu de production d'acide. Il est nécessaire de se rappeler que dans de nombreux cas des renseignements plus précis seront obtenus par l'emploi des méthodes de titration.

Lorsque la méthode des indicateurs est utilisée, ceux-ci doivent être ajoutés au milieu tout à fait au début, ou ajoutés seulement au moment où il s'agit de déterminer la réaction finale.

Dans ce cas, la couleur obtenue doit être comparée avec les couleurs standard.

L'utilisation des indicateurs mélangés au milieu est la moins sûre, mais c'est le procédé le plus rapide. Quand il s'agit de cultures sur gélose, c'est le seul procédé satisfaisant.

Pour l'utilisation du milieu contenant les indicateurs au départ, suivre les directives données à *Indicateur.*

L'indicateur le plus couramment utilisé est le pourpre de bromocrésol, mais quand il s'agit de microorganismes produisant des quantités considérables d'acide, on peut utiliser avantageusement le vert de bromocrésol ou le bleu de bromophénol.

Quand on étudie des séries d'organismes inconnus, il est souvent judicieux de les ensemencer dans des milieux sucrés contenant du pourpre de bromocrésol. Plus tard, ceux qui ont produit de l'acide sont réensemencés dans le même milieu avec du vert de bromocrésol; enfin, en dernier lieu, le réensemencement peut être fait dans les milieux contenant du bleu de bromophénol. Si l'on a décidé d'observer la production de l'alcalinité aussi bien que l'acidité, on peut employer le bleu de bromothymol ou mieux un mélange de pourpre de bromocrésol et de rouge de crésol.

Dans les milieux solides cette technique a souvent de la valeur parce qu'elle montre la production d'acide dans une partie du tube, la production de substances alcalines dans l'autre partie.

Il est parfois difficile de lire les réactions exactes des milieux auxquels on a primitivement mélangé l'indicateur en se référant simplement aux couleurs standard. Cependant, une bonne estimation de la concentration en ions hydrogènes peut être obtenue par l'examen de trois tubes ensemencés en même temps, chacun d'eux contenant l'un des trois indicateurs indiqués plus haut.

La table suivante donne des indications sur les variations des couleurs des indicateurs utilisés:

*p*H	7.0	6.0	5.5	5.0	4.0	3.0
Pourpre de bromocrésol	pourpre			jaune		
Vert de bromocrésol			bleu			jaune
Bleu de bromophénol				bleu		jaune

S'il n'est pas possible avec cette table de déterminer exactement la valeur du *p*H de la culture, on pourra noter l'intensité de la réaction dans chacun des indicateurs. Ceci peut être fait en notant la nuance de la culture, en comparaison avec son changement à partir de sa couleur de départ et de son orientation vers la couleur jaune. Les notations seront les suivantes: +, + +, + + +, + + + +. Le symbole o indiquera l'absence de changement de réactions.

Si on utilise le rouge de phénol pour dépister l'alcalinité, utiliser le symbole – pour indiquer le changement dans la direction de l'alcalinité.

Pour que les résultats aient une valeur, il est bien entendu que l'on doit indiquer l'indicateur qui a été employé.

La production de gaz dans un milieu liquide est habituellement mesurée par le volume de gaz dans la partie fermée du tube à fermentation de Smith et Durham. Le tube de Durham consiste en un petit tube (75 x 10 mm) renversé et soudé à un large tube (150 x 18 mm).

Dans le cas des milieux solides la présence ou l'absence de gaz est notée suivant la présence ou l'absence de bulles et de craquements dans la gélose.

Ce test (production de gaz) est surtout utilisé pour les organismes produisant primitivement de l'hydrogène. Si le gaz produit est du CO_2, une toute petite quantité peut venir s'accumuler dans le tube de fermentation. Dans certains cas aucun gaz n'est même décelable. Ceci est dû à la grande solubilité et à la grande diffusibilité du CO_2 dans l'air. La méthode la plus convenable pour dépister le dioxyde de carbone est celle de Elderedge et Rogers.

Interprétation des résultats. – Lorsqu'un organisme a produit des gaz ou des changements considérables dans l'acidité du bouillon ou de la gélose à l'extrait de viande en présence d'une substance fermentescible, qui n'a produit aucun changement dans les milieux de base non additionnés de la même substance, on peut en conclure qu'il y a eu dégradation de cette substance. Dans la pratique courante de tels résultats sont en général suffisants.

Pour comprendre la véritable action des microorganismes sur les substances fermentescibles, d'autres recherches sont nécessaires. *(S. A. B.)*

Fermentation anaerobie des sucres

On opère comme pour le test de fermentation aérobie, mais on ajoute 0.25 p. 100 de gélose et le milieu est distribué par 10 ml dans des tubes à essai sans tubes de Durham. *(W. I. C.)*

Fermentation—General

P. Hauduroy (1951). Techniques bactériologiques. Masson et Cie Éditeurs, Paris.

Etude des hydrates de carbone et des alcools

Comme milieu de base on se sert d'un bouillon a l'extrait de viande de Liebig.

Extrait de viande de Liebig	5 g
Peptone (Riedel)	10 g
NaCl	5 g
Eau de robinet	1 000 cm^3

Ajuster à *p*H 7,4.

Comme indicateur on ajoute 12 cm^3 par litre de la solution suivante :

Bleu de bromothymol	1 g
NaOH (n 1/10)	25 cm^3
Eau distillée	475 cm^3

On ajoute au milieu ainsi préparé 0,5 p. 100 des substances suivantes : adonite, *l*-arabinose, dextrine, *i*-dulcite, *d*-glucose, *i*-inosite, lactose, maltose, *d*-mannite, *l*-rhamnose, saccharose, salicine, *d*-sorbite, tréhalose, ou xylose.

On ensemence avec des cultures sur gélose âgées de 20 heures et on maintient à l'étuve à 37° pendant 30 jours. La réaction positive est marquée par un changement de couleur du milieu. *(C. I. E.)*

Fermentation—*Staphylococcus*

C. Shaw, J. M. Stitt, and S. T. Cowan (1951). J. gen. Microbiol. 5, 1010–1023.

Peptone water containing Andrade indicator and 1% (w/v) of "sugar" was used; final readings were made after 14 days at 37°C, 21 days at 30°C, or 28 days at 22°C. The "sugars" tested were glucose, xylose, galactose, lactose, sucrose, maltose, raffinose, starch, dextrin, glycerol, mannitol, dulcitol, and salicin.

Fermentation—*Streptococcus*

P. F. Swartling (1951). J. Dairy Res. 18, 256–267.

Peptone water (1% peptone, 0.5% NaCl) containing 0.5% of the test sugars, added after sterilization, with litmus as indicator, was inoculated with one loopful (4 mm) of an 18 hr broth culture and incubated at 30°C. Results were read after 1, 3 and 7 days. Acid production within 3 days is indicated by +, and acid production after 7 days by (+).

Fermentation—*Bacillus*

P. de Lajudie (1952). Annls Inst. Pasteur, Paris 82, 380–382.

Tous les germes poussèrent en présence d'azote minéral dans le milieu d'Ayers et coll., qui donna d'excellents resultats pour l'étude des fermentations.

Fermentation—*Bacillus*

N. R. Smith, R. E. Gordon and F. E. Clark (1952). U. S. D. A. Agr. Monograph No. 16.

For the fermentation studies a basal medium was used, a modification of one proposed by Ayers, Rupp and Johnson. The formula was as follows: $(NH_4)_2HPO_4$, 1.0 g; KCl, 0.2 g; $MgSO_4$, 0.2 g; yeast extract, 0.2 g; agar, 15 g; and distilled water, 1000 ml. Twenty milliliters of a 0.04 percent solution of bromcresol purple was added as the indicator, and the medium was tubed and autoclaved. A 10 or 15 per cent aqueous solution of the test carbohydrate, sterilized separately by autoclaving or filtration through a L2 Pasteur-Chamberland filter, was pipetted into each tube in such quantity as to result in a 0.5 per cent concentration of the carbohydrate. The tubes were incubated overnight, examined for sterility, inoculated, and observed for growth, acid and gas at intervals up to 10 or 20 days.

If a strain failed to hydrolyze glucose in this basal carbohydrate medium, it was inoculated on slopes of glucose-nutrient agar with an indicator, usually bromcresol purple. If acid was then formed, it was assumed that the strain could not utilize the ammoniacal nitrogen in the basal medium and required organic nitrogen. Unless otherwise indicated, the basal carbohydrate medium was used in this work.

S. H. Ayers, P. Rupp and W. T. Johnson, Jr. (1919). U. S. Dept. Agr. Bul. 782, 39 pp.

Fermentation—*Pseudomonas*

P. Villecourt, H. Blachère and G. Jacobelli (1952). Annls Inst. Pasteur, Paris 83, 316–322.

L'acidification de divers oses et alcoöls était étudiée en présence de rouge de phénol dans de l'eau peptonée contenant 5 p. 1000 de peptone et 20 p. 1000 de substrat hydrocarboné. L'aération facilitant la croissance, le milieu était réparti en tubes de 17 à raison de 2 cm^3 seulement par tube.

Fermentation—Peste "Sauvage" du Kurdistan

M. Baltazard and P. Aslani (1952). Annls Inst. Pasteur, Paris 83, 241–247.

La recherche de la fermentation de la glycérine et du rhamnose a été faite selon la technique utilisée par Chen [3] sur les indications de R. Pollitzer et de K. F. Meyer et recommandée par Pollitzer dans sa note [2] comme technique standard. Le milieu est l'eau peptonée (1 p. 100 de peptone (1) en eau distillée) à laquelle on ajoute glycérine ou rhamnose dans la proportion de 1 p. 100. L'indicateur est la fuchsine acide : indicateur d'Andrade (fuchsine acide ; 0.20 g, eau distillée ; 100 cm^3), ajoute au milieu dans la proportion de 1 p. 100. Après stérilisation (une heure, à 105°C seulement pour éviter la caramélisation) le pH de ces milieux est très soigneusement ajusté à 7,40.

Deux tubes au moins de chaque milieu sont ensemencés largement avec le produit de raclage (4 anses) d'une culture de quarante-huit heurs sur gélose ordinaire de la souche à examiner et portés à l'étuve à 37°C (2). Les tubes sont agités pour aération tous les jours et suivis pendant vingt et un jours. La pousse du bacille dans le milieu est estimée tous les jours par néphélométrie et notée en même temps que le virage de la couleur du milieu. Des subcultures sont faites chaque semaine pour vérifier la vitalité des germes.

(2) Nous avons respecté cette température pour raisons de standar-

disation, bien que la pousse du bacille pesteux soit moins rapide et abondante à 37°C qu'à 32°C.

(2) Pollitzer (R.). Note de documentation pour le Comite d'Experts de la Peste de l'OMS, 1950.
(3) Chen (R. H.). J. infect. Dis. 1949, 85, 97.

Fermentation—*Lactobacillus*

M. Rogosa, R. F. Wiseman, J. A. Mitchell, M. N. Disraely and A. J. Beaman (1953). J. Bact. 65, 681–699.

Seitz-filter sterilized sugars in 2% concentration (galactose, 1.4%) were added to the following medium: trypticase, 10 g; yeast extract, 5 g; KH_2PO_4, 6 g; ammonium citrate, 2 g; salt solution, 5 ml; sorbitan monooleate, 1 g; sodium acetate hydrate, 15 g; alizarine red S indicator, 0.04 g; distilled water to 1 liter. The salt solution comprises: $MgSO_4 \cdot 7H_2O$, 11.5 g; $MnSO_4 \cdot 2H_2O$, 2.4 g; or $MnSO_4 \cdot 4H_2O$, 2.8 g; $FeSO_4 \cdot 7H_2O$, 0.68 g; and distilled water to 100 ml.

The medium had a final pH after sterilization of 5.9.

Fermentation—*Mycobacterium*

R. E. Gordon and M. M. Smith (1953). J. Bact. 66, 41–48.

Acid production from carbohydrates

One-half ml amounts of a 10 per cent aqueous solution of the carbohydrate to be tested, sterilised by autoclaving, were added aseptically to tubes containing 5 ml of sterile, inorganic nitrogen base, a modification of one proposed by Ayers, Rupp, and Johnson (1919). The inorganic nitrogen base had the following composition: $(NH_4)_2HPO_4$, 1 g; KCl, 0.2 g; $MgSO_4$, 0.2 g; agar, 15 g; distilled water, 1,000 ml. The pH of the medium was adjusted to 7.0 before the addition of 15 ml of a 0.04 per cent solution of bromcresol purple. Cultures on this carbohydrate agar were observed for acid production after 7 and 28 days' incubation at 28°C.

S. H. Ayers, P. Rupp and W. T. Johnson (1919). U. S. Dept. Agr. Bull. 782.

Fermentation—*Pseudomonas (Malleomyces) pseudomallei*

P. de Lajudie and E. R. Brygoo (1953). Annls Inst. Pasteur, Paris 84, 509–515.

Action sur les hydrates de carbone: – Nous avons utilisé, pour la plupart des sucres, de l'eau peptonée, additionnée, après autoclavage, de 1 à 3 p. 100 de glucide en solution filtrée sur bougie, le rouge de phénol servant d'indicateur. L'attaque du glycérol fut recherchée en milieu de Stern, celle du mannitol, sur milieu semi-solide. Ces deux milieux étaient préparés selon la technique recommandée par Dumas. Les milieux liquides furent ensemencés avec I goutte d'une culture de vingt-quatre heures, le

milieu au mannitol ensemencé par piqûre avec une parcelle de culture sur gélose. La cloche du tube de Besson glucosé permettait la mise en évidence d'une éventuelle production de gaz.
J. Dumas. Bactériologie médicale, Flammarion, Paris, 1951.

Fermentation—*Actinomyces, Corynebacterium*

H. Beerens (1953). Annls Inst. Pasteur, Paris 84, 1026–1032.

Action sur les sucres. – Acidification constante (pH 3), sans production de gaz, des milieux contenant: glucose, galactose, lévulose, glycérol. Les autres sucres ne sont pas fermentés. Ces recherches ont été effectuées sur le milieu suivant: extrait de viande 3 g; peptone trypsique 10 g; chlorure de sodium 5 g; extrait de levure 4 g; cystéine (Chte) 0,30 g; gélose 0,5 g; sucre étudié 10 g; eau 1000 cm^3; ajusté à pH 7.2, stérilisé à 115°C pendant vingt minutes. L'acidification a été mise en évidence par addition d'indicateur de pH après culture à 37°C pendant sept jours. Un milieu non sucré, ensemence dans les mêmes conditions, servait de témoin.

Fermentation—*Listeria*

R. F. Jaeger and D. M. Myers (1954). Can. J. Microbiol. 1, 12–21.

Fermentation reactions were determined by using phenol red broth base (BBL) with 1% carbohydrate and pH 7.4. The fermentation tubes were inoculated from original cultures (isolated on blood agar) and incubated for 30 days at 37°C.

Fermentation—*Streptococcus*

G. Andrieu, L. Enjalbert and L. Lapchine (1954). Annls Inst. Pasteur, Paris 87, 617–634.

Les milieux sucrés : dans 5 cm^3 de bouillon ordinaire nous ajoutons une quantité suffisante de lactose, mannitol, raffinose, ou sorbitol pour avoir une concentration finale de 1 p. 100. Nous ajoutons ensuite II à III gouttes de rouge de phénol choisi comme indicateur coloré. Une réaction positive correspond à un virage du colorant au jaune citron. Une réaction négative laisse le colorant rouge orangé.

Fermentation—*Xanthomonas*

N. James (1955). Can. J. Microbiol. 1, 479–485.

In the basal medium of Ayers, Rupp and Johnson (1919, U. S. Dept. Agr. Bull. 782) plus 1% carbohydrate for 7 days and pH tested electrometrically.

Fermentation—*Lactobacillus*

D. M. Wheater (1955). J. gen. Microbiol. 12, 123–132.

The basic medium, consisting of Neopeptone 1.5%; Tween 80 0.1%; Yeastrel 0.6%; and agar 0.15%, had a final pH of 7.0–7.2 and was sterilized by autoclaving at 15 lb sq. in. for 20 min. The indicator was 0.04%

chlorophenol red. As the colour of an ethanolic solution was found to deteriorate when stored, the solid was dissolved in 1 ml of ethanol and added to the medium immediately before tubing and sterilization. Thirty "carbohydrates" were used and 2% solutions, sterilized by Seitz-filtration added to tubes of the basic medium to give a final concentration of 0.2%. Before inoculation the tubes were incubated for 24 hr to check sterility. Inoculated tubes were incubated for 7 days and readings taken on the 1st, 2nd, 4th and 7th days.

Fermentation—*Lactobacillus*

G. H. G. Davis (1955). J. gen. Microbiol. 13, 481–493.

Medium: casein partial-hydrolysate (Peters and Snell, 1954) was prepared as follows: 112 g of BDH light white soluble casein in 1600 ml of N-sulphuric acid was autoclaved for 5 hr at 15 lb. Approximately 180 g of barium hydroxide was mixed in thoroughly; the pH readjusted to *c.* 3.0 with N-sulphuric acid; 56 g of activated charcoal were added and mixed well for 30 min. The solution was filtered through paper pulp in a Büchner funnel.

A double strength basal medium was then prepared by adding 56 g of peptone, 28 g of yeast extract 5.6 ml of Tween 80, 28 ml of Salts A* and 28 ml of Salts B* to the casein hydrolysate solution and diluting to a final volume of 2800 ml. After steaming to dissolve the solids the pH was adjusted to 7.0; the medium re-steamed for 20 min and clarified by filtration through paper pulp. This medium was autoclaved for 15 min at 15 lb. and stored in 250 ml quantities. Fermentation tests were prepared by diluting 250 ml of the double strength medium to 450 ml adjusting the pH to 7.2, adding 1 ml of a 1.5% aqueous solution of bromcresol-purple, and, in the case of heat tolerant carbohydrates, 50 ml of 10% carbohydrate solution. The final medium was dispensed into bijou bottles and sterilized by steaming for 20 min on 3 consecutive days. With those carbohydrates which are degraded by heat, the basal medium was sterilized by autoclaving, cooled, and 50 ml of a 10% Seitz-filtered carbohydrate solution added with sterile precautions. The sterility was checked by incubation for 3 days at 37°C.

The following carbohydrates were employed: D-mannose, D-xylose, L-arabinose, L-rhamnose, D-melibiose, lactose, melezitose, sorbitol, mannitol, salicin and raffinose. The first three were sterilized by Seitz-filtration.

Tests were inoculated with 1 drop, incubated for 14 days and observed at 48 hr intervals for colour change from purple to brilliant yellow with accompanying turbidity.

*See A. C. Hayward (1957). J. gen. Microbiol. 16, 9–15.

V. J. Peters and E. E. Snell (1954). J. Bact. 67, 69.

Fermentation—*Streptomyces* and *Nocardia*

R. E. Gordon and M. M. Smith (1955). J. Bact. 69, 147–150.

Acid from Carbohydrates

The cultures were examined for acid production after 7 and 28 days' incubation at 28°C on inorganic nitrogen agar, a modification of the medium of Ayers *et al.* (1919) plus each carbohydrate. The medium contained $(NH_4)_2HPO_4$, 1 g; KCl, 0.2 g; $MgSO_4$, 0.2 g; agar, 15 g; distilled water, 1000 ml. The pH was adjusted to 7.0, and 15 ml of a 0.04 per cent solution of bromcresol purple (indicator solution) were added. The agar was tubed in 5 ml amounts and sterilized. One half ml of a 10 per cent aqueous solution of the carbohydrate to be tested, autoclaved separately, was added aseptically to each tube of agar.

S. H. Ayers, P. Rupp and W. T. Johnson (1919). U. S. Dept. Agr. Bull. 782.

Fermentation—various

C. Shaw and P. H. Clarke (1955). J. gen. Microbiol. 13, 155–161.

Medium consisted of peptone water + bromcresol purple + 1% (w/v) of "sugar" and final readings made after incubation of cultures for 14 days at 37°C.

Fermentation—*Bacillus*

E. Grinsted and L. F. L. Clegg (1955). J. Dairy Res. 22, 178–190.

The fermentation of glucose, arabinose and xylose followed the methods of Smith *et al.* (1946). For mesophiles the inorganic nitrogen basal medium was used, but this was unsuitable for thermophiles, which were cultured on nutrient agar slopes. One ml/L of a 1.6% alcoholic solution of bromcresol purple was added as an indicator since this was found to be more satisfactory than the aqueous solution originally recommended.

N. R. Smith et al. (1946). Misc. Publ. U. S. Dept. Agric. No. 559.

Fermentation—*Pseudomonas*

C. K. Williamson (1956). J. Bact. 71, 617–622.

Seitz-filtered "sugars" in a concentration of 1 per cent in nutrient broth, were used for these biochemical studies. The sugars were dissolved in 200-ml aliquots of the broth and the solutions were then filtered, incubated for 24 hr at 37°C to ascertain that the media were sterile, and finally inoculated with a definite number of smooth, mucoid, or filamentous cells. Flasks of plain nutrient broth were used as controls. Cell populations were quantitated by the poured plate technique. The initial pH of the media was 6.8, and pH readings were taken at the end of 24, 48, 96, 168 and 240 hr on a Beckman model H-2, line operated, hydrogen ion meter. All experiments were repeated at least three times.

Fermentation—*Bacillus*

K. L. Burdon (1956). J. Bact. 71, 25–42.

Fermentation tests were performed in bromcresol purple agar butt slants containing 1 per cent of "sugars." The basic medium was composed of L tryptose, 20 g; NaCl, 5 g; agar, 15 g; dissolved in 1 L of distilled water, adjusted to have a final pH of 7.0. To each 250 ml of this medium, after autoclaving and cooling to 50°C, were added 0.25 ml of a 1.6 per cent alcoholic solution of bromcresol purple indicator, and 12.5 ml of a 20 per cent Seitz-filtered solution of one of the carbohydrates. The medium was tubed aseptically in 18 by 180 mm test tubes and cooled so as to allow a 2-in deep butt and a slant approximately 2 in. long. Inoculation was made by stabbing the butt, then streaking the slant with the same inoculum. Uninoculated tubes of each sugar were incubated as controls. Readings were made after 12, 24, 36, 48 and 72 hr by comparison of the appropriate control tube with each inoculated tube under a shaded fluorescent light.

Fermentation—*Streptococcus*

M. Moreira-Jacob (1956). J. gen. Microbiol. 14, 268–280.

Seitz-filtered 0.5% (w/v) "sugar" solutions (except aesculin) were added to peptone water containing 1% Andrade's Indicator and 1% ox serum.

Fermentation—*Neisseria gonorrhoeae*

P. J. Mullaney (1956). J. Path. Bact. 71, 516–517.

Bovine albumin (Armour & Co.). A 5 per cent solution, in distilled water, was acidified with one-tenth of its volume of N HCl and sterilized by placing it for ten minutes in boiling water; it was then thoroughly cooled and adjusted to pH 7.2 with NaOH.

Sugar solutions. (15 per cent) were sterilized by steaming on three successive days.

Phenol red (0.2 per cent) in distilled water was sterilized, before use, by bringing to a boil over a bunsen flame.

Fermentation agar. 20 g of peptone and 5 g of NaCl were dissolved in 900 ml distilled water by heating in a steamer for 30 minutes; the solution was filtered through filter paper, adjusted to pH 7.6, and steamed for 30 minutes; after filtration the pH was adjusted to 7.6, and 25 g of agar powder were added and dissolved by autoclaving at 10 lb pressure for 20 minutes; the medium was then filtered through paper pulp.

Final media. Reagents were mixed in a Petri dish immediately before use. The rabbit serum agar medium contained 0.8 ml Seitz-filtered rabbit serum, 1 ml of sugar solution, 0.3 ml of 0.20 per cent phenol red, and 15 ml of fermentation agar. In the albumin-agar medium the rabbit serum was replaced by 1.1 ml of heated albumin solution. The media, when freshly made, were pink in colour. A heavy inoculum, taken from a pure culture on a blood-agar plate, was deposited on them and spread to the size of a shilling. Four strains can be tested on a 4-inch Petri dish. The plates were incubated in air containing approximately 10 per cent carbon dioxide, in a McIntosh and Fildes' jar, for 18 to 24 hours at 37°C; the lids of the Petri dishes were prevented from closing tightly by sterile aluminum clips. The carbon dioxide concentration was produced by placing a 6 × 1 in. test tube, containing a weighed quantity of sodium bicarbonate, in a jar alongside the culture plates. A measured quantity of N HCl was then run into the test tube and the lid of the jar quickly replaced. The jar was then incubated. The quantities of N HCl and sodium bicarbonate were calculated to yield a quantity of carbon dioxide approximately equal to 10 per cent. of the volume of the jar (1 g $NaHCO_3$ plus about 12 ml N HCl yields a little more than 250 ml CO_2). After incubation the entire medium was yellow. The plates were then reincubated for 6 to 24 hours at 37°C in air without the addition of carbon dioxide, and the medium regained its pink colour. Colonies producing acid still showed a yellow area at the site of inoculation and were surrounded by a narrow yellow zone.

Fermentation—*Zymomonas*

N. F. Millis (1956). J. gen. Microbiol. 15, 521–528.

Seitz-filtered carbohydrate solutions were added to sterile 1% Difco yeast extract at pH 5.5, to give a final concentration of 1% carbohydrate. Bromcresol green was used as an indicator.

Fermentation—carbohydrate indicator media

Society of American Bacteriologists *Manual of Microbiological Methods* – McGraw-Hill Book Co. Inc., New York 1957.

The common procedure is to add the carbohydrate or polyalcohol to be studied to a basal medium (either liquid or agar) to which an indicator has been added to detect changes in pH which develop during growth. Growth of some organisms, particularly the sporeforming anaerobes, may result in a marked reduction of the indicator, in which case the indicator must be added after rather than before growth; use of a spot plate or other methods of pH determination is then made at the time of observation. Early observation of fermentation results generally eliminates the difficulty due to reduction of the indicator, since acidity changes usually precede reduction. Production of gas is detected in liquid media by placing Durham tubes (small inverted vials which will fill with liquid during sterilization) in the tubes at the time the medium is dispensed. The tubes are unnecessary if a solid or semisolid medium is used. Semisolid agar is prepared by adding 0.3–0.5 per cent of agar to a satisfactory liquid medium and making stab inoculations in the column of medium with a straight inoculating needle. In such a medium (or in full-strength agar) gas production will be denoted by the appearance of gas bubbles and cracks in the medium; the same semisolid medium may also be used to determine motility. Full-strength (1.5 per cent) agar should be cooled in slanting position and inoculated on the surface of the slant.

The basal medium to be employed for fermentation tests should provide the necessary nutrients for the organism to be studied and must be free of fermentable carbohydrates. If good growth can be obtained on 2 per cent casein or gelatin-peptone solutions (or agar), these media are preferred. According to Vera (1949), some samples of beef and yeast extracts contain fermentable carbohydrate. In all instances, *control tubes* to which carbohydrate has not been added must be inoculated to check changes of pH due to the breakdown of carbohydrates or other substances.

The synthetic medium of Ayers, Rupp and Johnson (1919) may be used if a peptone-free medium is desired and if the organism being studied can utilize ammonium salts as a source of nitrogen. It is prepared as follows: $NH_4H_2PO_4$, 1.0 g; KCl, 0.2 g; $MgSO_4 \cdot 7H_2O$, 0.2 g; water 1,000 ml; carbohydrate, 10.0 g; adjust the pH by addition of 1 N NaOH.

The soluble carbohydrates or polyalcohols are added to the basal medium at a level of 0.5–1.0 per cent. The indicator (commonly 1 ml of a 1.6 per cent alcoholic solution per 1,000 ml of medium) is added before sterilization. Although litmus and Andrade's indidator (acid fuchsin decolorized with alkali) have been used widely, they do not give accurate results in terms of H-ion concentration; thus, except for special purposes it is recommended that sulfonphthalein indicators be employed. Select the appropriate indicator from the list given in Table 1, governing choice by the following considerations: phenol red, with pH range 6.9–8.5, is useful for indication of changes on the alkaline side of neutrality and slight changes to acid; brom-thymol blue has a sensitive range extending slightly in either direction from neutrality; brom-cresol purple, with a pH range of 5.4–7.0, is useful in synthetic media and for pronounced pH changes in more highly buffered media. Bromthymol blue is frequently the most useful of these, but must be used with caution in synthetic media, since it indicates even the minor pH changes due to CO_2 absorption.

Table 1. Acid-base indicators

Indicator	Concentration recommended, %*	Sensitive pH range	Full acid color	Full alkaline color
Bromphenol blue	0.04	3.1–4.7	Yellow	Blue
Bromcresol green	0.04	3.8–5.4	Yellow	Blue
Methyl red	0.02	4.2–6.3	Red	Yellow
Bromphenol red	0.04	5.2–6.8	Yellow	Red
Bromcresol purple	0.04	5.4–7.0	Yellow	Purple
Bromthymol blue	0.04	6.1–7.7	Yellow	Blue
Phenol red	0.02	6.9–8.5	Yellow	Red
Cresol red	0.02	7.4–9.0	Yellow	Red
Thymol blue (alk. range)	0.04	8.0–9.6	Yellow	Blue
Phenolphthalein	0.10	8.3–10.0	Colorless	Red

*Stock solutions in 95 per cent ethanol for the indicator acids or in water for the indicator salts.

S. H. Ayers, P. Rupp and W. T. Johnson (1919). U. S. Dept. Agr. Bull. 782.

H. D. Vera (1949). Soc. Am. Bacteriologists, Abstr. Papers, 49th gen. meeting, p. 6.

Fermentation—*Acetobacter*

R. Steel and T. K. Walker (1957). J. gen. Microbiol. 17, 445–452.

Difco yeast extract	0.5% (w/v)
Brom-cresol-purple	0.003% (w/v)
Distilled water	
Test sugars to	1.0% (w/v)

Incubate for 8 days.

Fermentation—*Actinomyces bovis*

S. King and E. Meyer (1957). J. Bact. 74, 234–238.

A comparison of different basal media indicated a semisolid trypticase agar base (no. 151 Baltimore Biological Laboratory) was preferable. This medium was sterilized by autoclaving at 15 lb pressure, 121°C for 20 min. Carbohydrates in 20 per cent concentrations, sterilized separately in the same manner, were added aseptically to this medium, making the final concentration of sugar in each tube 1 per cent. To our knowledge, this medium has not been used previously in studies covering *A. bovis* or "anaerobic diphtheroids."

Fermentation—thermophilic sulphate reducing bacteria

L. L. Campbell, Jr., H. A. Frank and E. R. Hall (1957). J. Bact. 73, 516–521.

Reactions on carbohydrates and alcohols were determined in the basal medium of Reed and Orr (1941). After incubation for 5 days at the appropriate temperature, acid production was measured with a Beckman pH meter.

G. B. Reed and J. H. Orr (1941). War Med. Chicago 1, 493–510.

Fermentation—*Lactobacillus*

J. C. de Man (1957). Antonie van Leeuwenhoek 23, 87–96.

The medium used for determining the fermentation of glucose and melibiose contained: Bactotryptone, 2%; yeast extract, 1%; K_2HPO_4, 0.2%; $MgSO_4 \cdot 7$ aq, 0.1%, and sugar, 2%, in distilled water. The pH was 7.0. After inoculation, tubes with this medium were incubated at 30°C for 14 days, then titrated with N/10 alkali with phenolphthalein as indicator.

Fermentation—enterobacteria

J. G. Heyl (1957). Antonie van Leeuwenhoek 23, 33–58.

Fermentation of carbohydrates.

Bacto peptone	10 g
Sodium chloride	5 g
Distilled water	1000 ml
pH adjusted to 7.6	

After dissolving the ingredients 10 ml of the Andrade-indicator is added. Sterilize for 15 minutes at 115°C. After sterilization 1% of the carbohydrate to be tested is added.

Andrade-indicator:

Distilled water	100 ml
N NaOH	16 ml
Acid fuchsin	0.5 g

The tests have to be read daily for 10–12 days.

Fermentation—*Microbacterium*

V. Bolcato (1957). Antonie van Leeuwenhoek 23, 351–356.

Acid production from carbohydrates: 0.1 g of the pure carbohydrates to be tested was dissolved in 10 ml of nutrient broth (peptone, 5 g; beef extract, 3 g; distilled water, 1000 ml; pH 6.5) placed in Durham's fermentation tubes. These were sterilized in the Arnold steamer by the intermittent method for 3 successive days and inoculated with three loops of a fresh culture of the organism. A first series of tubes was incubated at

37°C. and a second series at 50°C. for 7 days. Care was taken to avoid evaporation. At the end of the incubation period growth was determined measuring the optical density and the pH of the medium in each tube.

Fermentation—*Neisseria*

R. E. M. Thompson and A. Knudsen (1958). J. Path. Bact. 76, 501–504.

The "sugars" employed were tested for purity by paper chromatography. The sugars were prepared as 20 per cent solutions, sterilised by Seitz filtration and added to the sterile base media to final concentrations of 1%. Sterile rabbit serum was added to a final concentration of 5%.

Medium IV. New Zealand agar powder, 0.15%; phenol red (0.2% aqueous solution), 1%; Hartley's tryptic digest broth to 100%.

Media adjusted to pH 7.6 and sterilised in the autoclave for 15 min. at a pressure of 15 lb per sq. in. The "sloppy" medium was stored in bulk and distributed in 3 × ½ in. plugged tubes in 2 ml volumes for use. All tubes were incubated for 18 hr at 37°C before inoculation as a test of sterility.

The pure culture was emulsified in approximately 1 ml of 1% peptone-water to form a heavy suspension. The fermentation media, pre-heated to 37°C, were inoculated with 1 drop of this suspension from a Pasteur pipette.

The media were incubated at 37°C for 48 hr. Carbon dioxide did not improve the growth of the organism.

Fermentation—*Pediococcus*

J. C. Dacre (1958). J. Dairy Res. 25, 409–413.

In a double digest of casein supplemented with a yeast or tomato extract, growth occurred and acid was formed at 30°C from: arabinose, glucose, fructose, galactose, mannose, lactose, maltose and dextrin; glycerol was fermented slowly. No growth took place with raffinose, sucrose, trehalose, melibiose, xylose, sorbitol and mannitol while mannitol was not formed from fructose. The culture failed to hydrolyse starch, inulin and sodium hippurate; it hydrolysed aesculin and salicin and arginine with the production of ammonia. Acetate, succinate, lactate and citrate were not utilized as a sole carbon source; the organism utilized citrate in the skim milk and broth media only in the presence of a fermentable sugar.

Fermentation—*Bacillus*

E. R. Brown, M. D. Moody, E. L. Treece and C. W. Smith (1958). J. Bact. 75, 499–509.

Either infusion or tryptone broth was used as a basal medium for fermentation tests. Substrates for all tests were inoculated, incubated and read according to the procedures described by Smith *et al.* (1952). N. R. Smith, R. E. Gordon, and F. E. Clark (1952). U. S. Department of Agriculture Monograph No. 16.

Fermentation—*Streptococcus*

P. Morelis and L. Colobert (1958). Annls Inst. Pasteur, Paris 95, 568–587.

Pour l'étude des fermentations, les glucides ont été utilisés à la concentration de 0,5 p. 100 (1 p. 100 pour le mannitol) en eau peptonée, le bleu de bromothymol étant employé comme indicateur. Les résultats sont indiqués dans les tableaux.

Fermentation—*Aeromonas*

J. P. Stevenson (1959). J. gen. Microbiol. 21, 366–370.

Cultures were incubated at 37°C in peptone water containing 1.0% (w/v) of the "sugar," using Andrade's indicator (1.0% acid fuchsin) for the detection of acid production, and Durham tubes for the collection of gas. Examinations were made after 24 and 48 hr and 5, 10 and 21 days.

Fermentation—*Clostridium*

M. E. Brooks and H. B. G. Epps (1959). J. gen. Microbiol. 21, 144–155.

Media: Fermentation of carbohydrates was examined in a peptone water medium containing the carbohydrate under test at 1.0% (w/v) concentration, and 0.2% (v/v) of a 6.0% (w/v) solution of bromocresol purple. The medium was dispensed in ¼ oz screw-cap bottles, and iron strip added to promote reducing conditions.

Inoculation and Incubation: The peptone-water sugars were inoculated with 24 or 48 hr Robertson's meat broth cultures, and incubated at 37°C for 2 weeks.

Method of Testing: Cultures were examined for the production of acid and gas, Universal Indicator (British Drug Houses Ltd., Poole) being added where the bromocresol purple had been decolorized. The pH value of cultures in aesculin peptone water were checked potentiometrically and reading of less than pH 6.5 was recorded as positive for acid production.

Fermentation—*Pseudomonas fluorescens*

M. E. Rhodes (1959). J. gen. Microbiol. 21, 221–263.

"Sugars" were incorporated at 1.0% (w/v) in Dowson's (1949) ammonium + inorganic salts basal medium. Bromcresol purple (1 ml of a 1.6% (w/v) ethanolic solution/L medium) was added as an indicator.

Five ml volumes of the media were at first sterilized by steaming for 1 hr on each of 3 successive days; later it was sometimes found necessary to add the separately sterilized sugar solutions to the sterile basal medium because of the formation of toxic products when the complete medium was heat sterilized. The standard inoculum of one 4 mm loopful was retained because it had been shown that the use of this larger inoculum (as compared with the straight wire inoculum) did not give rise to falsely positive results. Control basal media not containing an added carbon source were also inoculated. The development of turbidity, acidity and gas was recorded.

W. J. Dowson (1949). *Manual of Bacterial Plant Diseases* 1st ed. London: A. and C. Black.

Fermentation—*Mycobacterium fortuitum*

A. J. Ross and F. P. Brancato (1959). J. Bact. 78, 392–395.

Acid Production from Carbohydrates

The inorganic nitrogen medium and carbon sources used by Gordon and Smith (1953) were employed with the following modifications. The carbohydrates to be tested were sterilized by Seitz filtration and an additional carbohydrate, levulose, was used.

R. E. Gordon and M. M. Smith (1953). J. Bact. 66, 41–48.

Fermentation—*Mycobacterium*

R. E. Gordon and J. M. Mihm (1959). J. gen. Microbiol. 21, 736–748.

Acid production from carbohydrates. One-half ml of a 10% (w/v) aqueous solution of each carbohydrate, sterilized by autoclaving, was pipetted aseptically to 5 ml of sterile, inorganic nitrogen agar base: $(NH_4)_2HPO_4$, 1 g; KCl, 0.2 g; $MgSO_4 \cdot 7H_2O$, 0.2 g; agar, 15 g; distilled water, 1000 ml (Ayers, Rupp and Johnson, 1919). The pH value of this medium was adjusted to 7.0 before the addition of 15 ml of a 0.04% (w/v) solution of bromcresol purple. Cultures on each carbohydrate agar were observed for the acid colour of the indicator after 7 and 28 days of incubation at 28°C.

S. H. Ayers, P. Rupp and W. T. Johnson, Jr. (1919). U. S. Dept. Agr. Bull. 782, 39 pp.

Fermentation—*Escherichia*

S. Schäfler and S. Benes (1959). Annls Inst. Pasteur, Paris 96, 231–237.

Les fermentations des glucides de ces mêmes souches ont été étudiées sur trois milieux de culture : 1° dans l'eau peptonée, additionnée de 0,5 p. 100 de chaque glucide étudié ; 2° dans le milieu de Koser dépourvu de citrate et additionné de 0,2 p. 100 de peptone Witte, 0,002 p. 100 de bleu de bromothymol et 0,5 p. 100 de chaque glucide (milieu de Koser modifié peptoné) ; 3° dans le même milieu gélosé (0,004 p. 100 de bleu de bromothymol et 2 p. 100 de gélose). Tous ces milieux ont été utilisés, tant pour mettre en évidence les fermentations primaires, que les fermentations mutatives.

Fermentation—general

V. B. D. Skerman, *A Guide to the Identification of the Genera of Bacteria* – Williams and Wilkins Book Co., Baltimore, 1959.

Production of acid or acid and gas from carbohydrates, sugar, alcohols and alkaloids.

Note: Production of acid may be the result of an oxidative process (e.g., glucose to gluconic acid) or a fermentative one. The great majority of descriptions of biochemical properties in which acid production is mentioned refer to the latter type of reaction. The fermentative reactions occur simply because the medium is dispensed in such a manner that solution of O_2 from air by diffusion rarely meets the demand of the growing cell population. Although tubes are incubated aerobically, processes occurring in the tubes are essentially anaerobic.

The necessity to discriminate between the oxidative and fermentative production of acid is becoming more apparent. The method suggested by Hugh and Leifson has considerable practical merit.

In any method used for the detection of acid production, the observation is usually restricted to a visible change in the colour of an acid-base indicator. If the test is performed in an inorganic basal medium, the degree of color change will be dependent on the buffer capacity of the medium. For this reason the concentration of buffering salts such as phosphates should be reduced to a minimum consistent with metabolic requirements.

In media which contain amino-nitrogen the liberation of ammonia may, in some cases, be sufficient to neutralize all the acid produced and yield a false negative reaction. If a negative reaction is thought to be due to this, retesting in a mineral salts medium (if growth occurs) or in a medium with a smaller amount of organic nitrogen as suggested by Hugh and Leifson may yield a positive result.

In all the media recommended for acid production in these notes the substrates are sterilized in concentrated (usually 10 per cent) solutions and added aseptically after the sterilization of the medium. Sterilization by filtration is recommended but, provided the solutions are prepared in distilled water and do not turn yellow, heat sterilization at 110°C for 25 minutes is satisfactory for other than di- and poly-saccharides which may undergo a degree of hydrolysis. Heat sterilization of the substrate in the peptone base is widely practised and, where positive results are obtained, is usually satisfactory. However, with some organisms negative reactions in such media are meaningless, because the solution is rendered toxic by the heat treatment.

Methods

Mineral Salts Base.

The basal mineral salts medium described by Pope and Skerman* is recommended for this purpose. Add 2.0 ml of a 1.6 per cent alcoholic solution of bromicresol purple per L before final adjustment of the pH. Distribute in 5-ml amounts in 150 by 13 mm tubes, insert a Durham tube, and sterilize by heating to 121°C for 20 minutes. Add the filtered substrates aseptically to a final concentration of 1.0 per cent.

Acid production is indicated by a change in the color of the indicator from purple to yellow. Gas, if produced, accumulates in the Durham tube.

Peptone Base.

Prepare as above with the 1 per cent peptone water used in place of the mineral salts base.

Hiss's Serum Water Base:

This base is employed for organisms which will not grow or which grow poorly in peptone water.

Peptone	5.0 g
K_2HPO_4	1.0 g
Water	1000 ml

Dissolve by steaming for 15 minutes. Adjust the pH to 7.4. Add 8 ml of Andrade's indicator (vide infra). Sterilize at 115°C for 10 minutes. Cool to 50°C and add 100 ml of sterile ox serum. Add the substrates aseptically to a final concentration of 1 per cent and dispense aseptically in 5-ml amounts in sterile 150 by 13 mm tubes.

Andrade's Indicator

Acid fuchsin	0.5 g
Distilled water	100.0 ml
N NaOH (4 g/100 ml)	1000.0 ml

The Durham tube is omitted from this basal medium, because acid production which changes the color from colorless to pink may be accompanied by a coagulation of the serum which renders the gas tube inoperative.

Other Bases.

Organisms which will not grow in any of the above bases must be tested in a base in which they will grow. In reporting such results a simultaneous report should be made of other bases tested and found unsuitable.

Fermentation—*Fusobacterium*

A. C. Baird-Parker (1960). J. gen. Microbiol. 22, 458–469.

A medium containing the following constituents gave reasonable growth of fusobacteria in the absence of added carbohydrate, and good growth in the presence of fermentable carbohydrate (%, w/v); proteose peptone no. 3, 1.0; Lab-Lemco, 0.3; yeast extract, 0.1; L-cysteine HCl, 0.05. All sugars were Seitz-filtered and added to the separately–sterilized basal medium to a final concentration of 1.0% (w/v) or in the case of

rare sugars, 0.5% (w/v). The complete medium was "flash-sterilized" by the method of Davis and Rogers (1939). Cultures were incubated for periods up to 14 days; acid production was determined electrometrically or by Johnson's narrow-range pH papers. The use of pH indicator in the medium was found to be completely unreliable.
J. G. Davis and H. T. Rogers (1939). Zentbl. Bakt. ParasitKde (2 Abt) 101, 102.

Fermentation—*Pasteurella septica*

J. M. Talbot and P. H. A. Sneath (1960). J. gen. Microbiol. 22, 303–311.

Fermentation of glucose, sucrose, mannitol, galactose, sorbitol, D-xylose, maltose, adonitol, inositol, L-arabinose, dulcitol, glycerol, glycogen, salicin, lactose, raffinose, dextrin, trehalose, mannose and fructose in 1% (w/v) solution in peptone water with Andrade's indicator was observed daily for 14 days.

Fermentation—*Fusobacterium*

R. R. Omata and R. C. Braunberg (1960). J. Bact. 80, 737–740.

Medium:

Trypticase (BBL)	1%
Yeast extract (Difco)	0.5%
L-cystine	0.02%
Sodium thioglycolate	0.5%
final pH 7.2	

Glucose, sucrose, maltose, or lactose was added separately to the sterile medium as sterile solutions to give a final concentration of 1% carbohydrate.

Fermentation—*Leuconostoc*

E. I. Garvie (1960). J. Dairy Res. 27, 283–292.

The ability to ferment different carbon sources has long been the basis of the classification of the leuconostocs. The basal medium selected was 1.5% peptone, 0.6% Yeastrel, and 0.5% NaCl. Brom-cresol purple was used as indicator. The carbon sources used were prepared as 2% solutions sterilized by Seitz filtration and 0.5 ml was added aseptically to each 5 ml of broth. The cultures were incubated for 14 days.

Fermentation—*Streptococcus* and *Lactobacillus*

C. W. Langston and C. Bouma (1960). Appl. Microbiol. 8, 212–222.

Fermentation reactions were carried out in the following basal medium: Trypticase 10 g ; yeast extract 5 g ; NaCl 5 g ; and distilled water to make 1 L. Substrates were used at the 1 per cent level and either autoclaved with the medium or sterilized separately and added to the autoclaved base. The medium was tubed in 7 ml quantities and incubated for 2 weeks. Fermentation reactions were detected through the use of a Beckman pH meter.

Fermentation—*Pediococcus*

H. L. Günther and H. R. White (1961). J. gen. Microbiol. 26, 185–197.

Yeast-extract peptone broth (containing, % w/v: peptone, 1.0; yeast extract, 0.5; NaCl, 0.5; $MgSO_4$, 0.05; $MnSO_4$, 0.05; at pH 7.0) was used as a basal medium for fermentation tests, and Seitz-filtered carbohydrate added to give 1% (w/v) final concentration. Acid and gas production were determined after 7 days of incubation [indicator, 0.04% (w/v) bromcresol purple, added after incubation], since preliminary results had shown that many isolates, especially fresh ones, were slow in producing acid. Acid, once produced, was not masked by subsequent production of alkaline substances. The carbohydrates tested were: arabinose, xylose, glucose, fructose, maltose, lactose, sucrose, trehalose, raffinose, inulin, dextrin, glycerol, mannitol, sorbitol, salicin.

Maintenance of stock cultures and methods of cultivation. For maintenance of stock cultures, preparation of inocula and in all experimental work, 'Oxoid' tomato juice (TJ) broth or tomato juice (TJ) agar, adjusted to pH 6.6, were used unless otherwise stated. The following were exceptions to this rule: for strain Tc. 1 sodium chloride (5%, w/v) was added to the medium; and for the aerococci glucose Lemco broth (Shattock and Hirsch, 1947) or glucose yeast extract (GY) agar (containing, as %, w/v; peptone, 1.0; Yeastrel, 0.3; glucose, 1.0; NaCl, 0.25; agar, 1.0; at pH 7.4) was used.

Cultures were incubated aerobically at 30°C with specified exceptions.
P. M. F. Shattock and A. Hirsch (1947). J. Path. Bact. 59, 495.

Fermentation—*Streptococcus*

O. G. Clausen (1961). J. Path. Bact. 82, 212–214.

Fermentation of carbohydrates, alcohols and glycosides was observed in cultures grown in peptone water containing 5 per cent horse serum, 1 per cent Andrade's indicator and 0.5 per cent of the substrate.

Fermentation—*Pasteurella*

G. R. Smith (1961). J. Path. Bact. 81, 431–440.

Fermentation reactions. Tests were carried out in the medium recommended by Bosworth and Lovell. This consisted of 1 per cent peptone water containing 10 per cent infusion broth, 7.5 per cent bromothymol blue (B. D. H. indicator solution) and 1 per cent of the fermentable substance. The final pH of the medium was 7.0–7.2. Each tube was inoculated with one drop of a 5-hr broth culture. The production of full yellow indicated a positive reaction; incomplete colour changes were

disregarded.
T. J. Bosworth and R. Lovell (1944). J. comp. Path. Ther. 54, 168.

Fermentation—*Neisseria*

B. W. Catlin and L. S. Cunningham (1961). J. gen. Microbiol. 26, 303–312.

Heart infusion broth (Difco) with or without 0.3% (w/v) yeast extract (Difco) was supplemented after sterilization with 250 μg ribonucleic acid (Nutritional Biochemicals Corporation)/ml, 0.00005 M- sodium glutamate and 0.0005 M- calcium chloride added separately as sterile solutions (Catlin, 1960).

Capacity to produce acid from carbohydrates was examined by using above media with 1% (w/v) agar, phenol red 0.015 mg/ml, and 0.5% (w/v) of either glucose, maltose, fructose, sucrose, mannitol or lactose. For the latter additions filter-sterilized 20% (w/v) solutions were added aseptically. Media were tubed with a butt and a short slope, and were inoculated by stabbing the deep agar and streaking the surface.
B. W. Catlin (1960). J. Bact. 79, 579.

Fermentation—*Escherichia aurescens*

H. Leclerc (1962). Annls Inst. Pasteur, Paris 102, 726–741.

b) Métabolisme des hydrates de carbone. – Il est étudié à partir du milieu des base défini par Mossel et Martin [34] dont voici la formule :

Trypticase (BBL)	10	g
Extrait de levure	1,5	g
Chlorure de sodium	5	g
Bacto agar	5	g (1)
Bromocrésol pourpre	0,015	g
Eau distillée	1000	ml

Ajuster à pH 7 ; couler 3 ml en tubes de 8 × 180 mm. Stériliser à 120° vingt minutes.

Les milieux sont désaérés au bain-marie bouillant pendant quinze minutes. Après refroidissement entre 45° et 50°, on les additionne de la solution sucrée stérilisée au préalable par filtration, en quantité définie pour obtenir une concentration finale de 1 p. 100.

Les milieux sont incubés vingt-quatre heures à 30° pour éliminer ceux qui auraient pu être accidentellement souillés. Ils sont ensuite ensemencés en piqûre centrale, incubés quatre jours à 30° et jusqu'à trente jours à 20–22° si nécessaire. Le virage du milieu (jaune) est la marque d'une acidification : oxydation s'il n'apparait qu'en surface, fermentation s'il apparaît sur toute la hauteur de la colonne de milieu. La présence de gaz se traduit par une fragmentation plus ou moins intense de la gélose.

(1) La concentration d'agar définie par Mossel et Martin est de 15 g/ 1000 ml. Nous l'avons diminuée afin d'augmenter la rapidité d'apparition

du gaz.
[34] Mossel (D. A. A.) et Martin (G.). *Ann. Inst. Pasteur Lille* (à paraître).

Fermentation—*Clostridium*

H. Beerens, M. M. Castel and H. M. C. Put (1962). Annls Inst. Pasteur, Paris 103, 117–121.

Fermentation des hydrates de carbone : Les milieux sucrés à 1 p. 100, préparés avec la base V. L., servent à étudier la fermentation des glucose, lactose, saccharose, salicine, mannitol, glycérol, amidon soluble. Un même milieu V. L. non sucré sert de témoin. La fermentation est révélée par addition de quelques gouttes d'un indicateur de pH après incubation à 37°C pendant deux à trois jours, selon l'intensité de la croissance. Une réaction positive s'accompagne souvent d'un dégagement intense de gaz et d'un développement microbien plus abondant que celui observé dans les tubes où l'hydrate de carbone n'a pas été utilisé par la bactérie.

Fermentation—*Actinobacillus actinomycetemcomitans* and *Haemophilus aphrophilus*

E. O. King and H. W. Tatum (1962). *Actinobacillus actinomycetemcomitans* and *Hemophilus aphrophilus.* J. infect. Dis. 111, 85–94.

The fermentation base consisted of Difco heart infusion broth with 0.2 ml of 1.5% aqueous bromcresol purple per liter of the medium. Carbohydrates were Seitz-sterilized and added to make up a 1% concentration.

Fermentation—*Lactobacillus*

R. E. Smith and J. D. Cunningham (1962). Can. J. Microbiol. 8, 727–735.

Hayward's medium was used for the routine carbohydrate fermentations, with 1.5% (w/v) concentrations of carbohydrates. It consisted of Peptone (Oxoid), 0.5% (w/v); yeast extract (Difco), 0.3% (w/v); salts A, 0.5% (v/v); salts B, 0.5% (v/v); Tween 80, 0.1% (v/v); sodium acetate (hydrated), 0.5% (w/v). The constituents were diluted in glass-distilled water and the pH value of the medium adjusted to 6.8–7.0. Solution A contained 10 g KH_2PO_4 and 10 g K_2HPO_4 in 100 ml of distilled water; solution B contained 11.5 g $MgSO_4 \cdot 7H_2O$, 2.4 g $MnSO_4 \cdot 2H_2O$ and 0.68 g $FeSO_4 \cdot 7H_2O$ in 100 ml of distilled water. Heat-labile sugars were Seitz-sterilized, and added to the basal media after autoclaving. Durham's fermentation tubes were used to detect gas formed. Negative gas production was confirmed as a general characteristic with the maltose broth [Hayward's medium with 2.5% (w/v) maltose], as recommended by Hayward.

A. C. Hayward (1957). J. gen. Microbiol. 16, 9–15.

Fermentation—*Streptococcus*

W. E. Sandine, P. R. Elliker, H. Hays (1962). Can. J. Microbiol. 8, 161–174.

Two milliliter quantities of lactic broth* medium without carbohydrate and containing 0.02% bromcresol purple were dispensed into 12 × 120 mm tubes and autoclaved at 15 lb pressure (121°C) for 10 minutes. After cooling, 2 ml of 1% filter-sterilized carbohydrate was added aseptically to each tube to provide a single-strength test medium containing 0.50% carbohydrate. The test medium was inoculated in duplicate with a loopful of inoculum (0.01 ml) from a 24-hour lactic broth culture and incubated at 30°C for 48 hours. Slow-growing cultures such as *Leuconostoc,* which showed no evidence of growth in 48 hours, were incubated 5 days longer at room temperature (25°C).

The ability of *Leuconostoc* organisms to produce dextran from sucrose was tested by streaking 48-hour nonfat milk cultures of *Leuconostoc* organisms on agar plates of medium composed of Tryptone, 5.0 g; yeast extract, 5.0 g; sucrose, 100.0 g; $CaCO_3$, 20.0 g; agar, 20.0 g; distilled water, 1000 ml. The medium was neutralized and sterilized as described for the citrate broth. Plates were incubated for 2 to 6 days at 21°C.

*P. R. Elliker, A. W. Anderson and G. Hannesson (1956). J. Dairy Sci. 39, 1611–1612 for "lactic agar."

Fermentation—*Vibrio*

G. H. G. Davis and R. W. A. Park (1962). J. gen. Microbiol. 27, 101–119.

Acid Production.

Basal Medium: Koser salt solution (Mackie and McCartney, 1953, p. 429) plus Oxoid yeast extract 0.3% (w/v). Initial pH adjusted to 7.0, 2 ml of 1.5% (w/v) solution of bromocresol purple added per litre and medium autoclaved. Carbohydrates were added as 20% (w/v) autoclaved solutions to give 1% (w/v) final concentration. The complete media were aseptically dispensed and tested for sterility by incubating for 48 hr at 30°C. Tests read daily over 14 days.

T. J. Mackie and J. E. McCartney (1953). *HandBook of Practical Bacteriology,* 9th ed. Edinburgh: Livingstone.

Fermentation—*Vibrio*

G. H. G. Davis and R. W. A. Park (1962). J. gen. Microbiol. 27, 101–119.

Fermentation tested in 1% (w/v) concentration in peptone water (Mackie and McCartney, 1953) and also in a dilute meat extract medium (Simon 1956) both media containing bromothymol blue as pH indicator. Arabinose was added as a Seitz-filtered solution to sterile base: gas production from this sugar was not tested for. All other carbohydrate media were steam-sterilized. Control tests without carbohydrate were also inoculated. Initial pH of all tests was 7.4, and tests were read after 24 hr and 7 days' incubation. As results were identical by both methods, only one set has been recorded.

T. J. Mackie and J. E. McCartney (1953). *Handbook of Practical Bacteriology.* 9th ed. Livingstone.
R. D. Simon (1956). Br. J. exp. Path. 37, 494.

Fermentation—*Butyrivibrio* and others of rumen flora.

T. H. Blackburn and P. N. Hobson (1962). J. gen. Microbiol. 29, 69–81.

Salts solution. Solution (a) contained (g/L): KH_2PO_4, 3.0; $(NH_4)_2SO_4$, 6.0; NaCl, 6.0; $MgSO_4$, 0.6; $CaCl_2$, 0.6. Solution (b) contained (g/L): K_2HPO_4, 3.0. Rumen fluid was prepared by straining rumen contents (freshly obtained from a hay and grass-fed sheep) through gauze and centrifuging at 62,000 *g* for 10 min. The clear liquid was kept for not more than a few days at 2°C before use.

Bicarbonate-dithionite solution. Dithionite was added, as a small volume of a concentrated solution (6%, w/v) in boiled water, to a sterile 10% (w/v) solution of $NaHCO_3$. The mixed solution was gassed with CO_2 and a portion added to the medium to bring the final concentrations to 0.5 and 0.003% respectively of $NaHCO_3$ and dithionite.

"Sugars" (including carbohydrates, alcohols and acids) were the usual commercially available reagent grades, the xylan was the xylan No. 3 used by Howard, Jones and Purdom (1960). Cellulose was prepared from Whatman No. 1 filter-paper ground in a small ball mill to a fine powder. Carboxymethyl cellulose was Cellofas B, low viscosity (I. C. I. Ltd.). For isolation of bacteria, cellulose, xylan, carboxymethyl cellulose and malt extract were incorporated into the medium before autoclaving. Other carbohydrates were added as concentrated sterile filtered solutions, except for starch which was an autoclaved solution.

Casein was Glaxo casein C (Glaxo, Ltd, Greenford, Middlesex). This was acid precipitated, washed with water and dissolved in dilute NaOH, neutralized and the solution freeze-dried to give an easily soluble powder, which was added as a solution to the medium before autoclaving.

Other constituents. Yeast extract was "Difco" brand (Bacto Laboratories, Detroit, U. S. A.). Tryptose was "Bacto" brand (Bacto Laboratories, Detroit, U. S. A.). Malt extract was "Bacto" brand (Bacto Laboratories, Detroit, U. S. A.). L-Cysteine hydrochloride was from L. Light, Ltd., Colnbrook, Bucks.

Media for isolations. The media used for the isolations described here were basically similar and the constituents and preparations of the medium for isolations from sheep 74 and 190 only will be described in detail.

This medium contained per 100 ml: salts solution (a) 15 ml; salts solution (b) 15 ml; phenosafranine, 0.0001 g; cysteine hydrochloride, 0.05 g; casein, 0.5 g; rumen fluid, 10 ml; tryptose, 0.3 g; agar, 2.5 g; water, 55 ml. These constitutents were added as solids or solutions, mixed and gassed with CO_2 and autoclaved at 120°C for 15 min under CO_2 in a flask fitted with a Bunsen valve and a stoppered side tube. For addition of other medium constituents or dispensing of medium the Bunsen valve was removed and a syringe inserted through the tubing to which the valve had been attached. A brisk stream of CO_2 was meanwhile passed through the side tube and flowed out around the syringe, ensuring anaerobic conditions. After addition of the bicarbonate-dithionite solution (5 ml; see above) and any carbohydrate solution (final carbohydrate concentration was usually 0.5% (w/v)) necessary, the medium at 50°C was dispensed into tubes under CO_2 by means of hypodermic syringes or a laboratory-made dispensing apparatus for use under CO_2. Tenfold dilutions of rumen fluid were made under CO_2 in a solution prepared exactly like the medium, but omitting the agar and carbohydrate. One half millilitre portions of the diluted rumen fluid were added to 4.5 ml of medium in 6 × $^5/_8$ in. test tubes under CO_2, the tubes tightly stoppered with rubber bungs and rolled under cold water. For liquid media the agar was omitted, but the cultures were still incubated in test tubes as above, at 38°C.

Media for biochemical reactions. For determination of fermentation and other reactions of the bacteria the isolation medium was appropriately modified. Unless the degree of hydrolysis of casein was to be determined this was omitted from the media and yeast extract (0.3%, w/v) was added. Growth without CO_2 was tested in media prepared and dispensed under oxygen-free nitrogen and with bicarbonate replaced by phosphate.

B. H. Howard, G. Jones and M. R. Purdom (1960). Biochem. J. 74, 173.

Fermentation—*Haemophilus*

P. N. Edmunds (1962). J. Path. Bact. 83, 411–422.

Fermentation reactions. Basic peptone broth was prepared as the basic peptone agar of Edmunds (1960) without the agar (see below). Five per cent of horse-flesh digest (Edmunds 1960) and 0.0025 per cent of bromothymol blue powder were added and the medium was dispensed in 100-ml bottles and sterilised by steaming at 100°C for 45 minutes. Sugar solutions (5–20 per cent) were sterilised by steaming for 20 min on three successive days, except for arabinose and xylose solutions, which were Seitz-filtered. Starch solution was made freshly each day. These solutions were added aseptically to the medium to give a final sugar concentration of 0.25 or 1.0 per cent and 3 ml amounts of the medium were then dispensed into 5 × $^5/_8$ in. tubes. Durham tubes were not used since preliminary work had shown that no gas was produced in fermentation. Sterile plasma was obtained from human blood specially collected by the Blood Transfusion Service into sterile disodium citrate solution without the usual addition of glucose (false-positive reactions occurred with plasma from blood collected in glucose citrate solution). Five per cent of plasma was added aseptically to the sugar medium, either before sterility testing (by incubation in air at 37°C for 48–72 hr) or simultaneously with the inoculum.

Two methods were used for inoculation. In *method* 1, the completed plasma sugar medium was inoculated with 6–8 drops of a 48-hr culture in the blood agar and broth medium of Edmunds (1960). In *method* 2, about a third of the growth on a blood agar plate was suspended in 1.5 ml plasma and 0.15 ml of the suspension was added to each 3 ml tube of plasma-free sugar medium. The tubes were incubated in air at 37°C for 12 days, and observed at 5 and 12 days. A control set of tubes contained sterile plasma added in the same way as the plasma was added to the other tubes. Comparison with control tubes was necessary because the plasma made the colour change difficult to see when fermentation was weak.

The *basic peptone agar* consisted of 20 g peptone (Evans), 2.5 g NaCl, 2.0 g sodium glycerophosphate, 11 ml 2 N Na_2CO_3 solution in water and 1000 ml distilled water. These were mixed and free-steamed for 1 hr at 100°C in an autoclave. After cooling and adjustment of the pH to 7.3, 0.8–0.9 per cent Davis New Zealand agar was added and allowed to soak overnight. The steaming was repeated to melt and sterilise the agar, and the medium dispensed aseptically into 300-ml, sterile flat bottles. These were then steamed for 20 min.

The *horse-flesh digest* (Levinthal, 1931) was prepared from 1000 g minced fat-free fresh horse flesh, 3000 ml distilled water, 50 ml 2N Na_2CO_3 solution, 5.0 g pancreatin and 35 ml chloroform. The pancreatin (British Drug Houses) was ground up in 100 ml water, the rest of the ingredients added, and the mixture incubated for 19 hr at 37°C with intermittent shaking. A 50-ml portion was then removed, boiled till the meat coagulated and filtered through Whatman's no. 1 paper. The filtrate was tested for its pH value, which should be 6.8–7.1, and for albumin, which should be present in large amounts. If these tests were satisfactory, the main bulk of the mixture was steamed for 35 min and left overnight to cool and settle. The supernatant fluid was decanted, filtered and dispensed with aseptic precautions into sterile 500 ml screw-capped bottles. These were steamed for 15 min and stored at room temperature. For current use, the digest was dispensed with sterile precautions into 28-ml vials and again steamed for 10 min.

P. N. Edmunds (1960). J. Path. Bact. 79, 273.
W. Levinthal (1931). Zentbl. Bakt. ParasitKde Abt. I. Orig. 121, 513.

Fermentation—*Pseudomonas odorans*

I. Malék, M. Radochová and O. Lysenko (1963). J. gen. Microbiol. 33, 349–355.

Investigated in peptone water (%, w/v; Bacto peptone 1.0; NaCl, 0.5; pH 7.2) containing 1% (w/v) of the appropriate carbohydrate, with bromothymol blue as pH indicator – 28°C.

Fermentation—*"Bacterium salmonicida" (Aeromonas)*

I. W. Smith (1963). J. gen. Microbiol. 33, 263–274.

The fermentation of carbohydrates was examined in 1% Difco peptone water with Andrade's indicator. The media were sterilized by Tyndallization and the gas production recorded as positive when more than one-tenth of the liquid in a Durham tube was displaced by gas. For a comparison of gas production in the presence of different peptones and from glucose sterilized by filtration and in buffered peptone, standard 50 mm Durham tubes were used. The displacement of gas was measured to the nearest 0.5 mm and the volumes expressed as a percentage.

The lactose sugars were read at 7 and 40 days: the remainder of the sugar tests were read only at 7 days.

Fermentation—*Corynebacterium vaginale ("Haemophilus vaginalis")*

K. Zinnemann and G. C. Turner (1963). J. Path. Bact. 85, 213–219.

The authors used Hiss's serum water medium with added carbohydrate. Tubes were incubated for 1 week at 37°C in a McIntosh and Fildes' Jar with half the normal pressure of oxygen and at least 10% CO_2.

Fermentation—*Shigella sonnei*

M. Nakamura (1963). J. Bact. 85, 487–488.

The paper reports fermentation of salicin by *Shigella sonnei* in the following medium: Beef extract (Difco), 3 g; peptone (Difco), 5 g; salicin (Difco), 5 g; bromothymol blue dye solution, 1 ml; and distilled and deionized water, 1,000 ml. The bromothymol blue solution was prepared by dissolving 0.4 g of bromothymol blue (Difco) in 500 ml of 95% ethyl alcohol and making a total volume of 1,000 ml with water. The solution was filtered through Whatman no. 1 filter paper and stored in a dark bottle at 4°C. The salicin fermentation broth was dispensed in screw-capped test tubes, 10 ml per tube. In addition, Phenol Red Broth Base (Difco) containing 0.5% salicin was used. The fermentation reactions were essentially the same in the two fermentation media employed. Ten tubes of fermentation broth were inoculated with each strain; the volume of the inocula was approximately 0.05 ml, obtained by washing 24-hr cultures in buffer and resuspending to yield a population of approximately 10^7 cells/ml. The tubes were incubated for 40 days at 37°C before being discarded as negative.

Fermentation—*Propionibacterium acnes*

W. E. C. Moore and E. P. Cato (1963). J. Bact. 85, 870–874.

Lyophilized cultures were revived and maintained on anaerobic peptone-yeast extract-glucose (PYG) medium prepared as follows. Peptone; 5.0 g; glucose, 2.5 g; yeast extract, 2.5 g; and resazurin solution, 1 ml (25 mg of resazurin/100 ml of water) were added to 250 ml of distilled water in a 250-ml Erlenmeyer flask. The flask was fitted with a stopper and a glass-tubing chimney (diameter, 1 in.; length, 12 in.). The solution was boiled vigorously until it turned yellow. When the flask was removed from the flame, the chimney was quickly replaced with a stopper containing two small holes through which two hypodermic needles (6 in., 15 gauge) were inserted. Carbon dioxide, which had been passed through hot reduced copper turnings, was then bubbled continuously through the medium. After the flask had cooled to room temperature in a cold-water bath, 0.125 g of cysteine was added, and the pH was adjusted to 7.0 with NaOH. After neutralization, the CO_2 was replaced with oxygen-free N_2 in the same line to prevent further pH changes in the medium. To tube the medium, one of the 6-in. hypodermic needles was removed from the flask and inserted to the bottom of plain rim test tubes (18 × 150 mm). Medium was transferred from the flask to the bottom of the tubes with a spring-actuated syringe fitted with a 6-in. hypodermic needle, which had been purged of air. The medium was bubbled to the top of the tube with N_2, while the gassing needle was withdrawn through a notch cut in a no. 1 stopper. The stopper was then firmly seated in the tube. This procedure insured rapid and positive removal of all air. Tubes were placed in the outside rows of racks with rubber mesh matting above and below them. A second rack was secured over the stoppers, by means of stiff wire hooks, to hold them in during sterilization at 121°C for 15 min. All inoculations and transfers to cooled media were made under jets of oxygen-free CO_2 by the methods of Hungate (1950) and of Bryant and Robinson (1961). Tests with indigo carmine showed that this method consistently produced oxidation-reduction potentials below −125 mv; however, resazurin (E'_0, −30 to −40 mv) was satisfactory for the detection of tubes which were not properly stoppered or gassed, and the end point was easier to see. The medium should be protected from light, since resazurin becomes inactive after a few hours in sunlight.

The organisms were also cultured in aerobic PYG medium made without boiling or the addition of cysteine and dispensed to a depth of 2 in. in 16-mm screw-cap tubes. Anaerobic lactate medium was prepared exactly as anaerobic PYG, but the glucose was replaced with sodium lactate. Aerobic lactate medium was prepared exactly as aerobic PYG with sodium lactate replacing the glucose.

Cultures for fermentation studies were inoculated with approximately

0.02 ml of broth cultures which had shown turbidity for no more than 24 hr. The fermentation cultures were routinely incubated for 3 weeks at 37°C or 39°C.

Fermentation acids were chromatographed according to the method of Bruno and Moore (1962) using the temperature-controlled column.

C. F. Bruno and W. E. C. Moore (1962). J. Dairy Sci. 45, 109–115.
M. P. Bryant and I. M. Robinson (1961). J. Dairy Sci. 44, 1446–1456.
R. E. Hungate (1950). Bact. Rev. 14, 1–49.

Fermentation—rumen bacteria

R. S. Fulghum and W. E. C. Moore (1963). J. Bact. 85, 808–815.

Media for the study of substrate fermentation consisted of SRP-basal medium* with 0.1% $(NH_4)_2SO_4$ and 0.5% of the substrate to be studied. Glucose fermentation tubes were incubated for one week at 39°C. Similar preparations without glucose were observed to determine whether growth and fermentation were in response to glucose or to the clarified rumen fluid. No growth or very scant growth occurred in the medium without glucose, whereas moderate to heavy growth occurred in the medium with glucose. The acid fermentation products of glucose metabolism were determined chromatographically by the method of Neish (1952). In a later study, fermentation tubes were prepared in the same way as those above, except that glucose was increased to 1%. These tubes were inoculated with lyophilized cultures of the isolates which had been stored at 2°C for 2 years and were then incubated at 37.5°C for 3 weeks. The acid fermentation products were determined by the modified chromatographic procedure described by Bruno and Moore (1962).

*Basal medium for physiological tests – *Butyrivibrio, Borrelia, Bacteroides, Selenomonas, Succinivibrio.*

The anaerobic methods of Hungate (Bact. Revs. 14, 1–49, 1950) were used in handling and incubation of the cultures.

C. F. Bruno and W. E. C. Moore (1962). J. Dairy Sci. 45, 109–115.
A. C. Neish (1952). Report no. 46–8–3 (2nd revision). Prairie Regional Laboratory, Saskatoon. National Research Council of Canada.

Fermentation—*Clostridium rubrum*

H. Ng and R. H. Vaughn (1963). J. Bact. 85, 1104–1113.

The criterion for fermentation of a carbohydrate was the ability to produce acid and gas in a basal medium of the following composition (per liter): test substrate, 10 g; proteose peptone, 10 g; Tryptone, 10 g; yeast extract, 5 g; sodium thioglycolate, 2 g; K_2HPO_4, 1 g; and agar, 2.5 g. The same medium minus the test substrate served as the control.

Fermentation—*Staphylococcus aureus*

R. L. Brown and J. B. Evans (1963). J. Bact. 85, 1409–1412.

Anaerobic metabolism of mannitol and glucose was detected as follows.

Quantities (6 ml) of a basal medium (consisting of Trypticase, 1%; yeast extract, 0.5%; NaCl, 0.5%; substrate, 1%; and bromcresol purple, 0.004%) were dispensed in 18-mm test tubes after the pH was adjusted to 7.2. The medium was sterilized, and each tube was inoculated with 0.05 ml of culture. Each tube was covered with a sterile vaspar seal, incubated, and observed after 24 and 48 hr for a color change.

For aerobic metabolism of glucose and mannitol, 12-ml samples of the same media were dispensed in 50-ml Erlenmeyer flasks. Inocula and other conditions were the same as above.

Fermentation—lactic acid bacteria

R. Whittenbury (1963). J. gen. Microbiol. 32, 375–384.

Soft agar medium and preparation of soft agar cultures: The basal medium contained : meat extract (Lab-Lemco), 0.5 g ; peptone (Evans), 0.5 g ; yeast autolysate (prepared as in Gibson, Stirling, Keddie and Rosenberger, 1958), 5.0 ml or yeast extract (Difco), 0.5 g ; Tween 80, 0.05 ml ; agar (Davis), 0.15 g ; tap water to 100 ml. Either bromocresol purple (BCP), 1.4 ml of a 1.6% (w/v) ethanolic solution, or bromocresol green (BCG), 2.8 ml of a 0.4% (w/v) aqueous solution, was added per L medium. Fermentable substrates, as distilled water solutions sterilized by autoclaving at 121°C for 15 min, or by Seitz-filtration, were added to the autoclaved (121°C for 15 min) basal medium to give a final concentration of 0.5% (w/v) or, in the case of slightly soluble substances, 0.25% (w/v). The medium used in salt tolerance tests was prepared at double strength and an equal volume of NaCl solution was added after sterilization. Soft agar containing BCP was adjusted to pH 6.8–7.0 ; that containing BCG to pH 5.4. The basal medium, 90 ml quantities in bottles, was liquefied by momentarily autoclaving at 115°C, allowed to cool to 48°C in a water bath and completed by adding 10 ml of a fermentable substrate solution. The completed medium was replaced in the water bath or kept in an incubator at 50°C until required, when it was distributed in 6–7 ml amounts to sterile 6 × $^5/_8$ in. test tubes, the amount being judged visually as the medium was poured into the tubes. A set of completed media was placed in the one rack in the water bath so that they could be inoculated with one culture. Immediately before inoculation the media were allowed to cool to 37–40°C, a temperature which allowed ample time for inoculation before the agar set. The liquefied medium was inoculated with a drop of a turbid culture added by capillary pipette and then tilted two or three times before being allowed to set. The subsequent growth indicated that the inoculum was uniformly distributed through the medium by this procedure. When for comparative purposes substrates were autoclaved in the medium the same method of inoculation was used.

Liquid medium. This was similar in composition to the soft agar medium, but without agar. The medium was distributed in 4.5 or 2.25 ml amounts in 5 × ½ in. tubes. Substrate solutions, 0.5 ml in the former instance and 0.25 ml in the latter, were added subsequently when not included in the medium originally. Inoculation was by capillary pipette.

Inoculum medium. This was similar to the liquid medium with the exceptions that glucose, 0.5% (w/v), was autoclaved in the medium, which was adjusted to pH 6.5, and the indicators were omitted.

T. Gibson, A. C. Stirling, R. M. Keddie and R. F. Rosenberger (1958). J. gen. Microbiol. 19, 112.

Fermentation—*Actinobacillus*

P. W. Wetmore, J. F. Thiel, Y. F. Herman and J. R. Harr (1963). Comparison of selected *Actinobacillus* species with a hemolytic variety of *Actinobacillus* from irradiated swine. J. infect. Dis. 113, 186–194.

The authors used 'phenol red base (Difco) with 10% horse serum and 1% carbohydrate solution (sterilized by filtration): readings were made during a 2-week period'.

Fermentation—*Corynebacterium pseudotuberculosis*

C. H. Pierce-Chase, R. M. Fauve and R. Dubos (1964). J. exp. Med. 120, 267–281.

Carbohydrate redi-discs were obtained from Pennsylvania Biological Laboratories, Inc., Philadelphia and used comparatively in three different media at pH 7.3, namely: (a) veal infusion broth, (b) PF broth previously fermented by 4 hour growth of *Escherichia coli* and (c) a standard carbohydrate free basal medium consisting of 1.0 per cent proteose peptone No. 3, 0.5 per cent bacto-beef extract, and 0.5 per cent NaCl. Fermentation tubes containing one redi-disc in 3.0 ml of medium (1.1 per cent carbohydrate) were inoculated with 0.1 ml PF broth culture or a saline suspension of growth from PF agar. Growth and acid production were determined after 24, 48, and 72 hours of incubation by withdrawing small aliquots and testing with bromthymol blue. The final acidity in tubes giving positive tests was pH 5.1, as tested by bromcresol purple and bromcresol green. Identical results were obtained with all three media.

Fermentation—*Lactobacillus casei* var. *rhamnosus*

W. Sims (1964). J. Path. Bact. 87, 99–105.

Fermentation reactions were carried out in sterile screw-capped bijou bottles containing 4.0 ml of Hiss's serum water with Andrade's indicator and 0.5 ml of a Seitz-filtered 10 per cent solution of the sugar to be tested.

Fermentation—*Streptococcus*

R. H. Deibel, J. Yao, N. J. Jacobs and C. F. Niven, Jr. (1964). Group E streptococci. I. Physiological characterization of strains isolated from swine cervical abscesses. J. infect. Dis. 114, 327–332.

Strains were maintained by daily transfer in the following medium: tryptone (Difco), 10 g; yeast extract (Difco), 5 g; sodium chloride, 5 g; dipotassium phosphate, 5 g; glucose, 2 g; beef extract (Difco), 5 g; distilled water, 1 L; pH 7.0 to 7.2. In fermentation studies the medium was altered in that the phosphate concentration was decreased to 0.2%, the glucose to 0.05%, and bromcresol purple (0.004%) was added. A 1% concentration of test substrates was employed.

Fermentation—*Actinomyces*

L. K. Georg, G. W. Robertstad and S. A. Brinkman (1964). J. Bact. 88, 477–490.

In the fermentation tests, the carbohydrates were added to the basal medium at a concentration of 0.5%. A drop in pH of less than 0.49 pH unit (as compared with control tubes) was considered to be a negative reaction; a drop of 0.5 to 0.9 unit, a plus-minus reaction; and a drop of one pH unit or more was considered positive.

The basal medium consisted of heart infusion broth, 25 g; pancreatic digest of casein, 4 g; and yeast extract, 5 g (per liter of distilled water at pH 7.0). The inoculum was taken from cultures made in sugar free heart infusion broth.

The cultures were incubated at 37°C under pyrogallol-carbonate seals and read at 3 and 10 days.

Fermentation—*Dermatophilus*

M. A. Gordon (1964). J. Bact. 88, 509–522.

The sugar fermentation media contained 1% peptone, 0.5% NaCl, 1% of the test carbohydrate, and Andrade's indicator at pH 7.4 to 7.5. The inoculum was pipetted from a 4 to 5 day broth culture. The medium was incubated at 36°C (± 1) and readings were made at 48 hours, 5 days, 1 week and 2 weeks.

Fermentation—*Vibrio marinus*

R. R. Colwell and R. Y. Morita (1964). J. Bact. 88, 831–837.

The authors used the method of R. Hugh and E. Leifson (1953), J. Bact. 66, 24–26 – modified by the addition to the media of the following salts to produce a synthetic seawater: – sodium chloride, 2.4%; potassium chloride, 0.07%; magnesium chloride (hydrated) 0.53% and magnesium sulfate (hydrated), 0.7%. A standard inoculum was 1 drop from a pasteur pipette (*c.* 0.05 ml) of a 24 hour artificial seawater broth.

Cultures were incubated at 18°C.

Fermentation—enterococci

C. G. Rogers and W. B. Sarles (1964). J. Bact. 88, 965–973.

All strains were tested for ability to ferment L(+)-arabinose, mannitol, sorbitol, melibiose, melezitose, and glycerol, each at a concentration of

0.5%, in sterile 1% peptone broth containing bromocresol purple as an internal indicator. Tubes were read for acid production after 7 days.
Cultures were incubated at 37°C.

Fermentation—*Bacteroides oralis*

W. J. Loesche, S. S. Socransky and R. J. Gibbons (1964). J. Bact. 88, 1329–1337.

Stock solutions (20%) of carbohydrates and carbohydrate derivatives were autoclaved separately, and were added aseptically to the basal medium to give a final concentration of 1%. A 2% starch solution was sterilized with ethylene oxide (Judge and Pelczar, 1955) and was added to the test medium in a 0.2% concentration. The pH after 6 to 8 days of incubation was measured with a glass electrode. A pH drop of 0.4 to 0.6 unit below control cultures grown without added sugar was interpreted as slight fermentation, and greater acidities were considered positive.

Thioglycolate Medium without Dextrose (BBL), supplemented with 0.2% yeast extract (Difco) and hemin, was used as a basal medium. All cultures were incubated at 35°C in Brewer jars containing an atmosphere of 95% H_2 and 5% CO_2.

L. F. Judge, Jr. and M. J. Pelczar, Jr. (1955). Appl. Microbiol. 3, 292–295.

Fermentation—*Clostridium*

K. Tamai and S. Nishida (1964). J. Bact. 88, 1647–1651.

Sugar-fermenting activity was determined by adding 0.2 ml of a 1% solution of methyl red-bromothymol blue reagent (Nishida *et al.*, 1964) to each of the broth cultures, which had been incubated for 7 days.

S. Nishida, K. Tamai, and T. Yamagishi (1964). J. Bact. 88, 1641–1646.

Fermentation—*Veillonella*

M. Rogosa (1964). J. Bact. 87, 162–170.

Fermentation of carbohydrates was tested in V17 medium modified by the omission of sodium lactate, Arabinose, galactose, fructose, maltose, mannose, rhamnose, sorbose, and xylose were sterilized by filtration and added aseptically. Inulin was always autoclaved in the medium. Adonitol, amygdalin, cellobiose, dulcitol, glucose, glycerol, inositol, lactose, mannitol, melibiose, melezitose, α-methyl D-glucoside, α-methyl D-mannoside, raffinose, salicin, sorbitol, sucrose, and trehalose were sterilized by filtration. Separate portions were also sterilized by autoclaving. The final concentrations of carbohydrates were 0.5 to 2% in different experiments. Tubed media were arranged in racks allowing direct contact of steam with each tube, and were autoclaved by quickly raising the temperature to 121°C and by immediately beginning the exhaust cycle. Cells from 24- to 48-hr V17 broth cultures were washed twice in freshly sterilized distilled water containing either 0.01% $Na_2S \cdot 9H_2O$ or 0.0125% sodium thioglycolate, and diluted to the original volume; 1-drop inocula were then made. Inoculated tubes of basal medium without added carbohydrate were always included. Anaerobiosis was effected within 10 min by repeated evacuation and flushing with 95% N_2 plus 5% CO_2 (at least four times) of McIntosh and Fildes aluminum jars. Slight positive pressure (100 mm of Hg) was maintained, and no results were accepted as valid unless positive pressure was still present after the usual 2-week incubation at 36°C. The pH of all tube contents was determined electrometrically.

For V17 medium see Basal medium—*Veillonella* (p. 114).

Fermentation—*Clostridium*

N. A. Sinclair and J. L. Stokes (1964). J. Bact. 87, 562–565.

Sugar fermentations were carried out in tubes containing the following medium: Trypticase, 2.0%; NaCl, 0.25%; K_2HPO_4, 0.15%; sodium thioglycolate, 0.05%; phenol red, 0.002%; specified carbohydrate, 0.5%; and agar, 0.075%; Carbohydrates tested included glycerol, erythritol, xylose, arabinose, glucose, fructose, galactose, mannitol, sucrose, maltose, lactose, raffinose, salicin, and cellulose,

Incubation temperatures were 0°C, 15°C, and 25°C.

Fermentation—motile marine bacteria

E. Leifson, B. J. Cosenza, R. Murchelano and C. Cleverdon (1964). J. Bact. 87, 652–666.

The tests for carbohydrate metabolism were made with the MOF medium of Leifson (1963) and the following carbohydrates: glucose, sucrose, lactose, xylose, maltose, and mannitol. The incubation temperature was 20°C, and the tubes were discarded after 7 days.

E. Leifson (1963). J. Bact. 85, 1183–1184.

Fermentation—*Streptococcus faecalis*

J. L. Goddard and J. R. Sokatch (1964). J. Bact. 87, 844–851.

The authors report studies on the fermentation of 2-ketogluconate.

Fermentation—*Moraxella*

W. J. Ryan (1964). J. gen. Microbiol. 35, 361–372.

Peptone water enriched with 10% (v/v) horse serum and containing 1% (w/v) of the appropriate carbohydrate was used with Andrade's indicator. Cultures were incubated for 3 weeks at 37°C before results were recorded as negative.

Fermentation—*Lactobacillus*

M. Gemmell and W. Hodgkiss (1964). J. gen. Microbiol. 35, 519–526.

Yeast autolysate (5.0%, v/v) prepared as described by Gibson, Stirling, Keddie and Rosenberger (1958) was substituted for yeast extract in the basal medium. Davis agar (0.15%, w/v) and bromocresol purple (2.8 ml of a 1.6%, w/v, ethanolic solution/L) or bromocresol green (5.6 ml of a 0.4%, w/v, aqueous solution/L) were added. Media containing bromcresol purple were adjusted to pH 7.0; those containing bromocresol green to pH 5.4. Fermentable substrates (0.5%, w/v, final concentration) were Seitz-filtered and added at 45–50°C to the autoclaved medium. A Pasteur pipette was used to inoculate 8–10 ml of the melted medium (Whittenbury, 1963).

The basal medium, which was used throughout with minor modifications, had the following constituents in 1 L tap water: meat extract (Lab-Lemco), 5 g; Evans' peptone, 5 g; Difco yeast extract, 5 g; Tween 80, 0.5 ml; $MnSO_4 \cdot 4H_2O$, 0.1 g; potassium citrate, 1 g; pH 6.5.

T. Gibson, A. C. Stirling, R. M. Keddie and R. F. Rosenberger (1958). J. gen. Microbiol. 19, 112.

R. Whittenbury (1963). J. gen Microbiol. 32, 375.

Fermentation—*Aerococcus catalasicus*

O. G. Clausen (1964). J. gen. Microbiol. 35, 1–8.

Cultures were incubated aerobically at 37°C unless otherwise stated.

Utilization and acid formation of certain sugars, alcohols and glycosides. 0.5% (aesculin, 0.2%) of the following sugars, etc., were added to peptone water + 5% normal horse serum + 1% Andrade's indicator: lactose, glucose, maltose, sucrose, raffinose, galactose, xylose, mannitol, dulcitol, glycerol, starch, dextrin, salicin, aesculin. The incubation period was 14 days.

Fermentation—*Pediococcus*

E. Coster and H. R. White (1964). J. gen. Microbiol. 37, 15–31.

The authors used the methods of H. L. Günther and H. R. White (J. gen. Microbiol. 26, 185, 1961). Their Tomato Juice (TJ) broth or agar with the addition of Tween 80 (0.1%, v/v), pH adjusted to 6.6, were used in the maintenance and preparation of inocula with certain specified exceptions. Transfers for preparing inocula were made every 24 hours except for *Pediococcus halophilus* (48 hr) and some brewing strains (fortnightly).

All cultures were incubated aerobically at 30°C except for "brewing strains," which were incubated in an atmosphere of 95% (v/v) hydrogen and 5% (v/v) carbon dioxide at 22°C. The basal medium was modified by the addition of Tween 80 (0.1%, v/v).

Fermentation—*Pseudomonas aeruginosa*

R. R. Colwell (1964). J. gen. Microbiol. 37, 181–194.

Production of acid from carbohydrate. Cultures were incubated for 4 weeks in a bromcresol purple broth basal medium (Difco Laboratories) with 1.0% filter-sterilized test carbohydrate added. Glucose, lactose, maltose, sucrose, galactose, mannitol, mannose, melibiose, melezitose, dextrin, glycerol, and sorbitol were used. In addition, the Hugh & Leifson (1953) basal medium without agar and with Durham tubes was used, with added filter-sterilized glucose, lactose, or maltose as test carbohydrate. Each culture was examined at 1, 7, 14 and 28 days for acid and gas production in all tests.

R. Hugh and E. Leifson (1953). J. Bact. 66, 24.

Fermentation

G. W. Wilson and A. A. Miles (1964). *Topley and Wilson's Principles of Bacteriology and Immunity,* 5th edition. – Arnold.

Tested in 1 per cent peptone water containing 1 per cent of the sugar and Andrade's indicator. A Durham's tube is included. Acid, or acid and gas production is noted. For certain groups of organisms which do not grow well in this medium 5 per cent of serum is added. Horse serum is generally employed, but in a maltose medium it is better replaced by human or rabbit serum, since it contains an enzyme, maltase, which may lead to a false reaction (Hendry, 1938). In testing for the fermentative ability of bacteria it is advisable to avoid as far as possible the addition to the medium of any substance such as yeast extract, serum, or ascitic fluid that contains enzymes (see Herrmann, 1944). The presence of nitrate should also be avoided, as this may interfere with gas production (Cook and Knox, 1950). To prevent the masking of weak acid production from the carbohydrate by alkali formed from the protein during growth, tests may be made with heavy suspensions containing preformed enzymes and chemically defined substrates as recommended by Clarke and Cowan (1952) and Cowan (1953). Alternatively, Hugh and Leifson's (1953) medium may be used. This is a glucose medium containing only a small amount of peptone and rendered semi-solid with 0.3 per cent agar. This medium may also be used to distinguish between *fermentation,* which is an anaerobic process, and *oxidation,* which is aerobic. For this purpose two tubes are inoculated by the stab method. One tube is covered with ¼–½ in of sterile melted petrolatum. In true fermentation, acid is formed in both tubes; in oxidation, only in the uncovered tube. For a modification of this medium, see Davis and Park (1962).

P. H. Clarke and S. T. Cowan (1952). J. gen. Microbiol. 6, 187.
G. T. Cook and R. Knox (1950). J. clin. Path. 3, 356.
S. T. Cowan (1953). J. gen. Microbiol. 8, 391.
G. H. G. Davis and R. W. A. Park (1962). J. gen. Microbiol. 27, 101.
C. B. Hendry (1938). J. Path. Bact. 46, 383.
W. Herrmann (1944). Zentbl.Bakt.ParasitKde 151, 427.
R. Hugh and E. Leifson (1953). J. Bact. 66, 24.

Fermentation—*Lactobacillus*

F. Gasser (1964). Annls Inst. Pasteur, Paris 106, 778–794.

La fermentation des oses et des polyalcoöls est recherchée dans le milieu suivant.

Peptone oxoïd L_{37}	15	g
Extrait de levure	5	g
Tween 80	1	ml
Phosphate bipotassique	1	g
Phosphate monopotassique	0.8	g
Sulfate de manganèse	50	mg
Sulfate de magnesium	200	mg
Rouge de chlorophénol	0.04	g
Eau	q.s.p. 1000	ml
pH 6.4.		

Répartir en tubes d'Ivan Hall, étroits, dépourvus de bille, jusqu'au niveau de l'étranglement (5 ml) et autoclaver trente minutes à 115°C. Les oses ou polyalcoöls sont stérilisés par filtration sur bougie L_3 et introduits extemporanément à la concentration finale de 0,5 p. 100. Aucune autre source de carbone utilisable par les *Lactobacillus* n'est introduite dans le milieu de culture et le résultat est apprécié autant sur la présence d'une culture permise par la source d'énergie introduite que sur le virage de l'indicateur coloré dont la stabilité n'est pas toujours constante.

Les tubes sont inoculés avec 2 gouttes d'une dilution au 1/10 en tampon phosphate M/20, pH 6,4, d'une culture de quarante-huit heures en bouillon M. R. S. M. Cette dilution évite un début de virage de l'indicateur que provoquerait la forte teneur en acide de bouillon non dilué et permet cependant de garder un inoculum de l'ordre de 10^7 bactéries.

Fermentation—*Vibrio comma*

J. C. Feeley (1965). J. Bact. 89, 665–670.

Fermentation tests were carried out in Purple Broth Base (Difco) containing 0.5% of the following carbohydrates: dextrose, lactose, sucrose, mannose, mannitol, arabinose, and salicin.

Fermentation—marine bacteria

R. M. Pfister and P. R. Burkholder (1965). J. Bact. 89, 863–872.

To determine the carbohydrate fermentation of these marine bacteria, a 1.0% carbohydrate medium was prepared according to Skerman (1959) with the seawater medium previously described, but with K_2HPO_4 added in the amount of 0.05 g per liter and use of agar at the rate of 5.0 g per liter. The solutions of carbohydrates were sterilized by Seitz filtration and added aseptically to 1-liter batches of melted base medium. All media were preincubated for a minimum of 72 hr to check for sterility. Each organism was inoculated in each test medium in duplicate with a butt stab. Samples were prepared for anaerobic fermentation by covering with sterile mineral oil. With the tropical bacteria, the test medium was covered immediately after inoculation, because a few of the warm-water organisms utilized the carbohydrates aerobically so rapidly that any anaerobic test was spoiled before leaving the inoculation area. All cultures used for inoculum were 48-hr transfers, actively growing on seawater agar. Incubation was at 25°C, except where otherwise specified.

The seawater medium contained (per liter): Trypticase (BBL) or N-Z-Case, 2.0 g; Soy-tone (Difco), 2.0 g; yeast extract, 1.0 g; vitamin B_{12}, 1.0 μg; aged seawater; agar (if desired), 16 g; The pH was adjusted to 7.0 prior to autoclaving.

V. B. D. Skerman (1959). *A guide to the identification of the genera of bacteria.* The Williams & Wilkins Co., Baltimore.

Fermentation—*Staphylococcus epidermidis*

O. Sandvik and R. W. Brown (1965). J. Bact. 89, 1201–1208.

The anaerobic fermentations were determined by stab cultures in Phenol Red Agar Base (Difco) with 1% carbohydrate; the cultures were incubated for 5 days at 37°C in a Brewer jar.

Fermentation—*Mycobacterium*

R. J. Jones and D. E. Jenkins (1965). Can. J. Microbiol. 11, 127–133.

Acid Production from Carbohydrates. The carbohydrates used were d-sorbitol, glucose, rhamnose, inositol, and l-arabinose. The basal medium, minus the carbohydrate and agar, contained: $(NH_4)_2HPO_4$, 1.0 g; KCl, 0.2 g; $MgSO_4 \cdot 7H_2O$, 0.2 g; bromocresol purple, 15 ml of 0.4% solution; and distilled water, 1 liter. The pH was adjusted to 7.0 with 1 N HCl before agar (1.5%) was added and the mixture was brought to a boil. Carbohydrates were sterilized separately and were added to give a final concentration of 1%. Medium was dispensed in tubes and allowed to solidify on a slant. Tubes containing each of the carbohydrates were inoculated with 0.1 ml of the inoculum suspension. Cultures were incubated at 37°C for 28 days before negative results were recorded.

Fermentation—*Salmonella typhi*

B. B. Diena, R. Wallace and L. Greenberg (1965). Can. J. Microbiol. 11, 427–433.

Sugar utilization was tested in peptone-water media with and without 1.5% glycine.

Fermentation—*Streptococcus*

R. Whittenbury (1965). J. gen. Microbiol. 38, 279–287.

Soft agar medium, which was used for showing oxygen relationships and fermentative activities, was prepared and inoculated as described previously (Whittenbury, 1963) with the exception that the bromcresol purple content was doubled. Sugars and polyhydroxy alcohols were prepared as Seitz-filtered distilled water solutions and added to give a final

concentration in the media of 0.5% (w/v).
R. Whittenbury (1963). J. gen. Microbiol. 32, 375.

Fermentation—*Bacillus*

G. R. F. Hilson (1965). J. gen. Microbiol. 39, 407–421.
See Basal medium for carbon compounds—*Bacillus* (p. 121).

Fermentation—*Alcaligenes*

R. G. Mitchell and S. K. R. Clarke (1965). J. gen. Microbiol. 40, 343–348.

Standard peptone water media (glucose, lactose, maltose, sucrose, salicin, mannitol and dulcitol) were incubated for 21 days. Acid production from glucose, mannose, maltose, arabinose, xylose and glycerol was also tested in the medium of Hayward and Hodgkiss (1961) incubated for 12 days (6 strains).
A. C. Hayward and W. Hodgkiss (1961). J. gen. Microbiol. 26, 133.

Fermentation—*Staphylococcus*

R. Zemelman and L. Longeri (1965). Appl. Microbiol. 13, 167–170.

One drop of bromothymol blue was added to the area from which a colony had been removed for the coagulase test, and resulting color was registered. Negative results were indicated by a blue color, and mannitol fermentation was indicated by a change to yellow. Another drop of indicator, added to a noninoculated plate, was used as control. The cultures were grown on Chapman's medium 110 (J. Bact. 51, 409–410, 1946), and incubated at 37°C for 48 hours.

Fermentation—*Mima, Herellea, Flavobacterium*

J. D. Nelson and S. Shelton (1965). Appl. Microbiol. 13, 801–807.

Aqueous solutions (10%) of sucrose, glucose, lactose, maltose, inositol, and galactose were sterilized at 10 psi for 10 min, and were added aseptically in 1% concentration to previously sterilized Cystine Trypticase semisolid medium at 45°C; the mixture was then cooled. The tubes were stabbed, incubated at 37°C, and observed for acidification at 24 and 48 hr, 7 days and 3 weeks. For a comparison of 1 and 10% lactose utilization, aqueous solutions of lactose were sterilized by Millipore filtration, and were added aseptically to Phenol Red Agar Base (BBL) at 45°C. Cultures were streaked on the surface and read at 24 hr, 48 hr, 6 days and 3 weeks.

Fermentation—*Shigella sonnei*

R. R. Gillies (1965). J. Path. Bact. 90, 345–348.

Media. The substrates xylose, raffinose and melibiose were prepared as 1 per cent solutions in peptone water according to Cruickshank *et al.* (1960, p. 200) and distributed in 5 ml amounts in ¼ oz (7 ml) screw-capped bottles. Inoculation was done with a standard loop made from

nichrome resistance wire SWG no. 24 with an internal loop diameter of 2 mm; this was loaded from an overnight culture grown at 37°C on nutrient agar (Cruickshank *et al.*, p. 195) and thorough emulsification of the bacterial suspension in the relevant substrate was ensured by rotating the loop in the sugar medium for 1 min. After inoculation, incubation was maintained at 37°C for 21 days and the cultures examined every 24 hr for evidence of acid production.
R. Cruickshank *et. al.* (1960). Mackie and McCartney's *Handbook of bacteriology,* 10th ed. Edinburgh and London.

Fermentation—*Mycoplasma*

J. G. Tully (1965). Biochemical, morphological, and serological characterization of *Mycoplasma* of murine origin. J. infect. Dis. 115, 171–185.

Carbohydrate fermentation. Fermentation of carbohydrates was determined in 5 ml of HSI broth containing horse serum that had been inactivated at 56°C for 30 minutes, 0.5% carbohydrate, and 1 drop of 1% alcoholic phenol red. Control tubes without carbohydrate were included in each series, and all tubes received 0.5 ml of a 24-hour mycoplasma culture in the same HSI broth base. Tubes were incubated at 37°C and read every 24 hours for 7 days. Subcultures of the control tubes were made at 2 days and 7 days to verify growth of mycoplasma. HSI broth consists of Difco PPLO broth (7 parts), fresh 25% yeast extract (1 part) and horse serum (Cappel Laboratories West Chester, Pa.) (2 parts).

Fermentation—anaerobes

P. Pohl (1965). Annls Inst. Pasteur, Paris 108, 140–145.

Nous préparons d'abord le milieu de base VL classique, à cette différence près que nous en portons le pH à 8,5 : peptone trypsique, 10 g ; chlorure de sodium, 5 g ; extrait de viande, 2 g ; extrait de levure, 5 g ; chlorhydrate de cystéine, 0,3 g ; eau distillée, 1000 ml ; bacto-agar, 20 g.

Ajuster à pH 8,5.

Porter à ébullition, filtrer, répartir 75 ml en flacons d'Erlenmeyer de 100 cm^3, stériliser vingt minutes à 115°C. Un pH de 8,5 est indispensable pour éviter le virage au moment de l'ensemencement à partir de cultures en milieu de Tarozzi.

Avant de couler ce milieu, nous y ajoutons un indicateur, le pourpre de bromocrésol, à raison de 1 ml d'une solution alcooiique à 1 p. 100 par ballon de 75 ml de milieu préalablement liquéfié, puis ramené à température de 45° à 50°C. Le choix de cet indicateur limite les réactions positives aux seules acidifications très nettes.

Nous préférons ajouter l'indicateur après la stérilisation, afin d'éviter les modifications dues aux trop hautes températures. Moyennant certaines précautions (flambage soigneux des pipettes et des cols de ballon, travail sous hotte), cette manière d'agir n'a jamais amené de contamination dans nos milieux. Nous coulons enfin en boîtes de Petri de 10 cm de diamètre. En moyenne, nous répartissons le contenu du ballon, soit 75 ml, entre cinq boîtes. Il importe que l'épaisseur de la couche de milieu ne dépasse pas 5 mm.

Une fois le milieu refroide et prêt à l'emploi, nous y disposons sept disques de papier imbibés chacun d'un sucre différent. Ces disques sont fournis par des industries spécialisées (disques taxo BBL). Nous en disposons 6 aux sommets d'un hexagone et le dernier au centre. Il paraît difficile d'augmenter le nombre des disques sans nuire à la netteté de la réaction du fait du chevauchement des réactions positives. On ensemence à partir d'une culture pure en milieu de Tarozzi. Nous en prenons le contenu d'une pipette Pasteur que nous diluons dans 5 ml de gélose ordinaire préalablement chauffée, puis refroidie à 45°–55°C et nous le coulons sur la surface du milieu. Cet ensemencement par l'intermédiaire d'une couche de gélose fondue (au lieu de l'ensemencement à l'anse de platine) assure une excellente répartition des germes. Les bulles emprisonnées dans la gélose permettent de se rendre compte d'un éventuel dégagement gazeux.

Pour obtenir l'anaérobiose, nous utilisons le mélange classique : carbonate de potasse, 3 g ; pyrogallol, 3 g ; terre d'infusoires, 15 g.

Le mélange est fait extemporanément au mortier, puis introduit dans un petit sachet de papier de 3 cm × 5 cm.

Cependant, avant de fermer le sachet, nous y disposons une mince feuille de buvard que nous imbibons de 1 ml d'une solution de soude à 40 p. 100 dans le but de neutraliser les produits acides volatils, qui, la boîte étant scellée peu après, ne peuvent s'éliminer. La soude ainsi ajoutée (sous forme de solution et non de cristaux) n'a jamais inhibé nos cultures. Nous avons calculé que 1 ml de NaOH à 40 p. 100 pouvait capter l'équivalent en CO_2 du volume libre dans la boîte de Petri. Le sachet est fixé sur le couvercle par trois bandes de ruban adhésif et la boîte elle-même retournée sur ce couvercle. On les solidarise enfin par une bande de ruban adhésif et on scelle en trempant le pourtour de l'ensemble dans un récipient contenant de la paraffine fondue.

Fermentation—*Streptomyces*

T. F. Fryer and M. E. Sharpe (1965). J. Dairy Res. 32, 27–34.

The authors used the methods of Gordon and Smith (1955) and Gordon and Mihm (1957).

One-twentieth ml of an 18–24 hr culture of the organism, grown in 5 ml Yeastrel nutrient broth, centrifuged and resuspended in 5 ml distilled water was used as inoculum.

R. E. Gordon and J. M. Mihm (1957). J. Bact. 73, 15.
R. E. Gordon and M. M. Smith (1955). J. Bact. 69, 147.

Fermentation—*Brevibacterium*

R. Chatelain and L. Second (1966). Annls Inst. Pasteur, Paris 111, 630–644.

Action sur les glucides: étudiée en eau peptonée (15) pour les souches fermentatives et en milieu macération-gélose pour les souches oxydatives. Ces milieux sont additionnés de 1 p. 100 des glucides suivants: xylose, arabinose, rhamnose, galactose, lévulose, mannose, saccharose, maltose, lactose, mannitol, glycérol, sorbitol, inositol, salicine et dextrine. La lecture est faite après 1, 2, 4, 8 et 16 jours de culture.

La température d'incubation des cultures est de 30°C.

(15) Olivier (H. R.). *Traité de biologie appliquée,* tome II, Maloine édit., Paris, 1963.

Fermentation—*Providencia, Rettgerella*

C. Richard (1966). Annls Inst. Pasteur, Paris 110, 105–114.

Pour des raisons de commodité, les milieux ont été incubés à 37°C. Fermentation des hydrates de carbone et polyalcoöls en eau peptonée blue de bromothymol pH 7,4.

Fermentation—*Bacteroides, Veillonella, Neisseria*

J. van Houte and R. J. Gibbons (1966). Antonie van Leeuwenhoek 32, 212–222.

Basal medium with 1% filter-sterilized carbohydrate was used for fermentations. After 7 days' growth, the pH was measured with a glass electrode. Inoculated tubes of basal medium without carbohydrate were used as a control.

The basal medium was BBL thioglycolate medium, without added dextrose, supplemented with 5% horse serum and hemin (5 μg/ml) and menadione (0.5 μg/ml).

Fermentation—*Mycobacterium*

M. Tsukamura (1966). J. gen. Microbiol. 45, 253–273.

Acid production from carbohydrates was observed on the following medium: $(NH_4)_2SO_4$, 2.64 g; KH_2PO_4, 0.5 g; $MgSO_4 \cdot 7H_2O$, 0.5 g; yeast extract, 0.2 g; 0.2% (w/v) bromthymol blue in 0.02 *M* NaOH, 20 ml*; purified agar, 20.0 g; distilled water, 980 ml. The medium was adjusted to pH 7.0 and sterilized by autoclaving at 115°C for 30 min. Carbohydrates were sterilized separately by heating in solution at 100°C for 15 min and added to the above medium aseptically to a final concentration of 0.5% (w/v). When necessary, the medium was readjusted to pH 7.0 aseptically and poured into tubes. Slopes were inoculated with one loopful of the stock cultures and incubated. Control medium without carbohydrate was always set up. Acid formation was observed after 2 weeks' incubation at 28°C (rapidly growing mycobacteria) or after 4 weeks' incubation at 28°C (slowly growing mycobacteria). The following carbohydrates were used: glucose, mannose, galactose, arabinose, xylose, rhamnose, trehalose, maltose, lactose, raffinose, inositol, mannitol and sorbitol.

*The author is now using 10 ml of 0.2% bromthymol blue in 0.02 M NaOH (the amount of bromthymol blue solution has been decreased to one half).

Fermentation—*Mycobacterium*

R. E. Gordon (1966). J. gen. Microbiol. 43, 329–343.

The ability of each strain to produce acid from a series of carbohydrates, including glucose, was determined by growing the cultures aerobically on slants of the inorganic nitrogen agar of Ayers, Rupp and Johnson (1919) containing 1% (w/v) of each carbohydrate (Gordon and Mihm, 1959). The cultures were also tested for the fermentation of glucose in the inorganic nitrogen agar: $(NH_4)_2HPO_4$, 1 g; KCl, 0.2 g; $MgSO_4 \cdot 7H_2O$, 0.2 g; agar, 3 g; distilled water, 1000 ml; 0.04% solution of bromcresol purple, 15 ml. Glucose was added to this medium in the same way as to the Hugh and Leifson medium; the cultures were inoculated, sealed with a vaspar seal, and observed at 7 and 28 days.

S. H. Ayers, P. Rupp and W. T. Johnson, Jr. (1919). Bull. U. S. Dept. Agr. no. 782.

R. E. Gordon and J. M. Mihm (1959). J. gen. Microbiol. 21, 736.

Fermentation—*Mycobacterium parafortuitum* n. sp.

M. Tsukamura (1966). J. gen. Microbiol. 42, 7–12.

Acid formation from sugars was tested in the following medium: NH_4Cl, 2.0 g; KH_2PO_4, 0.5 g; $MgSO_4 \cdot 7H_2O$, 0.5 g; 0.2% (w/v) bromothymol blue in 0.02 M NaOH, 20.0 ml; purified agar, 30.0 g; distilled water, 1000 ml (pH 7.2). Sugar solutions were sterilized separately by heating at 100°C for 5 min and added to medium aseptically to a final concentration of 0.5% (w/v). Acid formation was observed after 2 weeks' incubation at 37°C.

Fermentation—*Staphylococcus*

S. J. Edwards and G. W. Jones (1966). J. Dairy Res. 33, 261–270.

Strains were tested routinely for mannitol fermentation at 37°C for 7 days using media described by Cruickshank (1960).

R. Cruickshank (1960). *Mackie and McCartney's Handbook of Bacteriology.* Edinburgh and London: Livingstone.

Fermentation—*Clostridium*

A. V. Goudkov and M. E. Sharpe (1966). J. Dairy Res. 33, 139–149.

The basal medium used for many of the tests consisted of: BBL Trypticase, 1.5%; yeast extract, 1%; cysteine, 0.05%; pH of medium, 7.0. The methods of Beerens, Castel and Put (1962) Annls Inst. Pasteur, Paris

103, 117) were used.

Carbohydrate fermentation. To the basal medium, 0.5% Seitz-filtered carbohydrate was added. The extent of fermentation was determined after 10 days' incubation by measurement of the fall in pH and production of gas.

Lactate fermentation. The medium used was prepared by adding 0.5 g Ca lactate and 0.5 g Na acetate/100 ml to the basal medium and adjusting the pH to 6.8. Ca lactate was found preferable to Na lactate (Bryant and Burkey, 1956). Fermentation was judged to have occurred when the pH rose to 8.0 or higher after 10 days' incubation.

M. P. Bryant and L. A. Burkey (1956). J. Bact. 71, 43.

Fermentation—*Salmonella*

A. B. Gonzalez (1966). J. Bact. 91, 1661–1662.

The authors report a strain of *Salmonella tennessee* which ferments lactose and sucrose.

Fermentation—enterobacteria

R. E. Goodman and M. J. Pickett (1966). J. Bact. 92, 318–327.

The paper describes a study of lactose fermentation in a study of lactose uptake and β-galactosidase activity of several late-lactose fermenters.

The following media were used: – the 2% lactose medium was the tableted substrate described by Hoyt and Pickett (1957). EMB-lac was a mineral-eosin-methylene blue-agar medium containing only lactose as available carbon. Four fluid media, each containing 0.001% bromocresol purple, were designed to test the fermentation of lactose; these were: A, 1% lactose and 0.1% peptone; B, 5% lactose and 0.1% peptone; C, 1% lactose and 1% peptone; and D, 5% lactose and 1% peptone.

Each inoculum for six different media was 0.1 ml of a suspension obtained by harvesting the overnight growth on a Kligler Iron agar (KIA) slant with 1 ml of sterile water. The four broth media, A to D (4 ml per 13 by 100 mm test tube), were observed for 20 days before being discarded as negative; with all tubes incubated beyond 3 days, their steel caps (Morton) were replaced by rubber stoppers. The EMB-lac plates were examined after 3 days of incubation, and the number of colonies was estimated. The 2% lactose medium was examined at 6, 24, and 48 hr. When a broth medium showed delayed fermentation, a subculture was plated on EMB-lac or MacConkey agar, a colony from this was transferred to KIA and, after growth on this medium, was checked in 2% lactose medium for promptness (i.e., evidence of mutative) of fermentation. All incubation was done at 35°C.

R. E. Hoyt and M. J. Pickett (1957). Am. J. clin. Path. 27, 343–352.

Fermentation—anaerobes

A. -R. Prévot (1966). Techniqùes pour le diagnostic des Bactéries Anaérobes. Éditions de la Tourelle, St. Mandé.

Lecture de l'attaque des glucides.

Le 4e ou 5e jour après l'ensemencement des eaux peptonées additionnées de glucides, on étudie l'acidification de chacun des tubes par les indicateurs universels de pH et on note les divers degrés d'acidification des glucides, ainsi que le dégagement de gaz. Il faut savoir que certaines espèces telles que *Pl. tetani, Pl. putrificum,* douées d'un pouvoir intense de désamination, n'acidifient jamais les eaux peptonées additionnées de glucide car la production d'NH_3 neutralise l'acidité et même alcalinise le milieu. Dans ce cas la fermentation des glucides ne peut-être mise en évidence que par la méthode pondérale : on cherche par un titrage, selon Bertrand, par exemple, le poids de glucide disparu.

Eau peptonée

Faire dissoudre 20 g de peptone bactériologique dans 1 litre d'eau distillée.

Ajuster à pH 7,4 avec une solution de soude à 10 % (il en faut environ 10 à 12 cm^3). Filtrer sur papier Laurent. Répartir en tubes 13–14/180 mm. Stériliser à 110° pendant 30 minutes.

Cette eau peptonée sert par ailleurs à préparer les milieux glucidiques. On ajoute pour cela 10 gouttes par tube de solution des glucides suivants, préalablement stérilisées :

Glucose, lévulose, galactose, saccharose, maltose, glycérine — solutions à 30 %.

Lactose	17 %
Empois d'amidon	2 %
Mannitol	20 %

Les premières se stérilisent par tyndallisation au bain-marie à 100° pendant 1 heure, 3 fois de suite à 24 heures d'intervalle.

Le lévulose se stérilise par filtration sur bougie L_5.

La glycérine et l'amidon se stérilisent à 115°

L'extrait globulaire employé pour certains anaérobies est celui qui sert aux aérobies. De même pour l'extrait de cervelle.

Fermentation—anaerobes

A. -R. Prévot (1966). Techniqùes pour le diagnostic des Bactéries Anaérobes. Éditions de la Tourelle, St. Mandé.

Milieu de Rosenow cystéine : Beerens et Tahon-Castel préconisent le milieu de Rosenow modifié par Hayden et cystéine :

Tryptone	10g
Extrait de viande	3g
NaCl	5g

Chlorhydrate de cystéine	0.3g
Eau ordinaire	1000 cm^3

On fait bouillir quelques minutes ; on ajuste à pH 7,2. On reporte 10 minutes à l'ébullition et on filtre sur papier mouillé. On ajoute alors : indicateur d'Andrade 10 cm^3 (I) ; puis glucose : 2 g.

On répartit 20 cm^3 de ce milieu dans des tubes de 200 X 20 dans lesquels on a mis un petit morceau de marbre blanc, un fragment de cervelle de boeuf ou de mouton. On stérilise à 115°. Ce milieu s'altère rapidement. Il doit être préparé extemporanément. Il est préférable de le paraffiner.

La fermentation du glucose se traduit par le virage du rose au rouge de l'indicateur. La décoloration de l'indicateur indique la réduction due au germe et le virage au vert indique l'alcalinisation du milieu. Le dégagement gazeux se traduit par un soulèvement de la paraffine.

FIBRINOLYSIN

Fibrinolysin—*Streptococcus*

A. C. Evans (1936). J. Bact. 31, 423–437.

A number of strains of each species were tested for ability to dissolve human fibrin, according to the technique of Tillett and Garner (1933). Their technique was slightly modified, because when it was followed exactly, the diluted plasma did not always clot.

The tests were always carried out in duplicate, with blood from 2 normal human subjects. The plasma was prepared as follows: Twelve milliliters of blood were taken from a normal person and placed in a tube containing the powder obtained by evaporating 1 ml of a 2 per cent solution of potassium oxalate. The blood cells were separated from the plasma by centrifugation.

The streptococci to be tested for fibrinolysin were grown overnight in beef infusion broth containing 10 per cent rabbit serum. The cultures were centrifugated and the clear supernatant fluid was used for the test. In carrying out the tests, sterile conditions were maintained throughout. One-fifth ml of plasma was added to 0.8 ml of saline solution, 0.5 ml of supernatant fluid from the culture was added, the tube was shaken, then 0.25 ml of a 0.25 per cent solution of calcium chloride was added and the tube was shaken again. The tubes were incubated in a water bath at 37°C, and readings were made after 30 minutes, and after 1, 3 and 24 hours. Readings were recorded according to the following scheme:

Inhibition of clotting	++++
Complete lysis in 1 hour	+++
Complete lysis in 3 hours	++

Complete or partial lysis in
24 hours +
No lysis –

W. S. Tillett and R. L. Garner (1933). J. exp. Med. 58, 485–502.

Fibrinolysin—*Staphylococcus*

E. Neter (1937). J. Bact. 34, 243–254.

Organisms are cultured in plain infusion broth for 18 hours at 37°C, centrifuged and the supernatant serially diluted in 0.5 ml volumes. To each tube one volume of 1/5 dilution of plasma (10 ml human, rabbit or guinea pig blood + 1 ml 2% potassium oxalate) is added, followed by 0.25 ml of 0.25% $CaCl_2$ in normal saline. The tubes are shaken and held at 37°C. Fibrinolysin causes dissolution of the plasma clot. Results were read at various intervals.

Fibrinolysin—*Streptococcus*

P. L. Boisvert (1940). J. Bact. 39, 727–738.

Strains were tested for the ability to lyse human and rhesus monkey plasma clots.

Fibrinolysin—*Streptococcus*

J. D. Le Mar and M. F. Gunderson (1940). J. Bact. 39, 717–725.

One milliliter of a 1:5 dilution of oxalated plasma was well mixed with 0.5 ml of the culture (24 or 96 hours). To this mixture was added 0.25 ml of a 0.25% solution of $CaCl_2$; the tube was well shaken and incubated at 37°C in a water bath. Readings were taken at 30 minutes, 3 hours and 24 hours. The method was that of Tillett and Garner, 1933 (J. exp. Med. 58, 485–502).

Fibrinolysin—*Streptococcus*

Commission on Acute Respiratory Diseases (1947). J. exp. Med. 85, 441–457.

Fibrinolysin Assay:

Fibrinogen – The source of fibrinogen was a lyophilized preparation obtained from human plasma and was supplied by Dr. E. J. Cohn and Dr. S. Howard Armstrong, Jr. A solution was prepared containing 0.6 g of fraction I in 100 ml of buffered saline.

Buffered saline – A solution of 0.01 M phosphate in 0.85 per cent sodium chloride at pH 7.4 was employed throughout this study.

Thrombin – A commercial rabbit thrombin was employed in a 1 to 8 dilution.

Bacteriological media – Two different media were used to culture the beta hemolytic streptococcus:

1. Beef heart infusion broth containing 0.2 per cent dextrose.
2. Fibrinolysin assay broth prepared from a *single lot* of tryptose

phosphate broth (dehydrated). To each liter of broth were added 1.6 g of sodium bicarbonate in 40 ml of distilled water. The final pH was 8.4. *Technique of test* – Several colonies of beta hemolytic streptococci were picked from the blood-agar plate and inoculated into 5 ml of the beef heart infusion medium containing several drops of mule blood.

Following incubation in a water bath at 37°C for 18 to 24 hours, 0.1 ml was transferred to 5 ml of beef heart infusion broth containing 0.1 ml of mule blood. After 8 hours' incubation, 0.1 ml of the supernatant was transferred to 5 ml of the assay broth, and at the same time a blood-agar plate was inoculated to test for purity. At the end of 18 hours' incubation in the water bath all cultures showing visual turbidity after shaking were centrifuged at 2500 rpm for 20 minutes. One ml of the supernatant broth was then removed to 4 ml of buffered saline. Serial twofold dilutions in buffered saline were made in a series of nine chemically cleaned Wassermann tubes. The initial dilutions ranged from 1:5 to 1:1280 and were contained in a total volume of 1 ml. Automatic pipettes were then employed to add 0.5 ml of the fibrinogen solution, followed by 0.2 ml of thrombin. The racks containing the tubes were shaken immediately (gently). Clotting occurred regularly within 2 to 3 minutes. The tubes were then incubated in a water bath at 37°C for 60 minutes. The titer of fibrinolysin was taken as the highest dilution which lysed the clot at the end of this period. A clot was considered lysed if it poured or slid readily when the tube was inverted.

Fibrinolysin—*Staphylococcus*

J. Pillet, P. Mercier and R. Pery (1948). Annls Inst. Pasteur, Paris 75, 458–471.

a) Méthode de la gélose à la fibrine. – Cette méthode préconisée par Christie et Wilson [1] présente le grand avantage de permettre une étude systématique et rapide de la propriété fibrinolytique sur un grand nombre de souches.

On part de plasma de lapin oxalaté que l'on chauffe à 56°C au bain-marie pendant dix minutes, la fibrine précipite et est recueillie par centrifugation pour être reprise par une quantité d'eau physiologique égale à celle du plasma primitif; on dépose alors 1 cm^3 de cette émulsion dans une boîte de Petri de 10 cm de diamètre dans laquelle on coule ensuite 6 cm^3 de gélose à 3 p. 100 portée à 56°C. On mélange alors rapidement, puis on laisse refroidir. On obtient ainsi des boîtes de gélose opacifiée par la fibrine sur lesquelles on ensemence ponctuellement les souches à étudier. La lecture, très facile, se fait après douze heures d'étuve. On observe alors dans les cas positifs une auréole transparente due à la lyse de la fibrine tranchant sur le fond opaque de la boîte. Cette méthode se prête mal à l'étude quantitative de la fibrinolysine et nous pensons qu'il faut avoir recours à la méthode de Tillet et Garner [2] mise au point pour étudier la fibrinolysine streptococcique.

b) Méthode de Tillet et Garner. – Cette technique consiste à provoquer la coagulation d'un plasma oxalaté par recalcification et à faire agir sur le caillot ainsi obtenu le filtrat de la culture bactérienne à tester. En pratiquant des dilutions croissantes de la fibrinolysine employée on peut ainsi titrer son activité. Il est également possible d'obtenir un caillot en additionnant de thrombine une solution de fibrinogène purifiée et opérer sur ce caillot comme précédemment. Signalons que le liquide fibrinolytique peut être aussi bien ajouté au plasma ou au fibrinogène avant l'addition de calcium ou de thrombine, la lecture devenant ainsi à notre avis beaucoup plus facile grâce à la grande homogénéité du caillot.

[1] Christie (R.) et Wilson (H.). *Austr. J. exp. Biol.*, 1941, 19, 329.
[2] Tillet (W. S.) et Garner (R. L.). *J. exp. Med.*, 1933, 58, 485.

Fibrinolysin—*Staphylococcus*

P. Hauduroy (1951). Techniques Bactériologiques. Masson et Cie, Éditeurs, Paris.

Fibrinolysine (Streptocoques). – Recueillir 5 cm^3 de sang dans un tube à centrifuger stérile contenant 0.5 cm^3 d'une solution à 2 p. 100 d'oxalate de potassium. Agiter pour mélanger. Centrifuger. Recueillir à la pipette le plasma oxalaté qui peut être conservé à la glacière quelques heures.

Pour faire la réaction, opérer comme suit. Mélanger dans un tube ordinaire, en suivant l'ordre:

Eau physiologique	0.8 cm^3
Plasma oxalaté	0.2 ”
Culture de 18 heures en bouillon contenant 1 p. 100 de peptone et 10 p. 100 de sérum de cheval	0.5 ”
Solution stérile à 0.25 p. 100 de chlorure de calcium, eau contenant 0.85 p. 100 de NaCl	0.25 ”

Un caillot se forme après adjonction de chlorure de calcium.

Si le streptocoque possède une fibrinolysine, le caillot est dissous en 15 minutes environ.

Fibrinolysin—*Staphylococcus*

C. H. Lack and D. G. Wailling (1954). J. Path. Bact. 68, 431–443.

Fibrinolysin tests. A human plasma-agar mixture was heated at 56°C for 20 minutes before pouring. This produced an opaque medium and fibrinolysis was judged by the zone of clearing around colonies after 48 hours' incubation (Christie and Wilson, 1941). Separation of fibrinolysis by plasmin from that by protease was made by incorporated soya-bean trypsin inhibitor in the medium. This inhibited fibrinolysis by plasmin

but not staphylococcal protease.
R. Christie and H. Wilson (1941). Aust. J. exp. Biol. med. Sci. 29, 329.

Fibrinolysin—*Staphylococcus*

S. I. Jacobs, A. T. Willis and G. M. Goodburn (1964). J. Path. Bact. 87, 151–156.

Fibrin agar. Two hundred and fifty units of sterile human thrombin (Lister Institute) dissolved in 2 ml sterile distilled water were added to 400 ml of the cooled nutrient base. Approximately 400 mg of sterile human fibrinogen (Lister institute) dissolved in 20 ml sterile distilled water were then added, and, after thorough rapid mixing, plates were poured immediately. This medium, which contains finely precipitated fibrin free from other plasma proteins, is slightly opalescent. The development of a coarse precipitate of fibrin in the medium during preparation renders it unsatisfactory. Fibrinolysis is indicated by wide zones of clearing about areas of growth after 18 hours' incubation at 37°C.

The nutrient agar base was a heart infusion broth (Difco) solidified with 1.5 per cent Davis New Zealand agar. The medium was sterilized at 121°C for 15 min, and reagents were mixed with the sterilized base medium when it had cooled to 50–55°C.

The medium was inoculated from a fresh 18 hr broth culture with a sterile throat swab.

FINAL pH IN BROTH AND GLUCOSE BROTH

Final pH in glucose broth—anaerobic streptococcus

M. L. Stone (1940). J. Bact. 39, 559–582.

Final pH in 1 per cent glucose broth. The final pH in glucose broth was determined by inoculating beef-infusion broth containing 1 per cent glucose and incubating it anaerobically for four days. The pH was then determined directly by means of the glass electrode, thus avoiding the considerable error involved in the use of indicators. At least two tests were made on each culture and suitable controls of uninoculated medium and medium inoculated with cultures representative of low and high final pH were included.

Final pH in glucose broth—*Streptococcus*

P. L. Boisvert (1940). J. Bact. 39, 727–738.

The organisms were incubated for 4 days in 1.0% glucose broth (Avery and Cullen, 1919). The pH attained was determined colorimetrically using brom-cresol green as the indicator.
O. T. Avery and G. E. Cullen (1919). J. exp. Med. 29, 215–234.

Final pH in broth—*Corynebacterium*

R. F. Brooks and G. J. Hucker (1944). J. Bact. 48, 295–312.

Final pH in nutrient broth was determined at the end of the 2-week incubation period at 37°C. Three drops of 0.04 per cent bromthymol-blue indicator were added to each tube and to uninoculated controls which had also been incubated, and the final pH determined in a comparator apparatus using standard color disks.

Final pH in glucose broth—aerobic spore-forming bacteria

N. R. Smith, R. E. Gordon and F. E. Clark (1952). U.S.D.A. Agr. Monograph No. 16.

Before the 6- or 7-day old cultures were tested for acetylmethylcarbinol, about 1 ml was withdrawn and the pH determined by the colorimetric method. The medium used contained 7 g of proteose-peptone, 5 g of glucose, 5 g of NaCl and 1000 ml of distilled water.

Final pH in glucose broth—*Pediococcus*

J. C. Dacre (1958). J. Dairy Res. 25, 409–413.

The acids produced from glucose. The final pH in a 1.0% glucose broth medium was 4.0–4.2. Both inactive lactic acid and acetic acid were produced from glucose in the molecular ratio 100:12. Gas-liquid partition chromatographic analysis (after James and Martin, 1952) of the culture filtrate, after 3 and 7 days' incubation, indicated that acetic acid was the only volatile acid formed by the organism.

A. T. James and A. J. P. Martin (1952). Biochem. J. 50, 679.

Final pH in broth—*Pediococcus*

H. L. Günther and H. R. White (1961). J. gen. Microbiol. 26, 185–197.

Glucose (1%, w/v) yeast-extract liquid cultures were incubated for 18 days and the final pH values measured electrometrically. Some isolates grew poorly in this medium but the use of tomato juice broth was considered inadvisable because of its natural content of reducing sugar which might have resulted in the production of acids from compounds other than glucose.

Maintenance of stock and cultures and methods of cultivation. For maintenance of stock cultures, preparation of inocula and in all experimental work, 'Oxoid' tomato juice (TJ) broth or tomato juice (TJ) agar, adjusted to pH 6.6, were used unless otherwise stated. The following were exceptions to this rule: for strain Tc. 1 sodium chloride (5%, w/v) was added to the medium; and for the aerococci glucose Lemco broth (Shattock and Hirsch, 1947) or glucose yeast extract (GY) agar (containing, as %, w/v; peptone, 1.0; Yeastrel, 0.3; glucose, 1.0; NaCl, 0.25; agar, 1.0; at pH 7.4) was used.

Cultures were incubated aerobically at 30°C with specified exceptions.

P. M. F. Shattock and A. Hirsch (1947). J. Path. Bact. 59, 495.

Final pH in broth—rumen bacteria

R. S. Fulghum and W. E. C. Moore (1963). J. Bact. 85, 808–815.

Final pH was determined with a glass-electrode pH meter in a lightly buffered medium (SRP-basal medium* without bicarbonate plus 0.1% $(NH_4)_2SO_4$ and 0.5% glucose).

The anaerobic methods of Hungate (Bact. Rev. 14, 1–49, 1950) were used in handling and incubation of the cultures.

***Basal medium for physiological tests** – *Butyrivibrio, Borrelia, Bacteroides, Selenomonas, Succinivibrio* (p. 109).

Final pH in glucose broth—*Aerococcus catalasicus*

O. G. Clausen (1964). J. gen. Microbiol. 35, 1–8.

Cultures were incubated aerobically at 37°C unless otherwise stated. Determination of final pH in 1% glucose beef-infusion peptone phosphate broth (pH = 7.4) after 14 days. The pH was measured with a Beckman pH meter.

Final pH in broth—*Pediococcus*

E. Coster and H. R. White (1964). J. gen. Microbiol. 37, 15–31.

The authors used the methods of H. L. Günther and H. R. White (J. gen. Microbiol. 26, 185, 1961). Their Tomato Juice (TJ) broth or agar with the addition of Tween 80 (0.1%. v/v), and the pH adjusted to 6.6 were used in maintenance and preparation of inocula with certain specified exceptions.

Transfers for preparing inocula were made every 24 hours except for *Pediococcus halophilus* (48 hr) and some brewing strains (fortnightly).

To determine final hydrogen ion concentration all cultures were incubated aerobically at 30°C in Glucose Yeastrel (GY) broth except for "brewing strains," which were incubated in an atmosphere of 95% (v/v) hydrogen and 5% (v/v) carbon dioxide at 22°C.

Final pH in broth—*Pediococcus*

R. Whittenbury (1965). J. gen. Microbiol. 40, 97–106.

The final pH value reached in glucose (1%, w/v) liquid medium initially at pH 6.5 was measured electrometrically after 1 week at 30°C.

For "liquid medium" see "Fermentation – lactic acid bacteria" (p.000), or R. Whittenbury (1963). J. gen. Microbiol. 32, 375–384.

Final pH in broth—*Alcaligenes*

R. G. Mitchell and S. K. R. Clarke (1965). J. gen. Microbiol. 40, 343–348.

Final pH value in fluid culture was determined in Difco "Bacto" nutrient broth incubated for 7 days, using "Lyphan" multi-strip pH papers.

FLAGELLA STAINS

Flagella stains

P. H. H. Gray (1926). J. Bact. 12, 273–274.

Cultures: The best results are obtained with 24 hr to 3 day old slope cultures on nutrient agar of some other medium suited to active growth. (Nutrient agar in the case of the soil organisms on which the stain was successful is composed of peptone, 5 g; Lemco, 3 g; and agar, 15 g per litre of water; pH adjusted to 7.3.) Suspensions are made in sterile distilled water in a watchglass and left for 20 min to 30 min at room temperature, during which time the organisms, if active will be washed free of slime. They should be examined for motility in the fresh state, immediately before the film is made.

Slides: Clean slides are kept in ammonia-alcohol; for use they are dried with a clean duster and flamed by passing 24 times through the Bunsen at the level of the top of the inner cone. They are then put into an oven at 45° to 50°C to cool, flamed side uppermost.

Film: One large loopful of the suspension is placed near one end of the slide. A strip of unsized paper (e.g., typing paper) of a width less than that of the slide is lowered on to the loopful and drawn gently down the slide and off towards the operator. The slide is then put into the oven to dry. The paper strip is convenient in that it allows of a very thin film which dries quickly.

Mordant: This is made up of the following solutions which when stored separately have kept without deterioration for a year in this laboratory.

Solution 1

Potash Alum – saturated aqueous sol	5 ml
Tannic acid – 20 per cent aqueous sol (A few drops of $CHCl_3$ must be added to this if a large quantity is made up)	2 ml
Mercuric chloride – saturated aqueous sol	2 ml

Solution 2

Basic fuchsin – saturated alcoholic solution

For use 0.4 ml of the fuchsin solution is added to Solution 1; the solutions are mixed by rapid rotation of the test tube, when a precipitate will form. The solutions must be freshly mixed for each batch of slides as the mixture deteriorates after twenty-four hours.

About 0.5 ml is allowed to act on each slide for ten minutes at room temperature.

Stain: Wash off the mordant with a gentle stream of distilled water. When no more fine precipitate is removable apply a few drops of Ziehl's carbol fuchsin and leave for 5 or 10 minutes. Wash off with tap water.

Flagella stains

W. E. Maneval (1931). J. Bact. 21, 313–321.

Of 24 mordants tested the author recommends the following mixtures 1, 3 and 20.

	1	3	20
10% Tannic acid	20 ml	20 ml	20 ml
10% alcoholic fuchsin	10 drops	10 drops	10 drops
20% $AlCl_3$	4 ml	–	–
20% $CuCl_2$	–	–	–
30% $FeCl_3$	–	2 ml	–
$Fe(OH)_3$ sol	–	–	3.25 ml
20% $NiCl_2$	–	–	–
20% $ZnCl_2$	–	–	2 ml

The ferric hydroxide solution was prepared by adding slowly 27.5 ml of 30% ferric chloride solution to 100 ml of vigorously boiling water, the boiling continuing until all the ferric chloride has been added. The tannic acid solutions are always filtered as soon as made and used *fresh.* The filtered solutions of chlorides may be kept as stock solutions and used over a considerable period of time. Alcoholic basic fuchsin contained 10 g of dye in 100 ml of 95% alcohol.

Clean, flamed slides were used. A bit of bacterial growth is transferred to 2 drops of water on a slide, but not mixed. After 2–5 minutes, four separate drops of water are placed on a second slide and then one loopful of suspension from the first slide is touched on each of the four drops in succession to give 4 dilutions. The slide is then dried at room temperature in the absence of air currents.

Stains and Staining: Although several staining solutions were tested, no other, not even carbol fuchsin, proved as satisfactory as anilin-alcohol-fuchsin (Maneval, 1929). This stain contains:

	ml
Water, distilled	30
Basic fuchsin, 10 per cent alcoholic	10
Anilin oil (1 part) and 95 per cent alcohol (3 parts) mixed	5
Acetic acid, 4 per cent	1

Mix in the order given; filter once or twice and again before using. It is best after three to six days and may sometimes be kept several weeks, but it finally deteriorates. Good results have been obtained by substituting rosanilin or methyl violet for the basic fuchsin in this formula. Anilin fuchsin and anilin gentian violet were poorer.

The air-dried slides were placed on a level surface, treated with 4 to 8 drops of mordant and allowed to stand two to ten minutes. (One or two trials indicated the time to use.) They were washed with 6 to 10 changes of water, being careful to hold them level when the water was first poured

on. After the second change of water the preparations were wiped carefully with a clean towel to remove the edge of the film of mordant which may harden somewhat, and upon completing the washing treated with 3 to 5 drops of stain for three to five minutes. After washing thoroughly and drying they were examined with the immersion objective without a coverglass.
W. E. Maneval (1929). Stain Technol. 4, 21–25.

Flagella stains

H. J. Conn and G. E. Wolfe (1938). J. Bact. 36, 517–520.

Preparation of Glass Slides.

See Society of American Bacteriologists p. 339.

Preparation of Suspensions: Use young and actively growing cultures (e.g., 18–22 hr old) on agar slants. With a flamed but well-cooled loop transfer a small amount of growth to 5–10 ml of sterile distilled water, which has been held for several hours at room temperature. (Poor slides result from suspensions made in water that is too hot or too cold.)

Mix thoroughly in the distilled water and allow to stand 5–30 minutes according to the type of growth produced. Gum-forming bacteria require 30 minutes, as recommended by Hofer and Wilson; those that produce no gum must stand in the water only 5–10 minutes. Standing in the water should be just long enough to allow the flagella to become untangled; too long a time results in their breaking off.

With a loop that has been flamed and cooled, remove a drop from the top of the suspension and place it on a glass slide prepared as described above. (Material should be taken from near the surface of the suspension because the non-motile cells tend to settle, while the motile cells – at least in the case of aerobes – collect at the top.) Smear the drop over the slide with the use of a second slide, as in preparing blood films; the film should be thin enough to dry rapidly and thus minimize distortion.

Staining: Use the mordant recommended by Gray (1926): 5 ml saturated aqueous solution potassium alum; 2 ml saturated aqueous solution mercuric chloride; 2 ml 20 per cent aqueous solution tannic acid; 0.4 ml saturated alcoholic solution basic fuchsin (presumably about 6 per cent, which was the strength employed in the present work). For best results, filter just before using. The technique is essentially that of Gray. Apply cold for 8–10 minutes; 10 minutes is ordinarily best, but this varies with the organism studied. (More than 10 minutes of mordanting is apt to cause too much precipitate.)

After mordanting, wash slides about 10 seconds in running water. Dry in the air, without heating. Stain 5 minutes, without heating, with Ziehl's carbol fuchsin; wash in running water; dry and examine. (In the present work a recently certified batch of National Aniline basic fuchsin was employed.)

P. H. H. Gray (1929). J. Bact. 12, 273–274.
A. W. Hofer and J. K. Wilson (1938). Stain Technol. 13, 75–76.

Flagella stains

T. Y. Kingma-Boltjes (1948). J. Path. Bact. 60, 275–287.

Staining was done with the Peppler mordant and the Zettnow silver ethylamine solution. The mordant is prepared by dissolving 20 g of tannin in 80 ml of water to which is added, with continuous stirring, 15 ml of a 2.5 per cent solution of chromic acid in water. Before use the mordant is kept for 7 days at 20°C. The Zettnow silver ethylamine solution is prepared by dissolving 0.15 g of $AgSO_4$ in 60 ml of water and adding ethylamine, with continuous stirring, until the precipitate formed is just dissolved. Films of the bacterial suspension are made on freshly annealed coverslips, fixed by heat, flooded with the filtered mordant and gently heated. After 10–15 minutes the mordant is washed off and the preparation dried and flooded with the silver ethylamine. This is heated until steaming and after 1–2 minutes washed off and the preparation dried.

Flagella stains—general

P. Hauduroy (1951). Techniques Bactériologiques. Masson et Cie, Éditeurs, Paris.

COLORATION DES CILS

Les cils des bactéries sont en général difficiles à mettre en évidence si l'on n'a pas soin d'observer un certain nombre de précautions. Les cultures doivent autant que possible être des cultures jeunes (15 heures environ) sur milieu solide. L'émulsion sera faite en eau distillée. Elle sera à peine trouble et parfaitement homogène. Les lames doivent être parfaitement propres, lavées à l'acide et à l'eau.

Les méthodes de coloration des cils sont nombreuses. Nous n'en indiquerons que deux.

Méthode de Löffler (modifiée par M. Nicolle et Morax). – Sur la préparation non fixée, verser quelques gouttes du mordant de Löffler. Chauffer jusqu'à émission de vapeurs. Laver à l'eau distillée. Répéter cette opération trois ou quatre fois.

Verser sur la préparation de la fuchsine phéniquée de Ziehl. Chauffer jusqu'à émission de vapeurs pendant 10 à 15 secondes. Laver à l'eau distillée. Sécher.

Méthode de Casares-Gil. – Préparer la solution mère du réactif suivant.

a) Dissoudre dans un mortier 10 g de tanin et 18 g de chlorure d'ammonium hydraté dans 30 cm^3 d'alcoöl à 70 p. 100.

b) Dissoudre dans 10 cm^3 d'eau distillée, en triturant très soigneusement dans un mortier, 10 g de chlorure de zinc et 1.5 g de chlorhydrate de rosaniline.

Verser ensuite goutte à goutte en remuant la solution *b* dans la solution *a.* Cette solution mère peut se conserver très longtemps, à l'abri de la lumière.

Avant usage, mélanger une partie de la solution mère avec quatre parties d'eau distillée. Agiter. Laisser reposer une minute.

Pour colorer, verser à travers un filtre quelques gouttes du colorant dilué sur la préparation non fixée. Laisser agir à froid pendant 1 à 2 minutes environ.

Laver rapidement à grande eau sous le robinet.

Colorer à froid à l'aide de la fuchsine phéniquée non diluée pendant 1 à 2 minutes.

Laver à l'eau. Sécher à l'air.

(C. C. T. M.).

(Méthode de Seguin-Fontana-Van Ermengen) (Recommandée pour les anaérobies). – 1° Fixer la préparation à l'acide osmique.

2° Sécher.

3° Laver à l'alcoöl à 90°.

4° Mordancer avec du tanin à 40 ou 50 p. 100 pendant 10 minutes en chauffant légèrement.

5° Laver à l'eau.

6° Imprégner au nitrate d'argent ammoniacal pendant quelques minutes en chauffant légèrement.

Chasser le précipité avec un peu d'eau.

(A. I. P.).

COLORATION DE FONTANA

(Méthode de) – 1° Fixer les frottis par le *formol acétique.*

Eau distillée	100 cm^3
Formol neutre	20 "
Acide acétique crist	1 "

S'il s'agit d'un frottis de sang, fixer et déshémoglobiniser en même temps avec le *liquide de Rüge.*

Eau distillée	100 cm^3
Formol neutre	2 "
Acide acétique cuit	1 "

2° Laver à l'eau ordinaire

3° Mordancer par le *tanin phéniqué.*

Eau distillée	100 cm^3
Tanin	5 g
Acide phénique neigeux	1 g

Chauffer légèrement pendant quelques secondes.

4° Laver à l'eau ordinaire puis à l'eau distillée.

5° Impregner par le *liquide de Fontana* (Azotate d'argent ammoniacal) en en versant quelques gouttes sur le frottis. Chauffer légèrement pendant 20 à 30 secondes. On peut recommencer à plusieurs reprises.

Préparation du liquide de Fontana. – Prendre une solution de nitrate d'argent à 0.25 p. 100. Ajouter de l'ammoniaque. Un précipité se forme. Continuer à verser doucement de l'ammoniaque, goutte à goutte en agitant. Le précipité se dissout. A ce moment, ajouter à nouveau un peu de solution de nitrate d'argent. On doit en définitive obtenir un liquide légèrement opalescent.

Laver à l'eau distillée, sécher, examiner à l'immersion.

(C. C. T. M.).

Flagella stains—*Pseudomonas (Malleomyces) pseudomallei*

P. de Lajudie, J. Fournier and L. Chambon (1953). Annls Inst. Pasteur, Paris 85, 112–116.

La coloration des flagelles du b. de Whitmore est malaisée. Un choix judicieux doit intervenir sur les points suivants:

a) Souche appelée à fournir les préparations;

b) Culture de cette souche;

c) Préparation des frottis;

d) Coloration des frottis.

a) *La souche.* – Celle qui nous a fourni les meilleurs résultats (S. C141) est, dans notre collection, parmi celles qui paraissent avoir le moins subi la dissociation M.

b) *Culture de cette souche.* – Il nous a paru avantageux de faire un passage sur cobaye et de cultiver sur gélose molle et sur gélose au sang les germes provenant d'une ponction du coeur chez l'animal infecté.

c) *Préparation des frottis.* – L'extrémité d'un ensemenceur chargée sur une culture de huit heures est déposée avec précaution dans 5 ml d'eau distillée. On maintient l'ensemenceur dans le liquide, *sans agiter,* jusqu'à ce que l'amas bactérien se soit dissocié et ait diffusé en trainées opalescentes. On prélève alors le liquide à la pipette avec précaution et on dépose une grosse goutte sur chaque lame d'une série parfaitement propre et dégraissée. On étale cette goutte par simple inclinaison de la lame, on laisse sécher lentement à l'abri des poussières et on entoure d'un trait de crayon gras la surface recouverte par le liquide.

d) *Coloration.* – Nous avons comparé les résultats fournis par trois méthodes:

Celle de Novel (1933),

Celle de Leifson (1938) modifiée par l'utilisation du "flagella stain" des Baltimore Biological Laboratories;

Celle de Casares-Gil rapportée par Dumas (1951).

C'est cette dernière qui nous paraît la meilleure dans les conditions de notre travail. Elle a rendu possibles les microphotographies qui illustrent notre texte et qui sont dues à l'obligeance du Dr M. Piéchaud, de l'Institut Pasteur, que nous tenons à remercier ici.

J. Dumas. *Bactériologie Médicale,* 1951.

A. Whitmore. *J. Hyg.,* 1913, 13, 1.

Flagella stains—*Pseudomonas*

C. K. Williamson (1956). J. Bact. 71, 617–622.

Cellular characteristics of colonial variants: Hanging-drop preparations showed that cells of the smooth and mucoid states were actively motile. Accordingly, attempts were made to demonstrate flagella using Leifson's method (1951). A drop of the bacterial suspension was placed at one end of the slide which was then tilted to permit the drop to flow to the other end. It was found that the best preparations were those which dried rapidly. Invariably, the preparations which required several minutes to dry showed no flagella. Slides were flooded with the stain immediately after the preparations dried and it was permitted to act for 11 minutes on smooth cells. Flagella on cells of the mucoid state were best demonstrated after hydration, washing, and rehydration of cells harvested from agar plates. The preliminary hydration facilitated the removal of the capsular material which otherwise masked the flagella. The stain was permitted to act for 13 minutes on mucoid cells. While there may be some question about cellular and flagellar damage using the differential blood smear technique to prepare slides, it is interesting to note that excellent preparations were also achieved by this method.

E. Leifson (1951). J. Bact. 62, 377-389.

Flagella stains

Society of American Bacteriologists *Manual of Microbiological Methods* – McGraw-Hill Book Co. Inc., New York 1957.

Methods for preparing slides

Ordinary cleaning of glassware is not sufficient for the purpose. Various methods have been proposed, but the following directions seem to give as good results as any:

Use new slides if possible, preferably of Pyrex glass or with similar heat resistant properties. (This is because under the drastic method of cleaning to remove grease, old slides have a greater tendency to break.) Clean first first in a dichromate cleaning fluid, wash in water, and rinse in 95 per cent alcohol; then wipe with a clean piece of cheesecloth. (Wiping is not always necessary but is advisable unless fresh alcohol is used after every few slides.) Pass each slide back and forth through a flame for some time, ordinarily until the appearance of an orange color in the flame; some experience is necessary before the proper amount of heating can be accurately judged.

Unless heat-resistant slides are used, cool slides gradually in order to minimize breakage. An ordinarily satisfactory method of doing this is to place the flamed slides on a metal plate (flamed side up) standing on a vessel of boiling water and then to remove the flame from under the water so as to allow gradual cooling. (Too rapid cooling may result in breakage, sometimes as long as 2 weeks after the heating.)

Methods of handling cultures. Of various methods proposed, it is not possible to recommend any one as uniformly the best. As any laboratory worker becomes familiar with one particular method, he soon finds he can get better results with that than with any other. The following method, however, can be given as one of the most satisfactory, especially for students who have not had previous experience with some other method: Use young and actively growing cultures (e.g., 18-22 hr old) on agar slants. Before proceeding, check the culture for motility in hanging drop. If motile, wash off the growth by gentle agitation with 2-3 ml of sterile distilled water. Transfer to a sterile test tube, and incubate at optimum temperature for 10 min (30 min for those producing slime).

At this point again check motility under a microscope. Transfer a small drop from the top of the suspension (where motile organisms are most numerous) by means of a capillary pipette to one end of the slide prepared as described above. Tilt the slide and allow the drop to run slowly to the other end. (Two or three such streaks can be placed on a slide.)

Place the slide in a tilted position, and allow to dry in the air.

Staining Procedures—Flagella stain

(1) Casares-Gil

As published by Plimmer and Paine (1921).

Mordant:

Tannic acid	10 g
$AlCl_3 \cdot 6H_2O$	18 g
$ZnCl_2$	10 g
Basic fuchsin	1.5 g
Alcohol (60%)	40 ml

The solids are dissolved in the alcohol by trituration in a mortar, adding 10 ml of the alcohol first, and the rest slowly. This alcoholic solution may be kept several years. For use, mix with an equal quantity of water (Thatcher, 1926) or dilute with 4 parts of water (Casares-Gil), filter off precipitate and collect filtrate on the slide.

Staining schedule:

1. Prepare smears of young cultures, on scrupulously cleaned slides as directed above.
2. Filter mordant onto slide as directed above (preferably using Thatcher's 1:1 dilution); allow to act for 60 sec without heating.
3. Wash in tap water.

4. Flood slide with freshly filtered Ziehl's carbol fuchsin* and allow to stand 5 min without heating.
5. Wash with tap water.
6. Air-dry, and examine. Sometimes considerable search may be needed before finding a satisfactorily stained part of the smear.

Results: Flagella well stained (red) in the case of those bacteria (e.g., colon-typhoid group, aerobic sporeformers) that do not have extremely delicate flagella.

Ziehl's Carbol – fuchsin

Solution A

Basic fuchsin (90% dye content)	0.3 g
Ethyl alcohol (95%)	10 ml

Solution B

Phenol	5 g
Distilled water	95 ml

Mix solutions A and B.

H.G. Plimmer and S.G. Paine (1921). J. Path. Bact. 24, 286-288.
L.M. Thatcher (1926). Stain Technol. 1, 143.

(2) Gray, 1926.

Mordant:

Solution A

$KAl(SO_4)_2 \cdot 12H_2O$ (sat aqu solution)	5 ml
Tannic acid (20% aqu solution)	2 ml
(A few drops of chloroform must be added to this if a large quantity is made up)	
$HgCl_2$ (sat aqu solution)	2 ml

Solution B

Basic fuchsin (sat aqu solution)	0.4 ml

Mix solutions A and B less than 24 hr before using. Both solutions separately may be kept indefinitely, but deteriorate rapidly after mixing.

Staining schedule:

1. Prepare smears from young cultures as directed above.
2. Flood slide with freshly filtered mordant, and allow to act 8-10 min.
3. Wash with a gentle stream of distilled water, and follow steps 4-6 of above schedule (Casares-Gil method).

Results: Same as with Casares-Gil method.

P.H.H. Gray (1926). J. Bact. 12, 273-274.

(3) Bailey (1929). Modified by Fisher and Conn (1942).

This method is especially recommended for bacteria on which flagella are difficult to stain (as is frequently the case with soil and water non-sporeformers and with plant pathogens) because of slime production, unusually fine flagella, or flagella that are readily lost.

Mordant:

	Solution A	
Tannic Acid (10% aqu solution)		18 ml
$FeCl_3 \cdot 6H_2O$ (6% aqu solution)		6 ml
	Solution B	
Solution A		3.5 ml
Basic fuchsin (0.5% in ethyl alcohol)		0.5 ml
HCl, concentrated		0.5 ml
formalin		2.0 ml

Staining schedule:

1. Prepare smears of young cultures carefully following the procedure recommended on page 340 under "Methods of handling cultures."
2. Filter the above solution A onto the slide and allow it to stand 3.5 min without heating.
3. Pour off solution A, and without washing add solution B, also through a filter, and allow it to stand 7 min without heating.
4. Wash with distilled water.
5. Before the slide dries, cover with Ziehl's carbol fuchsin* allowing it to stand 1 min on a hot plate heated just enough for steam to be barely given off.
6. Wash in tap water.
7. Dry in the air and examine.

Results: Similar to the preceding methods, but the background precipitate is usually finer and less conspicuous, thus interfering less with the demonstration of unusually fine, delicate flagella.

*See p. 341.

Staining flagella of anaerobes. O'Toole (1942) calls attention to certain difficulties in staining the flagella of anaerobes and gives a modification of the above Bailey stain which is intended to overcome them. The method is not unlike that of Fisher and Conn, who had the O'Toole procedure in mind when working out their modification.

H. D. Bailey (1929). Proc. Soc. exp. Biol. Med. 27, 111–112.
P. J. Fisher and J. E. Conn (1942). Stain Technol. 17, 117–121.
E. O'Toole (1942). Stain Technol. 17, 33–40.

Flagella stains—motile Gram-positive cocci.

C. M. F. Hale and K. A. Bisset (1958). J. gen. Microbiol. 18, 688–691.

Actively motile cultures were examined at various ages from 8 to 24 hr. Suspensions of organisms were made in distilled water containing 0.5% (w/v) formaldehyde, and air-dried smears were mordanted for at least 18 hr at 20–25°C in a solution of 7.5% (w/v) tannic acid + 1% (w/v) ferric chloride in water, washed and mordanted for a further 5 min in the same solution, with the addition of 1.5% (v/v) of a second solution containing 10%

(v/v) aniline + 40% (v/v) ethanol, in water. After washing, the preparation was stained in 0.5% (w/v) crystal violet for *c.* 3 min, and allowed to dry without heating.

Flagella stains—*Salmonella*

C. Quadling (1958). J. gen. Microbiol. 18, 227–237.

A modification of Leifson's (1951) method was used to stain flagella on bacteria grown in broth. Cultures were fixed with formaldehyde to a final concentration of 1% (w/v), then sedimented by centrifugation for 40 min at 250 *g*. The bacteria were then washed twice by gentle resuspension in 0.01% (v/v) Teepol (Shell Chemicals, Ltd, 105/109 Strand, London, W.C.2) in distilled water, and recentrifuged. The bacteria were finally resuspended in distilled water. The detergent was used to ensure the uniform distribution of organisms in smears (Snieszko, 1942), and to assist in attaining the cleanliness necessary with this staining technique. Drops of the final distilled water suspension of organisms were transferred to acid-cleaned slides and stained as described in Leifson (1951).

Bacteria were usually well dispersed in the central part of each smear; counts of number of flagella/bacterium were made on bacteria in this central region. Counts were made at a magnification of x1200 by using a 2 mm x 95 oil immersion objective with phase contrast illumination and a green filter. The flagella were counted on each of the first five flagellated organisms seen in each of ten or more successive scans started at arbitrarily chosen points. The use of phase-contrast made counter-staining of the bacterial bodies unnecessary.

E. Leifson (1951). J. Bact. 62, 377.
S. F. Snieszko (1942). Science, N. Y. 96, 589.

Flagella stains—*Acetobacter* and *Acetomonas*

J. L. Shimwell (1958). Antonie van Leeuwenhoek 24, 187–192.

The author found Leifson's method unsuitable and used "a modification of the method of Fisher and Conn (1943 – reference not cited) in which 1% aqueous crystal violet replaced both 5% alcoholic basic fuchsin and carbol-fuchsin".

E. Leifson (1954). Antonie van Leeuwenhoek 20, 102.

Flagella stains—*Pseudomonas*

M. E. Rhodes (1958). J. gen. Microbiol. 18, 639–648.

A modification of Fontana's method of staining spirochetes (Mackie and McCartney, 1949, p. 108) was devised, the modifications consisting of slight differences in the reagents and in the methods of preparing bacterial films for staining. Two reagents were used:

(1) Ferric tannate mordant: To 10 ml 10% (w/v) tannic acid were added 5 ml saturated aqueous potash alum, followed by 1 ml saturated

solution of aniline in water; the curd which formed was redissolved by shaking. The addition of 1 ml 5% (w/v) ferric chloride gave a black solution which was allowed to stand 10 min before use.

(2) Ammoniacal silver nitrate solution. From 100 ml of a 5% (w/v) aqueous solution of silver nitrate *c.* 10 ml were set aside. Concentrated ammonia solution (sp.gr. 0.88) was slowly added to the 90 ml portion until the brown precipitate just redissolved. Drops of the 10 ml sample of silver nitrate solution were then added until the solution remained faintly cloudy, even after shaking. Stored in the dark, this solution remained stable for several weeks.

Method of making a preparation and staining to show flagella: Organisms were grown on yeast extract agar slopes in 6 x 5/8 in. tubes, to each of which was added 1–2 ml sterile distilled water because motile organisms were most readily obtained from the liquid at the base of such slopes. Flagella staining of these cultures was made routinely after 18–24 hr of incubation, or after 48 hr in the case of the few cultures which grew slowly, e.g. *Pseudomonas lachrymans.* After incubation, two 4 mm-loopfuls of the liquid at the base of a slope were inoculated into 5 ml sterile distilled water and incubated for 1 hr (25°C) to remove debris and excess mucilage from the organisms. A drop of this suspension was then transferred to a slide previously cleaned in chromic + sulfuric acid mixture. After allowing the drop to air-dry, the film was covered with the iron tannate mordant reagent (no. 1) for 3–5 min. After washing the film very thoroughly with distilled water, the silver reagent (no. 2), heated nearly to boiling, was applied and left in contact with the film for 3–5 min. After a second washing of the film with distilled water, it was blotted, and either examined directly or made permanent under a coverslip by sealing with neutral Canada Balsam. Exposure to air caused disintegration of these silver-plated preparations after about 1 week. A light microscope with 1/12 in. oil immersion objective and x 10 ocular was used to examine the preparations. The silver-plated flagella were found to be easily visible provided that the substage condenser had been carefully centered and focused.

T. J. Mackie and J. E. McCartney (1949). *Handbook of Practical Bacteriology,* 8th ed. Edinburgh; E. & S. Livingstone Ltd.

Flagella stains—Leifson's

V. B. D. Skerman, *A Guide to the Identification of the Genera of Bacteria.* 2nd ed. Williams and Wilkins Book Co., Baltimore, 1967.

Young liquid cultures are preferred for this procedure.

Transfer the broth cultures to 15 ml centrifuge tubes and centrifuge at 3000 rpm for 5 minutes.

Pour off the supernatant and gently resuspend the cells in about 7 ml of distilled water. Centrifuge. Pour off the supernatant and gently resuspend the cells in distilled water to give a just visible turbidity. Incubate at 37°C for 10 minutes and check for motility.

Select a perfectly clean slide. Do not use a cloth or wash in soap or detergent.

Flame the slide for a few seconds and before it has cooled use a grease pencil to mark a heavily defined band around an area of about 1.5 sq. in. at one of the flamed side of the slide. This acts as a retarding margin around the stained area.

When it is cool, incline the slide and place loopfuls of the washed suspension at two points at one end of the enclosed area and allow them to run to the other end. If this does not occur readily, the slide has not been correctly cleaned. Allow the slide to dry in air and then place on a staining rack.

Shake the bottle of Leifson's stain thoroughly. Using a pipette, slowly run 1.0 ml of stain into the marked area, taking care that the stain does not flow over the grease pencil marks.

Staining Time: Leifson states as follows: "The proper staining time is best determined by the formation of a fine precipitate in the stain on the slide. As soon as the precipitate is formed the flagella are stained. The formation of the precipitate is best observed by placing the slide on a black background and illuminating the stain with a beam of light. The precipitate has the appearance of rust coloured cloud which generally starts forming along one edge of the slide and then quickly spreads throughout the stain. As soon as the precipitate is formed throughout the stain, the time is up and slide should be washed. This technique eliminates all guess work as to the proper time to wash the slides."

Flood the stain off the slide with a gentle stream of water. Counterstain for 1 minute with 1 per cent methylene blue.

(Note: It has been found in this laboratory that some flagella stain more readily after fixation in formalin.)

Preparation of the Stain: Three solutions are prepared: (a) 1.5 per cent sodium chloride in distilled water (b) 3.0 per cent tannic acid in distilled water and (c) 1.2 per cent basic fuchsin (special, for flagella stain)* in ethyl alcohol. To effect complete solution, it is necessary to shake frequently for several hours.

Mix the three solutions in exactly equal proportions and store either at −10° to −20°C (stable indefinitely) or at 0°C (stable for a few weeks) for 1 day or more to clear before use.

*Pharmaceutical Laboratories, National Aniline Div. 40 Rector Street, New York.

E. Leifson (1951). J. Bact. 62, 377.

E. Leifson (1958). Stain Technol. 33, 249.

E. Leifson and L. Hugh (1953). J. Bact. 65, 263.

Flagella stains—*Flavobacterium*

J. Meyer, J. Malgras, Ch. Romond et C. Boeglin (1960). Annls. Inst. Pasteur, Paris 98, 28–34.

Notre méthode est une modification de celle de Zettnow.

1° Préparation du mordant. – Dissoudre 2 g d'antimonio-tartrate acide de potasse dans 50 cm^3 d'eau distillée. Chauffer doucement en agitant pour activer la dissolution. Verser 25 à 30 cm^3 de cette solution dans 200 cm^3 d'une solution de tanin à 5 p. 100 dans l'eau distillée, de façon qu'au refroidissement le mélange soit à peine lactescent.

2° Préparation de la solution d'éthylamine argentée. – Préparer une solution saturée de sulfate d'argent, puis mélanger:

Solution de sulfate d'argent	5 cm^3
Eau distillée	5 cm^3

Ajouter goutte à goutte une solution d'éthylamine à 33 p. 100 jusqu'à ce qu'il reste une légère opalescence. Ne préparer ce réactif qu'au moment de l'emploi.

Mode opératoire: Employer une culture sur gélose âgée de 20 à 24 heures. Introduire dans un tube à hémolyse 1 cm^3 d'eau distillée stérile. Déposer en surface, sur les bords du tube, une anse de culture. Laisser les bactéries se répartir uniformément dans le liquide placé à 37° pendant trente à soixante minutes. Après ce temps ajouter dans le liquide une goutte de la solution d'acide osmique à 2 p. 100 et laisser agir quinze à trente minutes, ou même davantage, à 37°. Préparer une pipette Pasteur effilée et légèrement recourbée à l'extrémité qui doit être franchement coupée.

Prélever un peu de liquide surnageant opalescent et le déposer en fines gouttelettes sur une lame parfaitement propre, en appuyant la pipette sur la lame (ne pas laisser tomber la goutte). Mettre à sécher à l'étuve à 37°. Fixer à la chaleur et placer la lame dans le mordant presque à l'ébullition. Laisser faire un bouillon et retirer de la flamme. Abandonner le tout jusqu'à début de trouble (ne pas laisser trop longtemps pour éviter la formation d'un précipité sur la lame; pour l'éviter il suffit de tourner le frottis vers le bas).

Sortir ensuite le frottis du mordant, le placer horizontalement et le recouvrir d'eau à l'aide d'un jet de pissette. Laver très soigneusement. La lame restant horizontale, la recouvrir de l'éthylamine argentée en couche épaisse. Chauffer doucement avec la veilleuse d'un bec Bunsen jusqu'à émission franche de vapeur, que l'on doit maintenir pendant cinq à dix minutes. Ne pas laisser la solution d'éthylamine se concentrer trop, en rajouter si nécessaire. Rincer soigneusement la lame restée en position horizontale; on chasse l'éthylamine argentée d'un jet de pissette, la lame étant encore chaude. On contrecolore pendant deux minutes avec le violet de gentiane phéniqué. On lave à l'eau. Sécher à l'air et examiner à l'immersion.

Flagella stains—agarolytic bacteria

A. E. Girard and R. C. Cleverdon (1961). Can. J. Microbiol. 7, 407–409.

The Fisher and Conn staining procedure was used to obtain stains of 10 cultures of marine agarolytic bacteria (tentatively placed in the genus *Alginomonas*). All had a single polar flagellum. Flagella on cells grown in synthetic medium were easily demonstrated with little stained background. A brief discussion of flagellation as a taxonomic criterion of the agarolytic bacteria is included.

Preparation of smears and staining procedure: When the organisms demonstrated motility, a loopful was taken from each tube and placed on a tilted slide, the drop being allowed to flow over the surface. (Fresh sonically cleaned slides from the Will Corporation were found to be quite as satisfactory as those cleaned with dichromate-sulfuric acid or with nitric-sulfuric acid mixtures.) The slide was then allowed to air-dry before being stained by the Bailey method as modified by Fisher and Conn [1]. The mordant and mordant-dye mixture proved to give better results when allowed to stand at least 1 day before use; fresh mordants, in most instances, were unsatisfactory.

The slide was flooded with solution A with filtering as recommended [1] and was allowed to stand for 3.5 minutes. After A was poured off, without washing, the slide was flooded with B, with filtering, and allowed to remain for 7 minutes, after which it was washed in distilled water. Ziehl-Neelsen carbol fuchsin was then applied for 1 minute while the slide was still wet; the slide was then washed in tap water and air-dried.

[1] P. J. Fisher and J. E. Conn (1942). Stain Technol. 17, 117–121.

Flagella stains—*Escherichia aurescens*

H. Leclerc (1962). Annls Inst. Pasteur, Paris 102, 726–741.

b) *Système ciliaire.* – Nous avons employé la méthode de Rhodes [41] décrite avec quelques modifications par Buttiaux et Gagnon [5].

[5] Buttiaux (R.) et Gagnon (P.). *Ann. Inst. Pasteur Lille,* 1958-1959, 10, 121.

[41] Rhodes (M. E.). *J. gen. Microbiol.,* 1958, 18, 639.

Flagella stains—motile marine bacteria

E. Leifson, B. J. Cosenza, R. Murchelano and C. Cleverdon (1964). J. Bact. 87, 652–666.

To 400 ml of selected samples of water was added 5% (v/v) of formalin (40% formaldehyde solution), and the bacteria were washed by centrifugation in preparation for flagellar staining by the method of Leifson (1960).

E. Leifson (1960). *Atlas of bacterial flagellation.* Academic Press, Inc., New York..

Flagella stains—*Lactobacillus*

M. Gemmell and W. Hodgkiss (1964). J. gen. Microbiol. 35, 519–526.

Flagella staining and electron microscopy. Strains 1 and 11 were grown on the basal medium* + 0.05% (w/v) glucose, which was used both as a liquid and as slopes containing 1.5% (w/v) agar. Bacteria were harvested from the slopes by washing with sterile, distilled water and from liquid medium by centrifugation. Harvested bacteria were then washed three times with sterile distilled water by centrifugation, and a final suspension prepared in distilled water.

Flagella staining was done by a modification of the method of Casares-Gil (*Manual of Microbiological Methods,* 1957). A drop of the bacterial suspension was streaked on the middle of a microscope slide and air-dried. As soon as drying was complete the preparation was treated for 5 min with a fixative (60 ml ethanol + 30 ml chloroform + 10 ml neutral formalin) and washed in distilled water. The mordant (10 ml) was added to distilled water (30 ml) agitated slowly for 1 min, and filtered on to the slide through a Whatman no. 12 filter paper. The mordant was allowed to act for 4–5 min. The slide was then washed, stained for 1 min with strong carbol fuchsin (Ziehl–Neelsen), washed again and air-dried.

The basal medium which was used throughout with minor modifications had the following constituents in 1 L tap water: meat extract (Lab-Lemco), 5 g; Evans peptone, 5 g; Difco yeast extract, 5 g; Tween 80, 0.5 ml $MnSO_4 \cdot 4H_2O$, 0.1 g; potassium citrate, 1 g; pH 6.5.

*See **Fermentation**—*Lactobacillus* (p. 314).

Society of American Bacteriologists *Manual of Microbiological Methods* – McGraw-Hill Book Co. Inc., New York 1957.

Flagella stains—*Pseudomonas*

G. L. Bullock, S. F. Snieszko and C. E. Dunbar (1965). J. gen. Microbiol. 38, 1–7.

Novel's (1939) staining method was used.

E. Novel (1939). Annls Inst. Pasteur, Paris 63, 302.

Flagella stains—*Leptospira*

D. D. Blenden and H. S. Goldberg (1965). J. Bact. 89, 899–900.

The silver impregnation technique described here is a modification of a technique described by D. S. Kim of the Veterinary Research Institute, Pusan, Korea (unpublished data). The modified technique is rapid, employs only two reagents, which are easily applied, and requires no special fixation or heating. Alteration of the morphology of leptospires and flagella is minimal. The spiral structure of the spirochetes can often be seen in a relatively normal state.

Reagent A is composed of 100 ml of distilled water containing 5 g of tannic acid, 1.5 g of ferric chloride, 2.0 ml of 15% formalin, and 1.0 ml of 1% sodium hydroxide.

Reagent B, ammoniated silver nitrate solution, is prepared by use of 100 ml of 2% silver nitrate. About 10 ml of this volume are removed and saved; to the remaining 90 ml, ammonium hydroxide is added dropwise until the heavy precipitate that is formed is dissolved. From the 10 ml previously removed, 2% silver nitrate is added dropwise until a slight clouding appears and persists. At this point, the pH is adjusted to 10.0 with the ammonium hydroxide and silver nitrate. Reagent B is relatively unstable, and must be used within 4 hr of preparation. Readjustment of the pH to 10.0 after it changes has not produced satisfactory results.

Smears are prepared in a routine fashion except that extreme caution must be employed in handling bacterial suspensions for flagella stain. The slides are alcohol-cleaned and a loopful of distilled water is placed on each. A loopful of culture or suspension is placed just touching the distilled water, so that the two diffuse together to achieve a gradation of dilution of the medium. Slides are then allowed to air-dry. Fixation by heat is unnecessary.

For the preparation of bacteria to produce and maintain maximal numbers of flagella, incubation at 20°C, harvesting organisms during the logarithmic or early stationary phase, preparation of a faintly cloudy suspension in distilled water, and extreme care in handling, all as described by Leifson (*Atlas of Bacterial Flagellation,* Academic Press, Inc., New York, 1960) are essential.

The smears are covered by reagent A for 2 to 4 min; they are then rinsed in distilled water. Distilled water and varying concentrations of ammonium hydroxide and ethyl alcohol have been tried, both as rinses and as 10-min washes, with no apparent difference. After the water rinse, reagent B (pH 10.0) is added for about 30 sec. The smears are immediately washed with distilled water, air-dried, and examined under oil immersion.

Flagella stains—*Rhizobium*

J. de Ley and A. Rassel (1965). J. gen. Microbiol. 41, 85–91.

Flagella staining. The organisms were grown on agar slants with the modified Ashby medium, containing 1 ml of distilled water in the bottom (Rhodes, 1958). After several days at room temperature a loopful of the liquid was suspended gently in water; 35% formalin was added to make a final concentration of 5–10%. The organisms were centrifuged for 5 min at 3000 rpm and repeatedly washed by suspension in water and centrifugation at the same speed. A drop of the final suspension (about 10^7 – 10^8 cells per ml) was applied on a perfectly clean microscope slide and dried (Leifson, 1960). The staining procedure of Rhodes (1958) was followed.

E. Leifson (1960). *Atlas of Bacterial Flagellation.* New York: Academic Press.
M. E. Rhodes (1958). J. gen. Microbiol. 18, 639.

GALACTOSIDASE

Galactosidase—enterobacteria

L. Le Minor et F. Ben Hamida (1962). Annls Inst. Pasteur, Paris 102, 267–277.

Une pleine anse de culture bactérienne de 18 heures à 37° (sauf pour les *Pseudomonas* et *Aeromonas* qui furent incubés à 30°), prélevée sur la pente du milieu lactose-glucose-SH_2, est mise en suspension dans 0.25 cm^3 d'eau physiologique. Pour favoriser la libération de l'enzyme, on y ajoute 1 goutte de toluène, et après agitation et séjour de quelques minutes à 37°, 0.25 cm^3 de la solution d'O. N. P. G. (1). Nous avons employé ces quantités réduites dans le but d'économiser la culture bactérienne, dont la partie restante sur la pente du milieu peut servir à effectuer d'autres réactions biochimiques rapides et à rechercher les caractères antigéniques. Il est important d'avoir une suspension dense pour obtenir une concentration élevée d'enzyme et ainsi une réaction plus rapide. Comme celle-ci se manifeste par une coloration, le volume total n'a pas besoin d'être important.

Les tubes sont placés au bain-marie à 37°. Dans notre protocole nous avons fait les lectures au bout de vingt minutes, une, deux, trois et vingt-quatre heures. Une réaction positive se manifeste par une couleur jaune franc. Dans la plupart des cas, elle apparaît rapidement. Dans nos résultats nous avons appelé "réaction fortement et rapidement positive" celle donnant une couleur jaune intense en moins de vingt minutes, identique à celle obtenue avec des *Escherichia coli* acidifiant les milieux lactosés habituels en dix huit heures. Les réactions plus tardives sont mentionnées.

La coloration jaune est stable. Si, pour des raisons techniques, une lecture précoce ne pouvait être faite, ceci ne conduirait donc pas à une conclusion erronée.

(1) *Préparation de la solution tampon* $PO_4H_2Na{\cdot}H_2O$ (phosphate monosodique cristallisé) M, pH 7.0.

Dissoudre 6.9 g de $PO_4H_2Na{\cdot}H_2O$ dans 45 cm^3 d'eau distillée environ. Ajouter environ 3 cm^3 de lessive de soude pure 36° Baumé de façon à ajuster le pH à 7. Compléter à 50 cm^3 avec de l'eau distillée. Conserver à +4°.

Préparation de la solution d'O. N. P. G. M/75 dans une solution tampon de $PO_4H_2Na{\cdot}H_2O$, M/4 de pH 7.0.

Dissoudre dans 15 cm^3 d'eau distillée à 37°C, 80 mg d'O. N. P. G.

cristallisé (Light and Co. Colnbrook-Bucks, Angleterre, représenté en France par Touzard et Matignon, 3, rue Amyot, Paris-5e). Ajouter 5 cm^3 de la solution tampon $PO_4H_2Na \cdot H_2O$, ajustée à pH 7.0 avec de la soude (voir préparation de la solution plus haut). La solution doit être presque incolore. On la conserve à +4°C.

Avant l'utilisation, laisser quelques minutes à 37°C pour remettre en solution le phosphate qui a pu cristalliser à froid.

See also H. Mollaret and L. Le Minor (1962) Annls Inst. Pasteur, Paris 102, 649–652. These authors describe the occurrence of β-galactosidase in *Pasteurella pestis* and *P. pseudotuberculosis* and support the proposal to place these two organisms in the genus *Yersinia.*

Galactosidase—*Escherichia aurescens*

H. Leclerc (1962). Annls Inst. Pasteur, Paris 102, 726–741.

c) *β-galactosidase.* – La présence de cette enzyme gouvernant la fermentation du lactose doit être recherchée chez toutes les bactéries où cette fermentation n'apparaît pas sur les milieux lactosés classiques, ou même sur les milieux hyperconcentrés en lactose comme celui décrit par Buttiaux [4] ou par Lowe et Evans [28].

La méthode est basée sur la détermination colorimétrique de l'O-nitrophénol libéré du substrat o-nitrophényl-β-D-galactopyranoside (ONPG) lorsqu'il est en contact avec une suspension de bactéries traitées par le benzène, conformément à Lederberg [25] et Rotman [44].

Les organismes sont cultivés sur le milieu défini par Buttiaux [4], mais sans indicateur de pH. Une suspension dense de bactéries est ensuite préparée en milieu tampon contenant 4.5 g de phosphate bipotassique et 0.5 g de phosphate monopotassique pour 100 ml d'eau.

A 1 ml de la suspension on ajoute 1 goutte de toluène et on porte au bain-marie à 37° durant trente minutes en agitant de temps en temps.

On ajoute 0.2 ml d'une solution contenant 4 mg d'ONPG pour 1 ml.

On observe après trente minutes de contact.

L'apparition d'une coloration jaune est la marque d'une activité β-galactosidasique, plus ou moins intense selon la teinte observée.

Des préparations identiques sont faites avec *Salmonella typhimurium* comme témoin négatif et *E. coli* comme témoin positif.

[4] Buttiaux (R.). *L'analyse bactériologique des eaux de consommation* Flammarion, Paris, 1951.

[25] Lederberg (J.). *J. Bact.* 1950, 60, 381.

[28] Lowe (G. H.) et Evans (J. H.). *J. clin. Path.* 1957, 10, 4, 318.

[44] Rotman (B.). *J. Bact.* 1958, 76, 1.

Galactosidase—*Citrobacter, Salmonella*

M. J. Pickett and R. E. Goodman (1966). Appl. Microbiol. 14, 178–182.

Based on tests of Le Minor and Ben Hamida (1962) and Lowe (1962).

For qualitative results, each strain was grown and harvested as for our other rapid biochemical tests*: the entire surface (approximately 6 cm^2) of a Kligler Iron Agar (KIA) slant (7 ml in a 16 by 125 mm screw-cap tube) was inoculated, the medium was incubated 20 to 24 hr at 35°C, the surface growth was harvested with 1.0 ml of sterile distilled water, and 0.1 ml of this suspension was used as inoculum for each test. A trace of phenol red occluded in the inoculum did not interfere with detection of *o*-nitrophenol (ONP) in positive qualitative tests, but for quantitative results Trypticase Soy Agar (BBL) supplemented with 0.1% glucose and 1.0% lactose was used. This medium resembles KIA but does not contain phenol red, which might interfere with photometric determination of ONP. After 24 hr at 35°C, growth on this medium was harvested as above, centrifuged, and resuspended to the desired concentration with sterile water.

Qualitative ONPG tests. Each tube (13 x 100 mm) received 1.0 ml of buffered (0.1 M potassium phosphate, pH 7.4) substrate (0.25 mg of ONPG, Calbiochem) containing 1.0 mg of sodium azide as bacteriostat, and 0.1 ml of inoculum from a KIA slant. Incubation was at 35°C, and readings were made visually. Most strains of *Citrobacter* were positive within 1 hr; tubes showing no color after 4 days of incubation were recorded as negative. Spot tests were also made to determine whether, in the routine processing of enteric specimens (particularly stools), it would be feasible to check single colonies for β-galactosidase. The test solution contained, per milliliter, 1.0 mg of ONPG and 2.0 mg of sodium azide in 0.1 M potassium phosphate, pH 7.4. For each spot test, one colony was emulsified in 2 drops (approximately 0.1 ml) of this solution in a depression of a ceramic plate. Incubation was at room temperature, and tests remaining colorless for 60 min were recorded as negative.

*See R. E. Hoyt and M. J. Pickett (1957). Am. J. Clin. Path. 27, 343–352.

L. Le Minor and F. Ben Hamida (1962). Annls Inst. Pasteur, Paris 102, 267–277.

G. H. Lowe (1962). J. med. Lab. Technol. 19, 21–25.

Galactosidase—*Brevibacterium*

R. Chatelain and L. Second (1966). Annls Inst. Pasteur, Paris 111, 630–644.

β-galactosidase: recherchée après 2 jours de culture selon la technique de Le Minor et Ben Hamida [13].

La température d'incubation des cultures est de 30°C.

[13] Le Minor (L.) et Ben Hamida (F.). Ann. Inst. Pasteur, 1962, 102, 267.

Galactosidase—*Providencia, Rettgerella*

C. Richard (1966). Annls Inst. Pasteur, Paris 110, 105–114.

Pour des raisons de commodité, les milieux ont été incubés à 37°C. Recherche de la β-galactosidase: teste à l'ONPG [16].

[16] Le Minor (L.) et Ben Hamida (F.). *Ann. Inst. Pasteur,* 1962, 102, 267.

Galactosidase—RM bacterium

A. J. Ross, R. R. Rucker and W. H. Ewing (1966). Can. J. Microbiol. 12, 763–770.

Tests for beta-D-galactosidase activity were made by means of O-nitrophenyl-beta-D-galactopyranoside (ONPG) procedure as employed by Le Minor and Ben Hamida (1962), Lubin and Ewing (1964), and Buelow (1964).

P. Buelow (1964). Acta path. microbiol. scand. 60, 376–402.
L. Le Minor and F. Ben Hamida (1962). Annls Inst. Pasteur, Paris 102, 276–277.
A. H. Lubin and W. H. Ewing (1964). Publ. Hlth Lab. 22, 83–101.

Galactosidase—*Mycobacterium*

A. Tacquet, F. Tison, B. Polspoel, P. Roos and B. Devulder (1966). Annls Inst. Pasteur, Paris 111, 86–89.

La β-D-galactosidase scinde le lactose en ses deux parties constitutives. Cette activité ne se manifeste, selon Lapage et Jayaraman [5], que si le lactose pénètre dans la cellule bactérienne. Cette pénétration est fonction d'un facteur, dit spécifique, présentant les caractères d'une enzyme, nommée "β-galactosidase-perméase."

Notre étude a porte sur 181 souches mycobactériennes de différentes espèces, dont un nombre très restreint fermentait le lactose. L'absence de fermentation du lactose est due, non à un manque d'activité β-D-galactosidasique, mais essentiellement à une imperméabilité de la membrane pour cette substance.

L'application du test ONPG permet la mise en évidence d'une β-D-galactosidase dans les cellules qui ne possèdent pas de perméase.

Cette méthode [1, 6, 8, 11] est fondée sur la détermination colorimétrique de l'O-nitrophénol libéré à partir du substrat, l'O-nitrophényl-β-D-galactopyranoside (ONPG), après mise en contact de ce dernier avec une suspension de bactéries traitées par le benzène.

Matériel et techniques

1° Réactif [2]. – a) *Solution tampon.* Dissoudre 6.9 g de phosphate monosodique ($PO_4H_2NaH_2O$) dans 45 ml d'eau distillée, ajouter la lessive de soude pour ajuster à pH 7 (environ 3 ml), compléter à 50 ml d'eau distillée.

b) *Solution d'ONPG M/75.* Dissoudre 80 mg d'ONPG dans 75 ml d'eau distillée à 50°C et laisser refroidir. Ajouter 5 ml de la solution tamponnée. Le mélange incolore doit être conservé à +4°C. Il est conseillé de réchauffer le réactif à 37°C avant l'usage.

2° Réaction. – Prélever sur une culture en phase logarithmique de croissance sur milieu de Löwenstein-Jensen, une grosse anse de colonies (15 mg) et mettre en suspension dans 0.25 ml d'eau distillée. Agiter avec 0.05 ml de toluène, porter dix minutes à 37°C, ajouter 0.25 ml de la solution d'ONPG et maintenir à 37°C. La lecture de la réaction se fait après dix huit heures d'incubation à 37°C.

3° Lecture. – L'apparition d'une coloration jaune traduit la présence d'une activité β-galactosidasique, plus ou moins forte selon la teinte.

Après les premiers essais techniques, notre étude a porté sur 181 souches de Mycobactéries appartenant à différentes espèces.

[1] Bulow (P.). *Acta path. microb. scand.*, 1964, 60, 376–386 et 387–402.
[2] Buttiaux (R.), Beerens (H.) et Tacquet (A.). *Manuel de techniques bactériologiques.* Flammarion, édit., Paris, 1963.
[5] Lapage (S. P.) and Jayaraman (M. S.). *J. clin. Path.*, 1964, 17, 117–121.
[6] Leclerc (H.). Thèse Doctorat d'Etat en Pharmacie, Lille, 1961.
[8] Le Minor (L.) et Ben Hamida (F.). *Ann. Inst. Pasteur,* 1962, 102, 267–277.
[11] Lubin (A. H.) and Ewing (W. H.). *Tuberc. Health Lab.*, 1964, 22, 83–101.

Galactosidase—anaerobes

A.-R. Prévot (1966). Techniques pour le diagnostic des Bactéries Anaérobes. Éditions de la Tourelle, St. Mandé.

Recherche de la β-galactosidase.

L'importance du métabolisme du lactose par les anaérobies a incité plusieurs auteurs à rechercher la β-galactosidase chez eux.

Réactifs nécessaires: 1) Solution tampon phosphate 1 M pH 7; 2) Solution ONPG (orthonitrophénylgalactoside) M/75.

On ajoute 5 cm^3 de 1 à 15 cm^3 de 2. Le mélange reste incolore à +4°. Avant de l'utiliser on l'amène à 37°.

Réaction: On cultive l'anaérobie sur milieu au lactose à 1% pour induire l'enzyme. On suspend ses corps microbiens (obtenus par centrifugation si le milieu est liquide, par raclage s'il est solide) dans 0.25 cm^3 d'eau physiologique. On ajoute une goutte de toluène pour rendre ses parois perméables. On place au bain-marie à 37° et après quelques minutes on ajoute la solution d'ONPG tamponnée.

La présence d'une β-galactosidase fait apparaître une coloration jaune-franc stable en moins de 2 heures.

GAS PRODUCTION

Gas production—various

L. A. Priem, W. H. Peterson, and E. B. Fred, (1927). J. agric. Res. 34, 79–95.

The cultures were inoculated into 10 ml of 1 per cent glucose-yeast water, sealed with sterile melted vaseline, and allowed to ferment at 28°C. They were observed from day to day for gas formation. If the vaseline plug was pushed up they were recorded as gas-positive. If the vaseline remained at the surface of the liquid and there was no apparent leakage of gas, they were called negative.

Gas production—*Lactobacillus*

H. R. Curran, L. A. Rogers and E. O. Whittier (1933). J. Bact. 25, 595–621.

Carbon dioxide was measured quantitatively by the following method: Twenty ml of unfiltered Swiss cheese whey were sterilized in Eldredge tubes. These were immediately cooled to 37°C and inoculated with 0.1 ml of a 48-hr milk culture. Ten milliliters of $Ba(OH)_2$ were then measured into each tube to absorb the CO_2, after which the cultures were incubated at 37°C. Absorption of atmospheric CO_2 was prevented by sealing the upright tubes after inoculation. At the end of five days the excess of $Ba(OH)_2$ was titrated with 0.1 N oxalic acid and the CO_2 calculated from the difference. The quantity of CO_2 was expressed in milliliters of N/10 $Ba(OH)_2$ neutralized.

Gas production—lactic acid bacteria

T. Gibson and Y. Abd-el-Malek (1945). J. Dairy Res. 14, 35–44.

A cultural method of detecting the formation of carbon dioxide from glucose.

The foregoing discussion shows that CO_2 is formed actively by the heterofermentative lactic acid bacteria in glucose (5%) milk or glucose (5%) nutrient gelatin (containing 10–12% gelatin) each supplemented with a preparation from yeast and the juice of tomato or cabbage. Either medium may be used to differentiate the heterofermentative lactic acid bacteria and *Bacillus licheniformis* from organisms with which they might be confused. For streptococci the plant juice is not essential, and for the sporeformer the yeast extract may also be omitted. If nutrient gelatin is prepared solely for the purpose of this test, it may with advantage be made with 20% gelatin, filtration being omitted. In order to detect the gas visually it is only necessary to ensure that it is adequately retained within the culture.

Much use was also made of straight-sided, V-shaped fermentation tubes about 1.2 cm in diameter with 10-cm closed limb and 20 cm open limb at an angle of about 30° to each other. This tube facilitates work in which constituents of the media must be mixed after sterilization, for the medium is easily introduced aseptically so as to fill the closed arm. Gases are also well retained, especially if the exposed surface of the medium is covered with a layer of sterile agar.

The V-tube already described may be used, but the following simple technique, employing the ordinary straight tube available in any laboratory, is equally reliable.

If milk is to be used, the glucose, yeast extract and plant juice are added to a mixture of four parts separated milk and one part nutrient agar, and the medium is distributed in tubes and sterilized by intermittent steaming. When gelatin is employed in place of milk there appears to be no advantage in adding agar even if a temperature above the melting point of gelatin is to be used for incubation.

To make a test the medium is liquefied, cooled to 45°C, inoculated and chilled; after the gel has formed, melted and cooled, nutrient agar is gently poured into the tube so that it forms a layer 2–3 cm deep above the gelatin or soft milk agar. The culture is then incubated at approximately the optimum temperature for the organism and observed for gas formation.

Where the method is used only occasionally it is convenient to make the necessary additions to tubes of litmus milk or gelatin by pipette. A 50% solution of glucose and the other special ingredients can be separately sterilized in the autoclave. Screw-capped bottles are suitable containers as they eliminate evaporation. Prepared in this way the medium is less acid than when sterilized complete, and it is slightly more favourable to gas formation, but the difference is unimportant. On no account should the completed medium be sterilized in the autoclave.

The amount of gas formed is proportional to the depth of the medium, and it would therefore be inadvisable to use shallow layers. In this work the tubes have been filled to a depth of 5–6 cm, exclusive of the agar seal.

An essential condition of the test is rapid growth of the organism, presumably in order to avoid a gradual dissipation of the gas. Consequently the temperature of incubation should not be greatly different from the optimum for the species. A relatively large inoculum results in earlier and more vigorous gas production than does a light seeding, and in the case of organisms which grow feebly on artificial media (e.g., certain lactobacilli) it is an advantage to inoculate heavily by pipette from a young active culture.

Gas production—*Streptococcus*

E. J. Hehre and J. M. Neill (1946). J. exp. Med. 83, 147–162.

Gas from sucrose: The basic medium was in test 4;* 5 per cent sucrose was added. After inoculation, the broth was sealed with a layer of sterile vaseline, incubated at 37°C, and observed at 2 day intervals for 10 days. Cultures were recorded as negative if no bubbles or other signs of gas developed under the seal; leuconostoc bacteria (incubated at 30°C and with glucose or fructose as well as with sucrose) invariably showed obvious signs of gas formation by this method, and the test seemed worthwhile to include in order to compare the present streptococci with the dextran-forming leuconostoc.

*See **Fermentation**—*Streptococcus* (p. 273).

Gas production—*Streptococcus*

Y. Abd-el-Malek and T. Gibson (1948). J. Dairy Res. 15, 233–248.

The formation of CO_2 from glucose: A test for this property permits the recognition of heterofermentative streptococci. The technique was described previously.*

*T. Gibson and Y. Abd-el-Malek (1945). J. Dairy Res., 14, 35.

Gas production—*Streptococcus*

P. F. Swartling (1951). J. Dairy Res. 18, 256–267.

Production of CO_2: The method of Gibson and Abd-el-Malek (1945) was followed, using the milk-yeast-tomato medium. Some strains were also tested in plain milk. In these cases CO_2-free nitrogen was passed through a 48-hour old milk culture, and the liberated carbon dioxide absorbed in an U-tube containing "Ascariet" (NaOH absorbed on asbestos) after having been dried in a similar tube containing "Dehydrite" (water-free magnesium perchlorate). The amount of CO_2 formed was estimated by weighing.

T. Gibson and Y. Abd-el-Malek (1945). J. Dairy Res. 14, 35.

Gas production

P. Hauduroy (1951). Techniques Bactériologiques. Masson et Cie, Éditeurs, Paris.

TECHNIQUE DE ELDREDGE

(Dépistage du CO_2). – Dans cette technique on se sert de tubes spéciaux ainsi construits. Un tube horizontal de 8–10 cm de long, fermé à ses deux extrémités, possède en son milieu une tubulure verticale de 8–10 cm de hauteur. Deux systèmes semblables sont placés parallèlement et reliés entre eux par un tube horizontal qui rejoint les deux tubulures verticales à peu près au niveau de leur milieu. L'ensemble de l'appareil repose en définitive sur les tubes horizontaux comme sur deux pieds. Les tubulures verticales sont ouvertes à leur extrémité supérieure et pourront être fermées à l'aide de bouchons de coton ou de caoutchouc.

Ces tubes ne se trouvent pas dans le commerce, mais on peut les trouver sur demande à la Will Corp., Rochester, N. Y.; à la Macalaster Bicknell Co. Washington, W. C. et au Moore Sts., Cambridge, Mass.

Pour utiliser les tubes de Eldredge, placer 20 cm^3 du milieu dans l'un des bras horizontaux et stériliser. Ensemencer et mettre alors dans l'-autre bras une quantité déterminée (dépendant de la quantité de CO_2 cherchée, habituellement 15 à 25 cm^3) d'une solution fraîchment préparée à N/10 d'hydroxyde de baryum (on peut utiliser NaOH ou KOH, mais l'insolubilité du $BaCO_3$ formé fait que $Ba(OH)_2$ donne une indication plus satisfaisante et mieux visible de la production de CO_2).

Immédiatement après avoir placé l'alcali, pousser le coton dans le tube et sceller (il semble que l'on puisse fermer le tube avec un bouchon de caoutchouc sans être obligé de le sceller). Après 2 semaines d'incubation, titrer l'hydroxyde de baryum avec N/10 HCl ou, ce qui est préférable, avec H_2SO_4 en utilisant la phénolphtaléine comme indicateur.

Calculer la quantité de CO_2 d'après l'équation ; cm^3 de $Ba(OH)_2$ $\times$ normalité de $Ba(OH)_2$ = grammes de CO_2 $\times$ 0.0022.

Par exemple, 1 cm^3 de N/10 $Ba(OH)_2$ transformé en carbonates représente 0.0022 g CO_2. *(S. A. B.)*

Gas production—*Streptococcus*

W. H. Oliver (1952). J. gen. Microbiol. 7, 329–334.

CO_2 produced during the growth of culture (of streptococci) was estimated by absorption in 0.1 N-baryta which was subsequently back titrated with 0.1 N oxalic acid, the amount of CO_2 being expressed in ml at N. T. P.

Gas production—*Lactobacillus*

M. Rogosa, R. F. Wiseman, J. A. Mitchell, M. N. Disraely and A. J. Beaman (1953). J. Bact. 65, 681–699.

Preliminary test for gas production: Immediately after the primary isolation and purification of the isolate, heavy inoculations from actively growing cultures were made into 10 ml of medium 1: Tryptone (Difco), 20 g; tryptose (Difco), 5 g; yeast extract (Difco), 5 g; tomato juice, 200 ml; water soluble liver extract (Wilson Laboratory), 1 g; glucose, 3 g; lactose, 2 g; Tween 80, 50 mg; distilled water to one liter; pH 6.5, supplemented with 2 per cent agar. The tubed medium had been melted previously and held at 45°C. After solidification, the medium was layered with an additional depth of approximately one-quarter inch of non-nutrient 2 per cent agar, over which a thin layer of heavy mineral oil was added. Gas production was noted by cracks or disturbance of the agar. Very often sufficient gas (CO_2) was produced to force the plug of agar out of the tube.

Gas production—*Lactobacillus*

G. H. G. Davis (1955). J. gen. Microbiol. 13, 481–493.

Gas production from glucose: Medium: peptone, 1% (w/v); yeast extract, 0.5% (w/v); glucose, 3% (w/v); Tween 80, 0.1% (v/v); Salts A,* 0.5% (v/v); Salts B*, 0.5% (v/v); pH 7.2, with 0.2% (v/v) of 1.5% aqueous solution of bromocresol-purple added. It was dispensed into 5 X ½ in. tubes containing inverted Durham tubes and steamed for 20 min on three consecutive days to sterilize. The casein digest fermentation test medium described below was also found to be effective as the basal medium for this test. After inoculation with 3 to 4 drops, the tests were sealed by a layer of *c.* 1 ml of sterile liquid paraffin. They were incubated for 6 days and observed every 48 hours.

*For Salts A and Salts B see A. C. Hayward (1957). J. gen. Microbiol. 16, 9–15.

Gas production—*Lactobacillus*

J. C. de Man (1957). Antonie van Leeuwenhook 23, 87–96.

Gas production from glucose or citrate was determined in the following medium: Bactotryptone, 1%; Oxo meat extract, 1%; Difco yeast extract, 0.5%; tomato juice, 4%; Tween 80, 0.1%; K_2HPO_4, 0.2%; and glucose, 2% or citric acid, 3%, in distilled water; pH 6.8. The vacuum tube technique described by van Beynum and Pette (1942) was used. The tubes were incubated for 7 days at 30°C before being opened under water.

J. van Beynum and J. W. Pette (1942). Versl. Landbouwk. Onderz. Ned. 48, 765.

Gas production—lactic acid bacteria

A. C. Hayward (1957). J. gen. Microbiol. 16, 9–15.

Tomato juice broth + 0.5% or 2.5% (w/v) glucose and sealed with paraffin oil resulted in consistent gas production in Durham tubes. Some strains produced more gas with maltose. Tomato juice (TJ) medium of Davis, Bisset, and Hale (1955).

G. H. G. Davis, K. A. Bisset and E. M. F. Hale (1955). J. gen. Microbiol. 13, 68.

Gas production—*Lactobacillus*

J. Naylor and M. E. Sharpe (1958). J. Dairy Res. 25, 92–103.

Production of gas from glucose: The method of Gibson and Abd-el-Malek was used, the inoculum being the deposit of a centrifuged 10 ml TDB culture. This large inoculum was found more satisfactory for *L. brevis* strains which otherwise often gave false negative results because of their very slow growth and consequent slow gas production.

T. Gibson and Y. Abd-el-Malek (1945). J. Dairy Res. 14, 35.

Gas production—*Pediococcus*

J. C. Dacre (1958). J. Dairy Res. 25, 409–413.

Gas produced was tested for by the methods of Gibson and Abd-el-Malek (1945) and Hayward (1957).

T. Gibson and Y. Abd-el-Malek (1945). J. Dairy Res. 14, 35.
A. C. Hayward (1957). J. gen. Microbiol. 16, 9.

Gas production—*Fusobacterium*

A. C. Baird-Parker (1960). J. gen. Microbiol. 22, 458–469.

Gas production: The formation of soluble and insoluble gases in the fermentation of glucose was investigated by the methods of Gibson and Abd-el-Malek (1946) and of Hayward (1957). F. M. medium* containing 5.0% (w/v) glucose was used.

*See A. C. Baird-Parker (1957). Nature, Lond. 180, 1056.
T. Gibson and Y. Abd-el-Malek (1946). J. Dairy Res. 14, 35.
A. C. Hayward (1957). J. gen. Microbiol. 16, 9.

Gas production—*Streptococcus* and *Lactobacillus*

C. W. Langston and C. Bouma (1960). Appl. Microbiol. 8, 212–222.

Production of gas: To detect gas, medium A was used with the following modifications: Bromocresol purple was deleted, glucose increased to 20 g, and 20 g of agar added. The medium was dispensed into test tubes in 7-ml quantities. An agar layer (2 per cent) and oil seal were used to prevent loss of gas. Heavy inoculations were made and the tubes were incubated for 2 weeks. It was not uncommon for some cultures to push the agar and oil layers to the top of the tubes.

Medium A:

Trypticase, 10 g; phytone, 5 g; yeast extract, 5 g; glucose, 5 g; sodium chloride, 5 g; potassium phosphate (dibasic), 1.5 g; Tween 80 (sorbitan monooleate), 0.5 ml; tomato juice, 200 ml; bromcresol purple, 0.016 g; and distilled water to make 1 L pH 7. Autoclave at 15 lb for 15 min.

Gas production—*Pediococcus*

H. L. Günther and H. R. White (1961). J. gen. Microbiol. 26, 185–197.

The method of Gibson and Abd-el-Malek (1945) was used; cultures were examined daily for gas production during a 2-week incubation period.

Maintenance of stock cultures and methods of cultivation. For maintenance of stock cultures, preparation of inocula and in all experimental work, "Oxoid" tomato juice (TJ) broth or tomato juice (TJ) agar, adjusted to pH 6.6, was used unless otherwise stated. The following were exceptions to this rule: for strain Tc. 1, sodium chloride (5%, w/v) was added to the medium; and for the aerococci, glucose Lemco broth (Shattock and Hirsch, 1947) or glucose yeast extract (GY) (containing, as %, w/v, peptone, 1.0; Yeastrel, 0.3; glucose, 1.0; NaCl, 0.25; agar, 1.0; at pH 7.4) was used.

Cultures were incubated aerobically at 30°C with specified exceptions.

T. Gibson and Y. Abd-el-Malek (1945). J. Dairy Res. 14, 35.

P. M. F. Shattock and A. Hirsch (1947). J. Path. Bact. 59, 495.

Gas production—*Vibrio*

G. H. G. Davis and R. W. A. Park (1962). J. gen. Microbiol. 27, 101–119.

Routine peptone water fermentation tests containing Durham tubes were observed over 14 days.

Gas production—*Staphylococcus epidermidis*

D. Jones, R. H. Deibel, and C. F. Niven (1963). J. Bact. 85, 62–67.

Carbon dioxide production from glucose was determined qualitatively by the method of Williams and Campbell (1951).

O. B. Williams and L. L. Campbell (1951). Fd Technol., Champaign 5, 306.

Gas production—rumen bacteria

R. S. Fulghum and W. E. C. Moore (1963). J. Bact. 85, 808–815.

The production of gas was determined by observations of splits or gas bubbles in SRP-A stab cultures or in other agar media used in this study. For SRP-A medium see **Protein digestion**—*Butyrivibrio, Borrelia, Bacteroides, Selenomonas Succinivibrio* (p. 759).

Gas production—*Bacteroides oralis*

W. J. Loesche, S. S. Socransky and R. J. Gibbons (1964). J. Bact. 88, 1329–1337.

Gas production was determined in the basal medium supplemented with 0.5% glucose by the appearance of bubbles in Durham tubes.

Thioglycolate Medium without Dextrose (BBL), supplemented with 0.2% yeast extract (Difco) and hemin, was used as a basal medium. All cultures were incubated at 35°C in Brewer jars containing an atmosphere of 95% H_2 and 5% CO_2.

Gas production—*Lactobacillus casei* var. *rhamnosus*

W. Sims (1964). J. Path. Bact. 87, 99–105.

Gas production from glucose was tested for in 5/8in. (16 mm) diameter test tubes containing 15 ml of MRS broth* under a paraffin wax seal.

*J. C. de Man, M. Rogosa and M. E. Sharpe (1960). J. appl. Bact. 23, 130.

Gas production—*Lactobacillus*

F. Gasser (1964). Annls Inst. Pasteur, Paris 106, 778–794.

Production de CO_2 à partir du glucose, du citrate du malate. La méthode utilisée est adaptée de celle de Gibson et Abd-el-Malek [19].

Le même milieu de culture sert pour ces trois tests. Il a la composition suivante:

Néopeptone Difco	12 g
Extrait de levure	6 g
Chlorure de sodium	2 g
Tween 80	1 ml
Macération de viande	500 ml
Sulfate de manganèse	50 mg
Sulfate de magnésium	200 mg
Gélatine	200 g
Eau q. s. p.	900 ml

Dissoudre en chauffant à 70-80°C. Ajuster le pH à 6.5 (soude N). Clarifier avec 75 ml de sérum de cheval. Précipiter à l'autoclave à 120°C pendant trente minutes. Le pH est alors à 5.5-5.7. Ajouter soit le glucose, soit la solution de citrate, soit la solution de malate à la concentration finale de 2 p. 100 et ajuster à 1000 ml avec de l'eau. Pour les tests de fermentation du citrate et du malate on ajoute au milieu 0.5 p. 100 de glucose dont la présence nécessaire sera discutée plus loin. Répartir 10 ml par tube de 17 × 170. Stériliser trente minutes à 115°C.

Après inoculation, on dépose à la surface du milieu une couche de vaseline stérile fondue de 1 cm d'épaisseur environ. La vaseline redevient solide à la température d'incubation et constitue un bouchon qui est repoussé vers le haut du tube lorsqu'il se dégage du CO_2.

Les solutions de citrate et de malate sont des solutions à 20 p. 100 d'acide citrique ou malique ajustées à pH 5.5 avec de la potasse.

[19] Gibson (T.) and Abd-el-Malek (Y.). *J. Dairy Res.* 1945, 14, 35.

Gas production (from glucose)—*Pediococcus*

E. Coster and H. R. White (1964). J. gen. Microbiol. 37, 15–31.

The authors used the methods of H. L. Günther and H. R. White (J. gen. Microbiol. 26, 185, 1961). Their Tomato Juice (TJ) broth or agar with the addition of Tween 80 (0.1%, v/v), and the pH adjusted to 6.6, was used in maintenance and preparation of inocula with certain specified exceptions.

Transfers for preparing inocula were made every 24 hours except for *Pediococcus halophilus* (48 hr) and some brewing strains (fortnightly).

All cultures were incubated aerobically at 30°C except for "brewing strains," which were incubated in an atmosphere of 95% (v/v) hydrogen and 5% (v/v) carbon dioxide at 22°C.

The basal fermentation media used for gas production was also modified by the addition of Tween 80 (0.1%, v/v).

Gas production—*Bacillus*

G. R. F. Hilson (1965). J. gen. Microbiol. 39, 407–421.

The formation of CO_2 gas from glucose under semi-anaerobic conditions was also tested for. The test was done as described by Gibson and Abd-el-Malek (1945) with the following modifications. The medium consisted of 5% (w/v) peptone (Oxoid) with 5% (w/v) glucose in distilled water, adjusted to pH 7.3, and distributed into U-tubes, each having one closed limb which was completely filled with medium. The surface of the medium in the open limb was sealed after inoculation with a layer of paraffin wax 1 cm deep. After incubation up to 5 days, the amount of bacterial growth and the size of any gas bubbles in the closed limb were noted; shrinkage of the latter after the addition of alkali indicated their CO_2 content. *Bacillus subtilis* and *B. pumilus* strains were included in this part of the investigation.

T. Gibson and Y. Abd-el-Malek (1945). J. Dairy Res. 14, 35.

Gas production—*Streptococcus*

R. Whittenbury (1965). J. gen. Microbiol. 38, 279–287.

To examine for gas production from citrate and malate, each tube was sharply tapped to initiate effervescence. In cultures above pH 6.5 bromcresol purple was frequently bleached except when the culture was in contact with the atmosphere (the top 2 mm of the agar seal) where it remained oxidized, allowing an estimate of pH change to be made in these circumstances. For citrate and malate see p. 000 and p. 000.

Gas from threonine—anaerobes

A. -R. Prévot (1966). Techniques pour le diagnostic des Bactéries Anaérobes. Éditions de la Tourelle, St. Mandé.

Milieu à la thréonine.

Protéose-peptone	2 g
Extrait de levure	0.1 g
Chlorhydrate de cystéine	0.06 g
Bacto-agar Difco	0.2 g
Eau distillée	100 cm^3

Ajuster à pH 7.2 - 7.4. Répartir. Steriliser. Ajouter 1 cm^3 d'une solution stérile de DL-thréonine à 2.5 %. Paraffiner. Le gaz formé aux dépens de la thréonine soulève la paraffine.

GAS VACUOLES

Gas vacuoles—*Halobacterium halobium*

A. L. Houwink (1956). J. gen. Microbiol. 15, 146–150.

Houwink discusses appearance in the electron microscope preparations.

The main argument was, and still is, that these "bodies" can be made to disappear by subjecting a suspension of pink bacteria to a pressure of a few atmospheres. The result is immediately visible: the suspension turns red. Microscopic examination shows that no "bodies" are left, so the opacity as well as the different colour is clearly due to the presence of these "bodies." Petter (1931 and 1932) named them "gas vacuoles," as she considered them comparable to the gas vacuoles which Klebahn had described in *Cyanophyceae.*

H. Klebahn (1919). Mitt. Inst. allg. Bot., Hamb. 4, 11.
H. F. M. Petter (1931). Proc. Acad. Sci., Amst. 34, 1417.
H. F. M. Petter (1932). Over rode en andere bacteriën van gezouten vis. Thesis, Utrecht.

GELATIN HYDROLYSIS

Gelatin hydrolysis—coliforms

L. A. Rogers, W. M. Clark, H. A. Lubs (1918). J. Bact. 3, 231–252.

The test for gelatin liquefaction was made by spreading about 0.5 ml of a twenty-four hour sugar free broth culture on gelatin held in a small test tube. This was incubated twenty days at 20°C. No gelatin liquefiers were found in this collection. This probably means that in these samples at least, liquefiers occurred in such small numbers that they were not isolated.

Gelatin hydrolysis—general

W. C. Frazier (1926). A method for the detection of changes in gelatin due to bacteria. J. infect. Dis. 39, 302–309.

The method to be described is that of detection of a change in composition of the gelatin due to bacteria rather than the detection of liquefaction, although thus far all liquefying organisms have responded to a new test.

Gelatin-agar medium for test: In 100 ml of distilled water are dissolved NaCl, 5.0 g; KH_2PO_4, 0.5 g; K_2HPO_4, 1.5 g. Four g of bacto-gelatin are dissolved in 400 ml distilled water, and to this solution are added dextrose, 0.05 g; bacto-peptone, 0.1 g; beef infusion, 5.0 ml.

The two solutions are poured together, heated in a steamer, and then mixed with 500 ml of 3% washed agar. The pH is adjusted to about 7.0. The medium is placed in tubes or flasks and sterilized in autoclave.

Plates of gelatin-agar medium are poured and allowed to harden. Duplicate plates are inoculated on the surface of the agar at the center so as to form a giant colony of the organism, and are incubated for 2–3 days at the optimum temperature of the organism. In general, incubation for 48 hours at 30°C was found to be satisfactory for all the organisms studied.

After incubation one plate is flooded with a 1% solution of tannic acid, while the duplicate plate is flooded with an acid solution of bichloride of mercury of the following composition: –

$HgCl_2$, 15 g; HCl (conc.) 20 ml and H_2O, 100 ml.

If the gelatin has been changed, a clear zone will appear about the giant colony on the plate flooded with acid bichloride of mercury solution, surrounded by the cloudy precipitate of unchanged gelatin. The reaction is slow, and 15–30 minutes should be allowed for its completion.

The plate flooded with tannic acid-solution will present an appearance that will vary with the amount and degree of action of the bacteria on the gelatin. An organism that acts on gelatin with little or no increase in amino-nitrogen gives a white precipitate about the colony, heavier than the precipitate of gelatin throughout the rest of the plate. With some groups of organisms this precipitate is very heavy and white, with a distinct edge. If there has been considerable decomposition of the gelatin, there will be a clear zone about the colony, surrounded by a distinct white ring. This reaction takes place rapidly.

Gelatin hydrolysis—*Xanthomonas beticola*

N. A. Brown (1928). J. agric. Res. 37, 155–168.

In beef-gelatin plates with a pH of 7.1, kept at a temperature of 20° to 22°C for 14 days.

Gelatin hydrolysis—various, in milk

W. C. Frazier and P. Rupp (1928). J. Bact. 16, 187–196.

The gelatin medium contained the following: Bacto-gelatin, 10 g; $NaH_2PO_4 \cdot 2H_2O$, 1.2 g; KCl, 5 g; peptone, 0.1 g; beef infusion, 10 ml; glucose, 0.1 g; distilled water to make 1000 ml. In one experiment the glucose, peptone and beef infusion were omitted. The gelatin agar plate method used is described in another paper (Frazier, 1926).

W. C. Frazier (1926). J. infect. Dis, 39, 302–309.

Gelatin hydrolysis—quantitative method—*Proteus*

A. T. Merrill and W. M. Clark (1928). J. Bact. 15, 267–296.

The authors describe a method of quantitative assay of gelatinase in *Proteus* cultures.

Gelatin hydrolysis—mustiness in eggs

M. Levine and D. Q. Anderson (1932). J. Bact. 23, 337–347.

Gelatin liquefaction was determined by growing the organisms in 14 per cent gelatin at 20°C and observing for five days.

Gelatin hydrolysis—*Bacteroides* of human feces

A. H. Eggerth and B. H. Gagnon (1933). J. Bact. 25, 389–413.

Gelatin medium was prepared by adding 10% gelatin and 0.15% glucose to phosphate infusion broth and adjusting to pH 7.8.

Gelatin hydrolysis—*Bacteroides*

A. H. Eggerth (1935). J. Bact. 30, 277–299.

Gelatin medium was prepared by adding 10 per cent gelatin and 0.15 per cent glucose to phosphate heart infusion broth and adjusting to pH 7.8. After sterilization 4 per cent sterile serum was added.

Gelatin hydrolysis—*Clostridium*

R. S. Spray (1936). J. Bact. 32, 135–155.

Iron gelatin: To make 1 liter, take –

128 g Difco nutrient gelatin
1 g dextrose
1 L distilled water

Dissolve in a double boiler to avoid scorching. Adjust the reaction to pH 7.3 to 7.4, and add water to make 1 L. Tube about 8 cm deep in 200 × 15 mm or 200 × 13 mm tubes. Add to each tube one strip of iron (as in Iron-Milk) and autoclave for 15 to 20 minutes at 15 pounds pressure.

Iron-gelatin liquefaction. This test is performed in Difco Nutrient Gelatin to which is added 0.1 per cent glucose and a strip of iron. Powdered iron reduced by hydrogen may be used, as advocated for milk by Hastings and McCoy (1932), although we prefer the iron strip, as suggested by H. G. Dunham of Difco Laboratories. The Standard Methods gelatin formula is to be avoided. In this simple gelatin solution many anaerobes fail to liquefy in any reasonable time, while in our formula the same organisms commonly liquefy completely in 24 to 48 hours. This is particularly striking in the case of *Cl. septicum* and *Cl. luciliae.*

All cultures are incubated at 37°C, and are tested daily in ice water until liquefied, or if negative, at least for 30 days. No time limit can be set for the negative cultures, but we have not observed any species showing liquefaction delayed beyond 10 days. If incubated beyond 10 days, the plugs should be paraffined.

The iron-gelatin shows a distinctive orange to wine-red color when inoculated with *Cl. histolyticum.* The orange color appears at about 24 hours deepens to red at about 48 hours, then slowly and variably fades in 3 to 5 days or more. The reaction appears specific for *Cl. histolyticum,* and establishes a valuable presumptive test for this organism, having the same significance as the reaction of *Cl. welchii* in iron-milk. The reaction is displayed in mixed as well as in pure cultures, but its particular value lies in its application to pure cultures.

Hastings and McCoy (1932). J. Bact. 23, 54.

Gelatin hydrolysis—*Xanthomonas*

S. S. Ivanoff, A. J. Riker and H. A. Dettwiler (1938). J. Bact. 35, 235–253.

Ten per cent gelatin in water at pH 7.0 was used as stabs and incubated for three weeks at 20°C.

Gelatin hydrolysis—*Lactobacillus, Bacteroides*

K. H. Lewis and L. F. Rettger (1940). J. Bact. 40, 287–307.

Examination over periods of 30 days (37°C) of gelatin cultures were made in an adapted glucose-cysteine medium* containing 12 per cent gelatin.

*J. E. Weiss and L. F. Rettger (1934). J. Bact. 28, 501–521.

Gelatin hydrolysis—*Streptobacillus moniliformis*

F. R. Heilman (1941). A study of *Asterococcus muris (Streptobacillus moniliformis).* II. Cultivation and biochemical activities. J. infect. Dis. 69, 45–51.

Nutrient gelatin containing 10% serum was incubated for five days at 37°C.

Gelatin hydrolysis—*Corynebacterium acne*

H. C. Douglas and S. E. Gunter (1946). J. Bact. 52, 15–23.

Incubated for 10 days at 37°C in peptone yeast extract phosphate glucose gelatin "deeps."

Gelatin hydrolysis—*Staphylococcus, Micrococcus*

Y. Abd-el-Malek and T. Gibson (1948). J. Dairy Res. 15, 249–260.

Stab cultures in nutrient gelatin were incubated at 22°C. In order to secure data which could be compared with reports by other workers, the tubes were closed with stoppers and, if liquefaction failed to occur, were incubated for several months. The usefulness of such long incubations, however, seems extremely doubtful.

Gelatin hydrolysis—*Proteus*

G. T. Cook (1948). J. Path. Bact. 60, 171–181.

Medium

Gelatin stab

Method

Inoculated with a straight wire dipped in peptone water culture incubated 24 hr at 37°C. The gelatin stabs were examined for liquefaction at 6 and 21 days after standing at bench temperature.

Gelatin hydrolysis—*Pseudomonas alboprecipitans*

A. G. Johnson, A. L. Robert, and L. Cash (1949). J. agric. Res. 78, 719–732.

On Bacto-nutrient gelatin plates at 25°C, there was definite liquefaction by all cultures after 5 days. Utilization of gelatin also was demonstrated by a modified method of Frazier (1926) in which 4 per cent gelatin was used in beef-peptone agar plates. In test tubes liquefaction was too slow to be detected.

W. C. Frazier (1926). J. infect. Dis. 39, 302–309.

Gelatin hydrolysis—*Diplococcus*

G. C. Langford, Jr., J. E. Faber, Jr., and M. J. Pelczar, Jr. (1950). J. Bact. 59, 349–356.

Tubes of nutrient gelatin were inoculated with 0.5 ml amounts of cystine trypticase agar stock cultures and incubated at 35°C in an anaerobic jar. They were observed 7 days at 24 hour intervals for evidence of liquefaction. Plates of modified Frazier's gelatin medium were also streaked and incubated at 37°C for 1 week in anaerobic jars. Tubes of gelatin were more valuable than plates, as these organisms grow more profusely in liquid medium.

Gelatin hydrolysis—*Streptococcus*

P. F. Swartling (1951). J. Dairy Res. 18, 256–267.

Nutrient gelatin was inoculated with a loopful of an 18 hr old broth culture, incubated at 30°C for 48 hr and then observed for liquefaction after cooling for 6 hr.

Gelatin hydrolysis—*Staphylococcus*

C. Shaw, J. M. Stitt and S. T. Cowan (1951). J. gen. Microbiol. 5, 1010–1023.

Stab inoculations were made in 20% nutrient gelatin; cultures were incubated for 30 days at 22°C.

Gelatin hydrolysis—general

P. Hauduroy (1951). Techniques Bactériologiques. Masson et Cie, Éditeurs, Paris.

HYDROLYSE DE LA GELATINE

Les cultures sont ensemencées en stries, sur des plaques de gélose nutritive contenant 0.4 p. 100 de gélatine. On place à l'étuve à 28°C, pendant 2 à 14 jours (suivant la rapidité de la croissance). Les plaques sont alors recouvertes avec 8 à 10 ml de la solution suivante :

Eau distillée	100 cm^3
HCl concentré	20 ml
$HgCl_2$	15 g

Cette solution forme un précipité opaque blanc si la gélatine n'est pas altérée, et laisse des zones claires là où la gélatine est décomposée (modification du milieu de Frazier : *Journal Infect. Dis.*, 1926, 39, 302).

(A. T. C. C.)

Gelatin hydrolysis

P. H. Clarke and S. T. Cowan (1952). J. gen. Microbiol. 6, 187–197.

The suspension is prepared from a 24 hr culture grown on any suitable medium. For each test 0.04 ml suspension is pipetted into each of two Durham tubes containing 0.4 ml gelatin-agar base. The tubes are placed in the air incubator at 37°C and one is tested at 4, and the other at 24 hr. Acid mercuric chloride ($HgCl_2$, 15 g; distilled water, 1000 ml; conc. HCl, 20 ml) is added, and the tubes read after 30 min at room temperature against a blank gelatin-agar tube treated in the same way.

Columns made up of 0.4% gelatin and 0.5% N. Z. agar were used.

Gelatin hydrolysis—*Bacillus*

N. R. Smith, R. E. Gordon and F. E. Clark (1952). U. S. D. A. Agr. Monograph No. 16.

Plates of nutrient agar with 0.4 per cent gelatin were streaked once across and incubated at a suitable temperature. After 3 to 5 days the plates were flooded with 8 to 10 ml of the following solution: $HgCl_2$, 15 g; concentrated HCl, 20 ml; distilled water, 100 ml. The extent of hydrolysis of the gelatin was indicated by a clear zone underneath and around the growth, in contrast to the white opaque precipitate of the unchanged gelatin. As in the case of the starch plates, the spreading of some cultures could be checked by storing the plates for several days before inoculation.

Gelatin hydrolysis—setting time method

D. G. Evans and A. C. Wardlaw (1952). J. gen. Microbiol. 7, 397–408.

Dilutions of culture filtrate differing by 10–20% were made in 1.7 × 9 cm tubes and to 1 ml of each dilution 1 ml of 10% (w/v) gelatin solution was added. The latter was prepared by dissolving flake gelatin in saline, adjusting to pH 7.8, and adding phenol to 0.5% (w/v) as preservative. A control tube of 2 ml of 1.5% gelatin in saline was included in each titration.

Incubate at 37°C for 1 hour with periodic shaking and then place in an ice bath.

The time taken for the mixtures to set to the point where they failed to flow when the tubes were tilted horizontally was determined. A titer was obtained from that dilution of the culture filtrate giving the same setting time as the control (5–9 minutes).

Gelatin hydrolysis—*Bacillus*

P. de Lajudie (1952). Annls Inst. Pasteur, Paris 82, 380–382.

Il en a été de même de la recherche de l'hydrolyse de la gélatine sur milieu de Hastings modifié; nous avons dû revenir à la technique classique de culture en bouillon gélatine en tubes de 18 mm et incubation à 37°C; l'hydrolyse est facilement appréciée par la persistance de la fluidité du milieu après séjour de quinze minutes à la glacière, un tube témoin se trouvant solidifié au bout de ce laps de temps.

Gelatin hydrolysis—*Actinomyces, Corynebacterium*

H. Beerens (1953). Annls Inst. Pasteur, Paris 84, 1026–1032.

Action sur la gélatine. – (Milieu V. F. glucosé à 2 p. 1 000, gélatiné à 15 p. 1 000.) En tube scellé sous vide: liquéfaction en dix jours.

Gelatin hydrolysis—*Mycobacterium*

R. E. Gordon and M. M. Smith (1953). J. Bact. 66, 41–48.

The cultures were streaked once across duplicate plates of nutrient agar plus 0.4 per cent gelatin and incubated at 28°C. One plate was examined for hydrolysis of the gelatin at 5 days, and the second at 10 days, by flooding with 8 to 10 ml of the following solution: $HgCl_2$, 15 g; concentrated HCl, 20 ml; distilled water, 100 ml (Frazier, 1926). A clear zone underneath and around the growth provided a measure of the hydrolysis, while the unchanged gelatin appeared as a white, opaque precipitate.

W. C. Frazier (1926). J. infect. Dis. 39, 302–309.

Gelatin hydrolysis

J. Kohn (1953). J. clin. Path. 6, 249.

Denatured gelatin does not melt, hence it can be sterilized and incubated at any desired temperature. In other respects it behaves like natural gelatin.

The medium is prepared from Difco nutrient gelatin which is mixed with tap water in the proportion of 15 g to 100 ml of water. To the melted gelatin, finely powdered charcoal is added in a proportion of 3 to 5 g per 100 ml of the gelatin medium. The mixture is thoroughly shaken, and poured into petri dishes or other suitable flat containers to form a layer about 3 mm thick. The mixture should be poured when quite cool so that it sets quickly before the charcoal can sediment. It is advisable to smear the bottom of the container very thinly with vaseline, as the charcoal gelatin mixture has a tendency to stick to glass.

After the mixture has thoroughly set, the whole sheet is lifted from the dish and placed in 10% formalin for 24 hours. The formalized gelatin sheet is then punched into discs about 1 cm in diameter or cut into strips about 2 cm long and 5 to 8 mm wide. The pieces of gelatin are then wrapped in gauze and placed in a basin under running tap water for 24 hours. The material is then ready for sterilization.

The pieces of prepared gelatin are best put into screw-capped bottles and covered with water. The bottles are then placed in a water bath for tyndallization or into a steamer. Sterilization is effected by steaming for 30 minutes or by repeated heating in a water bath at about 90 to 100°C for 20 minutes each time. Autoclaving is not recommended as it tends to soften the gelatin. After sterilization the water is decanted from the containers and the material is ready for use. The gelatin pieces can now be put under sterile conditions into any suitable fluid medium and incubated to check their sterility. The prepared media are then ready for use .

The test proper is performed by inoculating the medium in which the gelatin is suspended and incubating it at 37°C. The test can also be performed on solid media, such as nutrient agar, in which case the agar slope is inoculated first and a sterile gelatin disc or strip is placed on the surface of the agar, preferably near the bottom of the slope. Care should be taken that a certain amount of the condensation water is present. An already established culture, whether in a fluid or on a solid medium, can be tested in the same way as described above. There may even be some advantage in doing so, as a certain amount of gelatinases may be present in the culture to start with.

The first evidence of liquefaction is the appearance of free particles of charcoal sedimenting to the bottom of the medium. On shaking they become resuspended and form a clearly visible black cloud. Once the process has started it proceeds rapidly to total liquefaction. Mechanical factors such as shaking do not produce false-positive results.

Gelatin hydrolysis—various

N. C. T. C. Methods 1954.

Method 1. Inoculate gelatin stab and incubate at 22°C. Observe daily for growth and liquefaction (including type). Limit, 30 days.

Method 2. Inoculate gelatin stab and incubate at the optimal temperature. Each week (for 4 weeks) cool in the refrigerator for 4 hours before reading liquefaction. Control of uninoculated medium.

Method 3. (Frazier's method modified by Smith, Gordon and Clark). Streak culture onto a plate of nutrient agar containing 0.4% gelatin and incubate at optimal temperature for 3 days.

Test: Flood the plate with 8–10 ml of acid-corrosive sublimate solution. Clear zones indicate areas of gelatin hydrolysis.

Acid–corrosive sublimate solution

Mercuric chloride	15 g
Distilled water	100 ml
Concentrated HCl	20 ml

Make up in the order stated and shake well.

N. B. Spreading organisms must be inoculated onto well-dried plates. Both lid and plate must be glass, and after use the dishes should be boiled.

W. C. Frazier (1926). J. infect. Dis. 39, 302.
N. R. Smith, R. E. Gordon and F. E. Clark (1946). Aerobic mesophilic spore-forming bacteria. U. S. Dept. Ag. Misc. Publ., 559.

Gelatin hydrolysis—*Proteus-Providencia*

C. Shaw and P. H. Clarke (1955). J. gen. Microbiol. 13, 155–161.

Liquefaction was observed in nutrient gelatin stab cultures incubated for 30 days at 22°C.

Gelatin hydrolysis—*Bacillus*

K. L. Burdon (1956). J. Bact. 71, 25–42.

To test for decomposition of gelatin, a modified form of Frazier's gelatin agar medium (Frazier, 1926) was prepared as follows: Sodium chloride, 5 g; potassium phosphate, 1.5 g; and monopotassium phosphate, 0.5 g, were dissolved in distilled water, 100 ml; gelatin, 4 g, and tryptose (Difco), 20 g, were dissolved in water, 400 ml. The two solutions were then mixed and heated to about 50°C. Twenty grams of agar were dissolved in 500 ml distilled water and, after the solution was cooled to 50°C it was thoroughly stirred into the gelatin mixture. The medium was then sterilized by autoclaving, and plates were poured. The plates were inoculated in the centre only, so as to yield a single, giant colony. After incubation for 48 to 72 hr and observation of the colony characteristics, each plate was flooded for several minutes with a solution consisting of: $HgCl_2$, 15 g, and concentrated HCl, 20 ml, dissolved in distilled water, 100 ml. This fluid was then poured off and the plates were observed for clear zones developing around the colonies. The width of the zones of gelatin decomposition in relation to size of the colonies was recorded.
W. C. Frazier (1926). J. infect. Dis. 39, 302–309.

Gelatin hydrolysis—*Acetobacter*

R. Steel and T. K. Walker (1957). J. gen. Microbiol. 17, 445–452.

Gelatin	12% (w/v)
Difco yeast extract	0.5% (w/v)
Distilled water	

Incubation at room temperature for 4 weeks.

Gelatin hydrolysis—*Microbacterium*

V. Bolcato (1957). Antonie van Leeuwenhoek 23, 351–356.

The cultures were streaked once across duplicate plates of saccharose agar (peptone, 5 g; beef extract, 3 g; saccharose, 10 g; agar, 15 g; distilled water, 1000 ml; pH 6.5) plus 0.5 per cent gelatin and incubated at 35°C. One plate was for estimation of hydrolysis of the gelatin at 5 days and the second at 10 days by flooding with 10 ml of the Frazier (1926) solution.
W. C. Frazier (1926). J. infect. Dis. 39, 302.

Gelatin hydrolysis—*Leptotrichia buccalis*

R. D. Hamilton and S. A. Zahler (1957). J. Bact. 73, 386–393.

Medium:

Peptone (Difco)	20 g
Yeast extract (Difco)	2 g
Glucose	5 g
NaCl	5 g
K_2HPO_4	10 g
Sodium thioglycolate	1 g
Agar (Difco)	0.5 g
Distilled water	1 litre
0.2% aqueous methylene blue 1 ml/L	

pH adjusted to 7.2 before autoclaving.

Gelatin liquefaction was tested in the above medium with 12% gelatin added, incubated for 5 days at 37°C by the methods described in the Committee on Bacteriological Technic of the Society of American Bacteriologists (1953) *Manual of Methods for Pure Culture Study of Bacteria,* Williams and Wilkins, Co., Baltimore, Maryland, U. S. A.

Gelatin hydrolysis—*Nocardia*

R. E. Gordon and J. M. Mihm (1957). J. Bact. 73, 15–27.

Duplicate plates of the following medium were streaded once across with each culture: peptone, 5 g; beef extract, 3 g; agar, 15 g; gelatin, 4 g; distilled water, 1000 ml; pH 7.0. The gelatin was soaked in approximately 40 ml of cold water before mixing with the melted agar. After 5 days' incubation at 28°C, one plate was covered with 8 to 10 ml of the following solution: HCl, 20 ml (concentrated); $HgCl_2$, 15 g; distilled water, 100 ml (Frazier, 1926). Hydrolysis of the gelatin was measured by a clear zone underneath and around the growth; unchanged gelatin formed a white, opaque precipitate. The second plate was incubated for 10 days before testing.

W. C. Frazier (1926). J. infect. Dis. 39, 302–309.

Gelatin hydrolysis—thermophilic sulfate reducing bacteria

L. L. Campbell, Jr., H. A. Frank and E. R. Hall (1957). J. Bact. 73, 516–521.

Carried out in the medium of Reed and Orr (1941).

G. B. Reed and J. H. Orr (1941). War Med. Chicago 1, 493–510.

Gelatin hydrolysis—*Actinomyces bovis*

S. King and E. Meyer (1957). J. Bact. 74, 234–238.

Nutrient gelatin having been found inadequate to support the growth of the fastidious *A. bovis,* an enriched medium, Thiogel (no. 293, Baltimore Biological Laboratory), was employed. Although most of the organisms which liquified gelatin did so within 15 days, we kept all our cultures for 30 days before recording results as negative.

Gelatin hydrolysis—*Bacillus*

E. R. Brown, M. D. Moody, E. L. Treece, and C. W. Smith (1958). J. Bact. 75, 499–509.

Inoculated tubes of nutrient gelatin were observed each day for a period of 3 weeks. An arbitrary line was drawn after 4 days of incubation.

Gelatin hydrolysis—*Streptococcus*

P. Morelis and L. Colobert (1958). Annls Inst. Pasteur, Paris 95, 568–587.

La liquéfaction de la gélatine a été étudiée pour la totalité des souches, elle a été observée chez 10 entre elles : 5 provenant d'excréments animaux, 1 d'excréments humams, 3 de produits pathologiques, enfin la dernière : la souche *faecalis liquefaciens* VGB, nous a été procurée par Miss Barnes.

Gelatin hydrolysis—*Aeromonas*

J. P. Stevenson (1959). J. gen. Microbiol. 21, 366–370.

Examined at 4°C after incubation in nutrient gelatin for 5 days at 37°C.

Gelatin hydrolysis

V. B. D. Skerman, *A Guide to the Identification of the Genera of Bacteria.* 2nd ed. Williams and Wilkins Book Co., Baltimore, 1967.

This may be detected by two methods.

1. Gelatin Broth.

Add 25 per cent gelatin to meat infusion or papain digest broth. Steam until all the gelatin is dissolved. Add 80 ml of N NaOH per L and then adjust the pH to 7.4. Cool to 50°C. Clarify through paper pulp. Dispense in 5-ml quantities in 160- by 13-mm tubes and sterilize at 110°C for 25 minutes. Do not heat above this temperature or the gelling power will be lost.

Inoculate by stabbing the inoculum down the centre of the tube. Incubate at 22°C for 1 month and record the time at which liquefaction becomes evident if it occurs. For organisms which will grow at this temperature in this basal medium, this procedure is satisfactory.

Liquefaction of the gelatin occurs initially somewhere along the line of inoculation. The subsequent development varies with the organism.

For organisms which will not grow at 22°C, incubation must be carried out at the required optimum. At 37°C the gelatin melts and liquefaction can be observed only after placing the tube in an ice bath on removal from the incubator. If only partial liquefaction has occurred, it may not be detected.

2. Gelatin Agar.

This medium is more satisfactory for mesophilic organisms. It is recommended for use with all organisms. The test is limited, of course, to those which will grow on the medium. Other basal media may be employed provided they do not contain precipitable proteins.

Dispense 10 ml of sterile meat infusion agar or papain digest agar into a sterile petri dish.

Melt a second 10-ml quantity of the medium and a gelatin broth (as in method 1 above). Add 1 ml of the gelatin broth to the agar. Mix well and pour onto the surface of the solidified agar.

Inoculate the medium and incubate at the desired temperature.

To detect hydrolysis, flood the plates with a 15 per cent solution of $HgCl_2$ in HCl. The unhydrolyzed gelatin forms a white precipitate with the mercury salt. Always compare with a control plate. Some organisms hydrolyze the gelatin very rapidly and may completely clear the plate in 24 hours.

Acid $HgCl_2$ Solution

$HgCl_2$	15 g
H_2O	100 ml
HCl (concentrated)	20 ml

Mix in the order stated and shake well. This is a modification of the method of Frazier (1926).

W. C. Frazier (1926). J. infect. Dis. 39, 302.

Gelatin hydrolysis—*Pseudomonas fluorescens*

M. E. Rhodes (1959). J. gen. Microbiol. 21, 221–263.

Stab inoculations into nutrient gelatin (yeast extract broth + 12.0% (w/v) Oxoid gelatin) were made. Cultures were incubated at 22°C and examined at intervals, the shape and approximate amount of liquefaction, when present, being noted. The possibility that the peptone might be preferentially used was considered, and so a similar medium without peptone was used for comparison. The streak plate method of Frazier (1926) for detecting gelatin hydrolysis was also used, and the width of the hydrolysed zone was measured after 7 days of incubation.

W. C. Frazier (1926). J. infect. Dis. 39, 302.

Gelatin hydrolysis—*Clostridium*

M. E. Brooks and H. B. G. Epps (1959). J. gen. Microbiol. 21, 144–155.

Titration of gelatinase activity. The substrate used was a 10% (w/v) solution of gelatin in distilled water, with the addition of Congo red, and 0.5% phenol as preservative. The reagent was stored in the cold and melted by warming to about 40°C immediately before use.

Doubling dilutions of samples of 48 hr cultures were added in 1.0 ml amounts to 1.0 ml of the gelatinase reagent. After thorough mixing, as shown by the even distribution of the Congo red colour, the mixture was incubated at 37°C for 18 hr, then placed at 4°C for 1 hr, and the results read at this temperature.

Gelatin hydrolysis—various organisms

J. J. McDade and R. H. Weaver (1959). J. Bact. 77, 60–64.

Ninhydrin method

The best results were obtained with a 1.5 per cent gelatin medium containing 10 ppm of manganese sulfate. The medium was dispensed in 13 by 100 mm tubes and carefully sterilized to avoid overheating. Tubes of sterile medium were preheated in a 37°C water bath and then inoculated with 0.1 ml amounts of cell suspensions prepared by carefully harvesting the growth from 4-hr nutrient agar slant cultures in 1.5-ml amounts of physiological saline. It is necessary to avoid large clumps of organisms and particles of agar. After incubation periods of 2 to 24 hr in the water bath, gelatin hydrolysis may be determined by adding 0.5 ml of 0.1 per cent aqueous ninhydrin solution (freshly prepared or stored in the refrigerator for not more than 2 weeks) and heating the tube in an 80°C water bath for up to 3 hr. The appearance of a bluish-purple color indicates hydrolysis.

Positive tests have been obtained consistently with gelatin-hydrolyzing cultures but it has not been possible to avoid some false positive reactions. Although saline plus cell suspensions, and uninoculated gelatin medium gave consistently negative results, occasional cultures of non-gelatin-hydrolyzing organisms in the gelatin medium gave positive results. This may be explained by the extreme sensitivity or the lack of specificity of the ninhydrin reagent. These false positive reactions limit the usefulness of the procedure to that of a screening test.

Methods using gelatin-precipitating agents

Studies of the effects of varying different factors led to the development of four procedures that have given useful results. These procedures incorporate features of methods used by Frazier (1926), Chapman (1948, 1952), Oakley *et al.* (1948), Clarke and Cowan (1952) and Clarke (1953).

(1) Plate modification of the Frazier (1926) method: Plates are poured with 15-ml quantities of medium (tryptose, 20 g; beef extract, 3 g; agar, 15 g; $MnSO_4$, 10 ppm; gelatin, 10 g; distilled water, 1000 ml; pH 7.0). The plates are dried overnight under Coors porcelain tops and are then spot-inoculated with 3-mm loopfuls of actively growing cultures on tryptose-beef extract-agar slants. After suitable periods of incubation (at

37°C in this study), which may vary from 1 to 96 hr with different strains of organisms, the plates are developed by wiping off the site of inoculation with a cotton swab soaked in the acid mercuric chloride indicator of Frazier (1926) or 20 per cent sulfosalycylic acid (Chapman, 1952). Plates that are to be incubated for over 48 hr should have snugly-fitting sterile filter paper discs placed inside the tops of the plates and, after 48 hr, the discs should be moistened daily with sterile distilled water.

Inoculation of the plates from liquid-medium cultures or cell suspensions may lead to "halos" around the areas of inoculation because of reduction of the gelatin content of the medium from dilution with the liquid. These must not be confused with clearing due to gelatin hydrolysis.

Adjustment of the medium to pH 8.0 gave better results with a strain of *Staphylococcus aureus* but poorer results with some other organisms.

Manganese sulfate was added to the medium after it had been shown to give more rapid results, as suggested by the work of Levinson and Sevag (1954). The addition of cysteine or the substitution of cysteine and iron for the mangenese, as suggested by the works of Weil and Kochalty (1937) and Kochalty *et al.* (1938), did not result in a more rapid test. Likewise, the addition of 0.01 M calcium sulfate or its substitution for the manganese sulfate as suggested by the work of Lautrop (1956) did not result in a more rapid test.

(2) Tube modification of the Frazier (1926) method: The same medium is used as in the plate modification. It is dispensed in 1.0 ml amounts in 10 by 75 mm tubes. The tubes are preheated in a 37°C water bath and then inoculated by layering 0.2 ml amounts of dense cell suspensions on the surface. The cell suspensions are prepared by harvesting the growths from 4.0 to 4.5 hr slant cultures on tryptose-beef extract agar in 1.0 ml amounts of saline. After suitable periods of incubation in the water bath, 2 or 3 drops of the acid mercuric chloride "developer" are added to each tube to be tested. A clear area below the meniscus is indicative of gelatin hydrolysis. A slightly lighter area may result from reduction of the gelatin content at the surface of the medium by dilution with the liquid of the inoculum. To avoid misinterpretation, a control tube to which 0.2 ml of sterile saline has been added should be used. When the tubes are to be incubated beyond 48 hr, 0.1 ml quantities of saline should be added daily, after 48 hr, or the tubes should be closed with stoppers or screw caps.

If 0.1 per cent cysteine is added to the medium and lead acetate papers are inserted into the mouths of the tubes, hydrogen sulfide production can be detected in 1 to 8 hr without interference with the demonstration of gelatin hydrolysis.

(3) Plate method using an ammonium sulfate-sodium chloride medium: In studies with staphylococci, Chapman (1948) avoided the use of a developing solution by incorporating ammonium sulfate along with the

sodium chloride in Stone's medium. The medium was cloudy, and clear areas developed around the colonies of gelatin utilizers. By changing the basic medium and the proportion of ammonium sulfate and sodium chloride, we have been able to adapt the principle of this medium for use with organisms other than staphylococci.

The medium contains: gelatin, 5 g; agar, 10 g; NaCl, 9.0 g; $MnSO_4$, 10 ppm; $(NH_4)_2SO_4$, 120 g; and distilled water, 1000 ml. The gelatin, NaCl, $MnSO_4$ and agar are first put into solution by heating. The ammonium sulfate is added, whereupon the medium becomes opaque white because of the precipitated gelatin. The medium is adjusted to pH 7.0 and is dispensed into tubes in 11 to 12 ml amounts. It is sterilized by autoclaving for 10 min at 10 lb pressure. In pouring the plates, the agar should be decanted from any heavy precipitate that may have formed in the bottoms of the tubes during autoclaving.

The plates are inoculated as are the plates with the modified Frazier method. Apparently, hydrolysis of the gelatin is the result of preformed enzymes since the cultures fail to grow on the medium. Peculiarly, the addition of tryptose and beef extract to the medium produced a medium that gave some false negative results.

(4) Tube method using ammonium sulfate sodium chloride medium: The same medium is used as in the plate modification. It is dispensed into tubes and inoculated as in the tube modification of the Frazier method. Positive results are indicated by clear areas which extend down the tube below the inoculum. Again a control is necessary to avoid misinterpretation of the "halo" effect and measures must be taken to prevent excessive dehydration of the medium when the tubes are incubated beyond 48 hr. The medium may be used for the simultaneous determination of hydrogen sulfide production if 0.03 per cent sodium thiosulfate or 0.1 per cent cysteine is added and lead acetate papers were placed in the mouths of the tubes.

G. H. Chapman (1948). Fd Res. 13, 100–105.
G. H. Chapman (1952). J. Bact. 63, 147.
P. H. Clarke and S. T. Cowan (1952). J. gen. Microbiol. 6, 187–197.
S. K. R. Clarke (1953). J. clin. Path. 6, 246–248.
W. C. Frazier (1926). J. infect. Dis. 39, 302–309.
W. Kochalty, L. Weil and L. Smith (1938). Biochem. J. 32, 1685–90.
H. Lautrop (1956). Acta path. microbiol. scand. 39, 357–369.
H. S. Levinson and M. G. Sevag (1954). J. Bact. 67, 615–616.
C. L. Oakley, G. H. Warrack and M. E. Warren (1948). J. Path. Bact. 60, 495–503.
L. Weil and W. Kochalty (1937). Biochem. J. 31, 1255–1267.

Gelatin hydrolysis—*Pasteurella septica*

J. M. Talbot and P. H. A. Sneath (1960). J. gen. Microbiol. 22, 303–311.

Observed by stab inoculation into a lead acetate + gelatin medium (lead acetate, 0.25%, w/v; gelatin 12.5% w/v; nutrient broth to 100 ml) which was incubated at 22°C for 14 days.

Gelatin hydrolysis—*Fusobacterium*

A. C. Baird-Parker (1960). J. gen. Microbiol. 22, 458–469.

The breakdown of gelatin was tested by Frazier's technique (Frazier, 1926). Proteolytic activity was tested against coagulated egg-white protein, inspissated horse serum and whole casein. The breakdown of casein was tested by the method of Hastings (1904).

W. C. Frazier (1926). J. infect. Dis. 39, 302.
E. G. Hastings (1904). Zentbl. Bakt. ParasitKde (2 Abt.) 12, 590.

Gelatin hydrolysis—*Pediococcus*

H. L. Günther and H. R. White (1961). J. gen. Microbiol. 26, 185–197.

Stab cultures were incubated at optimum temperature and examined for liquefaction after chilling at 7, 14 and 28 days. The nutrient gelatin medium had the same formula as the nutrient broth, with the addition of 14% (w/v) gelatin.

Maintenance of stock cultures and methods of cultivation. For maintenance of stock cultures, preparation of inocula and in all experimental work, "Oxoid" tomato juice (TJ) broth or tomato juice (TJ) agar, adjusted to pH 6.6, was used unless otherwise stated. The following were exceptions to this rule: for strain Tc. 1, sodium chloride (5%, w/v) was added to the medium; and for the aerococci, glucose Lemco broth (Shattock and Hirsch, 1947) of glucose yeast extract (GY) agar (containing, as %, w/v, peptone, 1.0; Yeastrel, 0.3; glucose, 1.0; NaCl, 0.25; agar, 1.0; at pH 7.4) was used.

Cultures were incubated aerobically at 30°C, with specified exceptions.

P. M. F. Shattock and A. Hirsch (1947). J. Path. Bact. 59, 495.

Gelatin hydrolysis—*Xanthomonas*

A. C. Hayward and W. Hodgkiss (1961). J. gen. Microbiol. 26, 133–140.

The following agar medium was dispensed in 45 ml quantities in 2 oz bottles: peptone (Oxoid), 5.0 g; yeast extract (Difco), 3.0 g; agar, 20.0 g; distilled water, 1 L; adjusted to pH 7.2. Five ml quantities of 4.0% (w/v) gelatin (British Drug Houses, Ltd.) were added to the molten agar base from which three plates were poured. Three or four organisms were inoculated to each plate, into the centre of 0.5 cm diameter cavities made with a surface sterilized cork borer. After incubation for 6 days, gelatin plates were flooded with acid mercuric chloride solution (Frazier, 1926). Zones of hydrolysis were recorded.

W. C. Frazier (1926). J. infect. Dis. 39, 302.

Gelatin hydrolysis—rhizosphere bacteria

P. G. Brisbane and A. D. Rovira (1961). J. gen. Microbiol. 26, 379–392.

Gelatin agar. Gelatin (0.4%, w/v) in basal agar. Hydrolysis was observed by flooding the plates with saturated ammonium sulfate following incubation at 26°C for 7 days.

The Basal medium. The yeast-extract peptone nitrate broth of Sperber and Rovira (1959), with 15 g agar/L for solid medium, was used.
J. I. Sperber and A. D. Rovira (1959). J. appl. Bact. 22, 85.

Gelatin hydrolysis—*Pseudomonas*

M. Véron (1961). Annls Inst. Pasteur, Paris 100, Suppl. 6, 16–42.

À 20–22°C, on recherchera la liquéfaction de la gélatine nutritive ensemencée par piqûre centrale.

Gelatin hydrolysis—*Clostridium*

H. Beerens, M. M. Castel and H. M. C. Put (1962). Annls Inst. Pasteur, Paris 103, 117–121.

Gélatine: Choisir la méthode de Kohn [4] (disque de gélatine dénaturée introduit dans un milieu nutritif liquide).
[4] Kohn (J.). *J. Clin. Path.*, 1953, 6, 249.

Gelatin hydrolysis—*Escherichia aurescens*

H. Leclerc (1962). Annls Inst. Pasteur, Paris 102, 726–741.

a) *Protéolyse.* – La liquéfaction de la gélatine nutritive est recherchée après ensemencement, puis incubation à 20-22° durant trente jours.

La gélatine préparée selon Kohn [21] est également utilisée parallèlement selon la modification de Lautrop [24].
[21] Kohn (J.). *J. clin. Path.*, 1953, 6, 249.
[24] Lautrop (H.). *Acta path. microbiol. scand.*, 1956, 39, 357.

Gelatin hydrolysis—*Haemophilus*

P. N. Edmunds (1962). J. Path. Bact. 83, 411–422.

Nutrient gelatin was superimposed on blood-agar in a tube and incubated for 2 days at 37°C. It was then inoculated with 6 drops of a 48 hour blood agar and broth culture. After incubation for 3 weeks at 37°C the tubes were held overnight at 4°C and then observed.

Gelatin hydrolysis—*Actinobacillus* and *Haemophilus*

E. O. King and H. W. Tatum (1962). *Actinobacillus actinomycetemcomitans* and *Haemophilus aphrophilus.* J. infect. Dis. 111, 85–94.
The test was performed with 12% gelatin in heart infusion broth.

Gelatin hydrolysis—*Lactobacillus*

R. E. Smith and J. D. Cunningham (1962). Can. J. Microbiol. 8, 727–735.

Determined with Hayward's (1957) basal medium containing 12% (w/v) gelatin, after 2 weeks' incubation. The basal medium contains Peptone (Oxoid), 0.5% (w/v); yeast extract (Difco), 0.3% (w/v); salts A, 0.5% (v/v); salts B, 0.5% (v/v); Tween 80, 0.1% (v/v); sodium acetate (hydrated), 0.5% (w/v). The constituents were diluted in glass-distilled water and the pH value of the medium adjusted to 6.8–7.0. Solution A contained 10 g KH_2PO_4 and 10 g K_2HPO_4 in 100 ml of distilled water; solution B contained 11.5 g $MgSO_4 \cdot 7H_2O$, 2.4 g $MnSO_4 \cdot 2H_2O$ and 0.68 g $FeSO_4 \cdot 7H_2O$ in 100 ml of distilled water.

A. C. Hayward (1957). J. gen. Microbiol. 16, 9–15.

Gelatin hydrolysis—*Staphylococcus* and *Micrococcus*

D. A. A. Mossel (1962). J. Bact. 84, 1140–1147.

The suspension of the strain used for the determination of its oxygen tolerance was streaked as a 4 cm straight line on freshly dried plates of a slight modification of Frazier's (1926) gelatin agar (Mossel and de Bruin, 1957). Six strains were tested per plate (9 cm diam.).

After 48 hr of incubation at 30°C, plates were developed using Frazier's solution of mercuric chloride. A strain was considered gelatinase-positive if it had digested the gelatin around or under the streak.

W. C. Frazier (1926). J. infect. Dis. 39, 302–309.

D. A. A. Mossel and A. S. de Bruin (1957). Antonie van Leeuwenhoek 23, 218–224.

Gelatin hydrolysis—*Caryophanon*

P. J. Provost and R. N. Doetsch (1962). J. gen. Microbiol. 28, 547–557.

Medium:

Hy-Case SF	10.0 g
Bacto-yeast extract	5.0 g
Na Acetate (anh.)	0.5 g
Agar	15.0 g
pH 7.8 – 7.9	
+ 12% (w/v) gelatin in place of agar	

Gelatin hydrolysis—demonstration of rapid methods

W. H. Ewing (1962). *Enterobacteriaceae. Biochemical Methods for Group Differentiation.* U.S. Department of Health, Education and Welfare, Public Health Service Publication, No. 734 (revised).

Kohn (1953) method, modified by Lautrop (1956).

Preparation of denatured charcoal gelatin:

1. Dehydrated nutrient gelatin medium is mixed with water in the proportion of 15 g to 100 ml of water and the mixture is heated.
2. Finely powdered charcoal is added to the melted gelatin medium in

the proportion of 3 to 5 g to 100 ml of medium.
3. The charcoal gelatin medium is thoroughly mixed, and poured into Petri dishes or other suitable flat containers, to form a layer about 3 mm thick. It is advisable to apply a thin film of Vaseline to the containers before pouring the mixture to prevent it from sticking. Also the mixture should be cooled before it is poured and then chilled quickly so that the charcoal does not sediment.
4. After the medium has set thoroughly, the entire sheet is removed from the container and placed in 10 per cent formalin for 24 hours.
5. The formalin-gelatin sheet is then cut into pieces about 1 cm by 5 to 8 mm. The pieces are then wrapped in gauze and washed in running tap water for 24 hours.
6. The pieces are then placed in jars or wide-mouthed screw-capped bottles, covered with distilled water, and sterilized by treatment with flowing steam for 30 minutes on 3 successive days. After sterilization the water may be decanted. Sterility may be controlled by placing pieces from each container in nutrient broth, using aseptic technique.
Inoculation:
1. The growth from an 18- to 24-hour agar plate culture is suspended in 3 ml of 0.85 per cent sodium chloride solution containing 0.01 M calcium chloride in a small sterile test tube (e.g., 13 x 100 mm). The suspension should be very dense since the rapidity of the reaction in this test appears to be a function of density and of temperature of incubation. The agar plates should be thick; that is, they should contain 35 to 40 ml of infusion agar medium. Three or four drops of a broth culture of the strain to be tested should be spread over the entire surface of the agar medium so as to obtain maximum, confluent growth.
2. With aseptic precautions, add a piece of denatured charcoal-gelatin to each dense suspension prepared as outlined above.
3. Add 0.1 ml of toluene to each suspension and shake the tubes.
Incubation: 37°C. Examine after 5 to 6 hours, and daily for 14 days. Positive reactions are indicated by the release of charcoal particles which collect in the bottom of the tube.

Toluene is added because it appears to have an activating effect on the reactions of strains that ordinarily are slow liquefiers of gelatin. However, advantage may also be taken of its bactericidal effect: the suspending fluid (saline solution with 0.01 M calcium chloride) can be distributed in 3 ml amounts in small test tubes and sterilized at 121°C for 15 minutes. A piece of charcoal-gelatin and 0.1 ml of toluene are placed in each tube, after which the cotton stoppers are replaced with corks that have been soaked in hot paraffin. The tubes are then stored until needed.

J. Kohn (1953). J. clin. Path. 6, 249.
H. Lautrop (1956). Acta path. microbiol. scand. 39, 357.

Gelatin hydrolysis

W. H. Ewing (1962). *Enterobacteriaceae. Biochemical Methods for Group Differentiation.* U.S. Department of Health, Education and Welfare, Public health Service Publication, No. 734 (revised).

Nutrient gelatin:

Beef extract	3 g
Peptone	5 g
Gelatin	120 g
Distilled water	1000 ml

Sterilize at 121°C for 12 minutes.

Inoculation: Inoculate by stabbing the medium with a wire, using inoculum from an agar slant culture.

Incubation: 20°C, 30 days.

It is believed that the nutrient gelatin method should be employed as the "standard" method in taxonomic work, since the rate of gelatin liquefaction is important in the characterization of members of certain groups and subgroups within the family *Enterobacteriaceae.* Some of the more recently described rapid methods (see p. 381) are excellent for diagnostic work in which one is interested only in whether a culture liquefies gelatin or not, but they are of little or no value as differential tests in those areas of the family where the rate of gelatin liquefaction is of differential importance (e.g., within the *Klebsiella-Aerobacter-Serratia* division). In those areas where the rate of gelatin liquefaction is of differential value, positive tests obtained by means of rapid methods should be repeated using the conventional method in order to determine whether liquefaction is rapid or delayed. If the above-mentioned limitations are borne in mind, certain rapid methods can be recommended for preliminary work.

Gelatin hydrolysis—*Vibrio*

G. H. G. Davis and R. W. A. Park (1962). J. gen. Microbiol. 27, 101–119.

Using 0.5% (w/v) gelatin in nutrient agar and flooding 5-day growth with acid mercuric chloride (Frazier, 1926).

W. C. Frazier (1926). J. infect. Dis. 39, 302.

Gelatin hydrolysis—*Pseudomonas odorans*

I. Málek, M. Radochová and O. Lysenko (1963). J. gen. Microbiol. 33, 349–355.

Broth medium over 6 weeks at 28°C.

Gelatin hydrolysis—*Spirillum*

W. A. Pretorius (1963). J. gen. Microbiol. 32, 403–408.

Stab cultures were made in Difco nutrient gelatin and incubated at 20°C for 14 days.

Gelatin hydrolysis—*Aeromonas*

I. W. Smith (1963). J. gen. Microbiol. 33, 263–274.

Liquefaction of gelatin was recorded after 7 days in nutrient gelatin.

Gelatin hydrolysis—general

L. Le Minor (1963). Annls Inst. Pasteur, Paris 105, 792–794.

La méthode de Kohn [1] et sa modification par Lautrop [2] sont basées sur l'utilisation de cubes de gélatine à 15 p. 100 additionnée de 3 à 5 p. 100 de poudre de charbon et dénaturée par le formol. Cette dénaturation permet de stériliser au bain-marie bouillant les cubes de gélatine qui sont ensuite placés dans un milieu liquide, sur milieu solide ou, dans la modification de Lautrop, dans une suspension épaisse de bactéries. Le début de la protéolyse se traduit par la libération de particules de charbon. Cette méthode permet d'obtenir une réponse bien plus rapide que la méthode classique et elle est particulièrement intéressante dans l'étude de bactéries liquéfiant lentement la gélatine, telles que les *Arizona* [3].

La méthode que nous proposons est encore plus simple, puisqu'elle ne nécessite aucune préparation: les films photographiques sont constitués par un support transparent sur la surface duquel est déposée une fine pellicule de gélatine contenant les sels d'argent. Cette pellicule devient noire après exposition à la lumière et développement. Nous avons recherché si, en mettant des fragments de tels films dans une suspension de bactéries protéolytiques, la gélatine serait dissoute, mettant à nu le support transparent et libérant les particules noires d'argent.

Nous avons utilisé du film Kodak "Microfile" coupé dans le sens de la largeur en bandes de 5 mm de largeur environ. Divers moyens de stérilisation nous ont donné des résultats satisfaisants: dans des sacs scellés de polyéthylène traités par le peroxyde d'éthylène, ou même simple passage très rapide sur une flamme d'une bande tenue par une pince. La stérilisation dans des tubes à l'autoclave à 110° a l'inconvénient de faire brunir le film. Cette stérilisation n'a en réalité d'importance que lorsque les bandes sont placées dans un milieu de culture liquide ou à la surface d'un milieu solide. Elle est superflue quand on effectue les tests en plaçant la bande de film dans une suspension épaisse de bactéries additionnée d'une goutte de toluène.

Deux méthodes furent employées dans les essais: l'une "lente", en mettant une bande de film dans une culture en eau peptonée; l'autre "rapide", en laissant baigner la partie inférieure de la bande de film dans une suspension épaisse faite en eau physiologique ou encore en eau peptonée, de bactéries récoltées sur gélose, suspension qui peut être soit portée en bain-marie à 37°, soit, ce qui permet d'avoir des réponses encore plus rapides, placée sur un agitateur de Kahn. On peut l'employer aussi avec les milieux combinés pour le diagnostic rapide des entérobactéries: la bande du film est alors placée dans l'eau de condensation qui est en bas de la pente du milieu lactose-glucose-SH_2.

Le premier signe d'attaque de la gélatine se révèle par une teinte gris-noirâtre du liquide. Puis, quand toute la gélatine est protéolysée, la fraction immergée du support devient transparente contrastant avec la partie restée noire de la partie non immergée.

[1] Kohn (J.). *J. Clin. Path.*, 1953, 6, 249.
[2] Lautrop (H.). *Acta path. microb. scand.*, 1956, 39, 357.
[3] Le Minor (L.), Fife (M. A.) et Edwards (P. R.). *Ann. Inst. Pasteur*, 1958, 95, 326.

Gelatin hydrolysis—rumen bacteria

R. S. Fulghum and W. E. C. Moore (1963). J. Bact. 85, 808–815.

To determine gelatin liquefaction, SRP-basal medium* with 1% Trypticase and 5% gelatin was used. Incubation was at 39°C, and the tubes were transferred to a refrigerator (2°C) for 30 min before observing for liquefaction.

The anaerobic methods of Hungate (Bact. Rev. 14, 1–49, 1950) were used in handling and incubation of the cultures.

***Basal medium for physiological tests**—*Butyrivibrio, Borrelia, Bacteroides, Selenomonas, Succinivibrio.*

Gelatin hydrolysis—*Clostridium rubrum*

H. Ng and R. H. Vaughn (1963). J. Bact. 85, 1104–1113.

Liver-gelatin consisted of double-strength liver infusion diluted with an equal volume of distilled water. Then 12% (w/v) gelatin was added to the mixture.

Gelatin hydrolysis—*Micrococcus roseus*

R. C. Eisenberg and J. B. Evans (1963). Can. J. Microbiol. 9, 633–642.

Gelatin hydrolysis was tested by two methods: (1) pour plates of nutrient agar with 0.4% gelatin were inoculated and incubated for 72 hours and then flooded with 15% $HgCl_2$ diluted with five parts (v/v) of concentrated HCl; (2) liquefaction of a 12% gelatin stab was observed after 1 week of incubation at 20°C.

Gelatin hydrolysis—*Aerococcus catalasicus*

O. G. Clausen (1964). J. gen. Microbiol. 35, 1–8.

Cultures were incubated aerobically at 37°C unless otherwise stated.

Ability to liquefy gelatin was tested in stab cultures incubated at 22°C for 30 days.

Gelatin hydrolysis—*Pseudomonas aeruginosa*

R. S. Berk and L. Gronkowski (1964). Antonie van Leeuwenhoek 30, 141–153.

Gelatin liquefaction was determined by adding 0.4% gelatin to nutrient agar, inoculating with the cultures, incubating 24 hr, and adding 20% sulfosalicyclic acid. A white precipitate indicated the presence of unhydrolyzed gelatin, while a clear zone around the bacterial growth indicated gelatin hydrolysis.

Gelatin hydrolysis—motile marine bacteria and *Hyphomicrobium neptunium**

E. Leifson, B. J. Cosenza, R. Murchelano and R. C. Cleverdon (1964). J. Bact. 87, 652–666.

Tests for gelatin liquefaction and nitrate reduction were made using a single medium of the following composition: artificial seawater (full or half strength); Casitone (Difco), 0.2%; yeast extract, 0.1%; gelatin, 8.0%; sodium nitrate, 0.2%; and tris buffer, 0.05%. After heating to dissolve the gelatin, the pH was adjusted to 7.5 with HCl, and the medium was tubed and autoclaved. The stab-inoculated medium was incubated at 20°C up to 7 days, unless liquefaction was apparent sooner. The test for nitrite was made by adding two to four drops of reagent 1 (4 g of sulfanilic acid dissolved in 500 ml of 0.2 N acetic acid), followed by two to four drops of reagent 2 (2.5 g of α-naphthylamine acetate in 500 ml of 0.2 N acetic acid).

Reduction of nitrate with gas formation was apparent in the form of bubbles in the solid medium. In the few instances when the gelatin liquefaction was extremely rapid, gas formation was checked by using a liquid nitrate medium with an inverted vial to trap the gas.

*E. Leifson (1964). Antonie van Leeuwenhoek 30, 249–256.

Gelatin hydrolysis—*Dermatophilus*

M. A. Gordon (1964). J. Bact. 88, 509–522.

The authors used 'beef extract gelatin (0.03% beef extract, 0.05% peptone, and 12% gelatin, at pH 7.8). The inoculum was taken from Brain Heart Infusion agar slants.

The medium was incubated at 36°C (± 1) and readings were made at 2, 5, 7, and 14 days, or longer.

Gelatin hydrolysis—*Vibrio marinus*

R. R. Colwell and R. Y. Morita (1964). J. Bact. 88, 831–837.

The authors used the method of the Society of American Bacteriologists. 1957. *Manual of Microbiological Methods.* McGraw-Hill Book Co., Inc., New York – modified by the addition to the media of the following salts to produce a synthetic seawater:– sodium chloride, 2.4%; potassium chloride, 0.07%; magnesium chloride (hydrated), 0.53% and magnesium sulfate (hydrated), 0.7%.

A standard inoculum was 1 drop from a pasteur pipette (*c.* 0.05 ml) of a 24 hour artificial seawater broth. Cultures were incubated at 18°C.

Gelatin hydrolysis—enterococci

C. G. Rogers and W. B. Sarles (1964). J. Bact. 88, 965–973.

Hydrolysis of gelatin (Burnett *et al.*, 1957) was determined in a medium containing 2% tryptose, 0.5% sodium chloride, 0.25% K_2HPO_4, 0.3% yeast extract, 1.5% agar, and 0.4% gelatin (pH 7.2), incubated at 37°C.

G. W. Burnett, M. J. Pelczar, Jr. and H. J. Conn (1957). In Society of American Bacteriologists *Manual of Microbiological Methods.* McGraw-Hill Book Co., Inc., New York.

Gelatin hydrolysis—*Bacteroides oralis*

W. J. Loesche, S. S. Socransky and R. J. Gibbons (1964). J. Bact. 88, 1329–1337.

Gelatin liquefaction was determined in the basal medium to which 4% gelatin was added. Cultures were examined at 4°C after 3 to 15 days of incubation.

The test was performed in the presence and absence of 0.5% glucose.

Thioglycolate Medium without Dextrose (BBL), supplemented with 0.2% yeast extract (Difco) and hemin, was used as a basal medium. All cultures were incubated at 35°C in Brewer jars containing an atmosphere of 95% H_2 and 5% CO_2.

Gelatin hydrolysis—*Clostridium*

K. Tamai and S. Nishida (1964). J. Bact. 88, 1647–1651.

Proteolytic activity. Proteolytic activity was measured by Brooks and Epps' (1959) method. To eliminate the sterile procedure employed in this method, we modified it as follows. A 24-hr chopped-meat broth culture was centrifuged, and the supernatant fluid was diluted twofold. A 1-ml amount of each dilution was mixed with an equal volume of a 10% solution of gelatin in a small test tube (13 by 100 mm). After the mixture was incubated at 37°C for 4 hr, it was chilled to observe the presence or absence of gelatinolysis. The maximal dilution causing gelatinolysis was determined after a few hours.

M. E. Brooks and H. B. G. Epps (1959). J. gen. Microbiol. 21, 144–155.

Gelatin hydrolysis—*Moraxella*

W. J. Ryan (1964). J. gen. Microbiol. 35, 361–372.

Nutrient gelatin containing 10% (v/v) horse serum was used. Incubation was at 37°C and, when necessary, was prolonged for 2 weeks.

Gelatin hydrolysis—*Pediococcus*

E. Coster and H. R. White (1964). J. gen. Microbiol. 37, 15–31.

The authors used the methods of H. L. Günther and H. R. White (J. gen. Microbiol. 26, 185, 1961). Their Tomato Juice (TJ) broth or agar with the addition of Tween 80 (0.1%, v/v), pH adjusted to 6.6, was used in maintenance and preparation of inocula with certain specified exceptions.

Transfers for preparing inocula were made every 24 hours except for *Pediococcus halophilus* (48 hr) and some brewing strains (fortnightly). All cultures were incubated aerobically at 30°C, except for "brewing strains," which were incubated in an atmosphere of 95% (v/v) hydrogen and 5% (v/v) carbon dioxide at 22°C.

Gelatin hydrolysis

G. S. Wilson and A. A. Miles (1964). Topley and Wilson's *Principles of Bacteriology and Immunity*. 5th edition – Arnold.

This may be tested for by inoculating a gelatin stab culture, incubating at 22°C and observing the type of liquefaction, or at 37°C and transferring the tube without shaking to iced water every day to see if the gelatin will still set. Slight liquefaction of gelatin may not occur for 6 weeks or so, but for practical purposes observation should not be continued for more than 14 days. Alternatively, Barer's (1946) method may be used. Nutrient agar containing 2 per cent gelatin and 0.025 per cent glucose in a phosphate-buffered medium is distributed in petri dishes, and inoculated heavily with the organism. The plates are incubated at 37°C, and the results are read the following day. The growth is flooded with either a 1 per cent solution of tannic acid, which renders the medium opaque and produces zones of greater opacity around the gelatinase producing colonies, or an acid solution of 15 per cent mercuric chloride, which renders the medium opaque except around the gelatinase-producing colonies. This method is not quite as sensitive as the gelatin stab method, but has the advantage of rapidity and of allowing as many as 12 strains to be tested on a single plate. An even more rapid method is that described by Kohn (1953), in which a disc of formalin-denatured gelatin containing finely powdered charcoal is incubated in a fluid or on a solid medium at 37°C. As soon as liquefaction begins, the charcoal particles fall to the bottom. A slide test may also be used (Thirst, 1957).

G. Barer (1946). Mon. Bull. Minist. Hlth 5, 28.
J. Kohn (1953). J. clin. Path. 6, 249.
M. L. Thirst (1957). J. gen. Microbiol. 17, 396.

Gelatin hydrolysis—*Nocardia*

A. Gonzalez-Mendoza and F. Mariat (1964). Annls Inst. Pasteur, Paris 107, 560–564.

Les méthodes suivantes ont été appliquées, elles sont indiquées dans le tableau I sous les lettres A, B, C et D.

Méthode A. – C'est la technique de Frazier [3] telle qu'elle a été utilisée par Gordon et Mihm [5,6]. Le milieu employé (peptone 5.00 g; extrait de viande 3.00 g; gélose 15.00 g; gélatine 4.00 g; eau distillée, 1000 ml; pH 7.0) comprend des substances organiques complexes. L'action gélatinolytique éventuelle des organismes est mise en évidence en versant sur une culture en boîte de Petri, une solution acide de bichlorure de mercure ($HgCl_2$ 15.00 g; HCl concentré 20 ml; eau distillée 100 ml) qui précipite la gélatine non attaquée.

Méthode B. – Cette technique préconisée par Mariat [9] diffère essentiellement de la précédente par le milieu de culture qui ne contient comme source d'azote et de carbone que la seule gélatine (KH_2PO_4 0.18 g; $Na_2HPO_4 \cdot 12H_2O$ 1.11 g; $MgSO_4 \cdot 7H_2O$ 0.60 g; KCl 1.00 g; solution d'oligo-éléments: X gouttes; thiamine, amide de l'acide nicotinique, pantothénate de Ca, pyridoxine: 1.10^{-7}; biotine: 1.10^{-9}; eau: 1000 ml; gélatine: 4.00 g; gélose: 15.00 g; pH 7.0).

On recherche l'hydrolyse du substrat en couvrant la culture d'une solution de noir amide (acide acétique M: 450 ml; acétate de sodium M/10: 450 ml; solution à 1 p. 100 de noir bleu naphtol dans la glycérine: 100 ml). Après lavage, seul le milieu où la gélatine n'est pas hydrolysée reste intensément coloré en bleu.

Méthode C. – Décrite par Bojalil et Cerbón [1], cette méthode se résume en une culture de l'actinomycète à éprouver dans une solution aqueuse de gélatine à 0.4 p. 100 ajustée à pH 7.0. L'utilisation de la gélatine se traduit par un net développement de l'inoculum, une alcalinisation du milieu et une réaction positive de celui-ci à la ninhydrine.

Méthode D. – Cette technique récemment préconisée par Le Minor et Piéchaud [7] est une modification heureuse de la méthode de Kohn. A une suspension dense de l'actinomycète en eau physiologique, on ajoute quelques gouttes de toluène (1.5 ml de suspension et IV gouttes de toluène, dans un tube de 12 x 120 mm). On immerge ensuite dans ce mélange un fragment de pellicule photographique exposée et révélée (Kodak "Microfile"). Si une action gélatinolytique a lieu, les grains d'argent sont libérés du support et tombent au fond du tube. La pellicule apparaît alors transparente dans sa partie immergée. Cette méthode permet de suivre l'évolution de la réaction, jour après jour. Il est possible d'appliquer diverses variantes, comme par exemple, remplacer la suspension dans l'eau physiologique et le toluène par un milieu de culture normalement ensemencé. A titre de contrôle nous avons utilisé cette variante avec, comme milieu, une solution chimiquement définie: KH_2PO_4: 0.45 g; $Na_2HPO_4 \cdot 12H_2O$: 1.19 g; $MgSO_4 \cdot 7H_2O$: 0.60 g; KCl: 1.00 g; $(NH_4)_2HPO_4$: 1.00 g; solution d'oligo-élements: X gouttes; mêmes vitamines que pour la méthode B; glucose: 10.00 g; eau: 1000 ml.

Pour mettre en oeuvre les différentes méthodes nous avons suivi les indications fournies par les auteurs.

[1] Bojalil (L. F.) and Cerbón (J.). *J. Bact.*, 1959, 78, 852.
[3] Frazier (W. C.). *J. inf. Dis.*, 1926, 39, 302.
[5] Gordon (R. E.) and Mihm (J. M.). *J. Bact.*, 1957, 73, 15.
[6] Gordon (R. E.) and Mihm (J. M.). *J. gen. Microbiol.*, 1959, 20, 129.

[7] Le Minor (L.) et Piéchaud (M.). *Ann. Inst Pasteur,* 1963, 105, 792.
[9] Mariat (F.). *Arch. Inst. Pasteur Tunis,* 1962, 39, 309.

Gelatin hydrolysis—various

H. Lagodsky and Mlle. Jandard (1964). Annls Inst. Pasteur, Paris 107, 128–131.

En effet, le pouvoir gélatinolytique des microorganismes constitue un critère important de leur identification. Les méthodes utilisées à cette fin font appel à la culture sur bouillon gélatiné, ou mieux, sur agar gélatiné. On peut encore citer la mise en contact d'un microorganisme et de gélatine solidifiée, soit sous forme de cubes, soit de cylindres a l'intérieur de tubes semi-capillaires.

A côté de réponses nettes et rapides, la liquéfaction caractéristique de la gélatine est souvent laborieuse et il n'est pas rare de devoir s'imposer un mois d'observation sans aboutir à une réponse autre que douteuse.

La gélatine du film photographique, qu'il soit noir ou en couleurs, offre à cet égard la double supériorité d'un réactif standard et d'un substrat hydrolysable en couche mince.

On comprend qu'un pouvoir gélatinolytique, même faible, soit plus rapidement et plus nettement affirmable par l'attaque d'un substrat pratiquement à deux dimensions que par l'hydrolyse d'un solide volumineux. Autrement dit, il est plus facile d'apprécier une même perte de substance en transparence mince plutôt qu'à partir d'un volume massif.

Enfin, le film constitue une preuve documentaire directe, facile à conserver en archives.

Le recours au film en couleurs apporte peut-être un avantage de sensibilité par rapport au film en noir. En effet, d'une part, la couche n'est pas simple, mais composite et la minceur de chacune des trois couches de gélatine est avantageuse par rapport à la couche unique du film en noir. D'autre part, et surtout, l'attaque de la couche superficielle modifie aussitôt la transparence d'origine qui vire au pourpre, puis, l'attaque se poursuivant, progressivement au bleu, bleu-vert, vert, vert pâle pour faire place, finalement, à l'incolore du support plastique entièrement décapé.

Au lieu de lire une gamme de gris, on bénéficie d'une gamme colorée dont le départ et les transitions sont d'une extrême sensibilité.

En outre, à l'examen au microscope, la structure et la couleur des granulations du film hydrolysé offrent une sorte de préparation "pseudo-histologique" qui peut aider à l'interprétation recherchée.

Méthode.

En pratique, un fragment de film en couleurs est maintenu bien à plat dans un petit support à rainures.

La goutte (ou la série de gouttes) de culture ou d'émulsion microbienne à tester est déposée à la pipette fine, directement sur le film: on constate que la goutte ne s'étale pas. Pour éviter la dessiccation, le dispositif est mis dans une boîte de Petri dont le fond est garni d'un buvard humide.

L'ensemble peut être laissé à la température ordinaire pour une observation longue, ou porté à l'incubation à 25°, 32° ou 37° pour des épreuves plus courtes (en ce cas, ne pas excéder huit heures pour éviter un ramollissement du film).

Après le temps choisi, tremper et rincer le film dans une eau légèrement formolée, puis rincer abondamment à l'eau courante. Après séchage à plat, examiner en transparence.

Deux précautions s'imposent pour éviter les fausses réponses:

1° L'une consiste à essayer, simultanément, une goutte du véhicule seul. En effet, si on opère directement avec la culture dans son milieu nutritif, certains de ces milieux, en particulier les milieux préparés par digestion extemporanée, peuvent encore contenir des traces d'agent protéolytique qui causeraient une fausse réaction positive.

2° L'autre, destinée à éviter toute fausse réaction négative, consiste à disposer une goutte-témoin d'un microorganisme dont le pouvoir gélatinolytique, bien connu de l'expérimentateur, sert de référence.

Gelatin hydrolysis—*Pseudomonas*

G. L. Bullock, S. F. Snieszko and C. E. Dunbar (1965). J. gen. Microbiol. 38, 1–7.

Gelatinase was determined by Smith's modification of Frazier's technique as described in the Society of American Bacteriologists *Manual of Microbiological Methods* – McGraw-Hill Book Co. Inc., New York 1957. Cultures were incubated at 20–22°C for 1 week.

Gelatin hydrolysis—*Pseudomonas aeruginosa*

A. H. Wahba and J. H. Darrell (1965). J. gen. Microbiol. 38, 329–342.

Production of gelatinase was tested by the plate method of Clarke (1953) with nutrient agar No. 1 as base and inoculating 16 strains per plate. Results were read after 24 hr. The test was repeated with negative strains and read after 3 days.

Nutrient agar No. 1 (Oxoid No. 1) contained Lab Lemco beef extract, 1 g; yeast extract (Oxoid L20), 2 g; peptone (Oxoid L37), 5 g; sodium chloride, 5 g; agar, 15 g; distilled water to 1 L; pH 7.4.

S. K. R. Clarke (1953). J. clin. Path. 6, 246.

Gelatin hydrolysis—*Bacillus*

G. R. F. Hilson (1965). J. gen. Microbiol. 39, 407–421.

For gelatin hydrolysis, bottles (½-oz) containing 10 ml nutrient broth solidified with 10% gelatin were inoculated and incubated at 37°C up to

5 days. The cultures were tested daily by bringing them to 0–4°C in a melting ice bath. Failure to solidify at this temperature indicated gelatinase action. Control, uninoculated bottles of gelatin medium were tested in parallel.

Gelatin hydrolysis—*Alcaligenes*

R. G. Mitchell and S. K. R. Clarke (1965). J. gen. Microbiol. 40, 343–348.

Kohn's (1953) method with incubation for 14 days was used.

J. Kohn (1953). J. clin. Path. 6, 249.

Gelatin hydrolysis—*Thiobacillus*

M. Hutchinson, K. I. Johnstone and D. White (1965). J. gen. Microbiol. 41, 357–366.

"Liquefaction of charcoal gelatin discs, Oxoid".

Gelatin hydrolysis—marine bacteria

R. M. Pfister and P. R. Burkholder (1965). J. Bact. 89, 863–872.

Gelatin hydrolysis was carried out by use of seawater agar plates containing 0.4% gelatin. After incubation, the plates were flooded with a 15% acid $HgCl_2$ solution (Skerman, 1959) to make visible the zones of hydrolysis.

The seawater medium contained (per liter): Trypticase (BBL) or N-Z-Case, 2.0 g; Soy-tone (Difco), 2.0 g; yeast extract, 1.0 g; vitamin B_{12}, 1.0 μg; aged seawater; agar (if desired), 16 g. The pH was adjusted to 7.0 prior to autoclaving.

V. B. D. Skerman (1959). *A guide to the identification of the genera of bacteria.* The Williams & Wilkins Co., Baltimore.

Gelatin hydrolysis—*Streptomyces albus*

T. F. Fryer and M. E. Sharpe (1965). J. Dairy Res. 32, 27–34.

The authors used the methods of Gordon and Smith (1955) and Gordon and Mihm (1957).

R. E. Gordon and J. M. Mihm (1957). J. Bact. 73, 15.

R. E. Gordon and M. M. Smith (1955). J. Bact. 69, 147.

Gelatin hydrolysis—*Staphylococcus*

R. Zemelman and L. Longeri (1965). Appl. Microbiol. 13, 167–170.

Stone's reaction was conducted on Chapman's (1946) medium 110 after 48 hr of incubation at 37°C by flooding plates with 5 ml of a saturated solution of ammonium sulfate. After 10 min, clear zones around the colonies were observed in positive cases.

G. H. Chapman (1946). J. Bact. 51, 409–410.

Gelatin hydrolysis—*Malleomyces, Pseudomonas*

Bach-Toan-Vinh (1965). Annls Inst. Pasteur, Paris 109, 460–463.

Hydrolyse de la gelatine. – Méthode de Kohn [4] modifiée par Lautrop [5]; méthode de Le Minor et Piéchaud [6].

[4] Kohn (J.). *J. clin. Path.*, 1953, 6, 249.
[5] Lautrop (H.). *Acta path. microbiol. scand.*, 1956, 39, 357.
[6] Le Minor (L.) et Piéchaud (M.). *Ann. Inst. Pasteur,* 1963, 105, 792.

Gelatin hydrolysis—*Streptomyces*

T. F. Fryer and M. E. Sharpe (1965). J. Dairy Res. 32, 27–34.

The authors used the methods of Gordon and Smith (1955) and Gordon and Mihm (1957).

One twentieth ml of an 18–24 hr culture of the organism grown in 5 ml Yeastrel nutrient broth, centrifuged and resuspended in 5 ml distilled water was used as inoculum.

R. E. Gordon and J. M. Mihm (1957). J. Bact. 73, 15.
R. E. Gordon and M. M. Smith (1955). J. Bact. 69, 147.

Gelatin hydrolysis—*Brevibacterium*

R. Chatelain and L. Second (1966). Annls Inst. Pasteur, Paris 111, 630–644.

Hydrolyse de la gélatine: en gélatine nutritive ensemencée par piqûre centrale. L'importance de la liquéfaction est notée après 10 et 30 jours.

La température d'incubation des cultures en gélatine nutritive est de 20–22°C.

Gelatin hydrolysis—*Providencia, Rettgerella*

C. Richard (1966). Annls Inst. Pasteur, Paris 110, 105–114.

Pour des raisons de commodité, les milieux ont été incubés à 37°C. Protéolyse de la gélatine: méthodes de Kohn-Lautrop [14] et de Le Minor et Piéchaud [19].

[14] Lautrop (H.). *Acta path. microbiol. scand.*, 1956, 39, 85.
[19] Le Minor (L.) et Piéchaud (M.). *Ann. Inst. Pasteur,* 1963, 105, 792.

Gelatin hydrolysis—*Clostridium*

A. V. Goudkov and M. E. Sharpe (1966). J. Dairy Res. 33, 139–149.

The medium used was nutrient gelatin containing 0.1% glucose and 0.05% cysteine.

Gelatin hydrolysis—*Staphylococcus*

S. J. Edwards and G. W. Jones (1966). J. Dairy Res. 33, 261–270.

Strains were tested routinely for gelatin liquefaction at 37°C for 7 days using media described by Cruickshank (1960).

R. Cruickshank (1960). *Mackie and McCartney's Handbook of Bacteriology.* Edinburgh and London: Livingstone.

Gelatin hydrolysis—*Staphylococcus*

M. Kocur, F. Přecechtěl and T. Martinec (1966). J. Path. Bact. 92, 331–336.

Gelatin hydrolysis was studied by the method recommended by Clarke (1953).

S. K. R. Clarke (1953). J. clin. Path. 6, 246.

Gelatin hydrolysis—anaerobes

A. -R. Prévot (1966). Techniques pour le diagnostic des Bactéries Anaérobes. Éditions de la Tourelle, St. Mandé.

Gélatine nutritive.

Digestat VF acide	1 000 cm^3
Gélatine	120 g

Chauffer au bain-marie à 60° jusqu'à dissolution complète.

Ajuster à pH 7.4 ; puis refroidir à 55°. Coller avec un blanc d'oeuf ou avec 50 cm^3 de sérum frais. Remuer pour homogénéiser sans mousser. Chauffer 1 heure à l'autoclave à 100°. Filtrer sur papier Laurent. Ajouter 2 g de glucose. Répartir en tubes de 13-14/180 mm. Boucher, stériliser à 110° pendant 30 minutes.

On a intérêt dans de numbreux cas à utiliser les cubes de gélatine au charbon de Kohn. La liquéfaction est ainsi très visible quand le charbon tombe au fond du tube. La réaction n'a de valeur que si elle est rapide.

Digestat VF acide.

L'étude comparative de numbreux bouillons a permis de conclure que celui qui donnait les meilleurs résultats pour la culture des anaérobies était le bouillon VF, dérivé du milieu de Stickel et Mayer et modifié par Prévot et Boorsma en 1939.

Voici la technique de sa préparation :

Dans un digesteur en acier inox on verse :

Eau à 47°	12 litres
Viande de boeuf parée, dégraissée et hachée	3 kg 600
Foie de boeuf paré, haché	1 kg

Agiter pour bien mélanger et ajouter :

Eau à 60°	6 litres
Acide chlorhydrique pur	150 cm^3
Pepsine titre 500	10 g

Régler le bain-marie à 48° pendant 20 heures, en remuant de temps en temps. Arrêter la digestion par un chauffage à 85° pendant quelques minutes. Filtrer sur papier Laurent mouillé. Ce digestat acide peut être conservé en glacière pour le stockage. Toutefois il n'est pas rigoureusement stérile et parfois un champignon acidophile ou même une bactérie peut s'y développer accidentellement. Aussi a-t-on intérêt à l'utiliser soit immédiatement, soit sans délai trop long.

GROWTH AT pH 9.6 AND 9.2

Growth at pH 9.6—*Streptococcus*

J. M. Sherman and P. Stark (1934). J. Dairy Sci. 17, 525–526.

Limiting hydroxyl-ion concentration of growth. When seeded in poured lactose nutrient agar plates (not streaked), and adjusted to varying degrees of alkalinity, *Streptococcus fecalis* is more tolerant to hydroxyl-ion than is *Streptococcus lactis.* At a pH value of 9.6 all of the lactis cultures were entirely inhibited and all of those of the fecalis group grew rapidly. (The sterilized, melted and cooled agar was adjusted to the desired pH immediately before use.)

Growth at pH 9.6—*Streptococcus*

P. M. F. Shattock and A. Hirsch (1947). J. Path. Bact. 59, 495–497.

Preparation of medium—

Components

I. Dextrose Lemco broth
Dextrose, 10 g; Lab. Lemco, 10 g; peptone (Evans), 10 g; NaCl, 5 g; tap water, 1 litre. pH 7.0
Autoclave at 15 lb for 20 minutes.

II. Buffer solution (Clark, 1928): Glycocoll, 7.505 g; NaCl, 5.85 g; freshly boiled, glass-distilled water 1 litre. This solution also may be autoclaved and stored in a well-stoppered flask. Mix 6 parts with 4 parts of N/10 NaOH by volume.

To 900 ml of dextrose Lemco broth add 100 ml of the buffer solution and adjust to pH 9.8 with N NaOH. Store overnight in a stoppered flask in the cold to complete precipitation. Filter through a Seitz filter and immediately distribute with sterile precautions into ¼ oz McCartney bottles, leaving the minimum of air space at the top. This medium should be used within 48 hours. It is essential that the pH of uninoculated control bottles be checked electrometrically (glass electrode) immediately before and after incubation. The initial pH should be 9.60 ± 0.02 and should not drop during incubation by more than 0.04 units. Care must be taken to ensure that the temperature of the medium is approximately 18°C when checking the pH on the glass electrode.

Technique of test

The medium is inoculated with 2 loopfuls (4 mm diam.) of an 18 hour dextrose Lemco broth culture, incubated at 30°C ± 1.0°C for 24 hours and inspected for growth. Control bottles of dextrose Lemco broth +

buffer (I and II above) adjusted with N HCl to pH 7.0 should be inoculated with each strain examined. Know positive and negative controls are included in each test.

W. M. Clark (1928). The determination of hydrogen ions. 3rd ed., London. p. 206.

Growth at pH 9.6—*Streptococcus*

P. F. Swartling (1951). J. Dairy Res. 18, 256–267.

The test described by Shattock and Hirsch (1947) was used.

P. M. F. Shattock and A. Hirsch (1947). J. Path. Bact. 59, 495.

Growth at pH 9.6 and 9.2—*Streptococcus*

P. H. Clarke (1953). J. gen. Microbiol. 9, 350–352.

Shattock and Hirsch (1947) described a liquid glycine-buffered medium for testing ability of streptococci to grow at pH 9.6, one of the criteria found useful by Sherman and Stark (1934) and Shattock and Mattick (1943–1944) for distinguishing *Streptococcus faecalis* from *Strep. lactis.* It was thought that if suitable indicator could be added to the medium it might be possible to dispense with the electrometric measurement of pH value, and as streptococci readily produce acid from glucose, growth would be accompanied by a change in colour of the indicator. Phenolphthalein with a pK of 9.7 was selected.

Preparation of Medium

The medium was prepared as described by Shattock and Hirsch (1947) with the modifications discussed above. Only 100 ml were prepared at a time, sufficient for testing about 16 strains.

(1) Stock Lemco broth pH 7.0 + 1% w/v glucose.

(2) Glycine buffer: glycine, 0.75 g; NaCl, 0.585 g, were dissolved in 100 ml freshly boiled distilled water. To 60 ml glycine + NaCl were added 40 ml 0.1 N NaOH to make solution (2).

(3) Set of buffer solutions pH 9.5, 9.6, 9.7, 9.8, and 9.9. Clark and Lubs borate buffers were used, as they were less likely to become contaminated. Twenty ml were prepared at each pH value, to which 0.1 ml 0.2% phenolphthalein was added. After mixing, the solutions were poured into bijou bottles (*c.* 5 ml) and the caps screwed on tightly to exclude air. They remained unchanged for at least 2 weeks. To 90 ml solution (1) were added 10 ml solution (2) and 0.5 ml phenolphthalein, and the reaction adjusted to pH 9.9 with N NaOH by matching with the buffers in the bijou bottles, but any standard method for matching pH indicators may be used. No difficulty was experienced in matching the colours but the accuracy obtained was no greater than ±0.25 pH unit. The medium was kept overnight in a refrigerator and then passed through a Seitz filter. Occasionally precipitation continued after filtration, but this was ignored as it did not affect the results. The sterile medium was

distributed into bijou bottles, leaving as little air space as possible. The bottles were incubated overnight to check sterility and on the next day inoculated from 24 agar or serum agar slope cultures. The colour of the medium at this stage is a definite clear pink. Any bottles in which the pH value has fallen to 9.2 or less will appear a pale orange-pink and can be discarded. This makes possible a check on each individual bottle before inoculation.

Uninoculated bottles and known positive and negative organisms were included in each set of tests. Tests were carried out at 37°C and 30°C. Controls were prepared of the buffered medium adjusted to pH 7.0 with bromothymol blue as indicator, and without indicator.

Tubes of glucose peptone water as used for fermentation tests were incubated concurrently in a number of tests. Only a small number of strains have been tested; the group D and group N strains maintained in the N. C. T. C., strains of *Strep. lactis* kindly provided by Dr. Shattock from her collection, representative N. C. T. C. strains of the other serological groups of streptococci, and strains of *Aerococcus viridans* (Williams, Hirch and Cowan 1953).

P. M. F. Shattock and A. Hirsh (1947). J. Path. Bact. 59, 495.
P. M. F. Shattock and A. T. R. Mattick (1943–1944). J. Hyg., Camb. 43, 173.
J. M. Sherman and P. Stark (1934). J. Dairy Sci. 17, 325.
R. E. O. Williams, A. Hirch and S. T. Cowan (1953). J. gen. Microbiol. 8, 475.

Growth at pH 9.6—*Streptococcus*

N. C. T. C. Methods (1954).

Toleration of pH 9.6

Media: 90 ml lemco broth
10 ml glycine (NaCl) NaOH Buffer
0.5 ml phenolphthalein.

Use N NaOH to adjust pH to 9.9. Leave for 24 hours in a cold room. Filter (Seitz). This results in a drop of pH to 9.6. Distribute aseptically into bijou bottles. Inoculate. Use an uninoculated bottle as control. Incubate at 37°C for 20 hours.

Tolerance of pH 9.6 is indicated by heavy growth and decolorization of the indicator.

(For greater details see the method of P. H. Clarke, p. 396.)

Growth at pH 9.6—*Streptococcus*

P. Morelis and L. Colobert (1958). Annls Inst. Pasteur, Paris 95, 568–587.

Le milieu à pH 9.6, que nous avons utilisé, est constitué par du bouillon ordinaire tamponné par du glycocolle à raison de 5 g p. 1 000 [33].

Toutes les souches à l'exception de deux s'y sont développées, habituellement de manière plus abondante que sur le milieu hypersalé.
[33] Shattock (P. M. F.) et Hirsch (W.). *J. Path. Bact.*, 1947, 59, 495.

Growth at pH 9.6 and 9.2—*Streptococcus*

W. E. Sandine, P. R. Elliker and H. Hays (1962). Can. J. Microbiol. 8, 161–174.

Sterile 1 N sodium hydroxide was added aseptically to 10-ml volumes of lactic broth* so that pH values of 9.2 and 9.6 were obtained (about 0.2 and 0.3 ml, respectively, were required). The resulting media were inoculated with one drop of inoculum from a 24-hour lactic broth as described in tests on growth temperatures, incubated 48 hours at 30°C, and observed for growth.
*P. R. Elliker, A. W. Anderson and G. Hannesson (1956). J. Dairy Sci. 39, 1611–1612 for "lactic agar."

Growth at pH 9.6—*Aerococcus catalasicus*

O. G. Clausen (1964). J. gen. Microbiol. 35, 1–8.

Cultures were incubated aerobically at 37°C unless otherwise stated. Growth tests were carried out in beef-infusion peptone phosphate broth with a pH of 9.6; incubation period, 72 hr.

Growth at pH 9.6—*Streptococcus*

R. Whittenbury (1965). J. gen. Microbiol. 38, 279–287.

Ability to initiate growth at pH 9.6: The method of Chesbro and Evans (1959) was used. Additional experiments with the same medium converted to a soft agar were also done. Cultures were incubated at 30°C.
W. R. Chesbro and J. B. Evans (1959). J. Bact. 78, 858.

GROWTH WITH KCN

Growth with KCN—enterobacteria

R. Buttiaux, with the collaboration of J. Flament and N. Baylly (1952). Annls Inst. Pasteur, Paris 83, 156–166.

Le milieu de Braun, Unat et Delibeyoglu. – Voici la formule originale du milieu de base :

Peptone blanche (nous utilisons une peptone trypsique).	10 g
ClNa	5 g
Na_2HPO_4	5.637 g
KH_2PO_4	0.226 g
Agar	20 g

Eau distillée	1 000 cm^3
pH	7.4 à 7.6

Nous préparons le milieu final de la façon suivante :

2.2 cm^3 de la base gélosée précédente, fondue par chauffage à l'ebullition pendant quelques minutes, sont répartis dans des petits tubes de 80 mm × 10 mm, bouchés au coton cardé : on les stérilise à l'autoclave (vingt minutes à 120°). Le volume restant est alors de 2 cm^3. Ils doivent être soigneusement capuchonnés pour éviter toute concentration, s'ils ne sont pas utilisés immédiatement. Au moment de l'emploi, on fait fondre le contenu du tube au bain-marie bouillant. On laisse refroidir jusqu'à 50°. On ajoute, avec précision, 0.1 cm^3 de la solution :

Cyanure de potassium	0.5 g
Eau distillée	100 cm^3

On plonge le tube dans un bain-marie à 45° pendant quelques minutes. On mélange alors, à son contenu, 3 öses (boucle de fil de platine de 6 mm de diamètre) d'une culture de vingt quatre heures, en eau peptonée (incubation à 37°), du germe étudié. On refroidit rapidement sous l'eau du robinet.

Les tubes ensemencés sont portés à l'étuve à 37° durant vingt-quatre heures. Les microbes capables de se développer en présence de cyanure de potassium forment, à la surface du milieu, un anneau blanchâtre nettement et facilement visible. Les germes réfractaires ne cultivent pas.

Certains détails doivent retenir l'attention :

1° La solution de cyanure de potassium doit être de préparation immédiate ou récente. On peut la conserver pendant huit jours, dans un flacon bouché maintenu à + 4°. E. W. Taylor dit qu'elle est capable de résister ainse pendant un mois ; notre expérience personnelle nous a montré que ce délai était excessif pour l'étude des B. paracoli et *Salmonella.*

2° Les quantités respectives de cyanure de potassium et de culture microbienne doivent être scrupuleusement observées. Lorsqu'elles ne le sont pas tout à fait, on observera, cependant, une culture de germes résistants dans la portion superficielle du milieu ; celle-ci se présentera, alors, non sous la forme d'un anneau, mais de colonies isolées moins facilement décelables.

3° La lecture peut être faite parfois en moins de vingt-quatre heures d'incubation à 37° ; dix-huit heures suffisent souvent. L'anneau n'apparaît, pour certains germes (B. paracoli 29, 911 et certains *E. intermedium),* qu'après quarante-huit heures de séjour à cette température. On confirmera donc, quand il sera nécessaire, par une lecture supplémentaire faite à ce moment.

4° Il faut toujours utiliser trois témoins :

Le germe étudié doit être ensemencé sur milieu de base non additionné de cyanure. Il doit cultiver en surface.

Un milieu de base + cyanure est inoculé avec un *Aerobacter.* On doit constater un anneau.

Un milieu de base + cyanure est ensemencé avec un *Escherichia coli.* On ne doit pas observer de culture en surface.

Braun, Unat et Delibeyoglu, Istanbul Seririyati, 1941, 23, 11.
Taylor (E. W.). *Bull. Hyg.*, 1951, 26, 919.

Growth with KCN—enterobacteria

R. Buttiaux, J. Moriamez and J. Papavassiliou (1956). Annls Inst. Pasteur, Paris 90, 133–143.

5° Le test de Braun au KCN sera réalisé de la façon suivante :

On prépare le milieu :

Peptone trypsique	10 g
ClNa	5 g
PO_4HNa_2	5.63 g
PO_4H_2K	0.22 g
Agar en poudre	1 g
Eau distillée	1 000 ml

On dissout par chauffage et ajuste à pH 7.5. On répartit exactement 2 ml par tube de 80 mm × 12 mm. On autoclave vingt minutes à 120°. Au moment de l'emploi, on ajoute 0.1 ml d'une solution de KCN à 0.5 p. 100 dans l'eau, conservée au réfrigerateur. On agite doucement. On inocule 3 öses bouclées (6 mm de diamètre, environ) de la culture de dix-huit à vingt-quatre heures en eau peptonée obtenue en 2 ; elles sont émulsionnées dans la totalité du milieu. Celui-ci est alors placé à l'étuve à 37° et examiné dix-huit à vingt-quatre heures plus tard. Un disque blanc net à sa surface indique le développement de la bactérie étudiée ; on dit que le test au cyanure de potassium est positif.

Growth with KCN—enterobacteria

J. G. Heyl (1957). Antonie van Leeuwenhoek 23, 33–58.

Medium for the KCN test according to Mϕller.

Peptone	10 g
Sodium chloride	5 g
Monopotassium phosphate	0.225 g
Sodium phosphate	5.64 g
Distilled water	1000 ml.

This medium is autoclaved for 20 minutes at 115°C, and to the cold medium are added 15 ml of a 0.5% solution of KCN in distilled water. Divide the medium in quantities of 1 ml in small tubes, which are quickly stoppered with sterilized paraffinated corks or with rubber stoppers after adding a thin layer of sterilized paraffin to the medium. The inoculated tubes must be read daily for 4 days, incubated at 37°C.

Growth with KCN—enterobacteria

M. Gershman (1961). Can. J. Microbiol. 7, 286.

Use of a tetrazolium salt for an easily discernible KCN reaction.

Møller's KCN medium is a notable contribution to the differentiation of several groups of enteric bacteria [1,3]. KCN sensitivity is particularly useful in the differential diagnosis of *Salmonella* (KCN–negative) from the frequently occurring slow lactose, sucrose, and salicin fermenting strains of *Escherichia freundii* (KCN–positive).

KCN broth base medium is available commercially, which facilitates its routine use as a diagnostic tool. Tubes are inoculated with one loopful of a 24-hour broth culture, incubated at 37°C, and observed daily for growth for 4 days. On occasion growth may be so slight, however, as to necessitate comparison with a control tube for proper interpretation.

Colorless solutions of various tetrazolium salts, in the presence of viable cells, are reduced to insoluble pigments. Kelly and Fulton [2] made use of this property in correcting the difficulty in discerning growth in semisolids by incorporating tetrazolium salts into the test medium and in this way inducing a cellular formation of red pigments.

Positive KCN reactions, therefore, when they do occur, can be rendered easily discernible by the use of a sterile KCN broth base containing a 0.005%, Seitz-filtered, aqueous solution of 2,3,5-triphenyltetrazolium chloride. Growth is revealed by the presence of a red precipitate.

[1] P. R. Edwards and M. A. Fife (1956). Appl. Microbiol. 4, 46–48.
[2] A. T. Kelly and M. Fulton (1953). Am. J. clin. Pathol. 23, 512.
[3] V. Møller (1954). Acta path. microbiol. scand. 34, 115–126.

Growth with KCN—various

K. J. Steel and J. Midgley (1962). J. gen. Microbiol. 29, 171–178.

The authors used the Rogers and Taylor (1961) modification of the Møller method (1954). Incubation was for two days at 37°C or three days at 30°C or 22°C.

K. B. Rogers and J. Taylor (1961). Bull. Wld Hlth Org. 24, 59.
V. Møller (1954). Acta path. microbiol. scand. 34, 115.

Growth with KCN—*Escherichia aurescens*

H. Leclerc (1962). Annls Inst. Pasteur, Paris 102, 726–741.

Test au cyanure de potassium. – Il est effectué selon la méthode de Buttiaux [2, 3].

[2] Buttiaux (R.). *Ann. Inst. Pasteur,* 1952, 83, 156.
[3] Buttiaux (R.). *Ann. Inst. Pasteur,* 1956, 90, 133.

Growth with KCN—*Staphylococcus, Micrococcus*

D. A. A. Mossel (1962). J. Bact. 84, 1140–1147.

Following Gelosa's (1960) suggestion, a representative number of cultures were also examined for growth in the presence of potassium cyanide. The procedure developed by Buttiaux, Moriamez, and Papavassiliou (1956) was used for this purpose. Three loopfuls of a fresh culture of the strain were emulsified in 8-mm tubes containing freshly prepared 0.025% KCN-semi-solid agar and incubated for 24 hr at 30°C. KCN tolerance was indicated by the formation of a zone of intensive growth in the upper few millimeters of the tubes.

R. Buttiaux, J. Moriamez and J. Papavassiliou (1956). Annls Inst. Pasteur, Paris 90, 133–143.

L. Gelosa (1960). G. Batt. Immun. 53, 309–316.

Growth with KCN—enterobacteria

W. H. Ewing (1962). *Enterobacteriaceae. Biochemical Methods for Group Differentiation.* U. S. Department of Health, Education and Welfare, Public Health Service Publication No. 734 (revised).

Test for growth in the presence of KCN (Møller, 1954; Edwards and Fife, 1956; Edwards, *et al.*, 1956)

Peptone, Orthana special	10 g
Sodium chloride	5 g
Monobasic potassium phosphate	0.225 g
Dibasic sodium phosphate	5.64 g
Distilled water	1000 ml
Adjust to pH 7.6	

The basal medium is sterilized at 121°C, 15 minutes, then refrigerated until thoroughly chilled. To the cold medium are added 15 ml of 0.5% KCN solution (0.5 g KCN dissolved in 100 ml cold sterile distilled water). The medium is then tubed in approximately 1 ml amounts in sterile tubes (12 × 150 mm or 13 × 100 mm) and *stoppered quickly* with corks sterilized by heating in paraffin. The medium in such tubes can be stored safely for 2 weeks at 4°C. The final concentration of KCN in the medium is 1:13,300. It has been found that 0.3% Bacto Proteose Peptone No. 3 may be substituted for Orthana Peptone.

Inoculation: The tubes are inoculated with 1 loopful (3-mm loop) of a 24-hour broth culture grown at 37°C.

Incubation: Tests are incubated at 37°C and observed daily for 2 days. Positive results are indicated by growth in the presence of KCN.

P. R. Edwards and M. A. Fife (1956). Appl. Microbiol. 4, 46.

P. R. Edwards, M. A. Fife and W. H. Ewing (1956). Am. J. Med. Technol. 22, 28.

V. Møller (1954). Acta path. microbiol. scand. 34, 115.

Growth with KCN—*Pseudomonas odorans*

I. Málek, M. Radochová and O. Lysenko (1963). J. gen. Microbiol. 33, 349–355.

Modifications described in the Report (1958) were used. Report of the Enterobacteriaceae Subcommittee of the Nomenclature Committee of the I. A. M. S. (1958). Int. Bull. bact. Nomencl. Taxon. 8, 25.

Growth with KCN—enterobacteria

G. S. Wilson and A. A. Miles (1964). Topley and Wilson's *Principles of Bacteriology and Immunity.* 5th edition – Arnold.

This test is likewise used for distinguishing between different organisms in the group of enterobacteria. The final concentration of KCN in the buffered peptone liquid medium is about 1/13,000 (see Report 1958). Report (1958). Int. Bull. bact. Nomencl. Taxon. 8, 25.

Growth with KCN—*Alcaligenes*

R. G. Mitchell and S. K. R. Clarke (1965). J. gen. Microbiol. 40, 343–348.

Moeller's (1954) method was used.

V. Moeller (1954). Acta path. microbiol. scand. 34, 115.

HEAT RESISTANCE

Heat resistance—*Serratia*

Bartolomeo Bizio (1823) – Translated by: C. P. Merlino (1924). J. Bact. 9, 527–543.

I took, therefore, a little glass globe, and suspended my bit of colored polenta in its centre; then I closed it with a cork through which I passed a small thermometer, the bulb of which was very near the colored polenta. After having sealed the neck of the globe, I raised the temperature up to 80° Réaumur* and I left it at this temperature for ten minutes. Then I took this bit of colored polenta and repeated the ordinary experiments; but the red color appeared just as soon and just as bright as before.

Seeing then that if this phenomenon was the result of a very minute plant, even 80° Réaumur were not sufficient to deprive its seeds of vitality, thus I wanted to subject them to still greater temperatures to see if they would be made incapable of reproducing their kind. I buried the globe in sand, and at first raised the temperature to 100° Réaumur, but finding no sensible difference in the results, I made another attempt and raised the temperature to 120° Réaumur, and left the colored polenta at such a heat for only five minutes. Although in this last experiment I reduced the time to a half of that allowed in other experiments, yet it

was sufficient, and perhaps too much so, to kill entirely.
*80°R = 100°C.

Heat resistance—*Microbacterium*

Orla-Jensen (1919). *The Lactic Acid Bacteria.* Copenhagen.

Resists heating to 72°C for 15 minutes.

Mémoires de l'Académie Royale des Sciences et des Lettres de Danemark, Copenhague, Section des Sciences, 8me série, t.V. no. 2.

Heat resistance—*Streptococcus*

J. M. Alston (1928). J. Bact. 16, 397–407.

To test the heat resistance of the organisms, 2 ml of a young phosphate broth culture were pipetted into each of two sterile and stoppered Wassermann tubes. One tube was placed in a water bath at exactly 60°C for ten minutes, and the other for fifteen minutes. After this immersion the tubes were cooled in water at room temperature, and two large loopfuls of each tube were plated on blood-agar.

Heat resistance—*Streptococcus*

J. M. Sherman and P. Stark (1934). J. Dairy Sci. 17, 525–526.

Thermal death rates: When placed in sterile skimmed milk and heated for 30 minutes at 65°C, all of the *lactis* cultures were killed while all of those of the *fecalis* group survived. As would of course be expected, when subjected to lower heat treatments the *fecalis* group survived in much greater numbers than did the cultures of *Streptococcus lactis.*

Heat resistance—*Streptococcus*

C. E. Safford, J. M. Sherman, H. M. Hodge (1937). J. Bact. 33, 263–274.

Heat resistance tests were made in milk. Ten milliliters of sterile skimmed milk were added to 1 ml of culture, and the heavily seeded mixture was heated in a sealed tube in a water bath.

Heat resistance—*Mycobacterium*

R. E. Gordon (1937). J. Bact. 34, 617–630.

Survival at 60°C for one hour. Three milliliters of ten days' old cultures in glycerol-phosphate broth were pipetted into 5 ml ampoules. The ampoules were sealed, fastened in a wire rack, and immersed in a 60°C, constant temperature water bath for one hour. At the end of this time the ampoules were cooled, their tips broken, and the contents pipetted onto glycerol-phosphate agar slants. If growth appeared on the slopes, it was compared macroscopically and microscopically with the original culture.

Heat resistance—*Bacteroides* and *Lactobacillus*

K. H. Lewis and L. F. Rettger (1940). J. Bact. 40, 287–307.

Thermal resistance was tested in melted glucose-cysteine semisolid agar tubes inoculated with 0.1 ml of a 48 hour glucose-cysteine broth culture, and exposed to 60°C for 5 minute intervals over a period of 30 minutes. Upon removal, tubes were immediately water-cooled until solidified, then incubated at 37°C for two weeks.

Heat resistance—*Lactobacillus, Streptococcus*

H. J. Peppler and W. C. Frazier (1942). J. Bact. 43, 181–192.

The method of Peppler and Frazier (1941) in the preparation of cultures and determination of their activity at 37°C after heat treatment at 60° to 64°C for 30 minutes was followed.

H. J. Peppler and W. C. Frazier (1941). J. Dairy Sci. 24, 611–623.

Heat resistance—*Streptococcus*

P. M. F. Shattock and A. T. R. Mattick (1943–1944). J. Hyg., Camb. 43, 173.

Two drops of an 18 hr culture in yeast-dextrose broth (peptone 2 per cent, lemco 1 per cent, yeastrel 0.3 per cent, NaCl 0.5 per cent, dextrose 0.5 per cent, pH 7.0) were inoculated into 10 ml of the same medium previously heated to 60°C. The tubes were held in a water bath at 60°C for 30 mins: cooled and incubated at the optimum temperature. The presence or absence of growth, compared with unheated control tubes, was recorded after 24 and 48 hr.

Heat resistance—*Streptococcus*

K. E. Hite and H. C. Hesseltine (1947). A study of aerobic streptococci isolated from the uterus and vagina. J. infect. Dis. 80, 105–112.

The ability of strains to resist 60°C for 30 minutes was tested in 0.1% dextrose-veal infusion broth, the tubes being immersed in water bath adjusted to this temperature. Survival was determined by subculturing onto blood agar plates.

Heat resistance—*Staphylococcus* and *Micrococcus*

Y. Abd-el-Malek and T. Gibson (1948). J. Dairy Res. 15, 249–260.

The method used was the same as that previously described (Abd-el-Malek and Gibson, 1948, J. Dairy Res. 15, 233) for the examination of streptococci, except that the cells to be heated were grown in sugar-free broth. The essential features of the procedure are the heating in milk and the immediate plating of the heated suspension. A method which is at least semiquantitative is required in order to secure consistent results.

Heat resistance—*Streptococcus* (survival at 63°C)

Y. Abd-el-Malek and T. Gibson (1948). J. Dairy Res. 15, 233–248.

Resistance to heat is a useful differential characteristic among the streptococci, provided an adequate technique is used for its detection. Several factors are known to influence the result of a test of this property. Of special importance for the purpose of differentiation is the form of the test. Experience confirms the theoretical deduction that unless the organisms to be distinguished differ rather widely in the level of their resistance, the outcome of a test that depends on the complete sterilization of a bacterial suspension will show much variation. A carefully standardized quantitative method is too time-consuming for routine identifications, and the following procedure, which affords results of sufficient uniformity for the end in view, was used throughout this work.

In a sterile 5/8 in tube are placed 0.2 ml of a fresh, moderately turbid lactose broth culture and then 5 ml of skim milk (sterilized by intermittent steaming). Care is taken to avoid contaminating the side of the tube. The tube is placed in a water bath at 63°C, with the milk level well below the water level, and is heated for 30 min after the temperature of the milk has risen to 63°C. The tube is then cooled and a plate of glucose agar (the medium used in plating the milk samples) is poured with 0.2 ml of the milk. The plate is incubated at 30°C and is examined for growth after 3–4 days.

For every bacterium there is certain to be some heat treatment which permits the survival of a small proportion of cells, so that a few colonies will be produced in certain trials and none in others. Instances have been noted, for example, when *Str. lactis* was heated at 60°C and when *Str. kefir* was heated at 63°C. The choice for differential purposes of 63°C was made chiefly because this temperature is employed in the pasteurization of milk. An additional consideration is that resistance at 60°C, which has been more commonly used for specifying the thermoduric streptococci, is less clearly linked with the easily determined and valuable diagnostic character, the ability to grow at 45°C. Milk was adopted as the medium in which to heat the organisms partly because of the relative constancy of its composition and partly on account of the application of the results to dairy practice.

Heat resistance—*Streptococcus*

P. F. Swartling (1951). J. Dairy Res. 18, 256–267.

Two loopfuls (4 mm) of an 18 hr culture (D. L. B.) were transferred to D. L. B. held at 60°C for 30 min in a thermostatically controlled water bath. The time was reckoned from the moment when the broth in a control tube fitted with a thermometer had reached 60°C. After 30 min exposure the tubes were cooled in tap water, incubated at 30°C for 24 hr and examined for growth.

A second survival test, described by Demeter (1929), was carried out, using milk fortified with Yeastrel.

K. J. Demeter (1929). Milchw. Forsch. 8, 215.

Heat resistance—general

P. Hauduroy (1951). Techniques Bactériologiques. Masson et Cie, Éditeurs, Paris.

THERMORESISTANCE

(Anaérobies). – On ensemence une série de bouillon Vf glucosé à 2 p. 1000 que l'on soumet aux températures suivantes avant d'être placés :

à 70° pendant 15 minutes ;
à 80° pendant 5–10 minutes ;
à 100° à partir d'une minute.

Un germe qui résiste 15–20 minutes à 70° est présumé sporulé.

(A. I. P.).

Heat resistance—*Mycobacterium*

R. E. Gordon and M. M. Smith (1953). J. Bact. 66, 41–48.

Subcultures were prepared on glycerol agar slants, quickly preheated to 60°C and placed in a water bath at the same temperature. After 4 hours they were removed, immediately cooled, stored at 28°C for 2 weeks, and observed for growth.

Heat resistance—*Streptococcus*

G. Andrieu, L. Enjalbert and L. Lapchine (1954). Annls Inst. Pasteur, Paris 87, 617–634.

Pour l'étude de la résistance à la chaleur nous procédons de la façon suivante : en une ou deux ampoules contenant environ 2 cm^3 de bouillon sont ensemencées de II à III gouttes de la culture de vingt-quatre heures du streptocoque à étudier. Les ampoules sont scellées, puis immergées complètement, dans un bain-marie à 61–62°C pendant trente minutes. On met ensuite les ampoules à l'étuve à 37°C sans les ouvrir. La lecture se fait après vingt-quatre et quarante-huit heures. Si le germe résiste à la chaleur, la pousse est évidente et indiscutable.

Heat resistance—*Streptococcus*

N. C. T. C. Methods 1954.

Method 1. (Skadhauge)

Place 2 ml of 20 hour serum broth culture into a small test tube. Place in a water-bath adjusted to 60°C for 30 minutes precisely. Cool under a cold tap, and incubated at 37°C for 20 hours. Subculture to a serum agar slope and incubate for 20 hours.

Growth shows ability of the strain to survive heating at 60°C for 30 minutes.
K. Skadhauge (1950). *Studies on enterococci.* p62. Copenhagen: Einar Munksgaard.
Method 2. (Abd-el-Malek and Gibson)
Pipette 0.2 ml lactose broth culture into a sterile 5/8-inch tube and add 5 ml skimmed milk (sterilized by intermittent steaming), taking care to avoid soiling the side of the tube. Place in a water-bath at 63°C, with the milk level below the water level. Allow time for the milk to warm up to 63°C, and remove tube from the bath 30 minutes later. Cool under running water, and incorporate 0.2 ml in a glucose agar pour plate.
Y. Abd-el-Malek and T. Gibson (1948). J. Dairy Res. 15, 233.

Heat resistance—*Nocardia*

R. E. Gordon and J. M. Mihm (1957). J. Bact. 73, 15–27.

Survival of 60°C. The cultures were inoculated on slants of yeast dextrose agar, quickly heated to 60°C in a water bath, then transferred to a water bath at the same temperature inside a constant temperature incubator. After 4 hr they were quickly cooled, incubated at 28°C for 2 weeks, and examined for growth.

Heat resistance—*Listeria*

R. E. Bearns and K. F. Girard (1958). Can. J. Microbiol. 4, 55–61.

Each *L. monocytogenes* strain was inoculated onto a Difco tryptose agar slope and incubated at 37°C for 18–24 hours. The growth from each slope was suspended in distilled water, and aliquots of the suspensions were adjusted by dilution to yield 5×10^8 viable organisms per ml. For each strain a series of nine screw-capped test tubes (20 × 150 mm) in duplicate containing 9 ml of sterile skim milk was inoculated with serial dilutions of the standard suspensions from 1:10–1:10,000,000. The ninth tube was not inoculated and served as a control. The milk used was pasteurized skim milk obtained from a local milk supplier. This milk was divided into aliquots of 300 ml and placed in screw-capped quart bottles. The bottles were autoclaved at 115°C for 15 minutes, cooled in the refrigerator at 4°C for 2 to 3 hours, then placed in a water bath, and maintained at 62°C for 35 minutes. Tests on this milk showed the samples to be sterile.

Following addition of *L. monocytogenes* to the sterile milk samples, these were then pasteurized in a constant temperature water bath which brought the temperature of the milk in the tubes to 61.7°C within 2 minutes. The milk surface in the tubes was kept 3 to 4 cm below the water level in the bath, and during the entire course of the experiment the tubes were agitated. At the end of the pasteurization period the number of survivors was determined by a modification of the drop plate method (Reed and Reed, 1948). The sample was thoroughly mixed and 0.1 ml was pipetted onto the surface of a Difco tryptose agar plate. The fluid added was spread over the agar surface by gently tilting the plate and, after drying, was incubated at 37°C for 48 hours. Colony counts made at this time compared closely with drop plate and pour plate results on the same samples.

The water bath (temperature 61.7°C ± 0.5°C) used for pasteurization was covered with a sloping plexiglass-topped hood having a slot in one end. A Boerner shaker with an attached horizontal aluminum bar was placed end to end with the water bath so that the bar entered the slot in one end of the hood. A wire test tube holder was suspended from the bar inside the hood so that it was immersed in the water and would agitate freely. The tubes containing the milk samples were allowed to "bounce" in the holder to aid in the dispersion of the organisms throughout the milk samples. A control sample containing a specially fitted thermometer showed that the temperature of the milk sample inside the tubes was the same as that of the water outside the tubes ±0.5°C over the pasteurization period.

R. W. Reed and G. B. Reed (1948). Can. J. Res. 26, 317–326.

Heat resistance—*Streptococcus*

P. Morelis and L. Colobert (1958). Annls Inst. Pasteur, Paris 95, 568–587.

Les tests de Sherman. – L'épreuve de résistance à la chaleur a été pratiquée de la manière préconisée par Nyman [24]. Cinq souches seulement, qui, au demeurant, ont bien les caractères des entérocoques, n'ont pas résisté à l'épreuve 3 d'entre elles proviennent d'excréments de vaches, 1 d'excréments de cobaye, 1 de fèces humaines, 1 provient d'une hémoculture dans un cas d'endocardite. Il faut noter que, pour certaines souches provenant principalement de fèces animales, cette résistance n'a pu être mise en évidence qu'après répétition de l'épreuve. Cette inconstance a été signalées par d'autres auteurs [32, 36] et White [41] a observé des variations considérables de la résistance à la chaleur en fonction de l'âge des cellules et de la température d'incubation préliminaire.

La culture à 45° a été pratiquée par ensemencement sur 5 cm^3 de bouillon ordinaire, de II gouttes d'une préculture de 24 heures sur bouillon. Les tubes ont été placés dans un bain-marie chauffé à 45° et les résultats lus après quarante huit heures. Toutes les souches à l'exception de 7 se sont développées dans ces conditions. Les 7 souches défaillantes avaient été isolées d'excréments de porcs et de vaches.

[24] Nyman (O. H.). *Acta path. micr. scand.*, 1949, suppl. 83.
[32] Shattock (P. M. F.). *Ann. Inst. Pasteur de Lille,* 1955, 7, 95.

[36] Skadhauge (K.). *Studies on Enterococci.* Einar Munskgaard. édit., Copenhague, 1950.
[41] White (J. C.) et Sherman (J. M.). *J. Bact.*, 1944, 48, 262.

Heat resistance—*Mycobacterium*

R. E. Gordon and J. M. Mihm (1959). J. gen. Microbiol. 21, 736–748.

Survival at 60°C: Slants of glycerol agar were inoculated, quickly heated to 60°C in a water bath, then placed in another water bath at 60°C inside a constant temperature incubator. After 4 hr the cultures were rapidly cooled, incubated at 28°C for 14 days and inspected for growth.

Heat resistance—atypical ***Mycobacterium***

A. Beck (1959). J. Path. Bact. 77, 615–624.

Examined by exposure of a freshly inoculated Lowenstein culture in a water bath to 60°C for 4 hr and subsequent incubation at 37°C (Gordon and Smith, 1953).

R. E. Gordon and M. M. Smith (1953). J. Bact. 66, 41–48.

Heat resistance—*Leuconostoc*

E. I. Garvie (1960). J. Dairy Res. 27, 283–292.

The heat resistance of the strains was examined by inoculating 0.04 ml of a culture into YG citrate, heating for the required time at the temperature chosen, and then incubating the tubes for 3 days and examining for growth. The ability to grow at 37°, 39° and 40°C was tested in YG citrate and the cultures observed daily for growth for 7 days.

Heat resistance—*Sarcina* and ***Bacillus***

R. E. MacDonald and S. W. MacDonald (1962). Can. J. Microbiol. 8, 795–808.

Heat Sensitivity. In order to determine the heat sensitivity of vegetative cells, a 10 ml sample of an aerated 15 to 18 hour culture (*c.* 5×10^8 cells/ml) at 30°C in AC broth* was washed once and resuspended in 1 ml of gelatin-saline diluent. A screw-cap test tube containing 9 ml of diluent was immersed for 15 minutes in a water bath which had been equilibrated at the desired temperature for at least 30 minutes. The washed cell suspension (1 ml) was then added aseptically to the 9 ml of diluent at zero time. A duplicate sample was used to determine the viable cell concentration before heat treatment. Samples were removed at each of the chosen time intervals, diluted, and plated on AC agar† at pH 8.4. Plates were incubated at 30°C, and counts were recorded at the end of 5 days. A similar technique was used for spores, except that 0.1% soluble starch was added to the counting medium. This was found to

decrease the variation in counts on duplicate plates.
*AC broth: 1% glucose, 0.4% peptone, and 0.4% yeast extract, adjusted to pH 8.4 after sterilization.
†AC agar: AC broth with 2% agar and 0.1% soluble starch added.

Heat resistance—*Streptococcus*

W. E. Sandine, P. R. Elliker, and H. Hays (1962). Can. J. Microbiol. 8, 161–174.

Tubes were inoculated from 24 hr lactic broth* cultures by adding one drop of inoculum to 10-ml quantities of sterile lactic broth. Two 3 ml aliquots then were placed in small pyrex tubes to which strings were attached. The tubes were sealed and then completely submerged in a 60°C water bath for 30 minutes. The culture tubes then were chilled in ice water and incubated at 30°C for 24 hours. Nonheated controls were run simultaneously on each culture.
*See P. R. Elliker, A. W. Anderson and G. Hannesson (1956). J. Dairy Sci. 39, 1611–1612 for "lactic agar."

Heat resistance—*Azotobacter*

A. J. Garbósky and N. Giambiagi (1963). Annls Inst. Pasteur, Paris 105, 202–208.

L'objet du présent travail a été d'étudier la thermorésistance, et la morphologie cellulaire correspondante, de la famille *Azotobacteraceae,* en utilisant des espèces des genres *Azotobacter, Beijerinckia et Derxia.*

Le milieu de culture utilisé a été : glucose, 10 g ; PO_4HK_2, 0.5 g ; $SO_4Mg \cdot 7H_2O$, 0.2 g ; ClNa, 0.2 g ; SO_4Ca, 0.1 g ; $SO_4Mn \cdot 4H_2O$, traces ; $FeCl_3 \cdot 6H_2O$, traces ; $MoO_4Na_2 \cdot 2H_2O$ à 5 p. 100, 2 cm^3 ; eau distillée, 1 000 cm^3 ; gélose, 20 g.

Preuve de thermorésistance. – Dans un bain d'eau à une température requise et contrôlée rigoureusement avec thermomètre, on plonge simultanément des tubes stériles avec 1 ml de suspension de chacune des espèces considérées. Ces suspensions s'obtiennent en ajoutant 3 ou 4 ml d'eau stérile à une culture sur tube de gélose et en agitant avec l'anse. Après avoir soumis les suspensions à l'action de la température pendant le temps choisi, on refroidissait à la température ambiante. Ensuite, avec des pipettes Pasteur, on ensemençait dans des tubes de gélose pour constater la survivance de chaque souche et on mettait à l'étuve à 30°. Les lectures se faisaient au quatrième jour de culture.

Heat resistance—*Aerococcus catalasicus*

O. G. Clausen (1964). J. gen. Microbiol. 35, 1–8.

Cultures were incubated aerobically at 37°C unless otherwise stated.

Heat resistance was tested by heating a 20 hr beef infusion peptone phosphate broth culture at 60°C (in a water bath) for 30 min, cooling, and subculturing to new broth and blood agar. The media were incubated for 48 hr before the results were recorded. Each broth culture to be tested was drawn sufficiently far up a Pasteur pipette to enable the ends of the pipette to be sealed without heating the cultures. The pipette was then plunged so deeply into a water bath at 60°C that the entire section moistened with culture was submerged.

Heat resistance—*Moraxella*

W. J. Ryan (1964). J. gen. Microbiol. 35, 361–372.

Heavy bacterial suspensions in peptone water were placed in small screw-capped bottles and immersed in water baths at 56°C and 100°C. At intervals a large loopful was removed and plated on blood agar.

Heat resistance—enterococci

C. G. Rogers and W. B. Sarles (1964). J. Bact. 88, 965–973.

Ability to grow after exposure to 60°C for 30 min in tryptose yeast-extract broth was tested.

Each culture was also subjected to the heat tellurite tolerance test of Cooper and Ramadan (1955).

K. E. Cooper and F. M. Ramadan (1955). J. gen. Microbiol. 12, 180–190.

Heat resistance—*Clostridium*

C. T. Huang, K. Tamai and S. Nishida (1965). J. Bact. 90, 391–394.

A 24-hr culture in 10% chopped-meat broth was heated to 80°C for 10 min, 90°C for 30 min, or 100°C for 10 min, as prescribed. The heated culture was immediately transferred into a tube of 1% glucose chopped-meat broth and incubated at 37°C for 24 hr.

Heat resistance—*Streptomyces albus*

T. F. Fryer and M. E. Sharpe (1965). J. Dairy Res. 32, 27–34.

This was determined by a modification of the technique of Stern and Proctor (1954) described by Franklin, Williams and Clegg (1958). The temperatures used were 62.7°C (145°F) and 71.7°C (161°F), being those commonly used for the pasteurization of milk, with a range of times. A comparison of a number of media showed that recovery of the heat-shocked cells was most successful on nutrient agar containing 10% skim milk. The number of survivors was estimated by the Miles and Misra (1938) technique, the plates being incubated for 3 days in the case of the mycelium and 5 days for the conidia. A longer incubation period was required for the conidia, as some of them showed a considerable lag before germinating.

J. G. Franklin, D. J. Williams and L. F. L. Clegg (1958). J. appl. Bact. 21, 151.

A. A. Miles and S. S. Misra (1938). J. Hyg., Camb. 38, 732.
J. A. Stern and B. E. Proctor (1954). Fd Technol., Champaign 8, 139.

Heat resistance—*Bacillus* spores

G. R. F. Hilson (1965). J. gen. Microbiol. 39, 407–421.

Heat resistance. From each suspension before heating a standard loopful was taken and spread evenly over the surface of a nutrient agar plate. Samples (2 ml) of each suspension were placed in ¼ oz bijou bottles, which were totally immersed in a water bath already at the required temperature. The bottles were shaken gently during the first 5 min of each period of heating to ensure the even distribution of heat through the suspension. After the chosen period of time had elapsed, a loopful of the heated suspension was spread on a plate of nutrient agar as in the case of the unheated control. All the plates were incubated overnight and the amounts of growth on each compared. The following temperature time exposures were used: 70°C 30 min, 80°C 30 min, 90°C 20 min, 100°C 20 min.

Heat resistance—*Salmonella*

R. P. Elliott and P. K. Heiniger (1965). Appl. Microbiol. 13, 73–76.

This paper describes the construction of a temperature gradient incubator for testing temperature of growth and heat resistance.

Heat resistance—*Streptomyces*

T. F. Fryer and M. E. Sharpe (1965). J. Dairy Res. 32, 27–34.

Preservation of strains

Strains were stored at 4°C on agar slopes of the ammonium lactate medium (AL) of Gyllenberg, Eklund, Antila & Vartiovaara (1960). Growth on this medium resulted in an abundant yield of conidia which facilitated subculturing.

Cultural medium

At first the organism was grown in nutrient broth, but a comparison of the growth in 4 media, viz. nutrient broth, Yeastrel nutrient broth (YNB), glucose nutrient broth and Yeastrel glucose nutrient broth showed that YNB gave the best growth. This medium was used for general cultural work.

Preparation of mycelium free from conidia

Tubes containing 5 ml of YNB were inoculated with growth from the AL slopes and incubated in a sloped position for 24 hr, the increased aeration encouraging both growth of mycelium and germination of conidia. Three successive transfers were made in this manner, using 0·1 ml inoculum. It was necessary at each subculture to ensure that none of the inoculum ran down the side of the tube, in order to prevent the growth and sporulation thereupon of individual colonies of streptomyces.

One tube was selected in which no aerial mycelium was evident and the mycelial aggregates were broken up using a sterile Griffiths' tube. The mycelial fragments were centrifuged, washed once in saline and re-suspended in 5 ml of sterile skim milk. This culture was incubated in the sloped position for 4 hr, so that cells damaged by homogenization might have a chance to recover.

Preparation of conidial suspension

After 7 days' incubation, the growth of 5 AL agar slopes was harvested into 2 ml of skim milk containing a wetting agent 2% Tween 80. The suspension was homogenized in Griffiths' tube.

Heat resistance

This was determined by a modification of the technique of Stern & Proctor (1954) described by Franklin, Williams & Clegg (1958). The temperatures used were 62·7°C (145°F) and 71·7°C (161°F) being those commonly used for the pasteurization of milk, with a range of times. A comparison of a number of media showed that recovery of the heat-shocked cells was most successful on nutrient agar containing 10% skim milk. The number of survivors was estimated by the Miles & Misra (1938) technique, the plates being incubated for 3 days in the case of the mycelium and 5 days for the conidia. A longer incubation period was required for the conidia as some of them showed a considerable lag before germinating.

J. G. Franklin, D. J. Williams and L. F. L. Clegg (1958). J. appl. Bact. 21, 151.

H. Gyllenberg, E. Eklund, M. Antila and V. Vartiovaara (1960). Acta agric. scand. 10, 50.

A. A. Miles and S. S. Misra (1938). J. Hyg., Camb. 38, 732.

J. A. Stern and B. E. Proctor (1954). Fd Technol., Champaign, 8, 139.

Heat resistance—*Staphylococcus*

G. C. Walker and L. G. Harmon (1966). Appl. Microbiol. 14, 584–590.

This was measured in milk, whey and phosphate buffer.

The organisms were transferred daily in a broth composed of 3.7% Brain Heart Infusion, 2% mannitol, and 1% yeast extract. To prepare a solid medium for plate counts, 2% agar was added. After preliminary experimentation with different media and combinations of ingredients, including Plate Count Agar, Veal Infusion Agar, Beef Lactose Agar, Dextrose Tryptone Agar, and the several media recommended for growing the staphylococci, the above medium was selected as the best for propagation of cultures and recovery of sublethally heat-treated cells.

Preparation of heating media. Reconstituted skim milk containing 10% solids-not-fat was prepared from nonfat dry milk and distilled water. Reconstituted whole milk containing 3.6% fat and 9.4% solids-not-fat was prepared from nonfat dry milk, sweet cream, and distilled water, and was homogenized at 2000 psi. The *p*H of the milks was adjusted to 6.65. Cheddar cheese whey containing 0.4% milk fat and with a *p*H of 6.5 was obtained immediately after the curd was cut during manufacture of cheddar cheese. Phosphate buffer (0.067 M, *p*H 7.0) was prepared. The above media and buffer were dispensed in 200-ml quantities in 8-oz glass jars, sterilized at 121°C for 15 min, and stored at 0°C until used.

Preparation of organisms for thermal destruction trials. Active cultures were inoculated into 6-oz bottles containing 50 ml of broth and glass beads and incubated for 12 hr at 37°C. In cultures of *S. aureus,* the cells tend to clump. Satisfactory dispersion of these clumps, as determined by microscopic examination, was accomplished by agitating the cultures on a mechanical shaker for 5 min at 350 gyrations per min immediately prior to inoculation into the heating medium. This procedure facilitated determination of the number of cells inoculated.

Thermal destruction. Thermal death time studies were made in the heating media previously described. Two strains of *S. aureus* (161-C and S-1) were tested in all four media. Strains B-120 and S-18 were tested only in phosphate buffer and whole milk. Strain B-120 was used to determine the effect of age on the thermal resistance of cells.

The apparatus used for heating the inoculated medium was similar to the unit designed by Kaufmann and Andrews (1954). The 200 ml of heating medium, maintained under continuous uniform agitation and adjusted to the test temperature, was inoculated with 1 ml of the broth culture of *S. aureus* injected with a syringe and needle. The decrease in temperature caused by the inoculum was less than 0.25°F (0.14°C) as measured on a recording potentiometer.

At zero-time, the number of cells in the inoculated heating medium ranged from 6.2×10^6 to 19×10^6 per milliliter. Approximately 5 ml of the medium was withdrawn at 1- or 2-min intervals with a syringe and needle. The samples were promptly placed in sterile chilled test tubes and agitated in an ice bath for 20 sec to prevent further destruction or multiplication of the surviving cells. The survivors were enumerated on the agar medium previously described by use of the quantitative surface inoculation technique of Punch and Olson (Abstr. J. Dairy Sci. 44: 1160, 1961) with 1 ml spread on five plates, or approximately 0.2 ml per plate. The plates were incubated for 48 hr at 37°C.

O. W. Kaufmann and R. H. Andrews (1954). J. Dairy Sci. 37, 317–327.

Heat resistance—*Mycobacterium*

M. Tsukamura (1966). J. gen. Microbiol. 45, 253–273.

Survival at 60°C for 1 hr was observed on the Löwenstein-Jensen medium.

HEMOLYSIS

Hemolysis—*Escherichia*

M. L. Orcutt (1928). J. Bact. 16, 123–134.

Horse, sheep, cow, rabbit, guinea pig, and human corpuscles were used. The test was made by mixing 0.5 ml of a twenty-four-hour bouillon culture with 0.5 ml salt solution. The corpuscles were washed three times and suspended in salt solution to the original volume of the blood. One drop of this suspension was added to the 1 ml of culture dilution and the mixture incubated at 37°C for two hours and refrigerated overnight. The results showed strong or complete hemolysis of all corpuscles.

Hemolysis—*Staphylococcus*

G. H. Chapman, C. Berens, A. Peters and L. Curcio (1934). J. Bact. 28, 343–363.

Oxalated rabbit blood was used for hemolysis in agar plates incubated for 24 to 48 hours.

Hemolysis—(double zone)

J. H. Brown (1937). J. Bact. 34, 35–48.

The method for demonstrating the double zones is as follows:

To a tube of infusion agar, melted and cooled to 45 to 50°C, is added about 1 ml of defibrinated blood and a loop of suitably diluted culture. The dilution of the culture should be such that there may be fewer than 100 colonies in the plate. After thorough mixing the inoculated medium is poured into a Petri dish and allowed to harden. The plates are incubated at 37°C and then refrigerated. During the first 24 hours of incubation deep colonies with small zones of hemolysis appear and become larger as incubation is continued. Usually the zones of hemolysis have a rather poorly defined periphery, but they will be found to be clear immediately next to the colony and are therefore of the beta type. To determine this, the colony should be viewed in optical section under the 16 mm objective of the microscope. When the plate is refrigerated there appears a broad outer zone of partial hemolysis, resulting in the so-called "double zone." Between the inner clear zone and the outer partly hemolyzed zone is a fairly distinct ring or zone of unhemolyzed blood corpuscles. Throughout the zones there is no greenish or brownish discoloration such as appears in alpha zones.

Hemolysis—*Streptococcus*

C. E. Safford, M. M. Sherman and H. M. Hodge (1937). J. Bact. 33, 263–274.

Action on blood in poured agar plates was determined on subsurface colonies, the methods recommended by Brown (1919) being faithfully followed.

J. H. Brown (1919). Monogr. Rockefeller Inst. med. Res. No. 9.

Hemolysis—*Streptococcus*

J. B. Gunnison, M. P. Luxen, J. R. Cummings and M. S. Marshall (1940). J. Bact. 39, 689–708.

Cultures were grown for 12–18 hours in beef heart infusion broth containing particles of meat. Then 0.5 ml of the cultures was mixed with 0.5 ml of a 5% suspension of rabbit cells and incubated for 1 hour at 37°C.

Hemolysis (double zone)—*Streptococcus*

J. B. Gunnison, M. P. Luxen, J. R. Cummings and M. S. Marshall (1940). J. Bact. 39, 689–708.

Pour plates were made from cultures diluted so as to give not more than 100 colonies in beef-heart-infusion agar containing 1% Difco neopeptone and mixed with 5% rabbit blood. Plates were incubated at 37°C for 48 hours and were then placed in the ice box overnight as recommended by Brown (1939).

J. H. Brown (1939). J. Bact. 37, 133–144.

Hemolysis—*Corynebacterium acne*

H. C. Douglas and S. E. Gunter (1946). J. Bact. 52, 15–23.

On peptone yeast extract glucose phosphate agar containing 5.0 per cent by volume of citrated human blood, all strains produced beta-hemolysis.

Hemolysis—*Streptococcus*

E. J. Hehre and J. M. Neill (1946). J. exp. Med. 83, 147–162.

The tests for hemolytic capacity were made only on strains that produced dextran or levan. The capacity of the colonies to produce zones of hemolysis was studied on 5 per cent rabbit blood agar and on 5 per cent horse blood agar; one series of each medium was incubated aerobically and another series in a Brewer anaerobic jar; the plates were streaked on the surface with dilutions of the culture that gave well separated colonies. The aerobically incubated plates were examined after 1 and 2 days; the anaerobically incubated ones were examined after 2 days. In the anaerobic series on horse blood agar the hemolysis was much sharper and clearer than in any of the other three.

The capacity to produce soluble lysin was studied with cultures in an enriched broth that gave large amounts of lysin with control strains of groups A and C hemolytic streptococci. The fluids were tested, both before and after reduction with $Na_2S_2O_4$,* against 1 per cent suspensions of horse, rabbit, and sheep red blood cells. None of the dextran-forming strains gave any hemolysis.

Oxidation of hemoglobin

On the aerobically incubated plates, most of the dextran-forming endocarditis and group H throat streptococci caused greening around the colonies, which in some instances was pronounced. The question of oxidation of hemoglobin was investigated with the entire collection of streptococci by observations of the brown color that developed when equal amounts of 18 hour cultures grown in meat infusion broth were incubated at 37°C for 1 hour with 5 per cent suspensions of sheep red blood cells.

*J. M. Neill and T. B. Mallory (1926). J. exp. Med. 44, 241.

Hemolysis—*Streptococcus*

K. E. Hite and H. C. Hesseltine (1947). A study of the aerobic streptococci isolated from the uterus and the vagina. J. infect. Dis. 80, 105–112.

The hemolytic activity of strains was tested in 3.0% sheep blood agar pour plates, observations being made after successive incubation at 37°C, refrigeration for 24 hours and reincubation for 24 hours. Final observations were made with the aid of low power magnification to determine destruction of the red cell stromata. Only those organisms that destroyed the cell stromata were classified as hemolytic. Organisms that produced multiple zones of partial clearing and greening were classified as alpha-type, those that caused a slight greenish discoloration of the medium as greening; and those that caused no visible change as gamma-type streptococci.

Hemolysis—*Haemophilus, Alcaligenes*

P. Thibault, S. Szturm-Rubinsten and D. Piéchaud-Bourbon (1947). Annls Inst. Pasteur, Paris 88, 246–250.

Hémolyse. – Le pouvoir hémolytiques des cultures a été étudié en présence de globules humains et de globules de mouton en milieux solides et liquides. L'ensemencement des géloses au sang se faisait à la surface des plaques en stries, en points séparés et aussi dans la masse du milieu liquéfié à dilution convenable pour obtenir des colonies isolées incluses dans la gélose. L'épreuve en milieu liquide se pratiquait en eau peptonée suivant la technique de Kauffmann [6].

A 2 cm^3 d'une solution peptonée à 1 p. 100, on ajoute 0.2 cm^3 d'une suspension d'hématies lavées trois fois; on ensemence à l'aide du fil avec une culture de vingt-quatre heures sur gélose; on place les tubes à 37° pendant quarante-huit heures; on lit après vingt-quatre et quarante-huit heures.

Quelles que soient la méthode et l'origine des globules, quelle que soit la souche, aucune hémolyse n'était visible après vingt-quatre heures. Certaines cultures déterminaient en quarante-huit heures une hémolyse discrète dont l'interprétation est délicate.

[6] F. Kauffmann *Enterobacteriaceae,* Copenhague, Munksgaard, 1951, 152.

Hemolysis—*Corynebacterium diphtheriae*

L. F. Hewitt (1947). J. Path. Bact. 59, 145–157.

Hemolysis was observed in liquid cultures, since these are more amenable to quantitative study than cultures on solid media. The technique adopted as standard was the addition of 0.5 ml of a 5 per cent suspension of washed rabbit erythrocytes to the culture or other preparation and adjustment of the total volume to 2 ml with saline. Tubes were incubated in a water bath at 37°C for one hour and the degree of hemolysis was read immediately, since standing at room temperature had little effect on the result except to increase degrees of partial hemolysis slightly. Under these conditions 1 ml of culture of an actively hemolytic strain produced complete or nearly complete hemolysis; there was less hemolysis with 0.5 or 0.2 ml of culture, and none with 0.1 ml. Unless otherwise stated, readings throughout this communication refer to the degree of hemolysis produced by 1 ml of culture in 1 hour at 37°C.

Effect of age of culture. 0.5 ml of a rabbit red cell suspension was added to 1 ml of culture of a *mitis* strain of *C. diphtheriae* after varying incubation periods, and the degree of hemolysis was observed after further incubation for one hour.

Hemolysis—*Streptococcus*

Y. Abd-el-Malek and T. Gibson (1948). J. Dairy Res. 15, 233–248.

Poured plates of 5% horse-blood agar were incubated for 48 hr and were then placed in the refrigerator overnight. Ox-blood agar was used in a routine test for mastitis organisms.

Hemolysis—*Staphylococcus* and *Micrococcus*

Y. Abd-el-Malek and T. Gibson (1948). J. Dairy Res. 15, 249–260.

Plates of 5% sheep- or ox-blood agar, inoculated on the surface, were incubated for 48 hr at the optimum temperature of the organism and then placed in the refrigerator overnight.

Hemolysis—*Mycoplasma*

D. G. ff. Edward (1950). J. gen. Microbiol. 4, 311–329.

It was preferable to grow the culture on the surface of the ordinary horse-serum agar medium for 2 days and then to pour on top a thin layer of the same medium containing 5% of a horse red cell suspension. The results were read after incubation for a further 2 days. In the second method the strains were grown in horse-serum broth; each day 2 drops of a 5% suspension of horse erythrocytes were added to 1 ml of culture. The mixtures were incubated at 37°C for 5 hr, being shaken at about half-hourly intervals and examined for changes in the colour of the erythrocytes.

Hemolysis—anaerobic Gram-negative diplococci

G. C. Langford, Jr., J. E. Faber, Jr., M. J. Pelczar, Jr. (1950). J. Bact. 59, 349–356.

Five per cent rabbit blood in trypticase soy agar was used for the determination of hemolytic activity of these organisms. Plates were streaked with the organisms and incubated for 48 hours in anaerobic jars at 35°C.

Hemolysis—*Staphylococcus, Streptococcus*

P. Hauduroy (1951). Techniques Bactériologiques. Masson et Cie, Éditeurs, Paris.

HEMOLYSINES

(Recherche des) (Anaérobies). – Centrifuger une culture de 24 heures en bouillon Vf glucosé à 2 p. 100. Chercher la plus petite quantité de culture capable d'hémolyser 0.5 cm^3 de globules rouges de mouton à 5 p. 100.

(A. I. P.).

HEMOLYSINES TYPE α ET β

(Staphylocoques). –

Dépistage de l'hémolysine type α. – Semble se produire seulement dans les cultures faites en présence de CO_2. Agit surtout sur les globules rouges de lapin et de mouton, pas sur les globules rouges humains.

Pour la mettre en évidence, deux séries de tubes sont nécessaires: une série pour les globules rouges de lapin, une série pour les globules rouges de mouton. Mélanger des doses décroissantes de filtrat (par exemple: 0.5; 0.4; 0.1; 0.05; 0.025; etc.) avec 1 cm^3 d'une suspension en eau physiologique à 1 p. 100 de globules rouges de lapin lavés deux fois et, dans une autre série de tubes, avec une suspension semblable de globules rouges de mouton. Porter à l'étuve à 37° pendant deux heures. Lire les résultats et les interpréter suivant le tableau. Ne pas jeter les tubes qui vont servir à dépister l'hémolysine type β.

Dépistage de l'hémolysine type β. – Il peut n'y avoir aucune hémolyse dans des tubes préparés pour la recherche de l'hémolysine type α. Dans ce cas, à la sortie de l'étuve à 37°, porter ces tubes à la glacière à 5.8° C. Lire le résultat après 18 à 20 heures.

S'il existe de l'hémolysine type β, on observe l'hémolyse dans les tubes contenant des globules rouges de mouton, pas d'hémolyse dans les tubes contenant des globules rouges de lapin. L'hémolysine type β est active aussi sur les globules rouges de boeuf et d'homme.

	Hémolysine type a	*Hémolysine type β*
		Quelquefois légère hémolyse à 37°
Globules rouges de mouton	Hémolyse à 37°	Rapide hémolyse après refroidissement à 5.8°
Globules rouges de lapin	Hémolyse	Pas d'hémolyse à 5.8°

Hemolysis—*Streptococcus*

P. F. Swartling (1951). J. Dairy Res. 18, 256–267.

Brown's (1949) technique was followed, using nutrient agar and horse blood.

J. H. Brown (1949). Mongr. Rockefeller Inst. med. Res. No. 9.

Hemolysis—*Staphylococcus*

J. Marks (1952). J. Path. Bact. 64, 175–186.

Sheep blood agar

The medium on which swabs were inoculated consisted of Lemco peptone-agar with 2.5 per cent defibrinated sheep blood (obtained from the Serum Research Institute, Carshalton). Overheating of the basal medium during preparation may cause browning of the sheep blood before incubation; similar browning has been observed with some meat infusion bases. The concentration of blood was kept low to avoid the neutralization of weak toxins by normal antibody. Inoculation plates were read after 18 to 24 hours' incubation in air plus 20 per cent CO_2. Longer incubation increased the incidence of nonspecific hemolysis and rarely gave more information. For the investigation of nonspecific hemolysis, plates were also made containing 2 per cent of thrice-washed sheep cells in the same basal medium.

Hemolytic colonies of staphylococci grown on sheep blood agar were picked with a wire and stab-inoculated into a diagnostic medium: up to 25 strains were tested on a single plate. The medium consisted of sheep blood agar enriched with approximately 20 per cent of normal serum treated with acid (Hartley 1914; Dubos and Fenner, 1950) to reduce its antitoxin content by 90 to 95 per cent. The acid-treated serum inhibited nonspecific lysis but had little effect on lysis due to a-toxin. Horse and bovine sera were found suitable; slaughter house blood was a convenient source. The serum was collected as aseptically as possible but was frozen if stored. It was sterilized by the treatment, which was shown to be capable of killing *Proteus vulgaris* or spores of *Bacillus subtilis* previously added.

To prepare the diagnostic medium, two volumes of serum were diluted with two volumes of isotonic saline and one volume of N HCl was added. The mixture was heated in a water-bath at 65° to 70°C for 20 minutes (or at 56°C for 60 to 90 minutes), after which six volumes of a melted Lemco peptone base containing twice the normal concentration of agar were added at 56°C. The medium was neutralized with one volume of N NaOH and, when cool enough, received 2.5 per cent of sheep blood. A precipitate of denatured globulin appears when the nutrient agar is added to the acidified serum. The latter should not be neutralized before adding the agar, or the particles of precipitate may become too large. Clarification of acid-treated serum by treatment with ether or chloroform at pH 5 as described by Dubos and Fenner makes it ineffective for the present purpose. Plates of the medium may be stored for 3 weeks at 4°C. Readings on the medium are made after not more than 24 hours' incubation in air plus CO_2.

R. J. Dubos and F. Fenner (1950). J. exp. Med. 91, 261.
P. Hartley (1914). Mem. Dept. Agric. India vet. Ser. 1, 178.

Hemolysis—production of soluble hemolysins

N.C.T.C. Methods 1954.

S-hemolysin (Streptolysin S)

Cultures are grown in 5% serum broth for 12–15 hours (not longer, or S may be masked by O). A control tube of uninoculated medium must be set up in parallel. Centrifuge. To 2 ml of supernatant, add an equal volume of a freshly prepared 5% suspension of washed rabbit cells. Incubate at 37°C for 2 hours. Centrifuge.

If 'S' hemolysin is produced, the supernatant is wine-red and translucent; if 'S' is not produced, the supernatant is clear, yellowish, with a brownish sediment of centrifuged blood cells.

'O' hemolysin (Streptolysin O)

Grow strains in 0.2% glucose broth for 24–48 hours – not less, as it takes some time for 'O' to be produced. A control tube of uninoculated medium must be set up in parallel. Centrifuge. Reduce by adding 0.5 ml dilute thioglycollic acid (below) to 2 ml of supernatant. Shake well and allow to stand for 10 minutes. Add equal volumes of 5% red cell suspension. Incubate for 2 hours and read as for 'S'-hemolysin. 'O' hemolysin is produced by groups A, C and G.

Preparation of thioglycollic acid

Proportion: 0.1 ml thioglycollic acid to 5 ml saline. To the required amount of thioglycollic acid, add a few drops of saline and a few drops of indicator (bromthymol blue). Neutralize with 4% caustic soda until the colour turns from yellow to green. Make up the volume to 5 ml or the desired amount, with saline.

Hemolysis—*Streptococcus*

G. Andrieu, L. Enjalbert and L. Lapchine (1954). Annls Inst. Pasteur, Paris 87, 617–634.

Les caractères indiqués pour les colonies (β, α, α + h, ou γ) sont étudiés sur gélose au sang de lapin, en surface et en profondeur, avec lecture au faible grossissement du microscope [1].

Nos milieux sont à base de bouillon au coeur de boeuf digéré par la trypsine et enrichi d'extraits de levures.

[1] G. Andrieu, L. Enjalbert et L. Lapchine. *Ann. Inst. Pasteur,* 1954, 87, 553.

Hemolysis—*Staphylococcus*

C. H. Lack and D. G. Wailling (1954). J. Path. Bact. 68, 431–443.

α-Hemolysin. Two per cent nutrient agar containing 10 per cent (v/v) of trice-washed rabbit cells added at 50°C was poured into Petri dishes containing a bottom layer of 1.5 per cent saline agar. α-Hemolysin was indicated by complete lysis of rabbit cells around colonies when incubated in air for 18 hours at 37°C. That this was α-lysin was confirmed by streaking colonies up to filter paper strips containing approximately 10 units of antitoxin and observing inhibition (Elek and Levy, 1950). The antitoxin was prepared by the Lister Institute.

β-Hemolysin. Sheep cell plates were prepared in the same way, but using formalinized sheep blood from the Serum Research Laboratory, Carshalton. β-Hemolysin was indicated by partial lysis of sheep cells in air within 18 hours at 37°C, followed by complete hemolysis in air within 18 hours at 4°C. This lysis was not inhibited by the standard antitoxin.

γ-Hemolysin. Human red cells obtained from the blood bank were washed three times in saline before being added to the agar. After inoculation the plates were incubated in 20 per cent CO_2, and 80 per cent O_2. γ-Hemolysin was indicated by complete hemolysis around the colonies in 18 hours in this atmosphere. It was not inhibited by standard antitoxin. Lysis of human cells in air was inhibited by antitoxin and was therefore assumed to be due to α-lysin.

S. D. Elek and E. Levy (1950). J. Path. Bact. 62, 541.

Hemolysis—*Streptococcus*

M. Moreira-Jacob (1956). J. gen. Microbiol. 14, 268–280.

Tested by streaking 18 hour cultures on 5% horse blood agar with a basal medium of 2% (w/v) agar, 1% peptone (Evans), 1% Lab-Lemco and 0.5% NaCl. Plates incubated for 48 hours at 37°C.

Hemolysis—*Bacillus*

K. L. Burdon (1956). J. Bact. 71, 25–42.

The hemolytic action was tested on 7 per cent sheep-blood agar streak plates, and also by incubating the supernatant after centrifugation, or the filtrate of broth cultures, with a 2.0 per cent suspension of washed sheep or rabbit erythrocytes.

Hemolysis—thermophilic sulfate reducing bacteria

L. L. Campbell, Jr., H. A. Frank and E. R. Hall (1957). J. Bact. 73, 516–521.

Hemolytic activity was determined by using 5 per cent defibrinated rabbit blood in a blood-agar base medium (Difco). Streak plates were examined for hemolysis after incubation for 5 days in a Brewer anaerobic jar in an atmosphere of hydrogen.

Hemolysis—*Streptococcus*

P. Morelis and L. Colobert (1958). Annls Inst. Pasteur, Paris 95, 568–587.

Propriétés hémolytiques. – Pour la préparation de la gélose au sang, nous avons utilisé le sang de cheval défibriné (non citraté), exempt d'antistreptolysines, délivré par l'Institut Pasteur de Paris. La méthode suivie fut celle conseillée par Andrieu, Enjalbert et Lapchine [1].

L'examen pratiqué au microscope sur des colonies développées en aérobiose et en anaérobiose permet de distinguer trois types de réactions. Dans le type α la colonie s'entoure d'un halo plus ou moins large de verdissement, mais les hématies simplement décolorées ne sont pas lysées. Dans le type β le halo est très clairement hémolytique, incolore et transparent, l'examen microscopique montre la destruction des hématies; il arrive parfois que le halo est réduit à un fin listéré d'observation difficile. Dans le type γ on n'observe aucun changement de teinte du milieu à la périphérie des colonies: les hématies sont intactes et bien colorées.

[1] Andrieu (G.), Enjalbert (R.) et Lapchine (L.), *Ann. Inst. Pasteur,* 1954, 87, 554.

Hemolysis—*Pseudomonas*

M. E. Rhodes (1959). J. gen. Microbiol. 21, 221–263.

Streak cultures on 5% (v/v) horse blood yeast extract agar plates were examined for hemolyzed zones.

Hemolysis—*Vibrio*

P. V. Liu (1959). Studies on the hemolysin of *Vibrio cholerae.* J. infect. Dis. 104, 238–252.

So-called hemolysis of *Vibrio cholerae* on blood agar was actually produced by high alkalinity of the media as the result of growth of this organism and was not due to any enzymatic activity.

The hemolysis produced by mixing equal amounts of peptone-water culture with a suspension of erythrocytes appeared to be caused by the hypotonicity of the mixture.

Most of the strains of *V. cholerae* employed in this study, however, were able to produce some extracellular hemolytic substance in low titer when the cellophane plate technique was used with a suitable medium designed to keep the pH of the growths of these organisms lower than 7.0. These hemolytic substances were nonantigenic; were active in lysing human erythrocytes but not those of sheep or goats; did not produce any specific lesion in the skin of animals upon intracutaneous injection; and did not appear to be of any significance in the pathogenesis of *V. cholerae.*

The final criterion for the differential diagnosis of *V. cholerae* and El Tor vibrios should be the demonstration of the production by the latter organism of a heat-labile, antigenic hemolysin which is active in lysing both human and sheep erythrocytes and produces the characteristic hemorrhagic lesion on the skin of animals.

Hemolysis—*Pasteurella septica*

J. M. Talbot and P. H. A. Sneath (1960). J. gen. Microbiol. 22, 303–311.

Equal volumes of the supernatant fluid of an 18 hr broth culture and a suspension of red cells were incubated together for 1 hr at 37°C. Human group O, horse, rabbit and sheep cells were used.

Hemolysis—*Fusobacterium*

A. C. Baird-Parker (1960). J. gen. Microbiol. 22, 458–469.

Hemolysis was measured by plating on F. M. medium* containing 5% (v/v) suspension of horse blood.

*See A. C. Baird-Parker (1957). Nature, Lond. 180, 1056.

Hemolysis—*Haemophilus vaginalis*

P. N. Edmunds (1960). J. Path. Bact. 79, 273–284.

Blood agar half-plates were inoculated in duplicate from a 48-hr blood agar culture and incubated aerobically at 37°C for 48 hr. The resultant growths were scraped off into 0.8 ml sterile citrated human plasma. Tubes containing 2 per cent peptone broth with 5 per cent digest and 1 per cent maltose or arabinose (whichever sugar the strain was known to ferment) were each inoculated with 0.15 ml of the *H. vaginalis* suspension. After incubation at 37°C for 48 hr, three tubes of a culture of each strain were pooled, centrifuged and the deposit made up with 2 per cent peptone broth to 3 ml. This gave a dense suspension of *H. vaginalis.* Its pH was checked and found to be 7.0. The suspensions, in volumes of 0.5 ml, were pipetted into sterile tubes (3 in. x 3/8 in.) and 1 drop of a 10 per cent suspension of washed red cells was added to each. After shaking, the tubes were incubated in a 37°C water-bath for 2 hr, and the degree of hemolysis observed. The tubes were then shaken, reincubated for a further 4 hr and observed again.

Hemolysis—*Leptospira*

D. C. Bauer, L. N. Eames, S. D. Sleight, and L. C. Ferguson (1961). The significance of leptospiral hemolysin in the pathogenesis of *Leptospira pomona* infections. J. infect. Dis. 108, 229–236.

Production of hemolysin – Stuart's medium (Difco) was dispensed in 400-ml amounts into liter screw-cap bottles and autoclaved. Forty ml of sterile rabbit serum were added aseptically to each bottle and the complete medium was inactivated at 56°C for 30 minutes. The medium was inoculated with 2 ml of an actively growing culture and incubated at 30°C. The cultures were observed daily for evidence of growth, and 1 to 2 days after maximum turbidity was reached incubation was terminated. A sample was removed from each bottle and tested for contamination in thioglycolate broth (Difco). Any culture subsequently showing contamination was discarded. Leptospirae were removed from the cultures either by centrifugation at 12,000 rpm for 30 minutes or by passage through a Seitz filter pad. The supernatants or filtrates were stored at 4°C.

Hemolysin assay – For *in vivo* experiments the titration method of Russell (1956) was employed. The number of hemolytic units (HU) per ml was designated as the reciprocal of the highest dilution of hemolysin producing visible hemolysis of a 2% suspension of sheep erythrocytes after incubation for 4 hours at 37°C and 12 to 18 hours at 4°C.

A more rapid and accurate assay procedure was used in subsequent experiments. A series of hemolysin dilutions was made in saline to a total volume of 1.5 ml. For initial determinations five or six dilutions between 1:10 and 1:100 were usually employed. An equal volume of a 2% suspension of sheep erythrocytes was added. Incubation at 37°C for 30 minutes was followed by incubation for 60 min at 4°C. The tubes were then centrifuged at 1500 rpm to sediment the erythrocytes, and the optical density of the supernatant was determined in a Bausch and Lomb Spectronic 20 colorimeter at 540 mμ. In each test was included a tube containing saline in place of hemolysin. The supernatant from this tube served as a blank. Another tube contained 1.5 ml of a 1:1000 saponin solution and 1.5 ml of sheep red blood cell suspension. The optical density of the supernatant from this tube was assigned a value of 100% hemolysis, and the optical density readings of the hemolysin dilutions were converted to percentage hemolysis. The reciprocal of the hemolysin dilution producing 50% hemolysis was designated as the number of HU per ml of hemolysin. Increased accuracy could be achieved by closer spacing of the dilutions around the previously estimated 50% endpoint. The number of hemolytic units per mg of protein was used as an index of specific activity of the hemolysin preparations. Protein concentration was determined by the method of Lowry *et al.* (1951).

O. H. Lowry *et al.* (1951). J. biol. Chem. 193, 265–275.
C. M. Russell (1956). J. Immunol. 77, 405–409.

Hemolysis—*Streptococcus*

O. G. Clausen (1961). J. Path. Bact. 82, 212–214.

Demonstration of soluble hemolysin. Two methods were employed.

(a) Equal parts of a 5 per cent suspension of washed rabbit red blood cells and a 20-hr 5 per cent serum broth culture were mixed, incubated for 2 hr at 37°C, and then centrifuged and examined by Skadhauge's (1950) method. (b) The streptococcus was grown for 20 hr in a 1 per cent dextrose broth or a 5 per cent serum broth after addition of *c.* 20 per cent of a 5 per cent suspension of washed rabbit red blood cells. The cultures were centrifuged before the results were recorded. Control tests were made with uninoculated media and with cultures of a group A streptococcus strain that was a reliable producer of soluble hemolysin.
K. Skadhauge (1950). *Studies on enterococci,* p. 58. Einar Munksgaard. Kopenhagen.

Hemolysis—*Pediococcus*

H. L. Günther and H. R. White (1961). J. gen. Microbiol. 26, 185–197.

Horse blood (5%, v/v) agar streak plates were prepared and incubated both aerobically and anaerobically. Pour plates were also made and incubated similarly. Results were read after incubation for 48 hr and again after overnight storage at 4°C.

Maintenance of stock cultures and methods of cultivation. For maintenance of stock cultures, preparation of inocula and in all experimental work, "Oxoid" tomato juice (TJ) broth or tomato juice (TJ) agar, adjusted to pH 6.6, was used unless otherwise stated. The following were exceptions to this rule: for strain Tc. 1, sodium chloride (5%, w/v) was added to the medium; and for the aerococci, glucose Lemco broth (Shattock and Hirsch, 1947) or glucose yeast extract (GY) agar (containing, as %, w/v, peptone, 1.0; Yeastrel, 0.3; glucose, 1.0; NaCl, 0.25; agar, 1.0; pH 7.4) was used.

Cultures were incubated aerobically at 30°C, with specified exceptions.
P. M. F. Shattock and A. Hirsch (1947). J. Path. Bact. 59, 495.

Hemolysis—*Pseudomonas*

M. Véron (1961). Annls Inst. Pasteur, Paris 100, Suppl. 6, 16–42.

L'hémolyse des hématies de mouton ou de cheval sur une gélose au sang à 5 p. 100.

Hemolysis—*Moraxella (Acinetobacter)*

E. T. Cetin and K. Toreci (1961). Annls Inst. Pasteur, Paris 100, 509–523.

e) Nous avons recherché si les souches de *Moraxella lwoffi* pouvaient hémolyser les érythrocytes de l'homme, du lapin, du cobaye, du chat, du cheval, du mouton, du boeuf, et de la poule. Les échantillons sanguins ont été défibrines dans des flacons contenant des billes de verre et une fraction du sang défibriné a été soumise au lavage dans du sérum physiologique à trois reprises différentes. Ensuite, on a préparé des plaques de

20 cm^3 de gélose nutritive en boîtes de Petri contenant 5 p. 100 de sang défibriné ou des érythrocytes lavés. Puis, on a ensemencé à l'anse sous forme de stries ou de disques les cultures de dix-huit heures en bouillon sur ces milieux et l'on a contrôlé pendant cinq jours la production de l'hémolyse à 37°C et 22°C. En outre, on a ajouté à la fraction surnageante de la culture en bouillon centrifugée un antibiotique actif sur les bactéries et on a recherché si le liquide surnageant serait de nature à produire l'hémolyse sur milieu solide et s'il pourrait hémolyser les érythrocytes ajoutés dans les tubes après titrage du liquide surnageant.

Hemolysis—*"Bacterium salmonicida" (Aeromonas)*

I. W. Smith (1963). J. gen. Microbiol. 33, 263–274.

Hemolysis and caseinolysin were tested on Eddy's medium (1960).

B. P. Eddy (1960). J. appl. Bact. 23, 216.

Hemolysis—*Listeria monocytogenes*

A. N. Njoku-Obi, E. M. Jenkins, J. C. Njoku-Obi, J. Adams, and V. Covington (1963). J. Bact. 86, 1–8.

Hemolysin production. Brain Heart Infusion broth (Difco) containing an additional 0.5% glucose was used in most of the experiments. When a solid medium was desired, 1.5% agar (Difco) was added. This medium gave consistently higher yields of hemolysin than Liver Infusion, Veal Infusion, or Tryptose Phosphate Broth (Difco).

For hemolysin production in broth, 0.5 ml of an 18-hr Brain Heart Infusion broth culture was inoculated into 10 ml of medium contained in screw-cap tubes (16 x 125 mm).

When solid medium was used, the same inoculum was spread out on 20 ml of solidified medium in a petri plate.

After incubation at specified temperature and time, hemolysin was obtained from broth culture by centrifuging it at 3,020 x *g* for 30 min at 8°C in a Servall RC–2 centrifuge with an SS-34 rotor. The supernatant contained the hemolysin. For solid medium, the agar culture was chopped, extracted with normal physiological saline at 37°C for 45 min, and centrifuged as above. The saline extracted contained the hemolysin.

Titration of hemolysin. Hemolytic activity was titrated by twofold dilutions in 0.5 ml of Streptolysin O Buffer Solution (Cappel Laboratory, West Chester, Pa.). An equal volume of a 1% suspension of sheep red blood cells (Cappel) in Streptolysin O Buffer Solution was added to the tubes of diluted hemolysin. A reading was made after 2 hr of incubation in a water bath (37°C), and the final reading was made after overnight storage at 10°C.

Titers were expressed usually as minimal hemolytic units (MHU), defined as the reciprocal of the highest dilution of hemolysin showing lysis. When expressed as complete hemolytic units (CHU), the highest dilution showing complete hemolysis was taken as the endpoint.

Hemolysis—*Staphylococcus*

J. W. McLeod (1963). J. Path. Bact. 86, 35–53.

The author describes methods for preparation and determination of hemolytic activity of thermostable staphylococcal toxin.

Hemolysis—*Staphylococcus*

E. T. Cetin (1963). J. Bact. 86, 407–413.

The hemolyzing potentialities of 144 *S. aureus* strains, isolated from pathological specimens, were tested on blood agar plates prepared with blood or washed erythrocytes of humans, sheep, and rabbits. To prepare these plates, an agar medium containing 1, 1.5, or 2% Difco agar in beef infusion broth was dispensed in 16-ml volumes to tubes. These media, when needed, were melted at 100°C and transferred to a water bath at 52°C. Portions (0.75 ml) of defibrinated blood or washed erythrocytes of humans, sheep, or rabbits were put into petri dishes (9 cm in diameter). The melted agar content of each tube was then added to the dish, which was moved circularly and linearly to mix the blood and agar homogeneously. For other tests, blood agar plates were incubated at 37°C, then put at room temperature (24° or 28°C) or in a refrigerator at 2°C, or both. Blood agar plates prepared with washed rabbit erythrocytes became completely hemolyzed if kept for more than 48 hr at 37°C. The color of the plates prepared with washed sheep or human erythrocytes turned to dark brown. As the color of those prepared with human and sheep blood was not altered during long incubation periods, such plates were used in most of the tests. Inoculations on blood-agar plates were made from agar or broth cultures of strains grown for 24 to 48 hr. This inoculation was done with a loop, so as to form a round shape 0.5 cm in diameter.

Hemolysis—*Staphylococcus*

W. D. Foster (1963). J. Path. Bact. 86, 535–541.

The staphylococci were tested for alpha-hemolysin production *in vitro* by the method of Elek and Levy (1950).

S. D. Elek and E. Levy (1950). J. Path. Bact. 62, 541.

Hemolysis—*Streptococcus*

R. H. Deibel, D. E. Lake and C. F. Niven, Jr. (1963). J. Bact. 86, 1275–1282.

Cultures were pour-plated in blood-agar containing 5% sheep blood, employing the dilution method to obtain isolated colonies. Plates were observed after 24 and 48 hr at 37°C.

Hemolysis—*Escherichia coli*

H. W. Smith (1963). J. Path. Bact. 85, 197–211.

Erythrocyte agar medium for recognition of hemolytic Escherichia coli. Blood agar media made with sheep, ox and pig blood were unsuitable because the high antihemolysin content of these bloods occasionally made hemolysis difficult to discern. Instead, therefore, I used nutrient agar containing washed ox erythrocytes in a concentration equivalent to that in 5 per cent ox blood. Penicillin and bacitracin, 10 units of each per ml of medium, were added to inhibit many of the hemolytic faecal bacteria other than *E. coli.*

Tests for soluble (α), hemolysin. Ten ml of alkaline extract broth in a 1 oz screw-capped bottle were inoculated with approximately 5 x 10^7 viable organisms per ml of the strain to be tested and incubated aerobically at 37°C for 2¼–2½ hr. The culture was centrifuged on an M.S.E. anglehead centrifuge at 6000 rpm for 30 min, and 0.5 ml of the supernatant fluid was added to an equal volume of a 2 per cent suspension of washed ox erythrocytes in an agglutination tube. One drop of a solution of streptomycin sulfate (100 μg) was added to prevent the multiplication of any bacilli still present. The tubes were incubated in a water-bath at 37°C for 2 hr and then read. Many of the tests were repeated, and culture fluids of over 60 strains classified by this method were titrated for α-hemolysin content, often on more than one occasion, by the method described below; the results confirmed the reliability of this method of identifying α-hemolysin-producing strains of *E. coli.*

Bacteria-free preparations of α-hemolysin were obtained by filtering the supernatant fluids of centrifuged cultures through membrane filters with an average pore diameter of 250 mμ.

Alkaline extract broth. This was used for the production of soluble (α) hemolysin. One part of fresh veal was macerated in two parts of water and 1 per cent proteose peptone (Difco) and 0.5 per cent NaCl were added. The pH was adjusted to 7.5 and, after it had been allowed to stand for sufficient time to ensure that the pH was constant, the medium was autoclaved at 121°C for 10 min. The fluid was then passed through filter paper. If the filtrate was not clear it was refiltered, this time through a Seitz clarifying pad. The pH was then adjusted to 7.2–7.4, and 0.2 per cent of glucose was added. The medium was distributed into suitable containers and autoclaved at 115°C for 15 min.

Detailed methods are also given for –
Preparation of α-hemolysin-containing fluids
Titration of hemolysin in culture fluids
Titration of α-antihemolysin in serum
Preparation of antisera to hemolysins
Plate antihemolysin tests

Hemolysis—*Staphylococcus aureus*

R. D. Ekstedt (1963). Studies on immunity to staphylococcal infection in mice. I. Effect of dosage, viability, and interval between immunization and challenge on resistance to infection following injection of whole cell vaccines. J. infect. Dis. 112, 143–151.

No rabbit red blood cell hemolysin (α-toxin) could be detected in the supernatant from cultures grown for 18 to 20 hours at 37°C in brain heart infusion broth (Difco). Incubation was carried out in air. The organism was hemolytic on 5% rabbit, sheep and human blood agar.

Hemolysis—*Pseudomonas aeruginosa*

R. S. Berk and L. Gronkowski (1964). Antonie van Leeuwenhoek 30, 141–153.

The authors used the method described by Berk (1962 and 1963).

R. S. Berk (1962). J. Bact. 84, 1041–1048.

R. S. Berk (1963). J. Bact. 85, 522–526.

Hemolysis—*Staphylococcus*

Riaz-Ul Haque and J. N. Baldwin (1964). J. Bact. 88, 1442–1447.

Originally, strains were maintained on Brain Heart Infusion Agar (Difco). However, this medium favored the emergence of variants with altered patterns of hemolysis. Trypticase Soy (TS) agar (BBL) prevented the emergence of such variants, and all cultures were subsequently propagated on this medium.

The blood-agar plates for the detection of the hemolysins were prepared by incorporation of red blood cells from sheep, rabbit, human, and horse blood into TS agar. Erythrocytes from citrated whole blood were thrice washed with sterile saline, and 5.0 ml of packed cells were added to 195 ml of sterile and cooled TS agar; 20 ml of agar were poured into petri dish.

For the determination of the type of hemolysin by the radial streak method, the blood-agar plates were divided into six sectors, and a loopful of the culture to be tested was streaked onto one of the sections on each of the blood-agar plates. The plates were incubated at 37°C in an atmosphere containing 20% carbon dioxide (Wadsworth, 1927) for 24 hr, and were then refrigerated overnight. The reactions produced before and after refrigeration were recorded.

When hemolysin production was determined by tube titrations, the strains were inoculated into TS broth and TS broth containing 0.3% agar. The cultures were incubated in an atmosphere containing 20% carbon dioxide at 37°C for 48 hr, and the supernatant fluids were harvested by centrifugation. With the use of phosphate-buffered saline (pH 7.0) as diluent, these fluids were titrated against 1% suspensions of sheep, rabbit, human, and horse red blood cells. The tubes were incubated for 1 hr at 37°C, and were then refrigerated overnight. The titer of the supernatant fluids was recorded after incubation at 37°C and after overnight refrigeration.

For the determination of hemolysin production by individual cells, an

18-hr culture in TS broth was diluted 1:2 x 10^7 in sterile distilled water. By use of a glass spreader, 0.1 ml of this dilution was spread over the surface of a sheep blood-agar plate. The inside of the lids of these plates contained a sterile filter paper disc (11.0 cm) which absorbed the moisture evaporated from the surface of the plates. The plates were incubated at 37°C for 24 hr in an atmosphere containing 20% carbon dioxide. Plates which had 20 to 30 well-isolated colonies were replicated onto sheep, rabbit, human, and horse blood agar plates. Circular pieces of sterile wire gauze (mesh size, 1 mm^2) 3.5 in. (8.89 cm) in diameter mounted on wooden circular discs of the same diameter with thumb tacks were used for this purpose. A new device containing 7,500 stainless-steel pins [length, 0.25 in. (0.63 cm); diameter, 0.015 in. (0.038 cm)] spaced 0.030 in. (0.076 cm) apart into a stainless-steel disc [diameter, 3.5 in. (8.89 cm)] was later found to be more effective. Velveteen employed by Lederberg and Lederberg (1952) tended to spread the colonies on replication, which was undesirable for studying the hemolytic reactions. The blood-agar plates were incubated in 20% carbon dioxide at 37°C for 36 hr, and were then refrigerated overnight.

The interpretation of the hemolytic reactions obtained on blood-agar plates was made following the criteria outlined by Marks and Vaughan (1950), Marks (1951), and Elek and Levy (1950, 1954). Zones of complete lysis with hazy edges on sheep and rabbit blood-agar plates were considered due to α-hemolysin. Zones of darkening surrounding the colonies on sheep blood-agar plates, which cleared on subsequent refrigeration, were regarded as being due to β-hemolysin. Complete zones of lysis with sharply defined edges on sheep, rabbit, human, and horse blood-agar plates indicated the production of δ-hemolysin. γ-Hemolysin was recognized by the production of small zones of partial lysis on human and rabbit blood-agar plates, with no accompanying hemolysis on sheep and horse blood-agar plates.

In the tube titrations, δ-hemolysin was recognized by the lysis of sheep, rabbit, human, and horse red blood cells at 37°C. An increase in the titer of sheep blood cells after refrigeration indicated β-hemolysin. Production of α-hemolysin was indicated by complete lysis of sheep and rabbit red blood cells at 37°C, and possibly slight lysis of human red blood cells but not of horse cells.

S. D. Elek and E. Levy (1950). J. Path. Bact. 62, 541–554.
S. D. Elek and E. Levy (1954). J. Path. Bact. 68, 31–40.
J. Lederberg and E. M. Lederberg (1952). J. Bact. 63, 399–406.
J. Marks (1951). J. Hyg., Camb. 49, 52–66.
J. Marks and A. C. T. Vaughan (1950). J. Path. Bact. 62, 597–615.
A. B. Wadsworth (1927). *Standard methods of the division of laboratories and research of the New York State Department of Health.* The Williams and Wilkins Co., Baltimore.

Hemolysis—*Aerococcus catalasicus*

O. G. Clausen (1964). J. gen. Microbiol. 35, 1–8.

Cultures were incubated aerobically at 37°C unless otherwise stated. Hemolysis was tested on 5% citrated horse blood agar after 24 and 48 hr incubation.

Hemolysis—*Moraxella*

W. J. Ryan (1964). J. gen. Microbiol. 35, 361–372.

Ten per cent (v/v) of whole blood was added to Lemco nutrient agar. Layered plates were poured and incubated, after inoculation, under optimal growth conditions. Horse, human, sheep, rabbit and chicken blood was used. Evidence of hemolysis was sought after 24 and 48 hr.

Hemolysis—*Pediococcus*

E. Coster and H. R. White (1964). J. gen. Microbiol. 37, 15–31.

The authors used the methods of H. L. Günther and H. R. White (J. gen. Microbiol. 26, 185, 1961). Their Tomato Juice (TJ) broth or agar with the addition of Tween 80 (0.1%, v/v), pH adjusted to 6.6, was used in maintenance and preparation of inocula, with certain specified exceptions.

Transfers for preparing inocula were made every 24 hours except for *Pediococcus halophilus* (48 hr) and some brewing strains (fortnightly).

All cultures were incubated aerobically at 30°C, except for "brewing strains," which were incubated in an atmosphere of 95% (v/v) hydrogen and 5% (v/v) carbon dioxide at 22°C.

Brewing strains were incubated for 14 days.

Hemolysis—*Staphylococcus*

D. C. Smith, V. D. Foltz and T. H. Lord (1964). J. Bact. 87, 188–195.

The base for blood-agar was Proteose Peptone agar, consisting of Proteose Peptone, 20 g; NaCl, 5 g; beef extract (Difco), 3 g; and agar-agar (Difco), 20 g, per liter of distilled water, adjusted to pH 7.0. Blood-agar plates were made by adding 5% washed or unwashed sheep, human, or rabbit blood cells.

Blood. The use of blood in its normal state will hereafter be considered as unwashed, defibrinated blood. Washed cells were made from whole blood by aseptic washing of the cells three times with physiological saline until the supernatant fluid was clear. Saline was then added to the tube of packed cells to bring the level of solution to the mark of the original volume of blood.

Detection of induced hemolysis. Each hemolytic culture was inoculated in the center of the blood plates by filling a loop with the culture, gently touching the surface of the plate, and rotating the loop in a circular manner until the loop was empty. In a circle with a radius of about 1.5 cm from this inoculum, a nonhemolytic culture was similarly inoculated. Initially each hemolytic culture had two or three nonhemolytic cultures spotted around it. The nonhemolytic cultures were rotated so that each of the 13 nonhemolytic cultures was inoculated close to each hemolytic culture.

All plates were incubated at 37°C, and observed at 18, 48, 72 and 96 hours.

A second series of 5% blood plates were inoculated similarly and incubated in a 10% CO_2 atmosphere at 37°C for 72 hr, at which time the presence of the secondary zone encircling hemolytic colonies and induced hemolysis of nonhemolytic cultures was noted.

To observe the phenomenon of induced hemolysis of nonhemolytic cultures on other than sheep blood-agar, washed and unwashed rabbit and human blood was used for preparing blood plates. These plates were inoculated and incubated at 37°C for 24, 48, and 96 hr to demonstrate induced hemolysis of nonhemolytic cultures as described for sheep blood plates.

Hemolysis—synergistic reaction of various organisms

G. Fraser (1964). J. Path. Bact. 88, 43–53.

Culture medium. Blood agar plates were prepared with freshly drawn red cells of sheep, ox, goat, rabbit, horse and fowl. The red cells were thrice washed in sterile saline solution and resuspended in saline to the original volume of the blood. Four per cent of this suspension was incorporated in nutrient agar, which consisted of 1 per cent peptone, 1 per cent meat extract, 0.5 per cent NaCl, and 1.5 per cent agar, and had a pH value of 7.4.

Plate test for hemolytic phenomena. Forty-eight-hour inspissated serum slope cultures of the strains to be examined were streaked across and at right angles to each other on the surface of blood agar plates; both organisms were applied at the same time, except where stated otherwise. Duplicate plates were incubated for 24–48 hr at 37°C in air and in air containing 10 per cent CO_2.

Organisms tested included *Actinobacillus lignieresi, Bacillus anthracis, Brucella abortus, Escherichia coli, Corynebacterium pyogenes, C. equi, C. renale, C. diphtheriae, C. hofmanni, C. bovis, C. xerosis, C. murium, C. haemolyticum, Loefflerella pseudomallei, Erysipelothrix rhusiopathiae, Listeria monocytogenes, Pasteurella septica, P. haemolytica, P. pseudotuberculosis, Proteus mirabilis, Pr. vulgaris, Pseudomonas aeruginosa, Salmonella typhimurium, S. cholerae-suis, S. dublin, Vibrio fetus, Dermatophilus dermatonomus,* group-B streptococci and various staphylococci.

Hemolysis—*Staphylococcus*

S. I. Jacobs, A. T. Willis and G. M. Goodburn (1964). J. Path. Bact. 87, 151–156.

Fresh sheep blood agar. This medium was used to detect β-lysin production. Whole sheep blood (Burroughs Wellcome and Co.) was washed three times in sterile saline to remove the plasma. The washed blood cells were added to the cooled base to give a final concentration of cells equivalent to 10 per cent of whole blood. The inoculated plates were incubated anaerobically for 48 hr at 37°C and then cooled and held at 4°C for 1 hr before reading. β-lysin production was indicated by a hazy, but well-defined zone of lysis around the area of growth. Neither α- nor δ-lysins causes interference when this method is used.

The nutrient agar base was a heart infusion broth (Difco) solidified with 1.5 per cent Davis New Zealand agar. The medium was sterilized at 121°C for 15 min, and reagents were mixed with the sterilized base medium when it had cooled to 50–55°C.

The medium was inoculated from a fresh 18-hr broth culture with a sterile throat swab.

Hemolysis—*Pseudomonas aeruginosa*

K. Morihara (1964). J. Bact. 88, 745–757.

A convenient method for determining the extent of hemolysis was performed with bouillon-agar containing defibrinated rabbit blood. Cultures were incubated at 30°C.

Hemolysis—enterococci

C. G. Rogers and W. B. Sarles (1964). J. Bact. 88, 965–973.

Hemolysis in TYE agar containing 5% (v/v) human blood. Incubated at 37°C.

Hemolysis—*Staphylococcus*

D. C. Smith, V. D. Foltz and T. H. Lord (1964). J. Bact. 88, 1700–1704.

The authors describe the study of two lytic factors involved in induced synergistic hemolysis in normally non-hemolytic staphylococci.

Hemolysis (with staphylococcal β-toxin)—*Pasteurella*

G. Bouley (1965). Annls Inst. Pasteur, Paris 108, 129–131.

En 1944, Christie, Atkins et Munch-Petersen [1] faisaient connaître le phénomène qui est désigné aujourd'hui par les initiales de ces chercheurs : les streptocoques du groupe B produisent une substance qui hémolyse les hématies de mouton sensibilisées par la toxine staphylococcique β.

. . . .

Nous procédons comme pour l'étude des streptocoques, c'est-à-dire que nous utilisons de la gélose nutritive additionnée de 2 p. 100 d'hématies lavées de mouton, en boîtes de Petri. Un staphylocoque β-toxique, faiblement actif sur les globules rouges de mouton, est ensemencé en une strie selon un diamètre de la boîte ; perpendiculairement à ce diamètre, mais sans le couper, nous strions la gélose avec les souches de *Pasteurella* à identifier. Après une nuit à 37°C, un résultat positif (C. A. M. P. +) se traduit par l'apparition d'un important trapèze d'hémolyse totale à l'intérieur de la zone sensibilisée par la toxine staphylococcique, trapèze axé sur la strie d'ensemencement pasteurellique. Ce phénomène persiste même lorsque la souche étudiée a pratiquement perdu son caractère hémolytique propre.

[1] Christie (R.), Atkins (N.) and Munch-Petersen (E.). *Austr. J. exp. Biol.,* 1944, 22, 197.

Hemolysis—*Staphylococcus*

A. Kayser and M. Raynaud (1965). Annls Inst. Pasteur, Paris 108, 215–233.

These authors describe a new hemolysin produced by *Staphylococcus aureus.* Full details of specific identification procedures are given.

Hemolysis—*Malleomyces, Pseudomonas*

Bach-Toan-Vinh (1965). Annls Inst. Pasteur, Paris 109, 460–463.

Hémolyse – Recherche sur milieu solide (au sang de cheval) et sur milieu liquide (au sang de mouton).

Hemolysis—*Staphylococcus*

W. R. Chesbro, F. P. Heydrick, R. Martineau and G. N. Perkins (1965). J. Bact. 89, 378–389.

The authors give details for the production and study of the action of β-hemolysin on streptococcal cell walls.

Hemolysis—non-el Tor cholera vibrios

S. Rizvi, M. I. Huq and A. S. Benenson (1965). J. Bact. 89, 910–912.

The presence of hemolytic activity was determined by the method of Feeley and Pittman (Bull. Wld Hlth Org. 28, 347, 1963) and by the plate hemolysis test (Taylor, *Cholera Research in India 1934–1940,* Job Press, Cawnpore, India, 1941).

Hemolysis—*Vibrio comma*

J. C. Feeley (1965). J. Bact. 89, 665–670.

Tube hemolysis tests, done by use of sheep erythrocytes and heart Infusion Broth (Difco) cultures, were performed as recommended by Feeley and Pittman (1963). Blood-agar plates [Blood Agar Base (Difco) plus 5% defibrinated sheep blood] were incubated aerobically at 35°C, and also anaerobically as recommended by deMoor (1963) to avoid confusion with so-called "hemodigestion."

J. C. Feeley and M. Pittman (1963). Bull. Wld Hlth Org. 28, 347–356.
C. E. deMoor (1963). Trop. geogr. Med. 15, 97–107.

Hemolysis—*Mycoplasma pneumoniae*

N. L. Somerson, R. H. Purcell, D. Taylor-Robinson, and R. M. Chanock (1965). J. Bact. 89, 813–818.

Tests for hemolysin. The presence of hemolysin active against guinea pig RBC was demonstrated by overlaying mycoplasma colonies with an RBC-agar mixture (Somerson *et al.,* 1963). Guinea pig RBC were washed three times in Alsever's solution, diluted, and added to Difco PPLO agar medium (55 to 60°C) to give a final concentration of 3% blood cells. About 3 ml of the mixture were poured as overlay onto agar plates (5 cm in diameter) which contained surface colonies of mycoplasmas. Plates were inspected daily for plaques resulting from hemolyzed cells.

N. L. Somerson, D. Taylor-Robinson, and R. M. Chancock (1963). Am. J. Hyg. 77, 122–128.

Hemolysis—*Bacillus anthracis*

R. F. Knisely (1965). J. Bact. 90, 1778–1783.

The hemolytic activity of the organisms was determined by streaking cultures on Heart infusion agar (Difco) containing 2% fresh, defibrinated, washed sheep cells.

Hemolysis—*Staphylococcus*

R. Zemelman and L. Longeri (1965). Appl. Microbiol. 13, 167–170.

To determine this hemolysin, the technique described by Christie, Atkins, and Munch-Petersen (1944) was used. It is based on the rapid and strong hemolysis of sheep red blood cells at 37°C in the zone of the plates where β-hemolysin, if produced, encounters some substance liberated by Lancefield group B streptococci, and is known as the CAMP test. This technique has been employed in our country to investigate mastitis-producer streptococci in dairy cattle (Abel, 1953), but it can also be used to determine β-hemolysin if a strain of Lancefield group B streptococcus is employed as the known factor of the reaction. We have used a serologically identified strain of Lancefield group B streptococcus, which gives a positive CAMP test. This microorganism was streaked on the center of a sheep blood-agar plate in a line dividing the plate in two sections. Staphylococci under test were then streaked at right angles to the streptococcus, but keeping a distance of approximately 1 cm between different staphylococci. Plates were incubated at 37°C, and final results were registered after 24 hr. However, strong hemolysis in the zone of interaction was clearly visible after 6 to 8 hr of incubation. About 12 strains of staphylococci could be tested simultaneously on the same plate, and no refrigeration was required to detect β-hemolysin. In addition, any other hemolysin that might be produced could be easily distinguished from it.

R. Abel (1953). Agr. Tec. Mex. 13, 48–61.
R. Christie, N. E. Atkins and E. Munch-Petersen (1944). Aust. J. exp. Biol. med. Sci. 22, 197–200.

Hemolysis—*Mycoplasma*

J. G. Tully (1965). Biochemical, morphological, and serological characterization of *Mycoplasma* of murine origin. J. infect. Dis. 115, 171–185.

Hemolytic activity of murine mycoplasma strains was examined against horse, sheep, and guinea pig erythrocytes following the procedure of Somerson *et al.* (1963). Two-day broth cultures were diluted 10-fold to 10^{-4} in broth blanks, and 0.25 ml transferred to small agar plates (60 × 15 mm). Colonies usually appeared in 2 days, and each plate was overlaid with 3 ml of a 4% erythrocyte-Difco PPLO agar suspension cooled to 50°C. The occurrence of hemolysis around well-isolated and enlarging colonies was observed daily for 7 days at 37°C. Uninoculated agar plates were overlaid with RBC suspensions and incubated for control purposes.
N. L. Somerson, D. Taylor-Robinson and R. M. Chanock (1963). Am. J. Hyg. 77, 122–128.

Hemolysis—*Vibrio* (el Tor)

S. N. De, B. Mukherjee and A. R. Dutt (1965). Is it El Tor Vibrio in Calcutta? J. infect. Dis. 115, 377–381.

For testing the hemolytic property, a modified Greig technique was followed. A 24-hour culture in heart infusion broth (pH 7.8) was used. To 1 ml of culture was added 1 ml of 5% saline suspension of washed fresh sheep red blood cells. Readings were taken after the mixture was incubated for 2 hours at 37°C and again after it was left in the refrigerator overnight. The hemolytic test was repeated with 1-, 2-, and 3-day-old cultures of 30 hemagglutinating strains.

Hemolysis—*Staphylococcus aureus*

E. M. Hoffmann and M. M. Streitfeld (1965). Can. J. Microbiol. 11, 203–211.

The authors describe, *inter alia,* the preparation and assay of delta-hemolysin.

Hemolysis—*Streptococcus*

R. Whittenbury (1965). J. gen. Microbiol. 38, 279–287.

Brown's (1919) method was followed, with nutrient agar containing NaCl 0.5% (w/v) and defibrinated horse blood 3% (w/v). Poured agar plates were incubated for 2 days at 30°C and refrigerated overnight. Further tests were made with ox blood in the basal medium agar containing NaCl 1.0% (w/v).

The basal liquid medium contained meat extract (Lab-Lemco), 0.5 g; peptone (Evans), 0.5 g; yeast extract (Difco), 0.5 g; Tween 80, 0.05 ml; $MnSO_4 \cdot 4H_2O$, 0.01 g, in 100 ml tap water, adjusted to pH 6.5 and autoclaved at 121°C for 15 min.
J. H. Brown (1919). Rockefeller Inst. med. Res. Monogr. no. 9.

Hemolysis—*Streptococcus*

I. Ginsburg, T. N. Harris and N. Grossowicz (1965). J. exp. Med. 121, 633–645.

The paper describes studies on the oxygen stable hemolysins of group A streptococci.

See also Ginsburg, Bentwich and Harris (1965). J. exp. Med. 121, 633–645.

Hemolysis—*Clostridium welchii*

J. G. Collee (1965). J. Path. Bact. 90, 13–30.

Hemolysin tests. Serial two-fold dilutions of the test material were made in 0.5 ml amounts of physiological saline containing calcium chloride 0.01 – 0.001 per cent; 0.5 ml aliquots of 2 per cent washed human red cell suspension were then added. The mixtures were shaken, incubated for 1 hr in a 37°C water-bath and then chilled overnight at 4°C before final observations of the degree of hemolysis (complete, almost complete, definite, doubtful or negative) were made for each tube, in comparison with a negative control.

Hemolysis—*Staphylococcus*

G. M. Wiseman (1965). J. Path. Bact. 89, 187–207.

Determination of the hemolysin spectrum of strains. The method of Elek and Levy (1950), modified in some respects, was used. Instead of the medium described by these authors, that described by Dolman and Wilson (1940), containing 1.5 per cent Davis New Zealand agar, was employed. Antiserum was Wellcome commercial staphylococcal antiserum containing 920 International Units of anti-alpha-lysin per ml. Instead of the 48 hr incubation period of Elek and Levy, a 24-hr period was used.

The paper describes the study of the beta-hemolysin of *Staphylococcus aureus.*

C. E. Dolman and R. J. Wilson (1940). Can. publ. Hlth J. 31, 68.
S. D. Elek and E. Levy (1950). J. Path. Bact. 62, 541.

Hemolysis—*Staphylococcus*

S. J. Edwards and G. W. Jones (1966). J. Dairy Res. 33, 261–270.

The hemolytic pattern and colonial appearance of strains were examined on the blood-agar medium previously described. The effects of specific anti-α-hemolysin and of β-hemolysin on the hemolytic activity of coagulase-negative strains were examined by the plate techniques of Elek and Levy (1950 and 1954, respectively). Sheep and ox cells were employed in the plates, which were incubated in 20% (v/v) CO_2 for 24 hr and for a further 24 hr in air. The presence of ϵ-toxin was investigated by the plate method of Fraser (1962) employing a strain of *Corynebacterium haemolyticum* (N.C.T.C. H 8452).

S. D. Elek and E. J. Levy (1950). J. Path. Bact. 62, 541.
S. D. Elek and E. J. Levy (1954). J. Path. Bact. 68, 31.
G. Fraser (1962). Vet. Rec. 74, 753.

Hemolysis—*Pseudomonas aeruginosa*

R. A. Altenbern (1966). Can. J. Microbiol. 12, 231–241.

Hemolysin Production on Membrane Filters: This procedure was carried out essentially as described by Liu (1957) and Berk (1962). Standard (0.45 μ) porosity membrane filters were placed on the surface of previously poured and solidified Trypticase soy agar (Baltimore Biological Laboratories). The membrane surface was inoculated by the spreading method with 0.1 ml of an overnight Trypticase soy broth culture of the *Pseudomonas* strain to be investigated. After 48 to 72 hours of incubation at 37°C, the membranes were transferred to small beakers where the microbial growth was washed off into 5 ml of 0.1 M phosphate buffer, pH 6.0, containing 0.35% sodium chloride (referred to as "buffer-saline"). This suspension was then centrifuged (1640 × *g*) for 10 min and the supernatant fluid containing the hemolysin was decanted and stored at 4°C until used. The packed cells were discarded.

Assay of Hemolysin

Samples to be assayed were serially diluted in 1.5 fold dilution steps in the pH 6.0 buffer saline described above so that 0.5 ml of the dilution remained in each tube. Then 0.5 ml of a 1% suspension of washed sheep erythrocytes in 0.85% saline was added to each tube. The assay tubes were incubated in a 47°C water bath for 60 min. Subsequently each tube received 2 ml of 0.85% saline and the contents were centrifuged for 10 min to remove unlysed erythrocytes or erythrocyte debris. The supernatant fluid was decanted and the optical density at 550 mμ was determined in a Coleman junior spectrophotometer. The optical densities of the tube contents were plotted on linear graph paper against the dilution factors of the tubes and a 50% hemolysis dilution was found graphically. The reciprocal of this dilution was designated as the number of hemolysis units per milliliter of the original sample. Replicate series of assays showed that the maximum error of this procedure was ±20%.

R. S. Berk (1962). J. Bact. 84, 1041–1048.
P. V. Liu (1957). J. Bact. 72, 718–728.

Hemolysis—*Staphylococcus aureus*

J. A. Donahue and J. N. Baldwin (1966). Hemolysin and leukocidin production by 80/81 strains of *Staphylococcus aureus.* J. infect. Dis. 116, 324–328.

Hemolysis on blood agar plates. Heart infusion agar (Difco) containing 5% thrice washed rabbit, sheep, or human red blood cells was employed to determine patterns of hemolysis. Tubes of BHI broth were inoculated with the strains to be tested and incubated at 37°C for 18 hours. The cultures were diluted in distilled water to contain approximately 1000 cells per ml. One drop was then spread onto the surface of rabbit, sheep and human blood agar plates, which were incubated for 60 hours at 37°C in an atmosphere containing 20% carbon dioxide in air. The patterns of hemolysis were interpreted as described by Elek and Levy (1950, 1954).

S. D. Elek and E. Levy (1950). J. Path. Bact. 62, 541–554.
S. D. Elek and E. Levy (1954). J. Path. Bact. 68, 31–40.

Hemolysis—*Brevibacterium*

R. Chatelain and L. Second (1966). Annls Inst. Pasteur, Paris 111, 630–644.

Hémolyse: culture en bouillon nutritif additionné de 5 p. 100 de sang de cheval. Lecture après 2 jours.

La température d'incubation des cultures est de 30°C.

Hemolysis—*Staphylococcus*

M. Kocur, F. Přecechtěl and T. Martinec (1966). J. Path. Bact. 92, 331–336.

Hemolysins were demonstrated in crude filtrates of staphylococcal cultures. Filtrates were obtained by a modification of the method of Parish and Clark (1932). Cultures were grown on Beil's agar (0.75 per cent agar) in a jar containing air with 25 per cent (v/v) CO_2 for 18–20 hr at 37°C. Each culture was then flushed with 5 ml Martin's broth (Djačenko, 1962) and re-incubated for 24 hr at 37°C in a jar containing air with 25 per cent CO_2. After incubation, the broth was drawn off and centrifuged. As control, Johanovský's (1956) method of cultivation on cellophane was also used; it gave identical results.

Alpha-hemolysin was demonstrated in the following way. The filtrate of the culture was diluted in two-fold steps from 1 in 2 to 1 in 1024, and 0.5 ml of a suspension of washed rabbit blood cells was added to each 0.5 ml of the diluted filtrate. The tubes were placed in a water-bath at 37°C for 1 hr and then maintained at room temperature for 2 hr. The titer of hemolysin was the greatest dilution of the filtrate that caused complete hemolysis.

Beta-hemolysin was demonstrated by a similar technique with sheep instead of rabbit erythrocytes. Tubes with the filtrate, appropriately diluted, and the blood cells were held at 37°C for 1 hr and then at 4°C for 18–20 hr.

Delta-hemolysin was demonstrated by the same method, but human instead of rabbit cells were used.

S. S. Djačenko (1962). *Microbiological methods of diagnosis of infectious diseases* (in Russian), Kiev, p. 28.
J. Johanovský (1956). Čslká Epidem. Mikrobiol. Immunol. 5, 41.
H. J. Parish and W. H. M. Clark (1932). J. Path. Bact. 35, 251.

Hemolysis—*Pediococcus*

R. H. Deibel, J. H. Silliker, and P. T. Fagan (1964). J. Bact. 88, 1078–1083.

The authors note a correlation between requirement of oleate and hemolysis.

A correlation was noted between the time of obtaining the blood sample from the donor and the time of the donor's last partaking of food. Blood from donors who had partaken of food 1 to 3 hours prior to bleeding showed a higher incidence of hemolysis.

HIPPURATE HYDROLYSIS

Hippurate hydrolysis—*Streptococcus*

A. C. Evans (1936). J. Bact. 31, 423–437.

The strains were grown for 4 days in infusion broth containing 1 per cent sodium hippurate. Reduction of the hippurate to benzoic acid was detected by the addition of 12 per cent ferric chloride solution containing 2.5 ml concentrated hydrochloric acid per liter. To 1-ml amounts of the supernatant culture fluid, 0.5 ml of the ferric chloride solution was added, and the tubes were thoroughly shaken immediately. The uninoculated sodium hippurate broth control always gave complete clearing. A precipitate which failed to disappear when the tubes were shaken was regarded as evidence of the reduction of the hippurate by the growing organisms.

Hippurate hydrolysis—*Streptococcus*

P. L. Boisvert (1940). J. Bact. 39, 727–738.

Strains were grown for 4 days in infusion broth containing 1.0% sodium hippurate, and to the clear supernatant medium was added 0.25% concentrated HCl and 12.0% ferric chloride (Ayers and Rupp 1922). Controls were included.

S. H. Ayers and P. Rupp (1922). J. infect. Dis. 30, 388–399.

Hippurate hydrolysis—*Streptococcus*

J. B. Gunnison, M. P. Luxen, J. R. Cummings and M. S. Marshall (1940). J. Bact. 39, 689–708.

The medium described by Coffey and Foley (1937) was inoculated with 0.2 ml of culture and incubated 37°C for 72 hours and tested for hydrolysis with acidified ferric chloride.

J. M. Coffey and G. E. Foley (1937). Am. J. publ. Hlth 27, 972–974.

Hippurate hydrolysis—anaerobic streptococci

M. L. Stone (1940). J. Bact. 39, 559–582.

The test of the hydrolysis of sodium hippurate was made according to the method of Ayers and Rupp (1922). The cultures were grown anaerobi-

cally in the Brown jar for four days in a medium consisting of infusion broth to which had been added 1 per cent of sodium hippurate. At the conclusion of the test period, the cultures were centrifugated, and to 1 ml amounts of the clear supernatant were added 0.3, 0.4, and 0.5 ml, respectively, of a 12 per cent ferric chloride solution containing concentrated hydrochloric acid in the proportion of 2.5 ml per liter. Uninoculated sodium hippurate broth incubated for the same period of time was used as a control, and sufficient ferric chloride was used in the test to insure complete clearing of the uninoculated control. The tubes were shaken vigorously, immediately upon the addition of the ferric chloride. In the case of a positive reaction, a heavy precipitate of ferric benzoate, insoluble in an excess of ferric chloride, was formed. If the reaction was negative, the protein and hippurate precipitate formed on the addition of ferric chloride, redissolved in the excess, leaving a clear solution. Known positive and negative cultures were included as controls in each set of tests.
S. H. Ayers and P. Rupp (1922). J. infect. Dis. 30, 388–399.

Hippurate hydrolysis—*Corynebacterium*

R. F. Brooks and G. J. Hucker (1944). J. Bact. 48, 295–312.
Determined in a medium having the following composition:

	%
Peptone	1.0
Pepsin	0.5
Calcium chloride	0.003
Sodium hippurate	1.0
Aqueous ferric chloride	1 drop of 1% sol'n/liter
pH 7.1	

The medium was tubed in exactly 5.0 ml amounts.

Incubation was at 37°C for 7 days at the end of which period the cultures were tested by the addition to each culture of exactly 1.25 ml of a 7-per-cent aqueous ferric chloride solution, shaking, and letting stand for 5 to 10 minutes. A permanent precipitate indicates the hydrolysis of sodium hippurate, as a benzoic-acid ferric chloride precipitate does not redissolve, whereas a hippurate ferric-chloride precipitate does. About one-third of the collection hydrolyzed sodium hippurate, the remainder failing to produce any change. Twelve per cent of the cultures failed to grow.

Hippurate hydrolysis—*Streptococcus*

Y. Abd-el-Malek and T. Gibson (1948). J. Dairy Res. 15, 233–248.

Hydrolysis was determined by addition of ferric chloride after incubation for 5–7 days in a medium containing 0.1% aesculin, 1% sodium hippurate and 0.5% each of peptone, tryptone, meat extract and yeast extract. The two test substances were also used separately but there appeared to be no advantage in doing so. Strains that failed to grow were retested with 0.25% glucose in the medium.

Hippurate hydrolysis—*Streptococcus*

K. E. Hite and H. C. Hesseltine (1947). A study of aerobic streptococci isolated from the uterus and vagina. J. infect. Dis. 80, 105–112.

Hydrolysis of sodium hippurate – The ability of strains to hydrolyze sodium hippurate was tested by the method of Ayers and Rupp (1922) using a veal infusion medium.

S. H. Ayers and P. Rupp (1922). J. infect. Dis. 30, 388–399.

Hippurate hydrolysis—oral *Lactobacillus*

M. Rogosa, R. F. Wiseman, J. A. Mitchell, M. N. Disraely, and A. J. Beaman (1953). J. Bact. 65, 681–699.

For the hippurate hydrolysis tests Rogosa's medium 3 (see **Fermentation**—*Lactobacillus* p. 285) was modified by the addition of 1 per cent sodium hippurate and 2 per cent glucose. After 2 weeks' incubation at 37°C, 3 ml of 50 per cent H_2SO_4 were added to 2 ml of the medium. In positive reactions the characteristic crystals of benzoic acid were observed.

Hippurate hydrolysis—*Streptococcus*

N.C.T.C. Methods (1954).

Method 1: Grow the organisms for 4 days in infusion broth containing 1% sodium hippurate. Include a control tube of uninoculated medium in each series of tests.

Reduction of hippurate to benzoic acid is detected by the addition of a 12% ferric chloride solution containing 2.5 ml conc. hydrochloric acid per litre.

To 1 ml quantities of the control medium add rapidly, with shaking, 0.2, 0.3, 0.4, and 0.5 ml ferric chloride solution. With the smaller amounts a precipitate usually appears, soluble in excess; with the larger a clear solution is obtained. Add the smallest volume of ferric chloride solution giving a clear solution to 1 ml quantities of clear supernatant fluids from the broth cultures. A heavy precipitate is taken as evidence of reduction of sodium hippurate by the growing organisms.

R. Hare and L. Colebrook (1934). J. Path. Bact. 39, 429.

Method 2: This medium contains sodium hippurate as the sole source of carbon (NaCl, 5 g; $MgSO_4 \cdot 7H_2O$, 0.2 g; $(NH_4)H_2PO_4$, 1 g; K_2HPO_4, 1 g; distilled water, 1 litre; to which are added 3 g sodium hippurate).

Grow culture on solid medium; scrape off a little growth without touching the medium; suspend the growth in physiological saline; shake well.

Using a straight wire, inoculate the hippurate medium from the suspension.

A. A. Hajna and S. R. Damon (1934). Am. J. Hyg. 19, 545.

Hippurate hydrolysis—*Lactobacillus*

G. H. G. Davis (1955). J. gen. Microbiol. 13, 481–493.

Medium: peptone, 1% (w/v); yeast extract, 0.5% (w/v); glucose, 0.5% (w/v); sodium hippurate, 1% (w/v); sodium chloride, 0.5% (w/v); Tween 80, 0.1% (v/v); Salts A,* 0.5% (v/v); Salts B,* 0.5% (v/v); pH 6.8, bottled in *c.* 5 ml quantities in 1/4 oz bottles and sterilized by steaming for 20 min on 3 consecutive days.

Tests were inoculated with 1 to 2 drops, incubated for 10 days, and tested by removing two 1 ml samples of clear supernatant culture fluid (centrifuged if necessary) to two clean 5 x 1/2 in. tubes. To one, 0.5 ml ferric chloride reagent (12% ferric chloride solution plus 2.5 ml concentrated hydrochloric acid per litre) was added, and to the other, 1 ml of a 50% sulfuric acid solution. They were allowed to stand, being shaken occasionally, for 30–60 min. A fine crystalline precipitate indicated a positive reaction. The role of both reagents is to detect free benzoic acid; weak positive hydrolysis, however, may only be detectable by the sulfuric acid test. If allowed to stand for 24 hr, large needle-like crystals of unknown composition appeared in some of the sulfuric acid tests, but these did not indicate a true positive reaction.

*See A. C. Hayward (1957). J. gen. Microbiol. 16, 9–15.

Hippurate hydrolysis—*Klebsiella* and *Aeromonas*

M. L. Thirst (1957). J. gen. Microbiol. 17, 390–395.

Media: The defined medium (HD) was prepared by Hajna and Damon's (1934) method, and a modified form (A) was made up as follows: NaCl, 0.5 g; $MgSO_4 \cdot 7H_2O$, 0.02 g; $NH_4H_2PO_4$, 0.1 g; K_2HPO_4, 0.1 g; sodium hippurate, 0.3 g; agar (New Zealand), 1.0 g; 4% (w/v) aqueous phenol red, 0.25 ml; water, 100 ml. The medium, used in the form of slopes, was not cleared, and the final pH value (6.8–7.0) needed little or no adjustment. Sterilization was by autoclaving at 15 lb/sq. in for 15 minutes.

Detection of hydrolysis: Media HD and A were inoculated lightly from an overnight culture on nutrient agar. Medium HD was incubated for 5 days at the optimum temperature of the organism; after this it was examined for growth and two methods of testing for hydrolysis were used: (1) the ferric chloride test described by Hajna and Damon (1934); (2) the pyridine-copper sulfate test with Zwikker's (1931) reagent described by Munch-Petersen (1940).

Medium A was inspected at intervals during an incubation period of 7 days. In 1–2 days sodium hippurate decomposition was shown by slight growth and a pink colour which increased with further incubation until the entire medium was alkaline.

A. A. Hajna and S. R. Damon (1934). Am. J. Hyg. 19, 545.
E. Munch-Petersen (1940). Bull. Coun. scient. ind. Res., Melb. 34, 96.
J. J. L. Zwikker (1931). Pharm. Weekbl. Ned. 68, 975.

Hippurate hydrolysis

J. Naylor and M. E. Sharpe, (1958). J. Dairy Res. 25, 92–103.

The organisms were grown for 14 days in TDB + 1% sodium hippurate; then 1.5 ml 50% H_2SO_4 were added to 1 ml of the supernatant, giving white crystals of benzoic acid in positive reactions.

Hippurate hydrolysis—*Fusobacterium*

A. C. Baird-Parker (1960). J. gen. Microbiol. 22, 458–469.

Seitz-filtered sodium hippurate was added to the following medium to a final concentration of 1.0% (w/v): proteose peptone no. 3, 1.0; Oxoid yeast extract, 0.1; Lab. Lemco, 0.3; L-cysteine HCl, 0.05; glucose, 0.5; disodium hydrogen phosphate, 0.5. The formation of free benzoic acid was tested by the addition of 1.5 ml of 50% (v/v) conc. sulfuric acid which caused its precipitation as leaflike crystals after standing at room temperature for periods of up to 4 hr (Ayers and Rupp, 1922). The presence of adequate growth was always noted before carrying out the tests; negative results were repeated.

H. S. Ayers and P. Rupp (1922). J. infect. Dis. 30, 388.

Hippurate hydrolysis—*Aerococcus catalasicus*

O. G. Clausen (1964). J. gen. Microbiol. 35, 1–8.

Cultures were incubated aerobically at 37°C unless otherwise stated.

Hydrolysis of sodium hippurate (1%) in broth into sodium benzoate after 5 days' cultivation was determined by means of 12% $FeCl_3$ solution with 0.2% concentrated HCl (Roemer, 1948).

G. B. Roemer (1948). Zentbl. Bakt. ParasitKde Abt. 1, Orig. 152, 458.

Hippurate hydrolysis—*Lactobacillus*

M. Gemmell and W. Hodgkiss (1964). J. gen. Microbiol. 35, 519–526.

Cultures were grown in 5 ml basal medium lacking citrate, but containing sodium hippurate (1%, w/v) + glucose (0.5%, w/v). After a few days the cultures were centrifuged and 50% (v/v) H_2SO_4 was added to the supernatant fluid, which was allowed to stand overnight. The appearance of flat platelike crystals of benzoic acid with a melting point of 121°C indicated a positive result. Needle-like crystals with a melting point of 187°C were assumed to be hippuric acid.

The basal medium which was used throughout with minor modifications had the following constituents in 1 L tap water: meat extract (Lab Lemco), 5 g; Evans peptone, 5 g; Difco yeast extract, 5 g; Tween 80, 0.5 ml; $MnSO_4 \cdot 4H_2O$, 0.1 g; potassium citrate, 1 g; pH 6.5.

Hippurate hydrolysis—*Pediococcus*

R. Whittenbury (1965). J. gen. Microbiol. 40, 97–106.

The author used the method of M. Gemmell and W. Hodgkiss (1964, J. gen. Microbiol. 35, 519).

Hippurate hydrolysis—*Mycobacterium*

R. E. Gordon (1966). J. gen. Microbiol. 43, 329–343.

The modification of Ayers and Rupp's test (1922), described by Baird-Parker (1963), was used. The cultures were inoculated into hippurate broth (tryptone, 10 g; beef extract, 3 g; yeast extract, 1 g; glucose, 1 g; Na_2HPO_4, 5 g; Na hippurate, 10 g; distilled water, 1000 ml). After 6 weeks of incubation, the cultures were examined for benzoic acid by mixing 1 ml of the culture, as free of clumps of growth as possible, with 1.5 ml of 50% (v/v) H_2SO_4. The appearance of crystals in the acid mixture after 4 hr at room temperature was accepted as evidence of the hydrolysis of hippurate.

S. H. Ayers and P. Rupp (1922). J. infect. Dis. 30, 388.
A. C. Baird-Parker (1963). J. gen. Microbiol. 30, 409.

HUGH AND LEIFSON (H AND L) OXIDATIVE AND FERMENTATIVE BREAKDOWN OF CARBOHYDRATE

H&L test

R. Hugh and E. Leifson (1953). J. Bact. 66, 24–26.

For many years bacteriologists have observed that some bacteria produce acid from carbohydrates only under aerobic conditions while others produce acid both under aerobic and anaerobic conditions. The significance of these observations does not seem to have been appreciated generally by taxonomists. Studies of bacterial physiology have made it increasingly evident that the bacterial metabolism of carbohydrates may be accomplished by two apparently fundamentally different mechanisms (see, for example, Porter, 1946: Werkman and Wilson, 1951). By one mechanism, appropriately called fermentation, the glucose molecule first is phosphorylated and then split into two triose molecules which undergo further changes. This process is independent of oxygen. By the other mechanism, which we shall call oxidation, the glucose molecule is not split into two triose molecules, but the aldehyde group is oxidized to a carboxyl group, forming gluconic acid. Further oxidation may take place to form various products such as 2-ketogluconic acid. Several studies of this mechanism, summarized by Sebek and Randles (1952), have failed to detect phosphorylation of the glucose molecule preliminary to oxidation. In the absence of compounds such as nitrates, the oxidation of carbohydrates is a strictly aerobic process, whereas fermentation is an anaerobic process.

Experimental methods and results

The practical distinction between oxidation and fermentation of carbohydrates rests on the role played by atmospheric oxygen. The degree of acidity produced by oxidative metabolism is usually lower than that produced by fermentative metabolism. Since most bacteria produce alkaline substances from peptone, the small amount of acid which may be produced by oxidative metabolism may be completely neutralized in the ordinary carbohydrate-peptone media. These difficulties can be overcome by using a medium with a relatively high carbohydrate concentration, a low peptone concentration, and aerobic conditions.

The medium which we have adopted to detect oxidation of carbohydrates and distinguish it from fermentation has the following composition: peptone, 0.2 per cent; NaCl, 0.5 per cent; K_2HPO_4, 0.03 per cent; agar, 0.3 per cent; bromthymol blue, 0.003 per cent; carbohydrate, 1.0 per cent; pH 7.1.

For the peptone we recommend a pancreatic digest of casein. Other types of peptone may or may not be satisfactory. Some bacteria, such as species of the genus *Brucella,* seem to grow better with sodium chloride in the medium. The phosphate is added to promote fermentation and to stabilize the pH. The purpose of the agar is to prevent convection currents in the medium and consequent mixing of the acid produced at the surface with the bulk of the medium. The concentration of agar also is optimum for determination of motility.

The bromthymol blue is dissolved in water and 0.3 ml of a 1 per cent solution added to each 100 ml of medium. Alcoholic solution of indicator should not be used because acid may be produced from the alcohol added. The carbohydrate cannot be sterilized with the medium because of the chemical changes which occur. For critical work and with unstable carbohydrates, sterilization by filtration is recommended. Practically it seems that most carbohydrates may be sterilized satisfactorily by autoclaving a 10 per cent aqueous solution. The sterile carbohydrate solution is added aseptically to the sterile melted base.

The medium is tubed to a depth of about 1½ inches. The oxygen which diffuses into the medium on storage does not seem to alter appreciably the reactions obtained. Duplicate tubes of the solidified medium are inoculated by stabbing. After inoculation one of the pair of tubes is covered with a layer of sterile melted petrolatum to a depth of ¼ to ½ inch. Several types of reactions may be observed in the medium: Fermentative organisms will produce an acid reaction throughout in both tubes. Oxidative organisms will produce an acid reaction in the open tube only, leaving the petrolatum covered tube unchanged with little or no apparent growth. The acid reaction produced by the oxidative organisms is apparent first at the surface and extends gradually downward into the medium. Where the oxidation is weak or slow, it is usual to observe an initial alkaline reaction at the surface of the open tube. This may persist for a variable length of time, up to several days of incubation, before turning acid, and must not be mistaken for a negative reaction. Nonfermenters and nonoxidizers produce no change in the covered tube and only an alkaline reaction in the open tube. Organisms which oxidize glucose but do not ferment it have never been observed to ferment any carbohydrate, and the petrolatum-covered tube may be omitted in subsequent tests with other carbohydrates.

J. R. Porter (1946). *Bacterial chemistry and physiology.* John Wiley and Sons, Inc., New York, N. Y.

O. K. Sebek and C. I. Randles (1952). J. Bact. 63, 693–700.

C. W. Werkman and P. W. Wilson (1951). *Bacterial physiology.* Academic Press, Inc., New York, N. Y.

H&L test—luminous bacteria

R. Spencer (1955). J. gen. Microbiol. 13, 111–118.

In the determination of the type of dissimilation of carbohydrates, Hugh and Leifson's (1953) medium was used, but with 1.5% NaCl instead of 0.5%. Glucose was the only carbohydrate used, as Hugh and Leifson found that organisms which oxidize glucose, but do not ferment it, do not ferment other carbohydrates.

Unless otherwise stated, the temperature of incubation was always 20°C.

R. Hugh and E. Leifson (1953). J. Bact. 66, 24–26.

H&L test—*Pseudomonas*

M. E. Rhodes (1959). J. gen. Microbiol. 21, 221–263.

Aerobic *versus* anaerobic glucose utilization: Ten ml volumes of the basal peptone agar recommended by Hugh and Leifson (1953) were used; glucose solution sterilized by filtration (sintered glass) was added to make 1.0% (w/v). Stab inoculations were made in duplicate cultures. One set of cultures was incubated in the usual way and the duplicates were sealed with a 1 cm layer of paraffin wax + a 1 cm layer of mineral oil before incubation.

R. Hugh and E. Leifson (1953). J. Bact. 66, 24–26.

H&L test—general

V. B. D. Skerman, *A Guide to the Identification of the Genera of Bacteria.* 2nd ed. Williams and Wilkins Book Co., Baltimore, 1967.

Medium:

Peptone	2.0 g
NaCl	5.0 g
K_2HPO_4	0.3 g
Agar	3.0 g
Bromthymol blue (1% aqueous solution)	3.0 ml
Distilled water	1000.0 ml

Dissolve the ingredients and adjust the pH, if necessary, to 7.1. Sterilize at 121°C for 20 minutes.

Prepare 10 per cent aqueous solutions of the carbohydrates and sterilize by Seitz filtration. Add 10 ml of the sterile carbohydrates aseptically to every 100 ml of the sterile, melted medium and dispense in 5-ml amounts in sterile 150 by 13 mm tubes.

Inoculate two tubes of each carbohydrate with each organism by stabbing with inoculum from a fresh slope culture. Cover the surface of one tube with sterile paraffin (petrolatum).

Reading the tests: Fermentative organisms produce acid throughout both tubes. Oxidative organisms produce acid in the open tube only. In the latter, acid appears first at the surface and then progressively towards the base.

Slow oxidative reactions are sometimes preceded by a slight alkaline reaction.

Note: 1. Dr. P. H. A. Sneath (personal communication) has suggested the use of long narrow tubes and the elimination of the paraffin (petrolatum) seal to avoid the unpleasantness associated with washing the sealed tubes.

2. Although this test sharply differentiates between two groups of organisms, the suggestion that the acid produced by the oxidative organism is the result of an "oxidative" action of the sugar is open to question. Some organisms which give acid only in the open tubes have been shown to produce no acid at all if grown under strictly oxybiontic conditions in shallow liquid cultures. It is possible that the so-called oxidative organisms may, in some instances at least, actually ferment the carbohydrate but are unable to produce sufficient *growth* in the absence of oxygen. Once this initial growth is established and oxygen becomes limiting in the dense surface growth, acid *may* be produced *fermentatively*. This question needs further clarification but does not detract from the usefulness of the test.

R. Hugh and E. Leifson (1953). J. Bact. 66, 24–26.

H&L test—*Flavobacterium*

R. Buttiaux and J. Vandepitte (1960). Annls Inst. Pasteur, Paris 98, 398–404.

Hydrates de carbone. — Sur le milieu de Hugh et Leifson glucosé, un phénomène assez curieux retient l'attention. Le germe se comporte d'-abord comme un oxydant, puis attaque discrètement le glucose après quatre ou cinq jours sur le milieu recouvert d'huile de paraffine ; le virage du bleu de bromothymol est lent et léger. On peut admettre, après des expériences répétées, qu'un pouvoir fermentaire très peu développé existe. Nous avons remarqué des faits identiques chez certains *Xanthomonas.*

H&L test—*Pseudomonas*

R. Butiaux (1961). Annls Inst. Pasteur, Paris 100, Suppl. 6, 43–51.

Le milieu de Hugh et Leifson [20] donne de bons résultats pour leur recherche. Il est décrit par Véron [14].

Dans le tube recouvert d'huile, l'oxygène ne peut intervenir. Si l'indicateur vire au jaune, la transformation du glucose qu'il traduit ne peut être due qu'à une fermentation. Dans ce cas, d'ailleurs, les deux tubes inoculés présentent le même virage. Au contraire, la bactérie qui oxyde le sucre, agit seulement dans le milieu au contact de l'air.

Pour les travaux de routine, on peut utiliser le milieu suivant :

Bacto-peptone (Difco)	2	g
Protéose peptone n° 3 (Difco)	10	g
ClNa	5	g
PO_4K_2H	0.3	g
Rouge de phénol	0.025	g
Bacto-agar (Difco)	5	g
Eau distillée	1 000	ml

ajusté à pH 7.6. On répartit des culots de 85 mm de hauteur environ (4 ml) dans des tubes de 8 X 180 mm. On autoclave vingt minutes à 120°. Au moment de l'emploi, on régénère au bain-marie bouillant pendant dix minutes. On ajoute une quantité suffisante d'une solution de glucose (stérilisée par filtration) pour obtenir une concentration finale en sucre de 1 p. 100. Lorsque la gélification est suffisante, on inocule le culot par une longue piqûre centrale. On incube à la température optima de développement. L'oxydation se traduit par un virage au jaune dans la zone supérieure des 10 mm, les parties plus profondes du milieu restant rouges. Dans la fermentation, la coloration jaune s'étend à toute la hauteur du milieu et peut s'accompagner de bulles de gaz si la bactérie étudiée est aérogène. On peut ainsi observer, en même temps, la mobilité des germes anaérogenes.

Sur les deux milieux décrits, certaines bactéries n'attaquent pas le glucose ; dans ce cas, l'indicateur coloré ne vire pas ou vire vers la zone alcaline de pH. Ce dernier phénomène est dû à la libération de bases au cours de la transformation des matières protéiques. On dit alors que le germe est « inerte » ou « alcalinisant. »

[14] Mager (J.). *Biochim. Biophys. Acta,* 1959, 36, 529.
[20] Mac Quillen (K.). *Biochim. Biophys. Acta,* 1950, 6. 66.

H&L test—*Pseudomonas*

M. Véron (1961). Annls Inst. Pasteur, Paris 100, Suppl. 6, 16–42.

Mais une autre méthode, très simple et pratique, due à Hugh et Leifson [32], doit être bien connue, car elle est maintenant très largement employée.

Ces auteurs utilisent le milieu suivant : peptone pancréatique de caséine, 2 g ; NaCl, 5 g ; K_2HPO_4, 0.3 g ; gélose, 3 g ; bleu de bromothymol, 0.03 g ; eau distillée, q. s. p. 1 000 ml. Ajuster le pH à 7.1. Filtrer. Répartir par 10 ml en tubes 17 X 170 ou par 3 ml en tubes 12 X 120. Boucher au coton. Autoclaver à 115°C. Pour l'utilisation, faire fondre au bain-marie à 100° C (la régénération par ébullition est inutile), ramener à 50° C et ajouter le glucide à étudier (q. s. p. 1 p. 100). Faire solidifier en culot dans l'eau froide.

Pour étudier le métabolisme oxydatif ou fermentatif du glucose, Hugh et Leifson ont proposé d'utiliser pour chaque souche deux tubes de leur milieu, glucosé à 1 p. 100 ; ces tubes sont ensemencés par piqûre centrale avec le fil de platine trempé dans une culture de 24–48 heures en bouillon. L'un des tubes ensemencés est ensuite recouvert d'une couche de vaseline fondue (au bain-marie) sur une hauteur de 6 à 12 mm, et aussitôt plongé dans l'eau froide pour refroidir la vaseline.

Avec ces deux tubes, on peut distinguer trois catégories de germes :

a) Les germes fermentatifs qui donnent une réaction acide dans les deux tubes, avec ou sans production de gaz. Il est impossible, dans ce cas, de déceler le métabolisme oxydatif, la fermentation s'accompagnant d'un virage intense dans la totalité des deux tubes.

b) Les germes oxydatifs qui donnent une réaction acide seulement dans le tube ⟨⟨ ouvert. ⟩⟩ Très souvent, la culture provoque une légère alcalinisation transitoire en surface, puis la réaction acide apparaît à la surface du tube et s'étend graduellement vers le bas. Par contre dans le tube ⟨⟨ fermé, ⟩⟩ c'est-à-dire recouvert de vaseline, le pH ne change pas et la culture est faible, souvent limitée à un disque situé sous la vaseline. Ces bactéries oxydatives, souvent aérobies strictes en gélose profonde, sont donc incapables d'utiliser les glucides pour leur croissance en anaérobiose même relative.

c) Les germes non-oxydants et non-fermentants, encore appelés inactifs, qui ne modifient pas le pH dans aucun des deux tubes, ou bien donnent seulement une alcalinisation du tube ⟨⟨ ouvert ⟩⟩ dans le cas de variétés alcalinisantes.

. . . .

Quoi qu'il en soit, les acides provenant de l'attaque des différents glucides oxydés par les *Pseudomonas* sont toujours élaborés en quantités faibles et peuvent être masqués par les produits basiques du catabolisme protéique dans un milieu trop riche en protéines, tel que l'eau peptonée à 1 p. 100 habituelle, où de très numbreuses souches de *Pseudomonas,* y compris *Ps. aeruginosa,* ne produisent aucune acidité visible. Il convient donc d'utiliser un milieu relativement pauvre en protéines et assez peu tamponné.

Le milieu décrit par Hugh et Leifson [32], dont la formule est donnée plus haut, convient bien pour cette étude. Liu [50] avait utilisé, de son côté, une eau peptonée à 0.1 p. 100 et Simon [67] une dilution à 5 p. 100 de macération de viande dans l'eau. Mais nous avons montré (Véron et Chatelain [76]) qu'un milieu gélosé, dérivé de celui proposé par Simon, donnait des résultats positifs, plus précoces et plus nombreux que tous ces différents milieux. La formule du milieu macération-gélose que nous utilisons est la suivante :

Macération de viande (500 g de viande de boeuf par litre), 50 ml ; KCl, 5 g ; gélose, 3 g ; rouge de phénol, 0.02 g ; eau distillée, q. s. p. 1 000 ml. Ajuster le pH à 7.4. Filtrer. Répartir par 10 ml en tubes 17 × 170 ou par 3 ml en tubes 12 × 120. Boucher au coton. Autoclaver à 115° C. Ce milieu s'utilise de la même façon que le milieu de Hugh et Leifson cité plus haut ; il est parfois nécessaire de réajuster le pH à 7.4 (avec une solution 0.1 N de potasse) avant d'ensemencer, notamment lorsqu'on utilise certains glucides dont les solutions sont acides.

Dans ce milieu macération-gélose, 109 souches de *Ps. aeruginosa* ou *Ps. fluorescens* se sont toutes montrées oxydatives, sauf une souche de collection de *Ps. aeruginosa* qui était inactive sur les glucides. Sur 69 souches de *Ps. aeruginosa* isolées récemment, le résultat était acquis en vingt-quatre heures pour 60 souches, en trois jours pour 66 souches et en quatre jours pour 68 souches. Une seule souche (60 CN), très muqueuse, n'a oxydé le glucose qu'en trois semaines. Avec 19 souches de *Ps. fluorescens,* l'oxydation du glucose était manifeste après un jour ou moins (14 souches), deux jours (4 souches) et sept jours (1 souche).

[32] Hugh (R.) et Leifson (E.). *J. Bact.,* 1953, 66, 24.
[50] Liu (P. V.). *J. Bact.,* 1952, 64, 773.
[67] Simon (R. D.). *Brit. J. exp. Path.,* 1956, 37, 494.
[76] Véron (M.) et Chatelain (R.). *Ann. Inst. Pasteur,* 1960, 99, 253.

H&L test—*Pseudomonas*

O. Lysenko (1961). J. gen. Microbiol. 25, 379–408.

Iodoacetate test: –

Bacto tryptone	0.1% (w/v)
Difco yeast extract	0.1% (w/v)
NaCl	0.5% (w/v)
K_2HPO_4	0.03% (w/v)
glucose,	1.0% (w/v)
1.5% (w/v) bromthymol blue aq.	0.2% (w/v)
pH 7.2	

Steamed for 30 minutes on 3 successive days. Monoiodoacetate, sterilized by filtration, added to final concentration of 10^{-3} *M.* Control (– iodoacetate) used.

Inoculated in the normal way, incubated 28°C and read *for* 5 days.
+ve if acid in both
–ve if acid only in control.

H&L test—*Xanthomonas*

A. C. Hayward and W. Hodgkiss (1961). J. gen. Microbiol. 26, 133–140.

The medium of Hugh and Leifson (1953) was modified for use with the weakly oxidative plant pathogenic bacteria, as follows: peptone (Oxoid), 1.0 g; $NH_4H_2PO_4$, 1.0 g; $MgSO_4 \cdot 7H_2O$, 0.2 g; KCl, 0.2 g; agar, 3.0 g; bromthymol blue, 0.03 g; distilled water, 1 L; pH about 7.2. Five ml of a 10% (w/v) solution of glucose were added to 45 ml quantities of the molten agar base which was dispensed in sterile plugged test tubes (5 × 0.4 inch) to a depth of about 1.5 in. Sterile liquid paraffin (B. P. grade) was used as a seal, following stab inoculation of the test bacteria from agar cultures. Salicin (10.0 g/L) was sterilized with the medium. Inoculated tubes were examined for a period of 14 days.

R. Hugh and E. Leifson (1953). J. Bact. 66, 24–26.

H&L test—*Pseudomonas*

R. Hugh and E. Ryschenkow (1961). J. gen. Microbiol. 26, 123–132.

The effects of the bacteria on carbohydrates were determined by the OF (oxidative-fermentative) principle of Hugh and Leifson (1953) and Leifson (1958). The medium used had the following composition: Difco Casitone (pancreatic digest of casein), 5 g; agar, 3 g; bromthymol blue (2%, w/v, aqueous solution), 4 ml; carbohydrate, 10 g; distilled water, 1000 ml; final pH 7.1.

Seitz-filtered 10% (w/v) carbohydrate solution was added aseptically to the cool, autoclaved (15 min at 121°C), melted basal medium. The medium was then dispensed aseptically into 13 × 100 mm sterile tubes to a depth of 50 mm.

Maltose oxidation. The authors state that *Pseudomonas maltophila* does not produce acid from glucose but oxidizes maltose readily.

Note: Hugh and Leifson, (1963. Int. Bull. bact. Nomencl. Taxon. 13, 133–138) note that the unsealed surface of dextrose medium is generally alkaline after 24 hours but may slowly become weakly acid after 5 days' incubation. A few strains have recently been found to produce oxidative acidity promptly from glucose.

R. Hugh and E. Leifson (1953). J. Bact. 66, 24.
E. Leifson (1958). Zentbl. Bakt. ParasitKde (I. Abt. Orig.), 173, 487.

H&L test—(Oxidation – Fermentation medium)

W. H. Ewing (1962). *Enterobacteriaceae. Biochemical Methods for Group Differentiation.* U. S. Department of Health, Education and Welfare, Public Health Service Publication, No. 734 (revised).

This method aids in the differentiation of microorganisms that utilize carbohydrates oxidatively rather than fermentatively and therefore is helpful in the identification of pseudomonads and members of the tribe *Mimeae.* It also aids in the identification of microorganisms that do not utilize glucose in either way (e.g., *Alcaligenes*). Only a glucose medium is included here, but other carbohydrates may be substituted for special purposes (Hugh and Leifson, 1953).

Peptone	2 g
Sodium chloride	5 g
Dibasic potassium phosphate	0.3 g
Agar	3 g
Bromthymol blue (1% aqueous solution)	3 ml
Distilled water	1000 ml

Adjust pH to 7.1.

Distribute the basal medium in test tubes (e.g., 13 × 100 mm), 3 or 4 ml per tube, and sterilize at 121°C for 15 minutes. After sterilization, 1 percent of glucose is added to each tube. A 10 per cent solution of glucose in distilled water sterilized by filtration is convenient for this purpose.

Inoculation: Two tubes of medium are inoculated (stab) with the culture under investigation. The medium should be inoculated lightly, using a young agar slant culture as the source of inoculum. After inoculation a layer (0.25 to 0.5 cm) of sterile melted petrolatum or of sterile paraffin oil is added to one of the tubes.

Incubation: 37°C. Observe daily for 3 or 4 days. Acid formation in the open tube only indicates oxidative utilization of glucose. Acid formation in both the open and sealed tubes is indicative of a fermentation reaction. Lack of acid production in either tube indicates that the microorganism being tested does not utilize glucose by either method.

H&L test—*Vibrio*

G. H. G. Davis and R. W. A. Park (1962). J. gen. Microbiol. 27, 101–119.

Oxidation *versus* fermentation of glucose. Modified from Hugh and Leifson (1953). Basal medium plus 1% (w/v) glucose and 0.3% (w/v) agar. Initial pH, 7.0 and 2 ml of 1.5% (w/v) bromocresol purple added per litre. (Note: these modifications were recommended by Dr. A. C. Hayward, Commonwealth Mycological Institute, Kew, and Dr. A. C. Baird-Parker, Unilever, Bedford, and do appear to clearify the results of this test.)

R. Hugh and E. Leifson (1953). J. Bact. 66, 24–26.

H&L test (Dissimilation of D-mannitol)—*Staphylococcus* and *Micrococcus*

D. A. A. Mossel (1962). J. Bact. 84, 1140–1147.

Modification of Hugh and Leifson's (1953) test was used for this purpose. It consists of stabbing the strains under investigation into tubes filled to at least 8 cm in height with an agar of the following composition: Trypticase (BBL), 10 g; yeast extract (Difco), 1.5 g; NaCl, 5 g; D-mannitol (analytical reagent grade), 10 g; agar, 15 g; bromcresol purple (Difco), 15 mg; water, 1000 ml; pH 7.1. All tubes were first heated in a steaming water bath for 10 min, then rapidly cooled in ice water, and immediately inoculated.

Incubation was continued through 4 days at 36° ± 1°C; readings were carried out after 2 and 4 days. Results were recorded as fermentation of mannitol if the lower and upper parts of the tubes turned distinctly yellow; if only the upper part turned yellow the result was noted as oxidative attack of mannitol. If, after 4 days of incubation, either no color change was observed or a distinct intensification of the purple color occurred, it was concluded that D-mannitol was not attacked.

R. Hugh and E. Leifson (1953). J. Bact. 66, 24–26.

H&L test—*Vibrio*

G. H. G. Davis and R. W. A. Park (1962). J. gen. Microbiol. 27, 101–119.

Oxidation *versus* fermentation of glucose; test as described by Hugh and Leifson (1953) initial pH, 7.4.

R. Hugh and E. Leifson (1953). J. Bact. 66, 24–26.

H&L test—*Escherichia aurescens*

H. Leclerc (1962). Annls Inst. Pasteur, Paris 102, 726–741.

a) Oxydation, fermentation. – Le milieu de Hugh et Leifson [16] permet de définir le mode d'action des bactéries sur le glucose. D'après les résultats on classe les bactéries en oxydantes, fermentantes ou inactives.

[16] Hugh (R.) and Leifson (E.). *J. Bact.*, 1953, 66, 24.

H&L test—(iodoacetate test)—*Pseudomonas odorans*

I. Málek, M. Radochová and O. Lysenko (1963). J. gen. Microbiol. 33, 349–355.

According to Lysenko (1961) 28°C.

O. Lysenko (1961). J. gen. Microbiol. 25, 379.

H&L test—general

G. S. Wilson and A. A. Miles (1964). Topley and Wilson's *Principles of Bacteriology and Immunity*. 5th ed. Arnold.

Alternatively Hugh and Leifson's (1953) medium may be used. This is a glucose medium containing only a small amount of peptone and rendered semisolid with 0.3 per cent agar. This medium may also be used to distinguish between fermentation, which is an anaerobic process and oxidation, which is aerobic. For this purpose two tubes are inocu-

lated by the stab method. One tube is covered with ¼–½ in. of sterile melted petrolatum. In true fermentation, acid is formed in both tubes; in oxidation, only in the uncovered tube. For a modification of this medium see Davis and Park (1962).
G. H. G. Davis and R. W. A. Park (1962). J. gen. Microbiol. 27, 101.
R. Hugh and E. Leifson (1953). J. Bact. 66, 24–26.

H&L test—motile marine bacteria and *Hyphomicrobium neptunium**

E. Leifson, B. J. Cosenza, R. Murchelano and R. C. Cleverdon (1964). J. Bact. 87, 652–666.

The tests for carbohydrate metabolism were made with the MOF medium of Leifson (1963) and the following carbohydrates: glucose, sucrose, lactose, xylose, maltose, and mannitol. The incubation temperature was 20°C, and the tubes were discarded after 7 days.
E. Leifson (1963). J. Bact. 85, 1183–1184.
*E. Leifson (1964). Antonie van Leeuwenhoek 30, 249–256.

H&L test—*Vibrio marinus*

R. R. Colwell and R. Y. Morita (1964). J. Bact. 88, 831–837.

The Hugh and Leifson (1953) method for testing for the production of acid from carbohydrates was modified further; in addition to the substitution of synthetic seawater for the salts and distilled water, the agar was omitted and a Durham vial was added to each tube for checking gas production.

Fermentative attack on glucose was checked further by the addition of sodium iodoacetate, sterilized by filtration and at 0.001 M concentration, to the glucose-synthetic seawater-broth tubes.

Synthetic seawater containing sodium chloride, 2.4%; potassium chloride, 0.07%; magnesium chloride (hydrated), 0.53%; and magnesium sulfate (hydrated), 0.7%, was used as the base. The inoculum was prepared from a 24 hour artificial seawater broth. Cultures were incubated at 18°C.
R. Hugh and E. Leifson (1953). J. Bact. 66, 24–26.

H&L test—*Hyphomicrobium neptunium*

E. Leifson (1964). Antonie van Leeuwenhoek 30, 249–256.

The author used the test as published by him in J. Bact. 85, 1183–1184, 1963.

H&L test—*Pseudomonas aeruginosa*

R. R. Colwell (1964). J. gen. Microbiol. 37, 181–194.

The method of Hugh and Leifson (1953) was used to determine whether glucose was used oxidatively or fermentatively. The miniature tube method (Colwell and Quadling, 1962) was also used. Growth in glucose with and without added iodoacetate (Lysenko, 1961) was tested.
R. R. Colwell and C. Quadling (1962). Can. J. Microbiol. 8, 813.

R. Hugh and E. Leifson (1953). J. Bact. 66, 24.
O. Lysenko (1961). J. gen. Microbiol. 25, 379.

H&L test—*Pseudomonas aeruginosa*

A. H. Wahba and J. H. Darrell (1965). J. gen. Microbiol. 38, 329–342.

This was performed by the method described by R. Hugh and E. Leifson (1953. J. Bact. 66, 24–26).

H&L test—*Pseudomonas*

G. L. Bullock, S. F. Snieszko and C. E. Dunbar (1965). J. gen. Microbiol. 38, 1–7.

Oxidation of glucose was determined in O/F Basal Medium (Difco) containing 1% (w/v) glucose. This commercial medium is almost identical with medium of Hugh and Leifson (1953).

Cultures incubated at 20–22°C for 1 week.

R. Hugh and E. Leifson (1953). J. Bact. 66, 24–26.

H&L test—*Staphylococcus*

M. Kocur, F. Přecechtěl and T. Martinec (1966). J. Path. Bact. 92, 331–336.

Anaerobic utilization of glucose and mannitol was tested in a medium recommended by the Subcommittee on the Taxonomy of Staphylococci and Micrococci (1965).

Subcommittee on Taxonomy of Staphylococci and Micrococci (1965). Int. Bull. bact. Nomencl. Taxon. 15, 109.

H&L test—*Mycobacterium*

R. E. Gordon (1966). J. gen. Microbiol. 43, 329–343.

Fermentation or oxidation of glucose. Both the Hugh and Leifson (1953) test and a modification of it were used to determine whether the cultures utilized glucose fermentatively or oxidatively. The basal medium of Hugh and Leifson contained: peptone, 2 g; NaCl, 5 g; K_2HPO_4, 0.3 g; agar, 3 g; 1% (w/v) aqueous solution of bromthymol blue, 3 ml; distilled water, 1000 ml; pH 7.1. The semisolid agar was tubed (7.5 ml per tube, 16 mm in diameter), autoclaved, and quickly cooled. Before the agar solidified, 0.5 ml of 15% (w/v) aqueous solution of glucose, also sterilized by autoclaving, was added to each tube. As soon as the agar became firm, duplicate tubes were inoculated by stabbing from a 3-day-old culture in glucose broth; one tube was sealed with vaspar [paraffin, 60% (w/w). vaseline, 40% (w/w)]. The vaspar seal occasionally pulled away from the inside of the tube during the first 24 hr of incubation, and the culture was resealed by heating the glass around the vaspar in a small flame. The cultures were observed at 7 and 28 days. A culture that was growing and producing the acid colour of the indicator in both the aerobic and anaero-

bic tubes was recorded as fermenting glucose; a culture that was growing in both tubes but forming the acid colour of the indicator in only the aerobic tube was recorded as oxidizing glucose.
R. Hugh and E. Leifson (1953). J. Bact. 66, 24.

HUMIC ACID

Humic acid

H. Winogradsky (1950). Annls Inst. Pasteur, Paris 79, 354–356.

Voici, brièvement résumés, les deux modes d'extraction :

Extraction par la lessive de potasse à 10 *p.* 100. – A. 100 g de terreau (fumier consommé ou terreau de couches), ajouter 100 cm^3 de HCl à peu près N ;

Laisser macérer pendant vingt-quatre heures en remuant de temps en temps. Décanter ;

Ajouter 250 cm^3 de KOH à 10 p. 100. Bien remuer, puis laisser déposer. Décanter ;

Ajouter HCl (même concentration que ci-dessus) goutte à goutte jusqu'à réaction acide. Agiter. Il se forme un précipité brunâtre, floconneux, qu'on laisse déposer. On décante le liquide surnageant. On lave avec de l'eau distillée à plusieurs reprises ;

Dissoudre le précipité dans la lessive de potasse, reprécipiter par l'acide chlorhydrique comme ci-dessus à plusieurs reprises. On filtre sur Buchner, en lavant avec de l'eau distillée chaude.

On répète cette opération avec 7 ou 8 lots de 100 g de terreau et on dissout les humates obtenus dans 250 cm^3 de lessive de potasse N/10, car il est malaisé d'opérer avec plus de 100 g de terreau.

Pour avoir une idée de la concentration approximative de la solution de l'ensemble des précipités, on sèche un des lots jusqu'à poids constant et l'on en déduit le poids approché d'humate total dissous.

On stérilise la solution à l'autoclave. Nous avons constaté qu'elle se conserve très bien en flacon bouché hermétiquement.

Extraction des humates de calcium. – A 10 g de terreau (ou n'importe quel échantillon de terre riche en matières végétales décomposées) on ajoute 200 cm^3 de HCl à 1 p. 100 et on laisse macérer pendant vingt-quatre heures. Décanter et laver ;

Ajouter 100 cm^3 de NH_3 à peu près N sur le sédiment. Laisser macérer à l'étuve vingt-quatre heures. Filtrer (ou centrifuger). Chauffer le filtrat au bain-marie pour éliminer NH_3 en excès en ajoutant de l'eau distillée, si nécessaire, pour éviter que la solution se concentre. On étend avec de l'eau et on précipite par addition de 10 g de chlorure de calcium. On filtre sur Buchner et on lave jusqu'à disparition de l'excès de chlorure ;

Mettre le précipité en suspension dans 50 cm^3 d'eau distillée environ. On stérilise à l'autoclave et on conserve au frigidaire en flacon bouché hermétiquement.

Humic acid

H. de Barjac (1952). Annls Inst. Pasteur, Paris 83, 279–281.

Mode d'obtention des acides humiques. – 2 kg de terreau sont soumis à l'épuisement par 4 l d'ammoniaque en solution à 2 p. 100, en présence de chlorure de potassium à 0.2 p. 100 pour coaguler l'argile. L'épuisement est réalisé à l'abri de l'air (solution ammoniacale préparée avec de l'eau distillée bouillie, flacon d'extraction complètement rempli de liquide et soigneusement bouché), ce qui est indispensable pour empêcher l'oxydation de la lignine. Après douze heures de contact à froid, on porte douze heures à 60° en agitant à plusieurs reprises. Le liquide brun surnageant est filtré et le filtrat précipité par l'acide chlorhydrique jusqu'à obtention de pH = 2, point où l'on n'observe plus de précipitation. Le précipité est séparé par centrifugation puis soumis à différents lavages avec une solution de ClH à 2 p. 1 000, puis de l'eau acidifiée par quelques gouttes de ClH, puis enfin de l'eau distillée (trois à quatre fois). Le précipité lavé est alors remis en suspension dans l'eau distillée, et peut être purifié par redissolutions et précipitations successives. On obtient ainsi une suspension homogène et stable d'acides humiques dont le pH est de 2.5 et l'extrait sec est de 4.4 g par litre (dans les conditions de concentration où nous nous sommes placé).

Réalisation d'une gamme de pH acides en différents milieux. – Cette même suspension d'acides humiques ajoutée à diverses concentrations aux milieux usuels suivants nous a permis d'obtenir des pH échelonnés de 6.4 à 3.3.

HYALURONIDASE

Hyaluronidase

Lengthy descriptions of experimental methods are given in the following papers:

K. Meyer, G. L. Hobby, E. Chaffee and M. H. Dawson (1940). J. exp. Med. 71, 137–146.

G. L. Hobby, M. H. Dawson, K. Meyer and E. Chaffee (1941). J. exp. Med. 73, 109–123.

K. Meyer, E. Chaffee, G. L. Hobby (1941). J. exp. Med. 73, 309–326.

R. M. Pike (1948). J. infect. Dis. 83, 1–11.

A. P. MacLennan (1956). J. gen. Microbiol. 14, 134–142.

Hyaluronidase—anaerobes

A. -R. Prévot (1966). Techniques pour le diagnostic des Bactéries Anaérobes. Éditions de la Tourelle, St. Mandé.

Recherche de l'hyaluronidase.

On dissout de l'acide hyaluronique pur dans un tampon acéto-acétique 0.1 M de pH 6 dont la formule est :

Acide acétique 0.5 M	3	cm^3
Acétate de sodium 0.5 M	97	cm^3
NaCl	4.384	g
Eau distillée q.s.p.	500	cm^3

On mélange une partie aliquote de cette solution avec un filtrat de culture. Si ce dernier contient de l'hyaluronidase, l'acide hyaluronique est dépolymérisé et la turbidité de la suspension diminue. Une antihyaluronidase empêche cette dépolymerisation de façon spécifique et par conséquent maintient la turbidité du milieu.

HYDROGEN CYANIDE

Hydrogen cyanide—*Chromobacterium*

P. H. A. Sneath (1956). J. gen. Microbiol. 15, 70–98.

Nutrient or blood agar plates, heavily inoculated to give growth over most of the plate and incubated at 25°C for 2 days, were tested by placing the end of an indicator paper inside the plate (but not touching the medium) and replacing the lid. The indicator paper was made as follows: benzidine acetate was dissolved in boiling water and the saturated solution was cooled and filtered. To the filtrate was added one-tenth its volume of 3% cupric acetate solution. The tip of a strip of filter paper was dipped in the mixture, and this becomes blue in the presence of HCN. Other gases giving a positive test are chlorine, bromine and hydrogen chloride, while sulfur dioxide and hydrogen sulfide inhibit the reaction (Anonymous, 1938). The paper may turn brown after 10 min, apparently because of ammonia from the cultures.

Hydrogen cyanide—*Chromobacterium violaceum*

R. Michaels and W. A. Corpe (1965). J. Bact. 89, 106–112.

The paper describes a detailed study of cyanide production. The routine medium for cyanide production by growing bacteria contained 1% (w/v) peptone dissolved in distilled water and adjusted to pH 7.0. For solid media 2% (w/v) agar was added.

A chemically defined medium for the study of the effect of specific carbon and nitrogen sources is described.

The media were dispensed in 75 ml quantities in 500 ml rubber stoppered Erlenmeyer flasks, inoculated with 1 ml of inoculum, and incubated at 30°C on a rotary shaker for specified times.

Cyanide was estimated by the following procedures:

Qualitative and quantitative estimation of cyanide: The alkaline picrate test and the copper sulfide tests gave positive reactions for cyanide when performed on both standard cyanide solution and culture distillates. Aqueous solutions of potassium cyanide (Fisher Scientific Co., Pittsburgh, Pa.; reagent grade) were standardized by the Liebig silver nitrate titration method described by Kolthoff and Sandell (1952). Cyanide was estimated in culture distillates with colorimetric methods. A good linear response was obtained with both the Aldridge (1944) and the Epstein (1947) procedures. Direct quantitative determination of cyanide in the culture supernatant liquid was not possible with the colorimetric procedures, because of development of nonspecific color by noncyanide constituents. The Aldridge test was used routinely for making quantitative estimations.

W. N. Aldridge (1944). Analyst, Lond. 69, 262–265.
J. Epstein (1947). Analyt. Chem. 19, 272–274.
I. M. Kolthoff and E. B. Sandell (1952). *Textbook of quantitative inorganic analysis.* The Macmillan Co., New York.

HYDROGEN PEROXIDE PRODUCTION

Hydrogen peroxide production—*Streptococcus*

J. Gordon (1954). J. Path. Bact. 68, 645–646.

Heated blood-agar with added benzidine or other aromatic amines is commonly used for the detection of hydrogen peroxide by bacteria (Penfold, 1922; Gordon and McLeod, 1940; Hayward, 1942); a black colour develops around positive colonies. Some difficulty, however, is occasionally experienced in producing satisfactory media, as the low solubility of most of the amines used prevents even distribution through the medium. The only more soluble amine so far used, 2:7 diamino-fluorene hydrochloride (Holman and McLeod, unpublished), is not available commercially.

I have found that the addition in a final concentration of 0.1 or 0.2 per cent of *m*-amino-acetanilide hydrochloride (B.D.H.), adjusted to pH 7, to heated blood-agar gives characteristic blackening around colonies of pneumococci or *Streptococcus viridans.* The amine is very soluble, and even when green formation on heated blood-agar was very slight, blackening of the medium containing *m*-amino-acetanilide was easily detected.

J. Gordon and J. W. McLeod (1940). J. Path. Bact. 50, 167.
N. J. Hayward (1942). J. Path. Bact. 54, 379.
W. J. Penfold (1922). Med. J. Aust. 2, 120.

Hydrogen peroxide production—lactic acid bacteria

R. Whittenbury (1963). J. gen. Microbiol. 32, 375–384.

Manganese dioxide agar, based on the pyrolusite agar described by Kneteman (1947), was used. Glucose agar was poured as a plate to which was added a very thin layer of glucose agar containing pyrolusite or manganese dioxide (black tech.; Harrington Bros. Ltd.), 4.0% (w/v). The plates were dried and inoculated by streaking; H_2O_2 formation was indicated by clearing of the manganese dioxide under and around the growth.

Glucose agar was similar to the inoculum medium (see **Fermentation**—lactic acid bacteria p. 310), with the exceptions that agar, 1.5% (w/v), and $MnSO_4 \cdot 4H_2O$, 0.01% (w/v), were added.

Cultures were incubated at 30°C.

A. Kneteman (1947). Antonie van Leeuwenhoek 13, 55.

Hydrogen peroxide production—*Lactobacillus, Pediococcus, Aerococcus, Streptococcus* and *Leuconostoc.*

R. Whittenbury (1964). J. gen. Microbiol. 35, 13–26.

Plates of MDO-agar and HBD-agar were inoculated by streaking with a capillary pipette containing an 18 hr old inoculum. Cultures were examined daily for 7 days.

Manganese dioxide (MDO) agar was basal medium agar adjusted to pH 6.5 and containing 1% (w/v) of a separately sterilized sugar or polyhydroxy alcohol; this was poured in a plate, and then a very thin layer of the same medium (1–2 ml) to which had been added 4% (w/v) manganese dioxide (black tech. Harrington Bros. Ltd.), was poured on top. This medium is a variation of the pyrolusite agar described by Kneteman (1947). Clearing of the manganese dioxide under and around the bacterial growth indicates H_2O_2 formation.

Heated blood *o*-dianisidine (HBD) agar was made as follows. Basal medium agar (90 ml adjusted to pH 6.5) was melted; 5 ml of a 1 + 1 mixture of defibrinated ox blood and water added, and the whole heated at 100°C for 15 min. *o*-Dianisidine (0.1 g in 5 ml sterile water heated at 100° for 15 min) was transferred while still hot with a wide-mouthed pipette to the melted agar. The medium, cooled to 48°C, was completed by adding a separately sterilized sugar or polyhydroxy alcohol solution (final concentration, 1%, w/v) and poured in plates. This medium was a variation of the media used by Penfold (1922), Berger (1953) and Kraus, Nickerson, Perry and Walker (1957). Benzidine was replaced by *o*-dianisidine because of the carcinogenic property of the benzidine. Heated blood was used because many organisms produced a dark brown growth on media containing unheated blood, irrespective of the presence of *o*-dianisidine or the formation of H_2O_2. The production of H_2O_2 was indicated by the growth and the surrounding medium becoming dark brown or black, heme compounds having a peroxidase-like reaction in the oxidation of *o*-dianisidine by peroxide.

Cultures were incubated at 30°C.

U. Berger (1953). Z. Hyg. InfektKrankh 136, 94.
A. Kneteman (1947). Antonie van Leeuwenhoek 13, 55.
F. W. Kraus, J. F. Nickerson, W. I. Perry and A. P. Walker (1957). J. Bact. 73, 727.
W. J. Penfold (1922). Med. J. Aust. 9, 120.

Hydrogen peroxide production—*Streptococcus*

R. Whittenbury (1965). J. gen. Microbiol. 38, 279–287.

The method of R. Whittenbury (1964). J. gen. Microbiol. 35, 13 was used (see p. 463).

Hydrogen peroxide*—*Serratia marcescens*

J. E. Campbell and R. L. Dimmick (1966). J. Bact. 91, 925–929.

Unless noted, samples of a single, third-passage culture of appropriate age were centrifuged, and the cells were suspended in distilled water or supernatant fluid. To provide a relatively constant mass "demand," suspensions were standardized to a total cell mass of 230 Klett units (KU) with a Klett-Summerson photometer and no. 42 filter. This mass was equivalent to approximately 5×10^9 organisms per milliliter in a mature culture. The hydrogen peroxide used was a fresh 1:10 dilution (v/v) of "Superoxol" (30% H_2O_2; Merck and Co., Inc., Rahway, N. J.) in distilled water.

To test disinfectant activity, 1 ml of a standardized bacterial suspension was added to 99 ml of 3% H_2O_2 contained in a 250-ml flask immersed in a temperature-controlled water bath. At appropriate intervals, 1-ml samples were mixed with 9 ml of a 1:1,000 dilution of catalase (Nutritional Biochemicals; Corp., Cleveland, Ohio; technical grade) to neutralize residual peroxide. Catalase was shown not to affect viability of untreated cells at this concentration. Assay of colony-forming units was done by dropping 0.1 ml quantities of aqueous, 10-fold serial dilutions from a calibrated orifice onto the surface of plates containing either chemically defined medium agar or Trypticase Soy Agar (BBL). Plates were incubated at 31°C.

*Note: This abstract should have appeared on p. 90 under **Antibiotics and other inhibitory substances.**

HYDROGEN SULFIDE

Hydrogen sulfide—various

C. R. Fellers, O. E. Shostrom and E. D. Clark (1924). J. Bact. 9, 235–249.

The technique of this method is as follows: A strip of filter paper moistened with a 10 per cent solution of neutral lead acetate is suspended in the mouth of a flask in such a manner that a given length is exposed to the action of any hydrogen sulfide formed. After incubation the approximate number of millimeters of lead acetate paper blackened is used as the basis of comparison. Obviously this method gives only approximate quantitative results.

Hydrogen sulfide—*Pseudomonas*

C. Elliott (1924). J. agric. Res. 29, 483–490.

Strips of lead acetate paper were suspended over broth, agar, potato and rutabaga cylinders. Stabs were made (following the method recommended by the 1923 *Manual of Methods*) in media containing 5 ml of beer extract peptone agar and 5 ml of 0.1 per cent basic acetate. Growth at the surface gradually turned light brown and later medium brown; then the agar became light brown in the upper half and finally, after 2 months, light brown throughout. A small amount of H_2S was produced.

Hydrogen sulfide—*Clostridium* and others

M. C. Kahn (1925). J. Bact. 10, 439–447.

The formula, then, for the substance used to support growth and test for H_2S in these experiments was as follows: Beef heart infusion, 1000 ml; peptone (Difco), 10 g; casein digest fluid, 30 ml; agar, 5 g; sodium thiosulfate crystals c. p. (Baker), 2.5 g. The above ingredients were heated to 100°C for thirty minutes to dissolve; the reaction adjusted to pH 7.2, filtered through cotton and flannel, tubed in 10 ml amounts and sterilized in the autoclave at 15 pounds for twenty minutes.

Just prior to inoculation, the tubes were boiled for fifteen minutes to expel as much of the dissolved oxygen as possible, rapidly cooled to 46°C, and 0.1 ml of a sterilized 10 per cent solution of lead acetate added to each tube. Inoculations were made from 0.5 per cent casein digest agar cultures with the aid of a Pasteur pipette, employing about 0.25 ml of inoculum. In all experiments with anaerobic, spore-bearing organisms definite results are more likely to be obtained within a minimum of time if relatively large amounts of culture are used for transplanting.

The tubes were incubated at 37°C for 20 days and daily observations made during this period.

Hydrogen sulfide—*Salmonella* and *Proteus*

L. H. Almy and L. H. James (1926). J. Bact. 12, 319–331.

The authors describe the use of the methylene blue formation reaction in detecting H_2S which has been freed from acidified cultures with CO_2 and collected in zinc acetate to which p-aminodimethylaniline hydrochloride, hydrochloric acid and ferric chloride have been added.

Methods are described for separating H_2S from mercaptans.

Hydrogen sulfide—*Xanthomonas*

F. R. Jones and L. McCulloch (1926). J. agric. Res. 33, 493–521.

Lead acetate agar made with beef infusion and with a pH value of 6.8 was used in the tests. Other tests were made by suspending lead acetate paper over cultures in various media.

Hydrogen sulfide—modified Kligler's medium

S. F. Bailey and G. R. Lacy (1927). J. Bact. 13, 183–189.

Bacto-beef extract	5 g	Glucose	1 g
Peptone (P.D.)	10 g	Lead acetate*	0.5 g
Sodium Chloride (B. & A.)	5 g	0.02% phenol red	50.0 ml
Agar	15 g	H_2O	1000 ml
Lactose	10 g		

The glucose, lactose and lead acetate should preferably be added after the nutrient agar has been prepared and cooled to 50°C. The pH should be adjusted to 7.4 and the medium tubed and sterilised at 5 lb pressure for 20 minutes.

*$(PbC_2H_3O_2)_2$ Baker's analyzed.

Hydrogen sulfide—*Xanthomonas beticola*

N. A. Brown (1928). J. agric. Res. 37, 155–168.

The organism produces hydrogen sulfide. The different colonies grown on potato cylinders and in beef bouillon were tested by suspending lead-acetate paper in the culture tubes. The paper blackened in two to six days.

Hydrogen sulfide—mustiness in eggs

M. Levine and D. Q. Anderson (1932). J. Bact. 23, 337–347.

Production of H_2S was determined by blackening of lead acetate agar.

Hydrogen sulfide—*Bacteroides*

A. H. Eggerth and B. H. Gagnon (1933). J. Bact. 25, 389–413.

Lead acetate broth was prepared by adding 1 ml of sterile 1 per cent lead acetate solution to 200 ml of sterile beef infusion broth, then distributing in tubes.

Hydrogen sulfide

C. E. ZoBell and C. B. Feltham (1934). J. Bact. 28, 169–176.

The authors note that filter papers impregnated with lead carbonate or lead acetate have been used ever since Gagnon (C. r. hebd. Séanc. Acad. Sci., Paris 85, 1074, 1877) demonstrated the formation of hydrogen sulfide from albuminous matter by bacterial action.

They discuss several methods and recommend the use of these papers.

Hydrogen sulfide—*Clostridium*

S. E. Hartsell and L. F. Rettger (1934). J. Bact. 27, 497–515.

The lead acetate medium usually employed in the determination of hydrogen sulfide production by members of the typhoid-paratyphoid-dysentery group was found to inhibit the growth of *Cl. cochlearum* and "*Cl. putrificum.*" Though Hall (1924) advocated the use of brain medium to detect the ability of anaerobic bacteria to produce hydrogen sulfide, he explained that in the case of "*Cl. putrificum*" insufficient proteolysis usually takes place in this medium to liberate the iron present in the tissue and permit the formation of iron sulfide. His observations were corroborated in the present investigation; however, *Cl. cochlearum* and "*Cl. putrificum*" were found to produce considerable hydrogen sulfide when grown in egg cube gelatin, in egg cube broth, and in egg meat medium. Filter paper which had been soaked in a 10 per cent solution of lead acetate and then dried was suspended immediately beneath the cotton stopper of tubes of the above media, after inoculation with the test species. In some instances when egg cubes were employed, either in broth or in gelatin media, it was not necessary to use this paper, the presence of hydrogen sulfide being indicated by the blackening of the cubes.

I. C. Hall (1924). J. Bact. 9, 211–224.

Hydrogen sulfide—*Pasteurella tularense*

C. M. Downs and G. C. Bond (1935). J. Bact. 30, 485–490.

For this study a basic slant medium of semisolid glucose, 2 per cent Bacto-peptone and meat extract agar containing cystine or cystine hydrochloride was used. Ferrous sulfate was added as an indicator in some series; in others a strip of moistened lead acetate paper was introduced into the tube after seven days' incubation and the tubes tightly stoppered.

Hydrogen sulfide—*Bacteroides*

A. H. Eggerth (1935). J. Bact. 30, 277–299.

Lead acetate broth was prepared by adding 1 ml of sterile 1 per cent lead acetate to 200 ml of sterile broth containing 0.5 per cent of glucose. Duplicate tubes were prepared containing also 4 per cent of sterile serum.

Hydrogen sulfide—coliforms

R. Vaughn and M. Levine (1936). J. Bact. 32, 65–73.

The authors tested several media and found that only those containing cysteine yielded uniformly positive results. In the absence of cysteine none of the strains of *Aerobacter* and only some of *Escherichia* "intermediate" *coli* types were positive. The cysteine media were those of R. Patrick and C. H. Werkman (1933, Iowa St. Coll. J. Sci. 7, 404–418).

Media: (i) Proteose peptone, 20.0 g; cysteine, 1.0 g; agar, 5.0 g; glucose, 1.0 g; ferrous chloride, 1.0 g; K_2HPO_4, 0.3 g; water, 1000 ml (ii) Proteose peptone, 20.0 g; cysteine, 1.0 g; agar, 5.0 g; ferrous chloride, 1.0 g; K_2HPO_4, 0.3 g; water, 1000 ml. (iii) cysteine, 1.0 g; agar, 5.0 g; ferrous chloride, 1.0 g; K_2HPO_4, 0.3 g; $(NH_4)_2SO_4$, 0.3 g; $(NH_4)_2HPO_4$, 0.7 g; $CaCl_2$, 0.2 g; water, 1000 ml.

The ferrous chloride and the other constituents of all media have to be sterilized separately and the two sterile solutions mixed before distribution.

Hydrogen sulfide—*Clostridium*

R. S. Spray (1936). J. Bact. 32, 135–155.

Lead acetate semisolid agar. To make 1 L take 36 grams Difco lead acetate agar (2 per cent agar). Infuse this for 30 minutes, shaking frequently, in 1 L warm (37°C) distilled water to dissolve all ingredients except the agar. Filter through cotton, with slight suction, to remove the agar. Make the volume to 1 L by pouring warm water through the cotton filter. Add to this 2.5 grams agar flakes, and boil to dissolve the agar. Make the volume to 1 L, and adjust the reaction to pH 7.3 to 7.4. Tube about 8 cm deep, and autoclave as usual.

Note: Prepared by this method, the medium displays quantitatively distinctive reactions with the various anaerobes. We have been unable as yet to reproduce these distinctions by other formulas or other modes of preparation.

Hydrogen sulfide—some sulfur bacteria

R. L. Starkey (1937). J. Bact. 33, 545–571.

The nitroprusside test was found to give a strong, although quickly fading, violet color in 5 ml of a solution containing 0.05 mg of sulfide-sulfur, or 1 part in 100,000. A light violet test which persisted but a few seconds was detected in 5 ml of a solution containing slightly more than 1 part in 500,000. This was the limit of effective use of the test. The iodine titration as used would reveal 1 part of sulfide-sulfur in 100,000 with 5-ml samples, but this was close to the limit of its usefulness.

In making the nitroprusside test, 5 ml of the solution were made distinctly alkaline with ammonium hydroxide, 0.5 ml of 5 per cent sodium nitroprusside was added, and the color quickly noted. (Note: This test is also given by sulfhydryls.)

Lead acetate paper, a very sensitive indicator for sulfide, was used to test for evolution of sulfide from the culture media during growth of the bacterium. A strip of freshly prepared moist paper was held in place by the cotton plug so that the lower end was about 5 mm from the surface of the solution medium.

In another test, a 10 ml portion of the culture solution was boiled for 3 minutes with a lead acetate paper hanging to within 2 cm of the surface of the gently boiling solution.

Hydrogen sulfide—*Fusobacterium*

E. H. Spaulding and L. F. Rettger (1937). J. Bact. 34, 535–548.

Produced in

Proteose peptone	10 g
Liebig's meat extract	3 g

Cysteine HCl	0.5 g
Potato extract (aqueous)	100 ml
Distilled water	900 ml

pH adjusted to 7.6

Blackened lead acetate; method not specified.

In semisolid agar media (0.3% agar), having the above composition, the hydrogen-sulfide-positive strains produced a turbid zone at the surface of the medium after a few hours' exposure to the air. This opaque layer contained elemental sulfur.

Hydrogen sulfide—various

C. A. Hunter and H. G. Crecelius (1938). J. Bact. 35, 185–196.

The authors tested 650 variations of media and recommended the following medium which is prepared as slants.

Tryptone	7.0 g
K_2HPO_4	0.3 g
Agar	5.0 g
Water	1000 ml
Skim milk	50 ml

Dissolve and filter, then add:

Na_2SO_3 (20 per cent)	10 ml
Bismuth liquor (3 per cent)	5 ml
Mannitol	5 g

Tube and sterilize at 15 lb/15 min.

The Bismuth liquor is prepared as follows:

Bismuth liquor – 3%

Place 3 g bismuth citrate, Merck U.S.P. VIII, in a glass stoppered bottle and add about 10 ml of distilled water. Mix well and add approximately 1 ml of NH_4OH sp. gr. 0.90. The bismuth citrate dissolves quickly, forming a clear, colorless solution. If it does not clear readily, gentle heat will increase the solubility. Add distilled water to bring the volume up to 100 ml. One-half (0.5) ml of the bismuth liquor is added to each 100 ml of medium.

The bismuth liquor prepared according to the above directions will remain perfectly clear for several weeks and in that condition can be used with satisfactory results. Should the solution become turbid or opalescent, it should be discarded and a fresh solution prepared.

The color change of the medium is brownish with small quantities of hydrogen sulfide, and a deep brownish-black when larger amounts of sulfide are formed.

Hydrogen sulfide—*Pasteurella multocida*

C. T. Rosenbusch and I. A. Merchant (1939). J. Bact. 37, 69–89.

Kligler's lead acetate was used for the hydrogen sulfide test.

Hydrogen sulfide—*Lactobacillus, Bacteroides*

K. H. Lewis and L. F. Rettger (1940). J. Bact. 40, 287–307.

Tested in glucose-cysteine semisolid agar stab cultures containing 0.05 per cent basic lead acetate, the observations extending over periods of 33 days.

Hydrogen sulfide—wide range of organisms

W. P. Utermohlen, Jr. and C. E. Georgi (1940). J. Bact. 40, 449–459.

Difco Proteose Peptone	20.0 g
K_2HPO_4	1.0 g
Glucose	1.0 g
Cysteine HCl (Merck)	2.0 g
Agar (Bacto)	15.0 g
0.0050 M Co $(NO_3)_2$ solution,*	20.0 ml
0.0050 M Ni $(NO_3)_2$ solution,†	100.0 ml
Water	880.0 ml

pH adjusted to 7.2 ± 0.1. Sterilized 15 lb/20 minutes.
*0.727 g $Co(NO_3)_2 \cdot 6H_2O$ in 500 ml water.
†0.728 g $Ni(NO_3)_2 \cdot 6H_2O$ in 500 ml water.

The nickel and cobalt solutions can be made up as one solution if desired.

Observations were made at the end of 1, 2, 3 and 5 days following incubation at 37°C.

The method was reported to be more sensitive than – Bacto-Lead Acetate Agar, Bacto-Iron Peptone Agar, Bacto-Kligler Iron Agar and Hunter's bismuth sulfite tryptone agar except with *Salmonella typhi.*

Hydrogen sulfide—*Streptobacillus moniliformis*

F. R. Heilman (1941). A study of *Asterococcus muris (Streptobacillus moniliformis).* II. Cultivation and biochemical activities. J. infect. Dis. 69, 45–51.

Strips of filter paper impregnated with lead acetate were placed between the plug and glass in the top of tubes containing inoculated cultures. In serum veal infusion broth and on egg yolk agar slants . . . , in 2–3 days.

With the addition of small amounts of sodium thiosulfate to the medium there was greatly increased production of hydrogen sulfide.

The addition of various amounts of ammonium sulfate, sodium sulfate or cystine to the medium did not increase the production of hydrogen sulfide.

Hydrogen sulfide—*Escherichia coli* and *Phytomonas syringae*

M. A. Smith (1944). J. agric. Res. 68, 269–298.

Strips of lead acetate paper were hung over beef-extract broth cultures.

Lead acetate agar slants were also used with *Escherichia coli* as a test orgainsm.

Hydrogen sulfide—*Corynebacterium*

R. F. Brooks and G. J. Hucker (1944). J. Bact. 48, 295–312.

Slant-stabs of the bismuth citrate agar of Hunter and Crecelius (1938) were inoculated both on the surface and in the stab, incubated at 37°C and observed daily for 2 weeks.

C. A. Hunter and H. G. Crecelius (1938). J. Bact. 35, 185–196.

Hydrogen sulfide—enterobacteria

A. A. Hajna (1945). J. Bact. 49, 516–517.

A new triple-sugar iron agar (TSI) medium has been developed which gives more satisfactory reactions, that is, reactions which are more clear-cut for acid and gas, and more sensitive for H_2S. This is important since selection of carbohydrate test media to be used for preliminary identification of members of the *Salmonella,* or *Shigella* genera, should be based on the TSI reactions. The formula for the TSI medium is given below.

To each liter of distilled water, add

Agar (dry)	13.0 g
B.B.L. nutripeptone*	20.0 g
Sodium chloride	5.0 g

Dissolve in the Arnold sterilizer or in flowing steam in the autoclave. When melted, adjust to pH 7.5.

Admix –

Lactose	10.0 g
Sucrose	10.0 g
Glucose	1.0 g
Sodium thiosulfate	0.2 g
Ferrous ammonium sulfate	0.2 g

Check the pH, which should then be 7.4.

Lastly, add –

Phenol red (1% aqueous solution) 2.5 ml.

Dispense approximately 5 ml in 15 X 100 mm tubes. Autoclave at 12 pounds pressure for 15 to 17 minutes.

*Bacto-peptone, 15 g, plus bacto-proteose peptone, 5 g, gives equally good results.

Sodium sulfite and meat extract have not been found to increase the production of H_2S materially. The role of sucrose in the medium is to eliminate certain sucrose-fermenting, non lactose-fermenting bacteria of the *Proteus* and paracolon groups.

Hydrogen sulfide—*Proteus*

G. T. Cook (1948). J. Path. Bact. 60, 171–181.

This was indicated by the blackening of a strip of filter paper soaked in a saturated solution of lead acetate and placed over a 2 per cent peptone water (Eupeptone) culture. Readings were made after 24 hours' incubation and cultures failing to produce hydrogen sulfide were subsequently retested using tryptic digest broth instead of peptone water. Blackening confined to the edge of the paper was recorded as a negative result.

Hydrogen sulfide—*Pseudomonas alboprecipitans*

A. G. Johnson, A. L. Robert and L. Cash (1949). J. agric. Res. 78, 719–732.

The test strip method of ZoBell and Feltham and also the lead acetate agar test* were used.

*Committee on bacteriological technique of the Society of American Bacteriologists 1947. *Manual of methods for pure culture study of bacteria.* Biotech. Pub., Geneva, N.Y.

Hydrogen sulfide—anaerobic Gram-negative diplococci

G. C. Langford, Jr., J. E. Faber, Jr. and M. J. Pelczar, Jr. (1950). J. Bact. 59, 349–356.

Strips of lead acetate paper prepared according to the *Manual of Methods for Pure Culture Study* (1947) were placed in the tubes and extended down almost to the surface of the medium.

Fluid thioglycolate medium was used to indicate the production of hydrogen sulfide.

Committee on bacteriological technique of the Society of American Bacteriologists 1947. *Manual of methods for pure culture study of bacteria.* Biotech. Pub., Geneva, N.Y.

Hydrogen sulfide—micromethods

M. L. Morse and R. H. Weaver (1950). Am. J. clin. Path. 20, 481–484.

Materials and Methods: As a medium for the production of hydrogen sulfide, a 2 per cent solution of Thiopeptone (Baltimore Biological Laboratories) in distilled H_2O was selected. In preliminary work it was noted that other peptones, such as Bacto Peptone and Bacto Proteose Peptone (Difco), could be used in place of the Thiopeptone. Thiopeptone, however, gave stronger reactions than did other peptones. The medium is dispensed in 0.8-ml amounts in 75 x 10 mm tubes. It is unnecessary to sterilize either the medium or the test tubes. Using clean, dry tubes, the medium is prepared just before it is needed, by dissolving the correct amount of Thiopeptone in distilled water; this solution gives the desired reaction, pH 6.8. That sterilization is not necessary is proved by finding *consistently negative* controls in each batch of tests. In many instances, controls have been incubated for as long as 24 hour although micro tests have seldom been incubated beyond one hour.

The tubes of Thiopeptone medium should be preheated in a water bath at 37°C and then inoculated heavily with growth from logarithmic phase cultures. We have used inocula from six hour cultures on slants of rich media. At times we have transferred the inoculum with a large loop, but in the majority of cases we have used nonabsorbent cotton swabs, made with very small amounts of cotton and usually sterilized to "set" the cotton. The growth was swabbed from half of the surface of the slant. The swab was then introduced into the tube of test medium, rotated below the surface of the medium, and pressed against the side of the tube, thus yielding a heavy, even suspension of organisms. If the culture is suspected of being pathogenic, the swab is placed in the swabbed culture and autoclaved before it is discarded.

To indicate the production of hydrogen sulfide, a 5-mm strip of lead acetate paper was folded 1 cm from the end and inserted into the mouth of the tube. Incubation was at 37°C and the results were read at 15 minute intervals up to one hour. To obtain results that correspond with those of the majority of macrotechniques, final readings should be taken at the end of 45 minutes.

When a longer incubation period is used the microtechnique becomes more sensitive than the common techniques and will show many additional cultures to be producers of hydrogen sulfide.

Hydrogen sulfide—*Staphylococcus*

C. Shaw, J. M. Stitt, and S. T. Cowan (1951). J. gen. Microbiol. 5, 1010–1023.

Stroke and stab cultures were made on lead acetate agar. In the recheck lead acetate papers were exposed over a broth culture.

Hydrogen sulfide—general

P. Hauduroy (1951). Techniques Bactériologiques. Masson et Cie, Éditeurs, Paris.

HYDROGENE SULFURE

(Dépistage de l'). – Ensemencer le microbe dans un tube contenant 10 cm^3 de bouillon nutritif. Placer entre le coton et le bord du tube un fragment de papier-filtre trempé dans une solution saturée d'acétate de plomb et aussitôt desséché, de telle façon que l'extrémité du fragment de papier soit juste au-dessus de la surface du bouillon. Le tube ainsi préparé est placé 5 jours à 37°C. S'il y a production d'H_2S, le fragment de papier noircit. *(W. I. C.).*

(Dépistage) = Gélose à l'acétate de plomb

Hydrogen sulfide—general

P. Hauduroy (1951). Techniques Bactériologiques. Masson et Cie, Éditeurs, Paris.

GÉLOSE À L'ACÉTATE DE PLOMB

(Pour déterminer la production d'hydrogène sulfuré).

Tryptone (marque Bacto)	20 g
Agar-agar	15 g
Eau distillée	1000 ml

Préparer le milieu de culture de réserve et régler son *p*H entre 6.8 et 7.0; stériliser à l'autoclave. Le préparer pour l'emploi en ajoutant 0.4 cm^3 d'une solution stérile à 0.5 p. 100 d'acétate de plomb basique par 100 cm^3 de milieu de culture fondu. Mélanger et répartir aseptiquement dans les tubes et faire prendre très incliné. Incuber pour vérifier la stérilité. Vérifier la réaction du milieu avec des souches connues de *S. paratyphi* et *S. schottmuelleri.* *(M. T.).*

Hydrogen sulfide—general

P. Hauduroy (1951). Techniques Bactériologiques. Masson et Cie, Éditeurs, Paris.

GELATINE AU CHLORURE DE FER

Extrait de viande Liebig	7.5	g
Peptone (Park et Davis)	25	g
NaCl	5	g
Eau de robinet	1000	cm^3
$FeCl_2(4H_2O)$ Scherihg-Kahlbaum) à 10 p. 100	5	cm^3
Gélatine	100	g
*p*H 7.6.		

Répartir en petits tubes. Ensemencer à partir d'une gélose par piqûre avec un fil. Maintenir pendant 60 jours à 22°C.

Une formation importante d'H_2S se décèle en un ou deux jours.

La liquéfaction de la gélatine est plus ou moins longue suivant le type de bactérie étudiée. *(C. I. E.).*

Hydrogen sulfide—micromethods

P. H. Clarke and S. T. Cowan (1952). J. gen. Microbiol. 6, 187–197.

Lead acetate agar base (LAAB) was prepared by adding 1 ml of a 0.05% solution of basic lead acetate (containing just enough HCl to prevent precipitation) to 9 ml of 1% New Zealand agar in water.

Cystine is made up in 0.1 N HCl and adjusted before use to pH 7.4–7.6 with diluted NaOH. The test requires a very heavy suspension—at least 10^{12}/ml.

Method: Suspension is prepared from any substrate medium. To columns of 0.4 ml LAAB agar in Durham tubes, add 0.04 ml cystine (1% pH 7.4) and 0.04 ml suspension. Read to 24 hr at 37°C.

Hydrogen sulfide—*Pseudomonas*

P. Villecourt, H. Blachère, and G. Jacobelli (1952). Annls Inst. Pasteur, Paris 83, 316–322.

La production de SH_2 en eau peptonée à 1 p. 100 d'hyposulfite de soude était recherchée pendant quatre jours avec un papier imprégné d'acétate de plomb.

Hydrogen sulfide—*Enterobacteriaceae*

J. Brisou and P. Morand (1952). Annls Inst. Pasteur, Paris 82, 643–645.

A la suite d'échanges de souches avec un laboratoire étranger, nous nous sommes rendu compte que la production d'hydrogène sulfuré par *Escherichia coli* est essentiellement fonction de la technique de détection mise en oeuvre. Une expérience très simple le démontre : Une série de tubes d'eau peptonée ordinaire reçoit une quantité fixe de sels de métaux lourds. Le taux final en sel est fixé à 0.05 p. 100 (taux du milieu gélose-gélatine-chlorure ferreux de Kauffmann).

Un tube sans ions métalliques servira de témoin. Tous sont munis du classique papier au sous-acétate de plomb fixé entre le bouchon et la paroi de verre à quelques millimètres de la surface de culture. On ensemence avec des *Escherichia coli* dont le diagnostic d'espèce avait été confirmé par le laboratoire du Séruminstitut de Copenhague*.

On complète l'expérience par une autre série de milieux ensemencés avec une *Escherichia,* un *Proteus mirabilis,* et un germe du type *Ballerup,* récemment isolés.

Les réactions lues à la sixième heure étaient toutes positives dans les tubes témoins ; toutes négatives là où un ion métallique avait été ajouté. Il y a done une action nette de l'ion métallique sur la vitesse de la réaction.

Sels de métaux lourds utilisés :

Chlorure ferreux (fraîchement préparé) ;

Citrate de fer ;

Citrate de bismuth ;

Acétate de thallium ;

Acétate de cadmium.

Un tableau d'ensemble résume ces résultats. Le signe + correspond au noircissement du papier au sous-acétate de plomb. Signe X réaction faible ; CN = Noircissement du milieu ou du dépôt au fond du tube de culture ; O = pas de culture.

Production d'H_2S par *Escherichia coli.*
(Culture de vingt-quatre heures)

	Souches											
Produits	*Esch.* 34		*Esch.* 35		*Esch.* 36		*Esch. ga*		*Prot. mirab.*		*Baller.*	
	37°	20°	37°	20°	37°	20°	37°	20°	37°	20°	37°	20°
Fe Cl_2	X	–	–	–	CN	–	+CN	–	+	+	+	–
Fe Citrate	+	–	+	–	X	–	X	–	+	+	+	X
Bi Citrate	+CN	–	XCN	–	+CN	–	–CN	–	+	+	X	0
Tl Acétate	–	–	–	–	–	–	–	–	+	+	–	0
Cd Acétate	0	0	0	0	0	0	0	0	–	–	0	0
Témoins	++	+	++	+	++	+	++	+	++	+	++	+

On peut conclure de cette expérience :

1° Que la présence d'ions métalliques dans le milieu de culture empêche ou ralentit la production d'H_2S chez certains microbes. Cette action empêchante est particulièrement nette chez *Escherichia.* On ne peut en être étonné, car il s'agit là d'une notion banale de microbiologie générale.

2° Que la croissance et les réactions biochimiques sont encore plus lentes à 20°. La technique proposant la recherche de l'hydrogène sulfuré à 20° et en présence d'un ion métallique lourd place la bactérie dans des conditions où elle ne peut plus former son H_2S. On ne peut donc accepter cette technique de détection sans certaines réserves.

En fait : dans les conditions normales de culture, c'est-à-dire dans un milieu qui ne contient aucun produit hostile, et à 37°, il semble que toutes les *Escherichia,* ou presque toutes, soient capables de produire de l'hydrogène sulfuré. On peut alors se demander jusqu'à quel point il est permis de modifier les caractères biochimiques d'un germe sous prétexte de le faire entrer dans le cadre étroit et arbitraire de nos systématiques.

*Nous remercions M. le Dr. F. Kauffmann de ces identifications biochimiques et sérologiques.

Kauffmann (F.). *Enterobacteriaceae,* Einar Munksgaard, Copenhague, 1951, Tableaux 2, 22, et 23.

Hydrogen sulfide—*Pseudomonas (Malleomyces) pseudomallei*

P. de Lajudie and E. R. Brygoo (1953). Annls Inst. Pasteur, Paris 84, 509–515.

La production d'hydrogène sulfuré était mise en évidence par culture en gélose au sous-acétate de plomb.

Hydrogen sulfide—various

P. H. Clarke (1953). J. gen. Microbiol. 8, 397–407.

The author gives a review of all methods and recommends the

following: cysteine HCl (0.01%) added to Lemco broth. H_2S was detected with lead acetate papers.

Hydrogen sulfide—micromethods

P. H. Clarke (1953). J. gen. Microbiol. 6, 397–407.

Modified methods of Clark and Cowan (1952). J. gen. Microbiol. 6, 187–197.

Suspension, 0.04 ml, was mixed in 65 × 10 mm tubes with 0.1% cysteine hydrochloride (reaction adjusted to pH 7.4), 0.06 ml, and phosphate buffer (0.025 M, pH 6.8), 0.04 ml. A strip of lead acetate paper was held in the mouth of the tube by a cotton wool plug. The tubes were incubated at 37°C.

Results could usually be read after 15–30 min, but for routine tests 2 and 24 hr readings were taken.

A review of the literature is given.

Hydrogen sulfide—*Brucella*

M. J. Pickett, E. L. Nelson and J. D. Liberman (1953). J. Bact. 66, 210–219.

Tests. A slant (2 ml of medium per 12 by 100 mm tube) was inoculated with one 4 mm loopful of stock suspension. Sulfide production was detected by inserting a "lead acetate strip" under the cotton plug. A new strip was inserted daily through six days' incubation.

Media were inoculated from a suspension of 5×10^9 cells per ml from a 48 hour slant and incubated at 35°C under 10% CO_2.

Hydrogen sulfide—enterobacteria

A. L. Olitzki (1954). J. gen. Microbiol. 11, 160–174.

Some strains of *Proteus, Klebsiella, Salmonella, Arizona* and *Bethesda* groups produced H_2S when 10^{11} cells or more in washed suspension were used *without* any substrate.

Ten-fold greater amounts were required to produce H_2S from homocystine, sulfite, or thiosulfite compared with cystine or cysteine. *Brucella* strains which autoagglutinated did not produce H_2S. H_2S production was inhibited by penicillin and streptomycin.

One-half milliliter of suspension (washed twice from nutrient agar 48 hour cultures) was mixed with 0.5 ml of buffer and substrate solution (0.15% K_2HPO_4; 0.05% KH_2PO_4; 0.5% NaCl and substrate). Inorganic substrates, 0.5%; Organic substrates, 0.1% (or sat. at 37°C). pH adjusted to 7.2. Tests in 75 × 11 mm tubes with lead acetate papers and rubber stoppers were observed for 48 hours.

Hydrogen sulfide—*Pseudomonas (Malleomyces) pseudomallei*

L. Chambon and P. de Lajudie (1954). Annls Inst. Pasteur, Paris 86, 759–764.

Nous avons précisé ce dernier point en employant le milieu synthétique gélosé de composition suivante [Vaughn et Levine, (1936)] :

Chlorhydrate de cystéine	0.250
Agar	1.250
K_2HPO_4	0.075
$(NH_4)_2SO_4$	0.075
$CaCl_2$	0.050
Eau bidistillée	250 ml

Ce milieu est stérilisé par 3 chauffages à 100°C à vingt-quatre heures d'intervalle et réparti en culot à raison de 5 ml par tube. On y ajoute avant refroidissement du sous-acétate de plomb comme indicateur de réaction. Aucune des 25 souches de bacilles de Whitmore, ensemencées par piqûre, n'a produit le noircissement caractéristique du dégagement d'hydrogène sulfuré.
R. Vaughn et M. Levin. *J. Bact.,* 1936, 32, 65.

Hydrogen sulfide—various
N.C.T.C. Methods 1954.
Method 1. Inoculate as a stab a tube of lead acetate agar. Read at daily intervals for 7 days.
I. J. Kligler (1917). Am. J. publ. Hlth 7, 1042.
Method 2. Grow the organism in a Lemco broth + 0.01% cysteine hydrochloride, and insert a lead acetate paper into the tube, holding it up by the plug. Read daily for 7 days.
P. H. Clarke (1953). J. gen. Microbiol. 8, 397.
Method 3. Inoculate a liver extract agar slope and insert a lead acetate paper strip as in Method 2. Read and change the paper daily for 7 days.
C. E. ZoBell and K. R. Meyer (1932). J. infect. Dis. 51, 91.
Lead acetate papers
Soak strips of filter paper (100 X 8 mm) in a *hot,* saturated solution of basic lead acetate. Dry in the incubator at 37°C or 55°C.

Hydrogen sulfide—*Proteus*
C. Shaw and P. H. Clarke (1955). J. gen. Microbiol. 13, 155–161.
Lead acetate papers were inserted above cultures in Lemco Broth + 0.01% cysteine incubated at 37°C for up to 7 days (Clarke, 1953).
P. H. Clarke (1953). J. gen. Microbiol. 8, 397.

Hydrogen sulfide—general comments
Society of American Bacteriologists *Manual of Microbiological Methods* – McGraw-Hill Book Co. Inc., New York 1957.
Hydrogen sulfide is generally detected in bacterial cultures by observing the blackening it produces in the presence of salts of certain metals, such as lead, iron. or bismuth, owing to the dark color of the sulfide of these metals. Two methods have been utilized for employing these tests: one

incorporates the metallic salt in the medium, and the other uses a test strip of filter paper impregnated with the metallic salt in question.

In early editions of the manual four media containing either lead or iron salts were given. The lead salt media, however, were discredited some time ago because of the toxic properties of these salts, and Hunter and Crecelius (1938) have shown the superiority of bismuth media over iron media. Zobell and Feltham (1934), moreover, have shown distinct advantages in the use of lead acetate test strips, without any of these metallic salts in the media. The advantage of the test strip technique is that it is more sensitive and does not introduce the possibility of inhibiting the bacterial growth if the concentration of metallic salt in the medium is too great. It is important, as emphasized by Hunter and Crecelius, that the indicator and method employed be stated when results are given. Untermohlen and Georgi (1940) suggested the use of nickel or cobalt salts, but specially emphasized the variations in results with different media and indicators.

In the test strip technique the bacteria may be grown in ordinary broth, peptone solution alone, or a peptone agar suitable to the organism in question. One must be certain that the peptone contains available sulfur compounds. This can be determined by running a check tube inoculated with a slow hydrogen sulfide producer. For this procedure the test strip should be prepared by cutting white filter paper into strips approximately 5 by 50 mm, soaking them in a saturated solution of lead acetate, sterilizing them in plugged test tubes, and drying them in an oven at 120°C. One of these strips should be replaced in the mouth of the culture tube before incubation in such a position that one-quarter to one-half of the strip projects below the cotton plug. These tubes should be incubated at about the optimum temperature of the organism under investigation and examined daily to notice whether or not blackening of the test strip has occurred.

Because of the inconvenience of the test strip technique, media in which iron salts are incorporated are now generally preferred. A dehydrated medium of such composition is available and has been found quite satisfactory.

For a "quick" method, Morse and Weaver (1947) recommended using small tubes containing 0.8 ml portions of 2 per cent thiopentone solution preheated to 37°C in a water bath. These are inoculated from a 6-hr agar slant culture, using sufficient quantity to give a turbidity indicating 2,100 million per liter. A strip of paper saturated with lead acetate solution is inserted in the tubes, and they are reincubated for 30–45 min. Clarke (1953) proposed a slightly different test; she called attention to the fact that very few of the cultures she tested, using a micromethod, failed to produce H_2S from cysteine. Thiosulfate appeared to be a more useful substrate than cysteine, since not all the cultures studied were able to reduce the compound to H_2S. When micromethods are used, the importance of recording the conditions of the test cannot be overemphasized.

If the lead acetate paper tests recommended are not desired, use dehydrated media or consult the original papers of Vaughn and Levine (1936), Hunter and Crecelius (1938), and Untermohlen and Georgi (1940) for directions for preparation of media containing lead, bismuth, or iron salts, which will precipitate as the sulfides in the presence of H_2S.

P. H. Clarke (1953). J. gen. Microbiol. 8, 397–407.
C. A. Hunter and H. G. Crecelius (1938). J. Bact. 35, 185–196.
M. L. Morse and R. H. Weaver (1947). J. Bact. 54, 28–29.
W. P. Untermohlen and C. E. Georgi (1940). J. Bact. 40, 449–459.
R. H. Vaughn and M. Levine (1936). J. Bact. 32, 65–73.
C. E. Zobell and C. B. Feltham (1934). J. Bact. 28, 169–176.

Hydrogen sulfide—*Leptotrichia buccalis*

R. D. Hamilton and S. A. Zahler (1957). J. Bact. 73, 386–393.

Medium:

Peptone (Difco)	20 g
Yeast extract (Difco)	2 g
Glucose	5 g
NaCl	5 g
K_2HPO_4	10 g
Agar	0.5 g
Distilled water	1 litre

pH adjusted to 7.2 before autoclaving.

The test was made in this medium, using the lead acetate paper strip method as described by the Committee on Bacteriological Technic of the Society of American Bacteriologists. *Manual of Methods for Pure Culture Study of Bacteria* (1953). Williams and Wilkins, Baltimore, Maryland.

Hydrogen sulfide—*Clostridium perfringens*

A. R. Fuchs and G. J. Bonde (1957). J. gen. Microbiol. 16, 330–340.

The authors found lead acetate papers the better of two methods tested. The presence of mercaptoacetate will result in H_2S production in sterile media. *Clostridium perfringens* produces H_2S from sulfite, thiosulfite, cystine, cysteine and glutathione, but *not* from methionine.

Hydrogen sulfide—thermophilic sulfate-reducing bacteria

L. L. Campbell, Jr., H. A. Frank and E. R. Hall (1957). J. Bact. 73, 516–521.

Estimation of growth and hydrogen sulfide

The amount of growth in liquid media was determined by optical density measurement at 525 mμ with a Bausch and Lomb "spectronic 20" colorimeter.

Hydrogen sulfide—*Acetobacter*

R. Steel and T. K. Walker (1957). J. gen. Microbiol. 17, 445–452.

The authors used a medium containing (w/v) Evans' peptone, 2%; Difco yeast extract, 0.5%; pH 7.0.

H_2S was detected by the method of Morse and Weaver (see Conn, H. J. 1949, *Manual of Methods for Pure Culture Study of Bacteria,* Biotech).

Hydrogen sulfide—*Enterobacteriaceae*

J. G. Heyl (1957). Antonie van Leeuwenhoek 23, 33–58.

Medium for H_2S reduction

Beef broth	1000 ml
Agar	5.0 g
Sodium sulfite	40.0 g
Dextrose	10.0 g
Ferro-sulfate	0.8 g

pH at 7.6

Hydrogen sulfide—*Enterobacteriaceae*

T. M. Cook and M. J. Pelczar, Jr. (1958). Appl. Microbiol. 6, 193–197.

Paper disc methods for hydrogen sulfide tests

The paper discs used in this study were 12.7 mm in diameter by about 1 mm thick. These were impregnated with various reagents for hydrogen sulfide tests. Experimental discs containing sodium thiosulfate, ferric ammonium citrate and glucose were prepared by aseptically pipetting Seitz-filtered solutions onto sterile dry discs contained in Petri dishes. The discs were satisfactory for immediate use or could be used after drying. Discs containing sodium thiosulfate and ferric ammonium citrate but no glucose were impregnated by touching them to the surface of the solution with forceps and allowing them to become saturated by capillary action. After drying on aluminium pans overnight at 37°C, the discs were placed in loosely capped vials and sterilized by dry heat at 140°C for 3 hours. All solutions for impregnating discs were adjusted to contain the desired amount per disc in 0.08 ml, the absorptive capacity of a disc.

Content of disc: sodium thiosulfate, 1 mg; ferric ammonium citrate, 1 mg; glucose, 1 mg. Test organisms were transferred from stock slants to tubes containing 10 ml of phenol red broth base and incubated for 24 hr at 37°C. A single loopful of the broth culture was used routinely as inoculum for both tube and plate tests. For plate tests, 20 ml of melted, cooled (45°C) phenol red agar base were inoculated, mixed, and poured into a petri dish. On solidification of the seeded medium, one to four discs were placed aseptically on the surface with alcohol-flamed forceps. Tubes containing 10 ml of broth or 15 ml of solid or semisolid medium were inoculated with a single loopful of broth culture before addition of the test disc. Results were read after 8 and 24 hr at 37°C and daily thereafter through a 4 day period.

Hydrogen sulfide tests comparable to control media except for *Salmonella pullorum* were obtained using a disc containing 1 mg sodium thiosulfate and 1 mg ferric ammonium citrate on trypticase agar base stabs of enteric cultures. Glucose applied to the discs was essential for plate tests, but not for tube tests. Although 1 mg glucose plus 1 mg each of sodium thiosulfate and ferric ammonium citrate gave excellent results, such a disc could not be sterilized conveniently.

Hydrogen sulfide—*Pseudomonas fluorescens*

M. E. Rhodes (1959). J. gen. Microbiol. 21, 221–263.

All isolates were first tested for their ability to produce H_2S from Kligler's iron agar (Oxoid) medium. A few isolates gave a positive reaction; the H_2S might have arisen from S-containing amino acids or sodium thiosulfate in the medium, or from both. Therefore two chemically defined media were devised, one with cystine (0.01%, w/v) added to the Dowson-type ammonium + galactose + inorganic salts medium, and the other with 1.0% (w/v) sodium thiosulfate added to the same basal medium. H_2S was detected by sterile lead acetate papers held in place by the cotton wool plugs of the cultures. After 7 days of incubation the cultures were also tested for dissolved sulfides by the addition of lead acetate solution. Because some results were inconclusive, a microtest was also used; heavy suspensions of organisms grown on yeast extract-agar for 24 hr were suspended in 1 ml of each of the two chemically defined media and incubated at 37°C for 24 hr with lead acetate papers held in place by cotton wool plugs. No carbon source was added to the thiosulfate-containing suspending medium, because of the report by Leathen and Braley (1955) of the chemical decomposition of thiosulfate to sulfate, hydrogen sulfide and sulfur at acid pH values.

W. W. Leathen and S. A. Braley (1955). J. Bact. 69, 481.

Hydrogen Sulfide—*Aeromonas*

J. P. Stevenson (1959). J. gen. Microbiol. 21, 366–370.

Blackening of filter paper strips, soaked in lead acetate and suspended above nutrient broth enriched with cysteine, was taken to indicate H_2S production. Cultures were incubated at 37°C for a maximum of 10 days.

Hydrogen sulfide

V. B. D. Skerman, *A Guide to the Identification of the Genera of Bacteria.* 2nd ed. Williams and Wilkins Book Co., Baltimore, 1967.

(i) *Ferric ammonium citrate agar (for sulfide production)*

Black colonies on ferric ammonium citrate agar indicate the production of hydrogen sulfide by the organism.

Peptone	10.0 g
Yeast extract	2.0 g
Sodium chloride	5.0 g
Ferric ammonium citrate	0.3 g
Agar	14.0 g
Tap water	1000 ml

Dissolve by heating to 121°C for 15 minutes. Adjust the pH to 7.0 and filter if necessary. Dispense in 5 ml quantities in 150 by 13 mm tubes. Sterilize at 120°C for 15 minutes.

Inoculate the tubes by stabbing the medium with a wire. H_2S production is indicated by the production of black iron sulfide.

(ii) *Peptone-cystine-sulfate medium for sulfide production*

Peptone	10 g
Cystine	0.1 g
Na_2SO_4	0.5 g
Distilled water	1000 ml

Dissolve the ingredients. Adjust the pH to 7.0 and filter if necessary. Dispense in 5 ml amounts in 150 by 13 mm tubes and sterilize at 121°C for 15 minutes.

(iii) *Ferrous chloride gelatin for sulfide production and gelatin liquefaction*

Liebig's meat extract	7.5 g
Peptone (Parke, Davis)	25.0 g
NaCl	5.0 g
Tap water	1000 ml
Gelatin	120 g
10% $FeCl_2 \cdot 4H_2O$	5 ml

Dissolve all ingredients except the ferric chloride. Adjust the pH to 7.6 and filter if necessary. Sterilize at 110°C for 25 minutes.

Immediately before use, melt the medium and add the required amount of sterile ferrous chloride solution. Dispense aseptically in 150 x 13 mm tubes and cool promptly. Seal the tubes with sterile stoppers.

Inoculate by stabbing and incubate for 60 days at about 20°C.

F. Kauffmann, *The Enterobacteriaceae,* 2nd ed. 1954. Einar Munksgaard, Copenhagen.

Hydrogen sulfide—general comments

V. B. D. Skerman, *A Guide to the Identification of the Genera of Bacteria.* 2nd ed. Williams and Wilkins Book Co., Baltimore, 1967.

The production of sulfide (H_2S)

Microbial production of sulfide results from the reduction of sulfate or the decomposition of organic sulfur compounds such as cystine, methionine, and glutathione. Since the nutritional requirements of organisms which are potentially capable of producing hydrogen sulfide are very broad, any one medium will not suit all types. The recommendations are for this reason directed towards the test for the sulfide rather than media used for the growth of the organism. Nevertheless, for taxonomic purposes, comparative reactions on specific media are desired and suggestions have been made for consideration.

H_2S *per se* does not exist in a medium at pH 7.0. The sulfur is present predominantly as the –SH ion. The solubility of this determines the extent to which H_2S will be liberated from the medium. This is important in the application of the lead acetate strip technique.

H_2S production does not occur when an adequate supply of oxygen is available for cell metabolism. Therefore conditions of incubation which permit a high degree of aeration should not be used.

Growth "aerobically" on the surface of an agar plate does not necessarily constitute aerobic conditions within the colony. On the contrary, the availability of oxygen to cells within a colony more than one cell deep is very poor. This permits the production of H_2S and the formation of black colonies on iron-containing media, such as glucose-blood-cystine agar for *Pasteurella tularensis.*

Media

1. *Mineral Salts Media:* The essential feature of any mineral salts medium is that it contains at least one of the salts in the form of a soluble sulfate in a concentration not less than 500 mg per L. If an iron salt is present the production of sulfide will result in the production of black iron sulfide.

If the concentration of iron is too small it may be necessary to use the lead acetate paper method or to test the culture for sulfide by the methylene blue procedure.

2. *Organic Media:* Any organic liquid medium containing sufficient cystine may be used, but for comparative purposes the following media are recommended: (a) peptone water plus 0.01 per cent cystine and 0.05 per cent sodium sulfate; (b) ferric ammonium citrate-agar; and (c) ferrous chloride gelatin.

In the two iron-containing media production of sulfide is indicated by the production of black iron sulfide. Production of sulfide in the peptone medium may be detected by the lead acetate paper strip method or the methylene blue method.

(a) *The Lead-Acetate paper method*

Cut filter paper into 50- by 10-mm strips and immerse them in a 5 per cent solution of lead acetate. Dry in air. Sterilize in a suitable container at 121°C for 15 minutes.

Following inoculation of the liquid medium, insert a strip of the sterilized paper between the plug and the glass, with the lower end above the liquid level. Incubate.

If H_2S is liberated during the growth of the organism, the lower portion of the paper will turn black. If, after incubation, there has been no color change, remove the plug and add 0.5 ml of 2 N HCl; replace the paper and plug without delay. The addition of the acid will liberate any dissolved sulfide, which will react with the lead to yield the black lead sulfide.

If heat is applied to assist the reaction, use a control tube of uninoculated medium to eliminate any possibility that the reaction is caused by decomposition of the medium.

(b) *The Methylene blue method*

The test is based on the reaction in acid solution between *p*-amino-dimethyl aniline and sulfide in the presence of ferric chloride to form methylene blue. The test is extremely sensitive, and 0.01 ppm of sulfide may be detected. The color developed is proportional to the sulfide concentration.

Test

To the 5 ml of the culture add 0.5 ml of amine-sulfuric acid solution and 0.1 ml of ferric chloride solution. Spin the tube to mix.

Results

A blue color is positive for H_2S; pink is negative for H_2S.

Reagents

Stock Amine-HCl Solution: The commercial *p*-amino-dimethyl aniline may be received as a mixture of crystal and liquid. Drain the crystals and press dry between filter papers. If the original material is entirely liquid, cool until it has partially solidified. If the dry crystals are dark, melt them in a closed vessel and again partially solidify by slow cooling. Seeding may be necessary to prevent excessive supercooling. Again dry the crystals. Add a known amount of the amine gradually to concentrated HCl in a beaker surrounded with ice. The final solution should contain 20 g of amine in 100 ml of solution.

Amine-Sulfuric Acid Solution: Add 500 ml of concentrated H_2SO_4 to 480 ml of water and cool; add 20 ml of amine-HCl solution.

Ferric Chloride Solution: Dissolve 45 g of $FeCl_3$ (or $FeCl_3 \cdot 6H_2O$, 75 g) in enough water to make 100 ml of solution.

Reference: R. Pomeroy (1936). Sewage Wks J. 8, 572.

Hydrogen Sulfide—*Pasteurella septica*

J. M. Talbot and P. H. A. Sneath (1960). J. gen. Microbiol. 22, 303–311.

Observed by stab inoculation into a lead acetate + gelatin medium (lead acetate, 0.25%, w/v; gelatin, 12.5%, w/v; nutrient broth to 100 ml) which was incubated at 22°C for 14 days.

Hydrogen Sulfide—*Fusobacterium*

A. C. Baird-Parker (1960). J. gen. Microbiol. 22, 458–469.

Medium (%, w/v): proteose peptone No. 3 (Difco), 1.0; Lab. Lemco, 0.1; Oxoid yeast extract, 0.1; glucose, 0.5; L-cysteine HCl, 0.05; disodium hydrogen phosphate, 0.5. Tests were carried out in sealed bijou bottles containing strips of filter paper impregnated with lead acetate. The decomposition of specific inorganic and organic sulfur-containing compounds was investigated by means of washed suspensions of organisms, using the methods of Olitzki (1954). Tests were performed in 10 by 75 mm hard glass tubes fitted with rubber bungs to exclude air. Incubation was in a water bath at 37°C.

A. L. Olitzki (1954). J. gen. Microbiol. 11, 160.

Hydrogen sulfide—*Fusobacterium*

R. R. Omata and R. C. Braunberg (1960). J. Bact. 80, 737–740.

Medium:

Trypticase (BBL)	1%
Yeast extract (Difco)	0.5%
L-cystine	0.02%
final pH 7.2	

Glucose was added separately to the sterile medium as sterile solution to give final concentration of 1% glucose. No test was specified.

Hydrogen sulfide—*Xanthomonas*

A. C. Hayward and W. Hodgkiss (1961). J. gen. Microbiol. 26, 133–140.

Peptone water supplemented with casein hydrolysate (British Drug Houses Ltd., "laboratory reagent"), 1.0 g/L, and L-cysteine hydrochloride, 0.1 g/L was dispensed in 5 ml quantities in ½ oz screw-capped bottles. Lead acetate papers were held over the medium by the screw cap which was kept loose. At 6 days final observations were made for H_2S production.

Hydrogen sulfide—*Pseudomonas*

M. Véron (1961). Annls Inst. Pasteur, Paris 100, Suppl. 6, 16–42.

La production de H_2S en gélose nutritive molle au sous-acétate de plomb, ensemencée par piqûre.

Hydrogen sulfide—*Moraxella (Acinetobacter)*

E. T. Cetin and K. Toreci (1961). Annls Inst. Pasteur, Paris 100, 509–523.

La production d'H_2S par nos souches de *Moraxella lwoffi* a été étudiée dans différents milieux dont certains étaient des milieux synthétiques. Ces milieux synthétiques étaient préparés soit d'après la formule d'Audureau, additionnés de thiosulfate de sodium ou de cystine, soit comme le milieu synthétique précité avec, cependant, une diminution de la quantité d'alcool et de citrate de fer et l'addition de thiosulfate de sodium ou cystine en différentes quantités. Dans une partie de ces milieux synthétiques, les bactéries n'ont pas pu se multiplier. Lorsqu'elles se sont multipliées la production d'H_2S n'était pas constante. Ces constatations nous ont donc conduits à préparer des milieux non synthétiques. Dans ce but, nous avons ajouté 0.10 p. 100 de thiosulfate de sodium ou de cystine au bouillon ; mais le milieu au thiosulfate de sodium donnant de meilleurs résultats, dans les milieux suivants on n'a ajouté que du thiosulfate de sodium [4]. Nous avons employé au cours de nos recherches un milieu qui était préparé par addition de la solution de 1/10 d'acétate de plomb à 2 p. 100 et 0.1 p. 100 de thiosulfate de sodium au bouillon. En outre, nous avons aussi préparé des milieux par l'addition de 0.1 p. 100 de thiosulfate de sodium à du bouillon non peptoné, à du bouillon de Hottinger [5] et à la solution de tryptose Difco à 2 p. 100. Après l'ensemencement des souches dans les divers milieux, la production en H_2S a été vérifiée par la couleur foncée du résidu dans les milieux contenant de la solution d'acétate de plomb et par la coloration noire du papier à l'acétate de plomb placé entre le coton et la paroi du tube dans les autres milieux. Les souches ont été ensemencées dans huit tubes contenant les divers milieux et deux rubans de papier d'acétate de plomb ont été placés dans chaque tube. Quatre de ces tubes ont été portés à l'étuve à 37°C et les quatre autres dans une étuve à 22°C. Le lendemain, après vérification de la multiplication des bactéries, on a paraffiné l'orifice de deux tubes de chaque série maintenus à 37°C et à 22°C respectivement. Par la suite, on a contrôlé quotidiennement et pendant quatre semaines la production d'H_2S.

[4] Braun (H.) et Silberstein (W.). *Fen. Fak. Mec.*, 1942, 7, 1.
[5] Braun (H.) et Ozek (O.). *Tip. Fak. Mon. Serisi,* 1946, 1, 1.

Hydrogen sulfide—*Pseudomonas*

R. R. Colwell and C. Quadling (1962). Can. J. Microbiol. 8, 813–816.

In the course of taxonomic studies of bacteria, we have used thin-walled miniature test tubes of approximately 100 mm × 6 mm which hold about 2.5 ml of medium.

Two different media were used. (1) Proteose Peptone (Difco), 1 g; distilled water, 100 ml; cystine, 0.01 g. (2) Proteose Peptone (Difco), 1 g; distilled water, 100 ml; sodium thiosulfate, 1.0 g. The media were adjusted to pH 7.0 and dispensed into the miniature tubes in approximately 1.5 ml amounts, and the tubes then autoclaved at 15 lb pressure for 15 minutes. All tubes were inoculated with a small drop of a 24 – 48 hour broth (1.0% Proteose Peptone (Difco), 0.3% yeast extract; and 0.5% NaCl; at pH 7.2–7.4) culture which had been incubated at 30°C. After inoculation, tubes were plugged with strips of sterile lead acetate paper (Whatman No. 1 filter paper dipped into 5% lead acetate suspension, air-dried, cut into strips and autoclaved).

Hydrogen sulfide—*Escherichia aurescens*

H. Leclerc (1962). Annls Inst. Pasteur, Paris 102, 726–741.

Production d'H_2S.

Nous l'avons recherchée sur un milieu défini par Mossel*, de composition suivante :

Trypticase (BBL)	10	g
Extrait de levure	3	g
Thiosulfate de sodium	1	g
Sulfate ferreux	0.2	g
Agar (Difco)	15	g
Eau distillée	1 000	ml

Ajuster à pH 7.2. Stériliser à 120°C durant vingt minutes. On ensemence en piqûre centrale : une coloration noire apparaît le long de la strie en cas de réaction positive. On incube huit jours à 30°C.

Ce milieu doit fournir des résultats plus précis que le milieu de Kligler. Selon Mossel, en effet, il est particulièrement utile avec certaines souches à pouvoir fermentatif élevé.

*Nous remercions M. Mossel (Institut Central de la Nutrition, Utrecht) de nous avoir aimablement communiqué ces reseignements.

Hydrogen sulfide—*Clostridium*

H. Beerens, M. M. Castel and H. M. C. Put (1962). Annls Inst. Pasteur, Paris 103, 117–121.

Production d'H_2S : Employer un milieu V. L. semi-solide [1] non glucosé, additionné de 0.2 g p. 1 000 de sulfate ferreux et de 0.3 g p. 1 000 d'hyposulfite de sodium ; il est réparti en tubes de 160 × 16.

Formule du milieu V. L. de base :

Peptone trypsique	10	g
Chlorure de sodium	5	g
Extrait de viande	2	g
Extrait de levures	5	g
Chlorhydrate de cystéine	0.3	g
Agar	0.5	g
Eau distillée	1 000	ml

Ajuster le pH à 7.4–7.5. Autoclaver à 115–120°C pendant vingt minutes.

Les milieux semi-solides de ce genre s'ensemencent après régénération avec une culture de vingt-quatre heures en milieu liquide, à raison de 1 ml environ par tube. Incuber à 37°C à l'air libre.

[1] Beerens (H.). *Ann. Inst. Pasteur Lille,* 1953-54, 6, 36.

Hydrogen sulfide—*Staphylococcus* and *Micrococcus*

D. A. A. Mossel (1962). J. Bact. 84, 1140–1147.

Following the suggestion of Krynski *et al.* (1962), a representative selection of strains was tested for the capability of H_2S formation from cysteine or its peptides. The procedure described by Mossel (1961) was used for this purpose. Tubes containing brain heart infusion broth enriched with 0.05% cysteine HCl were inoculated and then equipped with a folded filter-paper band (*c.* 10 cm long and 5 mm diam), impregnated with a saturated solution of lead acetate, and dried subsequently at 55°C. After 48 hr of incubation at 30°C, any blackening of the indicator strips was noted.

S. W. Krynski, W. Kedzia, E. Becla and A. Krasowski (1962). Annls Inst. Pasteur, Paris 102, 231–235.
D. A. A. Mossel (1961). Arch. Lebensmittelhyg. 12, 180–182.

Hydrogen sulfide—*Vibrio*

G. H. G. Davis and R. W. A. Park (1962). J. gen. Microbiol. 27, 101–119.

A basal medium of Koser salt solution (Mackie and McCartney, 1953, p. 429) plus Oxid yeast extract, 0.3% (w/v), was used. The initial pH was adjusted to 7.0, plus either 0.05% (w/v) L-cysteine HCl or 0.1% (w/v) sodium thiosulfate. H_2S was detected by lead acetate paper over 7 days.

Also from peptone water, detected with lead acetate paper; examined after 1 and 7 days.

T. J. Mackie and J. E. McCartney (1953). *Handbook of Practical Bacteriology* 9th ed. Livingstone.

Hydrogen sulfide

W. H. Ewing (1962). *Enterobacteriaceae. Biochemical Methods for Group Differentiation.* U. S. Department of Health, Education and Welfare, Public Health Service Publication, No. 734 (revised).

Triple sugar iron agar (a modification of Kligler's iron or of Krumweide's triple sugar agar).

Beef extract	3 g
Yeast extract	3 g
Bacto peptone	15 g
Proteose peptone, Difco	5 g
Lactose	10 g
Sucrose	10 g
Glucose	1 g
Ferrous sulfate	0.2 g
Sodium chloride	5 g
Sodium thiosulfate	0.3 g
Agar	12 g
Phenol red	0.024 g
Distilled water	100 ml

Polypeptone (BBL) may be substituted for the Bacto Peptone and proteose peptone in the formula given above.

After sterilization at 121°C for 15 minutes, the medium is slanted with a deep butt (1-inch butt, 1.5-inch slant).
Inoculation: The butt of the medium is stabbed and the slant is streaked.
Incubation: 37°C. Observe daily for 7 days for blackening caused by hydrogen sulfide production.

It is probable that all *Enterobacteriaceae* produce at least small amounts of hydrogen sulfide from organic or inorganic substrates if sufficiently sensitive methods are used to detect its presence, but the writer believes that in this instance the most sensitive method is not necessarily the best for taxonomic purposes. Rather, it is felt that for a hydrogen sulfide test to be of value in taxonomic work within the family *Enterobacteriaceae,* it should be poised at a certain level of sensitivity. TSI agar, or a medium of similar composition but without lactose and sucrose, serves this purpose admirably and may be recommended because it gives results comparable to classical descriptions and permits easy primary group differentiation within the family.

If a lead acetate paper method is used at all it should be employed as it is with *Brucella* cultures, i.e., in a semiquantitative way. The papers should be replaced with new ones each day for 4 days and records kept as to weak, moderate, or marked hydrogen sulfide production. If this is done, the results obtained will be comparable to those obtained with TSI agar, or other medium of similar sensitivity, except that cultures giving weakly positive paper tests may be negative in TSI agar.

Hydrogen sulfide—*Actinobacillus* and *Haemophilus*

E. O. King and H. W. Tatum (1962). *Actinobacillus actinomycetemcomitans* and *Haemophilus aphrophilus.* J. infect. Dis. 111, 85–94.

The hydrogen sulfide test was performed with triple sugar iron agar with lead acetate paper inserted with the cotton plug.

Hydrogen sulfide—*Lactobacillus*

R. E. Smith and J. D. Cunningham (1962). Can. J. Microbiol. 8, 727–735.

A modification of motility sulfide medium (Difco) was developed, which gave superior growth of lactobacilli. The beef extract and sodium citrate were replaced by 0.3% (w/v) yeast extract and 0.2% (w/v) sodium acetate, respectively, and additions were made in the form of 0.5% (w/v) tryptone, 0.3% (w/v) sodium thiosulfate, 0.1% (v/v) Tween 80, and 0.5% (v/v) of each of Rogosa's salt solutions A and B.* The medium was inoculated by stab, and incubated for 2 weeks.

*J. Naylor and M. E. Sharpe (1958). J. Dairy Res. 25, 92–103.

Hydrogen sulfide—*Enterobacteriaceae*

M. Véron and F. Gasser (1963). Annls Inst. Pasteur, Paris 105, 524–534.

Comme milieu de référence pour l'étude de la production d'H_2S, nous avons utilisé une gélose nutritive molle additionnée de cystéine, de thiosulfate, et de citrate ferrique (milieu GCTF), réalisée avec les concentrations préconisées par Christensen [1]. Ces concentrations nous ont donné en effet les résultats les plus fidèles et les plus conformes à ceux qui figurent dans les descriptions classiques des caractères d'identification des bactéries.

La formule des milieux de culture utilisés est la suivante :

1° *Gélose-cystéine-thiosulfate-citrate ferrique (GCTF) :*

Macération de viande (500 g/l)	1 000 ml
Peptone trypsique de caséine	10 g
KCl	4 g
Agar purifié en poudre (Biomar)	3 g

Faire dissoudre à chaud, ajuster à pH 7.6, autoclaver à 120°C pendant vingt minutes filtrer sur papier et ajouter les composés suivants (chacun ayant été dissous dans 10 ml d'eau distillée) :

L-cystéine monochlorhydrate	100 mg
Thiosulfate de sodium ($Na_2S_2O_3 \cdot 5H_2O$)	100 mg
Citrate ferrique ammoniacal	400 mg

Répartir 10 ml de milieu en tubes de 17 × 170 mm et autoclaver à 110°C. Après stérilisation, agiter pour remettre le précipité en suspension et refroidir en culot dans l'eau froide. Le pH final est de 7.4.

Ce milieu est utilisable pendant plusieurs mois s'il est conservé à la température ordinaire, à l'obscurité et dans des tubes capuchonnés. Il est inoculé par quatre piqûres faites dans le culot de gélose selon des lignes parallèles à l'axe principal du tube et distantes de 2 ou 3 mm de la paroi ; la semence bactérienne est prélevée en une seule fois, avec le fil de platine, dans une culture de 24 heures en bouillon. Le tube ensemencé est capuchonné et incubé à 30°C.

La production d'H_2S dans ce milieu se signale par l'apparition, un peu au-dessous de la surface, d'un précipité noir de sulfure de fer qui diffuse entre les traits d'inoculation.

2° *Variantes du milieu précédent :*

a) Bouillon-cystéine-thiosulfate-citrate ferrique (BCTF) : même formule que GCTF, mais sans agar ; 5 ml en tubes de 17 × 170 mm.

b) Gélose-cystéine-thiosulfate (GCT) : même formule que GCTF, mais sans citrate ferrique ; 10 ml en tube de 17 × 170 mm.

c) Bouillon-cystéine-thiosulfate (BCT) : même formule que GCTF, mais sans agar et sans citrate ferrique : 5 ml en tube de 17 × 170 mm.

3° *Milieu glucose-lactose-H_2S :*

Eau distillée	1 000 ml
Peptone trypsique de fibrine	20 g
NaCl	5 g

Glucose	1 g
Lactose	10 g
Thiosulfate de sodium	
($Na_2S_2O_3 \cdot 5H_2O$)	0.2 g
Sulfate ferreux ammoniacal	0.2 g
Rouge de phénol, solution	
aqueuse à 0.2 p. 100	10 ml
Agar	13 g

Faire dissoudre à 95–100°C ; ajuster le pH à 7.6 ; répartir par 8 ml environ en tubes de 17 X 170 mm ; autoclaver à 115°C pendant trente minutes. Pour utiliser le milieu, faire fondre la gélose et incliner partiellement en laissant un culot haut de 2 cm environ. On ensemence le culot par piqûre et la pente par stries médianes en surface. Incuber à 30°C, sans capuchonner.

La production d'H_2S dans ce milieu détermine un noircissement débutant dans la zone où se joignent le culot et la pente du milieu. Les souches lactose + acidifient tout le milieu (culot et pente jaunes) ; les souches lactose – et glucose + n'acidifient que le culot, mais pas la pente du milieu (culot jaune, pente rouge). Ce milieu ne permet une interprétation correcte du résultat des fermentations que dans le cas des souches fermentatives, cultivant abondamment sur le milieu en vingt-quatre heures (Entérobactériacées, par exemple) ; les changements de pH doivent être interprétés uniquement après vingt-quatre heures de culture.

[1] Christensen (W. B.). *Res. Bull. Weld Co. Colo. Health Dept.*, 1949, 1, 3.

Hydrogen sulfide—*"Bacterium salmonicida" (Aeromonas)*

I. W. Smith (1963). J. gen. Microbiol. 33, 263–274.

Hydrogen sulfide production was tested with lead acetate strips and ZoBell and Feltham agar (1934, J. Bact. 28, 169).

Hydrogen sulfide—*Pseudomonas odorans*

I. Málek, M. Radochová and O. Lysenko (1963). J. gen. Microbiol. 33, 349–355.

Used microtest of Rhodes (1959). (All tests were done at 28°C unless otherwise stated).

M. E. Rhodes (1959). J. gen. Microbiol. 21, 221.

Hydrogen sulfide—*Escherichia coli*

D. C. Anderson and K. R. Johansson (1963). J. gen. Microbiol. 30, 485–495.

Assay: One volume of a freshly prepared concentrated buffered substrate solution (16 or 32 m*M* L-cysteine HCl, 0.1 or 0.2 M KH_2PO_4, 0.05 M $MgSO_4$; adjusted to pH 6.4 with 0.1 N NaOH) was mixed with four volumes of washed organisms resuspended from the packed state in deionized water to a density of 0.75–1.5 mg, dry weight. Replicate 5 ml quantities of the substrate-organism suspension were placed in Pyrex tubes (12 × 100 mm) with rubber stoppers to minimize entrance of air and incubated at 37°C in a water bath. At intervals tubes were removed, 0.25 ml of a 40% (w/v) solution of NaOH was added to terminate the reaction, and the stoppers were replaced. The tubes were gently agitated for 20 min at room temperature, and the cell-free supernatant fluid recovered after centrifugation for 10 min at 2800 *g*. The supernatant fluid was assayed for hydrogen sulfide within 2 hr by a modification of the method of Delwiche (1951): 1–4 ml of the supernatant fluid were mixed with 2 ml of 2 N NaOH, 4 ml lead acetate reagent (Pb acetate, 1.0 g; glacial acetic acid, 2.5 ml; gum arabic, 2.5 g; distilled water to 1 L) were added, and the resulting brownish-yellow color was measured promptly in a Klett-Summerson photoelectric colorimeter (filter 54) previously adjusted to 100% transmission with a reagent blank. The readings were compared with those of an alkaline sodium sulfide solution standardized by iodometric titration (Bethge, 1953). Over a limited range in concentration of sulfide, a plot of colorimeter readings against μM-sulfide yielded a straight line. The assay was simple and reproducible (replicates agreed within limits of ± 1% two-thirds of the time), being sensitive to 0.1 μg H_2S/ml substrate solution.

The results confirm that sugar in protein hydrolysate media suppresses H_2S production. Sugar may accelerate production by washed cells in a defined medium.

P. O. Bethge (1953). Analytica chima. Acta 9, 129.
E. A. Delwiche (1951). J. Bact. 62, 717.

Hydrogen sulfide—*Clostridium perfringens*

H. E. Hall, R. Angelotti, K. H. Lewis and M. J. Foter (1963). J. Bact. 85, 1094–1103.

Hydrogen sulfide production was observed in sulfadiazine-polymyxin-sulfite-agar (Angelotti *et al.*, 1962).

R. Angelotti, H. E. Hall, M. J. Foter and K. H. Lewis (1962). Appl. Microbiol. 10, 193–199.

Hydrogen sulfide—rumen bacteria

R. S. Fulghum and W. E. C. Moore (1963). J. Bact. 85, 808–815.

The ability of isolates to produce H_2S was studied by use of SRP-basal medium* with 2% Trypticase, 0.05% ferric ammonium citrate, and 1% agar. A blackening along the line of the stab indicated H_2S production.

The anaerobic methods of Hungate (Bact. Rev 14, 1–49, 1950) were used in handling and incubation of the cultures.

***Basal medium for physiological tests**—*Butyrivibrio, Borrelia, Bacteroides, Selenomonas, Succinivibrio* (p. 109).

Hydrogen sulfide—*Vibrio*

L. Ringen and F. W. Frank (1963). J. Bact. 86, 344–345.

i) Albimi Brucella broth containing 0.15% agar with lead acetate papers inserted in the tubes. Tests were read at 24 and 120 hours.

ii) 0.1% $FeSO_4 \cdot 7H_2O$ was added to a basal medium of Brucella agar (Albimi Laboratories, Inc., Flushing, N. Y.). Quantities (10 ml) were placed in screw-capped tubes and sterilized at 121°C for 15 min. The medium gave best results when used within 4 weeks. Inoculations were made by making several stabs to the bottom of the tube and incubating in candle jars at 37°C.

This method was less sensitive than method i, but differentiated more clearly between *V. fetus* and *V. bubulus.*

Hydrogen sulfide (and motility)—general

M. Gershman (1963). J. Bact. 86, 1122–1123.

By supplementing semisoft agar with a sulfide indicator, the hydrogen sulfide reaction can be exploited simultaneously with motility determinations. This modified medium is composed of the following components: Tryptose (Difco), 10 g; sodium chloride, 5 g; agar, 5 g; 2,3,5-triphenyltetrazolium chloride, 0.05 g; ferrous ammonium sulfate, 0.2 g; sodium thiosulfate, 0.2 g; and 1 liter of distilled water. After being boiled to ensure complete dissolution, the composite is distributed in tubes, autoclaved for 15 min at 15 psi (121°C), and allowed to cool in an upright position.

With semisolid media, motility is macroscopically manifested by a diffuse zone of growth spreading from the line of inoculation. The effects are cumulative, and localized outgrowths appear when only a small proportion of motile cells are involved. Growth and diffusion in semisolids, however, may be so slight as to necessitate comparison with a control tube for proper interpretation. Kelly and Fulton (Am. J. clin. Path. 23, 512, 1953) corrected the difficulty in discerning growth in semisolids by incorporating tetrazolium salts into the test medium and, in this way, inducing a cellular formation of red pigments.

Nonmotile, nonsulfide-producing organisms reveal a red line along the route of inoculation. Nonmotile, sulfide-producing cultures form a black line. Motile, sulfide-nonproducing cultures develop a diffuse pink cloud throughout the medium. Motile, sulfide-producing cultures blacken the entire tube.

Hydrogen sulfide—*Actinobacillus*

P. W. Wetmore, J. F. Thiel, Y. F. Herman and J. R. Harr (1963). Comparison of selected *Actinobacillus* species with a hemolytic variety of *Actinobacillus* from irradiated swine. J. infect. Dis. 113, 186–194. The authors used triple sugar iron agar (Difco) and lead acetate impregnated strips (Wilson and Miles, 1955) placed in tubes of liver slant cultures.

G. S. Wilson and A. A. Miles (1955). Topley and Wilson's *Principles of Bacteriology and Immunity,* 4th ed. Baltimore, The Williams and Wilkins Co.

Hydrogen sulfide—*Veillonella*

M. Rogosa (1964). J. Bact. 87, 162–170.

H_2S production was detected by the blackening of cultures grown in V17 and V23 broth media supplemented with 0.5 g per liter of ferric ammonium citrate as an internal indicator. This compound was nontoxic. Tests were performed under a variety of conditions and medium supplementations with 26 sulfur-containing compounds.

For V17 medium see **Basal medium**—*Veillonella* (p. 114).

The V23 medium consisted of Trypticase, 1%; yeast extract, 0.5%; sodium thioglycolate, 0.075%; Tween 80, 0.1%; 85% lactic acid, 1%; adjusted with solid K_2CO_3, while stirring, to pH 6.5 to 6.6.

Hydrogen sulfide—*Clostridium*

N. A. Sinclair and J. L. Stokes (1964). J. Bact. 87, 562–565.

For H_2S production, peptone-ferric ammonium citrate medium was used. Incubation was carried out at 0°, 15° and 25°C.

Hydrogen sulfide—*Streptomyces*

E. Küster and S. T. Williams (1964). Appl. Microbiol. 12, 46–52.

The ability of streptomyces to produce hydrogen sulfide is generally used for taxonomic purposes. It was found that the previously used method, the blackening of Peptone Iron Agar, does not clearly indicate formation of hydrogen sulfide. It was shown that the blackening of a lead acetate strip is the most accurate indicator for H_2S-producing streptomycetes. A great variety of organic and inorganic sulfur compounds were examined and compared, and the choice of the most suitable sulfur source and method for the detection of hydrogen sulfide is discussed.

No choice of a suitable sulfur source was made.

Hydrogen sulfide

G. W. Wilson and A. A. Miles (1964). Topley and Wilson's *Principles of Bacteriology and Immunity,* 5th ed., Arnold.

Tested on lead acetate medium (heart extract broth containing 4 per cent peptone and 2.5 per cent agar. Sterilized, after which an equal quantity of a sterile 0.1 per cent solution of basic lead acetate is added).

Brown or black coloration = positive
No coloration = negative

The lead acetate may be replaced by 0.05 per cent ferric ammonium citrate or 0.03 per cent ferrous acetate (ZoBell and Feltham, 1934). A higher proportion of positive reactions is obtained with some organisms by incubating at 30°C instead of 37°C (Tittsler, 1931). The most delicate method is to grow the organisms in a slope tube of liver extract agar, and to include between the cotton wool plug and the tube a slip of filter paper soaked in 10 per cent lead acetate solution and subsequently dried. The amount of browning or blackening of the paper is measured in millimetres. A fresh slip may be inserted daily. For references to different methods see Clarke (1953).

P. H. Clarke (1953). J. gen. Microbiol. 8, 397.
R. P. Tittsler (1931). J. Bact. 21, 111.
C. E. Zobell and C. B. Feltham (1934). J. Bact. 28, 169.

Hydrogen sulfide—*Actinomyces*

L. K. Georg, G. W. Robertstad and S. A. Brinkman (1964). J. Bact. 88, 477–490.

H_2S production was demonstrated by growing the organisms on BHI slants in cotton-plugged tubes with strips of lead acetate paper suspended above the medium, and incubating at 37°C in an anaerobic jar (95% N_2 plus 5% CO_2). (False reactions may be obtained with lead acetate paper if tubes are sealed with pyrogallol-carbonate seals.)

The inoculum was taken from 3 day old cultures in AM broth. See **Starch hydrolysis**—*Actinomyces* (p. 812) for composition of AM broth.

Hydrogen sulfide—*Pseudomonas aeruginosa*

K. Morihara (1964). J. Bact. 88, 745–757.

Bouillon containing 0.05% cysteine was used. Hydrogen sulfide was detected by use of a lead acetate paper strip. Cultures were incubated at 30°C.

Hydrogen sulfide—*Vibrio marinus*

R. R. Colwell and R. Y. Morita (1964). J. Bact. 88, 831–837.

The authors used the method of R. R. Colwell and C. Quadling (1962), Can. J. Microbiol. 8, 813–816 – modified by the addition to the media of the following salts to produce a synthetic seawater: – sodium chloride, 2.4%; potassium chloride, 0.07%; magnesium chloride (hydrated), 0.53% and magnesium sulfate (hydrated), 0.7%.

A standard inoculum was 1 drop from a pasteur pipette (*c.* 0.05 ml) of a 24 hour artificial seawater broth.

Cultures were incubated at 18°C.

Hydrogen sulfide—*Bacteroides oralis*

W. J. Loesche, S. S. Socransky and R. J. Gibbons (1964). J. Bact. 88, 1329–1337.

Hydrogen sulfide production was tested by use of Peptone Iron Agar

(Difco) and also lead acetate paper strips over Trypticase Soy Broth.
The test was performed in the presence and absence of 0.5% glucose.

Hydrogen sulfide—enterobacteria
J. M. Bulmash and M. Fulton (1964). J. Bact. 88, 1813.
The authors note a lack of correlation between Kligler's and Triple Sugar Iron Agar for the detection of hydrogen sulfide.

Hydrogen sulfide—*Aerococcus catalasicus*
O. G. Clausen (1964). J. gen. Microbiol. 35, 1–8.
Cultures were incubated aerobically at 37°C unless otherwise stated. Stab cultures in lead acetate agar were incubated for 14 days.

Hydrogen sulfide—*Pseudomonas aeruginosa*
R. R. Colwell (1964). J. gen. Microbiol. 37, 181–194.
Hydrogen sulfide formation from sodium thiosulfate and from cystine was tested by the miniature tube method of Colwell and Quadling (1962). Also used, for comparative purposes, were the Difco lead acetate agar slopes.
R. R. Colwell and C. Quadling (1962). Can. J. Microbiol. 8, 813.

Hydrogen sulfide—*Pseudomonas*
G. L. Bullock, S. F. Snieszko and C. E. Dunbar (1965). J. gen. Microbiol. 38. 1–7.
Hydrogen sulfide was determined in Motility Sulfide Medium (Difco; Difco Laboratories, Detroit, Michigan, U. S. A.). Cultures vere incubated at 20–22°C for one week.

Hydrogen sulfide—*Alcaligenes*
R. G. Mitchell and S. K. R. Clarke (1965). J. gen. Microbiol. 40, 343–348.
H_2S production was tested in peptone water containing 0.01% L-cysteine hydrochloride (lead acetate papers) incubated for 6 days.

Hydrogen sulfide—*Malleomyces, Pseudomonas*
Bach-Toan-Vinh (1965). Annls Inst. Pasteur, Paris 109, 460–463.
Procédé au sous-acétate de plomb utilisant un indicateur sur papier, durée d'observation: trois jours.

Hydrogen sulfide—*Mima, Herellea, Flavobacterium*
J. D. Nelson and S. Shelton (1965). Appl. Microbiol. 13, 801–807.
The Triple Sugar Iron (TSI)-Agar reaction was read after overnight incubation. All cultures which did not produce blackening of TSI Agar were tested for hydrogen sulfide production by the lead acetate paper method (Society of American Bacteriologists, 1957).
Society of American Bacteriologists (1957). *Manual of microbiological methods.* McGraw-Hill Book Co., Inc., New York.

Hydrogen sulfide—*Salmonella*

J. M. Bulmash, M. Fulton and J. Jiron (1965). J. Bact. 89, 259.

The authors describe a strain of *Salmonella tennessee* which produced lactose-fermenting variants. The lactose fermentation resulted in a negative hydrogen sulfide test in both Kligler's and Triple Sugar Iron Agar.

Hydrogen sulfide—*Vibrio comma*

J. C. Feeley (1965). J. Bact. 89, 665–670.

The author used Kligler's Iron Agar (Difco).

Hydrogen sulfide—marine bacteria

R. M. Pfister and P. R. Burkholder (1965). J. Bact. 89, 863–872.

H_2S production was determined by a technique modified from Skerman (1959), with the following medium (per liter): peptone, 10 g; cystine, 0.1 g; Na_2SO_4, 0.5 g; seawater; and vitamin B_{12}, 1 μg. The pH was adjusted to 7.0. Sterile lead acetate paper strips were placed at the top of the tube between the plug and the glass. Darkening of the paper strips at any time during incubation was considered a positive indication for the production of H_2S.

V. B. D. Skerman (1959). *A guide to the identification of the genera of bacteria.* The Williams & Wilkins Co., Baltimore.

Hydrogen sulfide—enterobacteria

S. T. Williams and M. Goodfellow (1966). J. Bact. 91, 907.

It was evident that the sulfide was unstable in the presence of atmospheric oxygen, and hence the results obtained were due to the chemical behavior of the sulfide rather than to the ability of the organisms to form hydrogen sulfide. Chemical tests on the medium and 0.05% solution of ferric ammonium citrate indicated that the sulfide formed was oxidized to sulfur and ferric compounds.

Therefore Peptone Iron Agar should not be used in slope cultures for detection of hydrogen sulfide.

Hydrogen sulfide—*Providencia, Rettgerella*

C. Richard (1966). Annls Inst. Pasteur, Paris 110, 105–114.

Pour des raisons de commodité, les milieux ont été incubés à 37°C. Recherche de l'H_2S: par noircissement du milieu de Hajna-Rolland [24], éventuellement d'une gélose au sous-acétate de plomb sensibilisée par addition de $S_2O_3Na_2$.

[24] Rolland (F.). *Bull. méd.,* 1946, 60, 397.

Hydrogen sulfide—*Brevibacterium*

R. Chatelain and L. Second (1966). Annls Inst. Pasteur, Paris 111, 630–644.

Production de H_2S: sur papier imprégné de sous-acétate de plomb et placé au-dessus d'une culture sur gélose nutritive [15]. Lecture après 4 jours.

La température d'incubation des cultures est de 30°C.
[15] Olivier (H. R.). *Traité de biologie appliquée,* tome II, Maloine édit., Paris, 1963.

Hydrogen sulfide—*Clostridium*
A. V. Goudkov and M. E. Sharpe (1966). J. Dairy Res. 33, 139–149.
The basal medium used for many of the tests consisted of: BBL Trypticase, 1.5%; yeast extract, 1%; cysteine, 0.05%; pH of medium, 7.0. The methods of Beerens, Castel and Put (1962, Annls Inst. Pasteur, Paris 103, 117) were used.
The basal medium containing 0.1% glucose and 0.05% ferric citrate was used. A positive reaction was indicated by blackening of the medium.

Hydrogen sulfide—anaerobes
A. -R. Prévot (1966). Techniques pour le diagnostic des Bactéries Anaérobes. Éditions de la Tourelle, St. Mandé.
Recherche de SH_2.
Pour compléter la liste des caractères biochimiques, il convient encore de noter la production ou non de SH_2 au cours du développement du germe en milieu liquide. Il suffira pour cela, en ouvrant le tube scellé contenant une culture du germe en bouillon VF glucosé, d'y introduire une fine bande de papier au sous-acétate de plomb préalablement mouillée. L'apparition sur le papier blanc d'un dépôt noir à reflets métalliques traduit la présence de SH_2 dans les produits du métabolisme bactérien.

INDOLE (AND SKATOLE)

Indole—coliforms
L. A. Rogers, W. M. Clark and H. A. Lubs (1918). J. Bact. 3, 231–252.
Indole was determined by incubation at 30°C in a medium containing, in 1000 ml of water, 0.3 g tryptophan, 5 g K_2HPO_4 and 1 g of Witte peptone. The test for indole was made by the *p*-dimethylamido-benzaldehyde-hydrochloric acid method.

Indole
H. F. Zoller (1920). J. biol. Chem. 41, 25–36.
Extracted culture with iso-amyl or iso-butyl alcohol and then used the Salkowski test.

Indole—bacterial spot of tomato
M. W. Gardner and J. B. Kendrik (1921). J. agric. Res. 21, No. 2, 123–156.
Cultures in beef-peptone bouillon, 6 days old, gave no test for indole

when tested with potassium nitrite and sulfuric acid. Cultures in 5 per cent peptone bouillon yielded no test for indole after 5 days' incubation at 22°C.

Indole—*Haemophilus influenzae*

W. L. Holman and F. L. Gonzales (1923). J. Bact. 8, 577–583.

Method:

The volatility of indole can be demonstrated on filter papers, strips of white tape, or even absorbent cotton plugs. These are dipped in a saturated watery solution of oxalic acid and allowed to dry. The filter paper should be folded four or five times to prevent it from lying against the side of the tube and to offer a greater surface to the rising indole. The tape may curve under the cotton plug, both ends being held in place. The absorbent plugs can be lightly dipped in the saturated solution and dried in situ, care being taken not to have an excess of crystals on the cotton. It is important to remember that the reaction does not occur if the papers are wet.

If indole is being produced, the oxalic acid papers or absorbent cotton plugs turn pink.

Indole—*Pseudomonas*

C. Elliott (1924). J. agric. Res. 29, 483–490.

Cultures were grown in a solution of 1 per cent peptone, 0.5 per cent disodium phosphate, and 0.1 per cent magnesium sulfate for 17 days. Tested with H_2SO_4 and sodium nitrite; results were negative.

The same test was made in broth cultures grown for 4 days. Broth cultures 2 weeks old were tested by the Ehrlich and Salkowski methods recommended in the "Manual of Methods" (1923).

Indole and **skatole**—various

C. R. Fellers and R. W. Clough (1925). J. Bact. 10, 105–133.

The authors found the Ehrlich test to give the best results. For good results the indole has to be steam distilled or distilled directly from the culture. The test is accurate to 1/25,000,000.

It is not safe to rely on a red color in the culture tube, irrespective of the method of testing, unless the color is pronounced and is soluble in chloroform.

Indole—various

W. L. Kulp (1925). J. Bact. 10, 459–471.

(The author emphasizes the need for tryptophane in the medium.)

Methods of testing indole production by bacteria:

Holman and Gonzales (1923) gave a good historical review of indole testing methods and brought to light an old test which had fallen into obscurity, although Holman has used it since 1911. This is the Gnezda oxalic acid paper test, a method which takes advantage of the extremely volatile nature of indole. Filter paper is saturated with 10 per cent oxalic acid solution and dried. It is cut into strips (as nearly aseptically as possible), and a strip is supported across the lower end of the plug in the mouth of the test tube containing the test solution. Indole volatilizes at ordinary room temperature and above, and reacts with the oxalic acid paper, imparting a distinct pink color to it. According to the report of these investigators, this test is specific for indole and checks up well with the Ehrlich-Böhme test. It is very sensitive, a one to one million solution of indole giving a positive reaction in twenty-four hours at 37°C. The paper may be placed in the culture tube when inoculated, and the first development of indole by the test organism noted. The paper must be kept dry in order to react.

Salkowski's test has gradually been discarded as a test for indole, because various workers have shown that other substances give the same reaction as indole does.

The Böhme-Ehrlich test (para-dimethyl-amido-benzaldehyde and potassium persulfate) has generally been accepted as the standard test by investigators, although there are a few who claim that it is not always specific. The preponderance of evidence, however, is in support of its specificity. This test cannot be used with solid or with highly colored media. Goré has modified it, however, in such a manner as to make it practical for such media. The test is made by wetting the cotton plug with the given reagents and volatilizing the indole by immersing the culture tube or flask in a boiling water bath. The volatilized indole imparts a characteristic cherry red colour to the moistened plug. This modification makes the test very sensitive, one part of indole in two million giving a positive reaction, according to Goré. The Böhme-Ehrlich method has been reported as giving a positive test in 1:300,000 dilution.

Goré (1921). Quoted from Holman and Gonzales (1923).
W. L. Holman and F. L. Gonzales (1923). J. Bact. 8, 577–583.

Indole—various

C. R. Fellers and R. W. Clough (1925). J. Bact. 10, 105–133.

These authors give an extensive review of numerous indole tests and tests for skatole. Only those which have been recommended elsewhere are reproduced here. The authors favour the Ehrlich test in a modified form.

Original Ehrlich Test

According to the Manual of Methods prepared by the Committee on Bacteriological Technic of the Society of American Bacteriologists (1923), the Ehrlich test is performed as follows. The reagent is a 2 per cent solution of paradimethylaminobenzaldehyde in 95 per cent alcohol.

One milliliter of this reagent is added to the culture, then drop by drop, concentrated hydrochloric acid is added until a red zone appears between the alcohol and the peptone solution. Not more than 0.5 ml of the acid is required. On standing the zone deepens and widens. The red color is soluble in chloroform, and the test may be confirmed by shaking the culture with chloroform to see if the pigment dissolves. If it proves soluble the test is considered positive.

Salkowski nitroso-indole reaction (Cholera-red reaction)

According to the Committee on Bacteriological Technic of the Society of American Bacteriologists in the Manual of Methods (1923), the test is performed as follows:

Mix 5 ml of the culture with about one-third its volume of 1:1 sulfuric acid. Then add on the surface a small amount of a 0.02 per cent solution of sodium nitrite. A positive reaction is indicated by a pink zone between the acidified culture fluid and the nitrite solution.

Indole-acetic acid gives a positive reaction by this method.

Steensma vanillin test

The Committee on Bacteriological Technic of the Society of American Bacteriologists (1923) recommends the following procedure in carrying out the vanillin test for indole. To 5 ml of the culture add 5 drops of a 5 per cent solution of vanillin in 95 per cent alcohol and 2 ml of concentrated sulfuric acid. Indole gives a clear orange by this test, which reaches its greatest depth in two or three minutes. Tryptophane, on the other hand, gives a reddish violet, which develops more slowly and deepens on standing or heating.

Gnezda (1899) Oxalic acid test

Blotting paper soaked in oxalic acid solution and suspended over the test tube culture reacts very sensitively to indole.

Gnezda (1899). C. r. hebd. Séanc. Acad. Sci., Paris 128, 1584.

Steensma test (modified by Fellers and Clough)

Make the test whenever possible on the distillate (direct or steam). To 5 ml of the solution being tested add 5 drops of 5 per cent solution of vanillin in 95 per cent alcohol, 2.5 ml concentrated hydrochloric acid and mix. Indole gives a clear orange color, while skatole gives a violet to violet blue color soluble in chloroform. The orange color produced by indole is insoluble in this reagent. Delicacy for indole and skatole was found to be about 1:3,000,000. The application of heat hastens the reaction. Only definite positive results should be recorded.

Technique recommended for indole and skatole determination in Bacteriol Cultures by Fellers and Clough Materials:

1. Dunham's peptone solution: 10 g peptone, 5 g NaCl, 1000 ml tap water. Reaction neutral (pH 7.0). Tube 10 ml per tube and sterilize.

Since peptone occasionally contains indole, checks should always be made on the media.

2. Concentrated C.P. HCl.
3. Ethyl ether U.S.P.
4. NaOH (2.5 per cent solution) for washing ether extracts in separatory funnels.
5. Dilute HCl (10 ml concentrated C.P. HCl in 200 ml H_2O) for washing ether extracts.
6. Paradimethylaminobenzaldehyde (2 g in 100 ml 95 per cent alcohol).
7. $CHCl_3$-U.S.P. For extracting the indole or skatole.
8. Indole color standards (containing 0.5, 1 and 5 and 10 micromilligrams indole in 5 ml H_2O). One micromilligram in 5 ml is equal to a dilution of 1:5,000,000 or 0.2 part per million.
9. HCl for final indole test (600 ml concentrated C.P. HCl plus 200 ml distilled H_2O).
10. Distilled water.

Method:

Incubated bacteria to be tested for indole production in Dunham's peptone solution, preferably at 37.5°C for five days. Transfer contents of the culture tube to a 250 ml Fry flask and wash with 40 ml H_2O. A current of steam is passed through the culture and 100 ml of distillate collected. If direct distillation is carried out make up to 100 ml and distill 75 ml. Acidify with 2 ml concentrated HCl (2) and extract once with 50 ml ethyl ether in a 300 ml separatory funnel and separate the ether layer. Then wash the ether extract in the same separatory funnel with 5 ml of 2.5 per cent NaOH (4) followed by 5 ml dilute HCl (5) separating the ether layer in each case. Usually these successive washings with alkali and acid are unnecessary, though they appear to remove phenols and other interfering substances. Add 10 ml H_2O and carefully evaporate the ether on a water bath. Divide the remaining water into two 5 ml portions and test for indole as follows.

Add to one 5 ml portion, in a small colorless test tube, 0.5 paradimethylaminobenzaldehyde (6), then 1 ml HCl (9). Place in boiling water for twenty seconds, shake vigorously, then place tube in ice water for one-half minute and extract with 1 ml $CHCl_3$ (7). Compare extracted red colour with that of indole standards prepared in exactly the same manner. This test is accurate to 1:25,000,000. It is also a good, though arbitrary, quantitative method.

Skatole standards are not usually prepared unless the qualitative test is positive, as it is so rarely found. Skatole is indicated in this modified Ehrlich test by a pale blue colour in the chloroform extract, which becomes deeper on standing. The regular dimethylaniline test for skatole has already been described and may be applied to the reserve 5 ml portion of the test solution.

Dimethylaniline test for skatole

To 5 ml of the solution to be tested for skatole, add a few drops of fresh dimethylaniline and shake vigorously. Add about 4 ml of concentrated sulfuric acid to form a layer at the bottom. A violet ring is formed in dilutions of 1:1,000,000 or more. The color is soluble in chloroform. Indole does not interfere.

Clough (1922) modified the test by using hydrochloric acid in place of sulfuric acid and heating to bring out the pink color. The dimethylaniline should be recently redistilled; else an interfering substance may obscure the color reaction. By this modified test a 1:5,000,000 dilution of skatole may easily be detected.

Clough (1922). Publs Puget Sound biol. Stn 3, 219.

Indole—*Xanthomonas*

F. R. Jones and L. McCullough (1926). J. agric. Res. 33, 493–521.

Tests were made on cultures in Dunham's solution, in peptonized Uschinsky's solution, and in a medium containing peptone, disodium phosphate, and magnesium sulfate.

Indole—*Xanthomonas beticola*

N. A. Brown (1928). J. agric. Res. 37, 155–168.

No indole was produced. Tests were made with six colonies grown in 1 per cent peptone water for 4 and 10 days. The sodium nitrite-sulfuric acid test was used.

Indole—mustiness in eggs

M. Levine and D. Q. Anderson (1932). J. Bact. 23, 337–347.

Tests for the production of indole were made after growing the organisms in tryptophan broth for various periods. Two methods were employed on duplicates. One set was tested by mixing equal parts of paradimethylaminobenzaldehyde and a saturated solution of potassium persulfate, moistening the end of the cotton plug with the mixture, and pushing it down to within an inch of the liquid, after which the tubes were placed in a steaming water bath for five to ten minutes. A positive result was indicated by the development of a pink coloration on the cotton plug. To each duplicate tube there was added 0.2 ml of the following mixture: 75 parts amyl alcohol, 25 parts concentrated hydrochloric acid and 5 parts (Eastman) para-dimethyl-amino-benzaldehyde. A red ring on the surface of the liquid showed the presence of indole.

Indole

P. M. Kon (1932–1933). J. Dairy Res. 4, 206–212.

The medium consisted of 1 per cent peptone with 0.5 per cent NaCl.

This test was also carried out after 5 days' incubation, it being found that reactions which were indefinite earlier became positive after this time. Ehrlich's reagent was used.

Indole—*Fusobacterium*

L. W. Slanetz and L. F. Rettger (1933). J. Bact. 26, 599–621.

Meat infusion potato broth cultures were used to test indole production.

Indole—*Bacteroides*

A. H. Eggerth and B. H. Gagnon (1933). J. Bact. 25, 389–413.

Peptone water for the indole tests consisted of 1 per cent of Parke Davis peptone and 0.2 per cent of disodium phosphate, set at pH 7.8. When growth on this medium was poor, duplicate cultures were made on beef infusion broth.

Tests: Twelve day cultures were used for the indole tests. We used two tests on each culture: the Ehrlich test as given by Kligler (1914) and the Zoller (1920) test. The two tests confirmed each other in every case of our series.

I. J. Kligler (1914). J. infect. Dis. 14, 81.
H. F. Zoller (1920). J. biol. Chem. 41, 31.

Indole—*Clostridium*

S. E. Hartsell and L. F. Rettger (1934). J. Bact. 27, 497–515.

Freshly prepared Difco tryptophane broth was inoculated with a well developed egg meat culture of the test organisms and the tubes incubated in anaerobic jars at 37°C. After three weeks incubation at this temperature all cultures were removed and the presence or absence of indole was determined by means of the Böhme-Ehrlich test.

Indole

F. C. Happold and L. Hoyle (1934). Biochem. J. 28, 1171–1173.

Technique recommended for estimation of indole: Ten milliliters of culture are acidified with one drop of concentrated HCl and extracted with two successive 10 ml quantities of light petroleum (B.P. 40–60°C). A third extraction can be made to ensure complete removal of indole. The pooled extracts are washed once with 10 ml of distilled water. The light petroleum is then extracted with successive 5 ml quantities of freshly prepared Ehrlich's reagent until no more rosindole is produced. The rosindole is completely insoluble in light petroleum and therefore collects exclusively in the layer of Ehrlich's reagent.

The addition of the reagent in 5 ml quantities at a time ensures that the correct amount is used, a point of considerable importance. If the amount of indole present is very small, of the order of 1:500,000, the reagent should be added 1 ml at a time.

A standard solution of indole is prepared in the same medium as the culture, a concentration of 1:50,000 being suitable, and extracted with light petroleum in the same manner. The rosindole from the culture is then compared with that from the standard in Nessler tubes. It is usually necessary to immerse the solutions in boiling water for a few minutes to remove milkiness; this does not affect the result provided both solutions are immersed for the same time.

The method has the advantage of simplicity, it can be used with small quantities of culture, and no difficulty has been experienced with substances interfering with the Ehrlich reaction, as such substances are apparently not extracted by light petroleum. The method is accurate with concentrations of indole down to 1:1,000,000.

Comparison of the method with estimations made by the steam distillation technique has shown that both methods are equally accurate if parallel controls are used in each procedure; that is, if standard solutions are distilled and extracted, respectively, to compare with the unknowns. If controls are not done, and the distillate and extract are compared directly with standard indole solutions, the extraction method is definitely the more accurate, since a greater proportion of the indole present is recoverable by extraction than by steam distillation.

Qualitative detection of indole in bacterial cultures:

The following simple application of the extraction technique has been found useful for detection of indole in cultures. Five milliliters of culture are shaken vigorously with 2 ml of xylene, and 1 ml of Ehrlich's reagent is run on to the surface of the mixture, the liquids being allowed to redistribute themselves by gravity. The xylene extracts the indole and, as the globules of xylene pass through the layer of Ehrlich's reagent on their way to the surface, the rosindole body is formed and a red ring appears at the lower surface of the xylene layer. Xylene is used in the test in preference to light petroleum, as Ehrlich's reagent mixes uniformly with light petroleum-culture mixture.

The test is between 20 and 50 times as sensitive as the usual method of mixing the culture and the Ehrlich's reagent (with or without the addition of persulfate) and extracting the rosindole with chloroform.

Many bacteriologists detect indole in cultures by the simple method of running a layer of Ehrlich's reagent onto the surface of the culture, when a red ring appears at the junction of the liquids. This test is exceedingly delicate and apparently owes its great sensitivity to the fact that indole, being a substance which lowers surface tension, becomes concentrated in the surface layer. In the xylene extraction test described above the intensity of the reaction depends on the concentration of indole in the whole culture, whereas in the ring test the intensity depends on the concentration in the surface layer. The xylene test gives more intense reactions than the ring test with concentrations of indole such as are usually present in bacterial cultures (e.g., 1:100,000) and is probably the preferable test for such cultures, but the ring test is more delicate, indole being detectable in concentrations with which the xylene test gives negative results. It was found impossible to determine accurately the end-point of the ring test, since when dilutions of indole of the order of 1:10,000,000 are reached the concentration of indole in the surface layer becomes an appreciable fraction of the whole amount, and further dilutions cannot be made with accuracy.

Indole—*Achromobacter delmarvae*

H. F. Smart (1935). J. agric. Res. 51, 363–364.

Indole was not produced in tryptophan broth in 10 days.

Indole—*Bacteroides*

A. H. Eggerth (1935). J. Bact. 30, 277–299.

For indole tests, heart-infusion broth was fermented with *Salmonella schottmuelleri* to remove the sugar.

Indole—*Clostridium*

R. S. Spray (1936). J. Bact. 32, 135–155.

This test is performed in the sugar-free fermentation control,* after reading the pH reaction. No discrepancies have been observed between the Ehrlich and vanillin tests. The vanillin test, however, is advocated because of a reaction described below as the "Vanillin Violet" test, which was shown by only three species in our series.

The minimum safe time of testing is about 72 hours. Tests may be made up to 7 days or more, with little change in reaction.

"Vanillin Violet" test. When 10 drops of 5 per cent vanillin in 95 per cent alcohol are added to the culture, followed by 10 drops of concentrated HCl, the orange indole reaction appears in degree according to the species tested. Certain cultures, however, including *Cl. sporogenes, Cl. tyrosinogenes* and *Cl. botulinum* (A and B), showed no orange, but rather a faint violet. On further addition of the reagents this color is intensified, diffusing as a deep violet color. The nature of this reaction is not known, but Hall suggests that it is probably due to skatole or some closely related body.

This reaction indicates, in our series, one of these three species, and it is interesting to note that it occurs in mixed as well as in pure cultures. Thus, it was observed in a culture of *Cl. welchii* containing *Cl. sporogenes* as a contaminant. This test may, then, be of value in checking cultures for freedom of *Cl. sporogenes,* probably the most common anaerobic contaminant.

**Sugar-free fermentation base*

Difco neopeptone, 10 g; Difco tryptone, 10 g; agar flakes, 2.5 g;

distilled H_2O, 1000 ml. Boil to dissolve and add water to make 1 L. Adjust the reaction to pH 7.3 to 7.4. Tube about 8 cm deep in 200 X 13 mm tubes. Autoclave for 15 to 20 minutes at 15 pounds pressure. Hall – personal acknowledgment.

Indole—*Fusobacterium*

E. H. Spaulding and L. F. Rettger (1937). J. Bact. 34, 535–548.

Produced in –

Proteose peptone	10 g
Liebig's meat extract	3 g
Potato extract (aq)	100.0 ml
Distilled water	900.0 ml
pH	7.6

The Ehrlich-Böhme reagent was used (no time stated).

Indole—*Xanthomonas*

S. S. Ivanoff, A. J. Riker and H. A. Dettwiler (1938). J. Bact., 35, 235–253.

Tested in –

Beef extract	3 g
Difco tryptophane peptone	10 g
Water	1000 ml
pH	7.0

Tested with Böhme's solution. No time specified.

Indole—*Pasteurella multocida*

C. T. Rosenbusch and I. A. Merchant (1939). J. Bact. 37, 69–89.

Tested in –

Difco Proteose peptone	5 g
NaCl	5 g
K_2HPO_4	0.2 g
$MgSO_4 \cdot 7H_2O$	0.1 g
H_2O	1000 ml
pH	7.2 – 7.4

Indole formation was tested on the basic broth medium after 4 or 5 days' incubation. A slight modification of Kovacs' test (1928) was used to detect the formation of indole. It consisted of the addition of 1 ml of ether to each tube, which concentrated the indole at the surface of the medium and allowed a rapid and definite purple indole ring to form. N. Kovacs (1928). Z. ImmunForsch. exp. Ther. 55, 311–315.

Indole—*Pseudomonas*

M. E. Caldwell and D. L. Ryerson (1940). J. Bact. 39, 323–336.

Bacto-tryptone broth was used with the Goré technique.

Indole—*Lactobacillus* and *Bacteroides*

K. H. Lewis and L. F. Rettger (1940). J. Bact. 40, 287–307.

In a medium composed of 2 per cent Bacto-tryptone, 1 per cent Bacto beef extract and 0.05 per cent cysteine hydrochloride. After 5 and 12 days of incubation under anaerobic conditions the cultures were tested by layering 0.5 ml of Kovacs' reagent (Ruchhoft, Kallas, Chinn and Coulter, 1931) over 5 ml of the cultures. The presence of indole was indicated by the appearance of a red ring at the surface.

C. C. Ruchhoft, J. C. Kallas, B. Chinn and E. W. Coulter (1931). J. Bact. 22, 125–181.

Indole—*Klebsiella, Escherichia, Enterobacter*

E. Osterman and L. F. Rettger (1941). J. Bact. 42, 721–743.

Determined by incubating for 48 hours in a 1 per cent Difco Bacto-tryptone broth and testing with the Bohme-Ehrlich test reagent.

Indole—*Clostridium*

A. R. Stanley and R. S. Spray (1941). J. Bact. 41, 251–257.

Preparation of standard curve. The authors describe a quantitative photoelectric method for determination of indole. Commencing with a series of 5 ml quantities containing decreasing amounts of indole (0.50 mg to 0.02 mg), they describe the following method.

Using a pipette made from 7-mm glass tubing and fitted with a 2 ml rubber bulb, approximately 2 ml of isoamyl alcohol are added to each tube. Rubber stoppers are fitted to the tubes and, after inverting about four times, the alcohol is allowed to form in a layer at the surface. This is drawn off with another pipette and transferred to a mixing graduate numbered to conform to the corresponding indole dilution. A separate rubber stopper and pipette are used for each tube. This extraction is repeated three times, pooling the alcoholic extracts in the corresponding mixing graduate. The volume is brought up to 10 ml with iso-amyl alcohol, then to 25 ml with ethyl alcohol (95 per cent), and finally, using a volumetric pipette, 10 ml of Quantitative Indole Reagent, described below, are added to each graduate and the contents thoroughly mixed. A control solution is prepared from 10 ml iso-amyl alcohol, 15 ml ethyl alcohol (95 per cent), and 10 ml Quantitative Indole Reagent.

These solutions are allowed to stand, after thorough mixing, for one hour and then readings are taken on the photoelectric colorimeter. The colorimeter is standardized at 100, using the control solution containing no indole but otherwise identical with the other solutions. Readings are then made on the solutions from each cylinder and the results plotted against the respective concentrations of indole.

Preparation of quantitative indole reagent. The amount of *p*-dimethylaminobenzaldehyde to be used in the procedure is very important, as increased quantities give deeper colors in high indole concentrations. As a result of experimentation the following formula has been derived. Exactly 35 grams of *p*-dimethylaminobenzaldehyde are dissolved in about 700 ml of 95 per cent ethyl alcohol in a 1000 ml volumetric flask. One hundred seventy milliliters of concentrated HCl are added, thoroughly mixed, cooled to 20°C, and the volume brought to 1000 ml with 95 per cent ethyl alcohol.

Indole determination in bacterial cultures. The indole is extracted from a measured portion (preferably 2 to 5 ml) of the culture medium by the same procedure as used in the preparation of the standard curve. If less than 5 ml is used, its volume is brought up to that with distilled water. The extract is made up to volume, the indole reagent added, and a numerical reading is taken on the photoelectric colorimeter. This reading is located on the standard curve and the milligrams of indole per unit of culture medium is either read directly or calculated from the size portion used in the test.

Indole—*Clostridium*

R. W. Reed (1942). J. Bact. 44, 425–431.

Bactotryptone	20 g
Na_2HPO_4	5 g
Glucose	1 g
Agar	1 g
Na thioglycolate	1 g
Water	1000 ml
pH	7.6

Autoclaved in deep tubes. Qualitative tests for indole were made on cultures after 1, 2, 5 and 10 days, using Feller and Clough's (1925, J. Bact. 10, 105–133) modification of the Ehrlich indole reagent.

Indole—*Corynebacterium*

R. F. Brooks and G. J. Hucker (1944). J. Bact. 48, 295–312.

Indole production was determined in 2-day cultures in a 1 per cent aqueous solution of tryptone, incubated at 37°C. The test was carried out by the para-dimethyl-aminobenzaldehyde, potassium persulfate method of Goré (1921, Indian J. med. Research 8, 505–507).

Indole—*Phytomonas syringae*

M. A. Smith (1944). J. agric. Res. 68, 269–298.

The isolates were tested for indole production on Bacto-tryptophane broth (pH 6.8) by the Ehrlich-Böhme test. Tests were made at the end of 7, 12 and 18 days.

Indole—*Corynebacterium*

M. Welsch, G. Demelenne-Jaminon and J. Thibaut (1946). Annls Inst. Pasteur, Paris 72, 203–215.

Production d'indol. – La recherche de l'indol, dans les cultures âgées de quatre jours, au moyen du réactif d'Ehrlich (p-di-méthyl-amino-benzaldéhyde), est restée négative avec les 85 souches examinées. Une culture d'*Escherichia coli,* dans les mêmes conditions, donnait une réaction très nettement positive.

La même recherche, pratiquée avec le réactif de Salkowski (acide sulfurique et nitrite de potassium), a donné un résultat positif avec la moitié des souches environ. On sait que Palmirski et Orlowski, 1895, Escallon et Sirce, 1908, etc., ont signalé le même résultat avec des cultures âgées de trois semaines, mais que Blumenthal, 1898, Zipfel, 1912, etc., n'ont obtenu que des résultats négatifs avec des cultures de cinq jours. On sait aussi que la réaction négative au réactif d'Ehrlich et positive avec celui de Salkowski indique, non pas la présence d'indol, mais bien celle d'acide indol-acétique (Hewlett, 1900, 1901 ; Frieber, 1921).

Indole—*Corynebacterium acne*

H. C. Douglas and S. E. Gunter (1946). J. Bact. 52, 15–23.

Medium:

Tryptose yeast-extract phosphate glucose thioglycolate broth. No method is given.

Indole—*Shigella*

W. H. Ewing (1946). J. Bact. 51, 433–445.

Two methods were used for indole tests:

(a) The Gnezda test, where filter paper strips were impregnated with a saturated solution of oxalic acid, dried, and suspended over 2 per cent peptone water.

(b) 1 per cent tryptone water culture tested after 24 hours incubation with Kovacs' reagent.

Indole—anaerobic micrococci

E. L. Foubert, Jr., and H. C. Douglas (1948). J. Bact. 56, 25–34.

Medium:

Difco tryptone	20 g
Difco yeast extract	2 g
Glucose	10 g
Sodium thioglycolate	1 g
Methylene blue	2 mg
distilled water	1000 ml

pH adjusted to 7.3–7.5 and drops to 6.9 to 7.1 during sterilization.

Indole production in the above medium was tested by the methods recommended by the Committee on Bacteriological Technique (1944). Com. Bact. Tech. Soc. Am. Bact. (1944). Leaflet V (9th ed.). Geneva, N. Y.: Biotech Publications.

Indole—*Proteus*

G. T. Cook (1948). J. Path. Bact. 60, 171–181.

Method:

Carried out in the usual way at 37°C.

Indole—*Escherichia coli*

Data taken from:

W. W. Arnold, Jr. and R. H. Weaver (1948). J. Lab. clin. Med. 33, 1334–1337. The C. V. Mosby Co. St. Louis, Missouri.

Medium:

I	tryptone	1 per cent
	beef extract	0.3 per cent
	distilled water	
	pH 7.4	
II	tryptophane	0.3 per cent
	peptone, bacto	0.1 per cent
	K_2HPO_4	0.5 per cent
	distilled water	
	pH 7.4	

Use 1 ml quantities of medium in 10 × 75 mm tubes. Preheat to 37°C in a water bath.

Inoculum:

Inoculation of a tube with all the growth obtainable from a colony of diameter 2 mm will give good results. Inoculation with a larger amount of growth obtainable from an agar slant culture will yield quicker indole production. Inoculation with growth from a culture which is in the logarithmic period of development will give quicker results than inoculation from an older culture.

Method:

Tubes are inoculated and incubated in a water bath at 37°C. In the test, four drops of fresh Kovacs' reagent are added to the 1 ml quantity of medium in the tube. The tube is shaken and the results are read after a few minutes. For the preparation of the reagent, various brands of amyl alcohol, isoamyl alcohol, and isobutyl alcohol were tried. Of these, isoamyl alcohol proved to be superior, some of the amyl alcohols giving false colorations and the isobutyl alcohol failing to separate sharply from the medium.

The period of incubation varies from 6 minutes to 2 hr.

N. Kovacs (1928). Z. ImmunForsch. exp. Ther. 55, 311–315.

Indole—*Pseudomonas alboprecipitans*

A. C. Johnson, A. L. Robert and L. Cash (1949). J. agric. Res. 78, 719–732.

The organism was grown in Bacto-tryptone broth. The Ehrlich-Böhme

and Gnezda methods* both gave negative tests.
*Society of American Bacteriologists Committee on Bacteriological Technique (1947). *Manual of Methods for Pure Culture Study of Bacteria.* Biotech Pub., Geneva, N. Y.

Indole—anaerobic Gram-negative diplococci

G. C. Langford, Jr., J. E. Faber, Jr., and M. J. Pelczar, Jr. (1950). J. Bact. 59, 349–356.

Fluid thioglycolate medium was used to indicate the production of indole. Strips of oxalic acid paper prepared according to the *Manual of Methods for Pure Culture Study* (1947) were placed in the tubes and extended down almost to the surface of the medium.
Society of American Bacteriologists Committee on Bacteriological Technique (1947). *Manual of Methods for Pure Culture Study of Bacteria.* Biotech Pub., Geneva, N. Y.

Indole—general

P. Hauduroy (1951). Techniques Bactériologiques. Masson et Cie, Éditeurs, Paris.

MILIEU POUR LA RECHERCHE DE L'INDOL (C. I. E.)

Comme milieu de base on utilise le bouillon à la caséine de Kristensen, Lester et Jurgens. A 1 litre de ce bouillon filtré on ajoute 2 litres d'eau de robinet.

L'ensemencement du milieu se fait au fil de platine à partir d'une culture sur gélose en boîte de Petri. Après 20 heures de séjour à 37°C on ajoute quelques gouttes (en surface) du réactif d'Ehrlich et Böhme. En cas de réaction positive, formation d'un anneau violet.

Réactif d'indol d'après Ehrlich-Böhme.

Paradiméthylamidobenzaldéhyde	8 g
Alcool à 96 p. 100	760 cm^3
Acide chlorhydrique concentré purifié	160 cm^3

Indole—general

P. Hauduroy (1951). Techniques Bactériologiques. Masson et Cie, Éditeurs, Paris.

INDOL

(Milieu) = Milieu pour la recherche de l'indol.

(Recherche de). – Les microorganismes sont ensemencés dans des tubes contenant 5 ml d'eau peptonée. On ajoute 0.5 ml de réactif pour dépister l'indol. Remuer le tube. Tout le réactif se réunit à la surface. Une réaction positive est indiquée par une couleur rouge ou rose. Le réactif consiste en :

p-diméthylaminobenzaldéhyde	5 g
Acide chlorhydrique pur	25 ml
Alcool amylique	75 ml

(W. I. C.).

Indole—*Staphylococcus*

C. Shaw, J. M. Stitt and S. T. Cowan (1951). J. gen. Microbiol. 5, 1010–1023.

Cultures were incubated for 5 days at 37°C, 10 days at 30°C or 14 days at 22°C. Indole was detected by the Holman and Gonzales Method (1923 J. Bact. 8, 577) based on the oxalic acid reaction (Gnezda, 1899, C. r. hebd. Séanc. Acad. Sci., Paris 128, 1584.)

Indole—micromethods

P. H. Clarke and S. T. Cowan (1952). J. gen. Microbiol. 6, 187–197.

Kovacs' reagent is preferred to the Böhme reagent. It was made up with iso-amyl alcohol which Arnold and Weaver (1948) found to give better results than other alcohols. It was found to be stable for several months on the bench in a dark bottle.

Organisms were grown on nutrient or serum agar. Positive results were obtained in ½–1 hour.

Growth in the presence of glucose suppresses the development of tryptophanase. Suspensions prepared from glucose agar recover after incubation in the presence of tryptophane for 1–3 hours.

Method:

Add 0.04 ml of suspension (from nutrient, serum, or 1% glucose agar) to a mixture of 0.1% DL-tryptophane, 0.06 ml and 0.025 *M*-phosphate buffer (pH 6.8), 0.04 ml. Incubate at 37°C for 1 hour (3 hours if from glucose-agar), and add 0.06 ml Kovacs' reagent (*p*-dimethylaminobenzaldehyde, 5 g; isoamyl alcohol, 75 ml; conc. HCl, 25 ml). Shake and read immediately.

W. M. Arnold, Jr. and R. H. Weaver (1948). J. Lab. clin. Med. 33, 1334.

Indole—*Pseudomonas*

P. Villecourt, H. Blachère, and G. Jacobelli (1952). Annls Inst. Pasteur, Paris 83, 316–322.

Après quatre jours l'indol était recherché.

Indole—*Bacillus*

N. R. Smith, R. E. Gordon, and F. E. Clark (1952). U.S.D.A. Agr. Monograph No. 16.

The cultures were inoculated into 1 percent Difco tryptone and 1 percent BBL trypticase broths (5 ml amounts in 18 mm tubes). After 6, 10 and 14 days' incubation at a suitable temperature, usually 32°C, the cultures were shaken with 2 ml of the following solution: Paradimethylaminobenzaldehyde, 5 g; amyl alcohol, 75 ml; concentrated HCl, 25 ml. A pink to red color in the alcohol layer indicated the presence of indole.

Indole—*Pseudomonas (Malleomyces) pseudomallei*

P. de Lajudie and E. R. Brygoo (1953). Annls Inst. Pasteur, Paris 84, 509–515.

Indol: – La mise en évidence de la production d'indol se fit par le réactif de Kowacz après deux et sept jours de culture sur eau peptonée.

Indole—*Actinomyces, Corynebacterium*

H. Beerens (1953). Annls Inst. Pasteur, Paris 84, 1026–1032.

Indole. – La recherche de l'indol effectuée sur les cultures, en milieu de base ci-dessus décrit pour l'étude des fermentations sucrées, s'est montrée positive pour toutes les souches. Le milieu non ensemencé donnait une réaction négative.

Indole—various

N.C.T.C. Methods 1954.

Medium for tests is Lemco broth; incubate at optimal temperature.

Method 1: Test after 2 days' incubation. Add 1 ml ether to the culture, shake well, and run down the side of the tube about 1 ml of Ehrlich's reagent.

Ehrlich's reagent:

Paradimethylaminobenzaldehyde	4 g
Absolute alcohol	380 ml
Concentrated HCl	80 ml

P. Ehrlich (1901). Berl. med. Wschr. 1, 151.

Method 2: Test after 2 days' incubation. Add 0.3 to 0.4 ml Kovacs' reagent to 5 ml culture; shake well, and read after a minute.

Kovacs' reagent:

Paradimethylaminobenzaldehyde	5 g
Amyl alcohol	75 ml
Concentrated HCl	25 ml

(The reagent should be bright yellow to light brown)

N. B. Some brands of amyl alcohol are unsatisfactory because they give a dark colour with paradimethylaminobenzaldehyde. For microtests use *iso*-amyl alcohol.

N. Kovacs (1928). Z. ImmunForsch. exp. Ther. 55, 311.

Method 3: Put an oxalic acid paper between the plug and tube of culture. Read daily for 7 days.

Oxalic acid papers

Soak strips of filter paper (100 × 8 mm) in a *hot* saturated solution of oxalic acid.

Allow to dry in an incubator at 37°C or 55°C.

J. Gnezda (1899). C. r. hebd. Séanc. Acad. Sci., Paris 128, 1584.

W. L. Holman and F. L. Gonzales (1923). J. Bact. 8, 577.

Indole—*Proteus*

C. Shaw and P. H. Clarke (1955). J. gen. Microbiol. 13, 155–161.

Oxalic acid paper — method of Holman and Gonzales (1923) J. Bact. 8, 577.

W. L. Holman and F. L. Gonzales (1923). J. Bact. 8, 577.

Indole—*Zymomonas*

N. F. Millis (1956). J. gen. Microbiol. 15, 521–528.

On 10 day cultures in 1 per cent Difco yeast extract containing 2 mg D. L. tryptophane per ml at pH 5.5. Indole reagent was that recommended by Mackie and McCartney (1946).

T. J. Mackie and J. E. McCartney (1946). *Handbook of Practical Bacteriology,* 7th ed. Edinburgh, Livingstone Ltd.

Indole—*Acetobacter*

R. Steel and T. K. Walker (1957). J. gen. Microbiol. 17, 445–452.

Peptone	2 per cent (w/v)
Yeast extract	0.5 per cent (w/v)
Fat free casein	0.5 per cent (w/v)

Incubation – 5 days.

Reagent not stated.

Indole—thermophilic sulfate-reducing bacteria

L. L. Campbell, Jr., H. A. Frank and E. R. Hall (1957). J. Bact. 73, 516–521.

Formation occurs in the medium of Reed and Orr (1941). Indole was determined after incubation for 48 hr, with Kovacs' reagent.

G. B. Reed and J. H. Orr (1941). War. Med., Chicago 1, 493–510.

Indole—*Leptotrichia buccalis*

R. D. Hamilton and S. A. Zahler (1957). J. Bact. 73, 386–393.

Medium:

Tryptone (Difco)	20 g
Yeast extract (Difco)	2 g
Glucose	5 g
NaCl	5 g
K_2HPO_4	10 g
Sodium thioglycolate	1 g
Agar (Difco)	0.5 g
Distilled water	1 litre
0.2% aqueous methylene blue	1 ml/L

pH adjusted to 7.2 and dispensed in screw-top tubes before autoclaving.

All cultures were incubated at 37°C.

The Goré test for the detection of indole was used as described in the *Manual of Methods for Pure Culture Study of Bacteria* (1953) Williams and Wilkins Co., Baltimore, Maryland.

Indole—*Actinomyces bovis*

S. King and E. Meyer (1957). J. Bact. 74, 234–238.

To determine the presence of indole, a buffered thioglycolate medium described by Douglas and Gunter (1946) was used, to which nitrate or tryptone was added as necessary. The tryptone broths were examined for the presence of indole within 5 to 7 days.

H. C. Douglas and S. E. Gunter (1946). J. Bact. 52, 15–23.

Indole—*Enterobacteriaceae*

J. G. Heyl (1957). Antonie van Leeuwenhoek 23, 33–58.

Indole Base

Bacto tryptone	10 g
Sodium chloride	5 g
Distilled water	1000 ml
pH	6.8–7.2

Sterilize for 15 minutes at 120°C.

After incubation for 24–48 hours at 37°C the test is carried out by adding about 0.2–0.3 ml of the reagent of Ehrlich-Böhme. The reagent is prepared by dissolving 5 g para-dimethyl-amido benzaldehyde in 75 ml N amyl alcohol and adding to this solution 25 ml concentrated (38%) HCl.

Indole

Society of American Bacteriologists *Manual of Microbiological Methods* – McGraw-Hill Book Co. Inc., New York 1957.

Use 1 per cent concentration of a peptone high in tryptophane, such as those prepared by enzymatic digestion of casein or lactalbumin.

The test for indole may be performed by the technique of Ehrlich-Böhme, by either the Goré or the Kovacs modification of the same or by the Gnezda technique. The Kovacs method is especially simple and convenient. These procedures are as follows:

Böhme (1905) called for the following solutions:

Solution 1

Para-dimethyl-amino-benzaldehyde	1 g
Ethyl alcohol (95%)	95 ml
Hydrochloric acid, concentrated	20 ml

Solution 2

Saturated aqueous solution of potassium persulfate ($K_2S_2O_8$).

To about 10 ml of the culture fluid add 5 ml of solution 1, then 5 ml of solution 2, and mix well by rotating between the hands; a red color appearing in 5 min indicates a positive reaction. This test may also be performed (and sometimes more satisfactorily) by first mixing the culture with ether and adding solution 1 (Ehrlich's reagent), dropping it down the side of the tube so that it spreads out as a layer between the ether and the culture fluid. After this method of applying, solution 2 seems to be unnecessary.

The Goré (1921) test uses these same solutions, but the method of application is as follows: Remove the plug of the culture tube (which must be of white *absorbent* cotton); moisten it first with 4–6 drops of solution 2, then with the same amount of solution 1. Replace the plug and push it down until 1 or 1½ in. above the surface of the culture. Place the tube upright in a boiling water bath, and heat for 15 min without letting the culture solution come in contact with the plug. The appearance of a red color on the plug indicates the presence of indole.

The Kovacs (1928) test is a simplification of that of Böhme, using only one solution; it is now the method of choice in many laboratories:

Para-dimethyl-amino-benzaldehyde	5 g
Amyl or butyl alcohol	75 ml
Hydrochloric acid, concentrated	25 ml

This reagent may be used as in the Böhme test, but no solution 2 is required. The red color appears in the alcohol layer.

The Gnezda (1899) oxalic acid test is done as follows: Dip a strip of filter paper in a warm saturated solution of oxalic acid; on cooling, this is covered with crystals of the acid. Dry the strip of paper thoroughly (sterilization by heat seems unnecessary), and insert into the culture tube under aseptic conditions, bent at such an angle that it presses against the side of the tube and remains near the mouth. Reinsert the plug, and incubate the culture. If indole is formed, the oxalic acid crystals take on a pink colour.

It is recommended that the Goré or the Kovacs test be used in a routine way. In interpreting the results obtained, it must be remembered that when the reagents are added directly to the medium, they react with alpha-methyl-indole as well as with indole itself, but, as the former compound is nonvolatile, it cannot react to the Goré or Gnezda tests. Hence the Ehrlich test, unmodified, is less specific for indole than the Goré modification or the Gnezda test.

Some samples of para-dimethyl-amino-benzaldehyde and of amyl and butyl alcohol have been found unsatisfactory for the indole test. It is well, therefore, to check new supplies of these chemicals against samples known to be satisfactory.

A. Böhme (1905). Zentbl. Bakt. ParasitKde I. Abt. Orig. 40, 129–133.
J. Gnezda (1899). C. r. hebd. Séanc. Acad. Sci., Paris 128, 1584.
S. N. Gore (1921). Indian J. med. Res. 8, 505–507.
N. Kovacs (1928). Z. ImmunForsch exp. Ther. 55, 311–315.

Indole—various organisms

H. D. Isenberg and L. H. Sundheim (1958). J. Bact. 75, 682–690.

Media:

All organisms were maintained on Eugon (BBL) agar slants, kept refrigerated after initial incubation at 37°C. Transfers were made to Eugon (BBL) broths which were incubated at 37°C for 18 hr. Such cultures served as inocula for all other media tested. Transfers were usually made with sterile Pasteur pipettes which delivered 0.05 ml per drop.

The substrate originally employed for the growth of and indole production by the organisms was trypticase nitrate broth (BBL) designated in this study as medium A. Media B and C contained, in addition to the ingredients of medium A, 0.5 g L-tryptophan, while medium C was further modified by the addition of 0.01 g pyridoxal per 100 ml. Medium D (in per cent) was composed as follows: glucose, 0.1 g; K_2HPO_4, 0.2 g; $FeSO_4 \cdot 7H_2O$, 0.001 g; $MgSO_4 \cdot 7H_2O$, 0.02 g; casein hydrolysate (Mallinckrodt), 0.01 g; yeast extract (Difco), 0.01 g; agar (Difco), 0.01 g; L-tryptophan, 0.5 g; pH 7.2 ± 0.1. In medium E the ingredients of medium D were supplemented with NH_4Cl, 0.02 g; and $NaNO_3$, 0.02 g per 100 ml.

Anaerobic cultures were set up with medium B in small ground glass stoppered bottles. These were filled, after dry sterilization, with sterile culture medium, boiled to remove excess air, inoculated after cooling, and filled with additional recently boiled and cooled medium to the rim of the bottle, care being taken to avoid introduction of air. The ground glass stopper was then jammed into place. The bottles were incubated at 37°C and indole determinations performed at various time intervals covering several days, a new bottle being assayed each time.

Reagents:

The reagent for the direct *p*-dimethylaminobenzaldehyde test was prepared according to the method of Gadebusch and Gabriel (1956). The quantitative and qualitative *p*-dimethylaminobenzaldehyde reagents were prepared according to Wood *et al.* (1947); the hydroxylamine hydrochloride reagents were made up as directed by Gallop and Seifter (1958, unpublished data) as follows: NaOH, 4 *N*; $NH_2OH \cdot HCl$, 0.25 *N*; H_2SO_4, 4 *N*.

Procedures:

(1) Direct *p*-dimethylaminobenzaldehyde: To 1.0 ml of an 18 hr or older broth culture, 0.5 ml of the reagent is added. The color developed is recorded immediately and after 5 min. A spot test modification of this test was also employed. To 4 to 5 drops of the culture in the depression of a porcelain or plastic plate, 2 drops of the reagent are added, and the color is recorded immediately and after 5 min.

(2) *p*-dimethylaminobenzaldehyde method with extracted indole:

(a) Quantitative procedure: to a 1 ml aliquot of the supernatant of an 18 hr or older broth culture, obtained by centrifugation for 10 min at

2000 rpm, 1 ml of toluene is added and the two are mixed gently. Five-tenths ml of the toluene layer is removed to a clean test tube. One ml *p*-dimethylaminobenzaldehyde in ethyl alcohol (95 per cent) is added, followed by 8.5 ml of acid alcohol. The tube is allowed to stand at room temperature and then its color is compared with a reagent blank at 540 mμ. Cultures or standards in excess of 15 μg were diluted with uninoculated broth to bring them into the 1 to 15 μg range.

(b) Spot test procedure: To 4 to 5 drops of culture medium in which the organism has grown for at least 18 hr, an equal number of drops of toluene is applied in the depression of a spot plate. Slight agitation with the tip of a Pasteur pipette is followed by the transfer of some of the toluene layer (2 to 3 drops) to an adjacent clean depression. One or two drops of the *p*-dimethylaminobenzaldehyde reagent are now added, as well as 5 to 6 drops if acid-alcohol. Color development is recorded immediately and after 20 min.

(3) Hydroxylamine method: (a) Quantitative procedure: one ml of an 18 hr or older broth culture supernatant, obtained by centrifugation at 2000 rpm for 10 min, and 1 ml of NaOH are mixed; 2 ml of the hydroxylamine reagent are added. This mixture is allowed to stand for 15 min. Five ml sulfuric acid are then added and the color developed read against a reagent blank at 530 mμ.

(b) Spot test procedure: This test was performed in a porcelain plate depression, using drops per ml of the quantitative procedure, without the waiting period.

Uninoculated control media and indole standards accompanied all determinations. All spectrophotometric determinations were carried out in a B and L Spectronic 20 spectrophotometer.

H. H. Gadebusch and S. Gabriel (1956). Am. J. clin. Path. 26, 1373–1375.

W. A. Wood, I. C. Gunsalus, and W. W. Umbreit (1947). J. biol. Chem. 170, 313–321.

Indole—*Aeromonas*

J. P. Stevenson (1959). J. gen. Microbiol. 21, 366–370.

Cultures were incubated in peptone water for 5 days at 37°C and tested with Böhme's reagents.

Indole—*Pseudomonas*

M. E. Rhodes (1959). J. gen. Microbiol. 21, 221–263.

Five ml volumes of peptone (Evans) water cultures were incubated for 7 days and then shaken with 1 ml ether to extract indole which was detected by means of Ehrlich's reagents. The oxalic acid test paper method was used with *Pseudomonas aeruginosa* (to test for false-positive results due to conversion of pyocyanine to its red salt in acid solution; Sandiford,

1937). Ability to produce indole from tryptophan was similarly investigated by using the Dowson (1949) galactose+ammonium+inorganic salts medium+0.01% (w/v) tryptophan.

B. R. Sandiford (1937). J. Path. Bact. 44, 567.

W. J. Dowson (1949). *Manual of Bacterial Plant Diseases,* 1st ed. London: A. and C. Black.

Indole—*Clostridium*

M. E. Brooks and H. B. G. Epps (1959). J. gen. Microbiol. 21, 144–155.

Twenty-four and forty-eight hour cultures in Robertson's meat broth were examined for indole production by means of Ehrlich's reagent and by the vanillin test (Spray, 1936).

R. S. Spray (1936). J. Bact. 32, 135.

Indole—*Pasteurella septica*

J. M. Talbot and P. H. A. Sneath (1960). J. gen. Microbiol. 22, 303–311.

Peptone water cultures incubated overnight at 37°C were extracted with xylene, and the xylene extract tested with Ehrlich's reagent.

Indole—*Fusobacterium*

A. C. Baird-Parker (1960). J. gen. Microbiol. 22, 458–469.

Medium (%, w/v); Oxoid Tryptone, 1.0; Oxoid yeast extract, 0.1; Lab. Lemco, 0.3; glucose, 0.1; L-cysteine HCl, 0.05; disodium hydrogen phosphate, 0.5. Cultures were incubated for 7 days. The indole was extracted by shaking 3 ml culture fluid with 1 ml toluene (Isenberg and Sundheim, 1958). The presence of indole was indicated by the formation of a reddish color in the organic phase when *c.* 0.5 ml of Ehrlich's reagent was added with a Pasteur pipette to form a layer at the interface between the culture fluid and toluene layer.

H. D. Isenberg and L. H. Sundheim (1958). J. Bact. 75, 682.

Indole—*Fusobacterium*

R. R. Omata and R. C. Braunberg (1960). J. Bact. 80, 737–740.

Medium:

Trypticase (BBL)	1%
Yeast extract (Difco)	0.5%
L-cystine	0.02%
final pH 7.2	

Glucose was added separately to the sterile medium as sterile solution to give a final concentration of 1% glucose.

Indole—marine bacterium

D. Pratt and F. C. Happold (1960). J. Bact. 80, 232–236.

Rate of indole formation by whole cells. Test suspensions were prepared by diluting 0.1 ml of the heavy suspension in 10 ml of the medium to be tested. After incubation for 5 min at 37°C, 3 ml of suspension were pipetted to a tube containing 0.25 ml of 0.4 per cent L-tryptophane. The mixture was incubated for 10 min; the reaction was then stopped by addition of 0.5 ml of 10 per cent trichloroacetic acid. The indole formed was determined on the entire contents of the tube using the method of Happold and Hoyle (1934). The color formed by the addition of Ehrlich's reagent was measured at 560 mμ in a Spectronic 20 (Bausch and Lomb). The incubation time and cell concentration were adjusted so that between 0 and 20 μg of indole were formed and the data fell on the linear portion of the calibration curve.

The rate of indole formation by cell-free extracts. Tested in a reaction mixture having a total volume of 2 ml; usually 0.5 ml of extract containing 0.7 to 0.8 mg of protein was used. Phosphate buffer, 0.05 *M*, pH 7.4, was present in all tests. The reaction was initiated by the addition of 0.2 ml of 0.4% L-tryptophane and was stopped after 10 min incubation at 37°C by the addition of 0.5 ml of 10% trichloroacetic acid. The indole formed was determined using the entire contents of the tube.

F. C. Happold and L. Hoyle (1934). Biochem J. 28, 1171–1173.

Indole—*Streptococcus, Lactobacillus*

C. W. Langston and C. Bouma (1960). Appl. Microbiol. 8, 212–222.

Indole test:

The test for indole was carried out in indole-nitrate medium (BBL) after 24 and 48 hr incubation. Kovacs' reagent was used to determine indole production.

Indole—*Flavobacterium*

R. Buttiaux and J. Vandepitte (1960). Annls Inst. Pasteur, Paris 98, 398–404.

Caractère important pour le diagnostic présomptif, les *Flavobacterium* des méningites produisent un peu d'indole. Pour le déceler constamment, il est nécessaire d'employer la technique recommandée par King [4] : le germe est inoculé dans une eau peptonée (Tryptone Difco, de préférence); après quarante-huit heures à 30°C, on ajoute à 3 ml du milieu, un peu de xylène; on agite et on recouvre ce dernier avec du réactif d'Ehrlich.

[4] King (E. O.). Am. J. clin. Path. 1959, 31, 241.

Indole—*Pseudomonas*

M. Véron (1961). Annls Inst. Pasteur, Paris 100, Suppl. 6, 16–42.

La production d'indole dans une culture de 16–24 heures en eau peptonée.

Indole—*Pseudomonas*

R. Hugh and E. Ryschenkow (1961). J. gen. Microbiol. 26, 123–132.

Tryptone broth and modified Kovacs' reagent were used according to Gadebusch and Gabriel (1956), Am. J. clin. Path. 26, 1373.

Indole—*Xanthomonas*

A. C. Hayward and W. Hodgkiss (1961). J. gen. Microbiol. 26, 133–140.

Peptone water supplemented with casein hydrolysate (British Drug Houses Ltd. Laboratory reagent), 1.0 g/L, and L-cysteine hydrochloride, 0.1 g/L was dispensed in 5 ml quantities in ½ oz screw-capped bottles. Lead acetate papers were held over the medium by the screw cap which was kept loose. At 6 days, final observations were made and indole was tested for by addition of ether and Ehrlich's rosindole reagent (Mackie and McCartney 1960, p. 609).

T. J. Mackie and J. E. McCartney (1960). *Handbook of Practical Bacteriology,* 10th ed. Edinburgh, Livingstone.

Indole—*Escherichia aurescens*

H. Leclerc (1962). Annls Inst. Pasteur, Paris 102, 726–741.

Production d'indole. – Elle est observée sur une culture de 24 ou 48 heures en eau peptonée, après addition de réactif nitrique nitreux en présence d'alcoöl iso-amylique.

Indole—*Lactobacillus*

R. E. Smith and J. D. Cunningham (1962). Can. J. Microbiol. 8, 727–735.

Indole-nitrite broth (BBL) was tested, after 48 hours' incubation, for indole with fresh Kovacs' reagent. A portion of this medium was examined for the presence of nitrite using Griess-Ilosva reagents, and negatives were confirmed with zinc dust.

Indole production was tested by adding Ehrlich's rosindole reagent to 7-day peptone water cultures.

Indole—*Haemophilus*

P. N. Edmunds (1962). J. Path. Bact. 83, 411–422.

Plasma digest broth was prepared by adding 5 per cent horse flesh digest and 5 per cent citrated human plasma to the basic peptone broth. It was inoculated heavily and incubated for 4 days at 37°C. Ehrlich's reagent was layered on the broth and the reaction observed after 10 min. A control culture of *Escherichia coli* grown in the plasma digest medium for 4 days gave a good reaction.

Indole

W. H. Ewing (1962). *Enterobacteriaceae. Biochemical Methods for Group Differentiation.* U.S. Department of Health, Education and Welfare, Public Health Service Publication, No. 734 (revised).

Peptone water	
Bacto peptone	20 g
Sodium chloride	5 g
Distilled water	1000 ml

Leave the reaction unadjusted and sterilize at 121°C for 15 minutes. One per cent Bacto tryptone may be substituted for the 2 per cent Bacto peptone, if desired. Or 1.5 per cent casitone (Difco) or trypticase (BBL) may be used.

Inoculation:

Inoculate lightly from a young agar slant culture (culture which has been incubated overnight or 16 to 20 hours)

Incubation:

37°C, 40 to 48 hours. Test with Kovacs' reagent. Add about 0.5 ml of reagent, and shake the tube gently. A deep red color develops in the presence of indole.

Kovacs' Reagent:

Pure amyl or isoamyl alcohol	150 ml
Paradimethylaminobenzaldehyde	10 g
Concentrated pure hydrochloric acid	50 ml

Dissolve aldehyde in alcohol and then slowly add acid. The dry aldehyde should be light in color. Kovacs' reagent should be prepared in small quantities and stored in the refrigerator when not in use.

Tests for indole production may be made after 24 hours' incubation but if this is to be done, 1 or 2 ml of culture should be removed aseptically from the tube and the test made on this sample. If the test is negative, the remaining portion of the peptone water culture should be reincubated for an additional 24 hours.

Indole—*Actinobacillus* and *Haemophilus*

E. O. King and H. W. Tatum (1962). *Actinobacillus actinomycetemcomitans* and *Haemophilus aphrophilus.* J. infect. Dis. 111, 85–94.

Indole test was performed with 2% tryptone broth or heart infusion broth.

Indole—*Caryophanon*

P. J. Provost and R. N. Doetsch (1962). J. gen. Microbiol. 28, 547–557.

Hy-Case SF	10.0 g
Bacto yeast extract	5.0 g
Na acetate (anh)	0.5 g
Agar	15.0 g

pH 7.8–7.9

+ 0.1% (w/v) DL tryptophane

Indole—*Vibrio*

G. H. G. Davis and R. W. A. Park (1962). J. gen. Microbiol. 27, 101–119.

Tested in 5 day old peptone (Oxoid) water cultures using Ehrlich reagent and ether extraction.

Indole—*Vibrio*

G. H. G. Davis and R. W. A. Park (1962). J. gen. Microbiol. 27, 101–119.

In 1% (w/v) Oxoid tryptone water, tested after 3 days by ether extraction and Ehrlich reagent.

Indole—*Pseudomonas odorans*

I. Málek, M. Radochová and O. Lysenko (1963). J. gen. Microbiol. 33, 349–355.

In tryptone broth by the Kovacs' method after incubation for 5 days at 28°C.

Indole—rumen bacteria

R. S. Fulghum and W. E. C. Moore (1963). J. Bact. 85, 808–815.

Determinations of indole production and nitrate reduction were made in SRP-basal medium* with 2% Trypticase and 0.1% KNO_3.

The anaerobic methods of Hungate (Bact. Rev. 14, 1–49, 1950) were used in handling and incubation of the cultures.

***Basal medium for physiological tests**—*Butyrivibrio, Borrelia, Bacteroides, Selenomonas, Succinivibrio* (p. 109).

Indole—*Actinobacillus*

P. W. Wetmore, J. F. Thiel, Y. F. Herman and J. R. Harr (1963). Comparison of selected *Actinobacillus* species with a hemolytic variety of *Actinobacillus* from irradiated swine. J. infect. Dis. 113, 186–194.

Kovacs' reagent was used.

Indole—*Veillonella*

M. Rogosa (1964). J. Bact. 87, 162–170.

Tests for indole were performed in V17 medium by the Kovacs method described in the *Manual of Microbiological Methods.* Positive reactions were detected by the immediate development of a red color when drops of reagent were introduced at the surface of cultures. Delayed reactions in which brown or greenish-brown colors occur are not caused by the presence of indole, but are the result of nonspecific reactions and decomposition of the reagent (Society of American Bacteriologists, 1957).

For V17 medium see **Basal medium**—*Veillonella* (p. 114).

Society of American Bacteriologists (1957). *Manual of microbiological methods.* McGraw-Hill Book Co., Inc., New York.

Indole—*Clostridium*

N. A. Sinclair and J. L. Stokes (1964). J. Bact. 87, 562–565.

For nitrate reduction and indole formation, indole-nitrate medium was used. In some cases the medium was enriched with 0.01% DL-tryptophan to increase indole formation.

Incubation was conducted at 0°, 15° and 25°C.

Indole

G. S. Wilson and A. A. Miles (1964). Topley and Wilson's *Principles of Bacteriology and Immunity* – 5th edition – Arnold.

Tested in 1 per cent peptone water after 5 days by adding Böhme's reagents. One ml of ether is added to the culture, which is shaken thoroughly and then allowed to stand until the ether collects on the surface. One ml of solution A is run down the side of the tube; if no color appears within a minute, 1 ml of Solution B is added. A positive reaction is characterized by a color varying from a faint pink to a deep magenta. According to Happold and Hoyle (1934) xylene is better than ether; so is petroleum ether, which is less volatile.

Solution A:

para-dimethylaminobenzaldehyde	4 g
96 per cent Ethanol	380 g
Concentrated HCl	80 g

Solution B:

Saturated watery solution of potassium persulfate.

As peptones vary in their composition, Ljutov (1959) recommended the addition of 0.1 per cent tryptophan to the medium. So also did Braun and Silberstein (1940), who preferred to grow the organisms on the surface of a solid medium, because indole is formed best aerobically. A piece of filter paper, previously soaked in a solution made up with 5 g *p*-dimethylamido-benzaldehyde, 10 ml of pure phosphoric acid and 50 ml of methanol, and subsequently dried, is inserted between the tube and the plug. The production of indole leads to a red coloration of the paper.

An alternative method, depending on the volatility of indole at 37°C, is recommended by Holman and Gonzales (1923). It consists in placing a strip of filter paper, soaked in a saturated watery solution of oxalic acid and subsequently dried, between the cotton wool plug and the tube. The paper should be carefully folded so as to present the maximum surface to the volatilizing indole, which turns it a pink colour.

Some organisms form indole, but break it down as rapidly as they produce it, and hence may give a false negative reaction (Reed, 1942). The presence of glucose in the medium should be avoided. According to Hugh and Leifson (1953) indole is formed only by strains that are able to ferment, as opposed to oxidize, glucose.

H. Braun and W. Silberstein (1940). İstanb. Üniv. Tib Fak. Mecm. 3, 1596.
F. C. Happold and L. Hoyle (1934). Biochem. J. 28, 1171.
W. H. Holman and F. L. Gonzales (1923). J. Bact. 8, 577.
R. Hugh and E. Leifson (1953). J. Bact. 66, 24.
V. Ljutov (1959). Acta path. microbiol. scand. 46, 349.
R. W. Reed (1942). J. Bact. 44, 425.

Indole—motile marine bacteria and *Hyphomicrobium neptunium**

E. Leifson, B. J. Cosenza, R. Murchelano and R. C. Cleverdon (1964). J. Bact. 87, 652–666.

The basic culture broth with 0.5% Casitone was satisfactory for the indole test with Kovacs' reagent.

Culture media: The culture broth and the plating agar were prepared with seawater taken from Noank Harbor and filtered through a 0.45-μ Millipore filter. To the water were added 0.2% Casitone (Difco) and 0.1% yeast extract; the pH was adjusted to 7.5; the mixture was then boiled, filtered through paper, and sterilized by autoclaving. This was the broth used for primary culture and for flagellar staining. For slants and plates, 1.5% agar was added to the broth. In studies of the isolated bacteria, artificial seawater was used in all media. Comparative studies showed that all the isolates grew equally well on media prepared with artificial seawater diluted with an equal quantity of distilled water, compared with media made with undiluted artificial seawater or with undiluted natural seawater. Since the acid-base buffer content of these media is low, it was advantageous to add additional buffer. Tris (hydroxymethyl)aminomethane (tris) buffer in 0.05% concentration proved to be satisfactory.

Note: The composition of the artificial seawater is not given or cited.

*E. Leifson (1964). Antonie van Leeuwenhoek 30, 249–256.

Indole—*Pseudomonas aeruginosa*

K. Morihara (1964). J. Bact. 88, 745–757.

Indole was detected by use of Kovacs' reagent after 1, 3, and 5 days of incubation in 1% Tryptone (Difco) water. Cultures were incubated at 30°C.

Indole—*Vibrio marinus*

R. R. Colwell and R. Y. Morita (1964). J. Bact. 88, 831–837.

The authors used the method of the Society of American Bacteriologists 1957, *Manual of Microbiological Methods* – McGraw-Hill Book Co., Inc., New York – modified by the addition to the media of the following salts to produce a synthetic seawater:– sodium chloride, 2.4%; potassium chloride, 0.07%; magnesium chloride (hydrated), 0.53%; and magnesium sulfate (hydrated), 0.7%.

A standard inoculum was 1 drop from a pasteur pipette (*c.* 0.05 ml) of a 24 hour artificial seawater broth.

Cultures were incubated at 18°C.

Indole—*Bacteroides oralis*

W. J. Loesche, S. S. Socransky, and R. J. Gibbons (1964). J. Bact. 88, 1329–1337.

Indole production was determined in the basal medium supplemented with 0.02% L-tryptophan.

The test was performed in the presence and absence of 0.5% glucose.

Thioglycolate Medium without Dextrose (BBL), supplemented with 0.2% yeast extract (Difco) and hemin, was used as a basal medium. All cultures were incubated at 35°C in Brewer jars containing an atmosphere of 95% H_2 and 5% CO_2.

Indole—*Clostridium*

K. Tamai and S. Nishida (1964). J. Bact. 88, 1647–1651.

Indole formation was examined by a routine method. A 0.5-ml amount of a 50% (v/v) solution of H_2SO_4 and 0.5 ml of amyl alcohol solution saturated with KNO_2 were added to 2.5 ml of the cultural fluid. The mixture, after sufficient shaking, was left standing for a short while. A positive reaction, corresponding to the strength of indole produced, was indicated by variation from a pinkish to a deep-red color.

Indole—*Moraxella*

W. J. Ryan (1964). J. gen. Microbiol. 35, 361–372.

Organisms were grown for 48 hr at 37°C in serum peptone water. Ehrlich's reagent was used to test for indole production after extraction of the culture with ether.

Indole—marine bacteria

R. M. Pfister and P. R. Burkholder (1965). J. Bact. 89, 863–872.

Indole production was determined (Skerman, 1959) by use of 1.0% tryptone (Difco) in the seawater medium.

The seawater medium contained (per liter): Trypticase (BBL) or NZ Case, 2.0 g; Soy-tone (Difco), 2.0 g; yeast extract, 1.0 g; vitamin B_{12}, 1.0 μg; aged seawater; agar (if desired), 16 g. The pH was adjusted to 7.0 prior to autoclaving.

V. B. D. Skerman (1959). *A guide to the identification of the genera of bacteria.* The Williams & Wilkins Co., Baltimore.

Indole—*Vibrio comma*

J. C. Feeley (1965). J. Bact. 89, 665–670.

Indole production was tested on 1% Trypticase broth with Kovacs' reagent.

Indole—*Pseudomonas*

G. L. Bullock, S. F. Snieszko, and C. E. Dunbar (1965). J. gen. Microbiol. 38, 1–7.

Indole production was determined in 1% (w/v) tryptone broth by the method of Kovacs, as described in the Society of American Bacteriologists' *Manual of Microbiological Methods* – McGraw-Hill Book Co. Inc., New York 1957. Cultures were incubated at 20–22°C for 1 week.

Indole—*Alcaligenes*

R. G. Mitchell and S. K. R. Clarke (1965). J. gen. Microbiol. 40, 343–348.

Indole production was tested with Ehrlich's reagent in peptone water cultures incubated for 6 days.

Indole—*Malleomyces, Pseudomonas*

Bach-Toan-Vinh (1965). Annls Inst. Pasteur, Paris 109, 460–463.

Recherche de l'indole – Procédé utilisant le réactif de Kovacs.

Indole—*Mima, Herellea, Flavobacterium*

J. D. Nelson and S. Shelton (1965). Appl. Microbiol. 13, 801–807.

Cultures were incubated for 24 hr in Indole-Nitrite Medium (BBL), extracted with ether, and layered with Ehrlich's reagent to test for indole production (Society of American Bacteriologists, 1957).

Society of American Bacteriologists (1957). *Manual of microbiological methods.* McGraw-Hill Book Co., Inc., New York.

Indole—*Providencia, Rettgerella*

C. Richard (1966). Annls Inst. Pasteur, Paris 110, 105–114.

Pour des raisons de commodité, les milieux ont été incubés à 37°C. Production d'indole: mise en évidence sur une culture en eau peptonée de vingt-quatre heures à l'aide du réactif de Kovacs.

Indole—enterobacteria

J. G. Johnson, L. J. Kunz, W. Barron and W. H. Ewing (1966). Appl. Microbiol. 14, 212–217.

In addition to Kovacs' test for indole, a screening test was used in which a paper strip impregnated with reagent (Weil and Saphra, 1953) was suspended from the cotton plug inside the tube of lysine-iron-agar. Because occasional paper strips gave false-negative reactions, Kovacs' method was also used whenever a test for indole was pertinent to the identification of cultures which were negative by the paper strip method.

A. J. Weil and I. Saphra (1953). *Salmonellae and shigellae,* p. 226, Charles C. Thomas, Publisher, Springfield, Ill.

Indole—*Bacteroides, Veillonella, Neisseria*

J. van Houte and R. J. Gibbons (1966). Antonie van Leeuwenhoek 32, 212–222.

Indole production was tested in 1-, 2- and 7-day cultures grown in the presence or absence of glucose in the basal medium.

The basal medium was BBL thioglycollate medium without added dextrose, supplemented with 5% horse serum, and hemin (5 μg/ml) menadione (0.5 μg/ml).

Indole and skatole—anaerobes

A. -R. Prévot (1966). Techniques pour le diagnostic des Bactéries Anaérobes. Éditions de la Tourelle, St. Mandé.

Indole : dans un tube à essais, verser 2 cm^3 de distillat* puis ajouter 1 cm^3 de réactif d'Ehrlich. Une réaction positive se traduit par l'apparition d'un anneau rouge cerise à la surface du liquide.

Scatole (méthyl-3-indole) : même processus que ci-dessus, mais la réaction se traduit par l'apparition d'un anneau de teinte bleu-violacé virant secondairement au bleu.

*Distillat Alcaline sous **Amines (volatile)** — anaerobes (p. 51).

INDOLYL-3-ACETIC ACID

Indolyl-3-acetic acid—*Agrobacterium tumefaciens*

G. Beaud and P. Manigault (1966). Annls Inst. Pasteur, Paris 111, 345–358.

Le dosage de l'acide indolyl-3-acétique (A. I. A.) s'effectue selon la méthode de Pilet [25], qui met en oeuvre le réactif de Salikowsky. On le prépare en dissolvant d'abord 6 g de chlorure ferrique dans 250 ml d'eau. Puis on ajoute lentement 300 ml d'acide sulfurique concentré, on laisse refroidir et on complète par addition de 250 ml d'eau. Le produit se conserve dans un flacon de verre brun. Pour le dosage, on ajoute 8 ml du réactif, 1 ml d'alcool ethylique, puis 1 ml de prélèvement (1ml de surnageant prélevé dans 3 ml de milieu de culture centrifugé). Vingt minutes plus tard, on lit la densité optique du liquide coloré (électro-colorimètre de Meunier, filtre vert, trajet optique 1 cm).

Le tryptophane est introduit dans le milieu de culture en utilisant une solution mère (5 g/l) stérilisée par filtration sur millipore HA. La courbe de référence s'établit à partir de solutions aqueuses d'A. I. A. (Prolabo) en présence de tryptophane.

[25] Pilet (P. E.). *Phytohormones de croissance.* Masson, édit., Paris, 1961.

LECITHINASE

Lecithinase—*Clostridium* (Nagler reaction)

N. J. Hayward (1943). J. Path. Bact. 55, 285–293.

The composition finally adopted for the Nagler plate was nutrient agar containing 20 per cent of human serum and 5 per cent of Fildes' peptic digest of sheep blood (Mackie and McCartney, 1938). Separate control plates containing 4 units of *welchii* antitoxin per ml of medium may be used; or, since the surface application of antitoxin inhibits Nagler reactions (Crook), considerable economy of materials can be effected by spreading over half the surface of the medium about 0.1 ml (50–100 international units) of antitoxin by means of a bent Pasteur pipette, and, after the antitoxin has been absorbed, sowing the inoculum equally on both halves. Great care must be taken to spread an even layer of antitoxin exactly to the central dividing line and to the periphery of the agar, and to leave uninoculated a few mm at the edge of the control half; otherwise colonies with zones may occur at the margin of the antitoxin treated area and confuse the reading of the test. Polyvalent gas gangrene antitoxin should not be used.

As the indicator for *welchii* toxin, human serum, being easier to obtain, is preferable to the egg yolk suspension described by Macfarlane *et al.* (1941). Yolk saline is more sensitive than human serum and varies less (Crook), but these qualities have no advantage under the conditions of the plate test, where the large yield of toxin makes a sensitive indicator unnecessary. With the human serum medium, overnight *welchii* colonies are surrounded by clearly defined opaque zones 1.3 mm in width.

E. M. Crook (1942). Br. J. exp. Path. 23, 37.
R. G. MacFarlane *et al.* (1941). J. Path. Bact. 52, 99.
T. J. Mackie and J. F. McCartney (1938). *Handbook of Practical Bacteriology,* 5th edition, Edinburgh, Livingstone. p. 44.

Lecithinase—*Clostridium* (Nagler reaction)

L. S. McClung, P. Heidenreich and R. Toabe (1946). J. Bact. 51, 751–752.

Hayward (1943) used nutrient agar containing 20% human serum and 5% peptic digest of sheep blood. Nagler (1944) used Weinberg's V. F. agar with 10% defibrinated sheep blood and 10% egg yolk suspension.

The authors suggest the following substitute:

Proteose peptone no. 2	40	g
Na_2HPO_4	5	g
KH_2PO_4	1	g
NaCl	2	g
$MgSO_4$	0.1	g
Glucose	2	g
Agar	25	g
Distilled water	1000	ml
pH	7.6	

Sterilize for 20 minutes at 240°F.

After autoclaving, add 10 ml of sterile egg yolk suspension to 100 ml of warm medium and pour approximately 15 ml in plates of 100 mm diameter. The addition of the blood to the medium, as suggested by Nagler, is unnecessary. To prepare egg yolk suspension aseptically, withdraw to a sterile rubber-stoppered tube, by aspiration, the yolk from a fresh hen's egg, after first removing the white. Add an equal amount of sterile 0.85 per cent NaCl to the yolk and invert the tube to mix the contents.

N. Hayward (1943). J. Path. Bact. 55, 285–293.
F. P. O. Nagler (1944). Nature, Lond. 158, 496.
F. P. O. Nagler (1944). Aust. J. exp. Biol. med. Sci. 23, 59–62.

Lecithinase—*Bacillus*

B. C. J. G. Knight and H. Proom (1950). J. gen. Microbiol. 4, 508–538.

Nutrient agar containing *c.* 3 g N/L was prepared by adding papain digest of horse muscle containing 1.5 g N/L to a water extract of fresh horse muscle and adding 1.2%/New Zealand agar. Tests for lecithinases and other enzymes visibly affecting egg yolk emulsion were made on this nutrient agar containing 10% (v/v) of egg yolk emulsion.

Two types of reactions were distinguished.

(1) a zone of opalescence extending well beyond the area of growth – because of C type lecithinases.

(2) a fainter reaction only immediately beneath the growth which may or may not be because of a lecithinase.

Lecithinase—*Staphylococcus*

W. A. Gillespie and V. G. Alder (1952). J. Path. Bact. 64, 187–200.

Yolk broth was prepared by the method of McGaughey and Chu: 5 parts by weight of fresh hen egg yolk were added to 100 parts of digest or infusion broth. Kieselguhr (2 g per 100 ml) was added after thorough stirring, and the mixture was stirred again and allowed to stand for about 30 minutes. It was then filtered through filter paper pulp. As opacity production was enhanced in the presence of glucose (probably because of the effect of lowered pH), one per cent of this sugar was added to the medium which was sterilised by filtration through a Seitz pad. It was an almost clear yellow fluid, of pH 7.2. Yolk saline was made by substituting 0.85 per cent NaCl for broth and omitting the glucose. These reagents were stored at 4–6°C and could be used up to at least 4 weeks after preparation.

The antiserum used in the experiments was Burroughs Wellcome therapeutic refined staphylococcus antitoxin, which was stated to contain 2500 units in 1.9 ml.

The egg yolk (EY) reaction

Two sterile Wassermann tubes (2 × 3/8 in) plugged with cotton wool

and containing 1 ml of glucose yolk broth, were inoculated with the staphylococcus; to one was added 1 drop (about 0.03 ml) of antitoxin. The tubes were incubated for 3 days at 37°C in a covered water bath to reduce evaporation. If the organism was egg yolk positive (EYP), the tube without antitoxin showed a distinct and usually dense white opacity, generally after one day. The opaque material often rose as a curd to the top of the fluid. The tube containing antitoxin remained clear, apart from opalescence due to organisms.
C. A. McGaughey and H. P. Chu (1948). J. gen. Microbiol. 2, 334.

Lecithinase—*Clostridium*

K. R. Johansson (1953). J. Bact. 65, 225–226.

The author used the egg yolk medium of McClung and Toabe (1947, J. Bact. 53, 139) with the addition of 0.02% sodium azide to enumerate intestinal anaerobes. Incubated anaerobically. Lecithinase positive colonies were those of true anaerobes. Aerobic spore formers failed to grow in the axide containing medium.

Lecithinase—*Clostridium*

A. G. Harbour (1954). J. Path. Bact. 67, 253–254.

Method. An egg yolk emulsion is made by mixing aseptically equal volumes of egg yolk and sterile saline; it is not filtered and it may be stored in the refrigerator. Lecithin plates are prepared by adding 10 ml of the emulsion to 100 ml nutrient agar cooled to 55°C and pouring the mixture into Petri dishes. A control plate is prepared by spreading 4 or 5 drops of antitoxin over the surface of a lecithin plate and drying the plate in the incubator.

To test an organism for lecithinase production, a control and a test plate are inoculated from an overnight culture on solid medium; the inoculated area is a spot 2.3 mm in diameter. This is covered with a sterile coverglass, care being taken that air bubbles are not trapped underneath. The plates are incubated at 37°C and examined at two, three and four hours. An opaque spot, easily visible by transmitted light, indicates lecithinase production.

Lecithinase

N.C.T.C. Methods 1954. – *In vitro* L.V. tests.

When bacterial fluids are to be examined for the presence of lecithinase, a convenient method is to take 1 ml fluid in an agglutination tube (2½ × ¼ in) and add 1 ml saline or sodium acetate buffer (0.2 *M* sodium acetate-acetic acid at pH 6.5) and 0.5 ml L.V.-saline. Mix by inversion, using small squares of nonabsorbent paper to cover the mouth of the tube. Incubate at 37°C for 1 hour. A control tube with 0.5 ml saline or buffer in place of L.V.-saline is also incubated. Read after 1 hour and again after 18 to 24 hours at room temperature. Opalescence appearing only in the "fluid" tube indicates the presence of lecithinase.

For the identification and measurement of lecithinases, supplies of sera containing known amounts of specific antilecithinases are required.

E. M. Crook (1942). Br. J. exp. Path. 23, 37.
R. G. Macfarlane, C. L. Oakley and C. G. Anderson (1941). J. Path. Bact. 52, No. 1, 99.
C. L. Oakley, G. H. Warrack and P. H. Clarke (1947). J. gen. Microbiol. 1, No. 1, 91.

Lecithinase—*Bacillus*

E. Grinsted and L. F. L. Clegg (1955). J. Dairy Res. 22, 178–190.

Yolk-agar plates were prepared by mixing equal parts of 5% yolk broth* and 4% nutrient agar at 45–50°C with aseptic precautions and pouring the mixture immediately into sterile Petri dishes. These were incubated overnight at 30°C to dry and harden the surface of the agar and to check for sterility. Cultures were diluted by mixing one drop in 9 ml of Ringer's solution and streaking one loopful of this over the surface of the agar to produce well-isolated colonies. The plates were then incubated for 24 hr at 30° or 55°C. The appearance of an opalescent zone beneath and around the colonies indicated the presence of lecithinase. The production of distinct zones was sometimes aided by standing the plates for a further 24 hr on the bench after their removal from the incubator.

*C. A. McGaughey and H. P. Chu (1948). J. gen. Microbiol. 2, 334.

Lecithinase—*Chromobacterium*

P. H. A. Sneath (1956). J. gen. Microbiol. 15, 70–98.

Egg yolk reaction

Used a medium containing 9 parts of nutrient agar and 1 part of a Seitz-filtered mixture of one egg yolk in 250 ml of saline.

The plates were incubated at 25°C for 4 days.

Lecithinase—*Bacillus*

E. R. Brown, M. D. Moody, E. L. Treece, C. W. Smith (1958). J. Bact. 75, 499–509.

The production of lecithinase by the organisms was determined by the method of McGaughey and Chu (1948). It is obvious that this test is unreliable for species differentiation to the extent that overlapping of results occurred between *B. anthracis* and *B. cereus.*

C. A. McGaughey and H. P. Chu (1948). J. gen. Microbiol. 2, 334–340.

Lecithinase—*Clostridium* (Nagler Reaction)

A. T. Willis and G. Hobbs (1958). J. Path Bact. 75, 299–305.

Media: Since egg yolk gives a stronger Nagler reaction than human serum (Hayward, 1941), and is now more readily available, it was used throughout. Furthermore, raw egg yolk media (Reed, 1948) not only give a stronger reaction than the L.V. preparation used by Macfarlene, Oakley and Anderson, but are simpler to prepare.

A nutrient agar base was prepared by dissolving 2 per cent agar in meat infusion broth and adjusting to pH 7.0. One per cent of lactose and 0.003 per cent of neutral red was added to this, and the mixture was autoclaved at 15 lb/sq in for 20 min. The egg yolk was separated from the white by the usual culinary technique and mixed with an equal volume of sterile 0.9 per cent saline; no particular aseptic precautions are necessary. The nutrient agar base was cooled to 45–50°C, and egg yolk suspension was added to a final concentration of 10 per cent. Plates were poured immediately.

It was found more convenient to keep the complete medium poured in plates, which, when kept at 4°C, were usable for several months, rather than to keep a merthiolate-preserved egg yolk emulsion, which keeps only about 10 days at this temperature.

To encourage the growth of some of the stricter anaerobes, 0.1 per cent sodium thioglycolate was added to the medium with the egg yolk suspension.

Inoculation of plates. The lactose-egg yolk-agar plates were dried for 20 min at 37°C before use. Half of each plate was then spread with 3 drops of *Cl. welchii* type A antitoxin, diluted if necessary to 600 units/ml, and the plates were then left at room temperature for a few minutes to absorb it. In all cases a loopful of a fresh chopped meat broth culture was streaked across the plate at right angles to the edge of the antitoxin area; the plates conveniently accommodated two organisms. The plates were incubated in anaerobic jars at 37°C, and examined after 24 and 48 hours.

N. J. Hayward (1941). Br. med. J. 1, 811 and 916.
R. G. MacFarlane, C. L. Oakley and C. G. Anderson (1941). J. Path. Bact. 52, 99.
G. B. Reed (1948). In Dubos, *Bacterial and Mycotic Infections of Man,* Philadelphia, p. 355.

Lecithinase—*Clostridium*

L. G. Jayko and H. C. Lichstein (1959). Nutritional factors concerned with growth and lecithinase production by *Clostridium perfringens.* J. infect. Dis. 104, 142–151.

Toxin assay. Lecithovitellin was prepared by adding the yolk of one egg to 500 ml of 0.9% saline. Twenty grams of Norite were then added and the suspension was agitated mechanically for half an hour. After removal of the Norite by centrifugation the lecithovitellin was passed through a Selas filter and stored at 4°C. Cultures to be assayed for lecithinase were neutralized to pH 7 with NaOH and centrifuged to remove the cells. Three ml of the culture supernate were mixed with an equal volume of the lecithovitellin diluted with 2 ml of 0.9% saline and incubated at 43°C in a water bath. The control for the assay was identical except that a few drops of polyvalent gas gangrene antitoxin (Lederle) were added to the culture supernate before it was mixed with the lecithovitellin. Lecithinase titers were measured as turbidity produced from lecithovitellin in a Coleman Junior Spectrophotometer at 650 mμ, with control tubes containing antitoxin employed as blanks.

Lecithinase—*Pseudomonas*

M. E. Rhodes (1959). J. gen. Microbiol. 21, 221–263.

Reaction on egg yolk medium. The appearance of the isolates when grown as surface streads on egg yolk plates (Knight and Proom, 1950, J. gen. Microbiol. 4, 508) was recorded.

Lecithinase—*Pasteurella septica*

J. M. Talbot and P. H. A. Sneath (1960). J. gen. Microbiol. 22, 303–311.

Egg yolk reaction was tested on the egg yolk medium of Knight and Proom (1950, J. gen. Microbiol. 4, 508) which was incubated at 37°C for 48 hours.

Lecithinase—*Fusobacterium*

A. C. Baird-Parker (1960). J. gen. Microbiol. 22, 458–469.

Measured by the method of McClung and Toabe (1947).

L. S. McCung and R. Toabe (1947). J. Bact. 53, 139.

Lecithinase—*Pseudomonas*

M. Véron (1961). Annls Inst. Pasteur, Paris 100, Suppl. 6, 16–42.

Un dernier caractère a été proposé pour différencier les *Pseudomonas* (Klinge et Gräf, Rhodes); il s'agit de la réaction au jaune d'oeuf, qui peut être recherchée sur gélose au jaune d'oeuf, selon la technique décrite par Knight et Proom pour l'étude des *Bacillus.*

Klinge (K.) et Graf (W.). Zbl. Bakt., I Abt. Orig., 1959, 174, 243.
Rhodes (M. E.). J. gen. Microbiol., 1959, 21, 221.
Knight (B. C. J. G.) et Proom (H.). J. gen. Microbiol. 1950, 4, 508.

Lecithinase—*Leptospira*

M. Füzi and R. Csóka (1961). J. Path. Bact. 82, 208–212.

Egg yolk reaction

Preliminary studies showed that the decomposition of egg yolk by leptospirae causes a reaction which is visible to the naked eye under proper experimental conditions. In fluid media, and especially when small inocula are used, the changes generally are inconstant or weak. If, however, a two-phase system is employed with solid egg yolk medium overlayered with Korthof's fluid medium containing numerous leptospirae, a turbidity and pellicle formation in the fluid can be observed after a period of incubation. A nutrient agar medium was prepared with 2 per cent agar fibre, 1 per cent Witte peptone and 5 per cent egg yolk emulsion. The egg yolk, unheated, was added aseptically to the melted agar medium held in a water bath at 50°C. Then 1 ml amounts of the medium were dispensed in 15 × 150 mm tubes, and these were closed with rubber plugs. Sterility was checked by incubating the tubes at 37°C for one day and at room temperature for an additional two days. Then 2 ml of 5–6 day old cultures containing about 10^6 leptospirae per ml were added to the egg yolk agar tubes, which were incubated in the upright position at 30°C for 4 weeks. Control tubes contained egg yolk medium overlayered with Korthof's medium without leptospirae. The tubes were observed daily with the naked eye and the number of living leptospirae was observed microscopically once a week. The whole investigation series was repeated five times over a period of 6 months.

Lecithinase—*Vibrio*

G. H. G. Davis and R. W. A. Park (1962). J. gen. Microbiol. 27, 101–119.

Egg yolk reaction tested as recommended by Willis (1960, J. Path. Bact. 80, 379) after 6 days.

Lecithinase—*"Bacterium salmonicida" (Aeromonas)*

I. W. Smith (1963). J. gen. Microbiol. 33, 263–274.

Production in egg-yolk medium was examined on Esselman and Liu's medium (1961).

M. T. Esselman and P. V. Liu (1961). J. Bact. 81, 939.

Lecithinase—*Staphylococcus*

D. B. Shah and J. B. Wilson (1963). J. Bact. 85, 516–521.

Preparation of the egg yolk factor. Staphylococcus aureus strain 10552 was grown in Heart Infusion broth (Difco) on a rotary shaker for 3 to 5 days. The enzyme had been purified from the cell-free culture filtrates by acid precipitation, two cycles of ethanol fractionation, and zone electrophoresis as described by Blobel, Shah, and Wilson (1961).

Because of the instability of the zone electrophoresis-purified egg yolk factor, the ethanol-fractionated enzyme preparation was used in all experiments. The lyophilized preparation was stored at –20°C. Using antiserum prepared against the ethanol-fractionated enzyme, gel diffusion precipitation analysis showed only one band with the zone electrophoresis-purified egg yolk factor and three bands with the homologous ethanol-fractionated enzyme. Staphylocoagulase was one of the three enzyme proteins present in this preparation (Blobel *et al.*, 1961).

Studies on lecithinases. Lecithinase C activity was measured by determining the rate of increase of acid-soluble phosphorus (Macfarlane and Knight, 1941). For use in the reaction mixtures, stable aqueous emulsions of partially purified egg lecithin and soybean lecithin (Nutritional Biochemicals Corp., Cleveland, Ohio) and purified egg lecithin (Hanahan, Turner, and Jayko, 1951) were prepared by treatment in a Raytheon 10-kc sonic oscillator for 10 min. After incubation of the enzyme, substrate, buffer, and varying amounts of $CaCl_2$ (10^{-1} to 10^{-4} *M*) at 37°C for 3 hr, the reaction was terminated by adding cold 20% trichloroacetic acid to a final concentration of 5% in the mixture. The mixture was allowed to stand at room temperature for 30 min and was filtered. The clear filtrate was assayed for acid-soluble phosphorus by the method of King (1932).

Lecithinase D activity was investigated according to the procedure of Davidson, Long, and Penny (1955). The discovery by Kates (1953, 1954) that degradation of lecithin by lecithinase D from cabbage and spinach is markedly stimulated by the presence of ether and $CaCl_2$ was also investigated. Free choline was measured by the enneaiodide method of Shapiro (1953).

Lecithinase A activity was measured by extraction of the fatty acids liberated by the enzyme in the reaction mixture (Fairbairn, 1945) and determined by titration with standard alkali with phenolphthalein as an indicator.

No lecithinase activity was found in the egg yolk factor of *Staphylococcus aureus* (see **Fat hydrolysis**—*Staphylococcus* p. 234).

H. Blobel, D. B. Shah, and J. B. Wilson (1961). J. Immun. 87, 285–289.
F. M. Davidson, G. Long and I. F. Penny (1955). In G. Popjak and E. LeBrenton (ed.), *Biochemical problems of lipids.* p. 253–262. Butterworths Scientific Publications, London.
D. Fairbairn (1945). J. biol. Chem. 157, 633–644.
D. J. Hanahan, M. B. Turner and M. E. Jayko (1951). J. biol. Chem. 192, 623–628.
M. Kates (1953). Nature, Lond. 172, 814–815.
M. Kates (1954). Can. J. Biochem. Physiol. 32, 571–583.
E. J. King (1932). Biochem. J. 26, 292–297.
M. G. Macfarlane and B. C. J. G. Knight (1941). Biochem. J. 35, 884–902.
B. Shapiro (1953). Biochem. J. 53, 663–666.

Lecithinase—*Clostridium perfringens*

H. E. Hall, R. Angelotti, K. H. Lewis and M. J. Foter (1963). J. Bact. 85, 1094–1103.

Their ability to produce lecithinase was determined on McClung-Toabe egg medium (McClung and Toabe, 1947).

L. S. McClung and R. Toabe (1947). J. Bact. 53, 139–147.

Lecithinase—*Pseudomonas aeruginosa*

R. R. Colwell (1964). J. gen. Microbiol. 37, 181–194.

Lecithinase was tested for by spotting a drop of a 24–48 hr YE broth culture on to an egg-yolk agar plate consisting of Difco Nutrient Agar containing 10% (v/v) egg-yolk suspension. The egg-yolk suspension was prepared as follows: an egg yolk was washed in 0.2% NaCl, placed in a Blendor jar with 200 ml 0.2% NaCl and the resulting suspension was centrifuged at 13,000 rev/min for 5 min in a high-speed refrigerated centrifuge. The supernatant fluid was sterilized by Seitz filtration and added to a cooled sterile Difco Nutrient Agar before plates were poured.

The standard inoculum for all tests was a single drop (1/20 ml) from a sterile disposable pipette (Fisher Scientific Company) of a 24–48 hr broth culture. The stock culture medium (YE) was a modification of that used by Rhodes (1959): Difco yeast extract, 0.3 g; Difco Bacto-Proteose Peptone, 1.0 g; NaCl, 0.5 g; agar, 1.5 g (agar omitted from liquid stock media); distilled water 1 L; adjusted to pH 7.2–7.4 with NaOH. Tests were carried out at room temperature (25°C) except where otherwise indicated.

M. E. Rhodes (1959). J. gen. Microbiol. 21, 221.

Lecithinase—*Pseudomonas*

G. L. Bullock, S. F. Snieszko and C. E. Dunbar (1965). J. gen. Microbiol. 38, 1–7.

Lecithinase was determined in egg yolk agar; method and medium were those described by the Society of American Bacteriologists *Manual of Microbiological Methods* – McGraw-Hill Book Co. Inc., New York 1957. Cultures were incubated at 20–22°C for 1 week.

Lecithinase—*Clostridium welchii*

J. G. Collee (1965). J. Path. Bact. 90, 13–30.

Tests of lecithinase activity. Serial doubling dilutions of 0.5 ml amounts of the test sample were made in 0.5 ml aliquots of saline containing calcium chloride 0.01 per cent. Volumes of 0.5 ml of egg yolk suspension were then added. The mixtures were shaken, incubated for 1 hr at 37°C in a water bath, and then chilled overnight at 4°C. An arbitrary visual estimation of turbidity was recorded as ++++, +++, ++, +, ⊥ or –, for each tube, in comparison with controls.

Lecithinase—*Pseudomonas aeruginosa*

P. V. Liu (1966). The roles of various fractions of *Pseudomonas aeruginosa* in its pathogenesis. J. infect. Dis. 116, 112–116.

The organisms were grown on tryptone-glucose-extract agar (Difco) enriched with 1% glucose and covered with a sheet of cellophane, as described previously (Liu *et al.,* 1961). The growth of the organism on the cellophane after 24 hours at 37°C was washed off with 3 ml of distilled water and centrifuged at 10,000 rcf for 1 hour. The supernatant fluid thus obtained will be referred to as "extract from the cellophane plate." This extract was

Millipore-filtered and dried to a powder in vacuum, resuspended to one-twentieth the original volume in distilled water, and dialyzed for 2 days with frequent changes of 0.02 *M* tris (tris[hydroxymethyl] aminomethane) buffer. The contents of the cellophane bag were centrifuged to remove the precipitate that formed during the dialysis, and the supernatant fluid was concentrated to one-fiftieth its original volume with Carbowax (polyethylene glycols, Union Carbide, N.Y.). This concentrated fluid was used in the column chromatography. For titration of lecithinase the preparations were diluted in saline containing 0.005% zinc sulfate, since this enzyme requires zinc ions for its maximum efficiency (Liu, 1964). The procedure was otherwise similar to that described previously (Liu *et al.*, 1961).

P. V. Liu (1964). J. Bact. 88, 1421–1427.

P. V. Liu, Y. Abe and J. L. Bates (1961). J. infect. Dis. 108, 218–228.

Lecithinase—anaerobes

A.-R. Prévot (1966). Techniques pour le diagnostic des Bactéries Anaérobes. Éditions de la Tourelle, St. Mandé.

Recherche des lécithinases

Certaines toxines hémolytiques ont pu être identifiées à des lécithinases. L'activité lécithinasique sera mise en évidence par l'action du filtrat de culture sur une émulsion d'ovolécithine préparée comme suit: un jaune d'oeuf est émulsionne dans 500 cm^3 d'eau physiologique contenant 0.8 g p. 1000 d'acétate de calcium. On ajoute 20 g de kaolin et l'on filtre sur papier, puis sur bougie ou sur membrane filtrante type Seitz. On peut conserver 8 à 10 jours en glacière.

On répartit dans un tube à hémolyse 0.5 cm^3 de cette émulsion et 1 cm^3 de filtrat de culture. On porte le mélange à l'étube à 37° pendant 4 heures. La présence de lécithinase se traduit par l'apparition d'un trouble. Ce trouble est empêché spécifiquement par l'antilécithinase homologue.

LEVAN PRODUCTION

Levan—*Streptococcus*

E. J. Hehre and J. M. Neill (1946). J. exp. Med. 83, 147–162.

Details of a chemical method and a serological method for the recognition of levan in the presence or absence of dextran in streptococcal cultures and methods for the recognition of dextran formation by streptococci are given. Details are as follows:

In the present investigation, the dextrans and levans were recognized by subjecting the supernatant fluids from sucrose and glucose broth cultures of the streptococci to the following physical, chemical, and serological tests.

Test 1. Gross Appearance.—The fluids were examined for gross evidence of gum formation; that is, for soluble gum (opalescent supernatant fluids) and for insoluble gum (gelatinous sediment or zooglea). The significance of the test rested upon the difference between the fluids from the sucrose and the glucose cultures of the same bacteria.

Tests 2 and 3. Alcohol-Precipitable Material.—The fluids were diluted 1:10 in 10 per cent sodium acetate, and 2.5 and 1.2 volumes of 95 per cent alcohol were added to separate portions of each sample. After they had been 3 hours at room temperature the mixtures were examined for cloudiness (controls with the uninoculated sucrose and glucose mediums gave practically no cloudiness). The test with 2.5 volumes was expected to be useful as an index for the presence of *either* dextran or levan, particularly since the interpretation could be based upon the presence of a precipitate in the sucrose but not in the glucose culture fluid. The test with 1.2 volumes was expected to have some limited use as a presumptive distinction between dextran and levan.

Test 4. Polyfructoside.—Chemical data on the presence of levan or levan-like material were obtained by the following procedure:—

The fluids were treated with 1.25 volumes of 95 per cent alcohol containing 10 per cent sodium acetate which would precipitate the major part of any dextran that was present but leave the levan material in solution. The mixtures were centrifuged, and after separation from the precipitates the supernatant fluids were treated with 5 volumes of 95 per cent alcohol, stored in the icebox overnight, and then centrifuged. The precipitates which would contain the levan, were carefully separated from the supernatant fluids which would contain any sucrose or fructose coming from the original culture fluid; as an additional precaution, the precipitates were dissolved in 10 per cent sodium acetate and the treatment with 5 volumes of alcohol was repeated. The final precipitates, after being carefully separated from the supernatant fluids, were heated in 0.8 N HCl for 15 minutes at 60°C, which conditions had sufficed for complete hydrolysis of the two purified levans we had previously studied (9).

The hydrolysates were treated with 9 volumes of absolute alcohol containing 1 per cent sodium acetate, stored in the icebox overnight, and then centrifuged. The sediments which would contain any traces of dextran not removed by previous procedures (and probably also the major part of any dextran degradation products) were removed. The supernatant fluids which would contain all of the fructose liberated by the acid hydrolysis of the levans, were analyzed for fructose by the Roe (21) procedure. The solutions of crystalline fructose used as standards were prepared in the same concentration of alcohol as that contained in the test solutions; the comparisons were made with a photoelectric colorimeter. The polyfructoside was calculated as nine-tenths of the fructose found in the Roe test, and is presented as milligrams per 1 cc.

Tests 5 to 10. Serological Reactions.—The bacterial culture fluids were neutralized, and then tested in dilutions ranging from 1:10 to 1:40,000. In the case of the fluids which showed no physical or chemical signs of dextran or levan formation, the serological tests were applied only to the 1:10 and 1:100 dilutions; all the others were tested in at least 3 different dilutions against all the serums. The text mixtures were observed for precipitation after incubation for 1 hour at 37°C and again after overnight storage in the icebox.

The antiserums used for the recognition of dextrans were from rabbits immunized with types 2, 20, and 12 pneumococci, and with sucrose-grown leuconostoc B bacteria; the type 20 sample was from a lot known not to react with levans (9) and the type 2 sample had been absorbed 3 times with R pneumococci. The antiserum used for recognition of the levans was from a rabbit immunized with a sucrose-grown spore-forming bacillus which we (9) have called bacillus N9; the chance of missing levans which might not react with that antiserum was well controlled by the preceding (No. 4) chemical test for polyfructosides. As a special control for the dextran reactions, a type 2 antipneumococcus serum that had been absorbed with sucrose-grown leuconostoc bacteria was included; this absorbed serum was known to have lost its original capacity to react with solutions of purified dextrans of leuconostoc bacteria although it retained its high capacity to agglutinate type 2 pneumococci and to react with the type 2 capsular S antigen (7). The serums were used in dilutions of from 1:10 to 1:15.

The antiserums were standardized by tests against solutions of purified dextrans prepared from sucrose broth cultures of group H streptococcus F90A and of leuconostoc A and B, and also against solutions of purified levans of *Streptococcus salivarius* S20B and of bacillus N9.* The types 2 and 20 and the leuconostoc antiserums reacted in about the same way with high dilutions (1:500,000 to 1:2 million) of all the dextrans; the type 12 antiserum reacted with high dilutions of the leuconostoc A dextran, and with low dilutions of the leuconostoc B dextran, but not at all with the streptococcal dextran; none of the four dextran-reactive antiserums reacted with either of the levans. The bacillus N9 antiserum reacted with 1:500,000 dilution of the *S. salivarius* levan and with 1:2,000,000 dilution of the bacillus N9 levan, which are the maximal dilutions detectable in tests with *S. salivarius* antiserum; the bacillus N9 antiserum did not react with any of the dextrans.

It should be noted that all of the antiserums utilized in the present study had been obtained by immunization with bacteria other than streptococci, which reduced to a minimum the possibility of confusion from streptococcal antigens other than dextran or levan.

Test 11. Reducing Sugars.—Neutralized samples of the fluids of the sucrose broth cultures were analyzed by the Hagedorn and Jensen method

(22). The values are calculated as glucose and are corrected for the slight reducing power of the uninoculated medium. This test served to supplement the others, because the accumulation of reducing sugar is a prominent feature in both dextran and levan formation. Streptococci that form neither dextran nor levan also may give reducing sugars, but the occurrence with them is infrequent in comparison to its regular occurrence with dextran- and levan-forming varieties.

Summary of the Tests.—The described series of tests take into account the general process of dextran and levan formation. These polysaccharides are synthesized through the action of bacterial enzymes (10, 11), by way of reactions which can be expressed as

$$n \text{ sucrose} \rightarrow \underset{\textit{dextran}}{(\text{glucose anhydride})_n} + n \text{ fructose}$$

$$n \text{ sucrose} \rightarrow \underset{\textit{levan}}{(\text{fructose anhydride})_n} + n \text{ glucose}$$

Tests 1 and 2 in this experiment are directed toward the detection of *either* of the polysaccharide polymers (dextran or levan); Nos. 4 to 10, toward the differentiation as well as the detection of the two polysaccharides; No. 11, toward the hexose "by-product" of the reactions.

*The chemical properties of the two levans and of the leuconostoc dextrans have been described (7, 9). The analysis of the dextran from the group H streptococcus which has not previously been published was as follows: Nitrogen: 0.10 per cent; ash (as Na): 0.17; $[\alpha]_D^{23°}$ +215° (c = 0.4 in 1 N NaOH), for the original product; and $[\alpha]_D^{23°}$ +53.4° (c = 2.0 in H_2O), and a reducing sugar content of 93 per cent for the acid hydrolysate; qualitative tests for fructose, pentoses, and uronic acid were negative. This streptococcus dextran was prepared by the method described (7) for the dextran of leuconostoc A; the yield (8 gm of the final purified product from 2 liters of culture) was much more abundant than would be obtained with most bacterial polysaccharides but was of the order of magnitude often gotten with bacterial dextrans.

(7) Sugg, J. Y., and Hehre, E. J., J. Immun. 1942, 43, 119. Hehre, E. J., Science, N.Y. 1941, 93, 237.
(9) Hehre, E. J., Genghof, D. S., and Neill, J. M., J. Immun. 1945, 51, 5.
(10) Hehre, E. J., Science, N.Y. 1941, 93, 237. Hehre, E. J., and Sugg, J. Y., J. exp. Med. 1942, 75, 339.
(11) Hestrin, S., Avineri-Shapiro, S., and Aschner, M., Biochem. J. 1943, 37, 450. Avineri-Shapiro, S., and Hestrin, S., Biochem. J. 1945, 39, 167. Hehre, E. J., Proc. Soc. exp. Biol. Med. 1945, 58, 219.
(21) Roe, J. H., J. biol. Chem. 1934, 107, 15.
(22) Hagedorn, H. C., and Jensen, B. N., Biochem. Z. 1923, 135, 46.

Levan—analysis of

J. R. Mattoon *et al.* (1955). Appl. Microbiol. 3, 321–333.

Details of procedures for the study of bacterial levans by analytical and physico-chemical methods are described.

Levan—*Corynebacterium laevaniformis*

F. F. Dias and J. V. Bhat (1964). Antonie van Leeuwenhoek, 30, 176–184.

For estimating levan in growing cultures the method of Hestrin *et al.* (1943) was employed.

S. Hestrin, S. Avineri-Shapiro and M. Aschner (1943). Biochem. J. 37, 450–456.

Levan—*Pseudomonas*

G. L. Bullock, S. F. Snieszko and C. E. Dunbar (1965). J. gen. Microbiol. 38, 1–7.

Levan production was determined on nutrient agar containing 4% (w/v) sucrose (Klinge, 1960). A thick slimy growth was considered positive for levan production. Cultures were incubated at 20–22°C for 1 week.

K. Klinge (1960). J. appl. Bact. 23, 442.

METHYL RED AND VOGES-PROSKAUER (ACETOIN) REACTIONS

Voges-Proskauer test—coliforms

G. C. Bunker, E. J. Tucker and H. W. Green (1918). J. Bact. 3, 493–498.

The use of Syracuse watchglasses, 1 ml of glucose-potassium phosphate broth – incubated for 48 hours at 30°C – and 0.5 ml of a 45 per cent solution of sodium hydroxide permits the development of a definite pink color reaction in this test in a maximum period of one and one-half hours after the addition of the latter solution.

The authors found that greater percentages of positive Voges-Proskauer reactions were obtained in Syracuse watchglasses than in test tubes. These glasses measure 5 cm in diameter and 0.8 cm deep. In them 1 ml of liquid spreads over an area of about 20 sq cm with an average depth of 0.05 cm.

Voges-Proskauer test

C. H. Werkman (1930). J. Bact. 20, 121–125.

The method to be described utilizes ferric chloride as a catalyst and avoids the disadvantages of the original V-P test or its modifications. The positive test is indicated by a deep copper coloration which appears at the surface after a few minutes and extends to the bottom of the tube. The color remains for several days and even after a week or longer it is clearly visible. The test may be positive for cultures of organisms belonging to the genus *Aerogenes* after three days' incubation at 30°C, but four-day cultures are recommended for standard procedure. Two drops of a 2 per cent solution of ferric chloride are added to 5 ml of the culture. Five milliliters of a 10 per cent solution of NaOH are now added, and the tube is shaken. The solution of ferric chloride must be added before the sodium hydroxide solution; addition after the alkali results in a marked flocculation.

In the positive tube two distinct layers of coloration may be observed. The upper is a copper color; the lower layer is the eosin pink of the standard V-P test. The copper color soon extends to the bottom of the tube. Acetylmethyl carbinol may be detected in three-day cultures when the standard technique fails to show its presence. Since the color of a positive test is that of bright copper, reading of the test is definite and certain.

Ferric chloride will not result in the oxidation of 2,3-butylene-glycol under the conditions of the test, and any organism which may have reduced the acetylmethyl carbinol to 2,3-butylene-glycol will not give a false Voges-Proskauer reaction.

Voges-Proskauer test—various

R. A. Q. O'Meara (1931). J. Path. Bact. 34, 401–406.

The composition of the culture medium arrived at is:

Sodium chloride	5.0 g
Magnesium sulfate	0.2 g
Ammonium phosphate $(NH_4)H_2PO_4$	1.0 g
Potassium phosphate K_2HPO_4	1.0 g
Sodium citrate (crystalline)	2.77 g
Glucose	5.0 g
Sodium fumarate	10.0 g
Distilled water to	1000 ml

The medium requires no adjustment of reaction. It is put up in wide tubes (¾ in. internal diameter), 5 ml being allowed to each tube, and may be sterilized in the autoclave. The object of putting a comparatively small amount of culture medium in so wide a tube is to ensure as liberal a supply of oxygen as possible, since increased aeration favours the production of acetoin. Organisms of the colon aerogenes type grow freely in the medium and may be tested for acetoin production as follows. A tube of the medium is inoculated with a loopful of a recent growth of the organism on solid medium. The culture is incubated at 37°C for 24 hours and is then tested for acetoin by adding a knife point of solid creatine (about 25 mg), followed by 5 ml of concentrated caustic soda. The caustic soda solution used should never be less than 40 per cent. After the addition of these reagents the tube is well agitated. If acetoin is present a red color appears in a couple of minutes, and, owing to the fact that the medium is colorless, may be appreciated very readily. Once the color has appeared it deepens rapidly on further agitation. The amount of acetoin capable of detection by this procedure is very minute. The presence of 1 part in 20,000 can be detected with great ease, and 1 part in 50,000 is just capable of recognition.

Voges-Proskauer test—*Escherichia coli*

G. A. Lindsey and C. M. Meckler (1932). J. Bact. 23, 115–121.

The authors used Werkman's $FeCl_3$ modification of the Voges-Proskauer reaction on 24 hour cultures at 37°C in standard glucose broth containing 5 g glucose, 5 g bacto-peptone, and 3 g bacto beef extract in 1000 ml water.

Two drops of a 1% solution of $FeCl_3$ were added to 5 ml of the culture, followed by 5 ml of 10% KOH.

C. H. Werkman (1930). J. Bact. 20, 121.

Voges-Proskauer test—mustiness in eggs

M. Levine and D. Q. Anderson (1932). J. Bact. 23, 337–347.

The production of acetylmethyl-carbinol was determined by making a qualitative test on glucose peptone cultures and in Clark and Lubs' medium. A 10 per cent KOH solution was added to an equal volume of a five-day old culture incubated at 25°C. The tubes were left exposed to the air and read at frequent intervals up to twenty-four hours. Tubes were considered as giving a positive Voges-Proskauer reaction if an eosin-pink coloration appeared.

Methyl red test

P. M. Kon (1932–1933). J. Dairy Res. 4, 206–212.

The medium for this test consisted of glucose (0.5 per cent), peptone (1.0 per cent) and water.

It was found that the examination was best made after 5 days' incubation at 37°C, since tests showed that some strains of *Bact. aerogenes,* if examined earlier, gave a positive reaction, i.e. the reversal to the negative reaction did not occur until the fifth day.

Voges-Proskauer test

P. M. Kon (1932–1933). J. Dairy Res. 4, 206–212.

O'Meara's (1931) modification, involving the addition of creatin, was used and found to be excellent, but instead of the Koser citrate medium plus glucose (0.5 per cent) used by O'Meara, it was found that good results could be secured by using the simple medium described for methyl red tests.

The tests were made after 2–5 days' incubation at 37°C, when the results were found to be clear-cut.

R. A. Q. O'Meara (1931). J. Path. Bact. 34, 401.

Voges-Proskauer test—various

M. M. Barritt (1936). J. Path. Bact. 42, 441–454.

It has been found advisable, as suggested by O'Meara, to use cultures grown in wide tubes. I therefore employed cultures in 6 x 5/8 in. tubes of glucose phosphate broth (Ministry of Health, 1934) made with Difco bactopeptone, which appears to be particularly suitable for this test. After 3 days' incubation at 37°C, 1 ml of culture was placed in a 6 x ¾ in. test tube, to which were added 0.6 ml of alpha-naphthol (5 per cent alcoholic solution) and 0.2 ml of KOH (40 per cent solution). Equally good results are obtainable with 0.5 ml of a 6 per cent solution of alpha-naphthol and 0.5 ml of a 16 per cent solution of KOH.

After the tubes have been shaken to mix the contents, positive reactions appear as a pink colour in 2–5 minutes. With well marked reactions this deepens to magenta or crimson in half an hour. Maximum colouration is reached within one hour when the stronger positives are deep crimson or ruby and the weaker are magenta or a rosy pink. Negatives are colourless, but very occasionally show a faint trace of pink, which is not usually noticeable unless the culture medium employed is colourless. After approximately one hour the negatives gradually develop a coppery colouration because of the action of KOH on the alpha-naphthol. The positives are readable for from 4 to 24 hours, depending on the strength of the reactions, and assume a coppery colouration on fading. The significance of the faint traces of pink observed in some cultures is uncertain. It is possible that they indicate minimal production of acetylmethyl carbinol under conditions which are not quite optimal. The question of the optimal conditions for the production of this substance is receiving attention. For the present such reactions are regarded as negative.

Addition of creatine slightly intensifies the reaction, which is also accelerated by heat.

Voges-Proskauer test—*Bacteroides*

K. H. Lewis and L. F. Rettger (1940). J. Bact. 40, 287–307.

Acetyl methyl carbinol formation in 4 day old glucose-cysteine broth* cultures, according to the method of Levine (1933).

M. Levine (1933). *An Introduction to Laboratory Technique in Bacteriology* 1st ed. MacMillan, New York.

*J. E. Weiss and L. F. Rettger (1934). J. Bact. 28, 501–521.

Voges-Proskauer test

M. Lemoigne, Mlle. B. Delaporte, and Mme. M. Croson (1944). Annls Inst. Pasteur, Paris 70, 65–79.

These authors discuss at length the determination of acetyl-methyl-carbinol qualitatively and quantitatively. The original Voges-Proskauer reaction and the O'Meara modification are cited for qualitative use.

Voges-Proskauer test—*Staphylococcus* and *Micrococcus*

Y. Abd-el-Malek and T. Gibson (1948). J. Dairy Res. 15, 249–260.

Cultures in peptone 1% and glucose 0.5% were incubated at 37°C and tested by Barritt's method after 3 days and, if negative, after 5 days.
M. M. Barritt (1936). J. Path. Bact. 42, 441.

Methyl red and **Voges-Proskauer tests**—*Proteus*

G. T. Cook (1948). J. Path. Bact. 60, 171–181.

Medium:

Glucose-phosphate peptone water (?)

Methods

Carried out in the usual way at 37°C. The inoculum was 2 or 3 drops of peptone water culture incubated 24 hours at 37°C.

Voges-Proskauer test—*Streptococcus*

P. F. Swartling (1951). J. Dairy Res. 18, 256–267.

A basal medium containing 1% tryptone (Difco), 1% beef extract (Lemco), 0.5% yeast extract (Lemco), 0.2% NaCl, 0.5% K_2HPO_4, 0.05% $MgSO_4$, and 0.05% $MnSO_4$ was steamed for half an hour and divided into two portions. To one of these was added 1% lactose; to the other, 1% lactose and 0.5% trisodium citrate. The media were distributed in 4 ml quantities in test tubes 24-mm wide and sterilized.

After inoculation (1 loopful of an 18-hr culture) the tubes were sloped to provide good aeration and incubated at 30°C. The Voges-Proskauer reaction was used to detect acetoin formation, and some strains were tested for diacetyl formation by den Herder's method (1947).
C. den Herder (1947). Ned. Melk- en Zuiveltijdschr.

Methyl red test—*Staphylococcus*

C. Shaw, J. M. Stitt and S. T. Cowan (1951). J. gen. Microbiol. 5, 1010–1023.

Cultures were grown in glucose phosphate peptone medium. The inoculum was incubated 5 days at 37°C, 10 days at 30°C or 14 days at 22°C.

Voges-Proskauer test

P. Hauduroy (1951). Techniques Bactériologiques. Masson et Cie, Éditeurs, Paris.

ACETYL-METHYL-CARBINOL

(Recherche de l'). — Préparer le milieu suivant:

Protéose-peptone	7 g
Glucose	5 g
NaCl	5 g
Eau distillée	1000 ml

On répartit en tubes de 18 mm de diamètre par 5 ml. Ensemencer. Placer à l'étuve à 32°C pendant 2, 4, 6, 20 jours. La présence d'acétyl-méthyl-carbinol est démontrée par l'apparition d'une coloration rouge, après addition de 5 ml d'une solution à 40 p. 100 de NaOH et de quelques milligrammes de créatine (O'Meara, R. A. Q., *J. Path. Bact.*, 1931, 34, 401).

(A. T. C. C.).

ACETYLMETHYLCARBINOL

(Recherche de l') (Lemoigne). — Ensemencer d'abord la souche étudiée dans du bouillon Vf glucosé à 1 p. 100, placé dans un Erlenmeyer.

Le milieu est désaéré par ébullition, l'ensemencement fait avec deux tubes de culture de 24 heures.

Lorsque la fermentation est terminée, ou que les germes se sont déposés (en moyenne après 7 jours), filtrer sur terre d'infusoire.

Le filtrat sert aux recherches.

On oxyde 100 cm^3 de culture au moyen de 5 cm^3 de perchlorure de fer pur. Distillation dans un ballon de 1 L en présence de talc pour éviter la mousse.

2 cm^3 de distillat sont recueillis dans un tube à essai contenant V gouttes de chlorhydrate d'hydroxylamine à 20 p. 100.

Réaction. — On tamponne avec XX gouttes d'acétate Na à 20 p. 100, puis on ajoute V gouttes de $NiCl_2$ à 10 p. 100. On porte à l'ébullition pendant 2 minutes. La présence d'acétylméthylcarbinol est caractérisée par un précipité rouge brique. En cas de doute, laisser reposer 2 heures.

(A. I. P.).

(Recherche de l'...) (Genre Colibacille) = Genre Colibacille.

(Recherche de l') (Réaction de Voges-Proskauer). — Préparer du bouillon phosphaté. Ensemencer avec une öse d'une culture en eau peptonée. Placer deux jours à 37°C. Ajouter une "pointe de canif" de créatin et 4 ml d'une solution à 40 p. 100 d'hydroxyde de sodium. Agiter pendant 2 à 5 minutes. Une réaction positive est révélée par l'apparition d'une couleur rose.

(W. I. C.).

Voges-Proskauer test—*Enterobacteriaceae*

P. Hauduroy (1951). Techniques Bactériologiques. Masson et Cie, Éditeurs, Paris.

Réaction de Voges-Proskauer et du rouge de méthyle.

K_2HPO_4 (Merck)	0.5 p. 100
Peptone de Witte	0.5 "
Glucose	0.5 "
Eau distillée	100 "

Mélanger K_2HPO_4, la peptone et l'eau distillée, chauffer jusqu'à ébullition, filtrer, autoclaver et ajouter 0.5 p. 100 de glucose.

Répartir en tubes étroits (1 cm de diamètre) sur une hauteur de 5 cm.

Porter à l'ébullition pendant 10 minutes.

Ensemencer et mettre à l'étuve à 37° pendant 4 jours.

Diviser alors le milieu en deux parties égales.

Dans l'une des parties, ajouter la moitié ou les deux tiers de son volume de NaOH à 20 p. 100. Remuer le tube et mélanger par renversement. Une fluorescence de couleur rougeâtre (comme de l'éosine) apparaissant pendant un couple d'heures indique une réaction positive.

Pour accélérer la réaction on peut ajouter quelques grains de créatinine.

A l'autre moitié du tube, ajouter quelques gouttes (2 à 3) d'une solution alcoolique à 0.25 p. 100 de rouge de méthyle.

Une réaction positive se manifeste par une couleur écarlate, une réaction faiblement positive par une couleur rouge-orange, une réaction négative par une couleur jaune.

Methyl red test—*Enterobacteriaceae*

P. Hauduroy (1951). Techniques Bactériologiques. Masson et Cie, Éditeurs, Paris.

ROUGE DE METHYLE.

(Réaction du). — Préparer du bouillon glucosé de la façon suivante:

Peptone Evan's	0.5
Phosphate monopotassique	0.5
Glucose	0.5
Eau	100

Ajuster à *p*H 7.5. Stériliser dans la vapeur fluente.

Ensemencer avec une öse un tube du milieu précédent en partant d'une culture en eau peptonée. Maintenir 3 jours à 37°C. Ajouter 5 gouttes d'une solution de rouge de méthyle contenant 0.04 p. 100 de rouge de méthyle. Une réaction positive est indiquée par l'apparition d'une coloration rouge, une réaction négative par l'apparition d'une coloration jaune, une réaction douteuse par l'apparition d'une coloration rose. *(W. I. C.).*

Voges-Proskauer test—*Staphylococcus*

C. Shaw, J. M. Stitt, and S. T. Cowan (1951). J. gen. Microbiol. 5, 1010–1023.

Two media were used, glucose phosphate peptone medium and a glucose peptone medium without salt or phosphate (Abd-el-Malek and Gibson, 1948). The test was carried out by adding 1 ml 40% (w/v) KOH to 4 ml culture, which was then incubated without a plug for 4 hr; a pink fluorescence was recorded as positive. Barritt's (1936) α-naphthol method was also used on 1 ml samples of each culture. In the recheck both the foregoing media and an additional medium—glucose peptone medium with NaCl but without phosphate (Smith, Gordon, and Clark, 1946)—were used; cultures were incubated at 30°C for 5 and 10 days, and the test was made by Batty-Smith's method (1941). Inocula were incubated

5 days at 37°C, 10 days at 30°C, and 14 days at 22°C.

Y. Abd-ed-Malek and T. Gibson (1948). J. Dairy Res. 15, 249.
M. M. Barritt (1936). J. Path. Bact. 42, 441.
C. G. Batty-Smith (1941). J. Hyg., Camb. 41, 521.
N. R. Smith, R. E. Gordon, and F. E. Clark (1946). Misc. Publ. U. S. Dept. Agric. No. 559.

Voges-Proskauer test—*Bacillus*

N. R. Smith, R. E. Gordon and F. E. Clark (1952). U. S. D. A. Agr. Monograph No. 16.

The medium recommended by the *Manual of Methods* (1951) and *Standard Methods* (1946) for the Voges-Proskauer reaction was modified after experiments had shown that phosphate inhibited the formation of acetoin by some of these organisms. The medium used contained 7 g of proteose-peptone, 5 g of glucose, 5 g of NaCl, and 1000 ml of distilled water. Five-milliliter portions in 18-mm tubes were inoculated and incubated at a suitable temperature for 2, 4 and 6 days or similar intervals, and occasionally for 10 and 20 days. The test was made by mixing the culture with an equal volume of a 40 percent solution of NaOH and adding 0.5 to 1.0 mg of creatine from a knife point. The appearance of a red color in the culture after it had stood 30 to 60 minutes demonstrated the presence of acetoin.

Amer. Pub. Hlth. Assoc. and Amer. Water Works Assoc. (1946). *Standard Methods for the Examination of Water and Sewage.* 9th ed., New York.
Soc. Amer. Bact. Comm. Bact. Tech. (1951). *Manual of Methods for Pure Culture Study of Bacteria.* Leaflet V. 12th ed., Geneva, N. Y.

Voges-Proskauer test—*Actinomyces, Corynebacterium*

H. Beerens (1953). Annls Inst. Pasteur, Paris 84, 1026.

Acétyl-méthyl-carbinol – *C. acnes* donne des traces d'acetoïne; les autres souches n'en produisent pas.

Cette dernière recherche a été effectuée en milieu V.F. glucose à 1 p. 100 selon la technique de Lemoigne.

Voges-Proskauer test—*Pseudomonas (Malleomyces) pseudomallei*

P. de Lajudie and E. R. Brygoo (1953). Annls Inst. Pasteur, Paris 84, 509–515.

Acétylméthylcarbinol: – L'acétoïne fut recherchée par la réaction de Voges-Proskauer, sur culture de trois jours en milieu de Clark-Lubs.

Methyl red test—micromethods

S. T. Cowan (1953). J. gen. Microbiol. 9, 101–109.

The author noted that adaptation of cells to glucose is essential but stated that the test is not really suitable for routine work.

Methyl red test

M. G. Jennens (1954). J. gen. Microbiol. 10, 121–126. (Originally by Clark and Lubs, (1915).

Defined medium RF glucose salt medium.

	% (w/v)
Glucose	0.5
NaCl	0.5
K_2HPO_4	0.5
KH_2PO_4	0.2
$(NH_4)H_2PO_4$	0.2
$MgSO_4 \cdot 7H_2O$	0.02
$MnSO_4 \cdot 4H_2O$	0.002
$FeCl_3$	0.00005

The author used 0.04% methyl red and, for the V.P. test, Batty-Smith's (1941) modification of Barritt's test.

Glucose phosphate peptone water: % (w/v)

Glucose	0.5
Peptone	0.5
K_2HPO_4	0.5
H_2O	1000 ml

Incubated 5 days at 30°C.

Variability was removed by use of the synthetic medium.

C. G. Batty-Smith (1941). J. Hyg., Camb. 41, 521.

Voges-Proskauer test—various

N.C.T.C. Methods 1954

Media:		
(a)	Glucose phosphate peptone	) All tubed in
(b)	Smith's glucose peptone	) 1.5 ml vols in
(c)	Gibson's glucose peptone	) 5½ × ½″ tubes.
(d)	Richard's glucose salt broth	)

Incubated cultures at 30°C; test after 5 days.

Method 1. Add 1 ml 40% KOH, shake well and incubate (without a plug) at 37°C for 4 hours.

Method 2. Add 0.6 ml 5% alcoholic α-naphthol solution, shake, and add 0.2 ml 40% KOH. Shake, slope tube, and read after 1 hour.

M. M. Barritt (1936). J. Path. Bact. 42, 441.

Method 3. Add 2 drops of 1% solution of creatine and 1 ml 40% KOH. Shake, slope and read after 4 hours.

R. A. Q. O'Meara (1931). J. Path. Bact. 34, 401.

Method 4. Add 1 ml alkaline copper sulfate [1.0 g $CuSO_4 \cdot 5H_2O$ dissolved in 40 ml concentrated NH_4OH (sp. gr. 0.88) added to 960 ml 10% KOH]. Shake; read after 1 hour.

E. Leifson (1932). J. Bact. 23, 353.

Method 5. Add 2 drops of 1% solution of creatine and 1 drop 2% ferric chloride; shake. Then add 0.6 ml 5% alcoholic α-naphthol, shake, and add 0.2 ml 40% KOH. Shake, slope, and read after 1 hour.
C. G. Batty-Smith (1941). J. Hyg., Camb. 41, 521.

Methyl red test

N.C.T.C. Methods 1954.

Medium: glucose phosphate peptone. Incubate cultures at 30°C and test after 5 and 10 days.

Method: Add methyl red solution (0.1 g dissolved in 100 ml absolute alcohol, and diluted to 250 ml with distilled water).

Red colour = +
Pale pink or orange = ±
Yellow colour = –
Controls
Positive = 86
Negative = 5,936

Methyl red test—*Proteus* and *Providencia*

C. Shaw and P. H. Clarke (1955). J. gen. Microbiol. 13, 155–161.

Cultures were grown in glucose phosphate peptone (Evans') medium and incubated for 5 days at 30°C.

Voges-Proskauer test—*Proteus* and *Providencia*

C. Shaw and P. H. Clarke (1955). J. gen. Microbiol. 13, 155–161.

Three media, each containing Evans' peptone, were used. Glucose phosphate peptone medium; glucose peptone medium with NaCl but without phosphate (Smith, Gordon and Clark, 1946); and glucose peptone medium without salt and phosphate (Abd-el-Malek and Gibson, 1948). Cultures were incubated at 30°C for 5 days and tested for the presence of acetoin by Batty-Smith's (1941) modification of Barritt's (1936) alpha-naphthol method.

Y. Abd-el-Malek and T. Gibson (1948). J. Dairy Res. 15, 249.
M. M. Barritt (1936). J. Path. Bact. 42, 441.
C. G. Batty-Smith (1941). J. Hyg., Camb. 41, 521.
N. R. Smith, R. E. Gordon and F. E. Clark (1946). Misc. Publ. U. S. Dept. Agric. No. 559.

Voges-Proskauer test—*Bacillus*

E. Grinsted and L. F. L. Clegg (1955). J. Dairy Res. 22, 178–190.

The medium used for the production of acetylmethyl carbinol (AMC) was that of Smith *et al.* (1946), but testing was done with the addition of alpha-naphthol and KOH to the incubated culture fluid.* One ml of the culture was placed in an 18 mm test tube and 0.5 ml of a 6% solution of alpha-naphthol in absolute alcohol and 0.5 ml of a 16% aqueous so-

lution of KOH were added. The misture was shaken vigorously to aerate it thoroughly and was also warmed slightly. The depth of colour was increased by allowing the tubes to stand on the bench for about 10 min.
*M. M. Barritt (1936). J. Path. Bact. 42, 441.
N. R. Smith, R. E. Gordon and F. E. Clark (1946). Misc. Publ. U. S. Dept. Agric. No. 559.

Voges-Proskauer test—*Lactobacillus*

D. M. Wheater (1955). J. gen. Microbiol. 12, 133–139.

Neopeptone	1.5%
glucose	2.0%
NaCl	0.5%
Tween 80	0.1%
Yeastrel	0.3% pH 6.8

Tubed in 5 ml amounts. Inoculated with 0.05 ml culture. Incubated at 30°C for 4 days.

Barritt's (1936) test used.
M. M. Barritt (1936). J. Path. Bact. 42, 441.

Voges-Proskauer test—*Lactobacillus*

G. H. G. Davis (1955). J. gen. Microbiol. 13, 481–493.

Acetylmethyl carbinol (acetoin) production from glucose

Medium: peptone, 0.5% (w/v); yeast extract, 0.3% (w/v); sodium acetate, 0.5% (w/v); glucose, 2% (w/v); Tween 80, 0.1% (v/v); Salts A*, 0.5% (v/v); Salts B*, 0.5% (v/v); at pH 6.8, dispensed in 3 ml quantities into bijou bottles and sterilized by steaming for 20 min on 3 consecutive days.

Tests were inoculated with 1 to 2 drops, incubated for 10 days, and tested by transferring the culture to a clean test tube, adding a few granules of creatine, 0.6 ml of 5% α-naphthol solution in absolute ethanol, and 0.2 ml of 40% potassium hydroxide solution (Barritt, 1936). Positive reactions varied in the intensity of the red colour produced with these reagents.
*See A. C. Hayward (1957). J. gen. Microbiol. 16, 9–15.
M. M. Barritt (1936). J. Path. Bact. 42, 441.

Methyl red and **Voges-Proskauer tests**—enterobacteria

J. G. Heyl (1957). Antonie van Leeuwenhoek 23, 33–58.

Bacto peptone	5 g
Potassium phosphate	5 g
Distilled water	975 ml

Boil to dissolve the ingredients, filter through paper and sterilize the filtrate for 20 minutes at 115°C. After sterilization 25 ml of a sterilized 25% solution of dextrose are added.

Tubes filled with about 10 ml of this substrate are inoculated and incubated for 3 days at 37°C. For the Voges-Proskauer test (Barritt's modification) 2 ml of the culture are placed in a second tube. To these 2 ml culture, 1.2 ml of a 5% alpha-naphthol solution in ethanol 96% and 0.4 ml of a 40% KOH solution are added. To the rest of the culture, 10 drops of a 0.25% alcoholic methyl red solution are added.

Methyl red and **Voges-Proskauer tests**

Society of American Bacteriologists *Manual of Microbiological Methoas* McGraw-Hill Book Co. Inc., New York 1957.

Prepare basal medium for these tests as follows: peptone, 7.0 g; glucose, 5.0 g; K_2HPO_4, 5.0 g; since all peptones are not suitable, use for peptone an enzymatic digest of casein.

Special tests for cleavage of glucose are commonly made in the differentiation of the organisms of the colon-aerogenes group. The medium ordinarily employed is as follows: 5 g of proteose peptone (Difco, Witte's, or some brand recognized as equivalent), 5 g of cp glucose, and 5 g of K_2HPO_4 in 1000 ml of distilled water. The dry potassium phosphate should be tested before using in dilute solution to see that it gives a distinct pink colour with phenolphthalein. According to Smith (1940), the K_2HPO_4 in this medium should be replaced with the same amount of NaCl, if the tests are to be carried out on aerobic spore formers. Tubes should be filled with 5 ml each, and each culture should be inoculated into duplicate (or triplicate) tubes for each of the two tests. Incubation should be at optimum temperature of the organism under investigation, and tubes should be incubated 2–7 days, according to the rate of growth of the organism in question. Although the same medium is used for both the methyl red and Voges-Proskauer tests, they must be performed in separate tubes. The latter test depends on the production of acetyl-methyl-carbinol from the glucose. Fabrizio and Weaver (1947) show the possibility of a rapid test for the production of this compound; Cowan (1953) agrees as to its practicability. A similar microtest for the methyl red reaction proves more difficult.

A positive methyl red reaction is regarded as being present when the culture is sufficiently acid to turn the methyl red (0.1 g dissolved in 300 ml of 95 per cent ethyl alcohol and diluted to 500 ml with distilled water) a distinct red; a yellow colour with the methyl red indicator is regarded as a negative reaction, while intermediate shades should be considered doubtful.

For the Voges-Proskauer reaction, according to the *Standard Methods* of the APHA (1946) to 1 ml of culture add 0.6 ml of 5 per cent α-naphthol in absolute alcohol and 0.2 ml of 40 per cent KOH. It is important to shake for about 5 sec after addition of each reagent. The development of a crimson to ruby colour in the mixture from 2 to 4 hr after addition of the reagents constitutes a positive test for acetyl-methyl-carbinol. Results should be read not later than 4 hr after addition of the reagents.

Various other tests have been suggested for this reaction, both to obtain results more quickly and because some organisms apparently give different results with different tests. In any case, weakly positive reactions may be obscured by the colour of the reagent. A procedure which has given excellent results with many thousand cultures run by a member of the committee (CAS) is the creatine test of O'Meara, as modified by Levine, Epstein and Vaughn (1934). In this procedure the test reagent added to the culture is 0.3 per cent creatine in 40 per cent KOH. This reagent deteriorates rapidly at temperatures over 50°C but may be kept 2 weeks at room temperature (22–25°C) or for 4 to 6 weeks in a refrigerator.

A recent modification by Coblentz (1943) is similar to the APHA method but uses a massive inoculum in broth from an infusion-agar slant culture, followed by incubation of the broth for 6 hr. Also, the 40 per cent KOH has 0.3 per cent of creatine added to it to intensify the reaction. After addition of the reagent (α-naphthol and KOH-creatine) the culture is shaken vigorously for 1 min; a positive reaction is characterized by an intense rose-pink color developing in a few seconds to 10 min.

The microtest of Fabrizio and Weaver (1947) calls for inoculation with a loopful of growth from a 6 to 12 hr infusion agar slant into 0.5 ml of infusion medium with 1 per cent trypticase and 0.5 per cent NaCl, placed in small tubes and preheated in a 30°C water bath. It is then incubated at 30°C in a water bath for 90 min and tested for acetyl-methyl-carbinol by the same method as given above, except that smaller quantities of the reagent (0.15 and 0.05 ml, respectively) are added.

American Public Health Association (1946). *Standard Methods for the Examination of Water and Sewage* 9th ed. publ. by the Assoc., N. Y.
J. M. Coblentz (1943). Am. J. publ. Hlth 33, 815.
S. T. Cowan (1953a). J. gen. Microbiol. 8, 391–396.
(1953b). J. gen. Microbiol. 9, 101–109.
A. Fabrizio and R. H. Weaver (1947). J. Bact. 54, 69.
M. Levine, S. S. Epstein and R. H. Vaughn (1934). Am. J. publ. Hlth 24, 505–510.
N. R. Smith (1940). J. Bact. 39, 575.

Voges-Proskauer test—*Acetobacter.*

J. de Ley (1959). J. gen. Microbiol. 21, 352–365.

The ordinary lactate medium contained, per litre of tap water:

0.067 *M* KH_2PO_4	125	ml
0.067 *M* Na_2HPO_4	125	ml
NH_4Cl	2	g
Yeast extract (Difco)	2	g
$MgSO_4 \cdot 7H_2O$	0.25	g
DL-lactic acid	10	g
Bromcresol purple	2	mg

Adjusted to pH 6.5 with NH_4OH.

Acetoin was determined by the modified Voges-Proskauer (VP) reaction of Westerfeld (1945).

W. W. Westerfeld (1945). J. biol. Chem. 161, 495.

Voges-Proskauer test—*Clostridium*

M. E. Brooks and H. B. G. Epps (1959). J. gen. Microbiol. 21, 144–155.

Grown for 24 and 48 hours on glucose phosphate broth and tested by O'Meara's method.

Methyl red and **Voges-Proskauer tests**—*Aeromonas*

J. P. Stevenson (1959). J. gen. Microbiol. 21, 366–370.

Cultures were incubated in glucose phosphate broth for 5 days at both 30°C and 37°C, and tested according to the details given by Vaughn, Mitchell and Levine (1939) and by Levine (1941). There was no difference in the results obtained at the different temperatures.

M. Levine (1941). Am. J. publ. Hlth 31, 351.

R. Vaughn, N. B. Mitchell and M. Levine (1939). J. Am. Wat. Wks Ass. 31, 993.

Methyl red and **Voges-Proskauer tests**—*Pseudomonas fluorescens*

M. E. Rhodes (1959). J. gen. Microbiol. 21, 221–263.

Fouad and Richards (1953) recommended a glucose-ammonium-inorganic salts medium for examining the VP and MR reactions by members of the *Enterobacteriaceae;* 7 ml volumes of this medium were used here. The VP and MR tests were carried out on 1 ml samples of the cultures after 5 and 7 days of incubation. The O'Meara (1931) method of testing for acetoin was used.

M. T. A. Fouad and T. Richards (1953). Proc. Soc. appl. Bact. 16, 35.

R. A. Q. O'Meara (1931). J. Path. Bact. 34, 401.

Methyl red test

V. B. D. Skerman, *A Guide to the Identification of the Genera of Bacteria.* 2nd ed. Williams and Wilkins Book Co., Baltimore, 1967.

This test depends on the ability of the organism to produce acid from glucose in amounts sufficient to reduce the pH to 4.2 or less and to maintain this low pH for at least 4 days. Some methyl red-negative organisms produce sufficient acid to lower the pH to 4.2 but subsequently metabolize the acid produced and give rise to neutral by-products within the 4-day period. The time interval at which the test is performed is therefore important. Four days is adequate for those enteric organisms which grow at 37°C, and for which this test has been extensively used. No comparable data are available for the more psychrophilic bacteria, but for comparative purposes a similar time interval is recommended at the lower optimal temperature.

In assessing the results of the test in relation to the Voges-Proskauer reaction, organisms which, at the 4-day interval, are both methyl red-positive and Voges-Proskauer-positive should be incubated for a longer period to determine whether complete conversion of the acid has taken place.

Methyl red Solution

Methyl red	0.1 g
Ethanol (95%)	300 ml
Distilled water	200 ml

Test

Add a few drops of the methyl red solution to the 4-day culture in glucose-phosphate-peptone water. A red color is positive and yellow is negative.

Note: It has been suggested that the phosphate should be omitted from the medium for sporing bacilli, because the buffering action of the phosphate prevents the development of the low pH. Nothing is gained by doing this. Mere acid production by species of the genus *Bacillus* is detectable in the usual sugar fermentation tests. To eliminate the phosphate is, in fact, to alter the conditions of the test as normally applied, and the two methods are no longer comparable.

W. M. Clarke and H. A. Lubs (1915). J. infect. Dis. 17, 160.

Voges-Proskauer test

V. B. D. Skerman, *A Guide to the Identification of the Genera of Bacteria.* 2nd ed. Williams and Wilkins Book Co., Baltimore, 1967.

The test depends on the ability of the organism to produce acid from glucose and subsequently to convert it to acetylmethylcarbinol or 2,3-butylene glycol, both neutral substances. On the addition of alkali, followed by vigorous shaking, both substances are oxidized to diacetyl, which reacts with the guanidine nucleus of arginine in peptone to produce a pink color. Creatine is added in the O'Meara modification of the test to provide an added source of the guanidine nucleus and thus accelerate the test.

A positive test is dependent primarily on the ability of the organism to bring about the conversion of the acid to acetoin and secondarily on the time of incubation. All Voges-Proskauer-positive organisms produce the acid first and hence give a positive methyl red test in the early stages of incubation. If the incubation period is too short, the Voges-Proskauer test may be negative or both it and the methyl red tests positive. On longer incubation the methyl red test becomes negative and the Voges-Proskauer strongly positive.

Notes on the time of incubation given under the methyl red test apply equally well here.

Method: Grow the organism in glucose-phosphate-peptone water for 4 days.

Original Test: Add 1 ml of a 10 per cent solution of KOH to the culture and shake vigorously with air. The color develops slowly, and the test should be read after 18 to 24 hours. A pink fluorescence is positive; no coloration, negative.

O'Meara Modification: Add a knife point of creatine to the culture, followed by 5 ml of 40 per cent NaOH. Shake the tube thoroughly. A pink color usually appears in about 2 minutes if positive; the development of the color may be delayed, however.

Another method in common use is *Barritt's Test.*

To the 3 ml culture add 1 ml of a fresh 10 per cent alcoholic solution of α-naphthol, and 1.0 ml of a 20 per cent aqueous solution of potassium hydroxide. Shake well. A bright cherry red color appears after 5 to 15 minutes, sometimes longer, and will fade after a few hours if acetoin is present in the culture.

The O'Meara test is recommended for general use. If the Barritt method is employed, the O'Meara method should be used for comparison and both tests reported.

R. A. Q. O'Meara (1931). J. Path. Bact. 34, 401.
M. M. Barritt (1936). J. Path. Bact. 42, 441.

Methyl red and **Voges-Proskauer tests**—*Pasteurella septica*

J. M. Talbot and P. H. A. Sneath (1960). J. gen. Microbiol. 22, 303–311.

Tested in the glucose-phosphate medium given in Topley and Wilson's *Principles,* 4th ed. (1955, p. 450) after growth at 37°C for 2 days. O'Meara's modification of the Voges-Proskauer test was used as described there.

Voges-Proskauer test—*Fusobacterium*

A. C. Baird-Parker (1960). J. gen. Microbiol. 22, 458–469.

Medium (%, w/v): proteose peptone no. 3, 1.0; Oxoid yeast extract, 0.1; Lab. Lemco, 0.3; L-cysteine HCl, 0.05; glucose, 2.0; disodium hydrogen phosphate, 0.5. Sterile glucose solution was added aseptically after heat sterilization of the other constitutents. Cultures were incubated for 8 days and acetoin tested for by the method of Barritt (1936).

M. M. Barritt (1936). J. Path. Bact. 42, 441.

Voges-Proskauer test—various

M. Fulton, D. Halkias and D. A. Yarashus (1960). Appl. Microbiol. 8, 361–362.

The paper describes the purification of 1-naphthol by steam distillation to yield white needle-like crystals in the distillate, for use in the Voges-Proskauer test. The purified 1-naphthol reagent is more stable and is free from interfering color.

Voges-Proskauer test—*Xanthomonas*

A. C. Hayward and W. Hodgkiss (1961). J. gen. Microbiol. 26, 133–140.

This was made at 6 days in standard glucose phosphate medium, using Barritt's modification (Mackie and McCartney, 1960).

T. J. Mackie and J. E. McCartney (1960). *Handbook of Practical Bacteriology,* 10th ed. Edinburgh: Livingstone.

Voges-Proskauer test—*Pediococcus*

H. L. Günther and H. R. White (1961). J. gen. Microbiol. 26, 185–197.

Tests were carried out in the medium of Swartling (1951), modified in one series of experiments by the substitution of glucose for lactose. Cultures were incubated for 6 days and tested for acetylmethylcarbinol by Barritt's (1936) modification of the Voges-Proskauer test.

Maintenance of stock cultures and methods of cultivation. For maintenance of stock cultures, preparation of inocula, and in all experimental work, "Oxoid" tomato juice (TJ) broth or tomato juice (TJ) agar, adjusted to pH 6.6, was used unless otherwise stated. The following were exceptions to this rule: for strain Tc. 1, sodium chloride (5%, w/v) was added to the medium; for the aerococci, glucose Lemco broth (Shattock and Hirsch, 1947) or glucose yeast extract (GY) agar (containing, as %, w/v, peptone, 1.0; Yeastrel, 0.3; glucose, 1.0; NaCl, 0.25; agar, 1.0; pH 7.4) was used.

Cultures were incubated aerobically at 30°C, with specified exceptions.

M. M. Barritt (1936). J. Path. Bact. 42, 441.

P. M. F. Shattock and A. Hirsch (1947). J. Path. Bact. 59, 495.

P. F. Swartling (1951). J. Dairy Res. 18, 256.

Voges-Proskauer test—*Vibrio*

G. H. G. Davis and R. W. A. Park (1962). J. gen. Microbiol. 27, 101–119.

Acetylmethylcarbinol (acetoin) production and final pH from 1% (w/v) glucose in the basal medium, initial pH 7.0. Tested after 4 days, using pH test papers and Barritt's (1936) method for acetylmethylcarbinol.

M. M. Barritt (1936). J. Path. Bact. 42, 441.

Voges-Proskauer test—*Vibrio*

G. H. G. Davis and R. W. A. Park (1962). J. gen. Microbiol. 27, 101–119.

Acetylmethylcarbinol production from glucose, using Barritt's (1936) method, but incubated for 5 days.

M. M. Barritt (1936). J. Path. Bact. 42, 441.

Methyl red and **Voges-Proskauer tests**—*Actinobacillus* and *Haemophilus*

E. O. King and H. W. Tatum (1962). *Actinobacillus actinomycetemcomitans* and *Haemophilus aphrophilus.* J. infect. Dis. 111, 85–94.

Difco MR–VP broth was used for determination of methyl red and Voges-Proskauer reactions.

Voges-Proskauer test

W. H. Ewing (1962). *Enterobacteriaceae. Biochemical Methods for Group Differentiation.* U.S. Department of Health, Education and Welfare, Public Health Service Publication, No. 734 (revised).

Test for the production of acetylmethyl carbinol (acetoin) and 2,3-butyleneglycol.

Buffered peptone glucose broth. The same medium used for the MR reaction (see p. 562) may be used for the VP test.

Inoculation: Inoculate lightly from a young agar slant culture.

Incubation: Standard methods generally recommend incubation at 30°C for 48 hours. However, incubation at 37°C for 48 hours is satisfactory in most instances. If equivocal results are obtained under these circumstances the tests should be repeated with cultures incubated at 25°C.

Test Reagent (O'Meara, modified)

Potassium hydroxide	40 g
Creatine	0.3 g
Distilled water	100 ml

Dissolve the alkali in distilled water and add creatine. Use the reagent in the proportion of 1 ml per ml of culture. Tests are left at room temperature and final readings are made after 4 hours. Tests may also be incubated at 37°C for 4 hours or at 48°C to 50°C for 2 hours. Aerate by shaking the tubes. A positive VP test is indicated by the development of a delicate eosin-pink color. The O'Meara reagent for V-P tests should be prepared frequently and should be refrigerated when not in use. If refrigerated the reagent may be used for 2 to 3 weeks but deteriorates rapidly thereafter (Levine, *et al.*, 1934. Suassuna *et al.*, 1961).

Alternate Test Reagent (Barritt).

Add 0.6 ml of 5% alpha-naphthol in absolute alcohol and 0.2 ml of 40% KOH to 1 ml of culture. Shake well after the addition of each reagent. Positive reactions occur at once or within 5 minutes and are indicated by the production of a red color. The development of a copper

color in some tubes should be disregarded.
M. Levine *et al.* (1934). Am. J. publ. Hlth 24, 505.
L. Suassuna *et al.* (1961). Publ. Hlth Lab. 19, No. 3; 38, No. 4, 67.

Methyl red and **Voges-Proskauer tests**—*Escherichia aurescens*

H. Leclerc (1962). Annls Inst. Pasteur, Paris 102, 726–741.

Test au rouge de méthyl et acétoïne.

Nous avons utilisé le milieu synthétique de Fouad et Richards [12] ensemencé et incube trois, puis cinq jours, à 30°. L'acétoïne est mise en évidence par la réaction d'O'Meara selon la technique d'Ewing [9].

[9] Ewing (W. H.) *Enterobacteriaceae, biochemical methods for group differentiation.* Public Health Service Publication, no. 734, U.S. Government Printing Office, 1960.

[12] Fouad (M. T. A.) and Richards (T.). *Proc. Soc. appl. Bact.,* 1953, 16, 35.

Methyl red test

W. H. Ewing (1962). *Enterobacteriaceae. Biochemical Methods for Group Differentiation.* U.S. Department of Health, Education and Welfare, Public Health Service Publication, No. 734 (revised).

Buffered peptone glucose broth. Several modifications of the Clark and Lubs formula are available. The two that follow are known to be satisfactory. The medium selected for MR tests may also be used for Voges-Proskauer (VP) reactions.

MR – VP medium (BBL)

Dipotassium phosphate	5 g
Polypeptone	7 g
Glucose	5 g
Distilled water	1000 ml

Suspend ingredients in water and heat slightly to dissolve them. Tube and sterilize at 118° to 121°C for 15 minutes.

MR – VP medium (Difco)

Buffered peptone	7 g
Glucose	5 g
Dipotassium phosphate	5 g
Distilled water	1000 ml

Tube and sterilize at 121°C for 10 minutes.

Inoculation: Inoculate lightly from a young agar slant culture.

Incubation: Standard methods generally recommend 5 days' incubation at 30°C. However, incubation at 37°C for 48 hours is sufficient for the determination of the MR reaction of the majority of cultures. Tests should not be made with cultures incubated less than 48 hours. If the results of tests incubated at 37°C for 48 hours are equivocal, the tests should be repeated with cultures that have been incubated for 4 or 5 days. In such instances duplicate tests should be incubated at 25°C.

Test Reagent

Methyl red	0.1 g
Ethyl alcohol (95 to 96%)	300 ml

Dissolve dye in the alcohol and then add sufficient distilled water to make 500 ml. Use 5 or 6 drops of reagent per 5 ml of culture. Reactions are read immediately. Positive tests are bright red, weakly positive tests are red-orange and negative tests are yellow.

Methyl red and **Voges-Proskauer tests**—*Haemophilus*

P. N. Edmunds (1962). J. Path. Bact. 83, 411–422.

Tests in conventional media supplemented with plasma were unsatisfactory owing to failure of growth. It was found necessary to use a massive inoculum of viable bacilli, sufficient to give the reaction without further growth. Twelve maltose digest broths (prepared as for fermentation tests) were inoculated with each strain. Heavy growth occurred in 48 hr at 37°C. The cultures were pooled and centrifuged, the deposit was resuspended in a small amount of broth and two drops were added to 0.5 ml of glucose phosphate medium (Cruickshank *et al.* 1960) the resulting suspension was equivalent to Brown's (1919–1920) opacity standard no. 5. The tests were incubated at 37°C for 4 days for the Voges-Proskauer and methyl red tests. The results were observed as recommended by Kauffmann (1954).

H. C. Brown (1919–1920). Indian J. med. Res. 7, 238.
R. Cruickshank *et al.* (1960). Mackie and McCartney *Handbook of Bacteriology* 10th ed., p. 218. Edinburgh: Livingstone.
F. Kauffmann (1954). *Enterobacteriaceae,* 2nd ed. Einar Munksgaard.

Methyl red and **Voges-Proskauer tests**—*Vibrio*

G. H. G. Davis and R. W. A. Park (1962). J. gen. Microbiol. 27, 101–119.

Tested after 5 days' incubation (Mackie and McCartney, 1953).

T. J. Mackie and J. E. McCartney (1953). *Handbook of Practical Bacteriology* 9th ed. – Edinburgh: Livingstone.

Methyl red and **Voges-Proskauer tests**—*Aeromonas*

I. W. Smith (1963). J. gen. Microbiol. 33, 263–274.

Methyl red and Voges-Proskauer tests were performed in 7-day glucose phosphate peptone water cultures. Acetylmethyl carbinol was detected by Barritt's modification.

Methyl red test—*Pseudomonas odorans*

I. Málek, M. Radochová and O. Lysenko (1963). J. gen. Microbiol. 33, 349–355.

Medium used: (%, w/v)

Bacto-peptone	0.5

K_2HPO_4	0.5	
Glucose	0.25	or galactose 0.25
Initial pH	7.0	

Results read during 14 days – 28°C.

Voges-Proskauer test—*Pseudomonas odorans*

I. Málek, M. Radochová and O. Lysenko (1963). J. gen. Microbiol. 33, 349–355.

Medium used: (%, w/v)

Bacto-peptone	0.5
K_2HPO_4	0.5
Glucose or galactose	0.25
Initial pH	7.0

Results read during 14 days and acetoin detected according to Barritt – 28°C.

Voges-Proskauer test—*Staphylococcus aureus*

R. L. Brown and J. B. Evans (1963). J. Bact. 85, 1409–1412.

Acetoin was determined in the glucose broth cultures after 3 days of incubation. A 1-ml amount of the culture was added to a solution containing 0.5 ml of α-naphthol reagent, 0.5 ml of 40% KOH, and a few crystals of creatine. Development of red color within a few minutes indicated a positive test.

Methyl red and **Voges-Proskauer test**

G. S. Wilson and A. A. Miles (1964). Topley and Wilson's *Principles of Bacteriology and Immunity* – 5th edition – Arnold.

Methyl red test

Tested by adding 5 drops of a 0.04 per cent solution of methyl red to a culture in glucose phosphate medium (peptone, 0.5 g; K_2HPO_4, 0.5 g; glucose, 0.5 g; water, 100 ml; pH 7.5). Cultures grown for 5 days at 30°C or 3 days at 37°C.

Red color = positive; Yellow color = negative.

Voges-Proskauer test

Tested by adding 1 ml of a 10 per cent solution of KOH to a glucose phosphate culture grown for 5 days at 30°C. Some workers recommend 2 days at 30°C. According to Smith, Gordon and Clark (1946), sodium chloride is preferable to K_2HPO_4 in the medium. The color develops slowly, and the test should be read after 18 to 24 hours.

Pink fluorescence = positive; no coloration = negative.

A higher proportion of positive reactions is obtained by the use of O'Meara's (1931) modification. A knife point of creatine is added to the culture, followed by 5 ml of 40 per cent sodium hydroxide. The tube is shaken thoroughly for 2 to 5 minutes. A positive reaction is characterized by the appearance of a pink color within about 2 minutes, unaccompanied by fluorescence; the development of the color may, however, be delayed for an hour or longer.

An even more sensitive test for acetylmethylcarbinol is that described by Barritt (1936). It consists in adding 0.6 ml of a 5 per cent ethanolic solution of α-naphthol and 0.2 ml of 40 per cent KOH solution to 1 ml of culture. In a positive reaction a pink color appears in 2–5 minutes, deepening to magenta or crimson in half an hour. In a negative reaction the mixture remains colorless for an hour or so, when it may become copper colored owing to the action of KOH on the α-naphthol. Traces of pink coloration are best neglected.

For taxonomic purposes the O'Meara method is preferred (Suassuna *et al.* 1961). For the chemistry of the reaction, see Eddy (1961).

M. M. Barritt (1936). J. Path. Bact. 42, 441.
B. P. Eddy (1961). J. appl. Bact. 24, 27.
R. A. Q. O'Meara (1931). J. Path. Bact. 34, 401.
N. R. Smith, R. E. Gordon and F. E. Clark (1946). U.S. Dept. Agric., Wash., Misc. Publ. No. 559.
I. Suassuna, I. R. Suassuna and W. H. Ewing (1961). Publ. Hlth Lab. 19, 67.

Voges-Proskauer test—*Enterobacteriaceae*

M. A. Benjaminson, B. C. de Guzman and A. J. Weil (1964). J. Bact. 87, 234–235.

Method A. Tubes containing 0.5 ml amounts of MR–VP (BBL) broth were prepared and sterilized as directed by the manufacturer. After inoculation, tubes were incubated at 37°C overnight. Next, three drops of α-naphthol (5% in 95% alcohol) plus one drop of 40% KOH were added. Tubes were read for definite pink or reddish coloration after standing for 5 min. (Shaking did not markedly improve the development of color. Color was stable for several hours.)

As mentioned before, this procedure loses reliability when incubation of the MR–VP medium is extended beyond 24 hr. If circumstances make it impractical to test within this time limit, the standard VP test should be used, or the following alternative method, which is positive with regularity after 24 hr of incubation and remains positive for 3 days, can be employed. It can also be used to advantage if no MR–VP medium is available.

Method B. After placing two drops of a 0.5% hydrous solution of creatine in a small test tube or on a spot plate, one heavy loopful of culture from the *acid* part of TSI medium was added. Next, three drops of α-naphthol and two drops of KOH (of solutions specified in method A) were added. After shaking, waiting 5 min, and shaking again, results were read. A positive test was pink to violet in color. No fading was observed for several hours. Bacteria from TSI slants incubated for up to 72 hr could be used, provided that material was taken from the acid part of the slant.

The two techniques were compared with the standard procedures of the VP test with both the Barritt and the O'Meara reagents on 567 strains of *Enterobacteriaceae,* including 169 *Escherichia,* 251 of the *Klebsiella-Aerobacter-Serratia* group, and 96 *Proteus.* Agreement was excellent, with the exception of *Proteus.* In this group we experienced large differences in the results with the standard procedures employing the Barritt and O'Meara reagents, and also between the standard procedures and the new modifications. As noted by Suassuna *et al.,* appearance of acetoin in cultures of *Proteus,* particularly of *P. mirabilis,* may vary greatly with conditions of medium and time of incubation. Accordingly, the VP test is of little taxonomic value for the *Proteus* group (Edwards and Ewing, *Identification of Enterobacteriaceae,* 1962, Minneapolis, Minn., Burgess Pub. Co.)

Dehydrated culture media were purchased from BBL with the exception of Triple Sugar Iron (TSI) medium, some of which was obtained from Difco, and Worfel-Ferguson medium, which was prepared according to Edwards and Ewing. The modified O'Meara and Barritt reagents were prepared and used according to Suassuna *et al.* (Publ. Hlth Lab. 19: 67, 1961).

Methyl red test—*Vibrio marinus*

R. R. Colwell and R. Y. Morita (1964). J. Bact. 88, 831–837.

The authors used the method of the Society of American Bacteriologists (1957), *Manual of Microbiological Methods* – McGraw-Hill Book Co., Inc., New York – modified by the addition of the media of the following salts to produce a synthetic seawater:– sodium chloride, 2.4%; potassium chloride, 0.07%; magnesium chloride (hydrated), 0.53%; and magnesium sulfate (hydrated), 0.7%.

A standard inoculum was 1 drop from a pasteur pipette (*c.* 0.05 ml) of a 24 hour artificial seawater broth.

Cultures were incubated at 18°C.

Voges-Proskauer test—*Vibrio marinus*

R. R. Colwell and R. Y. Morita (1964). J. Bact. 88, 831–837.

The authors used the method of the Society of American Bacteriologists (1957), *Manual of Microbiological Methods* – McGraw-Hill Book Co., Inc., New York – modified by the addition to the media of the following salts to produce a synthetic seawater:– sodium chloride, 2.4%; potassium chloride, 0.07%; magnesium chloride (hydrated), 0.53%; and magnesium sulfate (hydrated), 0.7%.

A standard inoculum was 1 drop from a pasteur pipette (*c.* 0.05 ml) of a 24 hour artificial seawater broth.

Cultures were incubated at 18°C.

Voges-Proskauer test—*Bacteroides oralis*

W. J. Loesche, S. S. Socransky and R. J. Gibbons (1964). J. Bact. 88, 1329–1337.

The authors used the methods in the *Manual of Microbiological Methods* (1957).

Society of American Bacteriologists *Manual of Microbiological Methods* – McGraw-Hill Book Co. Inc., New York 1957.

Methyl red test—*Aerococcus catalasicus*

O. G. Clausen (1964). J. gen. Microbiol. 35, 1–8.

Cultures were incubated aerobically at 37°C unless otherwise stated.

The methyl red test was carried out after 4, 5, and 7 days at 30°C.

Voges-Proskauer test—*Aerococcus catalasicus*

O. G. Clausen (1964). J. gen. Microbiol. 35, 1–8.

Cultures were incubated aerobically at 37°C unless otherwise stated.

The Voges-Proskauer test was carried out at 3 and 4 days at 30°C.

Voges-Proskauer test—*Pediococcus*

E. Coster and H. R. White (1964). J. gen. Microbiol. 37, 15–31.

The authors used the methods of H. L. Günther and H. R. White (J. gen. Microbiol. 26, 185, 1961). Their Tomato Juice (TJ) broth or agar with the addition of Tween 80 (0.1%, v/v), pH adjusted to 6.6, was used in maintenance and preparation of inocula with certain specified exceptions.

Transfers for preparing inocula were made every 24 hours except for *Pediococcus halophilus* (48 hrs) and some brewing strains (fortnightly).

All cultures were incubated aerobically at 30°C except for "brewing strains" which were incubated in an atmosphere of 95% (v/v) hydrogen and 5% (v/v) carbon dioxide at 22°C.

Voges-Proskauer test—*Vibrio comma*

J. C. Feeley (1965). J. Bact. 89, 665–670.

Cultures grown in MR–VP medium (Difco) for 48 hr at 35°C and 22°C were tested by the Barritt method (Ewing, 1962).

W. H. Ewing (1962). *Enterobacteriaceae: Biochemical methods for group differentiation.* U.S. Public Health Serv. Publ. 734 (revised).

Methyl red test—marine bacteria

R. M. Pfister and P. R. Burkholder (1965). J. Bact. 89, 863–872.

The methyl red test was tried according to Skerman (1959), with the substitution of seawater as the diluent, but the high percentage of K_2HPO_4 formed an insoluble precipitate and the phosphate had to be reduced to 0.05 g per liter.

V. B. D. Skerman (1959). *A guide to the identification of the genera of bacteria.* The Williams & Wilkins Co., Baltimore.

Voges-Proskauer test—marine bacteria

R. M. Pfister and P. R. Burkholder (1965). J. Bact. 89, 863–872.

Voges-Proskauer tests were tried according to Skerman (1959), with the substitution of seawater as the diluent, but the high percentage of K_2HPO_4 formed an insoluble precipitate and the phosphate had to be reduced to 0.05 g per liter.

V. B. D. Skerman (1959). *A guide to the identification of the genera of bacteria.* The Williams & Wilkins Co., Baltimore.

Methyl red test—*Pseudomonas*

G. L. Bullock, S. F. Snieszko and C. E. Dunbar (1965). J. gen. Microbiol. 38, 1–7.

This was determined in Difco MR–VP broth after 48 hr and 1 week at 20–22°C.

Voges-Proskauer test—*Pseudomonas*

G. L. Bullock, S. F. Snieszko and C. E. Dunbar (1965). J. gen. Microbiol. 38, 1–7.

The test was determined in Difco MR–VP broth after 48 hr and 1 week at 20–22°C.

Voges-Proskauer test—*Bacillus*

G. R. F. Hilson (1965). J. gen. Microbiol. 39, 407–421.

Cultures in M.R.V.P. medium (Oxoid) were incubated for 2 days and tested for the formation of acetylmethylcarbinol (*Mackie and McCartney's Handbook,* 1960).

Mackie and McCartney's Handbook of Bacteriology (1960), 10 ed. Ed. by R. Cruickshank, p. 610. Edinburgh and London: Livingstone.

Methyl red test—*Alcaligenes*

R. G. Mitchell and S. K. R. Clarke (1965). J. gen. Microbiol. 40, 343–348.

The methyl red test was performed on glucose phosphate broth cultures incubated for 6 days.

Voges-Proskauer test—*Alcaligenes*

R. G. Mitchell and S. K. R. Clarke (1965). J. gen. Microbiol. 40, 343–348.

(Barritt's method; Mackie and MacCartney, 1960) was used on glucose phosphate broth cultures incubated for 6 days.

T. J. Mackie and J. E. MacCartney (1960). *Handbook of Practical Bacteriology,* 10 ed., p. 610, Edinburgh, Livingstone.

Methyl red test—*Brevibacterium*

R. Chatelain and L. Second (1966). Annls Inst. Pasteur, Paris 111, 630–644.

Réaction du rouge de méthyle; culture de 4 jours en milieu de Clark et Lubs.

La température d'incubation des cultures est de 30°C.

Voges-Proskauer test—*Brevibacterium*

R. Chatelain and L. Second (1966). Annls Inst. Pasteur, Paris 111, 630–644.

Culture de 4 jours en milieu de Clark et Lubs: recherche de l'acétylméthylcarbinol par la technique de Barritt.

La température d'incubation des cultures est de 30°C.

Voges-Proskauer test—*Staphylococcus*

M. Kocur, F. Přecechtěl and T. Martinec (1966). J. Path. Bact. 92, 331–336.

Acetoin production was investigated on the Difco VP–MR medium; cultures were grown for 6–8 days and the presence of acetoin was shown by Barritt's method (1936).

M. M. Barritt (1936). J. Path. Bact. 42, 441.

Acetoin and 2,3-butylene glycol estimation

F. D. de Accadia (1947). Annls Inst. Pasteur, Paris 73, 1114–1116.

Pour doser ces deux composés, nous avons utilisé des méthodes nouvelles [3, 4] qui, sur une prise d'essai de 10 cm^3, permettent de déterminer à 5 p. 100 près, l'acétylméthyl-carbinol et le butylène-glycol dans des milieux qui en contiennent plus de 20 mg par litre, et qui permettent aussi de faire des dosages dans les milieux plus dilués mais alors avec une précision moindre. Ces méthodes sont spécifiques car elles sont basées sur une modification de la réaction de Lemoigne.

Techniques employées et résultats obtenus. – Le milieu est du bouillon de haricots contenant 0.25 p. 100 de phosphate di-ammonique, 0.25 p. 100 d'autolysat de levure, et 2.5 p. 100 de glucose.

Le milieu stérile réparti à raison de 120 cm^3 par boîte de Roux de 1 litre, est ensemencé puis porté à l'étuve à 30°C. Après vingt-quatre heures, on dose l'acétylméthylcarbinol et le butylèneglycol formés par les méthodes mentionnées ci-dessus.

[3] M. Hooreman, *C. R. Acad. Sci.*, 1946, 222, 1257.
[4] M. Hooreman, *C. R. Acad. Sci.*, 1947, 225, 208.

Acetoin estimation—*Welchia, Inflabilis, Clostridium*

H. Beerens and J. Guillaume (1951). Annls Inst. Pasteur, Paris 81, 93–96.

Le dosage de l'acétoïne a été effectué selon la méthode de Prill et Hammer [4].

5 ou 10 cm^3 de culture sont distillés à sec en présence de perchlorure de fer, et le distillat recueilli dans 1 cm^3 de la solution suivante: acétate

de Na, 70 g; hydroxylamine, 25 g; eau qs. p. 1000 cm^3, est complété à 15 cm^3. On chauffe une heure au bain-marie à 80°. On ajoute immédiatement 1 cm^3 de la solution: phosphate bipotassique, 144 g; acétone, 200 cm^3; eau, qs. p. 1000 cm^3. On laisse dix minutes à la température du laboratoire, on ajoute 0.3 cm^3 de solution d'ammoniaque, 2.2 cm^3 d'une solution de tartrate sodico-potassique à 90 g p. 50 cm^3 d'eau et 0.2 cm^3 de solution de sulfate ferreux à 5 p. 100 dans l'acide sulfurique à 1 p. 100. Il se développe une teinte rose dont l'intensité est proportionnelle à la quantité d'acétoïne présente.

Nous avons pratiqué le dosage par comparaison avec la teinte d'une solution formolée à concentration croissante de phénolsulfonephtaléine tamponnée à pH 7.7, conservée en tubes scellés. Cette gamme est elle-même étalonnée avec une solution de diméthylglyoxime.

Solution de P.S.P. à 2 p. 100 à introduire dans 10 cm^3 de solution à pH 7.7 formolée. .	0.011	0.027	0.083	0.15	0.23	0.52	0.69
Doses corrospondantes d'acétoïne en mg dans 15 cm^3 . . .	0.01	0.02	0.05	0.1	0.2	0.5	1

Cette méthode permet de doser facilement 0.02 mg à 0.5 mg d'acétoïne contenus dans 10 cm^3 de culture utilisés comme prise d'essai, soit 2 à 50 mg par litre. Le dosage, par la méthode de Lemoigne, n'est possible qu'à la concentration de 6 mg par litre, à la condition d'opérer sur 100 cm^3 de culture.

[4] Prill (E. A.) et Hammer (B. W.). J. Sci. 1938, 12, 385.

Acetoin estimation—*Staphylococcus aureus*

W. Kędzia (1962). J. Path. Bact. 84, 243–245.

Cultures were grown at 37°C in static Roux flasks containing 160 ml volumes of Clark's medium. This medium consisted of glucose, 5.0 g; peptone (product of 'Wytwornia Organopreparatow', Warsaw), 5.0 g; K_2HPO_4, 5.0 g; and distilled water to 1000 ml, and was sterilized by tyndallization. The quantity of acetoin in the culture was determined after 24, 48, 72 and 96 hr.

Acetoin was determined by a modification of the Voges-Proskauer reaction described by Sokatch and Gunsalus (1957). It was first collected by steam distillation of 25 ml of the culture. The collector, a 100 ml spherical flask containing 10 ml distilled water, was chilled in an ice bath. The tip of the delivery tube was kept under water to prevent loss of acetoin. Altogether 100 ml of distillate were obtained. A 5.0 or 2.5 ml sample of distillate was placed in a flask, and the colour reagents and enough water were added to bring the volume to 10 ml. The colour reagents were 1.3 ml α-naphthol (6 per cent in 95 per cent ethanol) and 0.3 ml arginine solution (300 mg in 10 ml 40 per cent KOH)—not creatine, as used by Sokatch and Gunsalus. The flask was placed in a shaker

for one hour at 20°C. The readings were then made in a Pulfrich colorimeter (4.35 mm tube) at 540 mμ in comparison with a blank. The acetoin production was calculated per 100 ml of culture.
J. T. Sokatch and I. C. Gunsalus (1957). J. Bact. 73, 452.

Acetoin estimation—*Staphylococcus*

S. Krynski, W. Kędzia, and M. Kaminska (1964). Some differences between staphylococci isolated from pus and from healthy carriers. J. infect. Dis. 114, 193–202.

Acetoin was measured according to the method of Sokatch and Gunsalus as modified by Kędzia (1962).
W. Kędzia (1962). J. Path. Bact. 84, 243–245.

Acetoin estimation—anaerobes

A.-R. Prévot (1966). Techniques pour le diagnostic des Bactéries Anaérobes. Éditions de la Tourelle, St. Mandé.

Recherche de l'acétoïne (*Fig. 4*).

Cette recherche s'effectue par la méthode de Lemoigne, modifiée par Kluyver, Donker et Vissert'hoff.

Mettre dans un ballon de 1 litre à fond rond:

Culture filtrée	100 cm^3
Solution de perchlorure de fer à 25 p. 100	5 cm^3
Talc (comme antimousse)	10 g

Monter le ballon avec un réfrigérant descendant; chauffer et recueillir 2 à 3 cm^3 de distillat dans un tube à essais contenant 1 cm^3 d'eau distillée environ et 5 à 10 gouttes d'une solution de chlorhydrate d'hydroxylamine à 20 p. 100. Arrêter la distillation et ajouter dans le tube:

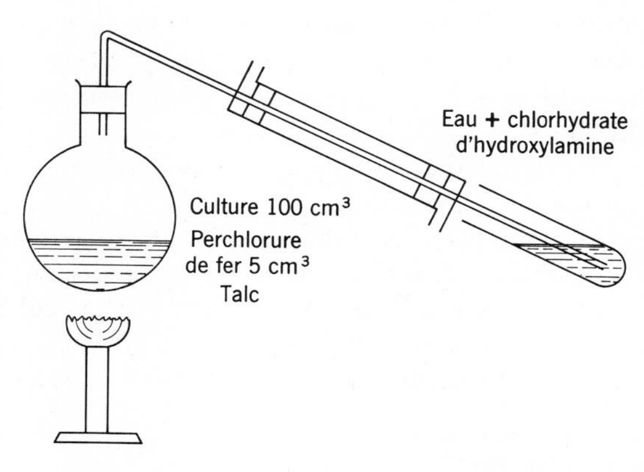

Fig. 4
Caractérisation de l'acétoïne

Solution d'acétate de soude à 20 p. 100 XX gouttes
Solution de chlorure de nickel à 10 p. 100 V gouttes

Faire bouillir le mélange 1 à 2 minutes puis le refroidir brusquement. Une réaction positive se traduit par l'apparition d'un précipité rouge brique au refroidissement; en cas de doute laisser reposer 24 heures.

MILK

Litmus milk—*Pseudomonas*

A. J. Riker, W. M. Banfield, W. H. Wright, G. W. Keitt and H. E. Sagen (1930). J. agric. Res. 41, 507–540.

Fresh skim milk containing 0.06 per cent of soluble litmus per liter was used for preparing the medium. Transfers were made with comparable suspensions of the bacterial growth from 48-hour-old agar slants, and the milk cultures were incubated at 28°C.

Iron milk—*Clostridium*

R. S. Spray (1936). J. Bact. 32, 135–155.

Mix fresh whole milk well and tube about 8 cm deep in 200 x 15 mm, or in 200 x 13 mm tubes. Add to each tube one strip of No. 26 gauge black stove-pipe iron, cut about 50 x 7 mm. Autoclave as usual, but reduce the pressure slowly when completed to avoid wetting or blowing of plugs.

Reactions:

1. Active gaseous fermentation with early coagulation (12 to 48 hr). The clot is violently disrupted, with no subsequent digestion or blackening.
2. Inactive gaseous fermentation, with coagulation quite uniformly delayed (4 to 6 days). No digestion or blackening of the clot is observed, although a dirty grey-brown discoloration may occur if evaporation exposes the end of the iron strip. Such changes are of no significance. Furthermore, cultures requiring prolonged incubation should have the plugs paraffined after the first 5 to 6 days.
3. Inactive gaseous fermentation, long continued, with late (if any) coagulation; no digestion or blackening.
4. Inactive gaseous fermentation, with digestion and blackening.
5. No gaseous fermentation, with no digestion or blackening. Coagulation is rather constant, but long delayed. It is not due to acid but to a weak coagulase. It appears at the earliest in 15 to 20 days. Some strains have failed to coagulate even after 5 months' incubation.

Litmus milk—*Mycobacterium*

R. E. Gordon (1937). J. Bact. 34, 617–630.

The cultures were inoculated into litmus milk and incubated at 25°C for one month.

Litmus milk—*Xanthomonas*

S. S. Ivanoff, A. J. Riker and H. A. Dettwiler (1938). J. Bact. 35, 235–253.

"90 grams of powdered skim milk, plus litmus, added to a liter of water". Cultures incubated 3 weeks at 28°C.

Iron milk—*Lactobacillus* and *Bacteroides*

K. H. Lewis and L. F. Rettger (1940). J. Bact. 40, 287–307.

In tubes of whole milk containing iron filings and incubated 18 days at 37°C.

Litmus milk—*Corynebacterium acne*

H. C. Douglas and S. E. Gunter (1946). J. Bact. 52, 15–23.

Standard litmus milk was exhausted by steaming, cooled, inoculated and layered with sterile vaspar. All strains produced a rennet curd after about 2 weeks' incubation at 37°C, followed by a slow peptonization of the casein.

Litmus milk—*Streptococcus*

K. E. Hite and H. C. Hesseltine (1947). A study of aerobic streptococci isolated from the uterus and the vagina. J. infect. Dis. 80, 105–112.

Reactions in milk – Both litmus milk and methylene blue (0.1%) milk were used. Only reduction occurring before coagulation was considered to be a positive reaction. Acid and coagulation were read from the litmus milk. When observed, peptonization occurred in both media.

Litmus milk—*Streptococcus*

Y. Abd-el-Malek and T. Gibson (1948). J. Dairy Res. 15, 233–248.

The main feature to receive attention was the reduction of the litmus. When it began before the indicator became pink, and was complete beneath the surface layer before the milk clotted, it was considered to be of the vigorous type. While this reaction is not entirely dependable, especially in the case of strains that have been cultivated artificially for long periods, it is useful for the detection of freshly isolated cultures of *Str. lactis, Str. cremoris* and many enterococci.

Litmus milk—anaerobic Gram-negative diplococci

G. C. Langford, Jr., J. E. Faber, Jr., and M. J. Pelczar, Jr. (1950). J. Bact. 59, 349–356.

Tubes of litmus milk were inoculated with 1 ml of the cultures and incubated in anaerobic jars at 35°C.

Litmus milk—*Staphylococcus*

C. Shaw, J. M. Stitt and S. T. Cowan (1951). J. gen. Microbiol. 5, 1010–1023.

Litmus milk was incubated for 14 days at 37°C, 21 days at 30°C, or 28 days at 22°C.

Litmus milk—*Streptococcus*

P. F. Swartling (1951). J. Dairy Res. 18, 256–267.

Special regard to reducing power was observed at 30°C.

Litmus milk—general

P. Hauduroy (1951). Techniques Bactériologiques. Masson et Cie, Éditeurs, Paris.

PETIT-LAIT TOURNESOLE

Opérer comme pour le petit-lait ordinaire (temps 1, 2, 3, 4, 5, 6).

Ajouter de la teinture de tournesol jusqu'à coloration du milieu en violet. Répartir. Stériliser à 100°, trois jours de suite pendant 30 minutes.

PETIT-LAIT TOURNESOLE (C. I. E.).

Le C. I. E. se procure le milieu tout prêt chez Schering-Kahlbaum à Berlin et on l'ensemence avec des cultures prises sur boîtes de gélose. Les tubes sont placés plusieurs jours à 37° et on note leurs variations de couleur. *(C. I. E.).*

Milk—*Pseudomonas (Malleomyces) pseudomallei*

P. de Lajudie and E. R. Brygoo (1953). Annls Inst. Pasteur, Paris 84, 509–515.

Ignorant les caractéristiques du lait utilisé par nos prédécesseurs, et désirant éliminer l'influence possible de la pasteurisation préalable, nous avons fait porter nos essais sur deux échantillons, l'un de lait pasteurisé, l'autre de lait non pasteurisé. Ils furent étudiés avec et sans addition de tournesol. Ils se comportèrent de façon identique.

Litmus milk—*Streptococcus*

G. Andrieu, L. Enjalbert and L. Lapchine (1954). Annls Inst. Pasteur, Paris 87, 617–634.

Pour le lait tournesolé nous ajoutons, à 6 à 7 cm^3 de lait écrémé stérile, II gouttes de teinture de tournesol R.A.L. Les cultures sont étudiées après quarante-huit heures et trois jours d'étuve. Les variations ne se produisent plus après trois jours d'étuve.

Litmus milk

N.C.T.C. Method (1954).

Incubate at optimal temperature; read daily for 2 weeks.

Litmus milk—*Lactobacillus*

D. M. Wheater (1955). J. gen. Microbiol. 12, 133–139.

Separated milk containing 0.3% yeastrel, 1% glucose, and 1.5% of a 1% solution of litmus. Incubated at 30°C for 48 and 72 hr.

Litmus milk—*Actinomyces bovis*

S. King and E. Meyer (1957). J. Bact. 74, 234–238.

Reactions in litmus milk medium, inoculated and overlaid with ½ in of Vaspar (50 percent Vaseline and 50 percent paraffin), were slow; generally an incubation period of 30 days was required.

Litmus milk

Society of American Bacteriologists *Manual of Microbiological Methods* – McGraw-Hill Book Co. Inc., New York 1957.

Prepare a saturated aqueous solution of litmus. Add a sufficient quantity of the solution to give a light lavender color to fresh skimmed milk (some grades of dried milk powder may be substituted, but many are unsatisfactory); sterilize 12–15 min at 120°C; cool tubes immediately by immersing in cold water. For anaerobic organisms, Spray's system of classification is based upon use of this medium (tubed in a deep column) to which is added 0.05 g of reduced iron or a thin strip of iron.
R. S. Spray (1936). J. Bact. 32, 135–155.

Litmus milk—*Bacillus*

E. R. Brown, M. D. Moody, E. L. Treece and C. W. Smith (1958). J. Bact. 75, 499–509.

Biochemical reactions. (1) The action of the cultures on litmus milk was determined, and reduction of methylene blue in the medium, as described by Smith *et al.* (1952), was recorded. The tubes of methylene blue and litmus milk were incubated at 37°C and examined for reduction.
N. R. Smith, R. E. Gordon and F. E. Clark (1952). U.S. Department of Agriculture Monograph No. 16.

Litmus milk—*Pediococcus*

J. C. Dacre (1958). J. Dairy Res. 25, 409–413.

A clot was formed in 18–20 days with some reduction of the litmus in 11–12 days at 30°C. The organism clots 0.2% yeast skim-milk media in 5–6 days at 22°C, in 3–4 days at 30°C, and in 3 days when incubated at 37°C.

Litmus milk—*Streptococcus*

P. Morelis and L. Colobert (1958). Annls Inst. Pasteur, Paris 95, 568–587.

La culture en lait tournesolé permet d'apprécier le pouvoir réducteur concurremment à celle en lait au bleu de méthylene; cependant, on a constaté que quelques souches, qui n'ont pas réduit ce dernier, ont décoloré le lait tournesolé; de plus la réduction du tournesol, chez les souches fortement réductrices, précède l'apparition de la coagulation et l'acidification.

Milk—*Streptococcus*

P. Morelis and L. Colobert (1958). Annls Inst. Pasteur, Paris 95, 568–587.

La coagulation du lait a été étudiée par la méthode de C. Gorini [14], plus sensible que le procédé habituel d'inoculation directe dans le lait. Elle consiste à verser le lait sur une culture bien développée sur gélose inclinée, puis à placer les tubes horizontalement dans une étuve à 37°. La coagulation s'est manifestée généralement entre douze et vingt-quatre heures. Nous avons constaté que seule une souche, la souche Ch4, n'a pas coagulé le lait. On a noté également que les souches qui liquéfiaient la gélatine digéraient secondairement le lait coagulé.

[14] Gorini (C.) *Le lait,* 1936, 16, 871.

Litmus and **methylene-blue milk**—*Pseudomonas fluorescens*

M. Rhodes (1959). J. gen. Microbiol. 21, 221–263.

Seven-ml volumes of litmus milk and of methylene-blue milk (methylene blue, 1/20,000) were inoculated and examined for colour changes, clotting, casein digestion and dye reduction. Because of the difficulty of detecting milk-protein digestion in strongly alkaline liquid cultures, the method involving spot inoculation of cultures onto skim-milk agar plates (skim milk, 20% (v/v); washed agar, 2.0% (w/v); adjusted to pH 7.2) was also used. After 7 days of incubation the plates were flooded with $HgCl_2$ solution (Frazier, 1926), and any clear zones of hydrolysis measured.

W. C. Frazier (1926). J. infect. Dis. 39, 302.

Litmus milk

V. B. D. Skerman, *A Guide to the Identification of the Genera of Bacteria* 2nd ed. Williams and Wilkins Book Co. Baltimore, 1967.

For the preparation of this medium use (a) fresh machine-separated milk, *or* (b) fresh milk, steamed for 20 minutes and allowed to stand for 18 hours for the fat to rise, after which the fat-free milk is siphoned off aseptically, *or* (c) good quality *spray-dried* powdered skim milk reconstituted by dissolving 100 gm per L of distilled water.

Add to the milk an alcoholic solution of litmus *(vide infra)* sufficient to give a distinct color (usually 40 ml per L). Adjust the pH to 7.0 by the addition of 1 N NaOH. Do not rely on visual adjustment. A glass electrode should be used. Steam for 15 minutes on 3 successive days with incubation at 37°C between steamings.

Alcoholic Litmus Solution: Grind 50 gm of litmus in a mortar with 150 ml of 40 per cent alcohol. Transfer to a flask, and boil gently on a steam bath for 1 minute. Decant the fluid and add another 150 ml of

40 per cent alcohol to the residue; boil again for 1 minute. Decant and mix the extracts. Allow to settle overnight and make up to 300 ml with 40 per cent alcohol.

Add N HCl drop by drop to adjust the pH to 7.0.

Litmus milk—*Streptococcus* and *Lactobacillus*

C. W. Langston and C. Bouma (1960). Appl. Microbiol. 8, 212–222.

Litmus milk was inoculated with test cultures and incubated 1 month. Tubes were examined and medium changes recorded at 24 hours, 48 hours, 1 and 4 weeks.

Litmus milk—*Pediococcus*

H. L. Günther and H. R. White (1961). J. gen. Microbiol. 26, 185–197.

Litmus milk cultures were examined for reduction of indicator, change in pH value or coagulation during 28 days of incubation.

Maintenance of stock cultures and methods of cultivation. For maintenance of stock cultures, preparation of inocula and in all experimental work, "Oxoid" tomato juice (TJ) broth or tomato juice (TJ) agar, adjusted to pH 6.6, was used unless otherwise stated. The following were exceptions to this rule: for strain Tc. 1, sodium chloride (5%, w/v) was added to the medium; and for the aerococci, glucose Lemco broth (Shattock and Hirsch, 1947) or glucose yeast extract (GY) agar (containing, as %, w/v; peptone, 1.0; Yeastrel, 0.3; glucose, 1.0; NaCl, 0.25; agar, 1.0; at pH 7.4) was used.

Cultures were incubated aerobically at 30°C with specified exceptions.
P. M. F. Shattock and A. Hirsch (1947). J. Path. Bact. 59, 495.

Milk—*Clostridium*

H. Beerens, M. M. Castel and H. M. C. Put (1962). Annls Inst. Pasteur, Paris 103, 117–121.

Lait cystéiné: A 1000 ml de lait écrémé, ajouter 0.8 g de chlorhydrate de cystéine. Ajuster le pH à 7.3-7.4. Répartir 12 ml environ par tube de 160 x 16. Stériliser à 115° pendant trente minutes. Ensemencer après régénération et refroidissement avec environ 1 ml d'une culture en milieu liquide. Incuber à 37°.

Lait acidifié à pH 5.8 (Weinzirl) par l'acide lactique 0.8 N. L'acide lactique est ajouté aseptiquement après autoclavage du lait. Distribuer en tubes de 160 x 16. Ensemencer avec 1 ml d'une dilution à 10^{-3} d'une culture en milieu liquide.

Litmus milk—*Pseudomonas odorans*

I. Málek, M. Radochová and O. Lysenko (1963). J. gen. Microbiol. 33, 349–355.

Examined over 14 days – 28°C.

Litmus milk—*Aeromonas*

I. W. Smith (1963). J. gen. Microbiol. 33, 263–274.

Litmus milk was examined at daily intervals up to 14 days for acid production, clotting, proteolysis and reduction of litmus.

Litmus milk

G. S. Wilson and A. A. Miles (1964). Topley and Wilson's *Principles of Bacteriology and Immunity,* 5th edition – Arnold.

No change, acid or alkali; clot disrupted by gas; peptonization; saponification. The term "clot" is unfortunately used for both an acid clot and a rennet clot. An acid clot results from the precipitation of the caseinogen; it is soft, gelatinous, does not retract, and can be completely dissolved in alkali. A rennet clot is due to the coagulation of the caseinogen under the influence of bacterial enzymes. A few hours after its formation it retracts with the expression of a clear greyish-coloured fluid called whey; the clot itself is firm and cannot be dissolved by alkali. Calcium caseinogenate is soluble in water. When lactic acid is produced from the lactose of the milk, the calcium combines with it, and the caseinogen, which is insoluble, is precipitated. This is the mechanism of formation of the acid clot. In coagulation by rennet the soluble calcium caseinogenate is converted into insoluble calcium caseinate, which forms the curd.

Litmus milk—*Actinomyces*

L. K. Georg, G. W. Robertstad and S. A. Brinkman (1964). J. Bact. 88, 477–490.

Litmus milk reactions were studied in litmus milk reinforced with 0.5% yeast extract and 0.3% glucose.

The inoculum was taken from 3 day old cultures in AM broth.

See **Starch hydrolysis**–*Actinomyces* (p. 812) for composition of AM broth.

The cultures were incubated at 37°C under pyrogallol-carbonate seals and read at 3 and 10 days.

Milk—*Dermatophilus*

M. A. Gordon (1964). J. Bact. 88, 509–522.

The author used 'bromocresol purple milk (1 ml of 1.6% alcoholic solution of indicator per liter of skim milk).'

The inoculum was pipetted from a 4 to 5 day broth culture.

The medium was incubated at 36°C (± 1) and readings were made at 48 hours, 5 days, 1 week and 2 weeks.

Litmus milk—*Vibrio marinus*

R. R. Colwell and R. Y. Morita (1964). J. Bact. 88, 831–837.

The authors used the method of the Society of American Bacteriologists, 1957, *Manual of Microbiological Methods.* McGraw-Hill Book Co., Inc., New York – modified by the addition to the media of the following salts to produce a synthetic seawater: sodium chloride, 2.4%; potassium chloride, 0.07%; magnesium chloride (hydrated), 0.53%; and magnesium sulfate (hydrated), 0.7%.

A standard inoculum was 1 drop from a pasteur pipette (*c.* 0.05 ml) of a 24 hour artificial seawater broth.

Cultures were incubated at 18°C.

The synthetic seawater litmus milk was prepared by first sterilizing the Litmus Milk (Difco) and double-strength synthetic seawater and mixing the components aseptically after cooling.

Litmus milk—enterococci

C. G. Rogers and W. B. Sarles (1964). J. Bact. 88, 965–973.

Reduction of litmus in skim milk (Burnett, Pelczar, and Conn, 1957; tubes were read after 24 hr and again at 7 days for acid production, reduction of litmus, clot formation, and proteolysis). Incubated at 37°C.

G. W. Burnett, M. J. Pelczar, Jr. and H. J. Conn (1957). In Society of American Bacteriologists *Manual of Microbiological Methods.* McGraw-Hill Book Co., Inc., New York.

Litmus milk—*Aerococcus catalasicus*

O. G. Clausen (1964). J. gen. Microbiol. 35, 1–8.

Cultures were incubated aerobically at 37°C unless otherwise stated.

Reduction prior to coagulation of litmus milk was observed after 12 and 24 hr, and daily for 5 days.

Litmus milk—*Pediococcus*

E. Coster and H. R. White (1964). J. gen. Microbiol. 37, 15–31.

The authors used the methods of H. L. Günther and H. R. White (J. gen. Microbiol. 26, 185, 1961). Their Tomato Juice (TJ) broth or agar with the addition of Tween 80 (0.1%, v/v), pH adjusted to 6.6, was used in maintenance and preparation of inocula with certain specified exceptions.

Transfers for preparing inocula were made every 24 hours except for *Pediococcus halophilus* (48 hr) and some brewing strains (fortnightly).

All cultures were incubated aerobically at 30°C, except for "brewing strains," which were incubated in an atmosphere of 95% (v/v) hydrogen and 5% (v/v) carbon dioxide at 22°C.

Litmus milk—*Moraxella*

W. J. Ryan (1964). J. gen. Microbiol. 35, 361–372.

Litmus milk. Ordinary litmus milk was an unsatisfactory growth medium; 10% (v/v) horse serum was incorporated and culture incubated up to 3 weeks.

Litmus milk—marine bacteria

R. M. Pfister and P. R. Burkholder (1965). J. Bact. 89, 863–872.

Litmus milk was prepared at 0.33% of the recommended concentration in seawater with the addition of 1 μg per liter of vitamin B_{12}. Sterilization of this medium was done by a 3-day successive tyndallization (5 min of free-flowing steam each time), with incubation at 25°C between steam treatments. Prepared medium was tested for sterility by preincubation for at least 72 hours prior to use.

Litmus milk—*Bacillus anthracis*

R. F. Knisely (1965). J. Bact. 90, 1778–1783.

The characteristics of growth in Litmus Milk (BBL) were observed and recorded.

Litmus milk—*Streptococcus*

R. Whittenbury (1965). J. gen. Microbiol. 38, 279–287.

A heavy inoculum, 3 capillary pipette drops, was added to the milk media which were incubated at 30°C and read at 8 hr.

Litmus milk—*Clostridium*

A. V. Goudkov and M. E. Sharpe (1966). J. Dairy Res. 33, 139–149.

The litmus milk contained 0.08% cysteine.

Milk—*Brevibacterium*

R. Chatelain and L. Second (1966). Annls Inst. Pasteur, Paris 111, 630–644.

Lait tournesolé: la modification du pH est notée après 4 jours et la peptonisation après 8 jours de culture.

La température d'incubation des cultures est de 30°C.

MOTILITY

Motility—enterobacteria

R. P. Tittsler and L. A. Sandholzer (1936). J. Bact. 31, 575–580.

Used a semisolid medium consisting of 0.3 per cent meat extract, 0.5 per cent peptone and 0.5 per cent agar; pH 6.8–7.2; dispensed in tubes and sterilized in the autoclave.

Stab inoculations were made with a wire and tubes incubated for up to 6 days at 37°C. Motility was manifested macroscopically by a diffuse zone of growth spreading from the line of inoculation.

Motility—*Aeromonas*

J. P. Stevenson (1959). J. gen. Microbiol. 21, 366–370.

Determined by microscopic examination of hanging-drop preparations after culture in nutrient broth for 2, 4, and 24 hr at 37°C.

Motility—enterobacteria

W. H. Ewing (1962). *Enterobacteriaceae. Biochemical Methods for Group Differentiation.* U.S. Department of Health, Education and Welfare, Public Health Service Publication No. 734 (revised).

Motility Test Medium

Beef extract	3 g
Peptone	10 g
Sodium chloride	5 g
Agar	4 g
Distilled water	1000 ml

Adjust reaction to pH 7.4 and tube, about 8 ml per tube, and sterilize at 121°C, 15 minutes. Inoculate by stabbing into the top of the column of medium to a depth of about 5 mm.

Incubation: 37°C for 1 or 2 days. If negative, follow with further incubation at 21° to 25°C for 5 days.

For special purposes such as enhancement of motility and flagellar development in poorly motile cultures, it often is advisable to passage cultures first through a semisolid medium containing 0.2 per cent agar tubed in Craigie tubes or in U-tubes. Subsequent passages may be made in the medium listed about (0.4 per cent agar).

Motility—*Actinobacillus* and *Haemophilus*

E. O. King and H. W. Tatum (1962). *Actinobacillus actinomycetem-comitans* and *Haemophilus aphrophilus.* J. infect. Dis. 111, 85–94.

Semisolid motility medium was observed for 1 week to detect any signs of motility.

Motility—*Escherichia aurescens*

H. Leclerc (1962). Annls Inst. Pasteur, Paris 102, 726–741.

a) *Mobilité.*—Nous l'avons étudiée par les deux techniques classiques: examen d'une culture jeune en eau peptonée, entre lame et lamelle; ensemencement en piqûre dans un culot de gélose nutritive à 5 g d'agar (Difco) pour 1000 ml. On incube vingt-quatre à quarante-huit heures à 30° puis dix jours à 20-22° si la première lecture n'est pas concluante.

Motility—*Moraxella*

M. Piéchaud (1963). Annls Inst. Pasteur, Paris 104, 291–297. Technique nouvelle pour la recherche de la mobilité des *Moraxella.*

Lautrop donne peu de détails sur la technique suivie par lui. Il dit simplement qu'il a fait ses observations en contraste de phase sur des plaques minces de gélose pauvre en éléments nutritifs, donc certainement selon le procédé habituel en chambre humide. Il n'a pas trouvé de conditions permettant une démonstration régulière.

Nous utilisions depuis plusieurs années pour des observations cytologiques une chambre à huile inspirée de celle de Fonbrune et modifiée de façon à permettre un examen en contraste de phase même à l'objectif à immersion avec un condensateur de focale normale. Dans cette cellule nous avions souvent observé la croissance des *Moraxella,* germes aérobies stricts.

La fabrication de cette cellule ne sera pas décrite en détail ici. Son fond est fait d'une grande lamelle couvre-objet 23 x 60 mm, sur laquelle sont collées (à l'aide de la cire spéciale en feuilles qui sert aux dentistes à prendre des empreintes de la cavité buccale, utilisée comme le serait la paraffine) des cales coupées au diamant dans une lame porte-objet de ±1 mm d'épaisseur. Les deux cales inférieures, coupées à 21 mm, supporteront une lamelle couvre-objet 22 x 22 mm dont la largeur est ramenée à 18-19 mm au diamant. Les deux cales supérieures coupées à 17 mm (une seule lame fournit les quatre cales nécessaires) sont collées en retrait de 3 mm par rapport aux précédentes et déterminent un espace de 24 mm au centre; elles empêchent le déplacement de la lamelle tout en lui laissant une latitude de 1 mm environ de chaque côté. Les verres doivent être parfaitement nettoyés avant le montage.

Avec l'extrémité droite d'un fil de fer chauffé et chargé de cire on borde, sur 2 mm environ de large, la face supérieure des bords du fond de la cellule, en veillant à bien fondre la cire aux quatre coins dans le dièdre des cales inférieures; la cire, ayant l'avantage de n'être mouillable ni par l'eau ni par l'huile, s'oppose aux fuites.

Cette chambre ou plus exactement la lamelle qui la ferme et qui porte la préparation immergée dans l'huile, se prête à des montages divers. Mais dans le cas qui nous intéressait ici nous avons eu l'idée, puisque les bactéries déjà connues comme ayant ce type de mobilité ont besoin d'un support pour se déplacer, de les coincer et de les faire se diviser dans un mince film de liquide entre une couche de gélose coulée sur la lamelle et l'huile de paraffine remplissant la chambre. Nous pensions qu'ainsi les bactéries pourraient prendre appui soit sur la surface de gélose soit sur l'huile à l'interface liquidehuile. Tout se passe comme s'il en était bien ainsi. Dans ce montage, en partant de la surface de la lamelle à travers laquelle regarde l'objectif, on rencontre successivement sa face inférieure, une couche de gélose, les bactéries dans une faible épaisseur de liquide et enfin l'huile de paraffine.

Montage de la préparation.

On prend une lamelle ramenée à 22 x 19 mm et parfaitement nettoyée. La saisissant par un coin avec une pince à pression inverse on la passe rapidement dans la flamme du bec, puis on étale une anse bien pleine de milieu nutritif gélosé fondu, sur environ 1 cm^2 au centre d'une de ses faces. On a ainsi une mince couche de gélose qui prend aussitôt, la lamelle maintenue horizontale. D'une main légère, on dépose en son centre une anse portant un film à peine bombé de culture liquide ou de suspension. Sans perdre de temps on verse rapidement dans la chambre 15 à 16 gouttes d'huile de paraffine stérile et on dépose tout de suite, en bonne place, la lamelle, sa face portant la préparation au contact de l'huile qui doit s'étaler en dépassant au moins un peu les limites de la gélose, si elle n'atteint pas les bords de la lamelle. On peut d'ailleurs ajouter ou enlever de l'huile avec une pipette Pasteur réeffilée.

Il faut aller vite, la gélose risquant de sécher, particulièrement avant d'avoir été attouchée d'un peu de liquide. Le volume de ce liquide doit être assez faible pour que, au contact de l'huile, il s'étale sans atteindre tout à fait les limites de la couche de gélose en donnant une lame fluide qui s'amenuise progressivement à partir du centre.

Il ne reste plus qu'à examiner la préparation au microscope. On peut immerger le condenseur à volonté. Si la gélose a été coulée en couche trop forte, l'épaisseur peut être plus grande que la distance frontale de l'objectif à immersion dont on ne pourra ajuster la mise au point.

Observations sur la mobilité des *Moraxella*.

1° Conditions de l'observation.

On peut faire une suspension légère en milieu liquide à partir d'une culture de 18-24 heures sur gélose ordinaire, gélose sérum, sérum coagulé, selon les exigences de la souche. On peut encore se servir d'une culture en bouillon ordinaire ou additionné d'un peu de sérum et âgée de quelques heures à 18-24 heures.

La gélose étalée sur la lamelle est préparée avec un culot de gélose nutritive (macération de viande, peptone 10g, NaCl 5 g, gélose 15-20g par litre) fondue au bain-marie et diluée au tiers soit avec de l'eau distillée, soit avec du milieu synthétique de base. Les suspensions ou cultures liquides servant d'inoculum sont de même diluées au tiers.

L'adjonction de sérum frais ou chauffé une heure à 56°, nous a paru présenter parfois un avantage pour les *Moraxella* du groupe II, et presque toujours pour celles du groupe I. De toute façon cette adjonction est indispensable pour *Moraxella lacunata* qui ne se développerait pas sans cela. Nous en ajoutons une goutte pour environ 2 ml soit de phase liquide, soit de gélose (ramenée alors à 55° après fusion et dilution).

Sans avoir étudié systématiquement les conditions de milieu, nous avons essayé les milieux purs ou dilués comme indiqué. Il semble qu'il y ait avantage à la dilution au moins de la gélose, et particulièrement avec du milieu synthétique pour toutes les *Moraxella.* L'incubation et l'examen ont été faits en général entre 20° et 25°, sauf pour *Moraxella lacunata* qui a été incubée entre 33° et 35° et examinée à la même température que les autres.

Le microscope est un statif Zeiss W, muni d'objectifs à contraste de phase, achromat x 40 et Néofluar x 100.

La préparation peut être regardée à différentes reprises depuis son montage, jusqu'à 24 heures et même souvent deux à trois jours. On doit examiner le centre et la périphérie en en faisant tout le tour à la recherche des zones favorables aux manifestations de mobilité.

Influence de la température.—Les mouvements sont faibles au-dessous de 15°. Une température de 20 à 25° est favorable. Une préparation placée à la glacière n'a repris ses mouvements qu'après un intervalle de l'ordre d'une demi-heure.

Influence de l'éclairement.—Nous ne savons pas si les germes sont mobiles dans l'obscurité. En tout cas nous avons vu souvent des mouvements, faibles ou nuls, s'exagérer et devenir très actifs plus ou moins vite après l'éclairement; ce temps de latence, très court, inappréciable quand la culture est jeune, devient nettement de quelques secondes à quelques dizaines de secondes pour une culture de vingt-quatre heures ou plus.

L'interposition d'un ou deux *filtres anticaloriques* n'empêche pas la mobilité; différentes régions du spectre conviennent, mais l'éclairage en lumière verte avec un filtre interférentiel nous a semblé particulièrement favorable.

Rôle éventuel de la polarisation de la lumière.—Dans certaines observations, des mouvements furent déclenchés, chez des bactéries jusque-là immobiles, après l'interposition d'un polaroïde ou sa rotation. Ce point mériterait d'être éclairci, d'autant plus que nous nous sommes aperçu qu'un bon quart de la lumière éclairant la préparation, mesurée en lux, était polarisé.

Eclairage du microscope par la lumière solaire.—Les mouvements d'une souche ont été aussi nets qu'avec la source lumineuse habituelle; ce qui montre que l'alimentation de la lampe par le courant alternatif n'a aucune influence sur le phénomène.

2° Mouvements observés.

Mouvements individuels.—Rotation comme autour d'un pivot, rappelant celle d'une aiguille aimantée attirée d'un côté ou de l'autre par du fer; complète, alternative, brusque, avec des périodes de repos.

Glissement selon l'axe longitudinal; la cellule semble tirée par un fil invisible auquel elle résisterait puis s'abandonnerait soudain. Parfois, elle revient en arrière aussi brusquement, comme retenue par un élastique, et le mouvement reprend avec des inflexions dans une direction ou une autre. Des cellules nombreuses mais séparées peuvent avoir l'air affairé d'abeilles sur un rayon de ruche.

Rotation d'une cellule fixée par un de ses pôles, ou oscillation.

Mouvements de groupe.—Dans un groupe formé par une microcolonie les germes les plus périphériques peuvent s'éloigner un peu et revenir vers la masse commune.

Un groupe de quelques cellules peut se diriger vers un autre groupe et rester collé à lui ou s'en éloigner ensuite comme après une visite.

Les groupes plus vastes sont généralement en mouvement continuel, addition des mouvements individuels, d'amplitude plus ou moins grande, qui donne souvent à l'ensemble l'air d'un protoplasme amiboïde.

La brusquerie de certains de ces mouvements de masse n'est pas sans rappeler celle des battements d'un explant de myocarde mis en culture.

Des chaînettes peuvent se déformer avec des contractions brusques.

Ces mouvements divers ont été retrouvés chez toutes les espèces de *Moraxella,* depuis *Moraxella lacunata* (bacille de Morax) jusqu'à *Moraxella glucidolytica.* Certaines souches étaient fraîchement isolées, d'autres conservées depuis longtemps.

Motility—*Bordetella*

M. Piéchaud and Mme. S. Szturm-Rubinsten (1965). Annls Inst. Pasteur, Paris 108, 391–395.

Mobilité. Il est indispensable de l'observer au microscope entre lame et lamelle et mieux encore sans lamelle avec un objectif x 40 en contraste de phase. Les deux espèces ne poussant qu'en aérobiose stricte ne donnent de culture qu'en surface d'une gélose molle de mobilité et rien le long de la piqûre d'ensemencement. L'examen microscopique seul peut révéler une différence entre les deux espèces.

NEUTRAL RED TEST

Neutral red test—*Mycobacterium*

A. J. Ross and F. P. Brancato (1959). J. Bact. 78, 392–395.

Two methods were used for this determination.

(a) the tube technique described by Dubos and Middlebrook (1948) was employed using 5-day-old cultures from heat infusion agar.

(b) Similar cultures were used in the molecular filter membrane procedure as outlined by Wayne (1955).

R. J. Dubos and G. Middlebrook (1948). Am. Rev. Tuberc. pulm. Dis. 58, 698–699.

L. G. Wayne (1955). J. Bact. 69, 92–96.

NITRATE AND NITRITE REDUCTION

Nitrate reduction—bacterial spot of tomato

M. W. Gardner and J. B. Kendrick (1921). J. agric. Res. 21, No. 2, 123–156.

In fermentation tubes containing 1 per cent potassium nitrate bouillon. Tests with Trommsdorf's reagent. Additional tests were made with test tube cultures in 2 per cent Difco peptone-water containing 1 per cent potassium nitrate. Examined at 4, 6, and 11 day intervals.

Nitrate reduction—*Pseudomonas*

C. Elliott (1924). J. agric. Res. 29, 483–490.

Two tests for nitrites were made in nitrate broth cultures with sulfanilic acid and alpha-naphthylamine in 5 N acetic acid. Cultures showed good growth. Results were negative during the first few days but showed definite reduction after 2 to 3 weeks. Two tests for nitrites in the synthetic nitrate medium (KNO_3, 1 g; K_2HPO_4, 0.5 g; $CaCl_2$, 0.5 g; glucose, 10 g; distilled water, 1000 ml) gave positive results. Reduction of nitrates is evident in this synthetic medium in 24 hr.

Nitrate reduction—*Xanthomonas*

F. A. Wolf (1924). J. agric. Res. 29, 57–68.

Nitrate broth consisting of 1 per cent peptone, 0.3 per cent beef extract, and 0.1 per cent potassium nitrate supports abundant growth. Tests at appropriate intervals were made with sulfanilic-acid solution or with naphthylamine acetate solution.

Nitrate reduction—*Xanthomonas beticola*

N. A. Brown (1928). J. agric. Res. 37, 155–168.

Tests were made with nitrate-bouillon cultures 9 days old, using the starch-iodine-sulfuric acid test.

Nitrate reduction—*Pseudomonas*

M. K. Bryan (1928). J. agric. Res. 36, 225–235.

There is no nitrite reaction when 10-day-old nitrate beef broth cultures are tested with the starch potassium iodide-sulfuric acid method, but a very decided though very moderate nitrite reaction when the alpha-naphthylamine-sulfanilic acid test was used. This would indicate very weak nitrate reduction.

Nitrate reduction—*Streptococcus*

R. A. McKinney (1934). J. Bact. 27, 373–401

The test of R. M. Greenthal (1930)* was used to detect the reduction of nitrates to nitrites. A few crystals of nitrite-free sodium nitrate were dropped in an eighteen to twenty-four hour broth culture of streptococci. A few drops of 10 per cent sulfuric acid, potassium iodide and dilute starch were added in succession. None of the pleomorphic streptococci thus tested gave a positive reaction.

*J. infect. Dis. 40, 569. 1930.

Nitrate reduction—*Achromobacter*

H. F. Smart (1935). J. agric. Res. 51, 363–364.

7 days at 26°C in medium containing 0.1 per cent Difco peptone and 1 per cent potassium nitrate was used to test for nitrate reduction.

Nitrate reduction—*Bacteroides*

A. H. Eggerth (1935). J. Bact. 30, 277–299.

Nitrate broth was prepared by adding 0.5 per cent of sodium nitrate to infusion broth.

Nitrate reduction—*Clostridium*

R. S. Spray (1936). J. Bact. 32, 135–155.

Nitrate semisolid agar:

To make 1 L, take –

5 g	Difco Tryptone
5 g	Difco Neopeptone
2.5 g	agar flakes
1 L	distilled water

Boil to dissolve and add water to make 1 L. Adjust the reaction to pH 7.3 to 7.4 then add –

1 g	potassium nitrate
0.5 g	glucose.

Dissolve and mix well. Tube about 8 cm deep in 200 x 15 mm or 200 x 13 mm tubes. Autoclave at 15lb for 20 min.

The test is usually performed after 72 hours incubation, although the reaction does not vary thereafter up to 7 days. We prefer the use of 0.6 per cent dimethyl-alpha-naphthylamine (Eastman), with sulfanilic acid, both in dilute acetic acid.

Nitrate reduction—*Fusobacterium*

E. H. Spaulding and L. F. Rettger (1937). J. Bact. 34, 535–548.

Used –

Proteose peptone	10.0 g
Liebig's meat extract	3.0 g
Potato extract (aq.)	100 ml
Distilled water	900 ml
Nitrate	not specified
pH	7.6

The usual naphthylamine-sulfanilic acid test was used for detection of nitrite and zinc dust (ZoBell, 1932) for residual nitrate. Examined after 5 days' incubation.

C. ZoBell (1932). J. Bact. 24, 273–281.

Nitrate reduction—*Mycobacterium*

R. E. Gordon (1937). J. Bact. 34, 617–630.

The cultures were grown in a beef-extract medium containing 1 per cent KNO_3. After ten days' incubation at 25°C they were tested for nirites with Trommsdorf's reagent.

Nitrate reduction—*Pasteurella multocida*

C. T. Rosenbusch and I. A. Merchant (1939). J. Bact. 37, 69–89.

Tested in –

Difco proteose peptone	5.0 g
NaCl	5.0 g
$MgSO_4 \cdot 7H_2O$	0.1 g
K_2HPO_4	0.2 g
$NaNO_3$	2.0 g
H_2O	1000 ml
pH	7.2–7.4

Tests were made on 5 day-old cultures in the above medium.

Nitrate reduction—*Lactobacillus, Bacteroides*

K. H. Lewis and L. F. Rettger (1940). J. Bact. 40, 287–307.

Nitrate reduction in glucose-cysteine broth* containing 0.5 per cent potassium nitrate, by the sulfanilic acid dimethyl-alpha-naphthylamine method, as recommended by the Society of American Bacteriologists (1937). Committee on Bacteriological Technique. *Manual of methods for pure culture study of bacteria, staining procedures,* Leaflet 4. Geneva, N.Y. Observations were made after 4 and 13 days' incubation at 37°C.

*J. E. Weiss and L. F. Rettger (1934). J. Bact. 28, 501–521.

Nitrate reduction—*Streptobacillus*

F. R. Heilman (1941). A study of *Asterococcus muris (Streptobacillus moniliformis).* II. Cultivation and biochemical activities. J. infect. Dis. 69, 45–51.

Serum veal infusion broth cultures containing 0.1% of potassium nitrate were tested for the presence of nitrites after 2 and 4 days of incubation. There was no evidence of the presence of nitrites as tested by the usual method in which acetic acid solutions of sulfanilic acid and alpha-naphthylamine are employed.

Nitrate reduction—*Clostridium*

R. W. Reed (1942). J. Bact. 44, 425–431.

Bacto tryptone	20 g
Na_2HPO_4	2 g
Glucose	1 g
Agar	1 g
KNO_3	1 g
Water	1000 ml
pH	7.6

Autoclaved in deep tubes.

Qualitative tests for nitrite were made with Tittsler's sulfanilic acid, dimethyl-α-naphthylamine reagent. When negative, tests were made for nitrate by adding zinc dust. (C. E. ZoBell, 1932, J. Bact. 24, 273–281). R. P. Tittsler (1930). J. Bact. 19, 261–267.

Nitrate reduction—*Corynebacterium acnes*

H. C. Douglas and S. E. Gunter (1946). J. Bact. 52, 15–23.

In peptone yeast extract phosphate glucose thioglycolate broth containing –

bacto peptone	2%
bacto yeast extract	0.5%
glucose	1.0%
KH_2PO_4	2%
B.B.L. sodium thioglycolate	0.10%
agar	0.05% or 1.5% (depending upon whether a fluid or solid medium is desired).

+ 0.2% KNO_3

The pH should be adjusted to about 7.1 before sterilization and will be about 6.8 after sterilization. Considerable darkening of the medium takes place during autoclaving, but this has no apparent adverse effect upon the growth of the organisms.

Nitrate reduction—*Staphylococcus* and *Micrococcus*

Y. Abd-el-Malek and T. Gibson (1948). J. Dairy Res. 15, 249–260.

Cultures in peptone 1%, KNO_3 0.1%, were tested for nitrite and, if negative, for nitrate.

Nitrate reduction—*Pseudomonas alboprecipitans*

A. G. Johnson, A. L. Robert and L. Cash (1949). J. agric. Res. 78, 719–732.

Bacto-beef-peptone agar with 0.1 per cent potassium nitrate added was used to test for nitrate reduction.

Nitrate reduction—anaerobic, Gram-negative diplococci

G. C. Langford, Jr., J. E. Faber, Jr., and M. J. Pelczar, Jr. (1950). J. Bact. 59, 349–356.

Fluid thioglycolate medium to which had been added 0.1 per cent potassium nitrate was used for the determination of nitrate reduction. The method of testing for nitrite was indicated in the *Manual of Methods for Pure Culture Study* of the Committee of Bacteriological Technique of the Society of American Bacteriologists (1947). Biotech Pub., Geneva, N.Y.

Nitrate reduction—*Staphylococcus*

C. Shaw, J. M. Stitt and S. T. Cowan (1951). J. gen. Microbiol. 5, 1010–1023.

Organisms were grown in nutrient broth containing 0.1% (w/v) KNO_3. In recheck tests cultures at 37°C were incubated for 10 days. Cultures were tested for nitrites by the Griess-Ilosvay method but, as Wallace and Neave (1927) have shown, excess nitrite may obscure a positive results by bleaching the red colour. In the recheck dimethyl-α-naphthylamine was used in place of α-naphthylamine. All negative tests were confirmed by the addition of zinc dust to reduce residual nitrate in the medium (ZoBell, 1932). In a true negative the red colour only appears after the addition of the zinc. (Inoculum incubated 5 days at 37°C, 10 days at 30°C or 14 days at 22°C.)

G. I. Wallace and S. L. Neave (1927). J. Bact. 14, 377.
C. E. ZoBell (1932). J. Bact., 24, 273.

Nitrate reduction—*Escherichia*

P. Hauduroy (1951). Techniques Bactériologiques. Masson et Cie, Éditeurs, Paris.

Milieu pour la réduction des nitrates.

KNO_3 libre de nitrites	0.2 g
Peptone	1 g
Eau distillée	1000 cm^3

Répartir en tubes de 5 cm^3. Autoclaver à 120°C pendant 5 minutes 3 jours de suite.

Ensemencer le milieu et mettre à l'étuve à 37° pendant 4 jours. A ce moment on mélange à parties égales immédiatement avant l'emploi les deux solutions suivantes et 0.1 cm^3 du mélange est ajouté à chaque tube.

Solution A. — Dissoudre 5 g d'acide sulphanilique dans 1000 cm^3 d'acide acétique (5 normal).

Solution B. —Dissoudre 5 g d'α-naphtylamine dans 1000 cm^3 d'acide acétique (5 normal).

Une réaction positive donne une coloration rouge en moins de 10 minutes.

Nitrate reduction—general

P. Hauduroy (1951). Techniques Bactériologiques. Masson et Cie, Éditeurs, Paris.

NITRATES

(Réduction des). — Ensemencer le microbe dans du bouillon nitraté (1.0 p. 100 de peptone de Fairchild's contenant 0.02 p. 100 de nitrate de potassium sans nitrite). Stériliser dans la vapeur fluente. Cultiver pendant 5 jours à 37°C. Ajouter alors 2 ml de réactif de Criers-Hosway. L'apparition d'une couleur rose indique la présence de nitrite. *(W. I. C.).*

(Réduction des) (Anaérobies). — On recherche les nitrites par la réaction de Tromsdorff. Préparer une culture de 24 heures en eau peptonée. Ensemencer avec cette culture de l'eau peptonée nitratée à 5 p. 100 et additionner de différents sucres à 2 p. 100 le milieu étant réparti en tubes de Hall.

Rechercher les nitrites après 24-48-72 heures. Pour ce faire, prélever à la pipette 0.5 ml de culture et ajouter une goutte d'empois d'amidon. La présence de nitrite est indiquée par l'apparition d'une coloration bleue. Il est important de rechercher l'action des germes sur les nitrates en présence des différents sucres, car l'attaque peut être plus ou moins rapide et en utilisant un seul sucre, le glucose par exemple, on pourrait manquer le stade nitrite de la réduction. *(A. I. P.).*

(Réduction des). — Cultiver les bactéries dans du bouillon nutritif contenant 0.1 p. 100 de KNO_3 pendant 3 à 5 jours. A 5 ml de cultures, ajouter 0.5 ml de la solution d'amidon de pomme de terre bouilli à 1 p. 100 et 0.5 ml d'une solution à 0.4 p. 100 de KI. Mélanger. Ajouter une goutte de SO_4H_2 concentré. Si des nitrates sont présents il se forme une couleur bleue. *(A. T. C. C.).*

NITRITES

Nitrites = Nitrates (Réduction des).

Nitrate reduction—microtest

P. H. Clarke and S. T. Cowan (1952). J. gen. Microbiol. 6, 187–197.

Method:

0.04 ml suspension is mixed in 65 x 10 mm tubes with 0.05% $NaNO_3$, 0.06 ml; phosphate buffer (0.025 M, pH 6.8), 0.04 ml. After 30 minutes at 37°C (water bath), 0.06 ml dimethyl-α-naphthylamine solution (6 ml/L of 5N-acetic acid) and 0.06 ml sulfanilic acid (8 g/L of 5N-acetic acid) is added and read 5 minutes later. A blank test on the nitrite and buffer is included.

Nitrate reduction—*Bacillus*

N. R. Smith, R. E. Gordon and F. E. Clark (1952). U.S.D.A. Agr. Monograph No. 16.

The method described in the Manual of Methods (1951) was adopted for this determination. The cultures were grown in nutrient broth containing 0.1 per cent KNO_3 and incubated at suitable temperatures. After 1, 3, 6 or more days several drops of each of the following solutions were added to a small portion of the culture; (1) sulfanilic acid, 8 g; 5 N acetic acid (1 part glacial acetic acid to 2.5 parts of water), 1,000 ml; (2) dimethyl-alpha-naphthylamine, 6 ml; 5 N acetic acid, 1000 ml. The appearance of a deep red color or a heavy yellowish precipitate indicated the presence of nitrite. If the reaction was negative, the rest of the culture was tested for nitrates by adding a few drops of diphenylamine and pouring concentrated H_2SO_4 down the side of the tube to form a layer underneath. A blue color at the junction of the two layers was indicative of nitrates.

Anaerobic production of gas from nitrate: A medium suggested by Gibson (1944) was modified for demonstrating anaerobic formation of gas from nitrate. It contained 1.0 percent tryptose, 0.5 percent K_2HPO_4, 0.3 percent beef extract, 0.2 percent yeast extract, and 1.0 percent $NaNO_3$. Approximately 8 ml of the broth were put into 16 mm tubes and autoclaved. The pH was about 7.6. Before use the medium was steamed to drive off the free oxygen, quickly cooled, inoculated, capped with 10 to 15 mm of sterile melted vaspar (equal parts of vaseline and paraffin) and incubated at 32°C. Observation for gas was made at 7 and 14 days.

For species requiring incubation temperatures of 45° or 50°C the technique had to be modified as follows: Melted vaspar was poured on top of the inoculated medium to a depth of 3 or 4 mm and allowed to harden. Melted 1.5 percent water agar was then added to a depth of 15 or 20 mm; the tubes cooled, and the culture finally incubated at the high temperatures. Results were usually obtained in 2 or 3 days before the agar developed cracks caused by drying. A layer of vaspar may also be placed on top of the agar.

T. Gibson (1944). J. Dairy Res. 13, 248–260.

Society of American Bacteriologists, Committee on Bacteriological Technic (1951). *Manual of Methods for Pure Culture Study of Bacteria.* Leaflet V. 12th ed. Geneva, N.Y.

Nitrate reduction—soil bacteria

H. de Barjac (1952). Annls Inst. Pasteur, Paris 83, 207–212.

1° Préparation du milieu. – Une solution contenant

NO_3K	2 g
Glucose	10 g
CO_3Ca	5 g
Milieu salin de Winogradsky à 1/20 Q.S.p.	1000 cm^3

est répartie, à raison de 1 cm^3, en tubes à hémolyse. On stérilise à 110°, vingt minutes.

2° Préparation des dilutions . — A partir d'une dilution au 1/10, effectuée au moriter stérile, avec 1 g de terre homogénéisée dans 9 g d'eau de Seine stérile, on réalise une gamme de dilutions allant de 10^{-1} à 10^{-8}.

3° Ensemencement. — Chaque dilution est répartie en 6 tubes de milieu à raison de 0.5 cm^3. On agite et l'on porte à l'étuve à 28°.

4° Lectures. — Les deuxième, quatrième, sixième, huitième, onzième et quatorzième jours, soit 6 lectures, une gamme de dilutions est testée:

A la diphénylamine sulfurique, après élimination de NO^{2-} par l'urée → NO^{3-}.

Au réactif de Griesz → NO^{2-}.

A l'iodure du potassium (10 p. 100) et l'extrait de Javel (le réactif de Nessler étant, en effet, inutilisable du fait de la présence de glucose, nous y substituons la recherche d'une formation d'iodure d'azote) → NH^3.

Ainsi, l'on peut suivre simultanément l'évolution respective des nitrates, des nitrites et de l'ammoniaque.

Nitrate reduction—Peste "Sauvage" du Kurdistan

M. Baltazard and P. Aslani (1952). Annls Inst. Pasteur, Paris 83, 241–247.

Pour la recherche de la production d'acide nitreux, et de la réduction des nitrates en nitrites, les techniques sont moins bien fixées. Pollitzer [2] recommande, pour la seconde de ces recherches, la technique de Petragnani [4], ce qui est inexact, cet auteur ayant seulement décrit une technique de recherche de la production par le bacille pesteux d'acide nitreux en bouillon de foie ordinaire, non additionné de nitrates. Cette technique de Petragnani étant par ailleurs décrite d'une manière peu précise nous avons, pour notre part, fixé les modes de recherche suivants d'après les indications de R. Pollitzer.

Le milieu utilisé est le bouillon de foie (foie de veau: 500 g, peptone Chapoteaut: 10 g, NaCl: 5 g), pH ajusté à 7.5, employé tel pour la recherche de la production d'acide nitreux. Pour la recherche de la réduction des nitrates en nitrites, le bouillon de foie est additionné de nitrate de potasse dans la proportion de 1 p. 100. Deux tubes au moins de chaque milieu sont ensemencés largement comme il a été dit précédemment. Après quarante-huit heures, la réaction suivante est pratiquée, après vérification de l'abondance de la pousse.

On ajoute à chaque tube 0.1 cm^3 d'une solution d'acide sulfanilique dans l'acide acétique (0.2 g pour 100 cm^3 d'acide acétique à 30 p. 100 [5N]. Poids spécifique 1041). Le mélange est soigneusement agité puis on ajoute goutte à goutte une solution d'α-naphtylamine dans l'acide acétique (0.5 g pour 100 cm^3 d'acide acétique comme ci-dessus). Si la réaction est positive, une coloration allant du rose franc au rouge clair doit apparaître dans le tube, après adjonction de V gouttes au maximum.

[2] Pollitzer (R.). *Note de documentation pour le Comité d'Experts de la Peste de l'OMS,* 1950.

[4] Petragnani (G.). *Bull Office internat. Hyg. publ.,* 1937, 29, 2522.

Nitrate reduction—*Pseudomonas*

P. Villecourt, H. Blachère and G. Jacobelli (1952). Annls Inst. Pasteur, Paris 83, 316–322.

En eau peptonée à 1 p. 100 de nitrate, la production de nitrite était recherchée par le réactif de Griess et la production de gaz avec une cloche.

Nitrate reduction—*Pseudomonas (Malleomyces) pseudomallei*

P. de Lajudie and E. R. Brygoo (1953). Annls Inst. Pasteur, Paris 84, 509–515.

Nous avons utilisé la technique de Griess-Ilosva sur culture de cinq jours en bouillon nitraté. La réaction fut également pratiquée sur culture en bouillon non nitraté, pour rechercher le pouvoir nitrifiant éventuel de bacille de Whitmore sur les rotéines. Devignat considère cette espèce comme "Griess-négative."

Nitrate reduction—*Haemophilus*

W. Smith, J. H. Hale, C. H. O'Callaghan (1953). J. Path. Bact. 65, 229–238.

For the nitrate-reduction tests, sodium nitrate was added to the medium to give a final concentration of 0.1 per cent (w/v).

Reduction:

This was detected by the demonstration of nitrite in the culture by means of the Greiss-Ilosvay reagent.

Nitrate reduction—*Mycobacterium*

R. E. Gordon and M. M. Smith (1953). J. Bact. 66, 41–48.

Tubes of nutrient broth plus 0.1 per cent KNO_3 were inoculated and incubated at 28°C. At 5, 10 and 14 days 1 ml of the broth culture was withdrawn with a sterile pipette and mixed with 3 drops of each of the following solutions: (a) sulfanilic acid, 8 g; 5 N acetic acid (1 part of glacial acetic acid to 2.5 parts of water), 1000 ml; (b) dimethyl-alpha-naphthylamine, 6 ml; 5 N acetic acid, 1000 ml. The appearance of a heavy, yellowish precipitate or red color within 10 minutes was considered proof of the presence of nitrites (Conn, 1951.) In the absence of nitrite, a qualitative determination for nitrate always was made by adding 4 to 5 mg of zinc dust to the tube previously tested for nitrites. Reduction of nitrate, if present, to nitrite by the zinc resulted in the formation of a red color. The use of an excessive amount of zinc dust should be avoided.

H. J. Conn, (1951). *Manual of Methods for the Pure Culture Study of Bacteria.* Biotech. Pub., Geneva, N.Y., Leaflet 5, 12th ed., 10–11.

Nitrate reduction—soil

H. de Barjac (1954). Annls Inst. Pasteur, Paris 87, 440–444.

Technique: Un milieu composé de NO_3K, 2 g; glucose, 10 g; CO_3Ca, 5 g, et solution de Winogradsky au 1/20 q. s. p. 1 000 est réparti en Erlenmeyer de 500 ml, à raison de 400 ml de milieu par fiole. L'ensemencement de chaque échantillon de sol est effectué sur une série de 8 fioles d'Erlenmeyer par la technique des dilutions de manière à obtenir une gamme de concentrations finales correspondant à celles utilisées dans la technique en tubes (soit 0.5 ml de dilution de 10 en 10 pour 1 ml de milieu). Les fioles ensemencées sont soumises au vide et portées à l'étuve à 28°C. Chaque jour un prélèvement est effectué dans chaque fiole, pour la recherche de NO_3^-, NO_2^-, et NH_3.

Nitrate reduction—*Brucella*

M. J. Pickett and E. L. Nelson (1954). J. Bact. 68, 63–66.

The procedure selected as most satisfactory for nitrate reduction tests was as follows: (1) Nitrate medium – 0.05 per cent sodium nitrate in 2 per cent of Albimi's C peptone, 1 ml in a 12 by 100 mm tube. (2) Inoculum – 0.1 ml of a stock suspension containing 5×10^9 bacteria per ml. (3) Incubation – 35°C, aerobic. (4) Nitrite reagent – 0.5 per cent sulfanilic acid and 0.5 per cent alpha-naphthylamine in 1 per cent acetic acid. (5) Nitrite detection – one 4 mm loop of nitrate broth culture was added to one drop of nitrite reagent. Both faintly pink and frankly red tests were recorded as positive. Tests were made after 6, 24, 30 and 48 hours of incubation; cultures which were nitrite negative at 48 hours were tested for nitrate by depositing a few mg of granular zinc and six drops of nitrite reagent directly into the culture tube.

Nitrate reduction—various

N.C.T.C. Methods 1954.

Method 1.

Medium: broth containing 0.1% KNO_3.

Incubate at 30°C and test after 3 and 14 days.

Reagents: *Solution A*

dimethyl-α-naphthylamine	6 ml
5 N acetic acid (1 part glacial acetic acid to 2.5 parts water).	1 litre

Solution B

Sulfanilic acid	8 g
5 N acetic acid	1 litre

Test: to culture add 1 ml solution A, followed by 1 ml solution B. Red colour = nitrites present.

Negatives: test by adding powdered zinc. Red colour = nitrates present

in medium (not reduced until zinc added); no colour = nitrates not present in medium (i.e., reduced to nitrites, and nitrites reduced further).
C. E. ZoBell (1932). J. Bact. 24, 273.

Method 2.
Medium: as in method 1. Incubate at 30°C.
Test: To 5 ml culture add 0.5 ml 1% solution of potato starch paste (boiled), 0.5 ml 0.4% solution KI, and, after mixing, 1 drop concentrated H_2SO_4. Blue colour = nitrites present.
Negatives: test by adding to 5 ml of culture a few drops of diphenylamine in concentrated H_2SO_4 to form a layer underneath: a blue colour at the junction of the two layers indicates presence of nitrate.

Method 3
Medium: broth containing 0.1% KNO_3 made semisolid with 0.3% agar; 8 ml per tube. Incubate at 30°C.
Test: as method 1, but leave for 1 hour before making final reading.
Negatives: Check with zinc dust and leave for 1 hour.

Method 4:

Medium:	Nitrite free KNO_3	0.2 g
	Peptone	1.0 g
	Distilled water	1 litre

Reagents: *Solution B* as for method 1
Solution C

Dissolve 5 g α-naphthylamine in 1 litre of 5 N acetic acid.

Mix solutions B and C in equal parts immediately before use and add to culture in proportion of 1 to 5.

Nitrate reduction—*Bacillus*

E. Grinsted and L. F. L. Clegg (1955). J. Dairy Res. 22, 178–190.

Reduction of nitrates to nitrites was tested for by the method described by the Society of American Bacteriologists (1934, Leaflet 5, 5th ed. p. 16). Three drops of the sulfanilic acid solution and of the alpha-naphthylamine solution were added to 1 ml of each broth culture after 1, 3 and 7 days incubation.

Nitrate reduction—*Lactobacillus*

G. H. G. Davis (1955). J. gen. Microbiol. 13, 481–493.

Nitrate reduction was tested in the following medium: Oxoid peptone, 0.5% (w/v); Difco yeast extract, 0.3% (w/v); glucose, 0.5% (w/v); agar, 0.1% (w/v); potassium nitrate (KNO_3), 0.2% (w/v); Salts A,* 0.5% (v/v); and Salts B,* 0.5% (v/v). The pH was adjusted to 6.8 and the medium dispensed into 5 x ½ in. tubes, each half-filled, and autoclaved at 15 lb for 15 min. The inoculum consisted of 2 or 3 drops of 24 hr TJ broth culture delivered from a capillary pipette. The tests were incubated for 6 days at 37°C and to each test was added one drop of dilute iodine solution and 1 ml of each of the Griess-Ilosvay reagents (Topley and Wilson's *Principles of Bacteriology and Immunity,* 1946, p. 368). The tests were allowed to stand at room temperature for 30 min and the results noted; false negative reactions were checked by the addition of powdered zinc and further observations of colour change. The presence of adequate growth was always noted before the test was made.

*See A. C. Hayward (1957). J. gen. Microbiol. 16, 9–15.

Nitrate reduction—*Bacillus*

K. L. Burdon (1956). J. Bact. 71, 25–42.

Cultures in nitrate broth (Difco) were tested for nitrite at intervals during incubation for 24 to 72 hr by adding the test reagents (sulfanilic acid and alpha-naphthylamine) to portions of the broth culture. If the reaction for nitrite remained negative for 72 hr, the original culture was tested for persistence of nitrate by addition of a pinch of zinc dust (after introduction of the nitrite test reagents) and observed for appearance of a pink colour.

Nitrate reduction—*Actinomyces bovis*

S. King and E. Meyer (1957). J. Bact. 74, 234–238.

To determine nitrate reduction, a buffered thioglycolate medium, described by Douglas and Gunter (1946), was used, to which nitrate was added. It is advisable to test for nitrate reduction within 5 to 7 days.

Occasionally, with the "anaerobic diphtheroids," nitrite had disappeared after this interval.

H. C. Douglas and S. E. Gunter (1946). J. Bact. 52, 15–23.

Nitrate reduction—*Leptotrichia buccalis*

R. D. Hamilton and S. A. Zahler (1957). J. Bact. 73, 386–393.

Medium:

Peptone (Difco)	20 g
Yeast extract (Difco)	2 g
Glucose	5 g
NaCl	5 g
Na thioglycolate	1 g
K_2HPO_4	10 g
KNO_3	1.0 g
Distilled water	1 litre

pH adjusted to 7.2 before autoclaving.

The medium was inoculated and tested for nitrite after 24 and 72 hours using the method recommended by the Committee on Bacteriological technic of the Society of American Bacteriologists *Manual of Methods for the Pure Culture Study of Bacteria* (1953). Williams and Wilkins Co., Baltimore, Maryland.

Nitrate reduction—*Nocardia*

R. E. Gordon and J. M. Mihm (1957). J. Bact. 73, 15–27.

Cultures were prepared in the following medium: peptone, 5 g; beef extract, 3 g; KNO_3, 1 g; distilled water, 1000 ml; pH 7.0. After 5, 10, and 14 days' incubation at 28°C, 1 ml of the broth culture was withdrawn aseptically and mixed with 3 drops of each of the following solutions: (1) Sulfanilic acid, 8 g; 5 N acetic acid (1 part of glacial acetic acid to 2.5 parts of water), 1000 ml; (2) Dimethyl-α-naphthylamine, 6 ml; 5 N acetic acid, 1000 ml (Conn, 1951). The presence of nitrite was indicated by a red color. In the absence of nitrite after 14 days' storage, 4 to 5 mg of zinc dust were added to the tube previously tested for nitrite. The zinc reduced the nitrate, if present, and produced a red color. Nitrate had to be demonstrated in the broth before a culture giving a negative reaction for nitrite was reported as unable to reduce nitrate to nitrite.

H. J. Conn (1951). Manual of Methods for Pure Culture Study of Bacteria, 12th ed., pp 10–11, Leaflet V. Biotech. Pub., Geneva, N.Y.

Nitrate reduction—thermophilic sulfate reducing bacteria

L. L. Campbell Jr., H. A. Frank and E. R. Hall (1957). J. Bact. 73, 516–521.

Determined in the medium of Reed and Orr (1941), after incubation for 14 and 48 hr. Sulfanilic acid and α-naphthylamine were used to determine nitrite.

G. B. Reed and J. H. Orr (1941). War Med. 1, 493–510.

Nitrate reduction

Society of American Bacteriologists *Manual of Microbiological Methods* – McGraw-Hill Book Co. Inc., New York 1957.

Nitrate reduction should be indicated by complete or partial disappearance of nitrate, accompanied by appearance of nitrite, ammonia, or free nitrogen. As quantitative nitrate tests are too time-consuming for routine pure culture work, one must ordinarily be satisfied with tests for the end products only.

The following routine procedure is recommended: Inoculate into nitrate broth and onto slants of nitrate agar (containing 0.1 per cent KNO_3 plus beef extract and peptone as usual). Test the cultures on various days. On these days examine first for gas as shown by foam on the broth or by cracks in the agar. Then test for nitrite with the following reagents.

1. Dissolve 8 g of sulfanilic acid in 1 liter of 5 N acetic acid (1 part of glacial acetic acid to 2.5 parts water), or in 1 liter of dilute sulfuric acid (1 part concentrated acid to 20 parts of water).

2. Dissolve 5 g of α-naphthylamine in 1 liter of 5 N acetic acid, or of very dilute sulfuric acid (1 part of concentrated acid to 125 parts of water). Or dissolve 6 ml of dimethyl-α-naphthylamine in 1 liter of 5 N acetic

acid. This latter reagent has recently been recommended by Wallace and Neave (1927), and by Tittsler (1930), as it gives a permanent red colour in the presence of high concentrations of nitrite.

Put a few drops of each of these reagents in each broth culture to be tested and on the surface of each agar slant. A distinct pink or red in the broth or agar indicates the presence of nitrite. It is well to test a sterile control which has been kept under the same conditions, to guard against errors due to absorption of nitrous acid from the air.

A rapid method, calling for less than an hour's incubation, has been devised by Bachmann and Weaver (1947), and a rapid microtechnique by Brough (1950) and by Clarke and Cowan (1952).

Among the micromethods, Brough's is detailed in a concise, easily followed form. The culture is grown 18–24 hr at optimum temperature on a nutrient agar giving good growth. The growth is washed off and sufficient to give high turbidity is added to 1 ml (in a small test tube) of 0.1 per cent KNO_3 in nutrient broth; Clarke and Cowan showed that the solution to which the KNO_3 is added may be pH 6.8 buffer instead of broth. The medium should be preheated in a water bath to 37°C before inoculation; after inoculation it needs to be incubated only 15 min at that temperature before adding the reagents of the nitrite test.

Presence of nitrite shows the nitrate to have been reduced, and the presence of gas is a strong indication that reduction has taken place. A negative result does not prove that the organism is unable to reduce nitrates; in such a case further study is necessary as follows:

In case the fault seems to lie in poor growth, search should be made for a nitrate medium in which the organism in question does make good growth by means of the following modifications: increasing or decreasing the amount of peptone, changing the amount of nitrate, altering the reaction, adding some readily available carbohydrate, adding 0.1–0.5 per cent of agar to a liquid medium to furnish a semisolid substrate. The appearance of nitrite in any nitrate medium whatever (while it is absent in a sterile control) should be recorded as nitrate reduction.

Absence of nitrite in the presence of good growth may indicate complete consumption of nitrate or its decomposition beyond the nitrite stage as well as *no* reduction at all. Test, therefore, for nitrate by adding a pinch of zinc dust to the tube to which the nitrite reagents have been introduced and allowing it to stand a few minutes. If nitrate is present, it will be reduced to nitrite and show the characteristic pink colour. Confirmation of the test may be obtained by placing a crystal of diphenylamine in a drop of concentrated sulfuric acid in a depression in a porcelain spot plate and touching with a drop of the culture (or of the liquid at the base of the slant if agar cultures are used). The test will be more delicate if the culture is first mixed with concentrated sulfuric acid and allowed to cool. A blue colour indicates presence of nitrate, provided nitrite is absent, but as nitrite gives the same colour with diphenylamine, this test must not be used when nitrite is present in the same or greater order of magnitude.

If none of these tests indicates utilization of the nitrate, the organism probably does not reduce nitrate, but to be certain of the fact further investigation is necessary. It must be understood, however, that for routine diagnostic work a determination of nitrite on standard nitrate broth or agar is ordinarily sufficient; this is because most descriptions in the literature containing the words "Nitrates not reduced" mean merely that no nitrite is produced on this medium. But in recording such results the student should be careful to state only the observed fact—i.e., that nitrite is or is not found in the nitrate medium employed.

Nitrate broth: Add 0.1 per cent of KNO_3 to a nutrient broth or agar. For obligate anaerobes, also add 0.1 per cent of glucose and 0.1 per cent of agar to basal medium and tube in deep columns. For organisms not reducing nitrates in a peptone medium the following synthetic medium of Dimmick (1947) is recommended:

K_2HPO_4	0.5 g
NaCl	0.5 g
$MgSO_4 \cdot 7H_2O$	0.2 g
$NaNO_3$	2.0 g
glucose	10.0 g
agar	15.0 g
distilled water	1000 ml

If the organism being studied requires more calcium, add 0.05 g of $CaCl_2$ to the above; in this case it is important that the final pH be 7.2; to assure this *after sterilization,* adjustment before sterilization should be to about 7.8.

B. Bachmann and R. H. Weaver (1947). J. Bact. 54, 28.
F. K. Brough (1950). J. Bact. 60, 365–366.
P. H. Clarke and S. T. Cowan (1952). J. gen. Microbiol. 6, 187–197.
I. Dimmick (1947). Can. J. Res., Sec. C 25, 271–273.
R. P. Tittsler (1930). J. Bact. 19, 261–267.
G. I. Wallace and S. L. Neave (1927). J. Bact. 14, 377–384.

Nitrate reduction

V. B. D. Skerman, *A Guide to the Identification of the Genera of Bacteria.* 2nd ed. Williams and Wilkins Book Co., Baltimore, 1967.

Note: This reaction is dependent on a number of factors.

1. It obviously will not occur if the medium in which the nitrate is incorporated does not support the normal growth of the organism concerned. The test need not be restricted to the use of the peptone basal medium

normally employed, but for comparative purposes the behavior of the organism in the nitrate-peptone water should be recorded.

2. It will not occur if the organism receives a supply of oxygen adequate for all respirational needs during growth, as may occur (a) in still cultures of slow-growing cells in which the medium is distributed in shallow layers which admit oxygen by diffusion as rapidly as it is utilized by the cells, or (b) in aerated cultures. This is due to one of two factors. The nitrate-reducing enzyme is not formed in the presence of an adequate oxygen supply and the enzyme, once formed, does not function under such conditions.

There are reports of strictly aerobic autotrophic organisms which will use nitrate as the sole source of nitrogen; this suggests that reduction is occurring in the presence of oxygen. In view of the generally slow growth of such organisms and the presence of adequate amounts of ammonia in the average laboratory air, such reports must be considered with caution. It is possible that some intracellular mechanism is involved in the reduction of nitrate for synthetic purposes.

Under anaerobic conditions, such as exist in deep still cultures incubated aerobically after the dissolved oxygen is utilized by the growing cells and under strictly anaerobic conditions, nitrate may be reduced to nitrite, ammonia, nitrous oxide, or gaseous nitrogen.

With organisms in which the reduction is limited to nitrite the latter is readily detected. With the others the detection of nitrite will depend on the degree to which it accumulates at any time during growth and the actual time at which the test is performed. For this reason simple testing of the medium for nitrite at a fixed time interval after inoculation is valid only if the test is positive. A negative reaction is inconclusive.

The use of suspensions of cells grown anaerobically in the presence of nitrate to adapt them to nitrate reduction, in tests employing a substrate and nitrate, is also subject to error, since in the presence of certain substrates some organisms will reduce nitrate to nitrogen without any detectable production of nitrite.

Procedure:

Select the medium most suited to the growth of the organism. Incorporate in the medium 0.2 per cent KNO_3, dispense in 5-ml amounts in 150- by 13- mm tubes, and sterilize. Allow the tubes to cool in air for 24 hours to permit the oxygen in the medium to equilibrate with air (except where required for anaerobic cultures). Inoculate the medium and cover it with a sterile paraffin seal. Incubate in a water bath at the optimal temperature. Observe the development of turbidity in the medium, and begin spot tests for nitrite as soon as the medium becomes obviously turbid to the naked eye. At this stage there is rarely any dissolved oxygen left in the medium. Remove the test samples with a warmed sterile dropping pipette, and reseal the paraffin after the removal of the sample. Continue the tests until maximal turbidity has been reached and for 12 hours beyond this time.

If the nitrite test is at any stage positive, record as "nitrite produced from nitrate" and specify the medium. If the latter is not nitrate-peptone water, the comparative reaction in nitrate-peptone water should be included and recorded even if no growth occurs.

If the nitrite tests are continually negative and there is no evidence of gas production, it is unlikely that nitrate is reduced.

Gas production may be due to evolution of nitrogen or nitrous oxide. The nature of the gas has not, as yet, any taxonomic significance. It should be remembered, however, that some organisms produce gas from peptone water alone under anaerobic conditions so that nitrate-free controls should be used.

1. *The Starch-Iodide Spot Test for Nitrite*

Reagents: Starch Iodide Solution:

Starch	0.4 g
$ZnCl_2$	2.0 g
H_2O	100 ml

Dissolve the $ZnCl_2$ in 10 ml of water. Boil and add the starch. Dilute to 100 ml, allow to stand for 1 week, and filter. Add an equal volume of 0.2 per cent solution of KI.

Hydrochloric Acid:

Concentrate HCl	16 ml
Water	84 ml

Test:

Using clean glass dropping pipettes, place 1 drop of each reagent in the depression of a white spot test plate. Add 1 drop of the culture. A blue color indicates the presence of nitrite.

The test depends on the formation of nitrous acid and its subsequent reaction with potassium iodide with the liberation of iodine which turns the starch blue. The test is not entirely specific. Control tests should be made with uninoculated media. Avoid the use of metal implements in taking samples.

2. *The Sulfanilic Acid-a-Naphthylamine method*

Reagents:

Sulfanilic Acid Solution: Dissolve 0.5 gm in 30 ml of glacial acetic acid. Add 100 ml of distilled water and filter. The reagent is stable for 1 month.

a-Naphthylamine Solution: This should not be more than 1 week old.

Dissolve 0.1 gm of *a*-naphthylamine in 100 ml of boiling distilled water. Cool and add 30 ml of glacial acetic acid. Filter. For use mix Solutions 1 and 2 in equal quantities.

Spot Test:

To 1 drop of the mixture of solutions 1 and 2, add 1 drop of the culture. The development of a pink color indicates the presence of nitrite.

Reference: The general procedures are the author's recommendation.

Nitrate reduction—*Gaffkya homari, Aerococcus viridans,* tetrad-forming cocci, *Pediococcus*

R. H. Deibel and C. F. Niven, Jr. (1960). J. Bact. 79, 175–180.

Nitrate reduction was tested in the semisolid indole-nitrite medium (BBL) supplemented with 0.5% yeast extract. Nitrite was detected using the sulfanilic acid-α-naphthylamine reagent.

Nitrate reduction—*Fusobacterium*

A. C. Baird-Parker (1960). J. gen. Microbiol. 22, 458–469.

Medium (%, w/v): proteose peptone no. 3, 1.0; Lab. Lemco, 0.3; Oxoid yeast extract, 0.1; glucose, 0.1; L-cysteine HCl, 0.05; disodium hydrogen phosphate, 0.5; 0.01, either sodium nitrate or sodium nitrite. Cultures were incubated for 10–14 days at 37°C; then to each culture which showed good growth, 2 or 3 drops of iodine were added followed by 1 ml each of the Griess-Ilosvay reagents (Society of American Bacteriologists, 1957). The tests were stood for 30 min at room temperature and the results noted. False negative, i.e. reduction of nitrate beyond nitrite to ammonia or nitrogen, were checked by adding powdered zinc to the broths. The production of ammonia from nitrite was tested with suspensions of washed organisms.

Nitrate reduction—*Escherichia*

F. Egami, Y. Hayase and S. Taniguchi (1960). Annls Inst. Pasteur, Paris 98, 429–438.

Mesure de l'activité de la nitrate-réductase en présence de divers donateurs d'électrons. — 1° Avec le formiate: un tube de Thunberg contenait en concentrations finales: KNO_3 0.01 M, le tampon phosphate pH 7.2 0.05 M, et le formiate 0.05 M, et dans la chambre secondaire, la suspension bactérienne ou les extraits. Volume total 1 ml. Le vide est réalisé à l'aide d'une pompe à eau; les tubes sont portés à 30°C; après dix minutes d'équilibre, la réaction est mise en route. Les nitrites formés sont mesurés avec le réactif de Griess-Ilosvey.

2° Avec le glucose ou le bleu de méthylène réduit par le glucose. Le formiate est ici remplacé par le glucose 0.05 M en présence ou en absence de bleu de méthylène 5×10^{-4} M.

3° Avec le méthylviologène réduit: méthylviologène (2×10^{-3} M) et moins de 1 mg de $Na_2S_2O_4 \cdot 2H_2O$. Le vide est rapidement effectué. Nous avons défini une unité de nitrate réductase comme étant la quantité d'enzyme nécessaire pour la formation d'une μmole de nitrite par heure dans les conditions indiquées.

L'activité spécifique est définie par le nombre d'unités par rapport à 1 mg d'azote de la préparation enzymatique.

Nitrate reduction—*Pediococcus*

H. L. Günther and H. R. White (1961). J. gen. Microbiol. 26, 185–197.

Incubation was carried out for 7 days in the medium of Davis (1955), from which salt solutions 'A' and 'B' had been omitted. Cultures were then tested for the presence of nitrite and of nitrogen gas as described in the *Manual for Pure Culture Study* (1954). The medium was tested for the presence of nitrite before incubation and for residual nitrate after incubation.

Maintenance of stock cultures and methods of cultivation. For maintenance of stock cultures, preparation of inocula, and in all experimental work, "Oxoid" tomato juice (TJ) broth or tomato juice (TJ) agar, adjusted to pH 6.6, was used unless otherwise stated. The following were exceptions to this rule: for strain Tc. 1, sodium chloride (5%, w/v) was added to the medium; and for the aerococci, glucose Lemco broth (Shattock and Hirsch, 1947) or glucose yeast extract (GY) agar (containing, as %, w/v, peptone, 1.0; Yeastrel, 0.3; glucose, 1.0; NaCl, 0.25; agar, 1.0; pH 7.4) was used.

Cultures were incubated aerobically at 30°C, with specified exceptions.

G. H. G. Davis (1955). J. gen. Microbiol. 13, 481.
Manual of Pure Culture Study of Bacteria (1954). Soc. Amer. Bact., Geneva, N.Y., Leaflet V., 54.
P. M. F. Shattock and A. Hirsch (1947). J. Path. Bact. 59, 495.

Nitrate reduction—*Pseudomonas*

M. Véron (1961). Annls Inst. Pasteur, Paris 100, Suppl. 6, 16–42.

La réduction des nitrates en bouillon-nitrate (0.1 p. 100 de NO_3K).

Nitrate reduction—*Mycobacterium*

H. Boisvert (1961). Annls Inst. Pasteur, Paris 100, 354–357.

La méthode préconisée avait été employée par Bratton et Marshall [4] pour le dosage des sulfamides. Les amines aromatiques donnent naissance en présence de NO_2Na et d'HCl, à un chlorure de diazonium. Celui-ci peut être copulé avec certains réactifs pour former un azoique coloré. Dans la technique de Bratton et Marshall, ce réactif copulant est le chlorhydrate de N-(1-naphtyl)éthylène-diamine. En France on utilise le chlorhydrate de naphtyl-diéthyl-propylène-diamine ou réactif IV de Tréfouël [5].

Dans le cas présent on utilise la méthode pour doser les nitrites en présence de sulfanilamide.

La réaction qualitative de Virtanen que nous employons consiste à introduire dans un tube à hémolyse contenant 1 goutte d'eau distillée pour faciliter le dépôt, une bonne anse de culture sur milieu de Jensen, soit 10 à 15 mg de bacilles.

On verse dans les tubes 2 cm^3 d'une solution M/100 de NO_3Na en tampon phosphate M/45 à pH 7 (NO_3Na: 0.085 g; PO_4H_2K: 0.117 g; PO_4H $Na_2 \cdot 12$ H_2O: 0.485 g; eau distillée: 100 cm^3).

Après deux heures d'incubation à l'étuve à 37° on ajoute dans chaque tube 1 goutte d'HCl concentré, II gouttes d'une solution de sulfanilamide à 0.2 p. 100, II gouttes de chlorhydrate de N·α-naphtyl-diéthyl-propylène-diamine (Prolabo) à 0.1 p. 100. Nous avons substitué ce dernier réactif, après essai, au bichlorhydrate de N-(1-naphtyl)-éthylènediamine (BDH) utilisé par Virtanen. Cette dernière solution doit être renouvelée toutes les semaines.

Dans les réactions positives le liquide est rouge-violet intense ou présente une coloration moyenne. Dans les réactions négatives le liquide reste incolore, ou l'on observe une légère teinte rosée. Tous les titrages doivent naturellement être accompagnés de témoins.

Observations. — Parfois la culture sur Jensen est pauvre, elle contient bien 10 mg de culture qu'il est évidemment impossible de prélever intégralement. On fait alors directement la réaction dans le tube qui est incubé en position couchée. Dans ce cas il ne faut tenir compte que des réactions positives. Dans les cultures en milieu liquide de Dubos ou de Youmans on introduit, pour 5 cm^3, II gouttes d'une solution à 4.25 p. 100 de NO_3Na et, après deux heures d'étuve, III gouttes d'HCl et V gouttes des autres réactifs.

La réaction doit se pratiquer sur des cultures en pleine croissance. Pour les souches de bacille tuberculeux humain, l'intensité de la réaction diminue après un mois.

[4] Bratton (A. C.) et Marshall (E. K.). *J. chem. Biol.,* 1937, 122, 263.
[5] Boyer (F.). *Thèse de sciences,* Paris, 1951.

Nitrate reduction—*Lactobacillus*

M. Rogosa (1961). J. gen. Microbiol. 24, 401–408.

Costilow and Humphreys's (1955) observation that certain strains of *Lactobacillus plantarum* reduced nitrates under certain conditions was confirmed. Two strains of *L. fermenti* also reduced nitrates. In static culture, agar and anaerobiosis were not essential for nitrate reduction, contrary to speculations in the literature. Nitrate reduction was possible only in media with restricted carbohydrate and with the pH value maintained at a relatively high value within the activity range of nitrate reductases. For good growth, lactobacilli for the nitrate test have been customarily grown in media with high carbohydrate content, with consequent low final pH values. This seems to be the essential reason why the genus *Lactobacillus* had previously been defined as unexceptionally nitratase-negative.

Until recent years the genus *Lactobacillus* has been described as unable to reduce nitrates (*Bergey's Manual,* 1948; Rogosa *et al.* 1953). However,

Costilow & Humphreys (1955) reported that 18 of 38 strains of *Lactobacillus plantarum* reduced nitrates. This result was achieved through the use of the BBL indole-nitrite medium (Baltimore Biological Laboratory, Inc.) which has an initial pH value of 7.2 and a composition of 2% (w/v) Trypticase, 0.2% (w/v) Na_2HPO_4, and 0.1% (w/v) each of glucose, KNO_3 and agar. Negative results, consistent with earlier general experience, were obtained with broth media, such as the BBL indole-nitrite medium minus agar, or Difco nitrate broth. To explain this, Costilow & Humphreys (1955) reasoned that agar would tend to decrease the oxygen tension of the medium and stated: 'It is obvious that the oxygen tension of the medium was the most important factor in nitrate reduction.....' From inspection of the formula of the BBL indole-nitrite medium it is evident that it contains very little fermentable carbohydrate (0.1 %, w/v) and is very highly buffered. Also the initial pH, 7.2, is high. This suggests that nitrate reduction may be a function of pH value and that the nitrate reductase enzymes may have a relatively high and narrow pH activity range. For example, Woods (1938) found with *Clostridium welchii* that, when the reduction rate at pH 6.8 was taken as 100%, it was 50% at pH 6.4 and only 10% at pH 6.1 after 25 min. After 35 min the reduction rate decreased further at pH 6.5 to only 10%. Similar results were obtained with a strain of *Escherichia coli.* Nason & Evans (1955) studied a purified enzyme from *Neurospora crassa* which catalysed the reaction $TPNH + H^+ + NO_3^- \rightarrow TPN^+ + NO_2^- + H_2O$ and the enzyme exhibited a sharp optimum for activity at pH 7.0 in phosphate buffer. Zucker & Nason (1955) found maximum activity between pH 8.0 and 9.0 for the intermediate enzyme, hydroxylamine reductase, in dentrification by *N. crassa.* Najjar (1955) obtained similar high pH optima with *Pseudomonas stutzeri* and *Bacillus subtilis* for the enzymic conversion of NO_2^- and NO to N_2. Because of the importance of nitrate reduction as a taxonomic criterion, a wide range of *Lactobacillus* species, and certain streptococci and pediococci, were tested in experiments designed to discover the appropriate conditions for nitrate reduction.

METHODS

The strains of lactobacilli were representative of 13 species in the collection of Dr. M. Elisabeth Sharpe (National Institute for Research in Dairying, NIRD, University of Reading), obtained through original isolations by Dr. Sharpe or other workers, and from such culture collections as the American Type Culture Collection (ATCC), the (British) National Collection of Type Cultures (NCTC), the (British) National Collection of Industrial Bacteria (NCIB) and the (British) National Collection of Dairy Organisms (NCDO maintained at NIRD). Each strain had been carefully studied by numerous tests (Rogosa *et al.,* 1953; Rogosa & Sharpe, 1959) including serological procedures (Sharpe, 1955) wherever specific group antisera could be prepared.

There were strains of *Lactobacillus plantarum.* The origins of the following 7 are of interest: *AR 1* NIRD 17/5; *P 4,* NCTC 6376 from ATCC 8014; *P 16,* ATCC 8014; *AR 5, L. arabinosus* V 743 Tittsler; *P 30,* V 322 Tittsler; *P 7,* NIRD 1–4; *P 8,* NIRD 1–8. Strains *AR 1, P 4, P 16* and *AR 5* are identical and are really only a single strain designated *L. arabinosus* 17/5 by Fred, Peterson & Anderson (1921), but ascribed different numbers in different culture collections.

Five strains of *Lactobacillus fermenti* were tested. Strain *F 1* is identical with ATCC 9338 and NCTC 6991, and strain *F 4* is NCTC 7230.

The remaining strains were distributed numerically as follows: *Lactobacillus acidophilus,* 5; *L. brevis,* 10; *L. buchneri,* 1; *L. bulgaricus,* 9; *L. casei,* 14; *L. cellobiosus,* 2; *L. delbrueckii,* 2; *L. helveticus,* 1; *L. jugurti,* 3; *L. lactis,* 11; *L. leichmannii,* 2; *L. salivarius,* 4.

In addition, one strain each of the following group D streptococci was tested: *Streptococcus bovis, S. durans, S. faecalis* var. *liquefaciens, S. faecalis* and *S. faecium.* Among group N streptococci were two strains of *S. cremoris,* one of *S. lactis* and two of *S. lactis* var. *diacetilacticus.* Three strains of *Pediococcus cerevisiae,* including ATCC 8081, were also tested.

The media used were: (1) the BBL indole-nitrite medium; (2) Difco nitrate broth; (3) various modifications of these such as the addition of agar, yeast extract, increased buffering with Na_2HPO_4, and changes in the concentration of glucose. The conditions for each experiment will be described in the tabulation of results.

Cultures were inoculated with one drop from good growth in MRS broth (de Man, Rogosa & Sharpe, 1960) and incubated for 6–7 days at 37° except for *Lactobacillus plantarum, L. brevis, L. casei,* the streptococci and the pediococci which were incubated at 30°.

When anaerobiosis was desired it was obtained by using McIntosh & Fildes jars filled with either H_2 or a 90% H_2 + 10% (v/v) CO_2 mixture, and combining residual traces of O_2 with H_2 by using the electrically heated catalyst.

The reagents for the detection of nitrite were: solution 1 containing 2 g sulfanilic acid in 250 ml 5 N-acetic acid; solution 2 containing 1.5 ml dimethyl-α-naphthylamine in 250 ml 5 N-acetic acid. The latter reagent was recommended by Wallace & Neave (1927) and Tittsler (1930) as superior to the α-naphthylamine reagent because the colour develops instantaneously, is more intense, and does not fade in a reasonable period of time. For the nitrite test, one drop of culture was deposited in a white porcelain spot plate followed by one drop each of solutions 1 and 2. The pink to red colour in positive tests developed at once and was unequivocal. The spot plate technique was often superior to tests made directly in the tubes. Some media contain substances (thioglycollate, for example) which inhibit colour development, but with the quantities of culture and reagents described for the spot plate test, colour development was uninhibited. Negative controls on the media and a positive control with *Micrococcus aureus*, phage type 80, were included routinely. All negative tests were confirmed by reducing residual nitrate to nitrite with Zn dust on the spot plate and testing for nitrite as described. Zn reductions in the tubes often gave ambiguous and false results.

RESULTS

The only species in which some strains reduced nitrates were *Lactobacillus plantarum* (7 of 12) and *L. fermenti* (2 of 5 tested). All other species of lactobacilli, and the streptococci and pediococci examined, were unable to reduce nitrates under any of the experimental conditions. Thus, Costilow & Humphreys's (1955) observation that certain strains of *L. plantarum* can reduce nitrates under appropriate conditions was confirmed.

When Difco nitrate broth (DNB) or DNB supplemented with agar and yeast extract was used at an initial pH 6.4 (a pH value favourable for growth) all strains repeatedly failed to reduce nitrates aerobically or anaerobically and the pH value decreased sharply to a final value of 3.4. At the relatively unfavourable value pH 7.4, however, the cultures reduced nitrates, the pH decrease was markedly less, and the final pH value was about 6.4, i.e. 3 units less than the decrease obtained when the initial value was pH 6.4.

In BBL indole-nitrite (IN) medium nitrate reduction was sometimes suppressed under anaerobic conditions. In IN medium supplemented with Difco yeast extract (YE) nitrate reduction was also occasionally absent or weak. From observation of growth it was obvious that these weak or negative responses were related to total growth and that growth in IN medium + YE was even poorer than in the relatively poor IN medium. Also, it is clear that nitrate reduction, both aerobically and anaerobically, took place in the absence of agar, which was irrelevent to the process. Yeast extract in broth media, contrary to the experience with agar-containing media, did not inhibit nitrate reduction. What is most germane to this discussion is that cultures in media such as IN medium without agar + YE supplemented with conventional quantities of glucose (0.9 %, w/v), had relatively low terminal pH values and also did not then reduce nitrates.

The above results strongly indicate that induced anaerobic conditions, yeast extract, and agar are not essential for nitrate reduction by these strains. Rather, it appears that the pH value, fermentable carbohydrate, and buffer capacity are the principal factors that effect nitrate reduction.

Anaerobiosis previously was achieved with 90% (v/v) H_2 + 10% (v/v) CO_2. Since CO_2 may decrease the pH value of the medium, anaerobiosis in further experiments was also maintained by an atmosphere of 100% H_2. As little as 0.1% (w/v) added glucose inhibited nitrate reduction by *Lactobacillus plantarum* P30. Nitrate reduction by this strain seemed most sensitive to slight acidity from the fermentation of glucose, but all nitrate positive strains in IN medium were, with one exception, unable to reduce nitrates when 0.3% (w/v) glucose was added and the pH value consequently decreased to 5.0 or less. With 0.5% (w/v) glucose, nitrate reduction was never observed. Again positive reactions were generally fewer or less intense in anaerobic as compared with aerobic cultures, and it is quite clear that agar had no enhancing effect. These repeated results with agar do not agree with those of Costilow & Humphreys (1955). In experiments which repeated their experimental protocol exactly as described, but in which the pH values of comparable broth and agar media were adjusted electrometrically exactly to the same initial value of 7.1, there was no difference in nitrate reduction in broth or agar containing media.

The strain of *Micrococcus aureus,* phage type 80, which was used as a positive control, reduced nitrate to nitrite at pH 7.0. But at pH 7.5 or above the organism consistently also reduced nitrite completely, with accumulation of N_2. In this case it would appear that the nitrite reductase enzyme had an even higher pH activity range than the nitrate reductase.

R. S. Breed, E. G. D. Murray and A. P. Hitchins *Bergey's Manual of Determinative Bacteriology.* Williams and Wilkins Book Co. Baltimore, 1948.
R. N. Costilow and T. W. Humphreys (1955). Science, N.Y. 121, 168.
E. B. Fred, W. H. Peterson and J. A. Anderson (1921). J. biol. Chem. 48, 385.
J. C. de Man, M. Rogosa and M. E. Sharpe (1960). J. appl. Bact. 23, 130.
V. A. Najjar (1955). *Methods of Enzymology,* 2, 420. Ed. S. P. Colowick and N. O. Kaplan, New York: Academic Press Inc.
A. Nason and H. J. Evans (1955). *Methods of Enzymology,* 2, 411. Ed. S. P. Colowick and N. O. Kaplan. New York: Academic Press Inc.
M. Rogosa, R. F. Wiseman, J. A. Mitchell, M. Disraely and A. Bearman (1953). J. Bact. 65, 681.
M. Rogosa and M. E. Sharpe (1959). J. appl. Bact. 22, 329.
M. E. Sharpe (1955). J. gen. Microbiol. 12, 107.
R. P. Tittsler (1930). J. Bact. 19, 261.
G. I. Wallace and S. L. Neave (1927). J. Bact. 14, 377.
D. D. Woods (1938). Biochem. J. 32, 2000.
M. Zucker and A. Nason (1955). *Methods in Enzymology,* 2, 416. Ed. S. P. Colowick and N. O. Kaplan. New York: Academic Press Inc.

Nitrate and nitrite reduction—*Xanthomonas*

A. C. Hayward and W. Hodgkiss (1961). J. gen. Microbiol. 26, 133–140.

A medium of the following composition was used: peptone (Oxoid), 10.0 g; K_2HPO_4, 5.0 g; yeast extract (Difco), 1.0 g; KNO_3, 1.0 g; or $NaNO_2$, 0.1 g; agar, 3.0 g; distilled water, 1 L.; adjusted to pH 7.0; 10 ml medium per 1 oz screw-capped bottle. At 5 days tests for nitrite formation or nitrite destruction were made by standard procedures.

Nitrate reduction—*Escherichia aurescens*

H. Leclerc (1962). Annls Inst. Pasteur, Paris 102, 726–741.

On la met en évidence à partir d'une culture de 24 heures sur gélose nitratée grâce au réactif de Griess.

Nitrate reduction—*Clostridium*

H. Beerens, M. M. Castel and H. M. C. Put (1962). Annls Inst. Pasteur, Paris 103, 117–121.

Nitrates: Le milieu V. L. (base décrite ci-dessus), glucosé à 2 p. 1 000 et additionné de 5 p. 1 000 de nitrate de sodium est recommandé. Rechercher les nitrites par la réaction habituelle à l'α-naphtylamine et l'acide sulfanilique.

Nitrate reduction—*Vibrio*

G. H. G. Davis and R. W. A. Park (1962). J. gen. Microbiol. 27, 101–119.

Tested in 5 day-old peptone water cultures containing 0.1% (w/v) KNO_3, using the Griess-Ilosvay reagents and zinc powder test for false negatives (ZoBell, 1932).

C. E. ZoBell (1932). J. Bact. 24, 273.

Nitrate reduction—*Vibrio*

G. H. G. Davis and R. W. A. Park (1962). J. gen. Microbiol. 27, 101–119.

Basal medium: Koser salt solution (Mackie and McCartney, 1953, p. 429) plus Oxoid yeast extract, 0.3% (w/v). Initial pH adjusted to 7.0, plus 0.1% (w/v) KNO_3 and 0.1% (w/v) agar.

T. J. Mackie and J. E. McCartney (1953). *Handbook of Practical Bacteriology,* 9th ed. Edinburgh: Livingstone.

Nitrate reduction—*Caryophanon*

P. J. Provost and R. N. Doetsch (1962). J. gen. Microbiol. 28, 547–557.

Medium:

Hy-Case SF	10.0 g
Bacto yeast extract	5.0 g
Na Acetate (anh)	0.5 g
Agar	15.0 g
pH 7.8 – 7.9	
+ 0.1% (w/v) KNO_3	

Nitrate reduction—*Haemophilus*

P. N. Edmunds (1962). J. Path. Bact. 83, 411–422.

Tests in conventional media supplemented with plasma were unsatisfactory owing to failure of growth. It was found necessary to use a massive inoculum of viable bacilli, sufficient to give the reaction without further growth. Twelve maltose digest broths (prepared as for fermentation tests) were inoculated with each strain. Heavy growth occurred in 48 hr at 37°C. The cultures were pooled and centrifuged, the deposit was resuspended in a small amount of broth, and two drops were added to 0.5 ml of nitrate medium (Kauffmann, 1954); the resulting suspension was equivalent to Brown's (1919–1920) opacity standard no. 5. The tests were incubated at 37°C for 5 days for the nitrate reduction test. The results were observed as recommended by Kauffmann.

H. C. Brown (1919–1920). Indian. J. med. Res. 7, 238.
F. Kauffmann (1954). *Enterobacteriaceae.* 2nd ed. Copenhagen: Einar Munksgaard. p. 357.

Nitrate reduction—*Actinobacillus* and *Haemophilus*

E. O. King and H. W. Tatum (1962). *Actinobacillus actinomycetemcomitans* and *Haemophilus aphrophilus.* J. infect. Dis. 111, 85–94.

Bacto peptone broth or heart infusion broth with 0.2% potassium nitrate.

Nitrate reduction—*Lactobacillus*

R. E. Smith and J. D. Cunningham (1962). Can. J. Microbiol. 8, 727–735.

Indole-nitrite broth (BBL) was tested, after 48 hours' incubation, for indole with fresh Kovacs' reagent. A portion of this medium was examined for the presence of nitrite, using Griess-Ilosva reagents, and negatives were confirmed with zinc dust.

Nitrate reduction—*Enterobacteriaceae*

W. H. Ewing (1962). *Enterobacteriaceae. Biochemical Methods for Group Differenciation.* U.S. Department of Health, Education and Welfare, Public Health Service Publication No. 734 (revised).

Although the nitrate reduction test is of little or no value for group differentiation within the *Enterobacteriaceae* it is included because a negative test is often of value in the exclusion of otherwise doubtful strains. The results of negative nitrate tests should always be confirmed. This may be done by adding a minute amount of zinc dust to the tube. The development of a red colour indicates the presence of unreduced nitrate.

Nitrate Reduction:

Tryptone	5 g
Neopeptone	5 g
Agar	2.5 g
Distilled water	1000 ml

Boil and adjust pH to 7.3–7.4

then add KNO_3 (nitrite-free) 1 g

Glucose 0.1 g

Sterilize at 121°C for 15 minutes.

Inoculation: The medium may be inoculated by stabbing into the column of semisolid agar medium.

Incubation: 37°C, 24 hours. The occasional culture that gives apparently equivocal results should be retested after 1, 2, 3 and 4 days' incubation.

Test Reagents:

A. Dissolve 8 g of sulfanilic acid in 1000 ml of 5 N acetic acid.

B. Dissolve 5 g of alpha-naphthylamine in 1000 ml of 5 N acetic acid.

Immediately before use, equal parts of solutions A and B are mixed, and 0.1 ml of the mixture is added to each culture.

Positive tests for reduction of nitrate to nitrite are indicated by the development of a red colour within a few minutes.

As might be expected, occasional strains that are otherwise typical *Enterobacteriaceae* may fail to reduce nitrate to nitrite. For example, one culture of *Shigella dysenteriae* 1 (Shiga) and three *Serratia* strains in the writer's collections failed to reduce nitrate and certain klebsiellae, particularly type 4 cultures, may be nitrate negative (P. R. Edwards, personal communication). However, failure to reduce nitrate to nitrite apparently is rare among members of the family *Enterobacteriaceae.*

Nitrate reduction—rumen bacteria

R. S. Fulghum and W. E. C. Moore (1963). J. Bact. 85, 808–815.

Determinations of indole production and nitrate reduction were made in SRP-basal medium* with 2% Trypticase and 0.1% KNO_3.

The anaerobic methods of Hungate (Bact. Rev. 14, 1–49, 1950) were used in handling and incubation of the cultures.

***Basal medium for physiological tests** – *Butyrivibrio, Borrelia, Bacteroides, Selenomonas, Succinivibrio* (p. 109).

Nitrate reduction—*Actinobacillus*

P. W. Wetmore, J. F. Thiel, Y. F. Herman and J. R. Harr (1963). Comparison of selected *Actinobacillus* species with a hemolytic variety of *Actinobacillus* from irradiated swine. J. infect. Dis. 113, 186–194.

The authors inoculated cultures into warm (liquid) cystine-trypticase agar (BBL) and on semisolid peptone agar containing 0.1% KNO_3.

Nitrate reduction—*Micrococcus roseus*

R. C. Eisenberg and J. B. Evans (1963). Can. J. Microbiol. 9, 633–642.

Nitrate reduction was tested on nitrate agar slants and in indole-nitrite medium after 3 days using materials and methods 'essentially those described in the *Manual of Microbiological Methods'.*

Society of American Bacteriologists *Manual of Microbiological Methods* – McGraw-Hill Book Co. Inc., New York 1957.

Nitrate reduction—*Staphylococcus epidermidis*

N. J. Jacobs, J. Johantges and R. H. Deibel (1963). J. Bact. 85, 782–787.

Organism and conditions of growth. The strain (AT2) used was a typical *S. epidermidis* (Breed, Murray and Smith, 1957) from our culture collection. The growth medium consisted of the following: Tryptone (Difco), 1 g; yeast extract (Difco), 0.5 g; NaCl, 0.5 g; K_2HPO_4, 0.5 g; glucose, 0.2 g; KNO_3, 0.3 g; and distilled water, 100 ml. Although higher concentrations of glucose would have yielded larger cell crops, the lack of pH control during growth would have interfered seriously with subsequent nitrite determination. The inoculum was grown for 1 to 2 days at 37°C in a medium similar to above but containing 0.5% of beef extract (Difco) and lacking nitrate. A 0.1 to 0.2% inoculum level was employed.

Anaerobic conditions were obtained by placing cotton-plugged Erlenmeyer flasks three-fourths filled with medium in a desiccator, evacuating three times, and replacing the atmosphere with helium after each evacuation. Static conditions of growth were obtained, except where indicated, by placing 200 ml of medium in a 2-liter Erlenmeyer flask and incubating without agitation. The flasks or desiccators were placed in a 37°C incubator for 24 hr.

For preparation of resting cells, approximately 600 ml of medium were centrifuged at 2 to 3°C. The cells were washed in 600 ml of distilled cold water (5 to 10°C), then in 40 ml of cold water, and finally resuspended in 10 to 40 ml of cold water. The static and anaerobic suspensions were adjusted to the same content according to dry weight determinations (110°C, 12 hr).

Analyses. Nitrite was determined by the method of Nason and Evans (1955), and nitrate by the method of Landmann *et al.* (1960). The nitrate method gave a precision of only about 5%, and, therefore, the data presented should be interpreted accordingly. (These chemical analyses were conducted after appropriate dilutions were made in cold water.)

If enough cells were present after dilution of samples to interfere with the colorimetric test, they were removed by centrifugation. Relative amounts of growth were determined by measuring optical densities of equally diluted cultures at 600 mμ in a Bausch and Lomb Spectronic 20 colorimeter. Nitrate reductase was assayed by a modification of the method of Taniguchi and Itagaki (1960). Experiments with resting suspensions were conducted anaerobically in Thunberg tubes. The tubes were evacuated twice and the atmosphere was replaced with helium. The tubes were then evacuated a third time and placed in a water bath (37°C) for the incubation period.

Chemicals. Benzyl viologen was purchased from Mann Research Laboratories, Inc., New York, N.Y. Hemin crystals (Armour Laboratories,

Chicago, Ill.) were dissolved in 0.01 N NaOH, sterilized by filtration, and diluted in sterile water before addition to sterile culture medium.

R. S. Breed, E. G. D. Murray, and N. R. Smith (1957). *Bergey's manual of determinative bacteriology,* 7th ed. The Williams and Wilkins Co., Baltimore.

W. A. Landmann, M. Saeed, K. Pih, and D. M. Doty (1960). J. Ass. off. agric. Chem. 43, 531–535.

A. Nason and H. J. Evans (1955). In S. P. Colowick and N. O. Kaplan (ed.), *Methods in enzymology,* vol. 2. Academic Press, Inc., New York. p. 411–415.

S. Taniguchi, E. Itagaki (1960). Biochim. biophys. Acta 44, 263–279.

Nitrate reduction—*Spirillum*

W. A. Pretorius (1963). J. gen. Microbiol. 32, 403–408.

Nitrate broth (Difco) with Durham fermentation tubes, was used. Cultures were tested with Griess-Ilosvay's reagents (with dimethyl-α-naphthylamine) after 5 days. Zinc metal was added to the negative cultures and when found again negative, tested for ammonia with Nessler's reagent.

Nitrate reduction—*Bacterium salmonicida (Aeromonas)*

I. W. Smith (1963). J. gen. Microbiol. 33, 263–274.

After incubation for 7 days, Griess-Ilosvay reagents were added to nitrate broth cultures (Topley and Wilson's *Principles of Bacteriology and Immunity* (1946). 3rd ed.)

Nitrate reduction—*Pseudomonas odorans*

I. Málek, M. Radochová and O. Lysenko (1963). J. gen. Microbiol. 33, 349–355.

Investigated in KNO_3 broth with Durham tubes after incubation for 5 days, using the Griess-Ilosvay reagent and checking with zinc dust for false negatives at 28°C.

Nitrate reduction—general

G. S. Wilson and A. A. Miles (1964). Topley and Wilson's *Principles of Bacteriology and Immunity* 5th edition – Arnold.

Tested on a broth culture containing 0.1 per cent KNO_3, grown for 5 days at 37°C, by the Griess-Ilosvay method.

Solution A:

α-Naphthylamine	1 g
Water	22 ml

Dissolve, filter, and then add 180 ml of dilute acetic acid (sp.gr. 1.04).

Solution B:

Sulfanilic acid	0.5 g
Dilute acetic acid	150 ml

Add 1 ml of Solution A, followed by 1 ml of Solution B.
Pink, red, or maroon color = positive
No coloration = negative.

A negative reaction may sometimes be due to the reduction of the nitrite to gaseous nitrogen almost as rapidly as it is formed, or to the production of hydroxylamine. The first possibility may be examined by growth in a gas fermentation tube, or by testing for nitrite in the way just described after adding 5 mg of zinc dust to reduce any residual nitrate (Steel 1961); the second by testing for nitrite after preliminary oxidation of the hydroxylamine with iodine (see Lindsey and Rhines 1932, Conn 1936, Reed 1942). A control tube inoculated with a strain known to reduce nitrates should always be tested.

The medium should be adjusted to the alkaline side of neutrality and should provide sufficiently anaerobic conditions for the action of the oxygen-sensitive enzyme nitratase, such as is formed by *Esch. coli.* The reducing enzyme of the denitrifying bacteria works satisfactorily in air (see Meiklejohn 1949).

An alternative test for nitrate reduction on a solid medium, based on the oxidation of hemoglobin to methemoglobin by nitrites, is described by Cook (1950).

H. J. Conn (1936). J. Bact. 31, 225.
G. T. Cook (1950). J. clin. Path. 3, 359.
G. A. Lindsey and C. M. Rhines (1932). J. Bact. 24, 489.
J. Meiklejohn (1949). Annls Inst. Pasteur, Paris 77, 389.
R. W. Reed (1942). J. Bact. 44, 425.
K. J. Steel (1961). Mon. Bull. Minist. Hlth 20, 63.

Nitrate reduction—*Mycobacterium buruli* n.sp.

J. K. Clancey (1964). J. Path. Bact. 88, 175–187.

The method of Bönicke (1962) was employed to detect nitrate-to-nitrite reduction.

R. Bönicke (1962). Bull. Un. Int. Tuberc. 32, 32.

Nitrate reduction—*Aerococcus catalasicus*

O. G. Clausen (1964). J. gen. Microbiol. 35, 1–8.

Cultures were incubated aerobically at 37°C unless otherwise stated.

Four-day cultures in nitrate broth were tested by adding Griess-Ilosvay reagent. Negative tests were checked by adding powdered zinc to prove that there was nitrate left in the culture medium. A surplus of nitrite in a culture can 'conceal' a positive reaction by decolorizing the red stain; for this reason the test was performed daily for up to 4 days (Shaw *et al.* 1951).

C. Shaw, J. M. Stitt and S. T. Cowan (1951). J. gen. Microbiol. 5, 1010.

Nitrate reduction—*Moraxella*

W. J. Ryan (1964). J. gen. Microbiol. 35, 361–372.

Strips of filter paper impregnated with potassium nitrate were laid across the surface of blood agar plates and inoculations made by the stab method (Cook, 1950). Browning of the medium around the inoculum indicated reduction of nitrate. Readings were made after incubation for 24 and 48 hr.
G. T. Cook (1950). J. clin. Path. 3, 359.

Nitrate reduction—*Photobacterium sepia*

D. J. D. Nicholas, W. J. Redmond and M. A. Wright (1964). J. gen. Microbiol. 35, 401–410.

Nitrate reductase. Method 1. The enzymic reduction of nitrate by whole bacteria was followed anaerobically in Thunberg tubes by using reduced benzylviologen (BVH) as a hydrogen donor. In the side-arm, 0.2 ml 10^{-3} M–benzylviologen (BV), 1 mg palladized asbestos and 0.3 ml 2.5×10^{-1} M-phosphate buffer (pH 7.5); in the tube 0.1 ml 10^{-1} M-KNO_3, 0.1 ml bacterial suspension (equiv. 2 mg N/ml), 0.3 ml 2.5×10^{-1} M-phosphate buffer (pH 7.5). The tube was evacuated and flushed with high purity hydrogen passed through a Deoxo-catalytic deoxygenator (Baker Platinum Division, Englehard Industries Ltd., 52 High Holborn Street, London) at which stage the dye was fully reduced. The tubes when finally evacuated were pre-incubated for 5 min at 30° before tipping in the reduced dye; the reaction was terminated after a further 10 min by adding 0.1 ml M-zinc acetate and 1.9 ml 95% (v/v) ethanol in water (Medina & Nicholas, 1957). After centrifuging at 4000 g for 5 min, nitrite was determined in a sample of the supernatant solution by the sulfanilamide method (Fewson & Nicholas, 1961).

Method 2. Reduced benzylviologen, prepared by the palladized asbestos-hydrogen method, did not function as a hydrogen donor for nitrate reductase in cell-free extracts. The dye reduced with sodium dithionite was, however, a suitable donor for the extracted enzyme (Sadana & McElroy, 1957). The specific activity of the enzyme is defined as μm-mole NO_2- formed/10 min/mg/bacterial total-N.

Nitric oxide (NO) reductase. The uptake of NO was measured in a Warburg apparatus. The main compartment contained 0.5 ml bacterial suspension (equiv. 2 mg N/ml), 1.5 ml 0.1 M-phosphate (pH 7.5); the side-arm contained 0.2 ml 10^{-2} M-$NADH_2$, 0.1 mg crystalline alcohol dehydrogenase, 0.1 ml 5% (v/v) ethanol in water, and the centre well 0.2 ml 20% (w/v) KOH. The vessels were flushed with oxygen-free nitrogen for 40 min and then with approximately 20% (v/v) NO in N_2 until damp litmus paper turned red when held in the exit gas stream from the side-arm stopper. The apparatus was equilibrated at 30° for 15 min and then the $NADH_2$-generating system tipped in to start the reaction. The nitric oxide was prepared by the reaction: $3\ Cu + 8\ HNO_3 = 2\ NO + 3\ Cu(NO_3)_2 + 4H_2O$. The NO evolved was displaced by a slow stream of oxygen-free nitrogen and after passing through Dreschel bottles containing boiled water the gas mixture was collected over water in a glass aspirator. The NO content of the gas mixture was determined in a gas burette connected to a reservoir containing saturated pyrogallol in 20% (w/v) KOH. The volume of gas was measured and oxygen introduced from a cylinder. After vigorous shaking the contraction in gas volume, due to the reaction $2NO + O_2 \rightarrow 2NO_2$, was measured. Nitrogen dioxide in the presence of excess O_2 dissolves readily in the alkaline pyrogallol. The gas mixtures prepared in this way contained between 20 and 25% (v/v) NO in N_2. Oxygen must be rigorously removed from the system since it reacts readily with NO to give NO_2 which dissolves in water to give a mixture of nitrous and nitric acids.

C. A. Fewson and D. J. D. Nicholas (1961). Biochim. biophys. Acta 49, 335.
A. Medina and D. J. D. Nicholas (1957). Biochim. biophys. Acta 23, 440.
J. C. Sadana and W. D. McElroy (1957). Archs Biochem. 67, 16.

Nitrate reduction—*Pediococcus*

E. Coster and H. R. White (1964). J. gen. Microbiol. 37, 15–31.

The method devised by Rogosa (1961) was employed to test for the reduction of nitrates. Cultures were grown in the indole nitrite medium (BBL) with $MgSO_4$, $MnSO_4$ (0.05%, w/v) and with and without added glucose (0.01%, w/v).

M. Rogosa (1961). J. gen. Microbiol. 24, 401.

Nitrate reduction—*Veillonella*

M. Rogosa (1964). J. Bact. 87, 162–170.

Nitrate reduction was determined in V17 broth containing 0.1% KNO_3 and adjusted to pH 7.5. Cultures were grown in this medium in the presence and absence of thioglycolate, and nitrite was detected with the reagents and the spot-plate technique described by Rogosa (1961).

For V17 medium see **Basal medium**—*Veillonella* (p. 114).

M. Rogosa (1961). J. gen. Microbiol. 24, 401–408.

Nitrate reduction—*Clostridium*

N. A. Sinclair and J. L. Stokes (1964). J. Bact. 87, 562–565.

For nitrate reduction and indole formation, indole-nitrate medium was used. In some cases, the medium was enriched with 0.01% DL-tryptophan to increase indole formation.

Incubation at 0, 15 and 25°C.

Nitrate reduction—motile marine bacteria

E. Leifson, B. J. Cosenza, R. Murchelano and C. Cleverdon (1964). J. Bact. 87, 652–666.

For the medium and incubation period see under **Gelatin hydrolysis** (p. 386).

The test for nitrite was made by adding two to four drops of reagent 1 (4 g of sulfanilic acid dissolved in 500 ml of 0.2 N acetic acid) followed by two to four drops of reagent 2 (2.5 g of α-naphthylamine acetate in 500 ml of 0.2 N acetic acid.

Reduction of nitrate with gas formation was apparent in the form of bubbles in the solid medium. In the few instances where the gelatin liquefaction was extremely rapid, gas formation was checked by using a liquid nitrate medium with an inverted vial to trap the gas.

Nitrate reduction—*Actinomyces*

L. K. Georg, G. W. Robertstad and S. A. Brinkman (1964). J. Bact. 88, 477–490.

Nitrate reduction was tested by growing the strains on the basal medium containing 0.1% KNO_3.

The basal medium consisted of: Heart Infusion Broth, 25 g; pancreatic digest of casein, 4 g; and yeast extract, 5 g (per liter of distilled water at pH 7.0).

The inoculum was taken from 3 day old cultures in AM broth. See **Starch hydrolysis**—*Actinomyces* (p. 812) for composition of AM broth.

The cultures were incubated at 37°C under pyrogallol-carbonate seals and read at 3 and 10 days.

Nitrate reduction—*Pseudomonas aeruginosa*

K. Morihara (1964). J. Bact. 88, 745–757.

Bouillon with 0.1% potassium nitrate was used for the test. Nitrite was detected by applying α-naphthylamine and sulfanilic acid after 1, 3, and 5 days of incubation. Cultures were incubated at 30°C.

Nitrate reduction—*Vibrio marinus*

R. R. Colwell and R. Y. Morita (1964). J. Bact. 88, 831–837.

The authors used the method of the Society of American Bacteriologists. 1957. *Manual of Microbiological Methods.* McGraw-Hill Book Co., Inc., New York – modified by the addition of the media of the following salts to produce a synthetic seawater:– sodium chloride, 2.4%; potassium chloride, 0.07%; magnesium chloride (hydrated) 0.53% and magnesium sulfate (hydrated) 0.7%.

A standard inoculum was 1 drop from a pasteur pipette (*c.* 0.05 ml) of a 24 hour artificial seawater broth.

Cultures were incubated at 18°C.

Nitrate reduction—*Bacteroides oralis*

W. J. Loesche, S. S. Socransky and R. J. Gibbons (1964). J. Bact. 88, 1329–1337.

Nitrate reduction was determined by use of the basal medium supplemented with 0.1% KNO_3, according to procedures described in the *Manual of Microbiological Methods* (Society of American Bacteriologists, 1957). The presence of residual nitrate was determined by the addition of zinc dust to acidified samples of the culture. Acidogenic strains were tested for nitrate reduction at pH 5.0 and 6.7 according to the suggestion of Rogosa (1961).

The test was performed in the presence and absence of 0.5% glucose.

Thioglycolate Medium without Dextrose (BBL), supplemented with 0.2% yeast extract (Difco) and hemin, was used as a basal medium. All cultures were incubated at 35°C in Brewer jars containing an atmosphere of 95% H_2 and 5% CO_2.

M. Rogosa (1961). J. gen. Microbiol. 24, 401–408.
Society of American Bacteriologists (1957). *Manual of Microbiological Methods.* McGraw-Hill Book Co., Inc., New York.

Nitrate reduction—*Lactobacillus*

F. Gasser (1964). Annls Inst. Pasteur, Paris 106, 778–796.

La réduction de nitrates en nitrites est recherchée en milieu M.R.S.M. additionné de 0.2 p. 100 de nitrate de potassium. Après deux jours et huit jours d'incubation la présence de nitrites est recherchée par la méthode de Wallace et Neave [59]. En l'absence de nitrites la présence de nitrates est contrôlée par la réaction de Zo Bell [66].

[59] Wallace (G. I.) and Neave (S. L.). *J. Bact.* 1927, 14, 377.
[66] Zo Bell. *J. Bact.* 1932, 24, 273.

Nitrate reduction—*Mycobacterium*

A. Tacquet, F. Tison and B. Devulder (1965). Annls Inst. Pasteur, Paris 108, 514–525.

The authors used the method of Tison, Gillaume and Devulder (Ann. Inst. Pasteur. 1964, 106, 797–801).

Nitrate reduction—*Bordetella*

M. Piéchaud and Mme. S. Szturm-Rubinsten (1965). Annls Inst. Pasteur, Paris 108, 391–395.

La réduction des nitrates en nitrites. La recherche de la réduction des nitrates en nitrites dans un bouillon nitraté à 1 p. 1000 est fortement positive pour toutes nos souches de *B. bronchiseptica,* qui cependant sont incapables de pousser en anaérobiose avec un nitrate comme accepteur d'électrons. Bien plus, nous avons remarqué que la réaction de Griess, effectuée sur une culture en eau peptonée à 10 p. 100 non nitratée, est également et constamment positive. Devignat [3], qui avait observé la même chose pour *Pasteurella pestis,* pense que ces germes peuvent former des nitrites à partir de certaines protéines des milieux de culture courants. *B. parapertussi* ne réduit pas les nitrates en nitrites et ne donne pas de nitrites en eau peptonée simple. Ce caractère différentiel se révèle de grande valeur pratique par sa constance.

[3] Devignat (R.). *Ann. Inst. Pasteur,* 1952, 82, 653.

Nitrate reduction—*Mima, Herellea, Flavobacterium*

J. D. Nelson and S. Shelton (1965). Appl. Microbiol. 13, 801–807.

Nitrate reduction was determined after 48 hr of incubation in Indole-Nitrite Medium. One drop of culture was spot-tested with one drop of sulfanilic acid in 5 N acetic acid (0.8 g per 100 ml) and one drop of D-α-naphthylamine in 5 N acetic acid (0.5 g per 100 ml).

Nitrate reduction—*Vibrio comma*

J. C. Feeley (1965). J. Bact. 89, 665–670.

Nitrate reduction was determined in semisolid motility-nitrate agar (1% Trypticase, 1% NaCl, 0.1% KNO_3, 0.4% agar).

Nitrate reduction—marine bacteria

R. M. Pfister and P. R. Burkholder (1965). J. Bact. 89, 863–872.

Nitrate reduction was detected by the method outlined in the *Manual of Microbiological Methods* (Society of American Bacteriologists, 1957), by use of seawater broth with 0.1% KNO_3.

Society of American Bacteriologists (1957). *Manual of Microbiological Methods.* McGraw-Hill Book Co., Inc., New York.

Nitrate reduction—*Corynebacterium nephridii*

L. T. Hart, A. D. Larson and C. S. McCleskey (1965). J. Bact. 89, 1104–1108.

Sodium nitrate was added in varying amounts to Trypticase Soy Broth (BBL) or to peptone broth. In the aerobic tests, cultures were grown in 500 ml Erlenmeyer flasks containing 100 ml of broth and incubated at 28°C on a shaker (180 rev/min). To obtain relatively anaerobic conditions, 4 ml of medium were introduced into 13 x 100 mm screwcapped tubes and autoclaved. After sterilization, the tubes were inoculated, then aseptically filled with sterile medium, and the caps were screwed down. One of the tubes was taken for analysis at each sampling time.

Anaerobic conditions were effected with the Parker (1955) method, except that nitrogen or nitrous oxide, admitted through a sterile cotton filter, was used to flush and fill the jar.

Nitrate utilization. Nitrate reduction was determined by the sulfanilic acid alpha-naphthylamine acetate method described by the Society of American Bacteriologists (1957). Ammonia nitrogen was determined by titration after distillation into 1% boric acid. Nitrite nitrogen was quantitated with the diazotization method described by the American Public Health Association (1955), and nitrate nitrogen by the method of Landman *et al.* (1960). Total nitrogen was measured by the micro-Kjeldahl method of Hiller, Plazin, and Van Slyke (1948).

Gas production from nitrate, nitrite, and nitrous oxide was determined by the manometric procedure described by Umbreit, Burris and Stauffer

(1959) with N_2 or N_2O as the gas phase. Sodium azide in a final concentration of 0.004 M was used as an inhibitor of nitrate reduction.

Analysis of gas. A mass spectrometer (Consolidated Electrodynamic Corp., Pasadena, Calif.) was used to analyze the gas produced by *C. nephridii* in a nitrate medium.

American Public Health Association (1955). *Standard methods for the examination of water, sewage, and industrial wastes,* 10th ed. American Public Health Association, Inc., New York.
A. Hiller, J. Plazin and D. D. Van Slyke (1948). J. biol. Chem. 176, 1401–1420.
W. A. Landman, S. Mohammed, P. I. H. Katherine and D. M. Doty (1960). J. Ass. off. Agric. Chem. 43, 531–535.
C. A. Parker (1955). Aust. J. exp. Biol. med. Sci. 33, 33–37.
Society of American Bacteriologists (1957). *Manual of Microbiological Methods* – McGraw-Hill Book Co. Inc., New York.
W. W. Umbreit, R. H. Burris and J. F. Stauffer (1959). *Manometric techniques,* 3rd ed. Burgess Publishing Co., Minneapolis.

Nitrate reduction—*Pseudomonas*

G. L. Bullock, S. F. Snieszko and C. E. Dunbar (1965). J. gen. Microbiol. 38, 1–7.

Nitrate reduction was determined in nutrient broth containing 0.1% (w/v) KNO_3; zinc powder (ZoBell, 1932) was used to detect false negatives. Tested after 1, 2, 7 days at 20–22°C.

C. E. ZoBell (1932). J. Bact. 24, 273.

Nitrate reduction—*Bacillus*

G. R. F. Hilson (1965). J. gen. Microbiol. 39, 407–421.

Reduction of nitrate to nitrite was tested by the method of Skerman (1959): samples were taken after 1, 2 and 3 days of incubation and tested for nitrite formation by a spot test on a white tile with starch + iodide indicator. Cultures were also tested for the production of gas from nitrate under anaerobic conditions (Gibson, 1944).

T. Gibson (1944). J. Dairy Res. 13, 248.
V. B. D. Skerman (1959). *A guide to the identification of the genera of bacteria,* Williams and Wilkins Book Co. Baltimore, Md.

Nitrate reduction—*Alcaligenes*

R. G. Mitchell and S. K. R. Clarke (1965). J. gen. Microbiol. 40, 343–348.

Nutrient broth containing 0.1% (w/v) KNO_3, incubated for 5 days, was tested with Griess-Ilosvay reagents, and with zinc dust for residual nitrate.

Nitrate reduction—*Streptomyces*

T. F. Fryer and M. E. Sharpe (1965). J. Dairy Res. 32, 27–34.

The authors used the methods of Gordon and Smith (1955) and Gordon and Mihm (1957).

0.05 ml of an 18–24 hr culture of the organism grown in 5 ml Yeastrel nutrient broth, centrifuged and resuspended in 5 ml distilled water was used as inoculum.

R. E. Gordon and J. M. Mihm (1957). J. Bact. 73, 15.
R. E. Gordon and M. M. Smith (1955). J. Bact. 69, 147.

Nitrate reduction—*Mycobacterium*

A. Tacquet, F. Tison and B. Devulder (1966). Annls Inst. Pasteur, Paris 110, 252–260.

L'étude de l'activité des mycobactéries sur les dérivés nitrés a permis de mettre au point de nouveaux tests pratiques de caractérisation des espèces *Mycobacterium tuberculosis, Mycobacterium bovis,* et des mycobactéries du groupe *avium.*

Selon que l'action des mycobactéries sur le nitrate et le nitrite de sodium est étudiée en suspension aqueuse ou en phase de croissance sur milieu de culture à l'oeuf, les résultats sont différents. En milieu solide, ils permettent de définir chaque espèce étudiée par des tests simples, facilement utilisables en pratique courante.

A. – RÉDUCTION DES NITRATES

La mise en évidence de la réduction des nitrates de sodium ou de potassium en nitrites est l'un des plus anciens procédés utilisés pour la différenciation des bactéries en général et de certaines espèces mycobactériennes en particulier.

Les travaux de Kumagai [9], de Edson [6], de Devignat [3, 4], de Gordon et Smith [8], de Hedgecock et Costello [7], et ceux qui sont colligés dans le *Bergey's Manual of Determinative Bacteriology* édit. 1957, nous apportent de précieuses indications au sujet de l'activité réductrice des nitrates exercée par certaines mycobactéries tuberculeuses ou ⟨⟨ paratuberculeuses ⟩⟩.

C'est cependant à Virtanen [15], puis à Boisvert [1], que revient le mérite d'avoir étudié et codifié cette réaction par l'étude systématique d'un grand nombre de souches de différentes espèces capables de jouer un rôle en pathologie humaine.

Techniques.

1. Méthode de Virtanen

La réaction de Griess permet de déceler la présence de nitrite de sodium et d'affirmer, par conséquent, la réduction des nitrates. Les nitrites sont en effet caractérisés par l'apparition d'une coloration violette, après addition successive du réactif A (acide sulfanilique : 0.8 g, acide acétique :

30 ml ; eau distillée : 100 ml) et du réactif B (α-naphtylamine : 0.5 g, acide acétique : 30 ml, eau distillée : 100 ml).

Cette réaction est pratiquée sur une suspension de 15 mg de mycobactéries laissée trois heures à 37° dans 2 ml de solution de tampon phosphatique ajustée à pH 7* et additionnée de 1 ml d'une solution de NO_3Na à 0.085 p. 100. La présence de nitrite se traduit par l'apparition immédiate de la coloration rouge violette.

Discussion.

a) Cette technique nécessite, après croissance des germes sur milieu de culture, leur mise en suspension en solution aqueuse ou tamponnée. Nous avons montré, en appliquant la méthode de Virtanen à des cultures de mycobactéries sur milieu solide, que cette manipulation pouvait être évitée.

b) D'autre part, en décelant la seule formation de nitrite, cette méthode ne permet pas d'apprécier la réduction du nitrate sans formation de nitrite. Nous avons mis en évidence une telle réduction du nitrate par les mycobactéries du groupe *avium.*

2. Méthode personnelle

1° Activité des mycobactéries sur le nitrate de sodium, en milieu solide

Nous avons adapté la technique de Virtanen au milieu solide à l'oeuf (Löwenstein-Jensen ou Coletsos), après qu'une étude préliminaire nous eut révélé que le nitrate de potassium et le nitrate de sodium pouvaient être utilisés indifféremment.

Technique.

1° Les tubes de milieu à l'oeuf sont imprégnés de nitrate de sodium, à raison de 0.30 mg/cm^2.

2° La souche à etudier est ensuite largement ensemencée, sur toute la surface du milieu.

3° Les tubes sont incubés à la température optimale de croissance jusqu'à apparition d'une végétation abondante.

4° La réaction de Griess est alors pratiquée, 0.25 ml des réactifs A et B étant successivement déposés à la surface du milieu de culture.

5° La présence de nitrite de sodium est décelée par l'apparition d'une coloration violette, foncée. En l'absence de nitrite, seul le verdissement du milieu par les réactifs acides est observé.

Résultats.

Les résultats obtenus par cette méthode, d'exécution très facile, sont identiques à ceux que donne la technique de Virtanen, bien qu'ils procèdent d'un comportement des souches à l'égard des nitrates différent de celui qu'on observe en suspension aqueuse, tamponnée ou non.

*Nous avons vérifié que cette réaction pouvait être faite aussi parfaitement sur une suspension de mycobactéries dans l'eau distillée.

2° Recherche du nitrate résiduel après action des mycobactéries.

Si la présence de nitrite de sodium, décelée par la réaction de Griess, traduit indubitablement la réduction du nitrate par les mycobactéries, le résultat négatif de la recherche des nitrites ne signifie pas, de façon formelle, la conservation du nitrate de sodium. Nous avons vérifié ce fait en déterminant le devenir du nitrate de sodium mis en présence de *M. bovis* ou de mycobactéries du groupe *avium,* incapables, d'après la méthode classique de Virtanen, de transformer le nitrate en nitrite de sodium.

Nous avons utilisé dans ce but la poudre de zinc qui réduit le nitrate de sodium en nitrite. L'addition de poudre de zinc à une solution aqueuse de nitrate ou à un milieu de culture à l'oeuf, préalablement imprégné en nitrate de sodium, détermine la formation de nitrite, lequel est mis en évidence par la réaction de Griess. Le même procédé, appliqué à une suspension de mycobactéries en solution aqueuse nitratée, ou à un milieu de culture imprégné de nitrate puis ensemencé avec des mycobactéries, précise la conservation ou la réduction de nitrate après croissance de ces germes.

La comparaison des colorations de Griess dans les tubes, témoin et ensemencé, permet d'apprécier l'intensité d'une éventuelle réduction du nitrate par des mycobactéries apparemment incapables de transformer ce nitrate en nitrite de sodium.

Pour sensibiliser cette méthode et la rendre applicable en pratique courante, nous avons déterminé, de façon empirique, la quantité *optimale* de nitrate de sodium, qui, sous l'influence de mycobactéries de différentes espèces, est soit totalement conservée, soit totalement dégradée en azote, soit réduite en une quantité de nitrite de sodium facilement décelable par la réaction de Griess. L'étude du comportement de 22 souches de mycobactéries de différentes espèces, ensemencées sur milieu à l'oeuf imprégné de nitrate de sodium à des concentrations variant de 0.05 à 0.50 mg/cm^2, nous a montré que la concentration optimale de nitrate est de 0.30 mg/cm^2 de milieu solide. En solution aqueuse, cette concentration est de 0.1 mg/ml.

Technique.

a) En milieu solide.

1° Trois tubes de milieu à l'oeuf sont imprégnés de nitrate de sodium, à raison de 0.30 mg/cm^2.

2° La souche à étudier est ensuite largement ensemencée sur toute la surface du milieu, dans 2 tubes sur 3.

3° Incubation à la température optimale de croissance jusqu'à apparition d'une végétation abondante.

4° A ce moment :

Premier tube (ensemencé) : Réaction de Griess (*Cf.* page 622) ; si la coloration violette n'apparaît pas, rechercher sur le deuxième tube, le nitrate de sodium résiduel.

Deuxième tube (ensemencé) et troisième tube (témoin) : 3 cg de poudre de zinc sont déposés à la surface du milieu. Après trois heures à l'étuve, la réaction de Griess permet de déceler le nitrite résultant de la réduction, par le zinc, du nitrate résiduel.

b) En solution aqueuse.

1° 2 ml d'une solution aqueuse stérile de nitrate de sodium à 0.1 mg/ml sont versés dans chacun des 3 tubes à hémolyse utilisés.

2° 15 mg de mycobactéries (poids humide) sont mis en suspension dans 2 de ces tubes.

4° Les 3 tubes sont laissés trois heures à l'étuve.

5° A ce moment :

Premier tube (ensemencé) : Réaction de Griess (*Cf.* page 622). Si la coloration violette n'apparaît pas, rechercher, sur le deuxième tube, le nitrate de sodium résiduel.

Deuxième tube (ensemencé) et troisième tube (témoin) : 1 cg de poudre de zinc est ajouté dans chaque tube. Après trois heures à l'étuve, la réaction de Griess permet de déceler le nitrite résultant de la réduction, par le zinc, du nitrate résiduel.

Nous avons comparé l'action sur le nitrate de sodium de mycobactéries (10 souches de *Mycobacterium tuberculosis,* 11 souches de *Mycobacterium bovis* et de BCG, et 20 souches de mycobactéries du groupe *avium*) en suspension aqueuse, et en phase de croissance sur milieu solide à l'oeuf.

Cette action a été appréciée en recherchant d'abord, par la réaction de Griess, une éventuelle réduction du nitrate en nitrite de sodium, puis, en cas d'absence de nitrite, en mettant en évidence la présence éventuelle de nitrate résiduel.

B. – RÉDUCTION DES NITRITES

L'action des mycobactéries du groupe *avium* sur le nitrate de sodium en milieu solide nous a amenés à rechercher si ces germes, qui transforment le nitrate en azote, sont également réducteurs du nitrite de sodium préalablement apporté au milieu de culture. Les mycobactéries appartenant aux autres espèces ont été étudiées dans les mêmes conditions.

La fragilité du nitrite, qui nécessite la stérilisation par filtration, et *l'action bactériostatique* de ce composé, nettement supérieure à celle du nitrate [12], imposent des techniques particulières.

Le principe de la recherche d'une réduction du nitrite de sodium par les mycobactéries est cependant simple : si l'on utilise la plus faible concentration de nitrite de sodium susceptible d'être facilement décelée par la réaction de Griess (0.005 mg/cm^2 de milieu solide à l'oeuf), la négativation de cette réaction traduit la réduction du nitrite.

Nous avons observé qu'une telle réduction n'apparaissait jamais lorsque les mycobactéries (de toutes espèces) étaient mises en suspension, même à forte concentration, dans une solution aqueuse ou tamponnée, préalablement nitritée.

En revanche, certaines espèces mycobactériennes sont capables de réduire le nitrite de sodium imprégnant le milieu solide (Löwenstein-Jensen ou Coletsos) sur lequel elles cultivent abondamment.

Nous avons montré [12] que les *facteurs de perméabilité* ne pouvaient pas expliquer cette constatation. De grosses quantités de germes broyés sont incapables de réduire les nitrites ; il ne s'agit donc pas d'une *endo-enzyme.* L'éventualité d'*enzymes adaptatives* ne peut être retenue. Enfin, les différents éléments des milieux de culture sont incapables de catalyser la réaction en l'absence de végétation. Il est donc possible de conclure que la réduction du nitrite de sodium par certaines souches de mycobactéries ne peut se produire que si les germes sont en *phase de croissance* et en pleine activité métabolique.

Technique.

1° Les tubes de milieu à l'oeuf sont imprégnés de nitrite de sodium, à raison de 0.05 mg/cm^2.

2° La souche à étudier est ensuite largement ensemencée sur toute la surface du milieu.

3° Les tubes sont incubés à la température optimale de croissance, jusqu'à apparition d'une végétation abondante.

4° La réaction de Griess est alors pratiquée, en même temps que sur le tube témoin nitrité mais non ensemencé.

5° La réduction du nitrite se traduit par une absence de coloration, tandis que la teinte violette caractéristique du nitrite apparaît sur le tube témoin.

[1] Boisvert (H.). *Ann. Inst. Pasteur,* 1961, 100, 352–357.
[3] Devignat (R.). *Ann. Inst. Pasteur,* 1952, 82, 650–652 et 653–658.
[4] Devignat (R.). *Ann. Soc. Belge Med. trop.,* 1964, 44, 439–452.
[6] Edson (N.). *Bact. Rev.,* 1951, 15, 147–152.
[8] Gordon (R.) and Smith (M.). *J. Bact.,* 1953, 66, 41–48.
[9] Kumagai (H.). *Sci. Rep. Res. Inst. Tohoku Univ. Serv.,* 1950, 2,
[12] Tison (F.), Guillaume (J.) et Devulder (B.). *Ann. Inst. Pasteur Lille,* 1963, 14, 125–138.
[15] Virtanen (S.). *Acta tuberc. scandin.,* 1960, supp. 48, 119.

Nitrate reduction—*Providencia, Rettgerella*

C. Richard (1966). Annls Inst. Pasteur, Paris 110, 105–114.

Pour des raisons de commodité, les milieux ont été incubés à 37°C. Réduction des nitrates: recherche des nitrites à l'aide du réactif de Griess (test confirmatif de Zo-Bell). Pichinoty [21, 22] en utilisant une méthode manométrique a montré qu'il existait deux types de nitratases:

nitrate-réductases A et B. Cet auteur a bien voulu nous communiquer les résultats de ses recherches concernant les nitrate-réductases de quelques souches de *Providencia* appartenant aux biotypes 1, 2, 5 et 6. Nous l'en remercions ici.

[21] Pichinoty (F.). *Ann. Inst. Pasteur,* 1963, 104, 219.
[22] Pichinoty (F.), Rigano (C.), Bigliardi-Rouvier (J.), Le Minor (L.) et Piéchaud (M.). *Ann. Inst. Pasteur,* 1966, 110, 128.

Nitrate reduction—*Mycobacterium*

H. Boisvert (1966). Annls Inst. Pasteur, Paris 111, 180–192.

Réduction des nitrates [37, 4]. Il est nécessaire d'utiliser une culture jeune, pratiquement dès que l'on peut prélever la quantité de germes suffisante.

Dans un tube à hémolyse contenant 0.1 ml d'eau distillée, on introduit environ 15 mg de bacilles et l'on ajoute 2 ml d'une solution à 0.085 p. 100 de nitrate de soude.

Après deux heures à 37° on laisse tomber dans le tube I goutte d'acide chlorhydrique pur, II goutes de sulfanilamide à 0.2 p. 100 et II gouttes de chlorhydrate de N-α-naphtyl-diéthyl-propylène-diamine à 0.1 p. 100. Si la réaction est positive, le liquide prend une couleur rouge violacé.

On peut employer, pour plus de sûreté, la méthode en tubes avec nitrate incorporé de Tacquet et Tison [34].

[4] Boisvert (H.). *Ann. Inst. Pasteur,* 1961, 100, 352.
[34] Tacquet (A.), Tison (F.) et Devulder (B.). *Ann. Inst. Pasteur,* 1966, 110, 252.
[37] Virtanen (S.). *Acta tub. scand.,* 1960, suppl. 48.

Nitrate reduction—yeast

F. Pichinoty and G. Méténier (1966). Annls Inst. Pasteur, Paris 111, 282–313.

Nous décrivons une méthode manométrique qui permet de mesurer l'activité de la nitrate-réductase de la levure *Hansenula anomala.* Chez l'organisme étudié, cette enzyme possède une fonction assimilatrice, mais pas de fonction respiratoire. Nous avons jugé nécessaire de lui conférer une appellation particulière et de la désigner par la lettre D car elle diffère, sur plusieurs points, des nitrate-réductases bactériennes A et B. Voici quelles sont ses principales propriétés : 1° Utilise probablement comme substrat un analogue structural du nitrate, le chlorate. 2° Dans les extraits bruts, fonctionne avec les formes réduites du benzyl-viologène, du méthyl-viologène et du FMN comme sources d'électrons. 3° Subit une inactivation importante de nature réversible lorsqu'elle se trouve placée en présence de l'un de ces donateurs réduits dans un milieu où ses substrats sont absents ; le nitrate et le chlorate protègent efficacement l'enzyme contre cette inhibition. 4° Est inhibée par le cyanure et l'azoture. 5°

Quel que soit le substrat, le pH correspondant à l'activité optimale est de 7. 6° Après centrifugation des extraits à grande vitesse, on retrouve une grande partie de l'activité dans la fraction particulaire ; ce fait suggère que l'enzyme D est mitochondriale.

Nitrate reduction—*Brevibacterium*

R. Chatelain and L. Second (1966). Annls Inst. Pasteur, Paris 111, 630–644.

Nitrate-réductase: après 4 jours de culture en bouillon nutritif additionné de 0.1 p. 100 de nitrate de potassium, la présence de nitrites est recherchée par la méthode de Wallace et Neave [26] ; la réaction de Zo Bell [28] permet ensuite de mettre en évidence la réduction au-delà des nitrites.

La température d'incubation des cultures est de 30°C.

[26] Wallace (G. I.) and Neave (S. L.). *J. Bact.*, 1927, 14, 377.
[28] Zo Bell (C. E.). *J. Bact.*, 1932, 24, 273.

Nitrate reduction—*Mycobacterium*

A. Tacquet, F. Tison, B. Polspoel, P. Roos and B. Devulder (1966). Annls Inst. Pasteur, Paris 111, 359–363.

L'activité réductrice des Mycobactéries sur le nitrate [1, 2, 7] et sur le nitrite de sodium ou de potassium varie selon les espèces. Son étude, dont nous avons souligné l'intérêt dans des travaux antérieurs [3, 4, 5, 6], nous a permis de constater, d'une part, que certains germes réduisant les nitrates étaient capables de dépasser le stade des nitrites, d'autre part que les résultats étaient différents selon que cette recherche était effectuée sur des Mycobactéries *en suspension aqueuse* ou sur des Mycobactéries *en phase de croissance.* L'objet de ce travail est d'étudier les variations de la réduction des *nitrites* en fonction des modalités expérimentales. Nous envisagerons successivement l'activité réductrice des Mycobactéries en suspension dans un milieu aqueux, en croissance sur milieu solide, en croissance sur milieu liquide. Quarante souches de Mycobactéries d'espèces diverses, en particulier du groupe *avium,* ont été étudiées.

1° Mycobactéries en suspension aqueuse. – Lorsque 15 à 60 mg de germes sont mis en présence de 2 ml de nitrite de sodium ou de potassium, à des concentrations variant de 2 à 50 μg/ml, on n'observe en aucun cas, même après quarante-huit heures de contact à 37°C, de réduction du nitrite, et la réaction de Griess reste positive avec la même intensité que dans les solutions témoins.

Nous avons constaté que cette activité n'était pas modifiée par l'existence de facteurs favorisant la perméabilité de la membrane cellulaire, ou l'existence d'endoenzyme ou d'enzyme adaptative [5].

2° Mycobactéries en croissance sur milieux solides. – Le comportement est tout différent lorsque les germes sont en *phase de croissance et de métabolisme.*

Des tubes de milieu à l'oeuf (Löwenstein-Jensen ou Coletsos base) sont imprégnés à raison de 10 μg de nitrite de sodium par cm^2, la solution utilisée étant stérilisée par filtration [5].

Après évaporation, ils sont ensemencés avec 0.20 ml d'une émulsion à 1 mg/ml de la souche à étudier.

Dès qu'il y a végétation, la réaction de Griess est pratiquée sur un tube témoin non ensemencé et sur le tube où une culture s'est développée.

Dans le cas de *M. tuberculosis* ou *bovis,* la même teinte violette est observée dans les tubes témoins et dans les tubes avec nitrite; dans le cas de *M. avium,* cette teinte n'apparaît pas, le nitrite étant réduit.

3° Mycobactéries en croissance en milieux liquides. – Cette étude a été faite en milieu liquide de Dubos, avec Tween 80 (500 mg par litre), qui a l'avantage de permettre des mesures quantitatives. Ce milieu liquide est enrichi de nitrite de sodium stérile à raison de 5 μg/ml et réparti en ballons de 25 ml. Ces derniers sont ensemencés avec 1 mg de la souche à étudier et portés à l'étuve en culture agitée.

Des prélèvements de 5 ml ont été faits aux troisième, sixième et douzième jours de végétation.

L'intensité de la végétation est notée en densité optique à 6 500 Å (spectrophotomètre Jouan Junior).

L'échantillon est ensuite centrifugé et le nitrite surnageant est dosé par photométrie à 5 200 Å après addition de 0.3 ml des réactifs :

A	Acide sulfanilique	0.8 g
	Acide acétique	30 ml
	Eau distillée	100 ml
B	α-naphtylamine	0.3 g
	Acide acétique	30 ml
	Eau distillée	100 ml

Reducing activity of Mycobacteria on sodium nitrite. Studies on the influence of culture media.

This activity varies according to the experimental conditions : culture on Loewenstein-Jensen' or liquid Dubos' medium.

Nitrite reduction is consistently observed with Mycobacteria belonging to the *avium* group cultivated on egg-media. This reduction does not always occur in liquid media.

Loewenstein-Jensen's media should therefore be used for the routine qualitative studies of this reducing agent.

[1] Boisvert (H.). *Ann. Inst. Pasteur,* 1961, 100, 352–357.
[2] Bonicke (R.). *Bull. int. Union Tuberc.,* 1962, 23, 13–68.
[3] Tacquet (A.). *Ann. Soc. Belge Med. trop.,* 1962, 42, 383–401.
[4] Tacquet (A.), Tison (F.) et Devulder (B.). *Ann. Inst. Pasteur,* 1966, 110, 252–260.

[5] Tison (F.), Guillaume (J.) et Devulder (B.). *Ann. Inst. Pasteur Lille,* 1963, 14, 125–138.
[6] Tison (F.) et Devulder (B.). *Path. Biol.,* 1965, 13, 458–462.
[7] Virtanen (S.). *Acta tuberc. scand.,* 1960, suppl. 48, 1–119.

Nitrate reduction (to nitrogen)—*Pseudomonas*

R. Y. Stanier, N. J. Palleroni and M. Doudoroff (1966). J. gen. Microbiol. 43, 159–271.

Tests were conducted in yeast extract (YE) medium, supplemented with 10 g glycerol, 10 g KNO_3 and 1 g Ionagar/L. Each culture was first grown for 24 hr in an unsealed tube containing 5 ml of this medium incubated without mechanical agitation. A loopful of this semi-aerobic culture was transferred to a tube containing 10 ml of the same medium, molten but cooled to about 40°C. The contents were mixed, and the tube was chilled briefly to allow the medium to become semi-solid. An overlayer of 2-3 ml of 1.0% (w/v) Ionagar in water provided an anaerobic seal and a trap for gas production. Vigorous denitrifiers like *Pseudomonas aeruginosa* produce dense turbidity in this medium after 18–24 hr of incubation at 30°C, soon followed by abundant gas production. Other denitrifying pseudomonads develop more slowly, and the tubes should be incubated 5 days before recording a negative result. With some strains, heavy anaerobic growth occurred without visible gas production; such strains were reported as positive. Pre-cultivation in the same medium under semi-aerobic conditions is important, since in some species prolonged aerobic cultivation in a nitrate-free medium seems to result in a loss or weakening of denitrifying ability, presumably as a result of mutation and selection. The precultivation step selects once more for denitrifying ability, and this ensures prompt growth of potential denitrifying strains when placed under rigorously anaerobic conditions.

We have also examined growth under denitrifying conditions in defined media.

Nitrate reduction—*Mycobacterium*

M. Tsukamura (1966). J. gen. Microbiol. 45, 253–273.

Nitrate reduction. Fifty mg (moist weight) of the organism were suspended in 5 ml 0.067 M-phosphate buffer (pH 7.1) containing 0.1% sodium nitrate, and incubated at 37°C for 15–16 hr. Formation of nitrite was examined by adding two drops of 2% (w/v) p-dimethylaminobenzaldehyde in 10% HCl + 1.0 ml 10% HCl.

Nitrate reduction—*Bacteroides, Veillonella, Neisseria*

J. van Houte and R. J. Gibbons (1966). Antonie van Leeuwenhoek 32, 212–222.

Nitrite (and residual nitrate) was determined according to the Society of American Bacteriologists' *Manual of Microbiological Methods* (1957).

The basal medium was BBL thioglycolate medium, without added dextrose, supplemented with 5% horse serum and hemin (5 μg/ml) menadione (0.5 μg/ml).

Society of American Bacteriologists (1957). *Manual of Microbiological Methods.* McGraw-Hill Book Co Inc. New York.

Nitrate reduction—*Clostridium*

A. V. Goudkov and M. E. Sharpe (1966). J. Dairy Res. 33, 139–149.

The basal medium used for many of the tests consisted of: BBL Trypticase, 1.5%; yeast extract, 1%; cysteine, 0.05%; pH of medium, 7.0. The methods of Beerens, Castel and Put (1962, Annls Inst. Pasteur, Paris 103, 117) were used.

The basal medium containing 0.2% glucose and 0.1% KNO_3 was used. Reduction was detected by methods described in the *Manual of Microbiological Methods* (1957).

Society of American Bacteriologists *Manual of Microbiological Methods* – McGraw-Hill Book Co. Inc., New York 1957.

Nitrate reduction—anaerobes

A. -R. Prévot (1966). Techniques pour le diagnostic des Bactéries Anaérobes. Éditions de la Tourelle, St. Mandé.

Réduction des nitrates en nitrites.

La première recherche se fait en utilisant soit les bouillons nitratés, soit l'eau peptonée nitratée, auxquelles on a ajouté divers glucides comme donateurs d'hydrogène. On prélève dans chacun des tubes un demi-cm^3 de milieu aux 1er, 2e, 3e, 5e jours d'étuvage et on met le prélèvement dans des tubes à essais bien propres ; on fait couler 1/2 à 1 cm^3 du réactif de Tromsdorff préparé extemporanément (empoi d'amidon additionné d'iodure de potassium et de 10 à 12 gouttes d'acide acétique). En cas de réduction des nitrates en nitrites, il se produit une coloration bleue. On note soigneusement les glucides donateurs d'hydrogène et le temps d'apparition de la réduction.

Eau peptonée avec nitrate.

Faire dissoudre 20 g de peptone bactériologique dans 1 litre d'eau distillée.

Ajuster à pH 7.4 avec une solution de soude à 10% (il en faut environ 10 à 12 cm^3). Filtrer sur papier Laurent. Répartir en tubes 13–14/180 mm. Stériliser à 110° pendant 30 minutes.

Cette eau peptonée sert par ailleurs à préparer l'eau peptonée nitratée à 5 p. 1000.

Bouillon VF ordinaire avec nitrate.

Neutraliser digestat acide*	1 litre
par lessive de soude pure (36° B)	9 cm^3

Ajuster à pH 7.4. Précipiter à 118°–120° pendant 30 minutes. Filtrer sur papier Laurent. Répartir dans les récipients adéquats (tubes de 13–14/180 mm ; tubes de Hall ; fioles de Jaubert et Gory ou d'Erlenmeyer). Boucher au coton cardé. Stériliser à 110° pendant 30 minutes.

Le même bouillon peut recevoir des additions diverses pour la mise en évidence des enzymes.

a) 5 p. 1000 de nitrate de soude pour l'étude de la réduction des nitrates.

*See **Gelatin hydrolysis** – anaerobes (p. 394).

NITROGEN FIXATION

Nitrogen fixation—*Clostridium*

J. Augier (1957). Annls Inst. Pasteur, Paris 92, 817–824.

Composition du milieu. – La source de carbone utilisée est le glucose, substance connue pour être favorable aux *Clostridium.* Il est probable que les hémicelluloses constituent également une bonne source de carbone. Nos essais à ce sujet sont encore en cours.

Le milieu est, en outre, légèrement tamponné pour éviter une baisse trop rapide du rH. Les cultures nous ont paru meilleures dans ces conditions.

La présence d'extrait de terre dans le milieu est indispensable, la sensibilité du milieu étant ainsi bien meilleure (en effet, Jensen [7] a précisé qu'en milieu purement synthétique un faible pourcentage seulement de *Clostridium* était capable de pousser). Le rôle des oligo-éléments est le même que pour les *Azotobacter,* aussi les avons-nous introduits de la même manière que pour ceux-ci [6].

Enfin, lorsqu'il est nécessaire d'introduire un indicateur de rH, nous prenons la phénosafranine, soit 8 ml d'une solution de phéno-safranine à 2 p. 1000, pour 1 l de milieu.

La formule est donc :

Solution de phosphate mono-potassique à 1 p. 100	75 ml
Soude décinormale	33 ml
Glucose	10 g
Extrait de terre	10 ml
Solution d'oligo-éléments [6]	1 ml
Solution de Winogradsky	50 ml
Carbonate de calcium	0.05 mg
Eau distillée	q.s.p. 1 000 ml

Répartir en tubes de 17, contenant chacun une cloche pour recueillir les gaz dégagés, 10 ml de milieu par tube ; stériliser à 100° pendant vingt minutes.

Méthode de culture. – Les tubes une fois ensemencés sont mis à l'étuve à 28°–30° :

1° Soit en les plaçant sous une cloche dans laquelle on fait le vide et ramenée à pression ordinaire par introduction d'azote.

2° Soit en les plaçant sous une cloche dans laquelle on dispose au fond d'une boîte de Petri un mélange de pyrogallol et de potasse. On fait le vide jusqu'à ce que la pression intérieure soit de 20 cm de mercure.

3° Plus simplement encore, on peut sceller l'extrémité des tubes de manière à ce que la flamme introduise des gaz de combustion dans le tube. Dans ce cas il est préférable de repartir le milieu

[6] Augier (J.). *Ann. Inst. Pasteur,* 1956, 91, 759.
[7] Jensen (H. L.). *Proc. Linn. Soc. N. S. W.* 1940, 65, 1 à 122.

Nitrogen fixation—*Rhizobium* (plant culture for)

E. Schiel, E. L. G. De Olivero, R. N. Diéguez, J. C. Pacheco and E. Enokida (1963). Annls Inst. Pasteur, Paris 105, 332–340.

Milieu de culture pour plantes. Récipients. – On a employé un milieu gélosé de pH 6, 5, selon la formule de Crone modifiée par Bryan [3], préparée selon Schiel et Ragonese [8], avec addition de 2 mg par litre d'un mélange d'oligo-éléments [7] auquel on a ajouté 1.6 g de Cl_2Ca 2 H_2O. On a transvasé en flacons de verre à large ouverture de 19.5 cm de hauteur et 6 cm de diamètre, à raison de 100 ml par récipient ; on a refroidi en agitant après la stérilisation, sans incliner. Culture de plantes totalement aseptique.

[3] Fred (E. B.) and Waksman (S. A.). *Laboratory manual of general microbiology.* McGraw-Hill Book Co, New York, 1928.
[7] Schiel (E.). de Olivero (E. L. G.) y Yepes (M.). *Rev. Inv. Agric.,* 1959, 13, 257–279.
[8] Schiel (E.) y Ragonese (A. E.). *Rev. Arg. de Agronomia,* 1942, 9, 114–169.

Nitrogen fixation—*Beijerinckia*

F. Hilger (1964). Annls Inst. Pasteur, Paris 106, 279–291.

Milieux de culture.

a) Milieu de base sans azote : glucose, 5 g ; $MgSO_4 \cdot 7H_2O$, 125 mg ; $Fe_2(SO_4)_3$, 2.5 mg ; $MnSO_4$, 2.5 mg ; solution d'oligoéléments selon Augier [1], 1 ml ; eau distillée, 1000 ml.

b) pH. La gamme suivante de pH est réalisée : 4.53 ; 4.98 ; 5.30 ; 5.60 ; 5.91 ; 6.24 ; 6.47 ; 6.81 ; 7.15. Les milieux sont fortement tamponnés aux phosphates (KH_2PO_4 et $Na_2HPO_4 \cdot 2H_2O$) qui y sont introduits dans les proportions indiquées par Sörensen et à la concentration molaire totale de 0.067. Cette concentration paraît élevée (environ 10 g par litre de phosphates au total), mais ne provoque cependant pas une dépression notable du développement des *Beijerinckia.*

Ceci découle d'essais préliminaires et est d'ailleurs prouvé par le fait que les quantités d'azote fixé, données plus loin, sont normales. Signalons d'ailleurs que Dommergues [7] utilise des concentrations analogues sans observer une dépression du développement et que Becking [3] constate une dépression relativement faible.

c) Ca. Les milieux à pH variables restent d'une part, dépourvus de Ca et reçoivent d'autre part, cet élément sous forme de $CaCl_2$, à la dose qui est normalement admise dans les milieux de culture pour *Azotobacter*, soit 0.5 g de $CaCl_2$ par litre ou 180 p.p.m. de Ca.

d) Précautions particulières. Tous les produits entrant dans la constitution des milieux sont de haute pureté. L'eau utilisée est bidistillée. Les récipients utilisés sont en verre pyrex ; ils sont nettoyés d'abord au bain sulfochromique, ensuite rincés soigneusement à l'eau bidistillée. Le glucose est stérilisé séparément en solution concentrée (12.5 p. 100) et est introduit aseptiquement en quantités exactement mesurées dans les récipients de culture.

3° Conditions de culture.

Les cultures sont faites en flacons d'Erlenmeyer de 300 ml contenant 50 ml de milieu.

La température d'incubation est de 28° C et les incubateurs sont pourvus de récipients contenant de l'acide sulfurique.

L'inoculum est constitué d'une suspension de cellules, obtenue au départ de cultures jeunes (six jours) sur gélose inclinée, dépourvue de Ca, tamponnée à pH 6 et contenant les éléments nutritifs aux concentrations indiquées plus haut. Les milieux sont ensemencés au moyen d'une goutte de suspension.

Nitrogen fixation—*Pseudomonas methanitrificans*

J. B. Davis, V. F. Coty and J. P. Stanley (1964). J. Bact. 468–472.

Atmospheric nitrogen fixation. Nitrogen fixation by the isolated methane-oxidizing bacteria was first tested on a small scale by inoculating triplicate 10-ml portions of nitrogen-free mineral-salts medium (< 2 μg of Kjeldahl N per ml) in 100-ml bottles. The medium had the following composition (in g per liter of nitrogen-free water): Na_2HPO_4, 0.3; KH_2PO_4, 0.2; $MgSO_4 \cdot 7H_2O$, 0.1; $FeSO_4 \cdot 7H_2O$, 0.005; and $Na_2MoO_4 \cdot 2H_2O$, 0.002. These systems were evacuated and filled with 30% pure-grade methane in air. One culture system was placed in the cold at 5°C to inhibit growth; the other two systems were incubated at 30°C. All systems were analyzed for fixed nitrogen content after an incubation period of 2 weeks. This analysis was performed by adding 3 ml of Kjeldahl reagent to each of the systems, thus acidifying the contents and trapping any ammonia that might be in the atmosphere or in the liquid. The contents were quantitatively transferred from each bottle to a

Kjeldahl digestion flask with washings of nitrogen-free distilled water. Larger scale tests of nitrogen fixation by the methane-oxidizing bacteria were performed by adding a small inoculum to 1- to 2-liter quantities of the nitrogen-free mineral-salts medium. A 50% solution of H_2SO_4 (200 ml) was employed in the gas-flow line preceding the culture system. This was for the purpose of absorbing ammonia that might possibly be in the methane-air mixture (1:9) which was bubbled at about 50 ml per min through the culture system. Turbidity due to bacterial growth was definite in these large systems in about 2 weeks and gradually increased. After an incubation period of 2 to 4 months, the growth systems were acidified to pH 2.0 with H_2SO_4, and the bacterial cells were harvested by centrifugation prior to analysis of their nitrogen content. The cell-free culture liquor was reduced in volume by evaporation to about 50 ml prior to Kjeldahl analysis. The H_2SO_4 solution was also tested for nitrogen content.

Nitrogen fixation—*Azotobacter, Clostridium*

P-C. Chang and R. Knowles (1965). Can. J. Microbiol. 11, 29–38.

Azotobacter was counted on nitrogen-free medium, according to the method of Brown *et al.* (1962) as mucilaginous colonies larger than 1 or 2 mm in diameter and, in addition, a count was made on the same plates of the total number of visible colonies. It was assumed that this would include those aerobic organisms which were able to fix nitrogen at a much slower rate than did *Azotobacter,* or which were able to grow in the presence of small quantities of nitrogen diffusing from the *Azotobacter* colonies or present as a contaminant in the agar medium.

Clostridium was counted on nitrogen-free medium (incubated in nitrogen with 5% carbon dioxide) according to the method of Ross (1958) but modified in the following way. To promote reducing conditions and to reduce spreading of colonies and splitting of the agar, 0.2% sodium thioglycollate and 0.15% sodium formaldehyde sulfoxylate were incorporated into the medium (Society of American Bacteriologists, 1957). After the flooding of the plates with iodine, the dark brownish blue-stained clostridia were counted. The total number of visible colonies was also counted on the same plates and was taken to be roughly indicative of the number of anaerobic or facultative organisms which were able to fix nitrogen at a much slower rate than did the clostridia, or which were able to grow in the presence of small quantities of nitrogen diffusing from the *Clostridium* colonies or present as a contaminant.

M. E. Brown, S. K. Burlingham and R. M. Jackson (1962). Pl. Soil 17, 309–319.

D. J. Ross (1958). Nature, Lond. 181, 1142–1143.

Society of American Bacteriologists. *Manual of Microbiological Methods* – McGraw-Hill Book Co. Inc., New York, 1957.

Nitrogen fixation—*Klebsiella*

M. C. Mahl, P. W. Wilson, M. A. Fife and W. H. Ewing (1965). J. Bact. 89, 1482–1487.

Growth. The growth medium and analytical procedures were essentially those described by Pengra and Wilson (1958) with the exception that mannitol (20 mg/ml) was used instead of sucrose as the carbon source. Ammonium sulfate (25 μg of N/ml) was added to insure good initial growth of all cultures.

The inoculum, an 18-to 20-hr culture grown in a test tube containing 10 ml of the medium, was transferred to a side-arm flask containing 90 ml of medium. The flasks were sealed and flushed with nitrogen as described by Pengra and Wilson (1958).

Growth was followed with a no. 66 filter in a Klett-Summerson colorimeter. Maximal growth (95 to 105 Klett units) was reached in 9 to 10 hr. An amount (1 ml) of a sterile yeast extract solution (4 mg/ml) was then added to each flask, and the flasks were again flushed with nitrogen. Yeast extract shortens the induction period of the nitrogen-fixing enzymes by providing a source of amino acids (Lindsay, 1963).

Nitrogen fixation. Fixation of N_2^{15} was determined by pipetting duplicate 5-ml samples of each culture into Warburg flasks after 20 to 22 hr of incubation and exposure to an atmosphere containing 10% N_2^{15} (33% atom N^{15} excess) and 90% helium. The samples were incubated at 30°C for 5 hr, digested, and analyzed for N^{15} in a Consolidated-Nier mass spectrometer (Burris and Wilson, 1957). Total nitrogen determinations were also made on cultures with use of a conventional Kjeldahl semimicro method. These cultures were incubated for 7 to 10 days in an effort to detect fixation by cultures that appeared to be negative in this respect.

R. H. Burris and P. W. Wilson (1957). In S. P. Colowick and N. O. Kaplan (ed.), *Methods of enzymology,* 4th ed., Academic Press, Inc. New York. p. 355–366.

H. L. Lindsay (1963). *Physiological studies of nitrogen fixation by cells and cell-free extracts of Aerobacter aerogenes.* Ph. D. Thesis, University of Wisconsin, Madison.

R. M. Pengra and P. W. Wilson (1958). J. Bact. 75, 21–25.

Nitrogen fixation—*Klebsiella pneumoniae*

D. C. Yoch and R. M. Pengra (1966). J. Bact. 92, 618–622.

The medium was prepared in two parts, which were autoclaved separately and mixed just before use. Solution I contained Na_2HPO_4, 6.25 g; KH_2PO_4, 0.75 g; and distilled water, 500 ml. Solution II contained: $MgSO_4 \cdot 7H_2O$, 0.2 g; Fe-Mo solutions* 1.0 ml; sucrose, 20 g; NaCl, 8.5 g; and distilled water, 500 ml. The inoculum, 1.5 ml of a culture actively fixing nitrogen in the late exponential growth phase, was transferred to 250-ml side-arm flasks containing 50 ml of the medium. The flasks were sealed and flushed with nitrogen as described by Pengra and Wilson. Unless otherwise indicated, the cultures were grown on a rotary shaker at 34°C.

R. M. Pengra, and P. W. Wilson (1958). J. Bact. 75, 21–25.

*P. W. Wilson and S. G. Knight (1952). *Experiments in bacterial physiology,* Burgess Publishing Co., Minneapolis, p. 53.

OPTICAL ROTATION OF LACTIC ACID

Optical rotation of lactic acid

M. Rogosa, R. F. Wiseman, J. A. Mitchell, M. N. Disraely, and A. J. Beaman (1953). J. Bact. 65, 681–699.

Medium:

Trypticase, 10 g; yeast extract, 5 g; KH_2PO_4, 6 g; ammonium citrate, 2 g; salt solution, 5 ml; glucose, 20 g; sorbitan monooleate, 1 g; (Tween 80), 0.3 g; sodium acetate hydrate, 10 g; acetic acid, 1.32 ml; distilled water, to 1000 ml. The salt solution contained $MgSO_4 \cdot 7H_2O$, 11.5 g; $MnSO_4 \cdot 2H_2O$, 2.4 g; $FeSO_4 \cdot 7H_2O$, 0.68 g; distilled H_2O to 100 ml.

Arabinose and xylose at concentrations of 0.5 per cent were added in the cases of certain heterofermentative strains which otherwise did not produce sufficient lactate for analysis. The medium was dispensed in 300 ml quantities into 500 ml flasks, 10 g of calcium carbonate added, and sterilized at 15 lb steam pressure for 15 minutes. The flasks were inoculated heavily and incubated for 2 weeks at 35°C. Volatile acids were distilled off and zinc lactates prepared from the ether extracts. The analyses were performed according to the method of H. R. Curran, L. A. Rogers and E. O. Whittier (1933, J. Bact. 25, 595–621).

Optical rotation of lactic acid

V. B. D. Skerman, *A Guide to the Identification of the Genera of Bacteria.* 2nd ed. Williams and Wilkins Book Co., Baltimore, 1967.

The culture should contain at least 2 g of lactic acid. Clarify by adding one-twentieth of its volume of 25 per cent zinc sulfate. Neutralize to a pH between 7.6 and 7.8 with 20 per cent sodium hydroxide, and centrifuge off the resulting zinc hydroxide precipitate.

Acidify the clarified solution to pH 2 with sulfuric acid and extract three times with ether (total volume of ether about 1 L). If appreciable amounts of volatile acids or succinic acid are present, they should be removed beforehand, but this is not necessary with homofermentative bacteria.

Add water (10 ml) to the ether extract, and distill off the ether. Make the remaining solution up to 50 ml, and titrate 1 ml to determine the total quantity of extracted acid. Then boil the solution for 10 minutes with a calculated excess of zinc carbonate; filter off the excess carbonate and wash. Evaporate the filtrate on a water bath until crystallization begins. Add alcohol to a concentration of 50 per cent and allow the mixture to stand overnight to crystallize. Filter off the product; wash with 95 per cent alcohol and then with ether, air-dry, and place in a desiccator over calcium chloride.

To test the optical form of the zinc lactate, the water of crystallization and the specific rotation are determined.

1. Weigh a 2 g sample of the salt accurately, dry to constant weight at 110°C, and calculate the amount of water of crystallization. Optically active salt contains 2 molecules (12.89 per cent) of water; the inactive salt contains 3 molecules (18.17 per cent).

2. Weigh 1 g of the anhydrous salt accurately, dissolve in distilled water, and dilute to 25 ml at 20°C. Determine the rotation of polarized light in a polarimeter, and calculate the specific rotation from the observed rotation by the formula

$$\alpha = \frac{100.a}{b.c}$$

where α is the specific rotation, a is the observed rotation, b is the length of the polarimeter tube in decimeters, and c is the concentration in g per 100 ml of solution.

The rotation of the active zinc salt is *opposite* to that of the free acid and varies slightly with concentration: a 4 per cent solution (w/v) has a specific rotation of ± 8.1 to 8.6.

C. S. Pederson, W. H. Peterson and E. B. Fred (1926). J. biol. Chem. 68, 151.

Also see: Biochem. J. 26, (1932) 846; Ind. Engng Chem. 27, (1935) 1492.

Optical rotation of lactic acid—*Pediococcus*

H. L. Günther and H. R. White (1961). J. gen. Microbiol. 26, 185–197.

The method of Pederson, Peterson and Fred (1926) was followed except that a continuous ether extraction apparatus was used, extracting the sample for 48 hr. The zinc content of the isolated zinc lactate was determined by the titrimetric method of Kolthoff and Sandell (1950) and the optical rotation determined polarimetrically, using the anhydrous salt in 1% (w/v) aqueous solution. Six pediococcus strains were examined.

Maintenance of stock cultures and methods of cultivation. For maintenance of stock cultures, preparation of inocula and in all experimental work, 'Oxoid' tomato juice (TJ) broth or tomato juice (TJ) agar, adjusted to pH 6.6, were used unless otherwise stated. The following were exceptions to this rule: for strain Tc. 1 sodium chloride (5%, w/v) was added to the medium; and for the aerococci glucose Lemco broth (Shattock and

Hirsch, 1947) or glucose yeast extract (GY) agar (containing, as %, w/v; peptone, 1.0; Yeastrel, 0.3; glucose, 1.0; NaCl, 0.25; agar, 1.0; at pH 7.4) was used.

Cultures were incubated aerobically at 30°C with specified exceptions.

I. M. Kolthoff and E. B. Sandell (1950). *Textbook of Quantitative Inorganic Analysis,* p. 577. London: Macmillan and Co., Ltd.
C. S. Pederson, W. H. Peterson and E. B. Fred (1926). J. biol. Chem. 68, 151.
P. M. F. Shattock and A. Hirsch (1947). J. Path. Bact. 59, 495.

Optical rotation of lactic acid—enterobacteria

M. C. Pascal and F. Pichinoty (1964). Annls Inst. Pasteur, Paris 107, 55–62.

Méthode de culture et extraction de l'acide lactique. – Les cultures sont effectuées en anaérobiose dans les ballons sous vide, à 32°, en milieu tamponné à pH 7 contenant du chlorure d'ammonium comme aliment azoté et du glucose (2 g/1000 ml) comme substrat fermentescible. Pour certaines bactéries (*Citrobacter, Klebsiella, Salmonella, Hafnia, Proteus* et *Providence*), on ajoute de l'extrait de levure Difco (50 mg/1000 ml). Après vingt-quatre à quarante-huit heures d'incubation, les cultures sont centrifugées afin d'éliminer les cellules. Le milieu (1000 ml) est acidifié par l'addition de quelques gouttes de H_2SO_4 concentré, puis placé dans un extracteur continu à l'éther. L'extraction est poursuivie durant deux jours. Le solvant est distillé et le résidu dissous dans quelques ml d'eau. L'échantillon est alors placé dans un dessiccateur sous vide en présence de NaOH (ou KOH) et de P_2O_5. On l'abandonne pendant plusieurs jours à la température du laboratoire jusqu'à déshydratation complète. Cette opération permet d'éliminer les acides volatils acétique et formique. Le résidu dissous dans 2 ml d'eau est neutralisé à pH 7 par NaOH à l'aide d'un papier-indicateur. La solution obtenue peut être conservée indéfiniment à – 10°; elle est utilisée directement pour l'identification du lactate.

Identification des antipodes. – Le L-lactate est recherché avec la L-lactique – déshydrogénase cristallisée de coeur (Sigma) de la façon suivante. Dans une cuve ayant un trajet optique de 1 cm, on place: 0.05 à 0.1 ml de l'échantillon, 2.6 ml d'un tampon glycine–NaOH pH 10, 0.3 ml d'une solution de NAD (2 μmoles), de l'eau pour compléter le volume total à 3 ml; au temps O, on ajoute 0.025 ml d'enzyme (préparation commerciale diluée cent fois). Les mesures sont effectuées à 340 mμ, toutes les trente secondes. La présence de L-lactate devrait se traduire par un accroissement continu de la densité optique.

Le D-lactate est identifié avec une préparation de D-lactique-déshydrogénase obtenue à partir d'une culture anaérobie de la souche 59 RL de *Saccharomyces cerevisiae* et purifiée selon la technique de Slonimski et Tysarowski [7]. On met dans une cuve de 1 cm: 0.1 à 0.2 ml de l'échantillon, 2 ml d'un tampon phosphates pH 7.3 (200 μmoles), 0.6 ml d'une solution de $K_3Fe(CN)_6$ (2.1 μmoles), de l'eau pour compléter le volume total à 3 ml; au temps O, on ajoute 0.025 à 0.25 ml d'enzyme suivant l'activité spécifique de la préparation. Les mesures sont faites à 420 mμ, toutes les trente secondes. La présence de D-lactate devrait se traduire par une diminution continue de la densité optique.

[7] Slonimski (P. P.) et Tysarowski (W.). *C. R. Acad. Sci.*, 1958, 246, 1111.

Optical rotation of lactic acid—*Lactobacillus*

F. Gasser (1964). Annls Inst. Pasteur, Paris 106, 778–796.

Détermination du pouvoir rotatoire de l'acide lactique.

a) *Principe.* La méthode utilisée est identique dans son principe à celle de Pederson, Peterson et Fred [42]. Celle-ci comprend deux temps:

1° Extraction de l'acide lactique par l'éther pendant un à trois jours dans un appareil continu.

2° Purification de l'acide lactique par cristallisation de son lactate de zinc. Celui-ci possède un pouvoir rotatoire spécifique $(\alpha)_D^{20} = \pm 7°4$ en solution à 5 p. 100, plus élevé que celui de l'acide lactique $(\alpha)_D^{20} = \pm 2°6$. L'activité optique du lactate de zinc est confirmée par son nombre de molécules d'eau de cristallisation: 3 pour le lactate racémique (18.18 p. 100) et 2 pour le lactate optiquement actif (12.89 p. 100).

L'extraction par l'éther, longue et parfois difficile à conduire dans un laboratoire de bactériologie, est remplacée par un passage sur résine échangeuse d'ions séparant du milieu de culture la totalité de sa fraction acide dont on extrait ensuite le lactate de zinc par cristallisation.

b) *Méthode.* Un bouillon M. R. S. M.* ne contenant ni citrate, ni acétate, ni glucose est réparti à raison de 400 ml dans de fioles de 500 ml bouchées à vis, permettant de garder une anaérobiose satisfaisante après autoclavage. Le glucose est stérilisé à part en solution à 32 p. 100, dont on ajoute 25 ml par fiole de bouillon. On évite ainsi pendant l'autoclavage la formation des composés bruns de la réaction de Maillard [33] qui traînent sur la colonne de résine. Chaque fiole est inoculée avec 20 ml d'une culture de vingt-quatre heures en bouillon M. R. S. M. de la souche à étudier. La culture obtenue après vingt-quatre heures d'incubation (quarante-huit heures pour certaines souches) a terminé sa phase exponentielle de croissance. Elle est centrifugée et le surnageant traité.

Les colonnes utilisées sont en verre, mesurant 27 cm de haut sur 5 cm de diamètre et ont un volume utile de 500 ml environ. Leur extrémité inférieure, munie d'un robinet rodé permettant le réglage du débit, est prolongée par un siphon dont le coude supérieur est au niveau de la surface de la résine contenue dans la colonne. On évite ainsi tout risque de dessiccation des résines.

L'une des colonnes contient 140 g (poids humide) de résine échangeuse d'anions faible (Amberlite IR_{45} analytical grade); l'autre, 170 g d'une résine échangeuse de cations forte (Permolite). Les résines sont maintenues à leurs parties supérieures et inférieures par un tampon de laine de verre. L'échangeur d'anions, traité par 400 ml de soude à 4 p. 100 et lavé jusqu'à pH 7 de l'effluent, est sous forme OH^-. L'échangeur de cations, traité par 400 ml d'HCl M et lavé jusqu'à pH 6.5 de l'effluent, est sous forme H^+. Au cours de toutes les opérations, le débit d'écoulement est réglé à 50 ml/minute environ.

Le surnageant de centrifugation est passé sur la colonne d'échangeur d'anions, qui est ensuite lavée avec 400 ml d'eau distillée (ou permutée). Des essais avec divers acides organiques en solution M/10 passés sur cette résine montrent que l'effluent de lavage reste légèrement acide, témoignant d'une lente hydrolyse, sauf avec l'acide lactique qui semble très fortement retenu. Il est donc avantageux ici de laver abondamment la colonne.

Les premiers essais d'élution a l'acide formique à 10 p. 100 ont été peu satisfaisants: l'acide lactique est difficilement élué et il faut utiliser près d'un litre de solution d'acide formique pour recouvrer la totalité de l'acide lactique fixé sur la colonne. La nécessité d'évaporer à sec pour éliminer ensuite l'acide formique devient alors très fastidieuse et provoque une concentration des résidus colorés en brun provenant du bouillon et entraînés au cours de l'élution, qui gênent ou même empêchent la cristallisation ultérieure du lactate de zinc. Des essais de fractionnement n'ont pas été non plus satisfaisants: compte tenu de la quantité d'acide à extraire, du débit rapide et du grand diamètre des colonnes, il n'est pas possible d'obtenir des fractions nettement séparées.

Pour éviter ces divers inconvénients, la résine est traitée par 200 ml de NaOH à 4 p. 100 décrochant la totalité de la fraction acide et l'éluat reçu directement sur la résine échangeuse de cations sous forme H^+. Il est donc nécessaire, avant toute opération, de contrôler si la quantité de résine échangeuse de cations est suffisante pour retenir la totalité des ions Na^+ contenus dans la quantité de soude utilisée pour l'élution de la résine échangeuse d'anions.

Après lavage suffisant des deux colonnes à l'eau distillée: 1 lavage avec 300 ml pour la première colonne, 2 lavages avec 300 ml pour la seconde et élimination des 200 premiers ml écoulés représentant le volume mort des deux colonnes, l'effluent total acide est de 900 ml environ.

Cet effluent est neutralisé à chaud par du carbonate de zinc en légei excès, laissé dix minutes à l'ébullition, filtré et évaporé sur bain-marie bouillant et sous courant d'air chaud jusqu'à 30 ml environ. Après refroidissement on ajoute 3 volumes d'alcool et on place au réfrigérateur à + 4°C afin de permettre aussi bien la cristallisation des formes racémiques que celle des formes optiquement actives de lactate de zinc [42].

Après un délai de vingt-quatre à quarante-huit heures les cristaux sont filtrés sur Büchner, lavés à l'alcool et séchés à l'air. Afin de mesurer le pourcentage d'eau de cristallisation, la dessiccation est effectuée dans une étude à 105°C pendant quatre heures après séchage du lactate de zinc jusqu'à poids constant dans un dessiccateur contenant du chlorure de calcium. Le pouvoir rotatoire est mesuré avec une solution aqueuse à 5 p. 100.

[42] Pederson (C. S.), Peterson (W. H.), Fred (E. B.). *J. biol. Chem.* 1926, 68, 151.

Optical rotation of lactic acid—*Pediococcus*

E. Coster and H. R. White (1964). J. gen. Microbiol. 37, 15–31.

The type of lactic acid produced from glucose by pediococci was examined using the medium of Camien and Dunn (1954). The medium was modified by the addition of $MgSO_4$ and $MnSO_4$ (0.05%, w/v) and Tween 80 (0.1%, v/v). Cultures were incubated for 1 month. The optical rotation was determined polarimetrically on a 5% (w/v) aqueous solution.

M. N. Camien and M. S. Dunn (1954). J. biol. Chem. 211, 593.

Optical rotation of lactic acid—*Lactobacillus*

E. P. Cato and W. E. C. Moore (1965). Can. J. Microbiol. 11, 319–324.

The authors describe an enzymatic method for the determination of optical rotation. It is a modification of methods proposed by Scholz *et al.* (1959) and Warburg *et al.* (1960), combined with the chromatographic method of Bruno and Moore (1962).

All organisms were cultured in broth containing 2% peptone, 1% yeast extract, and 1% glucose (PYG). Obligate anaerobes were grown in the same medium as modified by Moore and Cato (1963). Organisms were incubated 3 weeks at their optimum growth temperature. One milliliter of culture was deproteinized with 1 ml of 3.5% reagent grade perchloric acid. After 5 minutes at room temperature, the sample was clarified by centrifugation for 10 minutes at 1000 x *g* in tapered Pyrex tubes. A 0.1 ml sample of the supernatant liquid was used for the enzymatic determination. The remaining original broth culture was acidified below pH 2.0 with 50% H_2SO_4. Total concentrations of individual fermentation acids, including lactic acid, was determined chromatographically on 1 ml of this portion according to the method of Bruno and Moore (1962).

A 0.1 ml sample of the deproteinized portion was pipetted into a 12 x 75 mm test tube. To this was added 3.0 ml buffer (0.5 M glycine (Fisher Reagent), 0.4 M hydrazine sulfate (Fisher Certified Reagent) dissolved in redistilled water; pH adjusted to 9.0 with 2 N NaOH), 0.03 ml muscle lactic dehydrogenase (LDH) (Sigma, 41 mg protein/ml, diluted with an equal volume of redistilled water), and 0.2 ml nicotinamide adenine dinucleotide (NAD) (Sigma, assay 98%) solution (20 mg/ml redistilled water). Both the LDH and NAD solutions were freshly prepared and kept in an ice-water bath during preparation of the tubes. A series of standards was prepared with each set of unknown samples from 2.0 μmoles L(+)-lactate/0.1 ml PYG treated with perchloric acid (vol/vol). Dilutions were made with lactate-free PYG similarly treated to give 0.1, 0.2, 0.5, 0.8 and 1.0 μmoles lactic acid/0.1 ml solution. All tubes were incubated 60 minutes at 30°C. Each test solution was poured into a Bausch and Lomb ½ in. test tube and its absorbance read at 366 mμ against a reagent blank prepared and treated exactly as the standards but without lactate substrate. The blank is not stable and was used only once to bring the instrument to a zero reading. The reading of an unmatched tube containing medium and water was then used to restandardize the instrument periodically. The same colorimeter tube used for the unstable blank was used throughout the series.

It is extremely important that identical periods of time elapse between the addition of NAD and the reading of absorbance of each sample. NAD and hydrazine combine non-enzymatically to form a compound which absorbs at 300 mμ, the amount of absorbance a function of time (Pfleiderer and Dose, 1955). To obtain equal incubation periods all solutions except NAD were added to all the tubes to be tested. Then the NAD is added to the tubes in series at 30-second intervals. Under the conditions given, the change in optical density between blank and test solutions is essentially a linear function of lactate concentrations between 0.1 μmole/0.1 ml and 0.8 μmole/0.1 ml. The amount of L(+)-lactate per milliliter of culture is determined by multiplying μmoles lactate estimated from comparison with the standard curve by the sample dilution factor, 20. With the method described above, at least 20 unknown samples can be tested each day.

C. F. Bruno and W. E. C. Moore (1962). J. Dairy Sci. 43, 109–115.
W. E. C. Moore and E. P. Cato (1963). J. Bact. 85, 870–874.
G. Pfleiderer and K. Dose (1955). Biochem. Z. 326, 436–441.
R. Scholz, H. Schmitz, T. Bucher and J. O. Lampen (1959). Biochem. Z. 331, 71–86.
O. Warburg, K. Gawehn and A. W. Geissler (1960). Hoppe-Seyler's Z. physiol. Chem. 320, 277–279.

ORGANIC ACIDS (DETERMINATION OF)

Organic acids (determination of)—various

P. Hauduroy (1951). Techniques Bactériologiques. Masson et Cie, Éditeurs, Paris.

ACIDES FIXES

(Recherche des) (Anaérobies). – Opérer à partir d'un milieu de culture. Ensemencer d'abord la souche étudiée dans un bouillon Vf glucosé à 1 p. 100, placé dans un Erlenmeyer.

Le milieu est désaéré par ébullition, l'ensemencent fait avec deux tubes de culture de 24 heures.

Lorsque la fermentation est terminée, ou que les germes se sont déposés (en moyennes après 7 jours), filtrer sur terre d'infusoire.

Le filtrat sert aux recherches.

On place 100 cm^3 de culture acidifiée par SO_4H_2 1/10 dans un entonnoir à décantation. Réaliser une extraction en versant sans agitation une certaine quantité d'éther à 65° B. de façon à avoir une couche de 1 cm d'épaisseur. Après 24 heures de contact, recueillir l'éther, le distiller dans le vide. Reprendre le résidu dans 1 cm^3 d'eau distillée et chercher l'acide lactique et l'acide succinique.

Pour la recherche de l'acide lactique, ajouter à 2 gouttes du résidu 2 cm^3 de SO_4H_2 concentré. Porter au bain-marie à 100°pendant 2 minutes. Ajouter III gouttes de solution alcoolique de gaïacol à 5 p. 100 à la surface. L'apparition d'une coloration rouge indique la présence d'acide lactique.

Pour la recherche de l'acide succinique, ajouter à quelques gouttes du résidu, quelques gouttes d'eau ammoniacale à 10 p. 100. Évaporer à sec au bain-marie. Reprendre par quelques gouttes d'eau distillée. Ajouter quelques gouttes de perchlorure de fer dilué à 1/2 O. L'apparition d'un précipité gélatineux ocre indique la présence d'acide succinique. *(A. I. P.).*

ACIDES VOLATILS

(Recherche des) (Détermination de l'acidité volatile totale). – On prend le résidu de la distillation alcaline. On l'acidifie jusqu'à *p*H 3 avec de l'acide tartrique à 40 p. 100. Distiller par entraînement à la vapeur d'eau. Il faut obtenir environ 300 cm^3 de distillat. Arrêter la distillation avant l'obtention des 300 cm^3 si le distillat n'est plus acide. Noter le volume total du distillat obtenu.

Recherche des acides volatils *(Distillation fractionnée de Duclaux).* – Distiller à feu moyen 110 cm^3 de distillat acide. Doser à l'eau de chaux en présence de phénolphtaléine, 10 prises successives de 10 cm^3 chacune (rincer chaque fois les éprouvettes à l'eau distillée). Multiplier les 10 chiffres trouvés par un coefficient obtenu en divisant 100 par le chiffre de la dixième prise lu sur la burette. Établir une courbe sur papier millimétré. La comparer aux courbes étalons et aux tables de Duclaux.

Recherche de l'acide volatil total. – Prendre 10 cm^3 de distillat acide. Ajouter 3 gouttes de phénolphtaléine, titrer avec NaOH N/10.

$\frac{Ncm^3 \times 60}{1000 \times 10} \cdot \frac{V}{10}$ = g pour 200 cm^3 de culture. Diviser par 2 pour avoir un résultat en p. 100.

Ncm^3 = quantité de NaOH N/10 utilisée.

V = volume du distillat acide obtenu. *(A. I. P.).*

Organic acids (determination of)—anaerobes. (formic, acetic, propionic, n-butyric, n-valeric, n-caproic and β-oxybutyric)

A.-R. Prévot (1966). Techniques pour le diagnostic des Bactéries Anaérobes. Éditions de la Tourelle, St. Mandé.

Distillation acide: sans changer de ballon, amener le pH du résidu de la distillation alcaline aux environs de 3 par addition d'une quantité suffisante de solution d'acide tartrique à 40 p. 100 (acide que l'on ne risque pas de trouver comme produit acide de transformation du germe). Marquer au crayon gras, sur le ballon, le niveau du liquide au départ. Distiller pendant 1 heure à 1 heure et demie, à niveau constant et par entraînement à la vapeur d'eau, de façon à obtenir entre 300 et 350 cm^3 de distillat. Ne pas continuer à distiller si le distillat obtenu n'est plus acide et dans ce cas se contenter d'un minimum de 110 cm^3. Recevoir le distillat dans une éprouvette à pied graduée et bouchée à l'émeri.

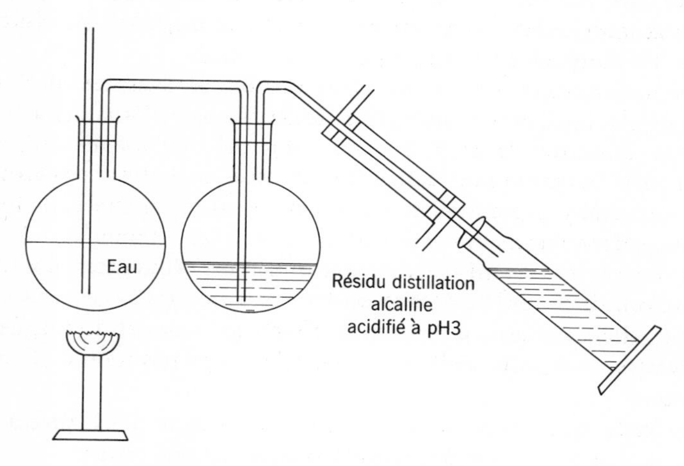

Fig. 5. *Distillation acide*

Quand la distillation est terminée, noter immédiatement la quantité totale de distillat obtenue, étiqueter et boucher l'éprouvette. Sur ce distillat acide deux opérations sont à effectuer:

1° La mesure de l'acidité volatile totale exprimée en acide acétique.

2° La caractérisation des acides volatils a été longtemps pratiquée suivant la méthode de Duclaux; mais les indéterminations qu'elle présente pour les mélanges de 2 acides dont l'un est isobutyrique, isovalérianique ou isocaproïque, et l'impossibilité de l'utiliser pour les mélanges de plus de 2 acides nous ont obligés à l'abandonner au profit de la chromatographie sur papier suivant la méthode de Guillaume et Osteux, modifiée par J. Blass.

Voici la technique que nous employons actuellement et qui donne d'excellents résultats:

– Le distillat est récolté dans un ballon et on arrête l'opération quand l'acidité totale du distillat correspond à 2 cm^3 d'une solution normale d'acide acétique, qui dans beaucoup de cas nécessite presque toujours une distillation pendant quarante minutes. Au bout de ce temps, on neutralise le produit par NaOH/N10 en présence de 1 goutte d'une solution alcoolique à 0.2 p. 100 de rouge de crésol, pour transformer les acides gras en sels de sodium, non volatils. Cette solution mise dans une capsule de porcelaine est portée sur un bain-marie à niveau constant, est concentrée à un volume de 4 cm^3. On met ces derniers dans une petite fiole de Fourneau et on ajoute environ 0.7 g de "permutite 50" traitée par HCl. On bouche la fiole et on agite pendant cinq minutes. La permutite fixe l'ion sodium, libérant ainsi les acides gras. On filtre la préparation sur coton de verre et on lave la petite fiole et le filtre deux fois avec 1 cm^3 d'eau distillée. On récupère une solution d'acides gras libres que l'on neutralise avec de la morpholine normale pour les rendre plus stables à la chromatographie, la morpholine ne gênant pas la révélation.

Avec une micropipette on dépose en quatre points espacés de 3 cm, situés sur une ligne horizontale d'une feuille de papier Whatmann 3 les quantités respectives de 2, 5, 5, 7, 5 et 10 mm^3. Le volume de 5 mm^3 permet déjà l'apparition de taches bien limitées, mais afin de pouvoir révéler un acide qui pourrait se trouver en quantités minimes, on doit recourir à des volumes plus grands soit 7.5 mm^3 et 10 mm^3.

Sur un point de la même feuille, on superpose VI gouttes, dont chacune d'un volume de 5 mm^3 est prélevée d'une solution N/10 des acides gras témoins. Il va sans dire qu'on attend le séchage complet de chaque goutte précédemment déposée avant de procéder à la superposition de la goutte suivante.

On fixe le haut du papier dans la rigole supérieure d'une cuve à chromatographie Jouan. On remplit la rigole avec le solvant suivant:

Butanol	180 ml
Cyclohexane	180 ml
Propanediol-1,2	60 ml
NH^3 à 22°	4.2 ml
Morpholine	0.42 ml
Eau distillée	21 ml

On ferme la cuve hermétiquement. Le solvant descend par capillarité, entraînant les différents acides. Dès que le front inférieur du solvant arrive assez bas, ce qui nécessite dix-huit heures on retire la feuille de papier de la cuve et on laisse à l'air libre pendant cinq minutes. Au bout de ce temps, on pulvérise rapidement avec un atomiseur à air comprimé le mélange des deux solutions suivantes:

Solution A:

O-crésol-sulfone phtaléine	0.2 g
Eau distillée	15 ml
NaOH N	0.3 ml

Après distillation ajouter:

Alcool à 96°	70 ml

Diluer cette solution au 1/5 avec de l'alcool à 96°.

Solution B

Véronal sodique	10 g
Eau distillée	100 ml
Alcool à 96°	150 ml

Mélanger extemporanément 3/4 de la solution A (diluée au 1/5) et 1/4 de la solution B pour la pulvérisation.

On a avantage à pulvériser le chromatogramme d'une certaine distance, la répartition des gouttelettes apparaissant ainsi beaucoup plus homogène, ce qui conduit à la meilleure révélation et délimitation des taches obtenues. On laisse sécher dix minutes après pulvérisation, puis on délimite les taches au crayon. On obtient des taches jaunes sur fond grenat, bien délimitées, qui persistent trois à quatre heures.

Les Rf obtenus avec les solutions témoins sont:

Acide formique	0.20
Acide acétique	0.26
Acide propionique	0.37
Acide n-butyrique	0.49
Acide n-valérianique	0.61
Acide n-caproïque	0.71

Le diamètre des taches permet une estimation grossière de la quantité de chacun des acides gras trouvés dans la goutte déposée sur le papier. C'est pourquoi on procède à la chromatographie simultanée de quantités croissantes du liquide de fermentation à examiner. En comparant le diamètre des taches obtenues avec celui des témoins, on peut apprécier avec une précision pratique satisfaisante la concentration de chacun des acides gras volatils dans la goutte et par un calcul simple en tenant compte des dilutions, dans le liquide à examiner. La teneur minima des révélations des différents acides est le millième de millimole.

Caractérisation des acides fixes.

Verser 100 cm^3 de filtrat de culture acidifié à pH3 avec de l'acide sulfurique au $1/10^e$, dans une ampoule à décantation. Ajouter ensuite de l'éther sulfurique, en opérant loin de toute flamme, de telle sorte que la couche recouvrant le filtrat mesure environ 1 cm d'épaisseur. Après 24 heures de contact, décanter le filtrat. Recueillir ensuite dans un verre l'éther restant dans l'ampoule à décantation. Le transvaser dans un ballon à fond rond de 150 cm^3 en prenant soin de laisser au fond du verre les quelques gouttes de filtrat qui peuvent y avoir été introduites avec l'éther. Monter sur ce ballon un réfrigérant descendant aboutissant à une fiole de Kitasato. Immerger, loin de toute flamme, la partie sphérique du ballon dans un récipient contenant de l'eau bouillante et distiller sous vide. Quand tout l'éther a distillé, ouvrir le ballon, chasser les dernières vapeurs d'éther par ventilation et reprendre le résidu par quelques gouttes d'eau distillée (1 cm^3 environ). Sur ce résidu rechercher:

a) la présence d'acide lactique: mettre dans un tube à essais 2 gouttes de résidu et ajouter 2 cm^3 d'acide sulfurique concentré. Porter au bain-marie à 100° pendant 2 minutes et refroidir. Déposer en surface 2 à 3 gouttes d'alcool gaïacolé à 5 p. 100. La présence d'acide lactique se traduit par l'apparition d'un anneau rouge;

b) la présence d'acide succinique: dans une petite coupelle en porcelaine, mettre quelques gouttes du résidu et 1 goutte d'ammoniaque pur. Évaporer à sec au bain-marie à niveau constant. Reprendre par quelques gouttes d'eau distillée et quelques gouttes d'une solution de perchlorure de fer à 10 p. 100. La présence d'acide succinique se traduit par l'apparition d'un précipité ocre gélatineux.

Acide β-oxybutyrique.

Il se caractérise par son lipide qu'on met en évidence par la technique de Lemoigne et Roukhelman. Il faut employer comme milieu de culture le bouillon de haricots, additionné de 20 g de saccharose, 2 g de peptone et 20 g de gélose par litre. La culture se fait à 20°-26°. On récolte les corps microbiens par raclage au 3e jour. On traite le culot microbien par HCl à 20 p. 100 qu'on maintient 3 minutes à 100°. On refroidit et on centrifuge. La masse microbienne est lavée 3 fois à l'alcool, 1 fois à l'éther, puis épuisée au chloroforme bouillant. La liqueur chloroformique filtrée est évaporée à sec et donne lieu à la formation d'un film de lipide β-hydroxybutyrique, se présentant en pellicules translucides, qui, par distillation donnent des cristaux d'acide α-crotonique fondant à 72°.

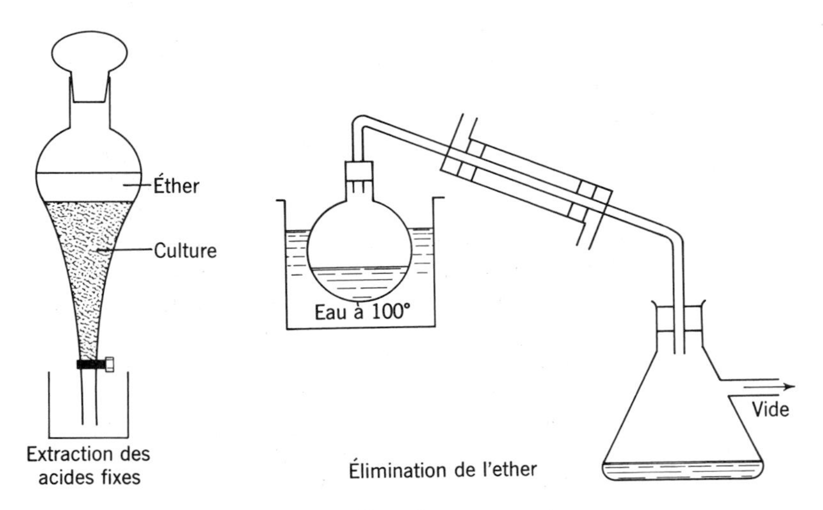

Fig. 6
Recherches des acides fixes

ORGANIC ACIDS–UTILISATION AND FERMENTATION*

Acetate—*Shigella* and *Escherichia*

I. D. Costin (1965). J. gen. Microbiol. 41, 23–27.

The sodium acetate agar medium (SAA) was prepared according to Trabulsi and Ewing (1962): distilled water, 100 ml; NaCl, 0.5 g; $MgSO_4 \cdot 7H_2O$, 0.02 g; $NH_4H_2PO_4$, 0.1 g; K_2HPO_4, 0.1 g; sodium acetate, 0.2 g; agar (washed 3 days), 2.0 g; bromthymol blue (1/500) aqueous solution, 4 ml.

L. R. Trabulsi and W. H. Ewing (1962). Publ. Hlth Lab. 20, 137.

Acetate—enterobacteria

J. G. Johnson, L. J. Kunz, W. Barron and W. H. Ewing (1966). Appl. Microbiol. 14, 212–217.

Acetate agar, used to test the ability of an organism to use acetate as the sole source of carbon, was prepared and used according to Trabulsi and Ewing (1962).

L. R. Trabulsi and W. H. Ewing (1962). Publ. Hlth Lab. 20, 137–140.

p-Aminosalicylic acid degradation—*Mycobacterium parafortuitum* n.sp

M. Tsukamura (1966). J. gen. Microbiol. 42, 7–12.

PAS degradation was observed on Lowenstein-Jensen medium or Ogawa egg medium containing 1 mg sodium p-aminosalicylate (PAS)/ml. Blackening of the medium after 7 days incubation at 37°C was recorded as a positive reaction.

p-Amino-salicylic acid degradation—*Mycobacterium*

M. Tsukamura (1966). J. gen. Microbiol. 45, 253–273.

Sodium p-aminosalicylate (PAS) degradation was observed according to a method described previously (Tsukamura, 1961; Tarshis, 1964; Tsukamura, 1965). Blackening of the Ogawa egg medium or Löwenstein-Jensen medium containing 0.2% (w/v) PAS after incubation at 37°C for 7 days was defined as a positive degradation.

M. S. Tarshis (1964). Tubercle 45, 267.

M. Tsukamura (1961). Jap. J. Tuberc. 9, 70.

M. Tsukamura (1965). J. gen. Microbiol. 41, 309.

p-Amino-benzoic acid degradation—*Mycobacterium*

M. Tsukamura (1966). J. gen. Microbiol. 45, 253–273.

p-Aminobenzoate degradation was recorded as a blackening of Sauton agar medium containing 0.2% (w/v) sodium p-aminobenzoate after incubation of 4 weeks (Tsukamura, unpublished test).

*See also **Basal media for carbon compounds** (pp. 101–128).

Citrate

S. A. Koser (1923). J. Bact. 8, 493–520.

The basal medium employed throughout this work consisted of:

Distilled water	1000 ml
NaCl	5.0 g
$MgSO_4 \cdot 7H_2O$	0.2 g
$(NH_4)H_2PO_4$	1.0 g
K_2HPO_4	1.0 g

This combination gives a colorless clear solution having a pH of 6.7 to 6.8. To this solution the various organic acids* were added and the reaction brought back to pH 6.8 by the addition of normal sodium hydroxide solution. Ammonium phosphate was used to supply the necessary nitrogen for development. An inorganic nitrogen containing salt must be employed since all carbon is to be excluded from the basal medium and to be derived solely from the test substances added.

*For the citrate medium 2 g anhydrous sodium citrate were added per litre.

Citrate—*Escherichia* and *Enterobacter*

S. A. Koser (1924). J. Bact. 9, 59–77.

Citrate medium: For the test of citrate utilization a chemically definite medium containing certain inorganic salts, an inorganic source of nitrogen, and a citrate as the only source of carbon was employed. Several different combinations may be used. Two of the most convenient are given here. (1) 5 g NaCl, 0.2 g $MgSO_4$, 1 g $(NH_4)H_2PO_4$, 1 g K_2HPO_4, and 2 g sodium citrate (2.77 g sodium citrate 5½ H_2O) in 1000 ml of distilled water. (2) 1.5 g $Na(NH_4)HPO_4 + 4H_2O$ (microcosmic salt), 1 g KH_2PO_4, 0.2 g $MgSO_4$ and 2 g sodium citrate in 1000 ml of distilled water. The hydrogen-ion concentration of both of these media is about 6.7 to 6.9. Potassium citrate may be substituted for the sodium citrate. Ammonium citrate constitutes a convenient combination of a source of nitrogen with the citrate radical, but has the disadvantage of being a very unstable salt.

The above media were tubed in 5 to 8 ml quantities, sterilized in the autoclave and inoculated directly from plain agar slants of the colon-aerogenes cultures. The temperature of incubation was 30°C. Observations of growth, as shown by visible turbidity, were made at intervals of one, two, three, four, and seven days and where negative results were obtained after two and three weeks.

Citrate—*Enterobacteriaceae*

J. S. Simmons (1926). A culture medium for differentiating organisms of typhoid-colon aerogenes groups and for isolation of certain fungi. J. infect. Dis. 39, 209–214.

Citrate agar medium:

Agar, clean and dry	20 g
NaCl	5 g
$MgSO_4$	0.2 g
$(NH_4)H_2PO_4$	1 g
K_2HPO_4	1 g
Sodium citrate (2.77 g sodium citrate 5½ H_2O)	2 g
Distilled water	1000 ml
Brom thymol blue (1.5% alcoholic)	10 ml

After dissolving the various salts in sterile distilled water add 20 g of clean, washed agar and sterilize the mixture at 15 pounds pressure for 15 minutes in an autoclave. After adjusting the reaction of the citrate agar to pH 6.8, add the indicator di-brom-thymol-sulphon-phthalein (brom-thymol blue). Variations have been observed in the intensity of the color reactions obtained with lots of brom-thymol blue distributed by different commercial firms. Uniformly satisfactory results have been obtained by using the indicator marketed by National Anilin and Chemical Company of Synthetical Laboratories, 10 ml of a 1.5% alcoholic solution in 1000 ml of citrate agar. The transparent olive green medium is then distributed in test tubes and allowed to solidify in a slanted position. Since brom-thymol blue is an indicator having a color range from yellow at pH 6 to blue at pH 7.6, the color of the citrate agar becomes yellow or orange when sufficient acid is produced, and turns Prussian blue when enough alkali is present.

A 0.4% aqueous solution of brom-thymol blue may be used in the proportion of 20 ml to 1000 ml of citrate agar, instead of the alcoholic indicator solution. The aqueous solution is prepared* by grinding one decigram (0.1 gram) of dry powdered brom-thymol blue in an agate mortar with 3.2 ml of N/20 NaOH and when solution is complete, dilute to 25 ml with water. The brom-thymol blue distributed by The Lamotte Chemical Products Co. gives uniformly satisfactory results when used in these proportions.

It is realized that the use of the agar, because of its organic nature, might be objected to in a test made solely to determine the ability of organisms to utilize citrate; and that some bacteria are able to reproduce on a simple jelly composed only of agar and water. However, trial has shown that this objection has no practical bearing on the use of citrate agar as a differential medium.

*W. M. Clark (1925). *The Determination of Hydrogen Ions;* Williams and Wilkins Co. Baltimore. p. 80.

Citrate—mustiness in eggs

M. Levine and D. Q. Anderson (1932). J. Bact. 23, 337–347.

The organic acid media were of the following compositions; dipotassium phosphate, 0.1 per cent; magnesium sulphate, 0.05 per cent; organic acid 0.2 per cent, adjusted to pH 6.0 with NH_4OH; 1.5 per cent agar added as a solidifying agent; 10 ml of 0.5 per cent aqueous solution of China blue per liter of solution added as an indicator. Inoculations were made from an agar slant on the surface of the organic acid agar. Utilization of the organic acid was indicated by decolorization of the China blue and growth on the surface.

Citrate

P. M. Kon (1932–1933). J. Dairy Res. 4, 206–212.

The method of Koser (1923) was used for this test, the results being read after 2-3 days' incubation at 37°C.

S. A. Koser (1923). J. Bact. 8, 493.

Citrate—*Haemophilus, Alcaligenes*

P. Thibault, S. Szturm-Rubinsten and D. Piéchaud-Bourbon (1947). Annls Inst. Pasteur, Paris 88, 246–250.

Utilisation du citrate: – L'utilisation du citrate de soude a été recherchée pour toutes les souches en milieu de Koser, pour certaines sur milieu de Simmons également. Les résultats ne furent pas significatifs: ils furent irréguliers pour les cultures de l'une et l'autre espèce.

Citrate—*Proteus*

G. T. Cook (1948). J. Path. Bact. 60, 171–181.

Medium: Koser's citrate medium.

Method: Inoculated with a straight wire which had been dipped in a peptone water culture incubated for 24 hours at 37°C. Incubated at 37°C.

Citrate—*Staphylococcus*

C. Shaw, J. M. Stitt and S. T. Cowan (1951). J. gen. Microbiol. 5, 1010–1023.

Koser's citrate medium incubated for 5 days at 37°C, 10 days at 30°C or 14 days at 22°C. All tubes showing turbidity were confirmed by needle reinoculation into citrate before being recorded as positive.

Citrate—general

P. Hauduroy (1951). Techniques Bactériologiques. Masson et Cie, Éditeurs, Paris.

GELOSE CITRATEE

Eau de robinet	1000	ml
Gélose	13	g

KH_2PO_4	0.5 g
NH_4NO_3	2 g
Citrate de soude	2 g
Solution de rouge de phénol à 0.4 p. 100	10 ml

Ajuster le *p*H à 6.8 avant l'adjonction de l'indicateur (modification de la solution de Koser. *J. Bact.*, 1924, 9, 59). *(A. T. C. C.).*

Citrate—*Vibrio*

C. Paris and J. Gallut (1951). Annls Inst. Pasteur, Paris 81, 343–346.

Techniques. – 1° *Composition des milieux :* Milieu de Koser :

Phosphate biammonique	1.5 g
Phosphate monopotassique	1.0 g
Sulfate de magnésium ($7H_2O$)	0.2 g
Citrate de sodium	3.0 g
Eau distillée	1 000 cm^3

Ce milieu est réparti en tubes de 17 et stérilisé à 115° vingt minutes.

Milieu de Simmons :

Chlorure de sodium	5 g
Sulfate de magnésium ($7H_2O$)	0.2 g
Phosphate monoammonique	1 g
Phosphate bipotassique	1 g
Citrate de sodium	2 g
Gélose	20 g
Eau distillée	1 000 cm^3

Chauffer pour faire fondre la gélose et ajouter 10 cm^3 de bleu de bromothymol en solution alcoolique à 1.5 p. 100. Répartir en tubes de 17 et stériliser vingt minutes à 115°. (La solution de bleu de bromothymol est obtenue en broyant au mortier 1.5 g de bleu de bromothymol R. A. L. dans 16 cm^3 de soude décinormale et en complétant à 100 cm^3 avec de l'alcool à 70°.)

2° *Ensemencement.* – Ces milieux ont été ensemencés largement en partant d'une culture de vingt-quatre heures sur gélose nutritive ordinaire.

L'alcalinisation du milieu mise en évidence par le virage au bleu du milieu de Simmons et du milieu de Koser, après addition d'une goutte de solution alcoolique de bleu de bromothymol, nous a permis d'identifier facilement les souches utilisant le citrate de sodium.

Koser (S. A.). *J. Bact.,* 1924, 9, 59.
Simmons (J. S.). *J. inf. Dis.,* 1926, 39, 208.

Citrate—*Bacillus*

N. R. Smith, R. E. Gordon and F. E. Clark (1952). U.S.D.A., Agr. Monograph No. 16.

Agar: NaCl, 1.0 g; $MgSO_4$, 0.2 g; $(NH_4)_2HPO_4$, 1.0 g; KH_2PO_4, 0.5 g;

Na citrate, 2.0 g; agar, 15 g; distilled water, 1000 ml and 20 ml of a 0.04 percent phenol red solution (pH adjusted to 7.0 before the addition of the indicator). This formula is a modification of Koser's citrate ammonium phosphate solution.
S. A. Koser (1924). J. Bact. 9, 59–77.

Citrate—*Bacillus*

P. de Lajudie (1952). Annls Inst. Pasteur, Paris 82, 380–382.

L'utilisation de citrate fut recherchée sur milieu de Koser-Simmons.

Citrate—*Pseudomonas (Malleomyces) pseudomallei*

P. de Lajudie and E. R. Brygoo (1953). Annls Inst. Pasteur, Paris 84, 509–515.

Utilisation du carbone du citrate de soude: – Cette propriété fut recherchée sur milieu de Simmons, ensemencé à partir d'une culture sur gélose.

Citrate, succinate and **malate**—*Mycobacterium*

R. E. Gordon and M. M. Smith (1953). J. Bact. 66, 41–48.

The cultures were inoculated on slants of each of the three following media, modifications of Koser's citrate agar (1924): NaCl, 1 g; $MgSO_4$, 0.2 g; $(NH_4)_2HPO_4$, 1 g; KH_2PO_4, 0.5 g; Na citrate (or Na succinate, or Ca malate), 2 g; agar, 15 g; distilled water, 1000 ml. The pH of each medium was adjusted to 7.0 before addition of 20 ml of a 0.04 per cent solution of phenol red. Observations for utilization of the citrate, succinate, and malate indicated by an alkaline reaction of the specific agar, were made after 7 and 28 days' storage at 28°C.
S. A. Koser (1924). J. Bact. 9, 59–77.

Citrate

N.C.T.C. Methods 1954.

Utilisation of citrate as the sole source of carbon

Make a light suspension in saline; all inoculations to the citrate media should be made with a straight wire. Incubated cultures at 30°C.
Method 1. Inoculation of Koser's citrate medium and daily observation for turbidity (up to 5 days). Check positives by subculture to Koser's citrate.
S. A. Koser (1923). J. Bact. 8, 493.
(1924). Ibid. 9, 59.
Method 2. Streak inoculation of Simmons' citrate agar, observe daily for 3 days. Check positives by subculture to Koser's citrate.
J. S. Simmons (1926). J. infect. Dis. 39, 209.

Citrate—*Pseudomonas (Malleomyces) pseudomallei*

L. Chambon and P. de Lajudie (1954). Annls Inst. Pasteur, Paris 86, 759–764.

L'utilisation du carbone du citrate de sodium est un test de grand intérêt qui est à la base de deux milieux différentiels, le milieu de Koser et le milieu de Simmons (Cf. Dumas (1951)).

Dans ces milieux nous avons constaté un développement rapide et abondant des cultures du bacille de Whitmore, fait déjà signalé par de Lajudie et Brygoo (1953).

J. Dumas. *Bactériologie Mèdicale,* Flammarion, édit., 1954.
P. de Lajudie et E. R. Brygoo. *Ann. Inst. Pasteur,* 1953, 84, 509.

Citrate—*Proteus*

C. Shaw and P. H. Clarke (1955). J. gen. Microbiol. 13, 155–161.

All tests were made by needle inoculation from a liquid culture. Koser's citrate medium (1923) was incubated for 7 days at 30°C. All tubes showing turbidity were regarded as positive when confirmed by needle reinoculation into citrate.

S. A. Koser (1923). J. Bact. 8, 493–520.

Citrate—enterobacteria

R. Buttiaux, J. Moriamez and J. Papavassiliou (1956). Annls Inst. Pasteur, Paris 90, 133–143.

Utilisation du citrate de sodium. On l'étudiera sur le milieu gélosé de Simmons en observant les précautions suivantes :

Le milieu sera réparti dans des tubes de verre neufs ou préalablement traités par du mélange sulfo-chromique ;

On ensemencera à sa surface, une très faible trace de la strie microbienne du milieu de Kligler, sans entraîner sur le fil de platine de particules de ce dernier ; ces substances nutritives étrangères fausseraient naturellement les résultats.

Citrate—enterobacteria

J. G. Heyl (1957). Antonie van Leeuwenhoek 23, 33–58.

Citrate medium according to Koser.

Sodium chloride	5.0 g
Sodium ammonium phosphate	1.0 g
Monopotassium phosphate	1.0 g
Magnesium sulfate	0.2 g
Sodium citrate	3.0 g
Distilled water	1000 ml
pH at 6.8	

Citrate

V. B. D. Skerman, *A Guide to the Identification of the Genera of Bacteria.* 2nd ed. Williams and Wilkins Book Co., Baltimore, 1967.

Koser's medium:

Organisms capable of utilizing citrate as the sole source of carbon will grow in the following medium.

NaCl	5 g
$MgSO_4 \cdot 7H_2O$	0.2 g
$NH_4H_2PO_4$	1 g
Sodium citrate	2 g
K_2HPO_4	1 g
Distilled water	1000 ml

Dissolve the salts in the water, adjust the pH to between 6.7 and 6.9. Filter, dispense, and sterilize at 121°C for 15 minutes. Inoculate from an agar culture and make at least three serial transfers before registering a definite positive result.

S. A. Koser (1924). J. Bact. 9, 59.

Citrate—*Aeromonas*

J. P. Stevenson (1959). J. gen. Microbiol. 21, 366–370.

Utilization was tested by inoculating three successive subcultures in Koser's citrate with the tip of a straight wire, and incubating at 37°C.

Citrate—*Pseudomonas*

M. E. Rhodes (1959). J. gen. Microbiol. 21, 221–263.

Growth in Koser's citrate medium. The ability of all isolates to grow in 5 ml of Koser's citrate medium (Koser, 1923) from a straight wire inoculum of a 24 hr yeast extract broth culture was tested. Falsely positive results were detected by a straight wire subculture into a second tube of Koser's medium from all the primary cultures which showed growth; further serial subcultures were carried out when necessary.

S. A. Koser (1923). J. Bact. 8, 493–520.

Citrate (and other organic acids)—*Mycobacterium*

R. E. Gordon and J. M. Mihm (1959). J. gen. Microbiol. 21, 736–748.

Modifications of Koser's (1924) citrate agar were prepared by adding 2 g of Na benzoate, Na citrate, Na lactate, Ca malate, mucic acid, Na oxalate, or Na succinate to NaCl, 1 g; $MgSO_4 \cdot 7H_2O$, 0.2 g; $(NH_4)_2HPO_4$, 1 g; KH_2PO_4, 0.5 g; agar, 15 g; distilled water, 1000 ml; 0.04% (w/v) solution of phenol red, 20 ml. To insure a slightly acid colour of the indicator in the sterilized agar, the pH value was adjusted before autoclaving as follows: benzoate, 7.0; citrate 6.8; lactate 6.8; malate 7.2; mucate 6.8; oxalate 6.7; and succinate 6.8. Use of the inorganic acid was indicated by the alkaline colour of the phenol red after 4 weeks of incubation at 28°C.

S. A. Koser (1924). J. Bact. 9, 59.

Citrate—*Enterobacteriaceae*

W. H. Ewing (1962). *Enterobacteriaceae. Biochemical Methods for Group Differentiation.* U. S. Department of Health, Education and

Welfare, Public Health Service Publication, No. 734 (revised)

Ammonium salts citrate medium

Test for the utilization of sodium citrate and ammonium salts.

Sodium chloride	5 g
Magnesium sulfate	0.2 g
Ammonium dihydrogen phosphate	1 g
Dipotassium phosphate	1 g
Sodium citrate	2 g
Agar (washed vigorously for 3 days)	20 g
H_2O	1000 ml

Add 40 ml of 1:500 bromthymol blue indicator solution.

Sterilize at 121°C, 15 minutes and slant so as to obtain a 1-inch butt and 1.5 inch slant.

Inoculation: Prepare a saline suspension from a young agar slant culture and inoculate the slant of the medium with a straight wire from the saline suspension. If desired, the butt of the medium may be stabbed.

Incubation: 37°C for 4 days.

If equivocal results are obtained as sometimes happens with members of the Providence group, for example, the test should be repeated and incubated at room temperature for 7 days. The above-mentioned medium is available from several commercial sources under the name of Simmon's citrate agar. These preparations are quite satisfactory.

Citrate—*Streptococcus*

W. E. Sandine, P. R. Elliker and H. Hays (1962). Can. J. Microbiol. 8, 161–174.

Previous studies with radioactive citric acid indicated that significant amounts of CO_2 were produced from this substrate by growing *S. diacetilactis* cells. Therefore a citrate-containing broth was inoculated with a loopful (0.01 ml) of a 24-hour nonfat milk* culture, incubated at 30°C for 24 hours, and visually observed for CO_2 production. The composition of citrate broth was as follows: Tryptone, 10.0 g; glucose, 10.0 g; sodium citrate dihydrate, 20.0 g; yeast extract, 5.0 g; dibasic potassium phosphate, 1.0 g; magnesium sulphate, 1.0 g; distilled water, 1000 ml. The broth was adjusted to pH 7.0 with HCl and autoclaved for 15 minutes at 121°C.

*10% nonfat milk sterilized by autoclaving at 121°C for 12 minutes.

Citrate—*Escherichia aurescens*

H. Leclerc (1962). Annls Inst. Pasteur, Paris 102, 726–741

On l'observe sur le milieu de Simmons incubé quatre jours à 30° puis à 20–22° jusqu'au trentième jour.

Citrate—*Enterobacteriaceae*

W. H. Ewing (1962). *Enterobacteriaceae. Biochemical Methods for*

Group Differentiation. U. S. Department of Health, Education and Welfare, Public Health Service Publication, No. 734 (revised).

Christensen's citrate agar (1949)

Test for citrate utilization in the presence of organic nitrogen.

Sodium citrate	3 g
Glucose	0.2 g
Yeast extract	0.5 g
Cysteine monohydrochloride	0.1 g
Ferric ammonium citrate	0.4 g
Monopotassium phosphate	1 g
Sodium chloride	5 g
Sodium thiosulfate	0.08 g
Phenol red	0.012 g
Agar	15 g
Distilled water	1000 ml

Tube and sterilize at 121°C for 15 minutes and slant (1 inch butt, 1.5 inch slant).

The ferric ammonium citrate and sodium thiosulfate may be omitted from the formula, if desired, since they do not affect the value of the medium as an indicator for citrate utilization.

Inoculation: The medium is inoculated over the entire surface of the slant.

Incubation: 37°C for 7 days. Positive reactions are indicated by alkalinization of the medium and development of a red color, particularly on the slant of the agar. This medium is of particular value in the differentiation of shigellae and anaerogenic, nonmotile *E. coli* biotypes (Edwards, Fife and Ewing, 1956).

P. R. Edwards, M. A. Fife and W. H. Ewing (1956). Am. J. med. Technol. 22, 28.

Citrate—*'Bacterium salmonicida' (Aeromonas)*

I. W. Smith (1963). J. gen. Microbiol. 33, 263–274.

Growth in citrate was recorded after 7 days in Koser's medium.

Citrate, malonate, benzoate and **formate**—*Pseudomonas odorans*

I. Málek, M. Radochová and O. Lysenko (1963). J. gen. Microbiol. 33, 349–355.

Tested in Koser base medium with phenol red and one of the following acids: (%, w/v) citric 1.0; malonic 1.0; benzoic 0.3; formic 0.1. Organic acid solutions were sterilised by filtration. The cultures were observed for 4 weeks at 28°C.

Citrate—*Streptococcus*

F. M. Ramadan and M. S. Sabir (1963). Can. J. Microbiol. 9, 443–450.

The authors used the method of P. Spaander and A. C. F. Roest (1959) Antonie van Leeuwenhoek 25, 169–178.

Citrate (Koser's)—*Vibrio marinus*

R. R. Colwell and R. Y. Morita (1964). J. Bact. 88, 831–837.

The authors used the method of the Society of American Bacteriologists. 1957. *Manual of Microbiological Methods.* McGraw-Hill Book Co., Inc., New York – modified by the addition to the media of the following salts to produce a synthetic seawater: – sodium chloride, 2.4%; potassium chloride, 0.07%; magnesium chloride (hydrated) 0.53% and magnesium sulfate (hydrated) 0.7%.

A standard inoculum was 1 drop from a pasteur pipette (*c.* 0.05 ml) of a 24 hour artificial seawater broth.

Cultures were incubated at 18°C.

Citrate (Simmon's)—*Vibrio marinus*

R. R. Colwell and R. Y. Morita (1964). J. Bact. 88, 831–837.

The authors used the method of the Society of American Bacteriologists. 1957. *Manual of Microbiological Methods.* McGraw-Hill Book Co., Inc., New York – modified by the addition to the media of the following salts to produce a synthetic seawater: – sodium chloride, 2.4%; potassium chloride, 0.07%; magnesium chloride (hydrated) 0.53% and magnesium sulfate (hydrated) 0.7%.

A standard inoculum was 1 drop from a pasteur pipette (*c.* 0.05 ml) of a 24 hour artificial seawater broth.

Cultures were incubated at 18°C.

Citrate (Koser's)—*Moraxella*

W. J. Ryan (1964). J. gen. Microbiol. 35, 361–372.

Koser's citrate medium (Gillies 1960) was used.

R. R. Gillies (1960). In *Mackie and McCartney's Handbook of Bacteriology,* 10th ed. Ed by R. Cruickshank. p. 609, Edinburgh and London, Livingstone.

Citrate—*Lactobacillus*

M. Gemmell and W. Hodgkiss (1964). J. gen. Microbiol. 35, 519–526.

The media used were (i) Basal medium + 3% (w/v) potassium citrate + 1% (w/v) glucose, and (ii) As (i) but without glucose. These media were distributed in 5 ml amounts in 6 X ½ in. test-tubes containing fermentation tubes. After inoculation the tubes were sealed with water agar.

The basal medium which was used throughout with minor modifications had the following constituents in 1 L tap water: meat extract (Lab Lemco), 5 g; Evans peptone, 5 g; Difco yeast extract, 5 g; Tween 80, 0.5 ml; $MnSO_4 \cdot 4H_2O$, 0.1 g; potassium citrate, 1 g; pH 6.5.

Citrate

G. S. Wilson and A. A. Miles (1964). Topley and Wilson's *Principles of Bacteriology and Immunity.* 5th edition – Arnold.

Brown (1921) drew attention to the usefulness of a medium containing citrate for distinguishing *E. coli* from *K. aerogenes.* Koser (1923, 1924) devised a fluid medium in which citrate provided the sole source of carbon. Ability to use this substance was indicated by growth and consequent turbidity. The composition of the medium now used is as follows:

Sodium chloride	5.0 g
Magnesium sulphate	0.2 g
Ammonium dihydrogen phosphate ($NH_4H_2PO_4$)	1.0 g
Dipotassium hydrogen phosphate (anhydrous)	1.0 g
Distilled water	1000 ml

This mixture forms a clear colorless solution with a pH of 6.8. Add 2 g of citric acid, and bring back the reaction to pH 6.8 with N/1 NaOH solution. Only a very light inoculum should be used, so as to avoid adding unnecessary organic matter to the medium and also to avoid obscuring the earliest signs of turbidity during growth.

Brown (1921) – no reference cited
Koser (1923, 1924) – no reference cited

Citrate—*Bordetella*

M. Piéchaud and Mme. S. Szturm-Rubinsten (1965). Annls Inst. Pasteur, Paris 108, 391–395.

Nous avons étudié l'utilisation du citrate de soude sur le milieu de Simmons. Les résultats sont plus nets et plus rapides que dans le milieu de Koser que nous avons surtout utilisé dans un travail précédent [9]. Toutes les souches de *B. bronchiseptica* ont alcalinisé en quarante-huit heures ou trois jours le milieu de Simmons. Aucune des dix souches de *B. parapertussis* n'a alcalinisé rapidement ce milieu ; On observe, après un délai de trois à six semaines, l'apparition de rares colonies citrate positives faisant alors virer l'indicateur. Moreno Lopez [7] trouve que *B. parapertussis* peut utiliser le citrate comme seule source de carbone ; ceci n'empêche pas l'usage du milieu au citrate de Simmons pour la différenciation, *B. bronchiseptica* l'alcalinisant très rapidement.

[7] Lopez (M. M.). *Microb. Española,* 1952, 5, 177.
[9] Thibault (P.), Szturm-Rubinsten (S.) et Piéchaud-Bourbon (D.). *Ann. Inst. Pasteur,* 1955, 88, 246.

Citrate—*Mycobacterium*

A. Tacquet, F. Tison and B. Devulder (1965). Annls Inst. Pasteur, Paris 108, 514–525.

The authors used ferric ammonium citrate in the method of Tisson, Tacquet and Devulder (Annls Inst. Pasteur, Paris, 1964, 106, 797–801.

Citrate—*Streptococcus*

R. Whittenbury (1965). J. gen. Microbiol. 38, 279–287.

Citrate dissimilation was examined in similar media to those used in the study of action on malate. Citrate (potassium citrate 3.0%, w/v) was added and the medium adjusted to pH 8.0 for routine use. Action on citrate in the absence of glucose was judged by gas production, acetoin formation, increase in pH measured electrometrically, and by increase of growth over that in the basal medium. In the presence of glucose (medium adjusted to pH 6.0) gas (CO_2) production indicated action on citrate.

The basal liquid medium contained meat extract (Lab-Lemco), 0.5 g; peptone (Evans), 0.5 g; yeast extract (Difco), 0.5 g; Tween 80, 0.05 ml; $MnSO_4 \cdot 4H_2O$, 0.01 g; in 100 ml tapwater, adjusted to pH 6.5 and autoclaved at 121°C for 15 min. (For malate see p. 663).

Citrate—*Bacillus*

G. R. F. Hilson (1965). J. gen. Microbiol. 39, 407–421.

Citrate utilization was tested by inoculating Koser's citrate medium (Oxoid), incubating for 2 days, and observing the development of turbidity; a loopful of any growth thus seen was subcultured to a new bottle of the same medium, and citrate utilization was assumed to occur only when the second subculture also showed growth of the strain under test.

Citrate—*Alcaligenes*

R. G. Mitchell and S. K. R. Clarke (1965). J. gen. Microbiol. 40, 343–348.

Koser's (1923) medium with incubation for 2 days was used.

S. A. Koser (1923). J. Bact. 8, 493.

Citrate—*Mima, Herellea, Flavobacterium*

J. D. Nelson and S. Shelton (1965). Appl. Microbiol. 13, 801–807.

Citrate utilization was determined on Simmon's citrate agar.

Citrate—marine bacteria

R. M. Pfister and P. R. Burkholder (1965). J. Bact. 89, 863–872.

Citrate utilization as a source of carbon for growth was determined with a medium containing synthetic seawater (Lyman and Fleming, 1940) containing the following (per liter): KH_2PO_4, 50 mg; $Fe(NH_4)_2(SO_4) \cdot 6H_2O$, 7 mg; N-Z-Case, 0.4 g; and sodium citrate, 3.0 g. To ensure that growth in the medium was due to citrate utilization and not the substitution of amino acids as a carbon source, the level of 0.4 g per liter of N-Z-Case was determined experimentally to be suitable for demonstrating increased growth with added citrate.

J. Lyman and R. H. Fleming (1940). J. mar. Res. 3, 134–146.

Citrate—*Providencia, Rettgerella*

C. Richard (1966). Annls Inst. Pasteur, Paris 110, 105–114.

Pour des raisons de commodité, les milieux ont été incubés à 37°C. Action sur les sels d'acides organiques. L'utilisation du citrate de sodium comme seule source de carbone est observée sur le milieu synthétique de Simmons [26].

[26] Simmons (J.). *J. inf. Dis.*, 1926, 39, 209.

Citrate—*Brevibacterium*

R. Chatelain and L. Second (1966). Annls Inst. Pasteur, Paris 111, 630–644.

Utilisation de citrate: sur milieu de Simmons. Lecture après 4 jours. La température d'incubation des cultures est de 30°C.

Citrate—RM bacterium

A. J. Ross, R. R. Rucker and W. H. Ewing (1966). Can. J. Microbiol. 12, 763–770.

Replicate tubes of sodium citrate were inoculated with each culture selected, and tests for utilization of the substrate was made with lead acetate solution after 1, 2, 7, and 14 days of incubation according to the method of Kauffmann and Petersen (1956).

F. Kauffmann and A. Petersen (1956). Acta path. microbiol. scand. 38, 481–491.

Lactate—*Clostridium*

H. Beerens, M. M. Castel and H. M. C. Put (1962). Annls Inst. Pasteur, Paris 103, 117–121.

Utilisation du lactate : Elle est étudiée sur le milieu de Bryant et Burkey [3].

Trypticase (B. B. L.)	15 g
Extrait de levures	5 g
Chlorhydrate de cystéine	0.50 g
Eau distillée	1 000 ml

Ajuster le pH à 7.00 et ajouter :

Acide lactique	5 g
Acétate de sodium, 3 H_2O	5 g

pH = 7.00.

Répartir 12 ml par tube de 160 × 16. Autoclaver vingt minutes à 115°. Ensemencer en introduisant au fond du tube 1 ml environ d'une culture de vingt-quatre heures en milieu V. L. liquide, glucosé à 2 p. 1 000. Recouvrir d'une couche de paraffine. Incuber vingt-quatre à trente-six heures à 37°. Mesurer le pH. Une alcalinisation par rapport à la réaction d'un tube témoin non ensemencé indique l'utilisation du lactate.

[3] Bryant (M. P.) and Burkey (L. A.). *J. Bact.* 1956, 71, 43.

Malate—*Lactobacillus*

M. Gemmell and W. Hodgkiss (1964). J. gen. Microbiol. 35, 519–526.

The media used were (i) Basal medium + 1% (w/v) glucose + 4% (w/v) malic acid, neutralized with KOH, and (ii) As (i) but without glucose. These media were distributed in 5 ml amounts in 6 X ½ in. test-tubes containing fermentation tubes. After inoculation the tubes were sealed with water agar.

The basal medium which was used throughout with minor modifications had the following constituents in 1 L tap water: meat extract (Lab Lemco), 5 g; Evans peptone, 5 g; Difco yeast extract, 5 g; Tween 80, 0.5 ml; $MnSO_4 \cdot 4H_2O$, 0.1 g; potassium citrate, 1 g; pH 6.5.

Malate—*Streptococcus*

R. Whittenbury (1965). J. gen. Microbiol. 38, 279–287.

Ability to dissimilate malate in the presence and absence of glucose was determined. Liquid medium was used containing DL-malic acid 4.0% (w/v); bromcresol purple, 2.8 ml of a 1.6% (w/v) ethanolic solution/L (unless acetoin was being tested for); glucose 1-2.0% (w/v); adjusted to pH 6.0 with KOH and distributed in 4 ml amounts in 5 X ½ in. test tubes containing Durham tubes. Water agar seals were added to the inoculated tubes; the function of the Durham tubes was to support the seals. Action on malate was judged by increase in pH value, gas (CO_2) production, and, in the absence of glucose, by increase in the amount of growth over that in the basal medium and by acetoin production.

The basal liquid medium contained meat extract (Lab-Lemco), 0.5 g; peptone (Evans), 0.5 g; yeast extract (Difco), 0.5 g; Tween 80, 0.05 ml; $MnSO_4 \cdot 4H_2O$, 0.01 g; in 100 ml tapwater, adjusted to pH 6.5 and autoclaved at 121°C for 15 min.

Malonate—See also under "Phenylalanine and Malonate"

Malonate—*Aerobacter* and *Escherichia*

E. Leifson (1933). J. Bact. 26, 329–330.

The composition of medium is as follows: $(NH_4)_2SO_4$, 2 g; K_2HPO_4, 0.6 g; KH_2PO_4, 0.4 g; NaCl, 2 g; Na malonate, 3 g; Indicator (0.5 per cent alcohol solution of B.T.B.), 5 ml; Distilled water, 1000 ml.

The phosphates adjust the pH to a medium green color with B.T.B. indicator. *Aerobacter* grows fairly well on medium and turns it blue. *Escherichia* does not grow appreciably and leaves the medium green.

Malonate—*Proteus-Providencia*

C. Shaw and P. H. Clarke (1955). J. gen. Microbiol. 13, 155–161.

Two drops suspension, 2 drops 0.03 M-sodium malonate, 2 drops 0.025 M-phosphate buffer, pH 6.0 + phenol red as indicator. Results were read up to 24 hr at 37°C and positive tests showed an alkaline reaction.

Malonate—*Aeromonas*

J. P. Stevenson (1959). J. gen. Microbiol. 21, 366–370.

The medium + methods of Shaw and Clarke (1955) were used.

C. Shaw and P. H. Clarke (1955). J. gen. Microbiol. 13, 155.

Malonate—*Enterobacteriaceae*

W. H. Ewing (1962). *Enterobacteriaceae. Biochemical Methods for Group Differentiation.* U. S. Department of Health, Education and Welfare, Public Health Service Publication, No. 734 (revised).

Sodium malonate broth (Leifson, 1953, modified)

Test for utilization of malonate.

Yeast extract	1	g
Ammonium sulfate	2	g
Dipotassium phosphate	0.6	g
Monopotassium phosphate	0.4	g
Sodium chloride	2	g
Sodium malonate	3	g
Glucose	0.25	g
Bromthymol blue	0.025	g
Distilled water	1000	ml

Sterilize at 121°C for 15 minutes.

Inoculation: Inoculate from a young agar slant or broth culture. (A 3-mm loopful of broth culture is preferred).

Incubation: 37°C for 48 hours.

Positive results are indicated by a change in the color of the indicator from green to Prussian blue.

Leifson's malonate broth, modified by the addition of a small amount of yeast extract and glucose is of considerable value in the differentiation of salmonellae and members of the *Arizona* group. The majority of *Salmonella* cultures do not utilize malonate, whereas the majority of strains belonging to the *Arizona* group do so (Schaub, 1948; Shaw 1956; Ellis *et al.,* 1957; Ewing *et al.,* 1957). Modified malonate medium is also of value in other areas of the family. For example, many cultures that belong to the *Klebsiella* and *Aerobacter* groups utilize malonate whereas *Serratia* strains, with rare exceptions, do not (Davis and Ewing 1957; Ewing and Davis 1959). Further, this medium may be used in conjunction with the organic acid media advised by Kauffmann and Petersen (1956) and Ellis, Edwards and Fife (1957) for the differentiation of *Citrobacter* as well as *Salmonella* and *Arizona* cultures (Ewing and Edwards, 1960).

B. R. Davis and W. H. Ewing (1957). Int. Bull. bact. Nomencl. Taxon. 7, 151.

R. J. Ellis, P. R. Edwards and M. A. Fife (1957). Publ. Hlth Lab. 15, 89.

W. H. Ewing and P. R. Edwards (1960). Int. Bull. bact. Nomencl. Taxon. 10, 1.
W. H. Ewing, B. R. Davis and R. W. Reavis (1957). Publ. Hlth Lab. 15, 153.
W. H. Ewing, B. R. Davis and R. W. Reavis (1959). Communicable Disease Centre Monograph.
F. Kauffmann and A. Petersen (1956). Acta path. microbiol. scand. 38, 481.
E. Leifson (1933). J. Bact. 26, 329.
I. G. Schaub (1948). Johns Hopkins Hosp. Bull. 83, 367.
C. Shaw (1956). Int. Bull. bact. Nomencl. Taxon. 6, 1.

Malonate—*Escherichia aurescens*

H. Leclerc (1962). Annls Inst. Pasteur, Paris 102, 726–741.

Utilisation des acides organiques. Le milieu au malonate décrit par Leifson [26] est utilisé d'après Ewing, Davis et Reavis [11].

[11] Ewing (W. H.), Davis (R.) and Reavis (R. W.). *Publ. Hlth. Lab.*, 1957, 15, 153.
[26] Leifson (E.). *J. Bact.*, 1933, 26, 329.

Malonate—*Moraxella*

W. J. Ryan (1964). J. gen. Microbiol. 35, 361–372.

The micromethod of C. Shaw and P. H. Clarke (J. gen. Microbiol. 13, 155, 1955) was used to detect utilisation of malonate.

Positive and negative controls were included.

Malonate—*Alcaligenes*

R. G. Mitchell and S. K. R. Clarke (1965). J. gen. Microbiol. 40, 343–348.

The combined medium of Shaw and Clarke (1955), incubation for 6 days was used.

C. Shaw and P. H. Clarke (1955). J. gen. Microbiol. 13, 155.

Organic acids

P. Hauduroy (1951). Techniques Bactériologiques. Masson et Cie, Éditeurs, Paris.

ETUDE DES SELS ORGANIQUES

Le milieu de base est ainsi composé :

Bactopeptone (Difco)	10	g
Eau distillée	1000	cm^3
NaOH (n 1/10)	8.5	cm^3

On ajoute comme indicateur une solution à 1 p. 5 000 de bleu de bromothymol à raison de 12 cm^3 par litre de milieu.

On utilise les substances suivantes aux doses indiquées :

1 p. 100 *d*-tartrate = Kalium-Natrium-tartrate, crist. pur (Merck).

0.5 p. 100 *l*-acide tartrique = *Acidum tartaricum*, lévogyre (Schering-Kahlbaum).

0.5 p. 100 *i*-acide tartrique = *Acidum tartaricum*, inactif (Schering-Kahlbaum).

1 p. 100 citrate de soude = *Natrium citricum* neutre, pulv. (Merck).

1 p. 100 acide mucique = *Acidum mucicum* (Schering-Kahlbaum).

Ajuster le *p*H à 7.4 environ en ajoutant NaOH (5 *n*).

Les milieux seront désignés sous les appellations abrégées de: *d, l, i*-tartrate, citrate ou mucate.

Les milieux sont ensemencés avec une öse d'un bouillon de culture âgé de 20 heures. On les place à l'étuve pendant 14 jours. On lit journellement.

La lecture se fait journellement : *a)* en observant le changement de coloration des tubes ; *b)* par adjonction à chaque tube contenant 3 à 4 cm^3 de milieu, 0.5 cm^3 d'une solution aqueuse et saturée d'acétate de plomb.

Pour les *d-, l-, i*-tartrate et citrate, il est nécessaire de pratiquer les deux réactions tandis que pour le mucate, seul le changement de couleur est pris en considération.

Les réactions sont positives pour tous les sels organiques lorsque les tubes primitivement bleu deviennent jaune-vert ou blanc et pour les *d-, l-, i*-tartrate et citrate on ne doit voir apparaître qu'un précipité très léger par adjonction de l'acétate de plomb.

Une réaction négative ne provoque pas de changement de couleur du milieu mais l'adjonction d'acétate de plomb provoque la formation d'un précipité abondant qui après 24 heures de repos, remplit les 2/3 du tube.

Il peut arriver que des tubes qui sont restés bleu donnent une réaction positive à l'acétate de plomb de telle sorte qu'on ne peut juger d'une réaction qu'après l'adjonction de ce sel.

Pour pouvoir suivre l'attaque des sels organiques il est nécessaire de préparer une série de tubes pour la même substance de façon à pouvoir faire des réactions à l'acétate et des lectures après 4, 5, 6, 7, 8 jours par exemple. *(C. I. E.).*

Organic acids

G. S. Wilson and A. A. Miles (1964). Topley and Wilson's *Principles of Bacteriology and Immunity.* 5th edition – Arnold.

The fermentation of organic acids, such as citric, tartaric, and mucic, must not be confused with their simple utilization as the sole source of carbon. In fermentation, the salt of the organic acid is broken down with the release of free sodium ions and the consequent production of an alkaline pH, which can be detected by a suitable dye. Kauffmann and

Petersen (1956), using the salts of these acids, found their fermentation to be very useful in the biochemical type differentiation of members of the *Salmonella* and *Klebsiella* groups, and in the group differentiation of the enterobacteria. For details of the test, see Ewing (1960).

Numerous micro-methods for testing for fermentation and other biochemical reactions are in use in different laboratories, mainly designed to shorten the period of observation (for references, see Clarke and Cowan 1952, McDade and Weaver 1959).

P. H. Clarke and S. T. Cowan (1952). J. gen. Microbiol. 6, 187.

W. H. Ewing (1960). *Enterobacteriaceae. Biochemical Methods for Group Differentiation.* Publ. Hlth Serv., Communicable Diseases Centre, Atlanta, Georgia.

F. Kauffmann and A. Petersen (1956). Acta path. microbiol. scand. 38, 481.

J. J. McDade and R. H. Weaver (1959). J. Bact. 77, 65.

Organic acids—*Mycobacterium parafortuitum* n.sp.

M. Tsukamura (1966). J. gen. Microbiol. 42, 7–12.

Utilization of organic acid as sole carbon source was observed in the following medium: $(NH_4)_2SO_4$, 2.64 g; KH_2PO_4, 0.5 g; $MgSO_4 \cdot 7H_2O$, 0.5 g; purified agar (Wako Chemical Co., Osaka), 30.0 g; distilled water, 1000 ml. This was adjusted to pH 7.0 by addition of 10% (w/v) KOH and the medium sterilized by autoclaving at 115°C for 30 min. Organic acids (sodium salts) were sterilized at 100°C for 15 min and added to medium at a concentration of 0.01 M. The medium was readjusted to pH 7.0 and sterilized at 100°C for 10 min. Growth was observed after 2 weeks incubation at 37°C. Organic acids tested were acetate, citrate, succinate, malate, pyruvate and benzoate.

Organic acids—*Providencia, Rettgerella*

C. Richard (1966). Annls Inst. Pasteur, Paris 110, 105–114.

Pour des raisons de commodité, les milieux ont été incubés à 37°C. Fermentations du d-tartrate, du mucate, du citrate et utilisation du malonate en milieux liquides selon Kauffmann et Petersen [12] et Leifson [15].

[12] Kauffmann (F.) et Petersen (A.). *Acta path. microbiol. scand.*, 1956, 38, 481.

[15] Leifson (E.). *J. Bact.*, 1933, 26, 329.

Organic acids—*Mycobacterium*

M. Tsukamura (1966). J. gen. Microbiol. 45, 253–273.

Utilization of organic acids as sole carbon sources was tested on the following medium: $(NH_4)_2SO_4$, 2.64 g; KH_2PO_4, 0.5 g; $MgSO_4 \cdot 7H_2O$, 0.5 g; purified agar (Eiken Co., Tokyo, or Wako Chemical Co., Osaka),

20.0 g; distilled water, 1000 ml. The medium was adjusted to pH 7.1 by adding 10% (w/v) KOH. The medium was supplemented with organic acid Na salt to a final concentration 0.02 M, and re-adjusted to pH 7.1. The medium was then sterilized by autoclaving at 115°C for 30 min and made up as slopes (sterilization by autoclaving gave the same results as sterilization by heating at 100°C for 15 min). Series of media containing different organic acids and control medium without carbon source were inoculated with one loopful of stock cultures. Growth of slow-growing mycobacteria was observed after incubation for 4 weeks, and growth of rapid-growing mycobacteria after incubation for 2 weeks. For incubation, a loop, 3.5 mm outside diameter and 2.0 mm inside diameter, was used throughout. After touching the loop slightly on the growth of test organisms, the inoculum was streaked on the slope of fresh media. Care was taken not to bring any visible bacterial mass on the fresh media. This method of inoculation gave nearly similar results as a semi-quantitative method in which one loopful of 3-day cultures of rapidly growing mycobacteria or 10-day cultures of slow-growing mycobacteria growing in the Dubos liquid medium were used for inoculation. The following organic acids were used for tests: acetate; citrate; succinate; malate; pyruvate; benzoate; malonate; fumarate.

Tartrate—RM bacterium

A. J. Ross, R. R. Rucker and W. H. Ewing (1966). Can. J. Microbiol. 12, 763–770.

Replicate tubes of D-tartrate were inoculated with each culture selected, and tests for utilization of the substrate was made with lead acetate solution after 1, 2, 7 and 14 days of incubation according to the method of Kauffmann and Petersen (1956).

F. Kauffmann and A. Petersen (1956). Acta path. microbiol. scand. 38, 481–491.

Salicylate—*Mycobacterium parafortuitum* n.sp.

M. Tsukamura (1966). J. gen. Microbiol. 42, 7–12.

Salicylate degradation was tested on Sauton agar medium containing 0.5 mg and 1.0 mg sodium salicylate/ml. Blackening of the medium after 7 days incubation at 37°C was recognized as positive degradation.

Salicylate—*Mycobacterium*

M. Tsukamura (1966). J. gen. Microbiol. 45, 253–273.

Salicylate degradation was observed as a blackening of Sauton agar medium containing 0.1% (w/v) sodium salicylate, after incubation at 37°C for 7 days (Tsukamura, 1965).

M. Tsukamura (1965). J. gen. Microbiol. 41, 309.

OXIDATION REACTIONS

See also "Basal media for carbon compounds"

Alkanes and ketones—*Mycobacterium*

H. B. Lukins and J. W. Foster (1963). J. Bact. 85, 1074–1087.

For method of growth see **Basal medium for carbon compounds** – *Mycobacterium* (p. 110).

Ketones were estimated photometrically by means of the colorimetric reaction with salicylaldehyde, acetone in alkaline solution (Neish, 1952), and the longer chain ketones in acidic solution (Mukherjee, 1951). Acetone and 2-butanone were distilled from the culture liquids into a solution of reagent for preparation of derivatives according to Vogel (1951). Longer chain ketones were extracted with three changes of ether. Acetol was assayed by periodate titration (Neish, 1952). Infrared-absorption spectra were obtained with a Baird instrument, and melting points with a Fisher block.

In addition to alkanes and ketones listed under the basal medium other ketones were oxidised.

S. Mukherjee (1951). Archs Biochem. Biophys. 33, 364–376.

A. C. Neish (1952). Report No. 46–8–3. Prairie Regional Laboratory, Saskatoon, Canada.

A. E. Vogel (1951). *A text-book of practical organic chemistry including qualitative organic analysis.* Longmans, Green and Co., Ltd., London.

Alkylamines—*Protaminobacter*

V. B. D. Skerman, *A Guide to the Identification of the Genera of Bacteria.* 2nd ed. Williams and Wilkins Book Co., Baltimore, 1967.

The original medium employed by den Dooren de Jong is prepared as follows:

$MgCl_2$, 2.0 g; K_2HPO_4, 10.0 g; tap water, 1000 ml: washed agar, 14 g.

Dissolve the ingredients, filter and adjust the pH to 7.0. Distribute in 20 ml quantities and sterilise at 121°C for 20 minutes.

Prepare a 5 per cent solution of the alkylamine. Neutralise with HCl. (Either monoethylamine or dimethylamine satisfies the requirements of the described species). Sterilise by filtration and store in screw-capped bottles or glass ampoules.

To prepare the medium melt the agar base and cool to 50°C. Add 1 ml of the alkylamine solution to 20 ml of the agar base and pour.

Note: Since the majority of agars available commercially contain quantities of metabolizable materials, growth of organisms on the above medium is not an indicator of use of the alkylamine. A liquid medium prepared with glass-distilled water should be employed as a final check

on the ability of the organism to grow with the use of the alkylamine as sole carbon and nitrogen sources.
den Dooren de Jong (1927). Zentbl. Bakt. ParasitKde Abt. II, 71, 218.

Ammonia—nitrifying organisms
R. N. Gowda (1924). J. Bact. 9, 251–272.
Isolation of the nitrite formers
Subculture method. Gibbs' (1919) nutrient solution modified, was employed.
Solution 1:
$(NH_4)_2SO_4$, 1.0 g; K_2HPO_4, 1.0 g; NaCl, 2.0 g; $MgSO_4$, 0.5 g; $Fe_2(SO_4)_3$, trace; water (conductivity), 1000 ml.
Solution 2:
$MgCO_3$, 5.0 g; water (conductivity), 100 ml.
Five millilitres of solution 1 with 0.5 ml of solution 2 were placed in flasks and inoculated with soil. Nitrites were first found after five to six days, with Trommsdorf's reagents. Complete oxidation of the $(NH_4)_2SO_4$ did not occur in cultures eight weeks old.
The nitrite formers may rapidly lose their power of oxidation of ammonia when grown in pure cultures.
Free CO_2 is essential for oxidation of ammonia by nitrate formers which grow most rapidly in washed agar containing soil extract.
W. M. Gibbs (1919). Soil Sci. 8, 427–481.

Ammonia—*Nitrosomonas*
V. B. D. Skerman, *A Guide to the Identification of the Genera of Bacteria.* 2nd ed. Williams and Wilkins Book Co., Baltimore, 1967.
NaCl, 0.3 g; $MgSO_4 \cdot 7H_2O$, 0.14 g; $FeSO_4 \cdot 7H_2O$, 0.3 g; H_2O, 90 ml; 0.1M KH_2PO_4*, 10 ml; + $(NH_4)_2SO_4$, 0.66 g.
Dilute to 1000 ml and add 10 g of powdered $CaCO_3$ and 0.4 ml of a trace element solution supplying Mn, 22 μg; B, 21 μg; Cu, 17 μg; Zn, 16 μg; and Co, 14 μg.
Dispense in layers not more than 1 cm deep in Erlenmeyer flasks. Sterilize at 121°C for 15 minutes.
J. Meiklejohn (personal communication).
*Previously boiled for 30 minutes, cooled, and made up to volume.

Carbon monoxide—*Carboxydomonas*
V. B. D. Skerman, *A Guide to the Identification of the Genera of Bacteria.* 2nd ed. Williams and Wilkins Book Co., Baltimore, 1967.
Medium for the Cultivation of Carboxydomonas
KNO_3, 2.0 g; K_2HPO_4, 1.0 g; $MgSO_4 \cdot 7H_2O$, 0.1 g; peptone, 0.2 g; H_2O, 1000 ml.
Dissolve the ingredients and adjust the pH to 7.2. Sterilize at 121°C for 20 minutes.

For an agar medium use only sufficient agar to make a moderately firm gel. An agar which is too hard inhibits growth.

Incubate under an atmosphere of 80 per cent CO and 20 per cent O_2.

A. Kistner (1953). Proc. K. ned. Akad. Wet., Sec. C 56, 443.

Dihydroxyacetone and hydroxymaleic acid (α-naphthol color test)

W. J. Turner, B. H. Kress, and N. B. Harrison (1942). J. Bact. 44, 249–250.

To 1 ml of the culture fluid add 0.5 ml of freshly prepared 6% alcoholic α-naphthol and then 0.2 ml 40% KOH. In the presence of more than 0.1 mg of dihydroxyacetone a yellow color soon changes to green and then to blue which is stable for hours. It has no characteristic absorption spectrum. It is soluble in polar organic solvents but not in benzene or ether.

The blue color is also given by 1 mg pyruvic acid or 25 mg of acetoacetic ester but not by glyceraldehyde, glucose, lactic acid, glucuronic acid, acetaldehyde or acetone.

Hydroxymaleic acid in concentration less than 0.5 mg per ml yields a pink color similar to that given by diacetyl, but it does not fluoresce. With concentrations greater than 1 mg per ml there is a preliminary color change through green, blue and purple to red. There is no characteristic absorption spectrum.

Ethanol—*Streptococcus*

E. C. Greisen and I. C. Gunsalus (1944). J. Bact. 48, 515–525.

Streptococcus mastitidis, a homofermentative lactic acid organism lacking the usual hemin catalysts, oxidizes ethyl alcohol to acetic acid in the presence of air without added carriers. The rate of oxidation is stimulated several fold by the addition of methylene blue. Contrary to expectations the oxidation requires but one mole of oxygen per mole of alcohol and hydrogen peroxide does not accumulate in the absence of added carriers.

With aldehyde fixative, alcohol is oxidized more rapidly with the utilization of 0.5 mole of oxygen per mole of alcohol.

Aldehyde is oxidized to acetic acid by 0.5 mole of oxygen. In the absence of oxygen, aldehyde is dismutated to acid and alcohol.

The alcohol oxidation, but neither the aldehyde oxidation nor dismutation, is inhibited by M/100 iodoacetate. Sodium cyanide does not inhibit the oxidation at the concentration effective against the usual hemin systems but does partially inhibit at M/100.

Ethanol—*Pseudomonas (Malleomyces) pseudomallei*

L. Chambon and P. de Lajudie (1954). Annls Inst. Pasteur, Paris 86, 759–764.

Utilisation de l'alcool éthylique et de l'alcool méthylique. –– Ces deux alcools n'assurent pas le métabolisme carboné du bacille de Whitmore, qui ne donne pas de culture dans le milieu synthétique proposé par Lwoff [*Cf.* Dumas (1951)] pour les *Moraxella,* milieu dont la formule est la suivante:

PO_4KH_2	4.5 g
$SO_4(NH_4)_2$	0.75 g
KCl	0.5 g
SO_4Mg	0.05 g
Eau bidistillée	1 000 ml

après stérilisation on ajoute:

Citrate de fer	0.01
$CaCl_2$	0.01
Alcool éthylique ou méthylique	10 ml

Dumas (J.). *Bactériologie Mèdicale,* Flammarion, édit., 1951.

Ethanol—*Acetobacter*

J. de Ley (1958). Antonie van Leeuwenhoek 24, 281–297.

All the strains produce acetic acid in a medium, containing 1% Difco yeast extract, 2% ethanol. After one week to ten days all the acid is consumed again.

Ethanol—*Acetobacter*

V. B. D. Skerman, *A Guide to the Identification of the Genera of Bacteria.* 2nd ed. Williams and Wilkins Book Co., Baltimore, 1967.

Dissolve 10 g of Bacto-tryptose in 1 L of distilled water and add 1 g of yeast extract and dissolve. Distribute in known quantities in small Erlenmeyer flasks to give a layer not more than 1 cm deep. Sterilize by autoclaving at 121°C for 20 minutes.

When cool add 10 per cent by volume of sterile absolute alcohol.

Inoculate the medium and incubate for 1 week or until obvious growth has occurred. To observe acid formation add a few drops of 1.6 per cent alcoholic solution of bromocresol purple.

NOTE: Numerous organisms may grow in this medium without oxidation of the alcohol, thus the test for acid is necessary.

Ethanol—*Acetobacter*

J. L. Shimwell, J. G. Carr and M. E. Rhodes (1960). J. gen. Microbiol. 23, 283–286.

Detected by the presence on agar plates of 1% (w/v) Difco yeast extract + 2% (w/v) ethanol + 2% (w/v) $CaCO_3$ of clear zones of dissolved $CaCO_3$ round the implanted (*c.* 1 cm^2) bacterial masses after 14 days at 26°C, or sooner.

Ethanol—*Vibrio*

G. H. G. Davis and R. W. A. Park (1962). J. gen. Microbiol. 27, 101–119.

Utilization of ethanol and production of excess acid was tested by growth within 7 days in two successive subcultures upon inorganic medium containing 1% (w/v) calcium carbonate and 1% (v/v) ethanol. Excess acid production was shown by clearing of $CaCO_3$ around growth.

Ethanol—*Vibrio marinus*

R. R. Colwell and R. Y. Morita (1964). J. Bact. 88, 831–837.

The authors used the method of J. L. Shimwell, J. G. Carr and M. E. Rhodes (1960). J. gen. Microbiol. 23, 283–286 – modified by the addition to the media of the following salts to produce a synthetic seawater:– sodium chloride, 2.4%; potassium chloride, 0.07%; magnesium chloride (hydrated) 0.53% and magnesium sulfate (hydrated) 0.7%.

A standard inoculum was 1 drop from a pasteur pipette (*c.* 0.05 ml) of a 24 hour artificial seawater broth.

Cultures were incubated at 18°C.

Ethanol—*Leuconostoc*

J. O. Mundt and J. L. Hammer (1966). Appl. Microbiol. 14, 1044.

Ethyl alcohol (95%) to a final concentration of 7% was added to heated and cooled Rogosa SL medium (Difco) containing 0.04% cycloheximide and adjusted to pH 5.4. The broth was tubed aseptically in 9-ml quantities if prepared in single strength for inocula of 0.1 ml. For larger quantities of inocula, the base medium was prepared with less water, and the final volume was obtained with inoculum alone or with inoculum and water. Plates were streaked from tubes with growth after incubation for 8 to 24 hr.

Prior to use of medium with ethyl alcohol, 82% of nearly 2,000 colonies taken from platings of various vegetables were *L. mesenteroides* whereas 4% were members of the genus *Lactobacillus*. With medium containing ethyl alcohol, 82% of several hundred colonies selected were lactobacilli, whereas 3% were *L. mesenteroides*. The percentage of pediococci remained the same.

Gluconate—*Pseudomonas aeruginosa*

W. C. Haynes (1951). J. gen. Microbiol. 5, 939–950.

Each strain was inoculated from a stock slant to a 300 ml Erlenmeyer flask containing 100 ml of sterile medium of the following composition: tryptone, 1.5 g; yeast extract, 1 g; K_2HPO_4, 1 g; potassium gluconate, 40 g; distilled water, 1000 ml; pH 7.0.

The inoculated flasks were then put on a rotary shaker (Gump-type) at 28–30°C. The shaker was set at 200 r.p.m. and each flask was rotated through an orbit of 2 ¼ in. diameter. At 2, 4, 7 and 14 days, samples were tested in the following manner.

To 1 ml of the culture in a 16 mm test-tube, 10 ml of a copper sulfate sugar reagent (Shaffer and Hartmann, 1921 – 'Carbonate-citrate reagent for cupric titration') were added. The contents of the tubes were thoroughly mixed, heated for 10 min in boiling water, rapidly cooled in cold water and then set aside until the next morning to settle the precipitate of reduced copper. The percentage conversion of potassium gluconate to potassium 2-ketogluconate was then estimated by comparison with previously prepared standards made with weighed quantities of calcium 2-ketogluconate. The conversion was recorded as 75, 50 or 25%, or none. Exact quantitative relationships were deemed unessential to establish the capability of these organisms to oxidize potassium gluconate to potassium 2-ketogluconate.

Slime is produced in the above medium when allowed to stand for 4 days after shaking.

Gluconate—*Proteus*

C. Shaw and P. H. Clarke (1955). J. gen. Microbiol. 13, 155–161.

The medium was a modification of that used by Haynes (1951) for pseudomonads. It consisted of: Evan's peptone, 1.5 g; Yeastrel, 1.0 g; K_2HPO_4, 1.0 g; potassium gluconate 40.0 g; distilled water, 1000 ml; adjusted to pH 7.0; sterilised 10 lb/sq in. for 10 min. After incubation of the culture for 48 hr at 37°C, 1.0 ml of Benedict's qualitative reagent for reducing sugars was added and the tubes placed in boiling water for 10 min.

A positive test showed a yellowish brown precipitate of cuprous oxide and there was a characteristic smell of decaying cabbage.

W. C. Haynes (1951). J. gen. Microbiol. 5, 939.

Gluconate—*Chromobacterium*

P. H. A. Sneath (1956). J. gen. Microbiol. 15, 70–98.

Tested in the modified Hayne's medium of Shaw and Clarke (1955) in shallow layers (3 ml in tubes 22 mm internal diameter).

After 4 days at 25°C, 1 ml of Benedict's qualitative reagent for glucose was added and after standing 10 minutes at room temperature the tubes were heated to 100°C for 10 minutes.

C. Shaw and P. H. Clarke (1955). J. gen. Microbiol. 13, 155.

Gluconate—*Pseudomonas*

W. L. Gaby and E. Free (1958). J. Bact. 76, 442–444.

Gluconate oxidation test (Haynes)

A practical laboratory method for the determination of gluconate oxidation by pseudomonads has only recently become available. This procedure is based on the ability of ketogluconate (oxidative product of gluconate by *Pseudomonas*) to reduce Benedict's reagent while gluconate is unable to do so. One gluconate substrate tablet (Key) is added to 1 ml of distilled water in a sterile test tube and inoculated heavily with the test organism. Following a 12 to 18 hr incubation period at 37°C the presence of reducing sugars was determined by adding the Benedict's reagent (one clinitest tablet, Ames) and comparing the resulting colour with a standard chart. It should be noted that any satisfactory test for reducing sugar can be employed.

W. C. Haynes (1951). J. gen. Microbiol. 5, 939–950.

Gluconate—*Pseudomonas*

M. E. Rhodes (1959). J. gen. Microbiol. 21, 221–263.

The defined medium of Koser (1923) but with the citric acid replaced by 0.5% (w/v) gluconic acid and neutralized with KOH, was used. Five ml volumes were inoculated, and after incubation for 7 days 1 ml Benedict's qualitative reagent for glucose was added. Then, after standing 10 min at room temperature, the cultures were heated at 100°C for 10 min; and examined for reduction (Sneath, 1956).

S. A. Koser (1923). J. Bact. 8, 493.

P. H. A. Sneath (1956). J. gen. Microbiol. 15, 70.

Gluconate—*Pseudomonas, Achromobacter*

H. B. Moore and M. J. Pickett (1960). Can. J. Microbiol. 6, 35–42.

The ability of organisms to oxidize gluconate to 2-ketogluconate was examined in a liquid medium of the following composition: potassium phosphate, 0.04 M, pH 6.5; potassium nitrate, 0.2%; and potassium gluconate, 2.0%. This medium was stored over chloroform at 4°C and tubed in 1 ml amounts for use. Tubes were steamed for 10 minutes and allowed to cool before being inoculated. After incubation for 12 to 16 hours, five drops of double-strength Benedict's solution were added to each tube. The tubes were steamed for 10 minutes, then observed for the typical red-brown precipitate of reduced copper.

Gluconate—*Vibrio*

G. H. G. Davis and R. W. A. Park (1962). J. gen. Microbiol. 27, 101–119.

Production of reducing compounds from gluconate (Haynes, 1951). Medium (%, w/v): Oxoid peptone, 0.2; Oxoid yeast extract, 0.1; K_2HPO_4, 0.1; pH 7.0. Basal medium autoclaved and Seitz-filtered 40% (w/v) aqueous potassium gluconate solution added to give final concentration of 4% (w/v). Dispensed aseptically, inoculated and shaken by hand twice daily during 5 days' incubation. Tested with Benedict's qualitative reagent.

W. C. Haynes (1951). J. gen. Microbiol. 5, 939.

Gluconate—*'Bacterium salmonicida' (Aeromonas)*

I. W. Smith (1963). J. gen. Microbiol. 33, 263–274.

Oxidation was tested by the method of P.H.A. Sneath (1956) J. gen. Microbiol. 15, 70).

Gluconate—*Pseudomonas odorans*

I. Málek, M. Radochová and O. Lysenko (1963). J. gen. Microbiol. 33, 349–355.

Tested in the medium of Sneath (1956) after incubation for 5 days at 28°C.

P. H. A. Sneath (1956). J. gen. Microbiol. 15, 71.

Gluconate—*Vibrio marinus*

R. R. Colwell and R. Y. Morita (1964). J. Bact. 88, 831–837.

The authors used the method of W. L. Gaby and E. Free (1958). J. Bact. 76, 442–444 – modified by the addition to the media of the following salts to produce a synthetic seawater:– sodium chloride, 2.4%; potassium chloride, 0.07%; magnesium chloride (hydrated) 0.53% and magnesium sulfate (hydrated) 0.7%.

A standard inoculum was 1 drop from a pasteur pipette (*c.* 0.05 ml) of a 24 hour artificial seawater broth.

Cultures were incubated at 18°C.

Gluconate—*Pseudomonas aeruginosa*

R. R. Colwell (1964). J. gen. Microbiol. 37, 181–194.

Production of 2-ketogluconic acid. Cultures were incubated for 7 days in Paton's broth medium (Paton, 1959); the presence of reducing sugars was determined by adding Benedict's reagent (Clinitest tablet, Ames Co. of Canada, Ltd., Toronto). The method of Haynes (1951) was also used and the production of slime and 'oyster' formation noted. The rapid test for gluconate oxidation (Gaby & Free, 1958) was done with gluconate substrate tablets (Key Scientific Products Inc., Los Angeles, U.S.A.) but the incubation time was increased to 72 hr at room temperature on a rotary shaker.

W. L. Gaby and E. Free (1958). J. Bact. 76, 442.
W. C. Haynes (1951). J. gen. Microbiol. 5, 939.
A. M. Paton (1959). Nature, Lond. 184, 1254.

Gluconate—*Pseudomonas*

G. L. Bullock, S. F. Snieszko and C. E. Dunbar (1965). J. gen. Microbiol. 38, 1–7.

The oxidation of gluconate was determined in the medium of Haynes (1951) with the exception that 2% (w/v) peptone was substituted for tryptone. The basal medium was autoclaved, and 0.5 ml of a 40.0% (w/v) Seitz-sterilized potassium gluconate solution was added to a final concentration of 4% (w/v). The inoculated medium was shaken twice by hand during incubation. Presence of reducing compounds was tested with Benedict's qualitative reagent. Cultures incubated at 20–22°C for 1 week.

W. C. Haynes (1951). J. gen. Microbiol. 5, 939.

Gluconate—*Pseudomonas aeruginosa*

A. H. Wahba and J. H. Darrell (1965). J. gen. Microbiol. 38, 329–342.

Oxidation of potassium gluconate and production of slime within 3 *days at* 37°C *and* 42°C. A modification of Haynes' (1951) original technique was used. The medium was dispensed in 5 ml amounts and every strain was inoculated in duplicate, one tube being incubated at 37°C and the other at 42°C (to investigate the possibility of combining three tests for identifying *Pseudomonas aeruginosa*). The cultures were not shaken during incubation. After incubation for 48 hr all tubes were examined for slime production by shaking. When slime was present, the cultures were then tested for reducing substances by adding half a 'Clinitest' tablet (Ames Co., Stoke Poges, Bucks, used for the estimation of sugar in urine) (Carpenter, 1961). When slime was absent, the culture was divided; one half was tested for the presence of reducing substance and the other incubated for a further 5 days at the appropriate temperature and re-examined. Strains not growing at 37°C were incubated at 22°C.

K. P. Carpenter (1961). J. gen. Microbiol. 26, 535.

W. C. Haynes (1951). J. gen. Microbiol. 5, 939.

Gluconate—*Alcaligenes*

R. G. Mitchell and S. K. R. Clarke (1965). J. gen. Microbiol. 40, 343–348.

Shaw and Clarke's (1955) method was used.

C. Shaw and P. H. Clarke (1955). J. gen. Microbiol. 13, 155.

Glucose (to 2-keto-gluconic acid)—*Pseudomonas* and *Phytomonas*

L. B. Lockwood, B. Tabenkin and G. E. Ward (1941). J. Bact. 42, 51–61.

Studies were conducted in Jena glass gas-washing bottles (Type 101-a), which are constructed with sintered glass false bottoms through which sterile air may be passed, thereby aerating and agitating the cultures. The basic nutrient solution has the composition:

Glucose, 100 g; corn steeping liquor, 5 g; KH_2PO_4, 0.6 g; $MgSO_4 \cdot 7H_2O$, 0.25 g; distilled water, 1000 ml.

Two hundred milliliters of this solution was placed in each Jena glass gas-washing bottle and sterilized at fifteen pounds pressure for one-half hour. After cooling, 2 ml of a sterile twenty per cent urea solution and

5 grams of calcium carbonate (sterilized dry) were added to each bottle. The cultures were incubated at 30°C for eight days, each culture being aerated constantly at the rate of 200 ml of air per minute.

At the conclusion of each experiment, a determination of calcium in solution was made by precipitation of calcium as the oxalate and subsequent titration with standard $KMnO_4$, in the usual manner.

The original nutrient solutions were analyzed for glucose and the fermented culture solutions for total reducing substances (glucose plus 2-ketogluconic acid) by the copper reduction method of Shaffer and Hartmann (1921). The optical activity of the fermented liquor was also determined after clarification; this value, together with the copper reduction value, permitted the calculation of the concentration of glucose and 2-ketogluconic acid in the liquors, according to a method described in an earlier publication (Stubbs *et al.* 1940). It should be mentioned here that calcium 2-ketogluconate has a very appreciable negative specific rotation (about −88°), in contrast to the considerable dextro-rotation of glucose and the very slight dextro-rotation of calcium gluconate. The exhibition of a levo-rotation by the fermented liquor is, therefore, indicative of 2-ketogluconic acid production. Agreement in the glucose values, as determined by polarimetric observations and copper reduction values, indicates that glucose is the only reducing material present in the cultures represented.

As a final check, calcium 2-ketogluconate was isolated from the culture liquors of representative species. In order to obtain this salt, the harvested liquors were concentrated at low temperatures to about one-third the original volume, cooled, and the crystalline material filtered off and recrystallized from water. Such purified samples were identified by the comparison of calcium content, copper reduction values and the optical rotation with pure known materials. The original identification of 2-ketogluconic acid was made by examination of the methyl ester (Ohle 1937) and the quinoxylin derivative (Ohle 1934). In later cases the methyl ester (mp 174–175, Ohle 1937) was prepared to check the identity of the 2-ketogluconic acid.

H. Ohle (1934). Ber. dt. chem. Ges. 67, 155–162.
H. Ohle (1937). Ibid. 70 B: 2153.
P. A. Shaffer and A. F. Hartman (1921). J. biol. Chem. 45, 365–394.
J. J. Stubbs, L. B. Lockwood, E. T. Roc, B. Tabenkin and G. E. Ward (1940). Ind. Engng Chem. 32, 1626–1631.

Glucose (to gluconic acid)—*Pseudomonas* and *Phytomonas*

L. B. Lockwood, B. Tabenkin and G. E. Ward (1941). J. Bact. 42, 51–61.

The production of gluconic acid by species of Pseudomonas and Phytomonas

200 ml nutrient solution in gas-washing bottles (see preceding abstract) contained 200 grams glucose, 5 grams $CaCO_3$, 1 gram corn steeping liquor, 0.4 gram urea, 0.12 gram KH_2PO_4, 0.05 gram $MgSO_4 \cdot 7H_2O$, and one drop oleic acid (antifoam agent). Temperature 30°C, air flow 200 ml per minute. Duration 8 days.

Gluconic acid is identified in the culture as the phenylhydrazide derative (Diemair *et al.* 1935).

W. Diemair, B. Bleyer and L. Schneider (1935). Unters. Lebensm, 69, 212–220.

Glucose (to gluconic acid)—*Acetobacter*

J. de Ley (1958). Antonie van Leeuwenhoek 24, 281–297.

The strains were grown in 1 liter flasks, containing 100 ml of a medium consisting of 1% Difco yeast extract, tap water and either 0.5, 2 or 10% glucose. The flasks were incubated at 30°C on a shaking machine. The glucose concentration was determined every second day over a period of two weeks by the method of Luff-Schoorl. Two strains formed much acid and after 2 days all glucose had disappeared. With one of the strains it could be shown that gluconic acid constituted the main endproduct, by several means: (1) by boiling with N HCl for 5 min a lactone was formed, since it develops the typical brown colour with the hydroxylamine reagent (Hestrin 1949); (2) this lactone behaved in the same way as the reference substance on paper chromatograms with the solvent n-butanol 3/ ethanol 1/ water 1; the position of the compound was revealed using the spray of Abdel-Akher and Smith (1951); (3) the free acid, obtained after treatment with Amberlite IR 120 (H+) behaved in the same way as the reference substance on paper chromatograms with the solvent methanol 6/HCOOH 1/ water 3 using bromcresolgreen as a spray reagent.

M. Abdel-Akher and F. Smith (1951). J. Am. chem. Soc. 73, 5859.
S. Hestrin (1949). J. biol. Chem. 180, 249.

Glycerol—*Acetobacter*

J. L. Shimwell, J. G. Carr and M. C. Rhodes (1960). J. gen. Microbiol. 23, 283–286.

Agar plates containing 1% Difco yeast extract and 2% (v/v) glycerol were used. Incubated 14 days at 26°C. If plates are flooded with Fehling's solution a quick (10 minutes) production of an aureole of yellow tored copper oxide(s) round the implanted masses indicates presence of dihydroxyacetone.

Glycerol—*Vibrio marinus*

R. R. Colwell and R. Y. Morita (1964). J. Bact. 88, 831–837.

The authors used the method of J. L. Shimwell, J. G. Carr and M. E. Rhodes (1960). J. gen. Microbiol. 23, 283–286 – modified by the addition to the media of the following salts to produce a synthetic seawater: – sodium chloride, 2.4%; potassium chloride, 0.07%; magnesium chloride (hydrated) 0.53% and magnesium sulfate (hydrated) 0.7%.

A standard inoculum was 1 drop from a pasteur pipette (*c.* 0.05 ml) of a 24 hour artificial seawater broth.

Cultures were incubated at 18°C.

Hydrogen—*Hydrogenomonas* and others

A. Schatz and C. Bovell, Jr. (1952). J. Bact. 63, 87–98.

see also A. Schatz (1952). J. gen. Microbiol. 6, 329–335.

Cultures were isolated, grown, and maintained on the following basal medium adjusted to pH 6.8 to 7.2: KH_2PO_4 0.1 g, NH_4NO_3 0.1 g, $MgSO_4 \cdot 7H_2O$ 0.02 g, $FeSO_4 \cdot 7H_2O$ 0.001 g, and $CaCl_2 \cdot 2H_2O$ 0.001 g, distilled water to 100 ml. Where desired, 1.5 per cent washed agar was incorporated. For autotrophic growth the base was supplemented with 0.05 per cent $NaHCO_3$. Stock solutions of the bicarbonate were autoclaved separately, flushed with CO_2, and added to the sterile medium prior to inoculation. Incubation was at 25°C in desiccators with the atmosphere initially adjusted to 10 per cent CO_2, 30 per cent air, and 60 per cent H_2.

Hydrogen—autotrophic medium for *Hydrogenomonas*

V. B. D. Skerman, *A Guide to the Identification of the Genera of Bacteria.* 2nd ed. Williams and Wilkins Book Co., Baltimore, 1967.

KH_2PO_4, 0.1 g; NH_4NO_3, 0.1 g; $MgSO_4 \cdot 7H_2O$, 0.02 g; $FeSO_4 \cdot 7H_2O$, 0.001 g; $CaCl_2 \cdot 2H_2O$, 0.001 g; Distilled water to 100 ml. Adjust the pH to between 6.8 and 7.2.

When desired, incorporate 1.5 per cent washed agar. For autotrophic growth, supplement the base with 0.05 per cent $NaHCO_3$. Autoclave stock solutions of the $NaHCO_3$ separately, flush with CO_2, and add to the sterile medium before inoculation.

Incubate under an atmosphere of 10 per cent CO_2, 30 per cent air, and 60 per cent hydrogen.

A. Schatz and C. Bovell (1952). J. Bact. 63, 87.

Indole—*Pseudomonas*

R. Y. Stanier, N. J. Palleroni and M. Doudoroff (1966). J. gen. Microbiol. 43, 159–271.

The oxidation of indole to the insoluble blue compound, indigotin, originally described by Gray (1928) as the salient property of *P. indoloxidans,* was tested on plates of the medium recommended by Gray, except that Na acetate was substituted for glycerol as the carbon source for strains unable to use glycerol. In positive strains, the patch and the adjacent agar turn blue, as a result of the deposition of crystals of indigotin.

Iron—*Ferrobacillus*

V. B. D. Skerman, *A Guide to the Identification of the Genera of Bacteria.* 2nd ed. Williams and Wilkins Book Co., Baltimore, 1967.

Liquid Medium: $(NH_4)_2SO_4$, 0.15 g; KCl, 0.05 g; $MgSO_4 \cdot 7H_2O$, 0.5 g; K_2HPO_4, 0.05 g; $Ca(NO_3)_2$, 0.01 g. Dissolve the salts in 1000 ml of distilled water. Sterilize at 121°C for 15 minutes. Prepare a stock solution consisting of 10 per cent $FeSO_4 \cdot 7H_2O$ in distilled water and sterilize by filtration; add 1 ml aseptically to each 100 ml of the above medium. The resultant medium is opalescent. Refrigerate the medium so that oxidation does not occur.

Iron—*Thiobacillus ferrooxidans*

V. B. D. Skerman, *A Guide to the Identification of the Genera of Bacteria.* 2nd ed. Williams and Wilkins Book Co., Baltimore, 1967.

$FeSO_4 \cdot 7H_2O$, 130.0 g; $MgSO_4 \cdot 7H_2O$, 1.0 g; $(NH_4)_2SO_4$, 0.5 g; Distilled water 1000 ml.

Dissolve the ingredients and adjust the pH to between 2.0 and 2.5 with sulfuric acid. Autoclave at 121°C for 15 minutes and allow to stand. A voluminous precipitate of ferric hydroxide settles out. Remove the supernatant aseptically and distribute as required, preferably in layers not more than 1 cm deep in Erlenmeyer flasks.

To prepare an agar medium dissolve the ferrous sulfate in 300 ml of the water and sterilize separately. Dissolve the other ingredients plus 20 g of agar in the remaining water and sterilize. Mix the two solutions just before plate pouring.

K. L. Temple and A. R. Colmer (1951). J. Bact. 62, 605.

Iron—*Thiobacillus ferrooxidans*

J. Landesman, D. W. Duncan and C. C. Walden (1966). Can. J. Microbiol. 12, 25–33.

The authors describe manometric studies on iron oxidation.

Isopropanol—photosynthetic bacteria

J. W. Foster (1944). J. Bact. 47, 355–372.

Oxidation of secondary alcohols

Cells harvested from a 10% yeast autolysate medium were resuspended in 0.1% $NaHCO_3$ containing 0.1% isopropanol and incubated 3 days anaerobically in the light. The cultures were then centrifuged in 10 ml portions of the supernatants slowly distilled in a microstill after acidification with H_2SO_4 and addition of a few drops of oleic acid as an antifoam. The first ml of distillate was caught in 0.5 ml of 1% 2,4-dinitrophenylhydrazine in 2N HCl. The appearance of a yellow colour indicated hydrazone formation. The foregoing method was used to obviate the disadvantages of the extremely slight growth on media containing secondary alcohols.

Lactate (to carbonate)—*Pseudomonas aeruginosa*

R. R. Colwell (1964). J. gen. Microbiol. 37, 181–194.

Oxidation of calcium lactate through acetate to carbonate and production and accumulation of dihydroxyacetone in media containing glycerol and the production of acetic acid from ethanol were tested by the methods given by Shimwell, Carr and Rhodes (1960).

J. L. Shimwell, J. G. Carr and M. E. Rhodes (1960). J. gen. Microbiol. 23, 283.

Lactate (to carbonate)—*Vibrio marinus*

R. R. Colwell and R. Y. Morita (1964). J. Bact. 88, 831–837.

The authors used the method of J. L. Shimwell, J. G. Carr and M. E. Rhodes (1960). J. gen. Microbiol. 23, 283–286 – modified by the addition to the media of the following salts to produce a synthetic seawater:– sodium chloride, 2.4%; potassium chloride, 0.07%; magnesium chloride (hydrated) 0.53% and magnesium sulfate (hydrated) 0.7%.

A standard inoculum was 1 drop from a pasteur pipette (*c.* 0.05 ml) of a 24 hour artificial seawater broth.

Cultures were incubated at 18°C.

Lactose (to 3-keto-lactose)—*Agrobacterium tumefaciens*

G. Beaud and P. Manigault (1966). Annls Inst. Pasteur, Paris 111, 345–358.

La production de 3 céto-lactose (3 CL) se fait qualitativement au moyen du test de Bernaerts et de Ley [7]. Des boîtes de gélose nutritive au lactose (20 g/l) sont ensemencées au centre, en un point. Après quarante-huit heures de séjour à l'étuve (30°C) la boîte est recouverte de liqueur de Fehling et gardée à la température de la pièce. Deux heures plus tard, on observe un anneau jaune d'oxydule autour des bactéries produisant le 3 CL. Une variante de cette méthode permet de déceler la production de 3 CL en milieu liquide.

[7] Bernaerts (M. J.) and De Ley (J.). *Nature,* 1963, 197, 406.

Methane—autotrophic medium for *Methanomonas*

V. B. D. Skerman, *A Guide to the Identification of the Genera of Bacteria.* 2nd ed. The Williams and Wilkins Book Co., Baltimore, 1967.

$NaNO_3$, 2.0 g; $MgSO_4 \cdot 7H_2O$, 0.2 g; $FeSO_4 \cdot 7H_2O$, 0.001 g; Na_2HPO_4, 0.21 g; NaH_2PO_4, 0.09 g; $CuSO_4 \cdot 5H_2O$, 200.0 μg; H_3BO_3, 60.0 μg; $MnSO_4 \cdot H_2O$, 30.0 μg; $ZnSO_4 \cdot 7H_2O$, 300.0 μg; MoO_3, 15.0 μg; KCl, 0.04 g; $CaCl_2$, 0.015 g; H_2O, 1000 ml;

Dissolve the salts and sterilize. Incubate under an atmosphere of 50 per cent methane and 50 per cent air.

J. W. Foster (personal communication)

Methanol—*Pseudomonas methanica*

A. A. Harrington and R. E. Kallio (1960). Can. J. Microbiol. 6, 1–7.

Two mineral media were used for routine culturing (1) of M. Dworkin and J. W. Foster (1956).
(2) $NH_4H_2PO_4$, 0.8 g; K_2HPO_4, 1.5 g; $MgSO_4 \cdot 7H_2O$, 0.2 g; Fe^{++}, 0.05 g; tap water, 10 ml (?); For routine cultures 100 ml mineral medium to which 1 ml of reagent-grade methanol had been added was inoculated and incubated on a gyratory shaker at 25°C.

Methanol was assayed by a modification of the method of Feldstein and Klendshoj (1954). A standard curve was prepared from weighed quantities of methanol.

M. Dworkin and J. W. Foster (1956). J. Bact. 72, 646–659.
M. Feldstein and N. C. Klendshoj (1954). Analyt. Chem. 26, 932–933.

Note: The medium of Dworkin and Foster referred to above was later replaced by the medium cited in the previous abstract.

Pentoses—*Pseudomonas*

L. B. Lockwood and G. E. N. Nelson (1946). J. Bact. 52, 581–586.

Oxidations of pentoses

d-arabinose $\longrightarrow$ d-arabonic acid
l-arabinose $\longrightarrow$ l-arabonic acid
d-xylose $\longrightarrow$ d-xylonic acid
d-ribose $\longrightarrow$ d-ribonic acid

Reactions catalysed by various species of *Pseudomonas* when growing in 100 ml aerated cultures in a corn steep liquor base + Ca CO_3, incubated at 30°C.

The corn steep liquor base contained 2 g urea, 0.6 g KH_2PO_4, 0.25 g $MgSO_4 \cdot 7H_2O$, and 5 ml of corn steep liquor per liter. Three drops of soybean oil were added to each culture to prevent excessive frothing. Sufficient sterile $CaCO_3$ was added to neutralize the pentonic acid which might be formed. $CaCO_3$ was sterilized dry. Culture media were sterilized by filtration.

The d-arabonate was identified by the production of the phenyl-hydrazide of the free acid (m p 213°C) (Glattfeld, 1913).

l-arabonic acid was identified by preparation of the brucine salt. (m p 152°C) and by the preparation of l-arabobenzimidazole (m p 235–236°C) (Nef, 1907 and Moore and Link, 1940).

d-xylonic acid was identified by the preparation of the brucine salt (m p 176°C) (Nef, 1914).

d-ribonic acid was identified as the benzimidazole derivative which melts at 191°C (Dimler and Link, 1943).

R. J. Dimler and K. P. Link (1943). J. biol. Chem. 150, 345–349.
J. W. E. Glattfeld (1913). Am. chem. J. 50, 137–157.

J. U. Nef (1907). Justus Liebigs Annln Chem. 357, 214–312.
J. U. Nef (1914). Ibid. 403, 204–383.
S. Moore and K. P. Link (1940). J. biol. Chem. 133, 293–311.

Nitrite—*Nitrobacter*

R. N. Gowda (1924). J. Bact. 9, 251–272.

The following nutrient solution was used (Gibbs, 1919): $NaNO_2$, 1.0 g; Na_2CO_3, 1.0 g; K_2HPO_4, 0.5 g; NaCl, 0.5 g; $MgSO_4$, 0.3 g; $Fe_2(SO_4)_3$, trace; water (conductivity), 1000 ml.

Flasks of this solution were inoculated with 1 gram of soil and the presence of nitrate tested by diphenylamine. In ten days all the nitrite was oxidized in practically all cases. Five millilitres of the NO_2^- solution were added to each culture when the nitrite had disappeared and this was repeated twenty-five times. The average time to oxidize the 5 ml of nitrite was four to five days.

W. M. Gibbs (1919). Soil Sci. 8, 427–481.

Nitrite—*Nitrobacter*

V. B. D. Skerman, *A Guide to the Identification of the Genera of Bacteria.* 2nd ed. The Williams and Wilkins Book Co., Baltimore, 1967.

NaCl, 0.3 g; $MgSO_4 \cdot 7H_2O$, 0.14 g; $FeSO_4 \cdot 7H_2O$, 0.3 g; H_2O, 90 ml; 0.1M KH_2PO_4*, 10 ml; + $NaNO_2$, 0.5 g.

Dilute to 1000 ml and add 10 g of powdered $CaCO_3$ and 0.4 ml of a trace element solution supplying Mn, 22μg; B, 21 μg; Cu, 17 μg; Zn, 16 μg; and Co, 14 μg.

Dispense in layers not more than 1 cm deep in Erlenmeyer flasks. Sterilize at 121°C for 15 minutes.

J. Meiklejohn (personal communication).

*Previously boiled for 30 minutes, cooled, and made up to volume.

Nitrite—*Nitrobacter*

G. W. Gould and H. Lees (1960). Can. J. Microbiol. 6, 299–307.

The medium G described by Lees and Simpson (Biochem. J. 65, 297–305, 1957) was diluted 10 times before the addition of sodium nitrite. The latter could be added at 100–200 μg/ml without depressing the oxidation rate. Max oxidation rate was 7 μg nitrite-N oxidized/ml per hour. For enrichment culture, 70 μg nitrite-N/ml was used.

Phenol—*Mycoplana*

V. B. D. Skerman, *A Guide to the Identification of the Genera of Bacteria.* 2nd ed. The Williams and Wilkins Book Co., Baltimore, 1967.

The original medium described by Gray and Thornton is prepared as follows.

K_2HPO_4	1	g
$MgSO_4 \cdot 7H_2O$	0.2	g

NaCl	0.1 g
$CaCl_2 \cdot 2H_2O$	0.1 g
$FeCl_3$	0.02 g
$(NH_4)_2SO_4$	0.5 to 1.0 g
Or KNO_3	0.5 to 1.0 g
Phenol	10 g
Distilled water	1000 ml

Dissolve the ingredients, adjust the pH to 7.0 and sterilize at 121°C for 20 minutes.

Serial transfers of the organisms must be achieved before growth in this medium is recorded as positive.

NOTE: The above medium yields a precipitate which is not desirable when turbidity is used as an index of growth. An equally satisfactory medium which is free of precipitate is described under 'Pope-Skerman Mineral Salts Media for the Cultivation of Autotrophs and Nonexacting Heterotrophs' in Skerman (1967, reference as for this abstract).

P. H. Gray and H. G. Thornton (1928). Zentbl. Bakt. ParasitKde Abt. II, 73,74.

Phenol—*Mycobacterium* (see p. 712).

Phenol—*Vibrio*

G. H. G. Davis and R. W. A. Park (1962). J. gen. Microbiol. 27, 101–119.

Utilization of tryptophane, phenylalanine, phenol, benzoic acid, and catechol as sole carbon sources. Compounds were supplied in 0.1% (w/v) concentration in the inorganic base. (NOTE: phenol was used in 0.05% (w/v) as 0.1% inhibited certain strains; autoxidation of catechol by light, heat or high pH was avoided by adjusting basal medium to pH 6.5 and sterilizing by Seitz-filtration).

Sulfur—*Thiobacillus thiooxidans*

F. W. Adair (1966). J. Bact. 92, 899–904.

Cells were grown in 1-liter Erlenmeyer flasks containing 500 ml of a sterile basal salts medium as described by Vogler and Umbreit with the addition of 1% (w/v) of sterile, sublimed elemental sulfur (Merck and Co., Inc. Rahway, N.J.). The cultures were allowed to stand for 3 days and then were placed on a rotary shaking machine at 23°C for 6 days.*

Sulfate production was determined qualitatively by the addition of $BaCl_2$ and quantitatively by adding a known amount of labeled elemental sulfur (S^{35}) (Volk Radiochemical Co., New York, N.Y.) to the reaction mixture and then, after a given time, stopping the reaction with 0.3 ml of 5% $HClO_4$. The reaction mixture was centrifuged at 5,000 x *g* for 10 min; the pellet was then discarded. The clear supernatant fluid was then removed and treated with excess 1.0 M $BaCl_2$. The labeled $BaSO_4$ precipitate was washed twice and resuspended in distilled water. A portion was placed in an aluminum planchet which was then heated to 600°C for 0.5 hr to remove any contaminating S^{35}. Radioactivity on the planchets was

measured, and the values were corrected for self absorption.
*T. M. Cook (1964). J. Bact. 88, 620–623.
K. G. Vogler and W. W. Umbreit (1941). Soil Sci. 51, 331–337.

Sulfur—*Ferrobacillus ferrooxidans*

P. Margalith, M. Silver and D. G. Lundgren (1966). J. Bact. 92, 1706–1709.

Growth on ferrous iron. The cells were grown under conditions similar to those reported previously by Silverman and Lundgren. The organism was propagated in 16-liter glass carboys on the ferrous sulfate-9K medium (9,000 ppm of Fe^{+++}, pH 3.3) under forced aeration and was harvested after 48 to 54 hr by use of a Sharples centrifuge.

Growth on elemental sulfur. The cells were grown in 2-liter Fernbach flasks containing 500 ml of the 9K salts solution (pH 3.3), 1.0 ppm of $FeSO_4$, and 5 g of precipitated sulfur. These flasks were autoclaved for 5 min at 121°C prior to inoculation and were cooled rapidly to prevent sulfur from coalescing. Flasks were agitated on a reciprocating shaker for 5 to 7 days at 28°C and were harvested, after the pH had dropped below 2.0, with a Sorvall RC-2 refrigerated centrifuge. The sulfur in the flasks was not depleted during this time.

Manometric studies were made on cells harvested from the two media.
M. P. Silverman and D. G. Lundgren (1959). J. Bact. 77, 642–647.

Sulfur—*Thiobacillus thiooxidans*

E. S. Kempner (1966). J. Bact. 92, 1842–1843.

Cultures of *T. thiooxidans* (ATCC 8085) were grown in Starkey's medium (R. L. Starkey, J. Bact. 10, 135, 1925) using 10 g of elemental sulfur per liter of medium. Cultures in cotton-plugged Erlenmeyer flasks were incubated at 26°C in a shaking water bath. Hydrogen ion activity was measured with an Expandomatic pH meter, (Beckman Instruments, Inc., Fullerton, Calif.), standardized at pH 2.0 with 0.05 M KCl-HCl buffer.

Sulfur—*Streptomyces*

K. T. Wieringa (1966). Antonie van Leeuwenhoek 32, 183–186.

Two solid media containing finely dispersed elemental sulphur and permitting autotrophic growth of sulphur-oxidizing organisms are described. On these media, colonies of sulphur-oxidizing cells are surrounded by a clear zone. A facultative S-oxidizing *Streptomyces* was isolated with the new technique.

Media with sulphur precipitated in situ from polysulphide. The following basal medium was used: K_2HPO_4, 500 mg; $(NH_4)_2SO_4$, 500 mg; $MgSO_4 \cdot 7H_2O$, 250 mg; $CaCl_2$, 100 mg; Na_2CO_3, 100 mg; Na silicate, 0.1 ml; Fe EDTA, 5 mg; Gaffron's microelement solution (Hughes, Gorham and Zehnder, 1958). 0.8 ml; dialysed agar (Oxoid No. 3), 15 g; glass-distilled water, 1 liter. The agar was dialysed for two days against distilled water. About 25 – 30 ml of molten, sterile basal medium at a temperature of *c.* 60°C were poured into a 10 cm Petri dish containing 1 – 1.5 ml of 0.1 N HCl solution. After mixing, the medium was allowed to solidify. A thin layer of 10 ml of the basal medium, supplemented with polysulphide, was poured on top of the first layer. The polysulphide solution was prepared by saturating a saturated solution of Na_2S in water with elemental sulphur, followed by sterilzation. Two ml of this solution per liter of basal medium sufficed to prepare the medium for the top layer. The HCl from the bottom layer diffuses through the top layer and precipitates the sulphur from the polysulphide as a very fine suspension. Any H_2S that may have been formed is removed by heating the plate on a water bath; at the same time the surface of the agar is dried.

Media with sulphur-covered bentonite. Na bentonite was rendered free from organic matter by ignition, thoroughly mixed with finely ground flowers of sulphur, and heated to 140°C to allow the sulphur to melt. On cooling, the bentonite particles became coated with elemental sulphur. The mixture was ground, suspended in water and sterilized. Ten ml of the basal medium containing *c* 20 mg of the S-coated bentonite mixture was poured on top of a primary layer consisting of solidified basal medium (20 – 25 ml) only. A good opalescent plate was obtained; in order to prevent settling of the S-coated bentonite particles, the top layer was cooled as rapidly as possible.

Solidification with silicic acid instead of agar. As Winogradsky's method for preparing silica gel plates proved to be time-consuming and laborious, the method of Giambiagi (1965) was followed later with some modifications. Silicic acid was prepared by running 50 ml of a sodium silicate solution with a density of 1.05 drop by drop through a 17 cm column of Amberlite resin IR-120(H) with a diameter of 19 mm. The retention water, 20 – 22 ml, was collected separately and, in order to wash out all the silicate solution from the column, was added again at the top. In this way 50 ml of a silicic acid solution was collected. The pH of the percolated solution was *c.* 3.5, a pH of *c* 4 being measured when first the retention water came through. In view of its low pH, the resulting silicic acid solution could easily be sterilized by boiling for a few minutes. For preparing plates, a tenfold concentrated basal solution containing polysulphide or S-coated bentonite was mixed with 9 parts by volume of silicic acid solution: then, if necessary, the pH was adjusted with a few drops of sterile Na_2CO_3 solution or with 0.1 N HCl. The plates solidified within 30 minutes.

An important difference between Winogradsky's and Giambiagi's methods is that in the first, by the dialyzing process, all soluble sulphur compounds are eliminated which is not the case with Giambiagi's method. The results, however, were identical.

Enrichment cultures were made in Erlenmeyer flasks containing the following liquid medium: K_2HPO_4, 1 g; $MgSO_4 \cdot 7H_2O$, 0.5 g; NH_4NO_3, 1 g; some flowers of sulphur; excess $CaCO_3$; tapwater, 1 liter. The flasks were inoculated with garden soil or sludge from a sewage treatment plant and incubated at 25°C.
N. Giambiagi, (1965). Bull. int. Soc. Soil Sci. N. S. No. 3, 24–28.
E. O. Hughes, P. R. Gorham and A. Zehnder (1958). Can. J. Microbiol. 4, 225–236.

Sulfur, thiosulfate, tetrathionate, trithionate and metallic sulfide minerals—*Thiobacillus ferrooxidans*

J. Landesman, D. W. Duncan and C. C. Walden (1966). Can. J. Microbiol. 12, 957–964.
Manometric methods are described.

Tetrathionate—*Providencia, Rettgerella*

C. Richard (1966). Annls Inst. Pasteur, Paris 110, 105–114.
Pour des raisons de commodité, les milieux ont été incubés à 37°C. Recherche de la tétrathionate réductase (TTR): selon la technique simplifiée de Le Minor et Pichinoty [18].
[18] Le Minor (L.) et Pichinoty (F.). *Ann. Inst. Pasteur,* 1963 104, 384.

Thiosulfate—*Thiobacillus*

R. L. Starkey (1934). J. Bact. 28, 365–386.
The medium commonly used in these studies was prepared of the following materials and distributed in 100 ml portions in 250 ml Erlenmeyer flasks:
Tap water, 1,000 g; $MgSO_4 \cdot 7H_2O$, 0.1 g; $CaCl_2 \cdot 2H_2O$, 0.1 g; $MnSO_4 \cdot 2H_2O$, 0.02 g; $FeCl_3 \cdot 6H_2O$, 0.02 g; $(NH_4)_2SO_4$, 0.1 g; $Na_2S_2O_3 \cdot 5H_2O$, 10.0 g; K_3PO_4, K_2HPO_4, or KH_2PO_4 or mixtures of these as stated for each experiment, 2.0 g.
The ammonium sulfate and sodium thiosulfate were sterilized separately. In this experiment, mixtures of phosphates were used to create a range of reactions from pH 9.0 to 4.8. Media at each reaction were inoculated with pure cultures of the five bacteria. Periodically the media were examined for extent of decomposition of thiosulfate and change in reaction. Residual thiosulfate was determined by titration with 0.01 N iodine solution. Reaction was measured colorimetrically.

Thiosulfate—*Thiobacillus*

V. B. D. Skerman, *A Guide to the Identification of the Genera of Bacteria.* 2nd ed. The Williams and Wilkins Book Co., Baltimore, 1967.
Autotrophic Medium for Nonaciduric Species of Thiobacillus
$(NH_4)_2SO_4$, 0.1 g; K_2HPO_4, 4.0 g; KH_2PO_4, 4.0 g; $MgSO_4 \cdot 7H_2O$,

0.1 g; $CaCl_2$, 0.1 g; $FeCl_3 \cdot 6H_2O$, 0.02 g; $MnSO_4 \cdot 4H_2O$, 0.02 g; $Na_2S_2O_3 \cdot 5H_2O$, 10 g; distilled water, 1000 ml.

Adjust the pH to 6.6, and steam for 1 hour on 3 successive days.

C. D. Parker and J. Frisk (1953). J. gen. Microbiol. 8, 344.

Thiosulfate—*Thiobacillus*

M. Hutchinson, K. I. Johnstone and D. White (1965). J. gen. Microbiol. 41, 357–366.

Oxidation of thiosulfate was tested in basal medium containing (g/L distilled water): Na_2HPO_4, 1.2; KH_2PO_4, 1.8; $MgSO_4$, 0.1; $(NH_4)_2SO_4$, 0.1; $CaCl_2$, 0.03; $FeCl_3 \cdot 6H_2O$, 0.02; $MnSO_4 \cdot 4H_2O$, 0.02. The concentration of thiosulfate was varied at 0.5, 1.0, 2.0, 4.0 and 6.0%. Cultures were incubated for 28 days and compared on the basis of titration of residual thiosulfate with 0.1 M iodine.

OXIDASE AND CYTOCHROME OXIDASE

Authors Note: The 'oxidase' and 'cytochrome oxidase' tests are, as clearly indicated by Véron *(vide infra; p. 693)* essentially identical reactions. The addition of α-napthol to the 'oxidase' test – if done rapidly – will convert it to a 'cytochrome oxidase' test.

Oxidase—*Neisseria gonorrhoeae*

P. J. Almaden (1938). The mucoid phase in dissociation of the Gonococcus. J. infect. Dis. 62, 36–39.

The base of the medium was Difco veal infusion prepared after the manner of Hitchens*. To each liter of infusion, there was added 2 g of potassium nitrate, 2 g of dextrose, and 300 ml of hydrolyzed egg albumin. The quantities of reagents employed for hydrolysis of the eggwhite were those suggested by Price**. However, instead of following his more involved method to hasten cleavage, the mixture was heated on a boiling water bath until the hydrolysis appeared to be complete. After hydrolysis, the albumin was adjusted to pH 7.4 and then autoclaved at 250°F for at least 20 minutes. The albumin was stored at room temperature until it was needed. Small quantities of the egg albumin veal infusion agar were prepared at one time since it was imperative to use fresh medium to obtain the best results. The necessary amount of egg albumin and 1.5% agar were added to the infusion and, after being heated to melt the agar, the medium was dispensed into test tubes measuring 1 × 8 inches and autoclaved. It was found that the use of large test tubes assured sufficient moisture being present on the surface of the slant – a factor of importance. This medium was satisfactory for use in Petri plates provided that the plates were inoculated on the same day that they were poured.

Oxidase test – A 1% aqueous solution of tetramethylparaphenylene-diamine hydrochloride was dropped on the surface of colonies and the production of color noted.

*J. infect. Dis. 29, 390, 1921.
**J. Path. Bact. 40, 345, 1935.

Oxidase—*Streptobacillus moniliformis*

F. R. Heilman (1941). A study of *Asterococcus muris (Streptobacillus moniliformis).* II. Cultivation and biochemical activities. J. infect. Dis. 69, 45–51.

Grown on the surface of serum veal infusion agar, the result of the oxidase test with 1% tetramethyl-paraphenylenediamine hydrochloride was negative.

Oxidase—various

S. E. Wedberg and L. F. Rettger (1941). J. Bact. 41, 725–743.

Indophenol oxidase

Three tubes were required for each strain and for each selected temperature. Into each tube were introduced 0.5 ml of di-potassium hydrogen phosphate-sodium hydroxide buffer (pH 7.3) and 0.5 ml bacterial suspension. Tube 1 was not heated; tube 2 was placed in flowing steam for 15 minutes; and tube 3 received 0.1 ml of a 1.5 per cent solution of potassium cyanide. When tube 2 was cool, 0.5 ml of a 1.0 per cent solution of paraphenylene-diamine-hydrochloride buffered at pH 7.3 and prepared immediately before use was added to each tube. Following thorough mixing, the tubes were allowed to remain undisturbed at room temperature in the dark. Observations were made over a period of 150 minutes, at the conclusion of which time tubes showing an intense brownish-purple color were recorded as 4 +. The other tubes were arbitrarily rated according to their degree of coloration. All of the tubes treated with potassium cyanide were negative, and the few steamed tubes which showed any coloration at all were never stronger than 2+ in their reaction.

Oxidase—anaerobic Gram-negative diplococci

G. C. Langford, Jr., J. E. Faber, Jr., and M. J. Pelczar, Jr. (1950). J. Bact. 59, 349–356.

The presence of oxidase was determined from colonies on trypticase soy agar plates flooded with a 1 per cent solution of *p*-amino-dimethyl-aniline monohydrochloride (Eastman Kodak Company) and observed for 15 minutes.

Oxidase

N. Kovacs (1956). Nature, Lond. 178, 703.

A piece of Whatman No. 1 filter paper about 6 cm square, is laid in a Petri dish. Two to three drops of 1% tetra-methyl-para-phenylenediamine dihydrochloride solution is dropped on the centre of the paper. The suspect colony is removed with a platinum rod and is smeared thoroughly on to the reagent-impregnated filter paper in a line 3-6 mm long. If the reaction is positive, the transferred colony turns dark purple in 5-10 sec. The reagent should be prepared every two weeks and kept in a dark, glass-stoppered dropping bottle in the refrigerator.

Oxidase—*Pseudomonas aeruginosa*

W. L. Gaby and C. Hadley (1957). J. Bact. 74, 356–358.

Cytochrome oxidase = Atmungsferment = indophenol oxidase = nadi oxidase = cytochrome c oxidase = cytochrome a_3 oxidase = cytochrome a oxidase.

It oxidises dimethyl-*p*-phenylene diamine in the presence of molecular oxygen and cytochrome c and upon the addition of α-naphthol, indophenol blue is formed.

Reagents used in the test were 1 per cent α-naphthol in 95 per cent ethanol and 1 per cent aqueous solution of *p*-aminodimethylaniline oxalate (Difco) both prepared fresh weekly.

0.2 ml of the α-naphthol and 0.3 ml of the *p*-aminodimethylaniline oxalate were added to each broth tube and the tubes shaken vigorously. A blue color indicates the presence of cytochrome oxidase.

For colonies on agar plates, several drops of a mixture of equal amounts of the two reagents are allowed to flow over isolated colonies.

Oxidase—*Pseudomonas*

W. L. Gaby and E. Free (1958). J. Bact. 76, 442–444.

To a 12 to 18 hr nutrient broth (2 to 5 ml) culture of the test organism 0.3 ml of a 1 per cent aqueous solution of *p*-aminodimethylaniline oxalate (Difco) and 0.2 ml of a 1 per cent ethanol solution of α-naphthol were added and the tube shaken vigorously to ensure thorough oxygenation of the culture. The appearance of a blue color is indicative of the presence of cytochrome oxidase. An immediate blue color indicates *Pseudomonas* while a slowly developing blue color is thought to indicate *Alcaligenes faecalis.*

W. L. Gaby and E. Free (1953). J. Bact. 65, 746.
W. L. Gaby and C. Hadley (1957). J. Bact. 74, 356–358.

Oxidase—*Pseudomonas*

A. Lutz, A. Schaeffer and Mlle. M. J. Hofferer (1958). Annls Inst. Pasteur, Paris 95, 49–61.

Nous résumons dans le tableau III les caractères des 100 souches qui ont fait lobjet de cette étude. La cytochrome-oxydase a été recherchée d'après la technique utilisée en 1957 par Gaby et Hadley [20] pour identifier *Pseudomonas aeruginosa,* à savoir: à 4.5 ml de bouillon de culture on ajoute 0.2 ml d'une solution d'α-naphtol à 1 p. 100 dans l'alcool éthylique à 95° et 0.3 ml de solution fraîche d'oxalate de *p*-aminodiméthylaniline à 1 p. 100 dans l'eau (Difco); les tubes sont agités vigoureusement aux fins de mélange et d'oxygénation de la culture; il apparait en quinze à trente secondes une coloration bleue.

[20] Gaby (W. L.) et Hadley (C.). *J. Bact.,* 1957, 74, 356.

Oxidase—*Pseudomonas*

W. L. Gaby and E. Free (1958). J. Bact. 76, 442–444.

A 6 cm square piece of Whatman's No. 1 filter paper was placed in a petri dish and 2 or 3 drops of a 1 per cent aqueous solution of tetramethylparaphenylenediamine dihydrochloride placed on the center of the paper. The test colony was removed with a platinum loop or rod and streaked onto the reagent-impregnated paper. The smeared colony turns dark purple in from 5 to 10 seconds if the reaction is positive (i.e. *Pseudomonas*) and it is assumed to be a member of the genus *Pseudomonas.*

N. Kovacs (1956). Nature, Lond. 178, 703.

Oxidase

V. B. D. Skerman, *A Guide to the Identification of the Genera of Bacteria.* 2nd ed. Williams and Wilkins Book Co., Baltimore, 1967.

Pour a freshly prepared 1 per cent aqueous solution of di- or tetramethyl-*p*-phenylenediamine hydrochloride over the culture after 24 hours of incubation. If oxidase is produced the colonies turn bluish purple.

J. Gordon and J. W. McLeod (1928). J. Path. Bact. 31, 185.

Oxidase—*Fusobacterium*

A. C. Baird-Parker (1960). J. gen. Microbiol. 22, 458–469.

The possession of an oxidase enzyme was tested by pouring a 1.0% (w/v) solution of tetramethyl-*p*-phenylenediamine over the colonies on an agar plate and examining for the development of a purple color within the colonies (Mackie and McCartney, 1953).

T. J. Mackie and J. E. McCartney (1953). *Handbook of Practical Bacteriology* 9th Ed. London. E. S. Livingstone Ltd.

Oxidase—catalase-producing strains of *Streptococcus*

C. W. Langston, J. Gutierrez, and C. Bouma (1960). J. Bact. 80, 693–695.

Oxidase activity was performed according to the method of Gordon and McLeod (1928).

J. Gordon and J. W. McLeod (1928). J. Path. Bact. 31, 185–190.

Oxidase—*Vibrio*

S. Szturm-Rubinsten, D. Piéchaud and M. Piéchaud (1960). Annls Inst. Pasteur, Paris 99, 309–314.

Test d'oxydase. – Nous l'avons recherché par la technique proposée par Kovacs [2] pour les *Pseudomonas*. Réactif: une solution de tétraméthyl-paraphénylène-daimine à 1 p. 100. Test facilement exécuté avec une anse de culture étalée en couche mince sur papier filtre. On ajoute une goutte de réactif préparé extemporanément. Une coloration rouge violacée dans les cas de réaction positive apparaît presque aussitôt en s'intensifiant après quelques secondes. La réaction était positive pour les souches étudiées comme pour les souches tests de vibrions que nous avons utilisées comme témoins. (Ces souches: *V. cholerae* Ogava, Inaba, souche *El Tor* et un vibrion des eaux proviennent de la collection de l'Institut Pasteur.) Cette réaction positive pour les vibrions permet d'éliminer non seulement les *Shigella* mais toutes les Entérobactéries [3].

Nous avons recherché des oxydases par la même méthode de Kovacs pour un certain nombre d'Entérobactéries, toujours avec un résultat négatif.

[2] Kovacs (N.). *Nature,* 1956, 178, 703.
[3] Gordon (J.) et Mc Leod (J. W.). *J. Path. Bact.,* 1928, 31, 185.

Oxidase—*Aeromonas*

W. H. Ewing, J. G. Johnson (1960). Int. Bull. bact. Nomencl. Taxon. 10, 223–230.

The authors first employed the method employed by Gaby and Hadley (1957), but subsequent tests indicated that nutrient broth tubes in 1.0 ml amounts gave results comparable to those obtained with larger volumes and long nutrient agar slants yielded superior results. Hence, nutrient agar slant cultures, incubated at 37°C, or at a lower temperature if required, for 18 to 20 hours were employed in the present studies. After incubation, two or three drops of each reagent were introduced and the tubes were tilted so that the reagents were mixed and flowed over the growth on the slants. Positive results were indicated by the development of an intense blue color in the growth within 30 seconds. Occasional cultures required up to one minute to give intense reactions. Rare strains required up to two minutes and produced less intense or spotty reactions. Any doubtful or very weak reaction that occurred after two minutes was ignored.

W. L. Gaby and C. Hadley (1957). J. Bact. 74, 356–358.

Oxidase—*Pseudomonas*

M. Véron (1961). Annls Inst. Pasteur, Paris 100, Suppl. 6, 16–42.

La cytochrome-oxydase est une enzyme ferroporphyrinique très largement répandue dans la nature, présente vraisemblablement dans toute cellule où fonctionne le système respiratoire des cytochromes. Dans ce système, la cytochrome-oxydase est la dernière enzyme de la chaîne, capable d'oxyder le cytochrome *c* réduit, en se transformant elle-même en enzyme réduite; la cytochromeoxydase réduite est alors oxydée directement au contact de l'oxygène moléculaire, elle reprend ainsi sa forme active.

Depuis très longtemps, la cytochrome-oxydase était mise en évidence dans les cellules grâce à la réaction de Nadi dont le principe est le suivant: en présence de cytochrome *c*, de cytochrome-oxydase, d'oxygène moléculaire, et dans un milieu tampon phosphate (pH: 7.8), le mélange équimoléculaire d'α-naphtol et de chlorhydrate de diméthyl-paraphénylènediamine donne une coloration bleue intense due à la formation d'un complexe, le bleu d'indophénol. Le plus souvent, le cytochrome *c* est supposé présent en quantité suffisante dans la cellule où l'on recherche la cytochrome-oxydase. Ne donnent une réaction colorée que les bactéries ayant un système cytochrome dont le potentiel de redox est supérieur au potentiel de redox du bleu d'indophénol formé.

Cette technique fut appliquée à la bactériologie par Gaby et Hadley [22], qui utilisent comme réactifs un mélange d'α-naphtol et d'oxalate de para-amino-diméthyl-aniline, ajouté soit dans la culture en bouillon, soit sur les colonies d'une gélose.

Si, dans les mêmes conditions opératoires que ci-dessus, l'on omet l'addition d'α-naphtol, on a alors une réaction un peu différente: il se forme dans ce cas une semi-quinone rouge, instable, qui, après quelques instants, s'oxyde en un dérivé noirâtre. Conventionnellement, l'enzyme bactérienne responsable de cette réaction est désignée d'un terme vague "oxydase", bien que, probablement, il s'agisse aussi de la cytochrome-oxydase; d'ailleurs, si, comme le fait Buttiaux [8], on ajoute de l'α-naphtol après avoir obtenu la coloration rouge, celle-ci vire rapidement à un bleu identique au bleu d'indophénol.

Cette technique de recherche de "l'oxydase" est utilisée en bactériologie depuis le travail de Gordon et Mac Leod [27] qui se servaient de solutions de chlorhydrate de diméthyl ou de tétraméthyl-paraphénylène-diamine versées directement sur les colonies bactériennes; les colonies oxydase-positives devenaient rouges, puis, en une à deux minutes, noires. Kovacs [45] utilisa la tétraméthyl-paraphénylène-diamine, mais avec une technique un peu différente: une goutte du réactif est versée sur un papier-filtre, et sur cette tache humide, la colonie éprouvée, prélevée à l'anse de platine, est ensuite étalée sur une petite surface.

Actuellement, on est donc en présence de plusieurs méthodes. Signalons cependant que la tétraméthyl-paraphénylène-diamine est à proscrire, car outre son prix élevé, c'est un corps beaucoup moins stable que le dérivé diméthylé, et qui est capable de donner de fausses réactions positives, comme l'a signalé Klinge [41].

On utilise donc actuellement un seul réactif: la diméthyl-paraphénylène-diamine (synonyme: para-amino-diméthyl-aniline), sous forme de chlorhydrate, ou mieux d'oxalate, en solutions aqueuses à 1 p. 100. Ces solutions se conservent quelques jours à la glacière dans le cas de l'oxalate, mais doivent se préparer extemporanément dans le cas du chlorhydrate. Pour faire une réaction de Nadi, on peut ajouter à l'une des solutions précédentes le même volume d'une solution à 1 p. 100 d'α-naphtol dans l'alcool à 95 p. 100. Cette solution se conserve aussi quelques jours à la glacière.

Il y a trois méthodes possibles de recherche: en milieu liquide (technique de Gaby et Hadley: 4.5 ml de bouillon + 0.2 ml d'α-naphtol + 0.3 ml d'oxalate de para-amino-diméthyl-aniline), sur gélose (addition sur les colonies de quelques gouttes de réactifs), ou sur papier, selon la technique de Kovacs.

Personnellement, nous utilisons indifféremment les deux sels du réactif, avec ou sans addition d'α-naphtol. Plutôt que la méthode en milieu liquide, dont les résultats sont parfois difficiles à interpréter avec certains germes donnant une réaction tardive, ou à cause de fausses réactions dues au bouillon, nous utilisons les cultures sur gélose, en recherchant la cytochrome-oxydase directement sur les colonies, ou par l'intermédiaire du papier-filtre.

Ces techniques sont équivalentes dans le cas de *Ps. aeruginosa* et de *Ps. fluorescens.* Les résultats sont absolument constants: toutes nos souches examinées (52 *Ps. aeruginosa* et 16 *Ps. fluorescens*) donnent une réaction de cytochrome-oxydase positive, la coloration rouge ou bleue apparaissant en moins d'une minute (et en deux à trois minutes en milieu liquide).

[8] Buttiaux (R.). *Ann. Inst. Pasteur,* 1961, Suppl. au numéro de juin, p. 43.
[22] Gaby (W. L.) et Hadley (C.). *J. Bact.,* 1957, 74, 356.
[27] Gordon (J.) et Mc Leod (J. W.). *J. Path. Bact.,* 1928, 31, 185.
[45] Kovacs (N.). *Nature,* 1956, 178, 703.

Oxidase—*Pseudomonas*

R. Buttiaux (1961). Annls Inst. Pasteur, Paris 100, Suppl. 6, 43–58.

Oxydase et cytochrome-oxydase. – La connaissance de ces systèmes a transformé l'étude des bacilles à Gram négatif. Elle met à notre disposition un test nouveau d'une grande utilité pour leur définition.

Différentes techniques peuvent être utilisées. Nous rappellerons, pour l'oxydase, celle de Gordon et McLeod [16] utilisant le chlorhydrate de para-diméthyl-phénylène-diamine qu'on peut avantageusement remplacer par l'oxalate beaucoup plus stable. Nous préférons la méthode de Kovacs [17] : au centre d'un petit fragment de papier-filtre (Whatman 1) placé sur une lame de verre on dépose deux à trois gouttes d'une solution aqueuse à 1 p.100 de tétra-méthyl-para-phénylène-diamine-hydro-chlore; on prélève un peu de la souche à étudier (colonie sur gélose) au moyen d'un fil de platine et on l'étale en strie sur le papier imbibé du réactif. Une coloration violet brunâtre survenant immédiatement prouve l'existence d'une oxydase. Il ne faut pas tenir compte des réactions plus tardives.

Gaby et Hadley [18] recherchent la cytochrome *c* oxydase de la facon suivante: à 5 ml d'une culture en bouillon de douze à dix-huit heures du germe à étudier, on ajoute 0.3 ml d'une solution aqueuse à 1 p. 100 d'oxalate de para-amino-diméthyl-aniline, puis 0.2 ml d'une solution alcoolique à 1 p. 100 d'α-naphtol. On agite vigoureusement. L'apparition rapide d'une couleur bleu outremer (bleu d'indophénol) indique l'existence d'une cytochrome oxydase. Ewing et Johnson [19] ont adapté ce procédé aux cultures en milieu solide. Le germe est cultivé sur gélose nutritive; après dix-huit à vingt-quatre heures d'incubation, on verse alors à sa surface deux à trois gouttes de chacun des réactifs suivants:

1° Solution à 1 p. 100 d'α-naphtol dans l'alcool à 95° ;

2° Solution à 1 p. 100 d'oxalate de para-amino-diméthylamine dans l'eau distillée.

On agite doucement pour qu'elles rocouvrent les colonies. Une réaction positive se traduit par une couleur bleu intense qui doit survenir en deux minutes, au plus. Cette technique donne d'excellents résultats.

Dans notre service, Mme Viarre a mis au point un procédé simple permettant de déceler successivement l'oxydase puis la cytochrome oxydase: dans un tube de verre de 12 x 80 mm, on place 0.5 ml (10 gouttes) d'une solution aqueuse à 1 p. 100 d'oxalate d'amino-diméthyl-aniline; on ajoute 2.5 ml environ d'une culture de 18-24 heures du germe en eau peptonée ou en bouillon. On agite. Une coloration rose apparaissant en quelques secondes, traduit la présence d'une oxydase. On ajoute alors 0.25 ml (5 gouttes) d'une solution alcoolique d'α-naphtol à 1 p. 100. Une coloration rapide, bleu vif, survient s'il existe une cytochrome oxydase.

[16] Gordon (J.) et McLeod (J. W.). *J. Path. Bact.,* 1928, 31, 185.
[17] Kovacs (N.). *Nature,* 1956, 178, 703.
[18] Gaby (W. L.) et Hadley (C.). *J. Bact.,* 1957, 74, 356.
[19] Ewing (W. H.) et Johnson (J. G.). *Int. Bull. Bact. Nom. Taxon.,* 1960, 10, 223.

Oxidase—*Xanthomonas*

A. C. Hayward and W. Hodgkiss (1961). J. gen. Microbiol. 26, 133–140.

Kovac's method (1956) was used.

N. Kovacs (1956). Nature, Lond. 178, 703.

Oxidase—*Escherichia aurescens*

H. Leclerc (1962). Annls Inst. Pasteur, Paris 102, 726–741.

Elle est recherchée d'après Kovacs [23].

[23] Kovacs (N.). *Nature,* 1956, 178, 703.

Oxidase—*Actinobacillus* and *Haemophilus*

E. O. King and H. W. Tatum (1962). *Actinobacillus actinomycetemcomitans* and *Haemophilus aphrophilus.* J. infect. Dis. 111, 85–94.

The oxidase reaction was determined by testing the growth on blood agar (5% rabbit blood) plates after 24 to 48 hours incubation using a 0.5% solution of tetramethyl *p*-phenylenediamine hydrochloride made up not more than 4 days previously.

Oxidase—*Pseudomonas odorans*

I. Málek, M. Radochová and O. Lysenko (1963). J. gen. Microbiol. 33, 349–355.

According to Kovacs (1956).

N. Kovacs (1956). Nature, Lond. 178, 703.

Oxidase

G. S. Wilson and A. A. Miles (1964). Topley and Wilson's *Principles of Bacteriology and Immunity* 5th edition – Arnold.

The oxidase reaction is believed to be due to a cytochrome oxidase which catalyses the oxidation of reduced cytochrome by molecular oxygen. In the original method introduced by Gordon and McLeod (1928) a 1–1.5 per cent solution of dimethyl-*p*-phenylenediamine is poured over the colonies in a Petri dish culture; oxidase-positive colonies become maroon, dark red, and black in 10-30 minutes. Steel (1961) prefers the method of Kovacs (1956). A small square of filter paper is laid in a Petri dish, and 2-3 drops of 1 per cent tetramethyl-*p*-phenylenediamine dihydrochloride are placed on it. The colony to be tested is smeared on the paper in a line about 5 mm long. In a positive reaction the colony turns dark purple in 5-10 seconds, and in a delayed reaction in 10-60 seconds. The solution should be prepared freshly every two weeks and kept in the dark. This reaction is of considerable taxonomic value. According to Steel (1961) all oxidase-positive strains also form catalase.

J. Gordon and J. W. McLeod (1928). J. Path. Bact. 31, 185.
N. Kovacs (1956). Nature, Lond. 178, 703.
K. J. Steel (1961). J. gen. Microbiol. 25, 297.

Oxidase—*Moraxella*

W. J. Ryan (1964). J. gen. Microbiol. 35, 361–372.

A few drops of a freshly prepared 1% (w/v) solution of tetramethyl-*p*-phenylenediamine hydrochloride were poured over a 24 hr culture on blood agar and the development of a purple color noted.

Oxidase—*Pseudomonas aeruginosa*

R. R. Colwell (1964). J. gen. Microbiol. 37, 181–194.

For cytochrome oxidase and oxidase formation the methods of Gaby and Free (1958) and Kovacs (1956), respectively, were used.

The standard inoculum for all tests was a single drop (1/20 ml) from a sterile disposable pipette (Fisher Scientific Company) of a 24-48 hr broth culture. The stock culture medium (YE) was a modification of that used by Rhodes (1959): Difco yeast extract, 0.3 g; Difco Bacto-Proteose Peptone, 1.0 g; NaCl, 0.5 g; agar, 1.5 g (agar omitted from liquid stock media); distilled water, 1 L; adjusted to pH 7.2 – 7.4 with NaOH. Tests were carried out at room temperature (25°C) except where otherwise indicated.

W. L. Gaby and E. Free (1958). J. Bact. 76, 442.
N. Kovacs (1956). Nature, Lond. 178, 703.
M. E. Rhodes (1959). J. gen. Microbiol. 21, 221.

Oxidase—*Pseudomonas aeruginosa*

K. Morihara (1964). J. Bact. 88, 745–757.

Cytochrome oxidase was determined by the method reported by Gaby and Hadley (1957). After 18 to 24 hr of culture, an α-naphthol solution and *p*-amino dimethyl aniline oxalate solution were added, and the mixture was shaken vigorously for 1 to 3 min. Cultures were incubated at 30°C.

W. L. Gaby and C. Hadley (1957). J. Bact. 74, 356-358.

Oxidase—*Veillonella*

M. Rogosa (1964). J. Bact. 87, 162–170.

Tests for oxidase enzyme were conducted by flooding the growth on V15 agar plates with 1 ml of a 1% solution of dimethyl-*p*-phenylenediamine hydrochloride or the oxalate salt, and were observed for a period of 10 min for usual color changes from pink to black.

The composition of the medium was as follows: trypticase (BBL), 5 g; yeast extract (Difco), 3 g; sodium thioglycollate, 750 mg; basic fuchsin, 0.002 g; 'tween 80', 1 g; sodium lactate (50 per cent), 25.0 ml agar, 15 g; distilled H_2O to make 1 L, pH adjusted to 7.5. The basic fuchsin was calculated on the basis of 100 per cent composition. Vancomycin from sterile vials was diluted in H_2O and added aseptically to a concentration of 7.5 μg/ml of medium just before pouring plates. (Rogosa, 1956; Rogosa *et al.* 1958).

M. Rogosa (1956). J. Bact. 72, 533–536.
M. Rogosa, R. J. Fitzgerald, M. E. MacKintosh and A. J. Beaman (1958). J. Bact. 76, 455–456.

Oxidase—*Vibrio marinus*

R. R. Colwell and R. Y. Morita (1964). J. Bact. 88, 831–837.

The authors used the methods of N. Kovacs (1956). Nature, Lond. 178, 703; W. H. Ewing and J. G. Johnson (1960). Int. Bull. bact. Nomencl. Taxon. 10, 223–230; W. L. Gaby and E. Free (1958). J. Bact. 76, 442–444 – modified by the addition to the media of the following salts to produce a synthetic seawater:– sodium chloride, 2.4%; potassium chloride, 0.07%; magnesium chloride (hydrated) 0.53% and magnesium sulfate (hydrated) 0.7%.

A standard inoculum was 1 drop from a pasteur pipette (*c.* 0.05 ml) of a 24 hour artificial seawater broth.

Cultures were incubated at 18°C.

Oxidase—*Pseudomonas*

G. L. Bullock, S. F. Snieszko and C. E. Dunbar (1965). J. gen. Microbiol. 38, 1–7.

The presence of cytochrome oxidase was examined by the method of Ewing and Johnson (1960).

W. H. Ewing and J. G. Johnson (1960). Int. Bull. bact. Nomencl. Taxon. 10, 223.

Oxidase—*Pseudomonas aeruginosa*

A. H. Wahba and J. H. Darrell (1965). J. gen. Microbiol. 38, 329–342.

Rogers' (1963) modification of Kovacs' (1956) method. Strains not growing at 37°C were incubated at 22°C.

N. Kovacs (1956). Nature, Lond. 178, 703.

K. B. Rogers (1963). Lancet 2, 682.

Oxidase—*Alcaligenes*

R. G. Mitchell and S. K. R. Clarke (1965). J. gen. Microbiol. 40, 343–348.

Kovacs's (1956) method was used.

N. Kovacs (1956). Nature, Lond. 178, 703.

Oxidase—*Bordetella*

M. Piéchaud and Mme. S. Szturm-Rubinsten (1965). Annls Inst. Pasteur, Paris 108, 391–395.

Nous l'avons recherchée par la technique de Kovacs [4]. Toutes les souches de *B. bronchiseptica* donnent une réaction rose nette et presque immédiate, contrairement aux souches de *B. parapertussis.*

[4] Kovacs (N.). *Nature,* 1956, 178, 703.

Oxidase—marine bacteria

R. M. Pfister and P. R. Burkholder (1965). J. Bact. 89, 863–872.

The Kovacs oxidase test was performed according to the method of Kovacs (1956). The cytochrome oxidase test of Gaby and Free (1958) was used as described, except for the addition of the reagents to an agar

slant in preference to broth culture. Catalase activity was qualitatively detected by adding a drop of H_2O_2 to an agar slant culture.
W. L. Gaby and E. Free (1958). J. Bact. 76, 442–444.
N. Kovacs (1956). Nature, Lond. 178, 703.

Oxidase—*Vibrio comma*

J. C. Feeley (1965). J. Bact. 89, 665–670.

This was determined by the method of Ewing and Johnson (1960).
W. H. Ewing and J. G. Johnson (1960). Int. Bull. bact. Nomencl. Taxon. 10, 223–230.

Oxidase—*Mima, Herellea, Flavobacterium*

J. D. Nelson and S. Shelton (1965). Appl. Microbiol. 13, 801–807.

Oxidase production was determined by flooding 24-hr BHI Agar culture plates with a 1% aqueous solution of *p*-aminodimethylaniline. To test for catalase production, 24-hr BHI Agar slants were layered with 3% hydrogen peroxide.

Oxidase—*Bacteroides, Veillonella, Neisseria*

J. van Houte and R. J. Gibbons (1966). Antonie van Leeuwenhoek 32, 212–222.

Oxidase was detected by flooding the plates with freshly prepared N,N′ dimethyl-*p*-phenylene diamine hydrochloride (Difco).

The basal medium was BBL thioglycollate medium, without added dextrose, supplemented with 5% horse serum and hemin (5 μg/ml) menadione (0.5 μg/ml).

Oxidase—*Pseudomonas*

R. Y. Stanier, N. J. Palleroni and M. Doudoroff (1966). J. gen. Microbiol. 43, 159–271.

The oxidase test was performed by soaking a strip of filter paper with 2-3 drops of a 1% (w/v) aqueous solution of N,N′-dimethyl-*p*-phenylenediamine, and immediately smearing a loopful of bacteria from a slope culture on the moist area. Oxidase-positive strains turn dark red within a few seconds. The test can also be performed with 0.2% (w/v) aqueous 2,6-dichlorophenolindophenol, first reduced with dithionite by careful addition of the reducing agent until the blue color just disappears. With this reagent, oxidase-positive strains turn blue.

Oxidase—*Providencia, Rettgerella*

C. Richard (1966). Annls Inst. Pasteur, Paris 110, 105–114.

Pour des raisons de commodité, les milieux ont été incubés à 37°C. Oxydase (méthode de Kovacs) [13].
[13] Kovacs (N.). *Nature,* 1956, 178, 703.

Oxidase—*Brevibacterium*

R. Chatelain and L. Second (1966). Annls Inst. Pasteur, Paris 111, 630–644.

Réaction dite de "l'oxydase": technique de Kovacs [9], modifée en utilisant une solution à 1 p. 100 d'oxalate de N-diméthylparaphénylène-diamine.

La température d'incubation des cultures est de 30°C.

[9] Kovacs (N.). *Nature,* 1956, 178, 703.

PECTIN HYDROLYSIS

Pectin hydrolysis—enterobacteria

J. E. Steinhaus and C. E. Georgi (1941). The effect of pectin, galacturonic acid and alphamethyl galacturonate upon the growth of *Enterobacteriaceae* J. infect. Dis. 69, 1–6.

The pectin broth, medium A, was prepared in the following manner. To sterile nutrient broth, containing 1% glucose and brom thymol blue indicator, pectin was dissolved by vigorous stirring. The resulting medium, which dropped to pH 3.5 upon addition of the pectin, was placed in an Arnold sterilizer for 10 minutes. Sterile K_2HPO_4 was added aseptically to buffer the medium. It was then adjusted to pH 7.0 with sterile NaOH. All hydrogen ion concentrations were determined with a glass electrode. The medium was then tubed aseptically using a modification of the method described by Riker and Riker (1936) and incubated at 37°C for 48 hours to check sterility. During this period, the pH dropped from 7.0 to 6.8; from the time of inoculation to the final observation, a further drop from pH 6.8 to 6.4 occurred.

Riker and Riker (1936). *Introduction to Research on Plant Diseases,* page 33, Swift, St. Louis.

Pectin hydrolysis

D. B. McFadden, R. H. Weaver and M. Scherago (1942). J. Bact. 44, 191–199.

Chemically pure apple pectin was found to hydrolyse readily on both the acid and alkaline side of pH 7.0 when autoclaved. The pectin was therefore sterilized as follows.

5 g of pectin were added aseptically to a sterile 1 litre flask. 20 ml of 75 per cent ethyl alcohol were added aseptically with rapid shaking. This was then allowed to stand for 5–8 days at room temperature to evaporate most of the alcohol. 500 ml of sterile distilled water was then added, the pectin dispersing immediately to give a clear 1 per cent solution.

This one per cent solution could be added to any basal medium aseptically and used for the determination of the ability of organisms to ferment pectin. In this investigation a basal medium containing 4 g of ammonium chloride and 4 g of dibasic potassium phosphate per liter of distilled water, and one containing 6 g of Bacto beef extract and 10 g of peptone per liter of distilled water, were used. It was necessary to prepare these basal media in the above double strengths and to so adjust the pH of each that, after the addition of an equal quantity of the alcohol-sterilized pectin solution, the desired concentration of the ingredients would be obtained and the final pH of each medium would be 7.1. A combination indicator containing brom-cresol-purple and cresol red was used. Two milliliter amounts of the basal medium were placed in Durham fermentation tubes and sterilized, after which 2 ml amounts of sterilized pectin solution were added aseptically. After the addition of the pectin, the tubes of media were placed in the incubator at 37°C for 48 hours and then at room temperature for 4 days, in order to detect contamination.

Since, throughout these studies, neither acid nor gas production was ever observed in the extract pectin broth, except for acid production in check *Erwinia* cultures, the results with this medium will be omitted from the tabulations. As acid was frequently produced in duplicate inoculations in the synthetic medium, it would appear that fermentation was probably masked in the extract broth.

Pectin hydrolysis—enterobacteria

S. C. Werch, R. W. Jung, A. A. Day, T. E. Friedemann and A. C. Ivy (1942). The decomposition of pectin and galacturonic acid by intestinal bacteria. J. infect. Dis. 70, 231–242.

The pectin used was the pure citrus variety employed in our previous work, and was obtained from the research department of the California Fruit Growers Exchange. This product is essentially free of such impurities as pentoses, pentosans, pigment and other materials frequently found in other pectins.

Before much work was done it became evident that a liquid, pectin-containing medium was necessary, and had to conform to the following requirements (a) that the pectin be unchanged (b) that the pH of the medium be readily set and adjusted within the range where pectin is relatively stable for the period of incubation and (c) that it be easily sterilized.

Heat or steam distillation resulted in decomposition of pectin, with acid production, lowered viscosity, and reduction in calcium pectate formation. Chemical sterilization would also alter its properties, besides complicating any chemical analysis, and thus was not tried. Desiccation proved inefficient.

Pectin powder could not be sterilized after 2 weeks of disiccation.

Berkefeld filtration proved satisfactory when certain precautions were observed. These precautions were included in a previous paper*. Lately we have found it better to make a water solution of pectin and to adjust the pH with normal alkali. Contamination is minimized by lengthening the arm of the suction flask, to accommodate a large quantity of cotton filter, and by adding a rubber diaphragm to its neck. This diaphragm is placed between the rubber stopper and mantle, after its centre is pierced by the metal end of the candle. It is then folded over the neck of the suction flask and tied.

A culture medium containing more than 1% of pectin can not be filtered successfully. If a greater percentage of pectin is included, filtration does not occur or is prolonged to the point where contamination can be expected even after every precaution is observed. Before use the medium is best incubated for at least 48 hours, in the flask in which it is to be used, for contaminating organisms often grow very slowly.

In our recent work we have used a medium, which for brevity we have designated "pectone" made up to contain 1% of each of pectin and peptone and 0.5% salt. To make up 2 liters of the medium, 20 g of pectin is dissolved in a liter of distilled water with the aid of an electric stirring machine, and then mixed with a liter of water solution of 20 g of peptone and 10 g of NaCl. This mixture is then adjusted to pH 7.2 with normal alkali and sterilized by Berkefeld filtration.

*Proc. Soc. expl Biol. Med. 46, 569, 1941.

Pectin hydrolysis—*Erwinia*-coliform relationship

R. P. Elrod (1942). J. Bact. 44, 433–440.

Inasmuch as the reducing sugars associated with pectin are soluble in 80 per cent alcohol, several soakings and washings with this solvent eventually removes any of these contaminating substances. The usual procedure was to place a weighed sample of the granulated pectin in a flask to which considerable 80 per cent alcohol was added. The mixture was shaken thoroughly and then incubated 12 hours with occasional shaking. The alcohol was filtered off, fresh alcohol added and the flask again incubated 6-8 hours. The mixture was filtered and the pectin on the filter paper washed 6-7 times with 80 per cent alcohol and twice with 95 per cent. The alcohol-moist pectin was then placed in a sterile petri dish and heated to dryness at 37°C. A 5 ml sample of the last 80 per cent alcohol filtrate was evaporated to dryness and a qualitative Benedict's test performed with the residuum; this invariably proved negative for reducing sugars. This procedure shortens the time of McFadden's method (1941) five or six days.

A basic synthetic medium was made up of 0.2 g magnesium sulfate, 0.1 g calcium chloride, 0.2 g sodium chloride and 0.2 g dipotassium phosphate per liter. This was sterilized by filtration through a Berkefeld N filter. The pectin was added so that the final concentration was approximately 0.5 per cent. Brom-cresol-purple was used as an indicator. On the addition of pectin, the medium became acid and it was necessary to adjust to neutral with sterile NaOH. After tubing aseptically, the tubes were incubated 48 hours at 37°C and 3-4 days at 20°C. The percentage of contaminated tubes was about 5 per cent; fairly effective sterilization of the pectin had taken place during the purifying process.

Four tubes of this medium were inoculated with each *Erwinia* and coliform culture, two incubated at 37°C and two at 20°C for one week.

Pathogenic action on carrot and turnip

As a means of testing the pathogenicity of the organisms, both carrots and turnips have been used. Jones' (1905) method was used to test the pathogenicity of the organisms against carrot and turnip. The vegetable tissue was sectioned aseptically and one piece (0.5 cm x 0.5 cm x 1.0 cm) added to a tube of nutrient broth and incubated several days to control for contamination. Tubes of both carrot and turnip were inoculated in duplicate; one set was incubated at 37°C and the other at 20°C. The progress of the rot was tested by probing with a stiff nichrome needle. Those showing active rot became soft and soon broke up. Unaffected pieces and controls in uninoculated broth were firm after weeks of incubation. This method proved far more satisfactory than inoculating sterile slices of carrot or turnip. This test is in reality a test of protopectinase production (Davison and Willaman, 1927).

F. R. Davison and J. J. Willaman (1927). Bot. Gaz. 83, 329–361.
L. R. Jones (1905). Zentbl. Bakt. ParasitKde II, 14, 910–940.
D. B. McFadden (1941). J. Bact. 42, 289–290.

Pectin hydrolysis—*Erwinia*

E. D. Garber, S. G. Shaeffer and M. Goldman (1956). J. gen. Microbiol. 14, 261–267.

Mature plants of nine varieties of radish and three varieties of turnip were harvested and the fleshy storage organs were washed with tap water, immersed in a 20% (v/v) solution of "Chlorox" for 3-5 min, and rinsed with sterile, distilled water. The fleshy organs were then sliced with a sterile knife, the slices being placed in a sterile Petri dish containing a saturated layer of filter-paper. The slices were inoculated by dropping the bacterial suspension on the upper surface and then incubated at 27°C for 24 hr. A slice from at least three different plants of each variety was used in routine tests of virulence and, in some experiments, as many as nine plants were used. After incubation, slices which had been attacked displayed a discolored, slimy, glistening surface which was easily penetrated by a blunt, glass rod; slices which had not been attacked remained white and firm. Suitable controls were routinely used.

Pectin hydrolysis—*Pectobacterium carotovorum*

W. J. Dowson (1957). Nature, Lond. 179, 682.

Wieringa'a Pectate Gel (Dowson's Modification) for Detection of Soft Rot Organisms*

Preparation of the Medium

1. Calcium Agar

Soil extract	500 ml
NaCl	0.5 g
$CaCl_2$	2.5 g
Yeast extract	0.5 g
Agar	10.0 g

Adjust the pH to 9.0 and autoclave at 15 lb for 15 minutes. Pour enough to cover the bottom of petri dishes and place in a desiccator to dry thoroughly (for about 2 days).

2. The 'Pectin' Solution.

Prepare a 2 per cent sodium polypectate solution by adding the powder slowly to hot distilled water, with continuous stirring. Add 4 per cent bromothymol blue, and adjust the pH to 8.0 by adding N sodium hydroxide drop by drop. Autoclave at 10 lb for 2 minutes. When cold pour just enough to cover the surface of the agar plates and leave to set and dry in a desiccator (about 8 days). The plates are ready for streaking if no liquid exudes from between the layers of pectate gel and agar when the plates are slightly tipped up. They keep well under a bell jar at laboratory temperature.

Note: Dowson (personal communication, September 1958) recommends the use of citrus sodium pectate (Product No. 24K12) obtainable from Sunkist Inc., of Ontario, California, U.S.A. or via London from S. and S. Services Ltd., 72 Victoria Street, London S.W.1.

This product will differentiate between the pectolytic pseudomonads of Paton (Nature, Lond. 181, 1958, No. 4601) and the pectolytic species of *Erwinia.* British pectate made from apples does not.

*K. T. Wieringa (1949). Rep. Proc. Fourth Internat. Cong. Microbiology, Copenhagen, 1947, p. 482.

Pectin hydrolysis—*Pseudomonas* and anaerobes

S. M. Betrabet and J. V. Bhat (1958). Appl. Microbiol. 6, 89–93.

Citrus pectin	1 g
Na_2HPO_4	0.08 g
KH_2PO_4	0.02 g
NaCl	0.005 g
$MgSO_4 \cdot 7H_2O$	0.05 g
$FeSO_4 \cdot 7H_2O$	0.001 g
$CaSO_4 \cdot 2H_2O$ (sat. sol.)	5 ml

$(NH_4)_2SO_4$	0.05 g
Micronutrient solution*	1.0 ml
H_2O	100.0 ml
pH	7.0

*Per 100 ml: $ZnSO_4 \cdot 7H_2O$, 1.1 g; $MnSO_4 \cdot H_2O$, 0.5 g; $CoSO_4$, 0.005 g; H_3BO_3, 0.005 g; Na_2MoO_4, 0.2 g; $CuSO_4 \cdot 5H_2O$, 0.0007 g.

Pectin hydrolysis—*Flavobacterium*

M. J. Dorey (1959). J. gen. Microbiol. 20, 91–104.

Used pectate gel medium of Richards and Fouad (1954) except that 2% (w/v) pectate was used in plates for streaking.

T. Richards and M. T. A. Fouad (1954). J. appl. Bact. 17, Proc. 19.

Pectin hydrolysis—*Pseudomonas*

M. E. Rhodes (1959). J. gen. Microbiol. 21, 221–263.

The method recommended by Dowson (1949, *Manual of Bacterial Plant Diseases,* 1st ed. London: A. & C. Black), was used. Stab inoculations were made into a gel of calcium pectate, prepared by layering a solution of sodium pectate (A.S.P. Chemical Co. Ltd.) over calcium chloride agar. Cultures were examined for growth and liquefaction.

Pectin hydrolysis

V. B. D. Skerman, *A Guide to the Identification of the Genera of Bacteria.* 2nd ed. Williams and Wilkins Book Co., Baltimore, 1967.

Carrot or potato plug medium for detection of soft rot organisms

Select young clean carrots (or new potatoes) preferably more than 1.5 inches at the widest part. Scrub the surface with a soft brush and then immerse in 0.2 per cent $HgCl_2$ for 2 minutes. Handling the carrot with aseptic precautions, rinse in several changes of sterile tap water in a sterile container. Cut across the center with a sterilized knife. Using a sterilized 6 mm cork borer fitted with a plunger, extract several cores of carrot by plunging the borer through the cut surface as far as possible without breaking through the outside surface. Discharge the plugs into a sterile petri dish. With a sterile knife, remove a few millimeters of the end of the core which formed the face of the original cut. Cut the plugs into 4-5 cm lengths and transfer them aseptically to tubes of sterile peptone water. Incubate for 3 days to check sterility and store.

When testing an organism for its ability to decompose the carrot, incubate a control tube with the test, and at intervals check the consistency of the carrot plug in the inoculated tube with a stiff mounted needle and compare with the control.

Pectin hydrolysis

W. H. Ewing (1962). *Enterobacteriaceae. Biochemical Methods for Group Differentiation.* U.S. Department of Health, Education and

Welfare, Public Health Service Publication, No. 734 (revised).

Certain bacteria presently classified as *Erwinia* and sometimes referred to as *Pectobacterium* occasionally may be seen and their differentiation from *Enterobacteriaceae* may be somewhat difficult. However, these bacteria produce pectinases and *Enterobacteriaceae* are not known to possess them. Hence, a medium containing pectin is of value in the differentiation of these bacteria. A suitable pectate medium (Starr, 1947) may be prepared as follows:

Place 100 ml distilled water in a beaker

Add 0.5 g yeast extract
0.9 ml N sodium hydroxide
0.6 ml 10% calcium chloride solution
($CaCl_2 \cdot 2H_2O$)
1.25 ml Bromthymol blue (0.2% solution)

Add 3 g sodium polypectate, #24, California Fruit Growers Exchange. (Add slowly and stir between additions). Stir to wet pectate.

Heat in a boiling-water bath to dissolve pectate (constant stirring). The color should now be blue-green (about pH 7.3). Further adjustment of pH cannot be made. Distribute in 3 to 4 ml amounts in small test tubes and sterilize at 121°C for 15 minutes. The color should now be yellowish-green (about pH 6.4). Allow tubes to cool in an upright position. This medium cannot be reheated.

Inoculation: Inoculate from a young agar slant culture by stabbing deep into the column of medium.

Incubation: 37°C for 7 days. Observe for evidence of liquefaction as with nutrient gelatin medium. (Changes in the color of the medium may occur during growth of *Enterobacteriaceae.* This apparently is of no importance so far as *Enterobacteriaceae* are concerned.)

M. P. Starr (1947). Phytopathology 37, 291.

Pectin hydrolysis—*Lactobacillus*

R. E. Smith and J. D. Cunningham (1962). Can. J. Microbiol. 8, 727–735.

Vaughn's medium, for detecting the hydrolysis of polypectate, was modified by substitution of 0.3% (w/v) malt extract for the beef extract, and by the addition of an equal quantity of yeast extract. 0.2% (w/v) sodium acetate, 0.1% (v/v) Tween 80 and 0.5% (v/v) of each of Rogosa's salt solutions A and B* were also added, using the concentrations outlined above. After 2 weeks' incubation streaked plates were checked for liquefaction.

*J. Naylor and M. E. Sharpe (1958). J. Dairy Res. 25, 92–103.

R. R. Vaughn *et al.* (1957). Fd Res. 22, 597–603.

Pectin hydrolysis—*Escherichia aurescens*

H. Leclerc (1962). Annls Inst. Pasteur, Paris 102, 726–741.

Seule la recherche de la polygalacturonidase est effectuée au cours de cette étude. Le milieu est celui proposé par Ewing [9] que nous avons modifié comme suit:

Extrait de levure	0.5 g
Chlorure de calcium 2 H_2O à 10 p. 100	0.6 ml
Polypectate de sodium*	1 g
Bleu de bromothymol à 0.2 p. 100	1.25 ml
Eau distillée	100 ml

L'extrait de levure est dissous dans l'eau. Ajouter le chlorure de calcium puis, lentement, le polypectate de sodium en agitant entre chaque addition. Ajuster à pH 7.3 avec de la soude normale.

Chauffer au bain-marie à 100° pour dissoudre, en agitant constamment. Distribuer 6 à 7 ml en tubes de 16 × 160 mm. Stériliser à 120°, quinze minutes.

On inocule en piqûre centrale à partir d'une culture de 24 heures. On incube quatre jours à 30° puis dix jours à 20-22°.

*Sunkists Growers, Inc Products Department, 616 East Grove St, Ontario, Calif., U.S.A.

[9] W. H. Ewing – *Enterobacteriaceae, biochemical methods for group differentiation.* Public Health Service Publication, No. 734, U.S. Government Printing Office, 1960.

Pectin hydrolysis—*Clostridium rubrum*

H. Ng and R. H. Vaughn (1963). J. Bact. 85, 1104–1113.

All media were adjusted to pH 7.0 with 1 N NaOH and sterilized by autoclaving at 120°C for 15 min. Solid media were prepared by addition of 2% agar to liquid media. These general procedures were adhered to, except where noted.

Sodium polypectate medium was prepared by blending to a creamy mixture in a Waring Blendor in 500 ml of cold distilled water, followed by the addition of 500 ml of boiling water: Tryptone, 5 g; proteose peptone, 5 g; soluble starch, 1 g; sodium thioglycolate, 0.5 g; Calgon (food grade), 2.5 g; sorbic acid, 1 g; and sodium polypectate, 60 g. The medium, which had a pH of about 6.5, was then sterilized in flasks filled to about one-fourth of capacity.

The sodium polypectate gel described above was used as the enrichment medium for the isolation of pectinolytic species of *Clostridium.* This medium contained soluble starch which served to overcome dormancy of spores (Wynne and Foster, 1948) and sorbic acid to inhibit catalase-positive organisms (York and Vaughn, 1954). To eliminate the nonsporeformers, the soil samples (1 g), in duplicate, were suspended in 1 ml of water in test tubes (25 by 200 mm), pasteurized in a water bath at 85°C for 5 min, and cooled. After addition of 0.6 ml of filter-sterilized, 10% $NaHCO_3$, about 30 ml of the polypectate medium, cooled to 60 to 70°C, were poured into the tubes containing the pasteurized soil suspensions. The tubes were then sealed with Vasper to inhibit the strict aerobes. When the tubes had cooled to room temperature and the gel had solidified, the cultures were incubated at 35°C. Liquefaction of the polypectate gel constituted a positive enrichment. All positive enrichments were carried through the enrichment procedure three times to assure the predominance of the pectinolytic flora.

E. S. Wynne and J. W. Foster (1948). J. Bact. 55, 61–68.

G. K. York II, and R. H. Vaughn, 1954. J. Bact. 68, 739–744.

Pectin hydrolysis—*Enterobacteriaceae*

B. R. Davis and W. H. Ewing (1964). J. Bact. 88, 16–19.

The pectate medium employed was that described by Starr (1947), to which 0.5% yeast extract was added (Ewing 1960).

Cultures were incubated for 21 days at 37°C.

W. H. Ewing (1960). *Enterobacteriaceae. Biochemical Methods for Group Differentiation.* U.S. Department of Health, Education and Welfare, Public Health Service Publication, No. 734.

M. P. Starr (1947). Phytopathology 37, 291–300.

Pectin hydrolysis—anaerobes

A. R. Prévot (1966). Techniques pour le diagnostic des Bactéries Anaérobes. Éditions de la Tourelle, St. Mandé.

Milieux à la pectine pour isolement des pectinolytiques*

Les enzymes responsables de la pectinolyse sont au nombre de trois: protopectinase, polygalacturonidase, et pectine-méthyl-estérase. Trois méthodes peuvent donc conduire à l'isolement et la culture des pectinolytiques.

1° Milieu pour la détection des protopectinases.

Bouillon de pomme de terre	500 cm^3
Eau de rivière	500 cm^3
Extrait de levure Byla	5 g

Dans chaque tube contenant 6 cm^3 de ce milieu, on immerge un petit cube de carotte. On stérilise 10 minutes à 120°. On ensemence avec des dilutions croissantes de terre. On étire les tubes, on les scelle sous vide et on incube à 30°. Dans les jours qui suivent on observe le ramollissement du fragment de carotte dans les tubes qui contiennent des protopectinolytiques. On les isole par les géloses profondes au jus de carotte — extrait de levure.

2° Milieu pour la détection des polygalacturonidases.

Pectine de pomme (Unipectine ruban rouge)	13 g
Extrait de levure Byla	5 g
Bouillon de pomme de terre	1 l

La pectine doit être traitée à part dans un récipient où elle est mouillée avec 35 cm^3 d'alcool à 96°. On verse alors le litre de bouillon à 80° d'un seul trait sur la pectine, qui se dissout presque instantanément. On répartit à chaud à raison de 6 cm^3 par tube. On stérilise à l'autoclave à 120°. Au moment de l'emploi on neutralise à chaud chaque tube par 0.1 cm^3 de lessive de soude au 1/20e. Les tubes ensemencés à 45° par des dilutions de terre, sont étirés et scellés sous vide. La pectinolyse se traduit par une liquéfaction du gel. On isole les pectinolytiques par la gélose profonde pomme de terre + pectine.

3° Milieu pour la détection de la pectineméthylestérase.

Bouillon de pomme de terre	1 l
Pectine de pomme (unipectine ruban brun)	7 g
Extrait de levure Byla	5 g

(Procéder comme plus haut).

Si une pectine-méthylestérase y a été élaborée, il y a acidification du milieu, qu'on évalue après 10 à 15 jours d'incubation (par exemple au moyen de l'indicateur de Smith).

(*) P. Kaiser et A. R. Prévot.—C. R. Acad. Sci., 1958, *247*, 1065.

PEROXIDASE

Peroxidase—*Haemophilus*

L. R. Anderson (1930). J. Bact. 20, 371–379.

Reagent. A large pinch of crystalline benzidine-hydrochloride was placed in a test tube with 3 ml of glacial acetic acid. After most of the crystals had dissolved, an equal volume of hydrogen peroxide was added. A peroxidase-containing substance in contact with this milky white reagent turns blue. A drop of the reagent is placed on the colony.

Peroxidase—*Clostridium*

J. Gordon and J. W. McLeod (1940). J. Path. Bact. 50, 167–168.

Colonies turn black when a peroxide producing strain (of *Clostridium*) "is grown on benzidine 'chocolate' agar under anaerobic conditions for 24 hours and then exposed for 30 minutes or longer to air. The medium used consists of 1 ml of benzidine solution (benzidine 0.25 g, N HCl 0.3 ml, distilled water 50 ml) and 11 ml of heated blood agar (i.e. 10 per cent defibrinated rabbit blood in agar heated for 15 minutes at 75°C). The mixture is poured into small plates immediately after the addition of the benzidine solution."

Peroxidase—various

S. E. Wedberg and L. F. Rettger (1941). J. Bact. 41, 725–743.

The test reagents employed in this investigation for the detection of peroxidase were (1) 0.5 per cent benzidine in 50 per cent alcohol, (2) 1.0 per cent Guaiac in 95 per cent alcohol, (3) 1.0 per cent "orthotolidine" in 50 per cent alcohol, and (4) 0.1 percent 2,7,diamino-fluorene-hydrochloride in water.

One-tenth milliliter of the color reagent was added to a mixture of 0.5 ml cell suspension, 0.5 ml potassium hydrogen phthalate-sodium hydroxide buffer (pH 4.5) and 0.1 ml of 3.0 per cent hydrogen peroxide. Appearance of a characteristic blue color has been generally accepted as a positive test for peroxidase with each of these indicators.

Peroxidase—*Mycobacterium*

A. Andrejew, Ch. Gernez-Rieux and A. Tacquet (1956). Annls Inst. Pasteur, Paris 91, 586–589.

On sait que l'activité catalasique des Mycobactéries INH-résistantes est affaiblie ou abolie [4]. Or, comme la catalase, la peroxydase (également un enzyme héminique) forme dans la cellule des systèmes subsidiaires d'-oxydation qui éliminent H_2O_2 résultant de l'activité d'autres systèmes. Rappelons que contrairement à la catalase la peroxydase ne décompose pas H_2O_2 mais catalyse, en présence de l'eau oxygénée, l'oxydation d'un grand nombre de phénols et d'amines aromatiques.

Méthodes. – Nous avons employé la méthode de Willstätter et Stoll [5], modifiée par Sumner et Gjessing [6], qui est basée sur la propriété de la peroxydase d'oxyder, en présence de H_2O_2, le pyrogallol en purpurogalline.

L'activité de la peroxydase est fonction de la quantité de purpurogalline formée. Le nombre de milligrammes de purpurogalline formés par milligramme d'enzyme en cinq minutes à 20° correspond à "P.Z." (Purpurogallin Zahl de Willstätter) et indique sa pureté. Pour la peroxydase cristallisée, Theorell [7] a trouvé un P.Z. de 1020 et Keilin et Hartree [8] un P.Z. de 1220.

Nous avons préparé la purpurogalline d'après la méthode de Willstätter et Stoll [5], modifiée par Sumner et Somers [9].

Les dosages de la purpurogalline sont effectués à l'aide d'un électrophotomètre Jobin et Yvon (λ = 430 mμ).

Notons que la souche de BCG a été obtenue par le Dr A. Tacquet à partir d'un seul bacille et dédoublée ensuite en BCG-CS (sensible à l'INH) et en BCG-CR (cultivée en présence d'INH et devenue résistante à 50 μg d'INH/ml).

[4] Middlebrook (G.). *Amer. Rev. Tub.,* 1954, 69, 471.
[5] Willstätter (R.) et Stoll (A.). *Untersuchungen über Enzyme,* 1928, 1, 414.
[6] Sumner (J. B.) et Gjessing (E. C.). *Arch. Biochem.,* 1943, 2, 291.
[7] Theorell (H.) et Maehly (A. C.). *Acta Chem. Scand.,* 1950, 4, 422.
[8] Keilin (D.) et Hartree (E. F.). *Biochem. J.,*1951, 49, 88.
[9] Sumner (J. B.) et Somers (G. F.). *Laboratory Experiments in Biological Chemistry,* Acad. Press Inc., édit., New York, 1944.

Peroxidase—*Mycobacterium*

A. Andrejew, Ch. Gernez-Rieux and A. Tacquet (1960). Annls Inst. Pasteur, Paris 99, 821–838.

Les résultats les plus rapides et les plus nets sont obtenus dans les conditions suivantes: on prépare une solution de benzidine à 1 p. 100 dans l'acide acétique normal (10 ml d'acide acétique pur + 163 ml d'eau). On agite cette solution et on filtre. On prépare extemporanément une solution de H_2O_2 à 1 p. 100 (0.5 ml de H_2O_2 à 110 vol., soit 30 p. 100 + 14.5 ml d'eau). Au dernier moment, on mélange ces deux solutions (benzidine et H_2O_2), à parties égales et on verse ensuite ce mélange sur le milieu de culture et en quantité suffisante pour recouvrir complètement toutes les colonies. Il est préférable de ne pas mélanger ces solutions sur le milieu même, afin de ne pas favoriser le décollement des colonies. La durée du test est de deux à quinze minutes. Au delà de cette limite,les bacilles tuberculeux INH-résistants prennent parfois une teinte grise ou faiblement bleutée (ou, à la longue, la couleur verte du milieu de Löwenstein), qui diminue le contraste avec le bleu foncé des bacilles tuberculeux sensibles à l'INH.

Note: The above article discusses the above under the name *Oxidase* and treats very extensively the interpretation of the oxidase test. It is listed here because benzidine is used in the test.

Phenol*—*Mycobacterium*

A. Andrejew, Ch. Gernez-Rieux and A. Tacquet (1961). Annls Inst. Pasteur, Paris 101, 754–770.

On sait que le brunissement ("darkening"), au contact de l'air, d'une pomme de terre coupée, par exemple, est principalement dû à l'activité polyphénol-oxydasique, très forte chez ce tubercule.

Test de brunissement.– Nous appellerons ainsi le test effectué dans les conditions suivantes: on introduit dans une série de tubes à hémolyse 0.3 ml de tampon phosphate 0.2M pH 7* [concentration finale de 0.03 M]; 0.1 à 0.2 ml d'une solution aqueuse de catéchol à 110 mg/ml (concentration finale 0.05 à 0.1 M); 0.5 ml (environ 5 mg en poids sec) d'une suspension (ou 1 ml de broyat) de bactéries. On complète à 2 ml avec l'eau bidistillée.

*Note: This abstract should have appeared on p. 685 under **Oxidation Reactions.**

Pour éviter toute contamination bactérienne, on peut ajouter au mélange 0.1 ml de pénicilline à 50 000 μg/ml et 0.1 de streptomycine à 5000 μg/ml, qui n'exercent aucune action sur l'activité étudiée. On complète ensuite à 2 ml avec de l'eau bidistillée.

Les témoins sont dépourvus de bactéries, ou additionnés de bactéries ou de leurs extraits, préalablement chauffés trente minutes à 100°.

Les tubes sont agités de temps en temps et laissés à 18°–20°.

Lorsque la réaction est fortement positive, le brunissement devient apparent dès les dix premières minutes. La couleur s'intensifie ensuite progressivement pour passer d'ambré-rose clair au brun-rouge moyen et, après vingt-quatre heures, au brun-rouge plus ou moins foncé. Il est nécessaire d'agiter les tubes avant de lire les résultats du test.

Dans les tubes témoins on n'observe, en vingt-quatre heures, qu'une couleur rose pâle due à l'auto-oxydation du catéchol (test négatif).

*Tampon phosphate 0.2 M pH 7: 390 ml de solution A + 610 ml de solution B. Solution A: 27.23 g de PO_4H_2K pour 1000 ml d'eau bidistillée. Solution B: 71.65 g de $PO_4HNa_2 . 12 H_2O$ pour 1000 ml d'eau bidistillée.

Peroxidase—*Haemophilus*

P. N. Edmunds (1962). J. Path. Bact. 83, 411–422.

A little material from a 48 hr culture on 5 per cent human-serum digest agar was emulsified in a drop of water on a slide and mixed with a drop of a solution of benzidine, acetic acid and H_2O_2. As a control, the medium was tested by placing a drop of solution on it and it gave neither a peroxidase reaction (blue colouration) nor a catalase reaction (bubbles of gas).

Peroxidase—*Veillonella*

M. Rogosa (1964). J. Bact. 87, 162–170.

The benzidine test, using both the benzidine base and the dihydrochloride for the presumptive detection of iron porphyrin compounds, was generally done as described by Deibel and Evans (1960) on plate cultures and cell suspensions from broth cultures. Since the benzidine base is ten times more sensitive than the hydrochloride salt, cell suspensions were washed three times in distilled water to remove interfering components (e.g., excess Fe and Cu), and were resuspended in 0.5 ml of the usual 5% H_2O_2.

For V17 medium see **Basal medium**—*Veillonella* (p. 114).

R. H. Deibel and J. B. Evans (1960). J. Bact. 79, 356–360.

Peroxidase—*Mycobacterium*

A. Tacquet, F. Tison and B. Devulder (1965). Annls Inst. Pasteur, Paris 108, 514–525.

Used the method of Andrejew, Gernez-Rieux and Tacquet (Ann. Inst. Pasteur, 1956, 91, 767–770.

Peroxidase—*Vibrio sputorum*

W. J. Loesche, R. J. Gibbons and S. S. Socransky (1965). J. Bact. 89, 1109–1116.

Iron-porphyrin compounds were sought by use of benzidine dihydrochloride reagent and 5% H_2O_2 (Deibel and Evans, 1960).

R. H. Deibel and J. B. Evans (1960). J. Bact. 79, 356–360.

Peroxidase—*Mycobacterium*

M. Tsukamura (1966). J. gen. Microbiol. 45, 253–273.

Peroxidase activity was tested according to the method described by Tiranarayanan and Vischer (1957).

M. O. Tiranarayanan and W. A. Vischer (1957). Am. Rev. Tuberc. pulm. Dis. 75, 62.

pH LIMITS OF GROWTH

pH limits of growth—*Pseudomonas*

A. J. Riker, W. M. Banfield, W. H. Wright, G. W. Keitt and H. E. Sagen (1930). J. agric. Res. 41, 507–540.

The effect of hydrogen-ion concentration on growth was studied by varying the reaction of the following medium: Glucose, 5.0 g; magnesium sulphate ($MgSO_4 \cdot 7H_2O$), 0.2 g; sodium chloride (NaCl), 0.2 g; calcium sulphate ($CaSO_4$), 0.1 g; dipotassium phosphate (K_2HPO_4), 0.2 g; 10 per cent yeast infusion, 100 ml; distilled water, 900 ml. This medium was prepared in quantities of several liters at a time and then divided into smaller quantities for each pH value to be used. Each lot of culture media was sterilized at pH 7.0 and then adjusted to the pH value desired by the addition of N/20 hydrochloric acid under aseptic conditions. The technic of seeding these cultures was standardized by making water suspensions of various strains, using organisms from 2-day-old cultures on potato-glucose agar slants. Cultures were first grown in solutions decreasing from pH 7.0 to pH 4.0 in steps of 0.5 pH. A second series was prepared with the following reactions: pH 4.6, 4.4, 4.2, 4.0, 3.8, and 3.6. Final results as to growth were recorded after 2 weeks at 21°C.

pH limits of growth—*Lactobacillus*

J. M. Sherman and H. M. Hodge (1940). J. Bact. 40, 11–22.

Medium:

Lactose, 1%; peptone, 1%; Bacto yeast extract, 1%; Na citrate, 0.5%; $NaHCO_3$, 0.065%.

The pH of this broth is about 7.2 before sterilization, rises to 7.8 to 7.84 after sterilization, and remains stable for one week at 37°C. All pH values were determined by use of the hydrogen electrode. Tubes of this medium were inoculated, incubated at 37°C and observed for growth and acid production.

pH limits of growth—*Microbacterium*

V. Bolcato (1957). Antonie van Leeuwenhoek 23, 351–356.

Optimum of pH for growth. 12 tubes containing 10 ml of sterile saccharose broth* to a pH of 4.5 to 9.0 were inoculated with some drops of a fresh culture of the organism. Upon incubation at 50°C for 10 and 24 hr, the tubes were examined for growth by direct microscopic observation and by determination of the final pH. The pH optimum for growth was from 6 to 7 and the limits for growth from pH 5 to 8.5. Final pH 4.5.

*peptone, 5 g; beef extract, 3 g; saccharose, 10 g; distilled water, 1000 ml.

pH limits of growth—*Acetobacter*

J. de Ley (1958). Antonie van Leeuwenhoek 24, 281–297.

Acid-resistance

It is characteristic of real acetic acid bacteria that they are able to grow at a pH of 4.5 and lower. Atkinson's medium (1956, J. Bact. 72, 189) was prepared, to which M/15 acetic acid and 2% ethanol were added. The medium was neutralized with NaOH, to pH 4.3, dispensed in 50 ml portions in 250 ml Erlenmeyer flasks, sterilized, inoculated and incubated for at least one week at 30°C. All the strains grew very well.

pH limits of growth—*Pseudomonas*

M. E. Rhodes (1959). J. gen. Microbiol. 21, 221–263.

Effect of the initial pH value of the medium. Five ml volumes of yeast extract broth were adjusted to pH 7.0, 6.0, 5.5, 5.0, 4.5 and 4.0 with N-HCl and used for growth tests in the usual way.

pH limits of growth—*Pediococcus*

H. L. Günther and H. R. White (1961). J. gen. Microbiol. 26, 185–197.

Growth at pH 9.0 and pH 4.2. In these experiments the technique was based on that described by Shattock and Hirsch (1947) for testing growth of streptococci at pH 9.6. The following modifications were made: tomato juice (TJ) broth was substituted for glucose Lemco broth; to obtain the medium at pH 9.0 suitable quantities of the 0.1 M-glycine buffer recommended by Shattock and Hirsch (1947) were added; for the medium at pH 4.2, sodium acetate + acetic acid buffer (Clark, 1928) at 0.04 M was selected, since some inhibitory effects were noted at higher concentrations.

Maintenance of stock cultures and methods of cultivation. For maintenance of stock cultures, preparation of inocula and in all experimental work, 'Oxoid' tomato juice (TJ) broth or tomato juice (TJ) agar, adjusted to pH 6.6, were used unless otherwise stated. The following were exceptions to this rule: for strain Tc. 1 sodium chloride (5%, w/v) was added to the medium; and for the aerococci glucose Lemco broth (Shattock and Hirsch, 1947) or glucose yeast extract (GY) agar (containing, as %, w/v, peptone, 1.0; Yeastrel, 0.3; glucose, 1.0; NaCl, 0.25; agar, 1.0; at pH 7.4) was used.

Cultures were incubated aerobically at 30°C with specified exceptions.

W. M. Clark (1928). *The Determination of Hydrogen Ions,* 3rd ed. London: Baillière, Tindall and Cox Ltd.

P. M. F. Shattock and A. Hirsch (1947). J. Path. Bact. 59, 495.

Temperature of growth*—*Xanthomonas*

A. C. Hayward and W. Hodgkiss (1961). J. gen. Microbiol. 26, 133–140.

Two drops of a light suspension of organisms in distilled water were added to glucose peptone agar slopes maintained in a water bath at 40° or 37°C for 2 days. Otherwise all cultures were incubated at 28°C for the period of test.

pH limits of growth—*Spirillum*

W. A. Pretorius (1963). J. gen. Microbiol. 32, 403–408.

pH value for growth. The defined medium of Giesberger (1936); NH_4Cl, 0.1%; K_2HPO_4, 0.05%; $MgSO_4$, 0.05%) with 0.2% (w/v) calcium lactate was used, and samples were adjusted with diluted HCl or KOH to give a range of pH values between 5.0 and 9.0 at 1 unit intervals.

pH limits of growth—enterococci

C. G. Rogers and W. B. Sarles (1964). J. Bact. 88, 965–973.

Ability to grow in tryptose yeast-extract broth at pH 9.6 was tested.

pH limits of growth—*Moraxella*

W. J. Ryan (1964). J. gen. Microbiol. 35, 361–372.

pH value of media. Gradient plates, prepared according to the method of Watson and Bennett (1957), were used to determine the optimum pH value. Nutrient agar + serum (10%, v/v) was the base used.

K. C. Watson and M. A. E. Bennett (1957). J. Lab. clin. Med. 50, 639.

pH limits of growth—*Pseudomonas aeruginosa*

R. R. Colwell (1964). J. gen. Microbiol. 37, 181–194.

YE broth, adjusted to pH 4.0, 4.5, 5.0, 5.5, 6.0, 7.0, 8.0 and 9.0, with HCl or NaOH as required, was inoculated and incubated for 14 days, growth being observed at 1, 2, 7 and 14 days.

The standard inoculum for all tests was a single drop (1/20 ml) from a sterile disposable pipette (Fisher Scientific Company) of a 24-48 hr broth culture. The stock culture medium (YE) was a modification of that used by Rhodes (1959): Difco yeast extract, 0.3 g; Difco Bacto-Proteose Pep-

*Note: This abstract should have appeared on p. 823.

tone, 1.0 g; NaCl, 0.5 g; agar, 1.5 g (agar omitted from liquid stock media); distilled water, 1 L; adjusted to pH 7.2 – 7.4 with NaOH. Tests were carried out at room temperature (25°C) except where otherwise indicated.
M. E. Rhodes (1959). J. gen. Microbiol. 21, 221.

pH limits of growth—*Pediococcus*

R. Whittenbury (1965). J. gen. Microbiol. 40, 97–106.

The ability to grow at pH 8.0 and above was tested by the method of Chesbro and Evans (1959), the final pH value of the medium being adjusted as required. The ability to grow lower than pH 8.0 was tested in glucose (0.5%, w/v) soft agar.

For "soft agar" see **Fermentation**—lactic acid bacteria (p. 310), or R. Whittenbury (1963). J. gen. Microbiol. 32, 375–384.
W. R. Chesbro and J. B. Evans (1959). J. Bact. 78, 858.

pH limits of growth—*Serratia marcescens* and yeasts

E. H. Battley and E. J. Bartlett (1966). Antonie van Leeuwenhoek 32, 245–255.

The following is the summary of the authors' paper to which the reader is referred for details.

Previously existing methods for determining the pH limits for the growth of microorganisms have involved (1), the setting up of individual cultures, each having a specific pH; (2), the pH gradient plate technique devised by Sacks (1956) in which a continuous pH gradient is established in a Petri dish by means of a buffer system; and (3), the pH gradient plate technique of Zak (unpublished), in which a continuous pH gradient is established by means of an electric current. The discontinuous pH gradient technique described here provides a convenient method of determining the maximum and minimum pH at which a microorganism can grow. The technique can be used aerobically or anaerobically, and has a precision of about ± 0.1 pH unit. Data are given for several yeasts and for *Serratia marcescens*. In all cases, the organisms tested continued to metabolize at pH values beyond those representing the limits for growth, sometimes by as much as 0.5 pH unit. The results suggest that pH limits are unsuitable criteria in microbial classification.

pH limits of growth—*Clostridium botulinum*

W. P. Segner, C. F. Schmidt and J. K. Boltz (1966). Appl. Microbiol. 14, 49–54.

This paper deals with the effect of pH on the outgrowth of spores. The term 'outgrowth' as used here indicates the first appearance of gas in Vaspar stratified tubes and a change in turbidity of the culture medium.

TPG medium was used to determine the effect of pH on the outgrowth of type E spore inocula. The medium was buffered to eliminate minor fluctuations of the adjusted pH levels. To each 100 ml of double-strength TPG (10% Trypticase, 1% peptone, and 0.8% glucose) was added 20 ml of 0.67 M potassium dihydrogen phosphate, plus 80 ml of deionized water. Therefore, the final medium consisted of 0.067 M phosphate-buffered single-strength TPG. Concentrated HCl was added to the approximate pH end point, followed by diluted HCl to the desired pH. A model 76 expanded scale pH meter (Beckman Instruments, Inc., Fullerton, Calif.) and a magnetic stirring device were used in the pH adjustments. The pH adjusted medium was transferred in 200-ml quantities to screw-cap bottles and sterilized. Media of pH 5.0 or lower were autoclaved for 5 min at 121°C, and those above pH 5.0 were autoclaved for 10 min at this temperature. Either sterilized sodium thioglycolate or L-cysteine hydrochloride was added to give respective concentrations of 0.02 and 0.05% just prior to the use of the medium.

In addition to TPG medium, liver broth was utilized to determine the effect of pH on spore outgrowth time. Liver infusion medium was prepared as recommended in the National Canners Association Laboratory Manual (1954). It was adjusted to various pH levels, bottles, and autoclaved as described above. At inoculation and at several periods during incubation, the pH of uninoculated TPG medium and liver broth medium was determined. All pH values cited are those obtained after autoclaving.

Inoculation and outgrowth techniques. Either 16 by 125 or 16 by 150 mm sterile screw-cap tubes were inoculated with 0.2 ml of a preheated (60°C, 13 min) spore inoculum equivalent to 2 million viable spores. With the use of a five-tube replicate test per variable, the inoculated tubes were poured with about 10 ml of medium, sealed with sterile melted Vaspar, and tempered in either cold tap water or ice water when incubation at low temperatures was desired. Each tube was dried and carefully examined for entrapped air bubbles. Converted to the nearest degree centigrade, tubes were incubated at 46°F (8°C), 50°F (10°C), 60°F (16°C), 70°F (21°C), 85°F (30°C), and 98°F (37°C). During incubation, tubes were examined on a Monday, Wednesday, and Friday schedule.

C. T. Townsend, I. I. Somers, F. C. Lamb and N. A. Olson (1954). *A laboratory manual for the canning industry.* National Canners Association Research Laboratories, Washington, D.C.

PHENOL (ESTIMATION OF)

Phenol, estimation of—*Clostridium*

A. R. Prévot and R. Saissac (1947). Annls Inst. Pasteur, Paris 73, 1125–1129.

On recherche couramment l'indol dans les cultures bactériennes, mais non les phénols. Ces derniers sont en effet plus difficiles à caractériser, la plupart de leurs réactions étant gênées par la présence des autres produits du métabolisme bactérien. En étudiant les anaérobies des sols d'Afrique, nous avons trouvé de nombreuses souches donnant les réactions du phénol, mais comme toutes ces souches produisent à la fois une grande quantité d'indol, il était nécessaire d'éliminer celui-ci. A cet effet nous avons adopté l'une des deux méthodes indiquées par Macé [9], méthode qui offre de plus l'avantage de réaliser une concentration de la solution dans laquelle on recherche les phénols: elle consiste à effectuer une première distillation en milieu acide, à alcaliniser par de la lessive de potasse le distillat obtenu, puis à le redistiller presque en totalité. On sature le résidu par un courant de CO_2, et on redistille presque jusju'à siccité. Si le dernier distillat contient des corps phénoliques, il donne une coloration rouge avec le réactif de Millon. Pour déterminer la nature de ces corps, nous avons effectué toutes les réactions suivantes, réactions classiques comme celles au Cl_3Fe: avec le phénol, coloration bleu violet, avec le *p.*crésol, coloration bleue, puis trouble; à l'eau de brome: avec phénol et *p.*crésol un précipité blanc, soluble dans la potasse (dans le cas du *p.*crésol, la solution obtenue est rose).

Par addition d'ammoniaque puis d'eau de brome avec le phénol; coloration bleue passant au rose par addition d'acides minéraux; avec le *p.*crésol, pas de coloration.

La réaction indiquée par Frieber [10], très sensible pour le phénol, négative pour le *p.*crésol (réaction n'ayant lieu que si la position en para est libre); elle consiste à oxyder en milieu alcalin, par l'hypochlorite, un mélange de phénol et de chlorhydrate de *p.*aminophénol, il y a formation d'indophénol de couleur bleue.

Par copulation en milieu alcalin avec le chlorure de l'acide diazobenzène sulfonique, coloration jaune pour le phénol, rouge pour le *p.*crésol.

Dans la réaction avec formol et SO_4H_2, coloration rose pour le phénol verte pour le *p.*crésol.

Dans la réaction de Melzer modifiée par Deichmann [11] avec SO_4H_2 et benzaldéhyde, le phénol produit à froid une coloration jaune qui devient brun rouge à chaud, puis par refroidissement et alcalinsation, violette; avec le *p.*crésol, pas de coloration.

Ces deux dernières réactions n'ont de valeur qu'après élimination de l'indol, qui donne, sauf dans la dernière partie de la réaction de Melzer Deichmann, les mêmes colorations que le phénol.

Enfin, les réactions de Ware [12] qui sont spécifiques du phénol et de l'orthocrésol. (Notons à propos de l'*o.*crésol que dans toutes les réactions précédentes, l'*o.*crésol donne les mêmes résultats que le phénol, sauf dans la réaction au Cl_3Fe: avec l'*o.*crésol, la coloration passe en quelques minutes au vert brunâtre). La réaction de Ware consiste à traiter en milieu chlorhydrique concentré la solution à identifier par un mélange de nitrite et de nitrate, ou de nitrite seul en quantité très petite. Les colorations obtenues avant et après addition du mélange à de l'eau ammoniacale sont tout à fait caractéristiques du phénol et de l'*o*.crésol; elles sont différentes pour ces deux corps et ne se produisent pas pour les autres phénols.

Toutes ces réactions sont extrêmement nettes quand on les effectue sur le dernier distillat, obtenu comme il a été dit plus haut. Notons que l'on ne peut pas employer l'autre méthode que Macé indique pour la recherche des phénols (méthode plus simple que la précédente: on distille en milieu acide et dans le distillat neutralisé par CO_3Ca on recherche le phénol par la réaction au Cl_3Fe) car même en utilisant comme milieu de culture un bouillon non glucosé, on est gêné par la présence dans le distillat d'acides organiques provenant de la fermentation des acides aminés, et donnant avec Cl_3Fe colorations et précipités. On est donc obligé d'utiliser la méthode comprenant les trois distillations successives décrites plus haut.

[9] Macé, *Traité de bactériologie,* 1, 340.
[10] Frieber, *Zentralbl. Bakt.*, 1921, 86, 58.
[11] Deichmann, *Ind. Eng. Chem. Anal.*, 1944, 16, 37.
[12] Ware, *Analyst.*, 1927, 52, 335.

Phenols and cresols—anaerobes

A.-R. Prévot (1966). Techniques pour le diagnostic des Bactéries Anaérobes. Éditions de la Tourelle, St. Mandé.

Phénols: dans un tube à essais, verser 1 cm^3 de distillat* et ajouter 1 goutte d'une solution de perchlorure de fer à 40 p. 100. Une réaction positive se traduit par l'apparition d'une coloration bleu-violet qu'il ne faut pas confondre avec un louche indiquant la présence d'acides aromatiques et autres produits du métabolisme bactérien.

Crésols: dans un tube à essais, à 1 cm^3 de distillat,* ajouter 4 gouttes d'alcool à 96°C, puis 4 cm^3 d'acide acétique et 3 gouttes de formol du commerce. Agiter. Ajouter 1 cm^3 d'acide sulfurique. La présence d'orthocrésol donne une teinte rouge, de métacrésol une teinte violette, de paracrésol une teinte verte. On peut également utiliser le réactif de Millon qui, ajouté à une quantité égale de distillat, donne une teinte rouge après chauffage en présence de crésols.

Crésols et phénols: les réactions indiquées ci-dessus pour la mise en évidence des crésols et des phénols ne sont que des réactions d'approche n'ayant pas une valeur absolue puisqu'elles peuvent être faussées par la présence de divers produits du métabolisme bactérien. Seule la technique décrite par A. R. Prévot et R. Saissac** permet d'affirmer avec certitude la présence de ces corps dans le distillat d'une culture microbienne. Mais en raison des opérations qu'elle nécessite, elle doit être réservée aux cas où une très grande précision est indispensable. La méthode à mettre en

oeuvre est celle de Macé qui réalise une concentration de la solution dans laquelle on recherche les phénols par deux distillations successives, la première en milieu acide, la seconde en milieu alcalin. Si le dernier distillat contient des corps phénoliques, il donne une coloration rouge avec le réactif de Millon. La nature de ces corps est alors déterminée par des réactions classiques (réaction au perchlorure de fer, à l'eau de brome, au formol et acide sulfurique, réaction de Frieber).
*Distillat Alcaline sous **Amines (volatile)** – anaérobes (p. 51).
**A. R. Prévot et R. Saissac. – Recherches sur la production des phénols par les bactéries anaérobies. Ann. Inst. Pasteur, 1947, 73, 1125.

PHOSPHATASE

Phosphatase—*Neisseria*

H. W. Leahy, H. E. Stokinger and C. M. Carpenter (1940). J. Bact. 40, 435–440.

Preparation of the bacterial cells

For the present investigation, 16 strains of Neisseria were employed. Of these strains, 10 were *Neisseria gonorrhoeae,* 2 *Neisseria intracellularis,* 2 *Neisseria catarrhalis,* and 2 *Neisseria sicca.* Each strain with the exception of two of *N. gonorrhoeae* strains (GNI and GNII) was grown for 6 or 7 days at 37°C in Douglas's broth containing 5 per cent of ascitic fluid. The cells were removed by centrifugation, washed once in a 0.85 per cent solution of NaCl (physiological saline solution), "lyophilized" by the technic of Flosdorf and Mudd (1935), placed in stoppered test tubes, and stored at 5°C until measurements were made. The preparations of dried cells from strains GN I and GN II were obtained from 6-day-old veal-infusion-broth cultures.

Preparation of the substrate solutions

The buffered substrate solutions employed for the measurement of pH optima were prepared as described by Leahy, Sandholzer and Woodside (1939). The 0.005 M disodium-phenyl-phosphate was buffered with a 0.05 M phthalate or veronal buffer at each pH unit from 4.0 to 9.0. After adjustment to the desired pH value had been made, it was checked by the glass-electrode method before the solution was used.

Measurement of optimal pH

The phosphatase activity of the "lyophilized" organisms was measured by the method of Leahy, Sandholzer and Woodside (1939). In brief, the procedure consisted in adding a known weight (from 2 to 5 mg) of "lyophilized" cells suspended in distilled water to a series of tubes containing buffered substrate solutions at each pH unit from 4.0 to 9.0. The mixtures were incubated for 24 hours at 37°C, ± 0.02°C, and the degree of hydrolysis was measured from the amount of phenol produced. Quantitative colorimetric estimation of the phenol was made by the use of Gibb's reagent (2,6-dibromo-quinonechloroimide), which is capable of measuring 0.1 gamma of phenol in 10 ml. The activity of the various preparations was expressed in gammas of phenol liberated per milligram of "lyophilized" cells in 24 hours at 37°C after a small correction, as indicated by the controls, had been made. In every instance, the pH of maximal activity was determined by plotting curves of the phenol values at each pH. Duplicate runs with different preparations of the same strain, and repetition of measurements, both in the absence and in the presence of magnesium ions, gave optima agreeing within ± 0.2 pH unit.

E. W. Flosdorf and S. Mudd (1935). J. Immun. 29, 389–425.
H. W. Leahy, L. A. Sandholzer and M. R. Woodside (1939). J. Bact. 38, 117.

Phosphatase—various

J. Bray and E. J. King (1943). J. Path. Bact. 55, 315–320.

Media: Eggs, either fresh or waterglass-preserved, are broken into a container and agitated with nutrient broth in the proportion of one part of broth to four of egg fluid. The mixture is then filtered through muslin. Phenol-phthalein-phosphate in the proportion of 0.1 g to 1 ml is ground up in a mortar with distilled water, plus a drop of caprylic alcohol to prevent frothing, and 1 ml of the creamy suspension is pipetted into each plate without allowing the powder to settle; 25 ml of the egg mixture are added and thoroughly mixed with the phenol-phthalein-phosphate. The plates are then placed in the inspissator with a disc of filter paper in the lid to absorb moisture, and the medium inspissated on two successive days in the usual manner. Plates are stored in the ice-chest.

Test: After inoculation and incubation at 37°C for 18 hours or longer the plates are examined for growth and then inverted over a Petri dish containing a concentrated solution of ammonia (which is renewed daily) and the results recorded before the colour fades. A rough indication of the intensity of the reaction may be obtained from the amount of alkalinisation required to bring up the colour. With strong reactors a mere "whiff" of ammonia makes the mass of organisms turn pink. Placing the plate down on the dish of ammonia for a few seconds brings up the weaker reactors, and exposure of the plate to the vapour for about thirty seconds shows up the organisms which only produce a faint pink colour. At this stage the strong reactors are stained intensely red. More prolonged exposure to the ammonia vapour causes the surface of the plate to become moist and difficult to read. As the colour begins to fade a few minutes after alkalinisation, partly due to the evaporation of the ammonia, it is necessary to read the results at once. The colour may, however, be partly revived by further exposure to ammonia vapour even after a week. A smear of each organism on the plate is made and examined microscopically to exclude contaminants. Where it has been necessary to subculture from the egg plate it has not appeared that the ammonia affects the viability of the organisms, although this point has not been thoroughly investigated.

Phosphatase

J. Tramer (1952). J. Dairy Res. 19, 275–287.

Simple and rapid methods for the estimation of bacterial phosphatases using di-sodium-*p*-nitrophenylphosphate as substrate.

The substrate

Di-sodium *p*-nitrophenylphosphate* (pNPP) is easily and rapidly hydrolysed by bacterial phosphatases. The salt is unstable at temperatures above 25°C and sensitive to light. It should be kept in the refrigerator in a dark stoppered bottle. In solutions, pNPP is also unstable at high temperatures and can, therefore, not be sterilized by autoclaving or steaming. Seitz-filtration, however, is satisfactory.

For streaking, pNPP plates were prepared in the following manner: 1 ml of a sterile unbuffered M/100 pNPP solution in distilled water was pipetted into a Petri dish followed by 10 ml of 1.5% milk agar and well mixed. When solid the surface was either streaked with the inoculum or inoculated from a peptone water culture by touching the surface with a straight needle. With *Bact. aerogenes,* for instance, yellow colonies were visible within 8 hr at 37°C. After 24 hr when the colonies were intense yellow in colour, diffusion of the *p*-nitrophenol from the colonies into the surrounding medium had occurred. Incubated controls remained unchanged, showing that there had been no auto-hydrolysis. Even quicker results can be obtained by transferring part of a colony from an agar plate with a loop to an agar plate prepared with pNPP as just described. *Bact. aerogenes* then produced a marked yellow colour within a few hours. These methods have been found useful as preliminary sorting tests.

*available from British Drug Houses Ltd., Poole, Dorset.

Phosphatase—*Staphylococcus*

F. S. Thatcher and W. Simon (1956). Can. J. Microbiol. 2, 703–714.

Detected by the method of Barber and Kuper (1951) which is based on the demonstration of free phenolphthalein around a colony grown on nutrient agar containing phenolphthalein phosphate (previously tested for absence of the free indicator). After incubation of the test cultures for 48 hr at 37°C, ammonia fumes were pumped into the tubed cultures with a hand pump. (The surface of the agar may alternatively be flooded with a dilute solution of ammonium hydroxide). This caused

any colony that had "split" the phosphate radicle from the colorless indicator complex to become surrounded by a red zone in the medium.
M. Barber and S. W. A. Kuper (1951). J. Path. Bact. 63, 65–68.

Phosphatase—*Pasteurella septica*

J. M. Talbot and P. H. A. Sneath (1960). J. gen. Microbiol. 22, 303–311.

Production observed after 18 hr incubation at 37°C on the medium of Bray and King (1943, J. Path. Bact. 55, 315).

Phosphatase—*Vibrio*

G. H. G. Davis and R. W. A. Park (1962). J. gen. Microbiol. 21, 101–119.

Production tested upon nutrient agar containing 0.006% (w/v) phenolphthalein phosphate, exposed after 5 days to vapour from 0.88 ammonia solution.

Phosphatase—*Aeromonas salmonicida*

I. W. Smith (1963). J. gen. Microbiol. 33, 263–274.

Activity was tested on Barber and Kuper's medium (J. Path. Bact. 1951, 63, 65) after 2 days.

Phosphatase—*Staphylococcus*

F. D. Cannon and C. V. Z. Hawn (1963). J. Bact. 86, 1052–1056.

To determine phosphatase activity, a modification of the procedure of Barnes and Morris (1957) was used. The substrate employed was phenolphthalein diphosphate, pentasodium salt (Sigma Chemical Co., St. Louis, Mo.). To 0.2 ml of the cell suspension in a Klett tube were added 2 ml of 0.05 M sodium acetate-acetic acid buffer (pH 6.0) and 0.5 ml of a 0.1% aqueous solution of phenolphthalein diphosphate. Phosphate buffers were avoided, since inorganic phosphate is known to inhibit the enzyme (Roche, 1950; Barnes and Morris, 1957). A blank was prepared by substituting 0.2 ml of sterile distilled water for the cell suspension. The tubes were incubated for 18 hr at 37°C. After incubation, 2 ml of 2.2 M sodium carbonate-bicarbonate buffer (pH 11) were added to each tube to stop the reaction and to develop the color of the phenolphthalein liberated; pH 11 was selected because it produced a stable red color with the phenolphthalein. A large excess of alkali causes formation of the trisodium salt which again is colorless. A Klett-Summerson photoelectric colorimeter with a number 54 filter was used. Phosphatase activity was expressed as Klett units. A standard 0.001% phenolphthalein solution at pH 11 read 375.

E. H. Barnes and J. F. Morris (1957). J. Bact. 73, 100–104.
J. Roche (1950). In J. B. Sumner and K. Myrbäck (ed.), *The enzymes,* vol 1, part 1. Academic Press, Inc., New York.

Phosphatase—*Staphylococcus*

S. I. Jacobs, A. T. Willis and G. M. Goodburn (1964). J. Path. Bact. 87, 151–156.

Phenolphthalein phosphate agar was used as described by Barber and Kuper. The medium differed from that described by these authors only in the composition of the nutrient base.

M. Barber and S. W. A. Kuper (1951). J. Path. Bact. 63, 65.

Phosphatase—*Staphylococcus*

S. Krynski, W. Kedzia and M. Kaminska (1964). Some differences between staphylococci isolated from pus and from healthy carriers. J. infect. Dis. 114, 193–202.

Acid phosphatase activity of the cells taken from a 24 hour 37°C broth culture was determined by the method of Barnes and Morris (1957).

E. H. Barnes and J. F. Morris (1957). J. Bact. 73, 100–109.

Phosphatase—*Moraxella*

W. J. Ryan (1964). J. gen. Microbiol. 35, 361–372.

The method used was essentially that described by Barber and Kuper (1951). The organisms were grown on serum agar plates containing 0.01% (w/v) phenolphthalein phosphate. When well grown, the culture was exposed to ammonia vapour.

M. Barber and S. W. A. Kuper (1951). J. Path. Bact. 63, 65.

Phosphatase—*Dictyostelium*

K. Gezelius and B. E. Wright (1965). J. gen. Microbiol. 38, 309–327.

Enzyme assays. The alkaline phosphatase activity was assayed in diluted extracts (about 0.1 mg protein/ml) with *p*-nitrophenyl phosphate (NPP) as substrate (Torriani, 1960) unless specified otherwise. The reaction was followed spectrophotometrically (Beckman DU or Zeiss PMQ II) at 23°C by the formation of nitrophenol (NP) at 420 mμ.

A. Torriani (1960). Biochim. biophys. Acta 38, 460.

Phosphatase—*Staphylococcus*

S. J. Edwards and G. W. Jones (1966). J. Dairy Res. 33, 261–270.

Phosphatase production on nutrient agar plates containing phenolphthalein diphosphate (Oxoid).

Phosphatase—*Staphylococcus*

M. Kocur, F. Přecechtěl and T. Martinec (1966). J. Path. Bact. 92, 331–336.

Phosphatase production was studied by means of Baird-Parker's (1963) modification of Barber and Kuper's (1951) method.

A. C. Baird-Parker (1963). J. gen. Microbiol. 30, 409.

M. Barber and S. W. A. Kuper (1951). J. Path. Bact. 63, 65.

Phosphatase—*Pseudomonas aeruginosa*

P. V. Liu (1966). The roles of various fractions of *Pseudomonas aeruginosa* in its pathogenesis. J. infect. Dis. 116, 112–116.

The organisms were grown on tryptone-glucose-extract agar enriched with 1% glucose and covered with cellophane. The extract was precipitated with ammonium sulfate (50% saturation) and dialyzed in 0.02 M tris buffer for 2 days. The contents of the cellophane bag were heated to 70°C for 20 minutes in order to destroy all the heat-labile products of the organisms. The alkaline phosphatase of *P. aeruginosa,* like that of *Escherichia coli* (Torriani, 1960), was resistant to this temperature. The preparation was then centrifuged to remove the precipitate and concentrated with Carbowax for chromatography. Titration of alkaline phosphatase was done with *p*-nitrophenylphosphate as substrate (Torriani, 1960). Because *P. aeruginosa* produces such a large amount of this enzyme, a unit equal to approximately 300 units as defined by Torriani was used.

A. Torriani (1960). Biochim. biophys. Acta 38, 460–469.

PIGMENTS – VISUAL REGISTRATION

Pigment—coliforms

L. A. Rogers, W. M. Clark and H. A. Lubs (1918). J. Bact. 3, 231–252.

Chromogenesis was determined in the human feces cultures by spreading on white paper the growth from an agar culture grown twelve days at 20°C and comparing with the plates in Ridgway's Color Standards. It was found that this series was almost entirely lacking in pigment. Nearly all of the culture gave a faint yellow color but this was so slight and showed so little variation that it was of no value. There were, however, a few exceptions to this statement.

Pigment—*Staphylococcus*

R. Thompson and D. Khorazo (1937). J. Bact. 34, 69–79.

The pigment was classified as orange, lemon or white by the appearance of the massed growth from a 72 hr agar plate placed upon white filter paper and pressed between two microscopic slides. No attempt was made to classify the various shades between orange and lemon. When any trace of orange color was present the strain was designated as orange.

Pigment—*Staphylococcus*

C. Shaw, J. M. Stitt and S. T. Cowan (1951). J. gen. Microbiol. 5, 1010–1023.

Strains were grown on potato and on nutrient agar at the optimal temperature for 24 hours and kept at room temperature for a week in diffuse daylight. Pigment was recorded as gold, lemon-yellow or pink.

Pigment—*Pseudomonas*

A. Lutz, A. Schaeffer and Mlle. M. J. Hofferer (1958). Annls Inst. Pasteur, Paris 95, 49–61.

Nous avons élaboré le milieu suivant (L1), qui nous a donné de bons résultats pour la biosynthèse quantitative de la pyocyanine par une grande proportion de nos souches, à savoir :

Acide L (+) glutamique		10 g
Extrait de levure Difco		1 g
Gélose Difco		13.5 g
Eau distillée	q. s. p.	1 000 ml

Le pH est ajusté à 7.5.

En moins de vingt-quatre heures, la coloration bleue des milieux apparaît et atteint un maximum en quarante-huit heures. Nous avons dans tous les cas extrait la pyocyanine en chloroforme alcalin et vérifié son virage au rouge en milieu aqueux acide.

Selon les souches productrices de pyocyanine, la gélose reste quelquefois teintée après extraction au chloroforme alcalin d'une couleur rouge violacée très pâle. D'autres fois, la couleur bleue tire tardivement un peu sur le bleu vert dû à un mélange de pyocyanine et d'un pigment jaune qui colore le chloroforme en jaune, la pyocyanine étant passée au rose rouge en phase aqueuse acide. Quelques souches ne donnent qu'un pigment rouge (pyorubine de Meader [25]) et d'autres plus nombreuses ne donnent pas de pigments.

Nous avons obtenu des pigments fluorescents en ajoutant au milieu ci-dessus du phosphate bi-potassique et du sulfate de magnésie dans les proportions suivantes (milieu L2) :

Acide L (+) glutamique		10 g
PO_4K_2H		7.5 g
SO_4Mg		2.5 g
Extrait de levure Difco		1 g
Agar (Difco)		13.5 g
Eau distillée	q. s. p.	1 000 ml

Le pH est ajusté à 7.

On inhibe ainsi ou on diminue fortement la biosynthèse de la pyocyanine et on observe aux rayons ultraviolets une fluorescence plus ou moins accentuée.

Signalons que Georgia et Poe [26] avaient obtenu en 1931 un bon rendement de pigments fluorescents sur le milieu suivant :

Asparagine	0.3 g
PO_4HK_2	0.05 g

$SO_4Mg \cdot 7H_2O$		0.05 g
Eau distillée	q. s. p.	100 ml

De son côté Turfitt [27] utilise pour la production d'un pigment vert fluorescent le milieu suivant où l'extraction chloroformique ne permet pas de déceler de pyocyanine :

NO_3NH_4		0.1 g
PO_4HK_2		0.025 g
$SO_4Mg \cdot 7H_2O$		0.025 g
Eau distillée	q. s. p.	100 ml

[25] Meader (P. D.), Robinson (J. H.) et Léonard (W.). *Am. J. Hyg.*, 1925, 5, 682.
[26] Georgia et Poe. *J. Bact.*, 1931, 22, 349.
[27] Turfitt (G. E.). *Biochem J.*, 1936, 30, 1323.

Pigment—*Staphylococcus* and *Micrococcus*

D. A. A. Mossel (1962). J. Bact. 84, 1140–1147.

Streak cultures (48 hr at 37°C on mannitol salt phenol red agar) were graded as: golden, weakly golden or white.

Pigment—*Vibrio*

G. H. G. Davis and R. W. A. Park (1962). J. gen. Microbiol. 21, 101–119.

Routine observation of all cultures in daylight; fluorescence under u. v. of growth upon a glucose-calcium carbonate medium.

Pigment—*Sarcina litoralis*

S. C. Nandy and S. N. Sen (1963). Can. J. Microbiol. 9, 601–611.

A solid medium was used for studying the growth of, and pigmentation by, *S. litoralis.* The medium was composed of the following ingredients: proteose peptone (Difco) 10 g; glycerine 10 g; magnesium sulphate heptahydrate 20 g; trisodium citrate 2.5 g; ferric chloride 0.02 g; potassium nitrate 1.0 g; sodium chloride 200 g; volume made up to 1000 ml with distilled water. The pH was adjusted to 6.7 – 6.8 using a Beckmann pH meter, and then 12 g agar (B. D. H. powder) was added to 1000 ml of the medium which was steamed for half an hour in an autoclave and poured into Petri dishes (4 in.) while hot. It was observed in preliminary experiments that steaming alone was sufficient to sterilize the medium, probably because of its high sodium chloride content.

Pigment was estimated according to the method suggested by Stahly and coworkers (1942a; 1942b) by using a spectrophotometer and a photoelectric colorimeter. While standardizing the method, the pigment of *S. litoralis* was extracted from the moist cells with cold methanol by triturating with sand. The methanol extract was then studied in a Beckman D. U. spectrophotometer, and the curve for the complete absorption spectra indicated that the pigment had its main absorption peak at 494 mμ and showed no absorption in the region of 660 mμ. A filter of 660 mμ was thus used in a Klett-Summerson photoelectric colorimeter to measure turbidity alone and a filter of 520 mμ for measuring the pigment and turbidity. The difference of these two readings gave the measure of the pigment. The results obtained were calculated on the basis of a standard turbidity and are expressed in terms of optical density per unit mass of cells.

B. Sobin and G. L. Stahly (1942a). J. Bact. 44, 265–276.

G. L. Stahly, C. L. Sesler and W. R. A. Brode (1942b). J. Bact. 43, 149–154.

Pigment—*Micrococcus roseus*

R. C. Eisenberg and J. B. Evans (1963). Can. J. Microbiol. 9, 633–642.

Pigment production was determined on slants of 'Staphylococcus medium' No. 110 (Difco) after 1 week of growth.

Pigment—*Staphylococcus*

A. T. Willis and G. C. Turner (1963). J. Path. Bact. 85, 395–405.

After 48 hours' incubation on monoacetate agar it was possible to determine whether pigmentation was yellow or 'orange', the latter category including deep orange, pale orange and cream-coloured pigmentation. This colour differentiation was observed in primary culture and all subsequent cultures on glycerol monoacetate agar, but was never seen on the other media used.

The monoacetate agar consists of brain-heart infusion agar containing 1 per cent of glycerol monoacetate (Willis and Turner, 1962).

A. T. Willis and G. C. Turner (1962). J. Path. Bact. 84, 337.

Pigment—*Clostridium rubrum*

H. Ng and R. H. Vaughn (1963). J. Bact. 85, 1104–1113.

The production of pigment was tested on potato, corn and pea infusion broths and agars.

Pea infusion broth was made by dissolving 5 g of yeast extract, 10 g of Tryptone, 1 g of soluble starch, 1 g of K_2HPO_4, and 0.5 g of sodium thioglycolate in 500 ml of distilled water and 500 ml of pea infusion (double strength) which was prepared according to the procedure given by the National Canners Association (1956).

Corn infusion broth was prepared by boiling 50 g of dried ground yellow corn in 1 liter of distilled water which was then filtered through cheese cloth to remove the larger particles. The filtrate was made up to its original volume to correct for evaporation. This medium had a pH of about 6.5 after autoclaving.

All media were adjusted to pH 7.0 with 1 N NaOH and sterilized by autoclaving at 120°C for 15 min. Solid media were prepared by addition of 2% agar to liquid media.
National Canners Association (1956). *A laboratory manual for the canning industry.* Washington D. C.

Pigment—*Streptococcus faecalis*

D. Jones, R. H. Deibel and C. F. Niven, Jr. (1963). J. Bact. 86, 171–172.

When into the basal medium (Tryptone, 1%; yeast extract, 0.5%; NaCl, 0.5%; K_2HPO_4, 0.5%; sodium citrate, 0.5%; glucose, 0.05%; with and without 1.5% agar; pH 7.0), the sulfate or chloride salts of manganese, iron, or zinc were added at concentrations of 0.01 to 0.05%, growth of the organisms on agar plates was brown in color; the color intensity varied with cation added. If citrate was omitted, or the concentration of glucose in the basal medium was increased to 0.5%, no coloration occurred. Pyruvate or serine (0.5%) could replace citrate in the basal medium.

Pigmentation developed only in aerobically grown cultures, either on agar plates or in liquid medium incubated on a reciprocal shaker. However, if 24-hr cultures incubated anaerobically on agar plates were allowed to stand in contact with air, pigmentation developed after about 1 hr.

The color may be due to the precipitation of metallic hydroxide.

Pigment—*Staphylococcus*

L. M. Carantonis and M. S. Spink (1963). J. Path. Bact. 86, 217–220.

The authors observed this on a new selective salt egg medium of the following composition.

A solution of salts is made up by mixing 170 ml 30 per cent NaCl, 44 ml 20 per cent $CaCl_2$, and 22 ml 20 per cent LiCl. This is sterilized by autoclaving and allowed to cool. One egg yolk is broken sterilely into this salt soltuion, and mixed well with it, making about 260 ml of stock salt egg fluid. This keeps well at 2–4°C. For use, 33 ml of this stock is pre-warmed to 45–50°C and added to 100 ml nutrient agar (1.0 per cent each of Lab-Lemco meat extract and peptone, and 1.2 per cent Davis New Zealand agar), melted and allowed to cool to about 50°C. The final concentrations of the constituents are approximately (g per 100 ml) NaCl, 5.0; $CaCl_2$, 1.0; LiCl, 0.5; egg yolk, 1.5. The medium sets with a moderate opalescence against which pigmentation shows up well.

Pigment—*Xanthomonas*

M. P. Starr and W. L. Stephens (1964). J. Bact. 87, 293–302.

The authors described in detail methods for the extraction and identification of a carotenoid "alcohol" with absorption maxima at 418, 437 and 463 mμ in petroleum ether.

Pigment—*Pseudomonas aeruginosa*

Y. Azuma and L. D. Witter (1964). J. Bact. 87, 1254.

A number of media have been formulated to enhance pyocyanin production, i.e., Klinge's medium, (Arch. Mikrobiol. 33:1, 1959). Pseudomonas Agar P (Difco Supplementary Literature, p. 281, 1962), and Sellers' medium (Sellers, Wynne and Graber, Bact. Proc., p. 129, 1961). Despite these media, there are still apyocyanogenic strains of *P. aeruginosa.*

An apyocyanogenic psychrophilic mutant recovered the ability to produce pyocyanin on Klinge's medium when it was previously grown in a medium consisting of 0.2% carbobenzoxy-DL-alanine, 0.9% DL-alanine, 0.5% glucose, 0.09% K_2HPO_4, 0.09% KH_2PO_4, 0.04% $MgSO_4 \cdot 7H_2O$, 0.002% $FeSO_4 \cdot 7H_2O$, 0.002% NaCl, and 0.002% $MnSO_4$, in distilled water for 4 days at 20°C and stored for 3 weeks at 5°C. This organism again became apyocyanogenic after three transfers on Klinge's medium or on Trypticase Soy Agar, but could consistently be made pyocyanogenic again by repeating the above procedure. The carbobenzoxy-DL-alanine could be replaced by carbobenzoxy-L-lysine or carbobenzoxy-DL-valine, but not by carbobenzoxychloride.

Pigment—*Vibrio marinus*

R. R. Colwell and R. Y. Morita (1964). J. Bact. 88, 831–837.

Growth and production of pyocyanine or fluorescin, or both, on Sabouraud maltose agar, Sabouraud dextrose agar, Burton's medium (Burton, Campbell, and Eagles, 1948), and King's medium (King, Campbell, and Eagles, 1948).

Synthetic seawater containing sodium chloride, 2.4%; potassium chloride, 0.07%; magnesium chloride (hydrated) 0.53% and magnesium sulfate (hydrated) 0.7% was used as the base. The inoculum was prepared from a 24 hour artificial seawater broth. Cultures were incubated at 18°C.

M. O. Burton, J. J. R. Campbell and B. A. Eagles (1948). Can. J. Res. 26C, 15–22.

J. V. King, J. J. R. Campbell and B. A. Eagles (1948). Can. J. Res. 26C, 514–519.

Pigment—*Micrococcus violagabriellae*

J. L. Nichols and J. N. Campbell (1964). Can. J. Microbiol. 10, 633–640.

The authors describe detailed methods for the investigation of a melanin-like pigment.

Pigment—*Micrococcus violagabriellae*

K. J. Steel (1964). J. gen. Microbiol. 36, 133–138.

The author reports a study of the pigment from cultures grown on 22 different media at 4 temperatures (37°C, 30°C, 22°C and ambient temperature), in light and darkness, in air, in an atmosphere of CO_2 and anaerobically.

Pigment—*Pseudomonas aeruginosa*

R. R. Colwell (1964). J. gen. Microbiol. 37, 181–194.

Production of pigment. Several media, designed for, or described as enhancing, pigment production by *Pseudomonas aeruginosa* were tested. All the media used were scored for production of a green water-soluble pigment. Media designed for production of pyocyanine (Burton, Eagles & Campbell, 1947; Burton, Campbell & Eagles, 1948; King, Ward & Raney, 1954), pyorubin (Meader, Robinson & Leonard, 1925; King, Ward & Raney, 1954), and fluorescin (King, Campbell & Eagles, 1948; King, Ward & Raney 1954; Paton, 1959), were prepared as described by these authors, inoculated and incubated up to 7 days, being routinely examined for the production of visible pigment or pigment fluorescence under ultraviolet radiation in the case of fluorescin. Pigment production was tested for on Sabouraud maltose agar (Martineau & Forget, 1958). Pyocyanine and pyorubin were identified by the presumptive tests outlined by Wetmore & Gochenour (1956). Finally, sporadic production of oxychlororaphin and chlororaphin crystals was noted but, as observed by Haynes & Rhodes (1962), no single medium served as indicator for consistent production of these compounds.

M. O. Burton, J. J. R. Campbell and B. A. Eagles (1948). Can. J. Res. 26, 15.
M. O. Burton, B. A. Eagles and J. J. R. Campbell (1947). Can. J. Res. 25, 121.
W. C. Haynes and L. J. Rhodes (1962). J. Bact. 84, 1080.
J. V. King, J. J. R. Campbell and B. A. Eagles (1948). Can. J. Res. 26, 514.
E. O. King, M. K. Ward and D. E. Raney (1954). J. Lab. clin. Med. 44, 301.
B. Martineau and A. Forget (1958). J. Bact. 76, 118.
P. D. Meader, G. H. Robinson and V. Leonard (1925). Am. J. Hyg. 5, 682.
A. M. Paton (1959). Nature, Lond. 184, 1254.
P. W. Wetmore and W. S. Gochenour (1956). J. Bact. 72, 79.

Pigment—*Staphylococcus*

S. I. Jacobs, A. T. Willis and G. M. Goodburn (1964). J. Path. Bact. 87, 151–156.

Glycerol monoacetate agar was as described by Willis and Turner (1962, 1963). The nutrient agar base containing 1 per cent glycerol monoacetate (B.D.H.) was autoclaved at 121°C for 15 min. (It should be noted that the use of the heart infusion base is essential for satisfactory results, and that overheating of the medium should be avoided.) After sterilization the medium was allowed to cool to 50–55°C when it was thoroughly mixed to ensure even distribution of the glyceride. Plates were poured immediately. Cultures on this medium were incubated at 37°C in the dark and read after 48 hr.

The nutrient agar base was a heart infusion broth (Difco) solidified with 1.5 per cent Davis New Zealand agar.

The medium was inoculated from a fresh 18 hr broth culture with a sterile throat swab.

A. T. Willis and G. C. Turner (1962). J. Path. Bact. 84, 337.
A. T. Willis and G. C. Turner (1963). J. Path. Bact. 85, 395.

Pigment—*Streptomyces*

A. J. Lyons, Jr. and T. G. Pridham (1965). J. Bact. 89, 159–169.

A detailed statement on the colorimetric determination of color is given.

Pigments—*Pseudomonas piscicida*

A. J. Hansen, O. B. Weeks and R. R. Colwell (1965). J. Bact. 89, 752–761.

Pigments were extracted from 48-hr populations grown on peptone-seawater-agar, by use of the procedures described by Weeks *et al.* (1962) except that initial extraction was done with methanol. Methanol extracts were evaporated to dryness *in vacuo,* and were redissolved in 95% ethyl alcohol for resolution by thin-layer chromatography (Brinkmann Instruments Inc., Westbury, N. Y.) by use of silica gel G (E. Merck, Darmstadt, West Germany) developed with concentrated NH_4OH-*n*-propanol (23:77) at 25°C. Pigment extracts also were separated by column chromatography. Columns were packed with a slurry of Permutit (Hartman-Leddon, Philadelphia, Pa.), and a concentrated methanolic extract was put on the column; the column was then developed progressively with: (i) water, (ii) NH_4OH-water (2:98 and 5:95, v/v), and (iii) NH_4OH-methanol (20:80, v/v), followed by a wash with water which removed a final fraction.

O. B. Weeks, S. M. Beck, M. D. Thomas and H. D. Isenberg (1962). J. Bact. 84, 1118.

Pigment—*Staphylococcus epidermidis*

O. Sandvik and R. W. Brown (1965). J. Bact. 89, 1201–1208.

The paper describes the mass production, extraction and examination of the pigment.

Pigment—*Achromobacter*

J. A. Duerre and P. J. Buckley (1965). J. Bact. 90, 1686–1691.

An unnamed species assigned to the genus *Achromobacter* produced a red pigment when grown on a medium containing yeast extract and tryptophane at an optimal pH of 8.0 and optimal temperature of 25°C.

The pigment acts as an electron acceptor to formic dehydrogenase. The oxidised form has absorption peaks at 506 and 304 mμ. The former disappears on reduction with sulfite.

Pigment—streptomycetes

T. G. Pridham (1965). Appl. Microbiol. 13, 43–61.

This is a report of the International workshop on determination of the color of streptomycetes.

Pigment—*Staphylococcus*

R. Zemelman and L. Longeri (1965). Appl. Microbiol. 13, 167–170.

As the high sodium chloride content of Chapman's (1946) Medium 110 enhances pigmentation of staphylococci, the color of the pigment was observed after 48 hr of incubation at 37°C on this medium and after exposing the plates to room temperature. Strains were classified in three different groups according to the pigment they produced: those showing golden (ranging from light-cream to deep-orange), white, and yellow-green pigment.

G. H. Chapman (1946). J. Bact. 51, 409–410.

Pigment—*Pseudomonas*

A. E. Wasserman (1965). Appl. Microbiol. 13, 175–180.

The authors report procedure for the extraction and spectro-photometric examination of water soluble pigment of *Pseudomonas aeruginosa, P. geniculata, P. ovalis* and *P. fluorescens.*

Three media were used to grow the organisms. Asparagine broth, as described by Elliott (1958), contained 0.1% asparagine, 0.05% $MgSO_4 \cdot 7H_2O$, and 0.05% K_2HPO_4 in distilled water. Malate-phosphate medium (M-P) contained 0.34% KH_2PO_4, 0.67% Na_2HPO_4, 0.5% DL-malic acid, 0.04% $MgSO_4 \cdot 7H_2O$, and 0.2% $(NH_4)_2HPO_4$ in distilled water; the pH was adjusted to 6.8 before sterilization. The third medium was Difco Pseudomonas Agar F.

R. P. Elliott (1958). Appl. Microbiol. 6, 241–246.

Pigment—*Pseudomonas*

A. H. Wahba (1965). Appl. Microbiol. 13, 291–292.

The authors report the production of pyrorubin by strains of *Pseudomonas aeruginosa* which did not produce pyocyanin on "stock nutrient agar plates."

Pigment—*Pseudomonas*

G. L. Bullock, S. F. Snieszko and C. E. Dunbar (1965). J. gen. Microbiol. 38, 1–7.

Production of fluorescent pigment was tested on Pseudomonas F Agar

(Difco) slopes. This commercial medium is designed to enhance fluorescin production by *Pseudomonas.* Fluorescence was tested by exposing inoculated slopes to ultraviolet radiation after 24-48 hr incubation. Lyophilized cultures and the same cultures which had been kept for 10 years in stock culture were tested. Test cultures incubated at 20-22°C for 1 week.

Pigment—*Pseudomonas aeruginosa*

A. H. Wahba and J. H. Darrell (1965). J. gen. Microbiol. 38, 329–342.

Enhancement of pyocyanin production on a medium modified from Sierra (1957) and containing (g): Bactopeptone, 10; sodium chloride, 5; $CaCl_2 \cdot H_2O$, 0.2; $MgSO_4 \cdot 7H_2O$, 0.1; agar, 18; distilled water to 1 L. Tween 80 solution, autoclaved separately, was added in a final concentration of 1%; pH 7.4. For comparative purposes, strains were also inoculated on nutrient agar No. 2 plates, as this medium improved the production of pyocyanin in some strains.

Nutrient agar No. 2 (Oxoid) contained Lab Lemco beef extract, 10 g; peptone (Oxoid L 37), 10 g; sodium chloride, 5 g; 'Ionagar' No. 2, 10 g; distilled water to 1 L; pH 7.4.

G. Sierra (1957). Antonie van Leeuwenhoek 23, 15.

Pigment—*Alcaligenes*

R. G. Mitchell and S. K. R. Clarke (1965). J. gen. Microbiol. 40, 343–348.

Fluorescin production was tested with ultraviolet radiation on the medium B of King, Ward and Raney (1954).

E. O. King, M. K. Ward and D. E. Raney (1954). J. Lab. clin. Med. 44, 301.

Pigment—*Mycoplasma*

S. Razin and R. C. Cleverdon (1965). J. gen. Microbiol. 41, 409–415.

Estimation of carotenoids. Part of the membrane suspension (usually one-fifth) was centrifuged at 37,000 *g* for 30 min. The resulting pellet was extracted with 7 ml of boiling ethanol for 10 min in the dark and under a nitrogen atmosphere (Rothblat, Ellis and Krtichevsky, 1964). The extracted membranes were removed by centrifugation and the extinction at 442 mμ of the supernatant fluid was measured in a Beckman DB spectrophotometer. Extraction of the carotenoids from wet membrane material with boiling ethanol was found superior to extraction of freeze-dried membranes with chloroform+methanol (2 + 1, by vol). The absorption spectrum of *Mycoplasma laidlawii* carotenoids measured in ethanol showed absorption maxima at 418, 442 and 472 mμ; the peak at 442 mμ was the highest. The amount of carotenoid pigments in membranes was expressed therefore as extinction at 442 mμ $\times$ 1000

per mg membrane protein or per mg dry-weight of membrane material.
G. H. Rothblat, D. S. Ellis and D. Kritchevsky (1964). Biochim. biophys. Acta 84, 340.

Pigment—*Pseudomonas aeruginosa*

V. Hurst and V. L. Sutter (1966). Survival of *Pseudomonas aeruginosa* in the hospital environment. J. infect. Dis. 116, 151–154.

The authors used Seller's agar slants.

W. Sellers (1964). J. Bact. 87, 46–48.

Pigment—*Staphylococcus aureus*

A. T. Willis, J. J. O'Connor and J. A. Smith (1966). J. Path. Bact. 92, 97–106.

The authors prepared a number of media from natural fats, prepared and purchased soaps, and fatty acid salts using a nutrient agar base to test for pigment production.

Procedures were as follows: –

The nutrient base used in all media was heart infusion broth (Difco) solidified with 1.5 per cent of Davis New Zealand agar (pH 7.4). Sterilisation was at 121°C for 10 min.

Water-soluble substances, such as salts and carbohydrates, were prepared as sterile stock solutions for addition to the sterile agar base.

Fat media. These were prepared by emulsifying the fat into the hot melted nutrient agar base with a Silverson mixer. The emulsions were sterilised and the plates poured immediately.

Fatty acid media. The sodium salt of each fatty acid was neutralised with 1.5 N-HCl and the volume made up with distilled water to give a final concentration of 20 per cent of the salt. This stock solution was "sterilised" at 100°C for 30 min, before it was added to the sterile melted nutrient base.

Water-insoluble methyl esters were incorporated as described above for fats.

Commercial soap and TEM 4T media. The soaps and TEM 4T were neutralised with 1.5N-HCl and 1.5 N-NaOH respectively, and were then dealt with as described for fatty acids.

Butter soap and cream-soap media. (1) After separation from butter and cream, the lipid was saponified by boiling with alcoholic NaOH for about 2 hr. The fatty acids, which were then obtained by acidification, were washed with water and neutralised with 1.5N-NaOH. (2) The fat was treated with excess of aqueous lipase at 37°C for 48 hr, after which digestion was complete. The fatty acids were separated from the mixture and were washed and neutralised with 1.5 N-NaOH. Various concentrations of these additives were tested.

Inoculation of media. Inocula were taken from 24-hr nutrient agar cultures. The test media were spot-inoculated with charged sterile swabs, six or more cultures being accommodated on each plate. Incubation was at 37°C for 24 and 48 hr. Control cultures on the nutrient base alone were included in each series of experiments.

Butter oil was prepared from dairy cream and from commercial butter by (1) extraction with ether, and (2) holding at 60°C, which allowed the lipid fraction to layer out. These preparations of butter oil were emulsified into the hot melted nutrient base in concentrations ranging from 0.2 to 4.0 per cent.

The following proved the most useful in the concentrations indicated – cream, 10%; butter oil, 3.5%; human fat, 3.0%; glycerol monoacetate, 1.0%; TEM 4T (diacetyl tartaric acid ester of tallow monoglycerides, Hachmeister Inc.), 5.0%; sodium stearate, 2.0% and methyl caproate, 2.0%.

Pigment—*Staphylococcus*

M. Kocur, F. Přecechtěl and T. Martinec (1966). J. Path. Bact. 92, 331–336.

Pigmentation was examined in cultures grown at 37°C for 48 hr on egg yolk agar, tryptone agar, and blood agar and then stored at room temperature in the light for another two days.

Pigment—*Staphylococcus*

A. T. Willis, J. A. Smith and J. J. O'Connor (1966). J. Path. Bact. 92, 345–358.

Culture media. The nutrient base used in all the following media was heart infusion broth (Difco), solidified as required with 1.5 per cent of Davis New Zealand agar. Sterilisation was at 121°C for 15 min.

Glycerol monoacetate agar was prepared as described by Willis and Turner (1962, 1963) and Jacobs, Willis and Goodburn (1964).

Cream agar was as described by Willis (1960).

Inoculation of media. The agar media were inoculated from fresh 18-hr broth cultures with sterile throat swabs. This method ensured even inoculation, and rapid multiple subcultures could be made without recharging the swab. A 9 cm petri dish conveniently accommodated 9-12 cultures.

The authors found the cream agar to be superior to the glycerol monoacetate agar for pigment production.

S. I. Jacobs, A. T. Willis and G. M. Goodburn (1964). J. Path. Bact. 87, 151.

A. T. Willis (1960). J. Path. Bact. 80, 379.

A. T. Willis and G. C. Turner (1962). J. Path. Bact. 84, 337.

A. T. Willis and G. C. Turner (1963). J. Path. Bact. 85, 395.

Pigment—*Staphylococcus*

J. J. O'Connor, A. T. Willis and J. A. Smith (1966). J. Path. Bact. 92, 585–588.

Culture media. Unless otherwise stated the nutrient base used in all media was heart infusion broth (Difco) solidified with 1.5 per cent of Davis New Zealand agar (pH 7.4). Sterilization was at 121°C for 10 min. Cream agar was prepared as described by Willis (1960). Salt agar was prepared by adding 6.5 per cent of sodium chloride to the nutrient base before sterilization. Sodium chloride cream agar was cream agar containing 6.5 per cent of sodium chloride.

Inoculation of media. Inocula were taken from 24-hr nutrient agar cultures. The test media were spot-inoculated with charged sterile swabs, six cultures being accommodated conveniently on each plate. Incubation was at 37°C for 24 and 48 hr unless otherwise stated.

A. T. Willis (1960). J. Path. Bact. 80, 379.

Pigment—*Staphylococcus*

R. W. Brown (1966). J. Bact. 91, 911–918.

The authors describe the determination of color of cultures of *S. epidermidis* by spectral reflectance colorimetry.

Pigment—*Nocardia corallina*

O. R. Brown and J. B. Clark (1966). J. Bact. 92, 1844–1845.

The paper describes the crystallisation and study of the major pigment which apparently is not a carotenoid.

The authors note that the Carr-Price test (P. Karrer and E. Jucker, *Carotenoids,* Elsevier, New York, 1950) "may be misleading as it is not absolutely specific for carotenoids."

Pigment —*Micrococcus roseus*

J. J. Cooney and O. C. Thierry (1966). Can. J. Microbiol. 12, 83–89.

A defined medium has been developed which supports good growth and pigment synthesis of *Micrococcus roseus* ATCC 516. The medium contains fructose, adenine, alanine, arginine, glutamic acid, glycine, isoleucine, methionine, proline, serine, and inorganic salts. The medium is not a minimal medium, but omission of any component decreases growth or pigment content, or both. Pigment synthesis parallels culture development. Addition of leucine or mevalonic acid decreases pigment content. Diphenylamine (10^{-7} M) decreases pigment content 27%, suggesting that the carotenoids are principally xanthophylls. Absorption spectra of extracted pigments differ when glucose or fructose is the carbon source, and when fructose-containing medium is supplemented with mevalonic acid.

Pigment—*Bacillus cereus* var. *alesti*

R. L. Uffen and E. Canale-Parola (1966). Can. J. Microbiol. 12, 590–593.

Bacillus cereus var. *alesti (Bacillus thuringiensis* var. *alesti)* Toumanoff and Vago, 1951 (see also Heimpel and Angus, 1963), a pathogen of silkworms, produces a red pigment when grown on nutrient agar at 15°C, but forms no pigment on the same medium at 28°C (W. A. Smirnoff, personal communication).

The absorption spectrum of this pigment in 2 N NaOH exhibited maxima at 242, 282, and 410 mμ, identical with those characteristic of pulcherrimin purified from other aerobic sporeforming bacteria (Canale-Parola, 1963).

E. Canale-Parola (1963). Arch. Mikrobiol. 46, 414–427.

A. M. Heimpel and T. A. Angus (1963). *Insect pathology,* Vol. 2, Academic Press, New York. pp. 21–73.

C. Toumanoff and C. Vago (1951). C.r. hebd. Séanc. Acad. Sci., Paris, 233, 1504–1506.

Pigment—*Pseudomonas aeruginosa*

J. C. MacDonald (1966). Can. J. Microbiol. 12, 771–774.

Medium, Growth Conditions, and Determination of Pyocyanine. The compositions of the media are presented in Table 1, K_2HPO_4 was autoclaved separately from other components of the media, and quinic acid was neutralized with NaOH and sterilized by filtration. Media (10 ml for stationary cultures and 25 ml for shake cultures) was contained in 125 ml Erlenmeyer flasks, and was adjusted to pH 7.0 to 7.2 with sterile, dilute NaOH or HCl just before inoculation. After inoculation of the media, the flasks were incubated for 4 days at 30°C either without shaking or on a rotary shaker operating at 110 rpm with a radius of motion of 2.9 cm. The cultures were then acidified to pH 1 to 3 with HCl, water was added to replace that lost by evaporation, and a portion of the cultures was centrifuged. Part (0.5 ml) of the supernatant liquid was made slightly basic with 1.0 M $NaHCO_3$, and the blue pyocyanine was extracted by thorough mixture with 10 ml of chloroform. The chloroform was clarified by centrifuging and its optical density determined at 695 mμ. (Spectronic 20 spectrophotometer) and corrected for a blank. A calibration had been made for the instrument and cuvettes, relating optical density to concentration of pyocyanine. The extraction and determination of optical density were done rapidly because of the reported instability of pyocyanine (Kurachi, 1958).[7] To obtain blank values (which were always low), the pyocyanine was extracted from the chloroform with 1.5 M HCl, and the optical density of the chloroform determined again.

TABLE I

Production of pyocyanine on various media

Constituents of medium, g/100 ml	Medium number and reference						
	1 Burton *et al.* (2)	2 Hellinger (5)	3 Blackwood and Neish (1)	4 Ingram and Blackwood (6)	5 MacDonald (8)	6 Sheikh and MacDonald (9)	7 Frank and DeMoss (4)
Glycerol	1.0	1.0	1.0	1.0	1.0	1.0	2.5
L-Leucine	0.6	0.8	0.8	0.8			
DL-Alanine		0.4	0.4	0.8			1.0
L-Alanine					0.6	0.8	
Glycine	0.6						
Quinic acid						1.0	
Ferric citrate							0.01
$FeSO_4 \cdot 7H_2O$	0.001	0.001	0.001	0.001	0.001	0.001	
$MgSO_4 \cdot 7H_2O$	2.0	0.2	0.2	0.2	0.2	0.2	
$MgCl_2 \cdot 6H_2O$							0.41
Na_2SO_4							1.42
K_2HPO_4	0.04	0.04	0.05	0.01	0.01	0.01	0.014
$CaCO_3$		0.1	0.1	0.1			
Yield of pyocyanine,* mg/ml of culture							
Stationary culture							
ATCC 9027	0.11–0.25	0.23–0.26	0.22–0.26	0.23–0.30	0.07–0.08	0.08–0.11	0.21–0.23
PRL F20	0.27–0.28	0.20–0.24	0.16–0.19	0.14–0.22	0.01–0.02	0.07–0.14	0.13–0.16
Shake culture							
ATCC 9027	0.02–0.05	0.01–0.02	0.01	0.26–0.32	0.12–0.16	0.14–0.19	0.29–0.30
PRL F20	0.02–0.05	0.05–0.06	0.02–0.05	0.29–0.32	0.10–0.14	0.16–0.22	0.31–0.35

*The range of results for two experiments with duplicate flasks in each is given.

1. A. C. Blackwood and A. C. Neish (1957). Can. J. Microbiol. 3, 165–169.
2. M. O. Burton, J. J. R. Campbell and B. A. Eagles (1948). Can. J. Res. C. 26, 15–22.
3. M. O. Burton, B. A. Eagles and J. J. R. Campbell (1947). Can. J. Res. C. 25, 121–128.
4. L. H. Frank and R. D. DeMoss (1959). J. Bact. 77, 776–782.
5. E. Hellinger (1951). J. gen. Microbiol. 5, 633–639.
6. J. M. Ingram and A. C. Blackwood (1962). Can. J. Microbiol. 8, 49–56.
7. M. Kurachi (1958). Bull. Inst. Chem. Res. Kyoto Univ. 36, 174–187.
8. J. C. MacDonald (1963). Can. J. Microbiol. 9, 809–819.
9. N. M. Sheikh and J. C. MacDonald (1964). Can. J. Microbiol. 10, 861–866.

Pigment—*Erwinia*

M. P. Starr, G. Cosens and H-J. Knackmuss (1966). Appl. Microbiol. 14, 870–872.

A blue water insoluble pigment is produced by some species of *Erwinia (vide infra).* It has been identified as indigoidine, 5,5′-diamino-4,4′-dehydroxy-3,3′-diazadiphenoquinone-(2,2′).

Conditions for production were as follows:

The three cultures intensively studied were ICPB-EM108, *Erwinia maydis,* isolated from corn stalk rot by Kelman in 1957; ICPB-EC176, *Erwinia chrysanthemi,* isolated from slow wilt of carnations by Bakker and Scholten in Holland; and ICPB-EC207, *Erwinia cytolytica,* isolated by Lazar from dahlia in Roumania in 1962.

E. maydis was grown on YDC agar poured in thick layers into 100 mm petri dishes. *E. chrysanthemi* and *E. cytolytica* were grown on potato-glucose-peptone-agar prepared as follows. Sound potatoes were scrubbed, peeled, sliced, and diced; 200 g of the diced potatoes was placed in 1,000 ml of distilled water and cooked in autoclave for 1 hr at 5 psi of steam pressure. The potato infusion was filtered through cheesecloth, and the volume was restored to 1,000 ml. To 1,000 ml of potato infusion were added 10 g of glucose, 10 g of peptone (Difco), and 20 g of agar (Difco). The pH was adjusted to 7.2, and the medium was sterilized by autoclaving. The medium was then poured in thick layers into 100 mm petri dishes.

Selected pigmented colonies of the three organisms were suspended in sterile distilled water in test tubes. Six to eight streaks were made on each plate with a transfer loop.

Plates inoculated with *E. maydis* were incubated at room temperature (*c.* 23 to 25°C) for 4 days. (Longer incubation resulted in production of a gritty material which centrifuged in a layer under the pigment and made separation of the pigment difficult). *E. chrysanthemi* and *E. cytolytica* were incubated at room temperature for 3 or 4 days, and the plates were then stored in a refrigerator at 4°C for 24 hr. The plates were scraped, allowed to incubated at room temperature, and then scraped a second time. This procedure tended to increase the meager yield with these two strains.

The pigment was separated from the cells by centrifugation, lyophilised and then dissolved in dimethylformanide, filtered and the concentration estimated spectrophotometrically at 602 m μ (authentic λ max, 602 mμ; log ϵ , 4.37).

The pigment was purified by a modification of the method of Kuhn *et al.* (1965).

To minimize the time the pigment stays in the unstable leucoform, a large excess of the sodium dithionite must be avoided.

The lyophilized, pigment-containing material (*c.* 500 mg) was suspended in 100 ml of saturated sodium bicarbonate solution. The sodium dithionite was added in small quantities until most of the indigoidine was dissolved. The solution was quickly filtered through silica gel; the pigment was reoxidized with oxygen, separated by centrifugation, and then washed with 1 N acetic acid.

The pigment so obtained were examined spectrophotometrically.

R. Kuhn, M. P. Starr, D. A. Kuhn, H. Bauer and H. J. Knackmuss (1965). Arch. Mikrobiol. 51, 71–84.

Pigment—*Pseudomonas*

R. Y. Stanier, N. J. Palleroni and M. Doudoroff (1966). J. gen. Microbiol. 43, 159–271.

The authors used Medium A and Medium B of King, Ward and Raney (1954). In addition, to elicit the blue pigment of *Ps. lemonnieri* they used a medium containing (%, w/v) Bacto peptone, 0.5; glucose, 1.0; Bacto agar, 2; in tap water.

The original should be consulted for the lengthy statement on pigmentation.

E. O. King, M. K. Ward and D. E. Raney (1954). J. Lab. clin. Med. 44, 301.

Pigment—*Azotobacter*

C. Hardisson and Mme M. Robert-Gero (1966). Annls Inst. Pasteur, Paris 111, 486–496.

Hardisson and Pochon (Annls Inst. Pasteur, Paris 1966, 111, 66–75) describe the production of the dark brown-black pigment by strains of *Azotobacter* growing on a synthetic medium containing benzoate. This pigment was analyzed in comparison with similar humic pigments.

The extracts were fractionated by dialysis and precipitation in three main fractions: fulvic, humic and hymetomelanic acids. U. V. and I. R. absorption spectra, approximative molecular weights (on Sephadex column), the presence of amino acids, carbohydrates and other chemical properties were determined. The results of the studies reveal a close analogy between the natural humic extracts and the *Azotobacter*-synthetized humic-like substance.

They used the following methods of extraction and examination.

Extraction. – Pour les trois sols, nous avons utilisé la méthode d'extraction au pyrophosphate de sodium décrite précédemment [23]. Les extraits bruts lyophilisés dans un lyophiliseur Lyoboy C. S. D. (Secfroid) ont été repris dans de l'eau distillée, puis dialysés afin d'éliminer les sels et les petites molécules. L'eau de dialyse est changée fréquemment jusqu'à ce que son pH soit égal à celui de l'eau distillée et qu'elle soit incolore.

Les eaux de dialyse sont concentrées sous vide, puis lyophilisées (fraction D).

La fraction non dialysable est également lyophilisée et gardée dans un dessiccateur sous vide (fraction ND).

L'obtention de la substance para-humique synthétisée par *Azotobacter chroococcum* a été décrite précédemment [10] ; cette substance est également divisée en fraction dialysable et non dialysable.

Fractionnement. – Les parties dialysables ont été fractionnées par passages successifs sur résines échangeuses d'ions : cationique (Dovex 50 H^+) et anionique (Amberlite $IR_{45}OH^-$). Comme éluants, nous avons utilisé NH_4OH 2 N pour obtenir la fraction basique et ClH à concentrations croissantes jusqu'à 2 N pour l'obtention de la fraction acide. Dans ces différentes fractions, nous avons dosé les acides aminés, les glucides et les acides organiques.

Détermination des poids moléculaires approximatifs par passage sur gel de Sephadex G-75, G-100 et G-200 – Comme éluant, nous avons utilisé une solution 0.01 M de ClNa. Les fractions ont été recueillies, par volumes de 5 ml, à l'aide d'un collecteur de fractions Seive. La densité optique des fractions a été mesurée à 450 mμ à l'aide d'un photocolorimètre Spectronic 20 [20].

Etude spectrométrique. – L'examen dans l'U. V. a été fait en utilisant un spectrophotomètre Beckman DK-2A.

Les fractions non dialysables ont été mises en suspension dans le nujol et passées dans un appareil Perkin-Elmer infracord, avec prisme de ClNa, pour enregistrer le spectre I. R.

a) *Etude chimique.* La méthode de Burges [3] a été utilisée pour le fractionnement classique en acides humique, fulvique et hymétomélanique des fractions non dialysables.

b) La *matière organique* a été determinée par perte au feu à 450°C dans un four électrique.

c) Le *carbone organique* a été dosé par la méthode d'Anne modifiée [5] et l'azote par la méthode dite microkjeldahl.

d) *P total* a été déterminé par la méthode de Fiske et Subbarow modifiée par Faure [7].

e) NH_2 *total* a été déterminé par la méthode de Rosen [25] avant et après hydrolyse par ClH-6 N dans des tubes scellés, à 105°C pendant seize heures.

f) Les *acides aminés* ont été mis en évidence par chromatographic couche mince sur plaque de cellulose M N 300 (Macherey et Nagel) ; les solvants utilisés ont été les suivants :

n-propanol : eau	7 : 3	[33]
n-propanol : ammoniaque 8.8 p. 100	8 : 2	[33]
n-butanol : acide formique : eau	6 : 1 : 2	[1]
n-butanol : acide acétique : eau (phase organique)	4 : 1 : 5	[33]

Les acides aminés ont été révélés par une solution 0.2 p. 100 de nynhidrine dans l'éthanol.

Les *hexoses* ont été déterminées par la méthode à l'anthrone [28].

g) Les *glucides* ont été mis en évidence par chromatographie couche mince sur plaque de cellulose et sur papier Whatman n° 1, avant et après hydrolyse acide dans HCl 2N dans tubes scellés à 105°C pendant quatre heures. Les solvants utilisés ont été les suivants :

acétate d'éthyle : pyridine : eau	2 : 1 : 2	[13]
n-butanol : acide acétique : eau	2 : 1 : 1	[12]
n-butanol : éthanol : eau	4 : 1.1 : 1.9	[12]

Révélateurs utilisés : phtalate d'aniline [22] et solution acétonique de NO_3Ag.

h) Les *acides organiques* ont été mis en évidence par chromatographie couche mince sur cellulose, sur gel de silice G (Merk), et sur papier Whatman n° 1. Les solvants utilisés ont été les suivants :

n-butanol : acide formique : eau	6 : 1 : 2	[1]
propanol : ammoniaque	6 : 4	[1]
éthanol 96° : NH_3 25 p. 100 : eau	25 : 4 : 3	[2]

Comme révélateur, nous avons utilisé le réactif de Linskens [19] et le vert de bromocrésol.

[1] Bancher (E.) et Scherz (H.). *Mikrochim. Acta,* 1964, 6, 1159.

[2] Bram (D.) et Geenen (H.). *J. Chromatog.,* 1962, 7, 56.

[3] Burges (A.). *Sci. Proc. Roy. Dublin Soc.,* 1960, Ser. A, 1, 53.

[7] Faure (M.). in J. Loiseleur, *Techniques de Laboratoire,* 2e édit., Masson, édit., Paris, 1954.

[10] Hardisson (C.) et Pochon (J.). *Ann. Inst. Pasteur,* 1966, 111, 66.

[12] Hough (L.), Jones (J. K. N.) and Wadman (W. H.). *J. chem. Soc.*, 1950, 1072.
[13] Isherwood (F. A.) and Jermyn (M. A.). *Biochem. J.*, 1951, 48, 515.
[20] Martin (F.), Dubach (P.), Mehta (N. C.) und Deel (H.). *Pflanzenernähr.*, 1963, 103, 27.
[23] Pignaud (G.), Milkowska (A.), Chalvignac (M. A.), Robert-Gero (M.) et Pochon (J.). *Ann. Inst. Pasteur,* 1966, 111, 76.
[25] Rosen (H.). *Arch. Biochem. Biophys.*, 1957, 67, 10.
[33] Wollenweber (P.). *J. Chromatog.*, 1962, 9, 369.

PIGMENT FROM TYROSINE

Pigment from tyrosine—organism resembling *Haemophilus pertussis*
W. L. Bradford and B. Slavin (1937). Am. J. publ. Hlth 27, 1277–1282. (Copyright 1937, by the American Public Health Association, Inc.)

1. Medium: Growth of the organism upon slants of Bordet medium produced a definitely darker discoloration of the medium in 48 hours than did typical strains of *H. pertussis.*

2. Medium: Growth in Douglas broth was abundant and after 5-7 days produced the characteristic copy – mucoid type with a brownish-red discoloration of the broth.

3. Medium: Upon potato, the organism grew well and produced a brownish black color in 48 hours. A similar color resulted from the growth of 2 strains of *Brucella bronchisepticus.*

4. Medium: On Levinthal's medium a light brown discoloration of the medium occurred.

5. Medium: All of the atypical strains were then cultured on peptone-iron agar slants and in each a definite brownish-red change occurred along the slant but very little along a stab into the butt.

Method of determination of nature of change in color:

When 48 hour slant growths were extracted with water, alcohol, or acetone for 24 hr, the solutions became light brown and resembled in color known solutions of iron compounds. It was not soluble in ether or chloroform. Aqueous extracts of the agar mixture upon which typical strains had grown and of uninoculated agar mixture remained colorless. The pH of the aqueous extracts determined with the La Motte Comparator were:

atypical strains	7.4
typical strains	7.0
agar control	6.8

Absorption curves were made with a Bausch and Lomb Quartz Spectroscope on the aqueous extracts of the agar growth of 3 atypical strains, of 1 typical strain, of uninoculated agar medium, and of solutions of ferric ammonium citrate, of ferric chloride and of ferric alum. The solutions of the known iron compounds were made approximately to match in color those of the atypical strain extracts.

In each instance curves were obtained in the ultraviolet with peaks at 2,670 Å.

Pigment from tyrosine—*Rhizobium*

H. Z. Gaw (1944). J. Bact. 48, 483–489.

Tested by using the carrot extract medium of Fred and Waksman (1928)* plus 0.15% tyrosine.

**Laboratory Manual of General Microbiology.* McGraw-Hill, N. Y. 1928 p. 145.

Pigment from tyrosine

P. Hauduroy (1951). Techniques Bactériologiques. Masson et Cie, Éditeurs, Paris.

GELOSE A LA TYROSINE

Ajouter 0.05 p. 100 de tyrosine à de la gélose nutritive. Si une pigmentation apparaît, elle se produit entre le 2e et 14e jour. *(A. T. C. C.).*

Pigment from tyrosine—*Haemophilus parapertussis*

P. W. Ensminger (1953). J. Bact. 65, 509–510.

1. Medium: Medium proposed by Cohen and Wheeler (1946) was modified in order to eliminate as much iron as possible. Instead of Difco Casamino Acids, Technical, individual amino acids were added according to the Difco analyses of Casamino Acids, Technical (personal communication). $FeSO_4 \cdot 7H_2O$, $CuSO_4 \cdot 5H_2O$, and KH_2PO_4 were left out of the media, as they would interfere with quantitative iron determinations. To this medium made with triple glass distilled water, were added 2.0% agar and 0.1% L-tyrosine. Using the methods of Sandell (1950) and Woiwood (1947), no iron was found in this medium. Results were read with a Coleman, Jr. spectrophotometer using a 508 mμ filter.

Method. When grown on the above agar all eight strains of *H. parapertussis* produced the red-brown pigment. The strains then were carried for eight sub-cultures on this medium with the same pigment production throughout the eight transfers.

Thus *H. parapertussis* pigment is formed in the absence of any iron.

2. Medium: Bordet-Gengou agar with or without blood.

0.1% L-tyrosine added.

Method. A dark brown pigment was observed in 4-5 days when *H. parapertussis* was grown on Bordet-Gengou agar with or without blood.

However, when 0.1% L-tyrosine was added, *H. parapertussis* rapidly discolored the medium. Within 18 hr there was a deep red-brown color running the length of the slant about 7 mm deep, which spread within 48 hr throughout the entire agar butt. The color gradually turned to a very dark brown-black pigment when the tube was kept at 37°C.

S. Cohen and M. W. Wheeler (1946). Am. J. publ. Hlth 36, 371–376.
E. B. Sandell (1950). *Colorimetric determination of traces metals* 2nd Ed. Interscience Publ. Inc. New York.
A. J. Woiwood (1947). Biochem. J. 41, 39–41.

Pigment from tyrosine—*Aeromonas*

P. J. Griffin, S. F. Snieszko, S. B. Friddle (1953). J. Bact. 65, 652–659.

Some factors determining the production of a melano-pigment by *Bacterium salmonicida* were investigated. Oxygen, hydrogen ion concentration, temperature, and amount of substrate were found to be important in pigment formation. In the absence of tyrosine and phenylalanine, pigmentation of the medium did not occur. The addition of either glycyl-L-phenylalanine, L-phenylalanine, or D-phenylalanine resulted in the production of a brown or amber-to-black pigment. A bright salmon pink color developed when β-2-thienylalanine was tested.

Compounds which contained an acetyl or chloracetyl substituent in place of a hydrogen atom of the α amino group, or a hydroxyl group in place of a hydrogen atom of the β carbon atom in the alanine side chain of phenylalanine were inactive. Saturation of the benzene ring also produced an inactive substance. Glycyl-L-tyrosine, L-tyrosine, and L-tyrosine ethyl ester were chromogenic. Substitution of the hydrogen atom of the α amino or the carboxyl group of the alanine side chain of tyrosine and substituents of the 3 position on the benzene ring yielded inactive compounds. *B. salmonicida* is proposed as a useful tool in studies of pigmented products from phenylalanine and tyrosine.

For full details of methods, see the original.

Pigment from tyrosine—*Bordetella parapertussis*

E. Rowatt (1955). J. gen. Microbiol. 13, 552–560.

Inoculum and medium:

C. W. 1. medium – this was a liquid medium similar to that of Cohen and Wheeler (1946) except that cysteine and yeast extract were sterilized by filtration and added to the medium after it had been autoclaved. Yeast extract was made by extracting dried yeast (Standard Yeast Co.) for 15 min with an equal weight of water at 100°C and used at a concentration of 1 in 400. The amino acid source was Difco vitamin-free Casamino acids containing 36% amino acid (dry weight). The pH value of C. W. 1 medium after autoclaving was 6.9–7.0.

Washed suspensions: Organisms were grown in 250 ml quantities of C. W. 1 medium contained in penicillin flasks for 24 hours. Used in inoculum of about 1 mg dry weight organisms per 250 ml. The organisms were spun from the medium, washed in phosphate saline (pH7) and re-suspended in an appropriate buffer.

Method. Bradford and Slavin (1937) showed that *Bordetella parapertussis* formed a brown pigment during growth. Ensminger (1953) found that tyrosine was necessary for the formation of this pigment. Washed suspensions of *B. parapertussis* incubated with tyrosine formed a pink pigment which turned brown on prolonged incubation. Glutamate increased pigment formation and the reaction was more rapid at pH 7.4 than at the lower pH values.

S. M. Cohen and M. W. Wheeler (1946). Am. J. publ. Hlth 36, 371.
W. L. Bradford and B. Slavin (1937). Am. J. publ. Hlth 27, 1277.
P. W. Ensminger (1953). J. Bact. 65, 509.

Pigment from tyrosine—*Dermatophilus*

M. A. Gordon (1964). J. Bact. 88, 509–522.

The authors used tyrosine agar plates prepared and read as described by Gordon and Smith (1955) and Gordon and Mihm (1957).

The inoculum was pipetted from a 4 to 5 day broth culture.

The medium was incubated at 36°C (± 1) and readings were made at 48 hours, 5 days, 1 week and 2 weeks.

R. E. Gordon and J. M. Mihm (1957). J. Bact. 73, 15–27.
R. E. Gordon and M. M. Smith (1955). J. Bact. 69, 147–150.

Pigment from tyrosine—*Streptomyces*

K. F. Gregory and J. C. C. Haung (1964). J. Bact. 87, 1281–1286.

Tyrosinase production is revealed by the formation of a dark-brown pigment surrounding colonies grown on media containing either protein or free L-tyrosine. The similar pigment formed by *S. lavendulae* was shown to be melanin (Mencher and Heim, 1962).

A glucose-asparagine-mineral salts medium (Gregory and Vaisey, 1956), prepared with washed agar and double-distilled water in acid-washed flasks, served as the "minimal" medium. The same medium supplemented with 0.1% casein hydrolysate (Casamino Acids) and 0.3% yeast extract (Difco) served as the "complete" medium, except in experiments with strain A26-59, where the medium of Hickey and Tresner (1952) was required to achieve good sporulation. To permit recognition of tyrosinase production by individual colonies, 0.04% L-tyrosine was usually added to the media.

All incubations were at 30°C for 4 to 7 days.

K. F. Gregory and E. B. Vaisey (1956). Can. J. Microbiol. 2, 65–71.
H. J. Hickey and H. D. Tresner (1952). J. Bact. 64, 891–892.
J. R. Mencher and A. H. Heim (1962). J. gen. Microbiol. 28, 665–670.

Pigment from tyrosine—*Pseudomonas aeruginosa*

R. R. Colwell (1964). J. gen. Microbiol. 37, 181–194.

Melanin production from tyrosine and from phenylalanine was observed by the method of Brisou and Menantaud (1957).

J. Brisou and Menantaud (1957). Bull. Ass. Dipl. Microbiol., Nancy 67, 3.

POLY-β-HYDROXY BUTYRIC ACID

Poly-β-hydroxy butyric acid—*Bacillus*

M. Lemoigne, Mlle. B. Delaporte and Mme. M. Croson (1944). Annls Inst. Pasteur, Paris 70, 224–233.

Cette technique a été déjà exposée [4] et complétée ultérieurement [5]. Nous la rappelons: on peut opérer, soit en milieu gélosé, soit en milieu liquide. Dans les deux cas, le milieu nutritif est le même et a été décrit dans notre premier mémoire relatif au test de l'acétylméthylcarbinol [6]. En milieu solide, ce bouillon de haricots, contenant 20 g de saccharose, 2 g de peptone et 20 g de gélose par litre, est réparti en tubes inclinés ou, mieux, en tubes de Legroux. 3 tubes de gélose inclinés suffisent pour l'essai. La culture est faite à 30°C. Peu avant la sporulation, en général le troisième jour, la culture est mise en suspension dans une solution d'acide chlorhydrique à 20 p. 100.

Si l'on opère en milieu liquide, on répartit par 75 ou 100 c. c. dans des boîtes de Roux de 1000 c. c. La culture est faite à 30°C, la boîte placée horizontalement, pour que le liquide ait une faible épaisseur. 3 boîtes de Roux suffisent pour l'essai. On centrifuge le quatrième jour, après avoir prélevé 2 c. c. qui servent au test de l'acétylméthylcarbinol.

Le culot microbien est mis en suspension dans l'acide chlorhydrique à 20 p. 100.

Dans les deux cas nous avons une émulsion chlorhydrique que l'on maintient trois minutes à l'ébullition.

Après refroidissement, on centrifuge.

La masse microbienne est lavée trois fois à l'alcool et une fois à l'éther et enfin épuisée par le chloroforme bouillant.

La liqueur chloroformique filtrée est évaporée à sec.

La réaction est positive si l'on obtient une pellicule translucide, mince ou épaisse mais en tour cas cohérente, facilement détachable, tout à fait caractéristique. Examinée au microscope, elle présente le plus souvent un aspect qui a été décrit antérieurement [5]. Cette pellicule est un lipide β-hydroxybutyrique qui, distillé, donne des cristaux d'acide α-crotonique fondant à 72°C.

Cette réaction a une sensibilité assez faible, mais cependant suffisante pour un test bactériologique. En effet les bacilles les moins riches donnent facilement un film de 6 à 9 mg pour les prises d'essais que nous avons indiquées. Or, un film de 2 mg, obtenu par évaporation de la solution chloroformique dans une petite capsule, est très visible puisqu'il a un diamètre de 2.5 cm.

[4] M. Lemoigne et N. Roukhelman, *Ann. Ferment.*, 1940, 5, 527.
[5] M. Lemoigne, G. Sanchez et H. Girard, *Ann. Inst. Pasteur,* 1943, 69, 187.
[6] M. Lemoigne, Mlle B. Delaporte et Mme M. Croson, *Ann. Inst. Pasteur,* 1944, 70, 65.

Poly-β-hydroxy butyric acid—*Micrococcus*

G. Sierra and N. E. Gibbons (1962). Can. J. Microbiol. 8, 249–253.

Isolation and Extraction of PHBA: As a reference material, PHBA was isolated from *Bacillus megaterium* (NRC strain B-506). Cells were grown on a rotary shaker at 30°C in the Macrae and Wilkinson medium [1] with 2% of glucose. The washed cells were disrupted with Ballotini beads (No. 12) in a Waring blendor, dried, and extracted with chloroform-dioxane (80:20, v/v) in a Soxhlet. The extract was evaporated to dryness, the residue redissolved in hot chloroform, and the polyester precipitated from the solution by the addition of an acetone-ether (1:2, v/v) mixture. The PHBA was reprecipitated from the chloroform solution 3 times and finally dried overnight *in vacuo* over P_2O_5.

The above method of extraction could be used with cells of *M. halodentrificans.* However, these cells were usually digested by shaking them overnight at 35°C with alkaline hypochlorite (prepared from commercial chlorinated lime, 12% available chlorine), essentially as described by Williamson and Wilkinson [2]. Dried cells were used since much better digestion was obtained and less hypochlorite was required than with wet cells. The insolubel material was sedimented by centrifugation, washed three times with distilled water, dried at 100°C for 24 hours, and weighed. A white amorphous powder, readily soluble in chloroform, was obtained. As it contained very little acetone or ether-soluble material, it was designated as "lipid-free inclusions". For chemical characterization and elementary analysis, the "lipid-free inclusions" were further purified by fractional precipitation from chloroform with acetone-ether (1:2 v/v), and identified as PHBA by elementary analysis.

[1] R. M. Macrae and J. F. Wilkinson (1958). J. gen. Microbiol. 19, 210–222.
[2] D. H. Williamson and J. F. Wilkinson (1958). J. gen. Microbiol. 19, 198–209.

Poly-β-hydroxy butyric acid—*Pseudomonas*

R. Y. Stanier, N. J. Palleroni and M. Doudoroff (1966). J. gen. Microbiol. 43, 159–271.

The accumulation of poly-β-hydroxybutyric acid as an intracellular reserve material is an extremely valuable taxonomic character among aerobic pseudomonads. It was detected by microscopic examination of bacteria under phase contrast, and confirmed where necessary by staining-smears with Sudan Black or by chemical extraction (Williamson and Wilkinson, 1958). The accumulation of cellular organic reserve materials is favoured under conditions of nitrogen starvation (Doudoroff and Stanier, 1959); and in some purple bacteria, which can form both glycogen and poly-β-hydroxybutyrate reserves, the formation of poly-β-hydroxybutyrate is dependent on the provision of an organic substrate metabolically closely related to it. These considerations led us to determine poly-β-hydroxybutyrate accumulation on bacteria grown in a special chemically defined medium, consisting of standard mineral base with $(NH_4)_2SO_4$ at a low concentration (0.2 g/L), furnished with DL-β-hydroxybutyrate (5 g/L) as carbon and energy source. The medium was used in liquid form (5 ml/tube); cultures were mechanically agitated, and examined after 24–48 hr, when growth had become nitrogen limited. Under these circumstances, polymer-accumulating species regularly contained extensive intracellular deposits of the polymer, readily observable by microscopic examination of wet mounts. Although β-hydroxybutyrate is a carbon and energy source for most aerobic pseudomonads, a few cannot use it. For such organisms, either Na acetate of Na succinate (5 g/L) was used instead.

The standard mineral base contained, per liter: 40 ml of Na_2HPO_4 + KH_2PO_4 buffer (M; pH 6.8); 20 ml of Hutner's vitamin-free mineral base (Cohen-Bazire, Sistrom and Stanier, 1957); and 0.2 g of $(NH_4)_2SO_4$. This basal medium is easy to prepare, and provides all necessary minerals, including trace elements. It is heavily chelated with nitrilotriacetic acid and EDTA, and forms a copious precipitate upon autoclaving. The precipitate redissolves as the medium cools, to form a water-clear solution.

G. Cohen-Bazire, W. R. Sistrom and R. Y. Stanier (1957). J. cell. comp. Physiol. 49, 25.
M. Doudoroff and R. Y. Stanier (1959). Nature, Lond. 183, 1440.
D. H. Williamson and J. F. Wilkinson (1958). J. gen. Microbiol. 19, 198.

Poly-β-hydroxy butyric acid esterase—*Pseudomonas*

R. Y. Stanier, N. J. Palleroni and M. Doudoroff (1966). J. gen. Microbiol. 43, 159–271.

The production of an extracellular esterase capable of depolymerizing poly-β-hydroxybutyric acid was determined by patching strains on mineral agar plates overlayered with a suspension of poly-β-hydroxybutyric acid (0.25%, w/v) in mineral agar. Positive strains showed growth of the patch and clearing of the surrounding medium. This is a test of great taxonomic value among aerobic pseudomonads, but unfortunately the substrate, a

natural product not commercially available, is tedious to prepare. We used polymer granules extracted from *Bacillus megaterium* and prepared as described by Delafield *et al.* (1965).
F. P. Delafield, M. Doudoroff, N. J. Palleroni, C. J. Lusty and R. Contopoulou (1965). J. Bact. 90, 1455.

Other papers dealing with poly-β-hydroxy butyric acid are:

W. G. C. Forsyth, A. C. Hayward and J. B. Roberts (1958). Nature, Lond. 182, 800–801.

A. C. Hayward, W. G. C. Forsyth and J. B. Roberts (1959). J. gen. Microbiol. 20, 510–518.

A. C. Hayward, (1959). J. gen. Microbiol. 21, (ii).

P. J. Provost and R. N. Doetsch (1962). J. gen. Microbiol. 28, 547–557.

J. M. Merrick and M. Doudoroff (1964). J. Bact. 88, 60–71.

J. M. Merrick, D. G. Lundgren and R. M. Pfister (1965). J. Bact. 89, 234–239.

D. G. Lundgren, R. Alper, C. Schnaitman and R. H. Marchessault (1965). J. Bact. 89, 245–251.

F. P. Delafield, M. Doudoroff, N. J. Palleroni, C. J. Lusty and R. Contopoulos (1965). J. Bact. 90, 1455–1466.

PROTEIN DIGESTION

Protein digestion—*Xanthomonas*

F. A. Wolf (1924). J. agric. Res. 29, 57–68.

Digestion of casein.

1 per cent casein was added to the stock agar in poured plate cultures. After a week's incubation wide halos, in which the casein was entirely dissolved, had formed around the colonies, thus demonstrating the ability of these organisms to form erepsin.

Protein digestion—various organisms

W. C. Frazier and P. Rupp (1928). J. Bact. 16, 57–63.

Casein

The casein agar which has finally been adopted as most satisfactory for the isolation of caseolytic organisms is made as follows: 3.5 g of casein (according to Hammarsten) are soaked for fifteen minutes in 150 ml of distilled water and 72 ml of saturated lime water are then added. The mixture is shaken until the casein is almost dissolved, 0.35 g of potassium citrate is added, and the shaking is continued until the casein is dissolved, after which 10 ml of double strength beef infusion are added and the solution is made up to 300 ml. To this solution are added 100 ml of a 0.15 per cent calcium chloride solution and 100 ml of a phosphate solution which contains 0.105 per cent of $Na_2HPO_4 \cdot 2H_2O$ (Sorensen's phosphate) and 0.035 per cent of KH_2PO_4 and which has a pH value of 7.4. The resulting 500 ml of casein solution are divided into 50 ml portions in 200 ml flasks and autoclaved at 20 pounds pressure for fifteen minutes. The resulting pH value should be 7.0. The medium will be markedly opalescent after sterilization. Likewise 50 ml portions of 3 per cent washed agar are sterilized in separate flasks. When the medium is to be used the casein solution and agar are heated in the steamer and while hot the agar is poured into the flasks of casein solution and thoroughly mixed. Plates are poured with the milky solution which results. The casein or agar solutions may be steamed separately a number of times without apparent change; but after the solutions are mixed they should be used without further steaming. The mixture will withstand a few meltings but gradually becomes more flocculent and finally, on further steaming, goes into larger clots.

Protein digestion

S. A. Waksman and R. L. Starkey (1932). J. Bact. 23, 405–428.

The medium generally employed in these studies contained the following mineral constituents:

K_2HPO_4	1.0 or 3.0 g
$MgSO_4 \cdot 7H_2O$	0 or 0.5 g
NaCl	0.1 or 0.5 g
$FeSO_4$	0 or 0.01 g
Distilled water	1,000 ml

The proteins were used as the sole sources of energy and nitrogen in most cases. In a few instances glucose was used as an accessory source of energy. The liquid medium was distributed in 100 ml amounts in 250 ml Erlenmeyer flasks. In the presence of magnesium sulfate the large amounts of ammonia which were produced by the microorganisms led to the formation of crystals of magnesium-ammonium phosphate which were relatively insoluble in the medium and consequently interfered with the complete recovery of the ammonia in the analyses. In most cases the magnesium salt was left out of the medium and no influence on growth was observed as a result.

The proteins were added to the media in amounts of either 0.5 or 1.0 per cent. Where fibrin, albumin, zein, gliadin or edestin were used, the desired amounts of the air-dried proteins were weighed out and added to the flasks of liquid medium. The casein was brought into solution before addition to the medium, by dissolving each gram in 8 ml of 0.1 N NaOH. The gelatin was also dissolved in the liquid medium before being added to the flasks. The reaction of the media was adjusted to neutrality. Incubation of cultures was carried out at 27 to 28°C.

The analyses of the cultures were carried out as follows: The cultures were filtered through weighed filter papers and the residual protein and the cellular material on the paper were washed with distilled water. The filtrate and washings were added together and made up to a definite volume. The residual protein and cellular material was dried at 75 to 85°C to constant weight, and total nitrogen determined by a modified Kjeldahl method. The filtrate was analyzed for total nitrogen, amino nitrogen and ammonia nitrogen. The ammonia nitrogen was determined by a modified Folin aeration method. To 25 ml of the filtrate was added 5 ml of 40 per cent Na_2CO_3 solution, 5 ml of 20 per cent NaCl solution and a small amount of crude oil to prevent foaming. The tube containing the liquid was placed in a water bath at 65°C and attached to an aeration system and the volatilized ammonia was trapped in standard 0.05 N H_2SO_4. Aeration was continued for four to six hours. After the ammonia had been removed, the liquid was neutralized with acetic acid, made up to volume, and the amino nitrogen determined by the Van Slyke micro method. The hydrogen-ion concentration of the culture was determined by the colorimetric method. Casein nitrogen was determined on portions of the filtrate by the method of Sherman and Neun (1916). The solution was poured into 25 ml of 20 per cent potassium sulfate solution and then 5 ml of 0.2 N HCl was added. After standing for one hour the precipitated casein was separated from the liquid by filtration and the nitrogen content was determined. Glucose was determined by the Bertrand method.
H. C. Sherman and E. D. Neun (1916). J. Am. chem. Soc. 38, 2199–2216.

Protein digestion—*Bacteroides*

A. H. Eggerth (1935). J. Bact. 30, 277–299.

Coagulated egg-albumen broth was prepared by cutting cubes of coagulated egg white into tubes of infusion broth and sterilizing.

Protein digestion—*Staphylococcus* and *Micrococcus*

Y. Abd-el-Malek and T. Gibson (1948). J. Dairy Res. 15, 249–260.

Casein digestion: Plates of (a) nutrient agar + 20% skim milk, and (b) a casein agar* were inoculated on the surface. After incubation (a) was flooded with 10% HCl and (b) with 1% tannic acid.
*Stand. Spec. Brit. Stand. Inst. (1940). No. 895.

Protein digestion—general

P. Hauduroy (1951). Techniques Bactériologiques. Masson et Cie, Éditeurs, Paris.

SERUM DE LOFFLER

Trois parties de sérum stérile de cheval sont mélangées avec une partie de bouillon glucosé à 0.5 p. 100. Le mélange est distribué dans des récipients stériles. La coagulation est obtenue en plaçant les tubes dans un appareil à coaguler à une température de 75°C pendant 6 heures. La stérilisation se fait dans le même appareil par un chauffage à 90°C pendant 1 heure deux jours de suite. *(W. I. C.).*

Protein digestion—general

P. Hauduroy (1951). Techniques Bactériologiques. Masson et Cie, Éditeurs, Paris.

HYDROLYSE DE LA CASEINE

Préparer des boîtes de gélose au lait en mélangeant en quantités égales du lait "skim" stérile et de la gélose stérile à 2 p. 100. Les deux substances ont été chauffées à 45°-50°C avant le mélange. Après solidification, les cultures sont ensemencées en stries, et on les observe pour voir le développement de la culture et l'éclaircissement de la caséine (modification du milieu de Hasting: *Cent. Bakt.,* 1903, 2. Abt., 10, 384).

(A. T. C. C.).

Protein digestion—*Staphylococcus*

C. Shaw, J. M. Stitt and S. T. Cowan (1951). J. gen. Microbiol. 5, 1010–1023.

Digestion of serum: Organisms were grown on slopes of Loeffler's medium.

Protein digestion—*Bacillus*

N. R. Smith, R. E. Gordon and F. E. Clark (1952). U.S.D.A. Agr. Monograph No. 16.

Hydrolysis of casein

Milk agar was prepared by mixing hot sterile 2.5 per cent water agar with half as much cool sterile skim milk. Plates were quickly poured and, after solidification, were streaked once across. At various intervals they were observed for growth of the organism and zone of clearing of the casein. This is a modification of Hastings' method.

E. G. Hastings (1903). Zentbl. Bakt. ParasitKde Abt. 2, 10, 384.

Protein digestion—*Staphylococcus*

J. Marks (1952). J. Path. Bact. 64, 175–186.

Medium for demonstrating proteolysis

Two volumes of serum, two of saline and one of N HCl were heated in a water bath at 65°C to 70°C for 20 minutes, twenty volumes of melted nutrient agar (of normal agar content) added at 56°C, and the mixture neutralised with one volume of N NaOH. The resulting medium was poured into plates. It contained a dispersed fine precipitate of denatured globulin, the digestion of which produced zones of clearing.

Protein digestion—*Pseudomonas (Malleomyces) pseudomallei*

P. de Lajudie and E. R. Brygoo (1953). Annls Inst. Pasteur, Paris 84, 509–515.

Activités protéolytiques – Sérum coagulé.

Protein digestion—*Mycobacterium*

R. E. Gordon and M. M. Smith (1953). J. Bact. 66, 41–48.

Hydrolysis of casein: Skim milk or skim milk powder suspended in the proper amount of water was sterilized by autoclaving at 121°C for 20 minutes; cooled to 47°C; mixed with an equal volume of sterile 2 per cent water agar, also at 47°C; and poured into sterile plates. These were streaked, incubated at 28°C, and observed for clearing of the casein at 7 and 14 days (Hastings, 1903).

E. G. Hastings (1903). Zentbl. Bakt. ParasitKde II, 10, 384.

Protein digestion—Loeffler's inspissated serum

N.C.T.C. Methods 1954.

Loeffler slope: incubated 5 days at optimal temperature.

Protein digestion—*Streptomyces* and *Nocardia*

R. E. Gordon and M. M. Smith (1955). J. Bact. 69, 147–150.

Hydrolysis of casein: Skim milk or skim milk powder suspended in the proper amount of water and an equal volume of 2 per cent water agar were autoclaved separately at 121°C for 20 minutes, cooled to 47°C, mixed, and poured into sterile plates. Each culture was streaked on a plate, incubated at 28°C, and observed for clearing of the casein at 7 and 14 days (Hastings, 1903).

E. G. Hastings (1903). Zentbl. Bakt. ParasitKde 2 Abt., 10, 834.

Protein digestion—thermophilic sulphate reducing bacteria

L. L. Campbell, Jr., H. A. Frank and E. R. Hall (1957). J. Bact. 73, 516–521.

Tests for proteolysis were carried out as follows – (1) liquefaction of coagulated egg albumin (2) proteolysis of alkaline egg medium (3) blackening of brain medium (4) reddening of beef heart tissue – using methods recommended in the *Manual of Methods for the Pure Culture Study of Bacteria* (1944) Biotech. Publ., Geneva, N. Y. were used (5) action on litmus milk.

Protein digestion—*Mycobacterium*

R. E. Gordon and J. M. Mihm (1959). J. gen. Microbiol. 21, 736–748.

Decomposition of casein: Skim milk or skim-milk powder suspended in the required amount of water and an equal volume of 2% (w/v) water agar were sterilized separately by autoclaving, cooled to 47°C, mixed, and poured into sterile plates (5 plates from 100 ml). Each culture was

streaked on a plate and inspected for clearing of the casein after incubation at 28°C for 7 and 14 days (Hastings, 1903).
E. G. Hastings (1903). Zentbl. Bakt. ParasitKde 2 Abt. 10, 384.

Protein digestion—Loeffler's inspissated serum

V. B. D. Skerman, *A Guide to the Identification of the Genera of Bacteria.* 2nd ed. Williams and Wilkins Book Co., Baltimore, 1967.

Glucose	2.0 g
Peptone	2.0 g
NaCl	1.0 g
Tap water	200 ml

Dissolve the ingredients and filter if necessary. Add 400 ml of blood serum and mix. Dispense in 5 ml amounts in 150 by 13 mm tubes and place them in a vertical position in an inspissator. Heat to 80°C very slowly and hold at this temperature for 6 hours to coagulate the serum. Sterilize by heating to 85°C for 20 minutes on 3 successive days.

If the glucose-peptone water is sterilized separately and the serum is collected and added aseptically, only the initial heating is necessary to coagulate the serum.

To detect liquefaction inoculate by stabbing and incubate for 1 to 2 weeks.

Protein digestion—*Micrococcus*

I. J. McDonald (1961). Can. J. Microbiol. 7, 111–117.

Proteinase activity was measured at 37°C by incubating 5.0 ml of supernatant, obtained by centrifugation of cultures (10,000 x *g* at 0°C) with 5.0 ml of casein substrate and 0.1 ml of toluene. The casein substrate [1] was modified by adding 30 ml of 0.05 N monopotassium phosphate after boiling and then by adjusting to pH 5.7 with hydrochloric acid. Where necessary, tubes were adjusted to contain the same amount of salt.

Samples of these reaction mixtures, taken at 0 and 4 hours, were mixed thoroughly with an equal volume of 0.6 N trichloroacetic acid (TCA) and filtered through Whatman No. 2 filter paper. Tyrosine was determined in 1.0 ml of the TCA filtrate by the method of Anson [2]. One unit of enzyme was defined as the amount of enzyme which liberated 1.0 μg of tyrosine under these conditions.

[1] I. Husain, I. J. McDonald (1958). Can. J. Microbiol. 4, 237–242.
[2] M. L. Anson (1938). J. gen. Physiol. 22, 79–89.

Protein digestion—rhizosphere bacteria

P. G. Brisbane and A. D. Rovira (1961). J. gen. Microbiol. 26, 379–392.

Casein agar. Ten ml of sterile 9% (w/v) solution of powdered milk were added to every 100 ml basal agar immediately before pouring plates.

Hydrolysis was observed as a clearing round the colony.

The Basal medium. The yeast-extract peptone nitrate broth of Sperber and Rovira (1959), with 15 g agar/L for solid medium, was used.
J. I. Sperber and A. D. Rovira (1959). J. appl. Bact. 22, 85.

Protein digestion—*Xanthomonas*

A. C. Hayward and W. Hodgkiss (1961). J. gen. Microbiol. 26, 133–140.

The following agar medium was dispensed in 45 ml quantities in 2 oz bottles: peptone (Oxoid), 5.0 g; yeast extract (Difco), 3.0 g; agar, 20.0 g; distilled water, 1 L; adjusted to pH 7.2. Five ml quantities of 4.0% (w/v) casein (Judex Ltd., light white soluble) were added to the molten agar base from which three plates were poured. Three or four organisms were inoculated to each plate, into the centre of 0.5 cm diameter cavities made with a surface sterilized cork borer. After incubation for 6 days, zones of hydrolysis were recorded.

Protein digestion—*Vibrio*

G. H. G. Davis and R. W. A. Park (1962). J. gen. Microbiol. 27, 101–119.

Serum hydrolysis, observed after 24 hr and 7 days growth upon inspissated serum.

Casein hydrolysis, using Hastings' (1904) method with 10% (v/v) skimmed milk in nutrient agar; examined after 5 days.
E. G. Hastings (1904). Zentbl. Bakt. ParasitKde (2 Abt.) 12, 590.

Protein digestion—*Pseudomonas aeruginosa*

J. D. Mull and W. S. Callahan (1962). J. Bact. 85, 1178–1179.

By use of the method of Lansing *et al.* (Anat. Rec. 1952, 114, 555), elastin was prepared from the aortas of young adults obtained at the time of necropsy. The elastin was oven dried at 100°C and ground to a powder filtrable through a 40-mesh filter. Quadruplicate nitrogen determinations showed 14% nitrogen. To tubes containing 4 ml of Brain Heart Infusion (Difco) were added 200 mg of desiccated elastin. Inoculations were made and the tubes were incubated at 37°C with controls for 5 days. Slime factor was removed with 0.1 N sodium hydroxide, and the elastin was recovered by filtration, dried, and weighed. The weight loss was taken as an expression of the elastolytic activity.

Protein digestion—*Clostridium rubrum*

H. Ng and R. H. Vaughn (1963). J. Bact. 85, 1104–1113.

The test for ability to digest coagulated albumen was carried out in a liver infusion broth to which was added approximately 1 cm^3 of heat-coagulated egg white.

Protein digestion—*Butyrivibrio, Borrelia, Bacteroides, Selenomonas, Succinivibrio*

R. S. Fulghum and W. E. C. Moore (1963). J. Bact. 85, 808–815.

The SRP–A medium was made up of 50% stock salts solution, 30% clarified rumen fluid, and 10% CO_2-equilibrated, oxygen-free, distilled water. Cysteine (0.08%, w/v) and, except where noted otherwise, 0.05% (w/v) glucose and 0.05% (w/v) cellobiose were added. The pH was adjusted to 6.8. Portions (9 ml) were dispensed into tubes containing 0.2 g of agar. All additions were made under oxygen-free CO_2 (Hungate, 1950). The stoppered tubes were placed in tightly covered racks and sterilized at 121°C for 20 min. Since protein suspensions were found to coagulate when sterilized in the presence of the salts solution, 1 ml of sterile skim milk was added aseptically to tubes of medium held at 45°C, immediately preceding inoculation. Rehydrated commercial powdered skim milk was prepared with distilled water and sterilized at 116°C for 12 min. The sterile milk was always cooled to and held at 45°C after sterilization, to minimize the re-entry of oxygen into the solution when the milk was rapidly added to the medium.

Clarified rumen fluid was prepared from rumen fluid expressed through two layers of cheesecloth. The liquor was heated at 121°C for 20 min under an oxygen-free CO_2 atmosphere in a sealed container. Particulate debris was then removed by centrifugation at 22,000 x *g* for 20 min.

Several variations of SRP medium were evaluated to determine their suitability for obtaining total viable counts and for enumerating the proteolytic flora of the rumen. Variations included: SRP–A which contained powdered skim milk as described above; SRP–B was the same as A, with 10% dialyzed skim milk (3.5% solids, w/v) replacing the skim milk; SRP–C was the same as A, with 1% thioglycolate replacing cysteine; SRP–D was the same as A but with 40% rumen fluid. Incubation was at 39°C for 120 hours.

R. E. Hungate (1950). Bact. Rev. 14, 1–49.

Protein digestion—*Streptococcus*

R. H. Deibel, D. E. Lake and C. F. Niven, Jr. (1963). J. Bact. 86, 1275–1282.

Proteolysis was determined in the following medium: Tryptone (Difco), 10 g; yeast extract, 5 g; K_2HPO_4, 5 g; NaCl, 5 g; glucose, 0.5 g, and distilled water, 1 liter. Various concentrations of gelatin (Difco) were used, and agar plates were prepared with 2.0% agar. Both the tube and plate methods were employed to detect proteolysis. The plates were inoculated by a central streak and developed after incubation by the addition of acidified mercuric chloride (Burnett, Pelczar, and Conn, 1957).

Cultures were incubated at 37°C.

G. W. Burnett, M. J. Pelczar, Jr., and H. J. Conn (1957). In Society of American Bacteriologists, *Manual of Microbiological Methods.* McGraw-Hill Book Co., Inc., New York.

Protein digestion—*Pseudomonas*

I. J. McDonald, C. Quadling and A. K. Chambers (1963). Can. J. Microbiol. 9, 303–315.

The authors described detailed analytical procedures for the estimation of proteinases active on casein, plasma albumin, β-lactoglobulin and hemoglobin.

Reaction mixtures consisting of 3.0 ml of casein and 3.0 ml of culture supernatant; or 3.0 ml of solutions of albumin, β-lactoglobulin, or hemoglobin, 2.0 ml of buffer (0.1 M boric acid and 0.1 M KH_2PO_4 at pH 7.5) and 1.0 ml of culture supernatant, were incubated at 35°C (except where otherwise stated). Samples of the reaction mixture, removed immediately after the addition of supernatant and after incubation for 4, 6, or 20–24 hours, were mixed thoroughly with an equal volume of 0.6 N trichloroacetic acid and filtered through Whatman No. 2 paper. The liberated tyrosine was determined in the filtrate by Anson's (1938) method.

M. L. Anson (1938). J. gen. Physiol. 22, 79–89.

Protein digestion—*Clostridium*

N. A. Sinclair and J. L. Stokes (1964). J. Bact. 87, 562–565.

The cultures were tested for proteolysis by growing them in media containing litmus iron milk, cubes of coagulated egg white, ground beef heart, and gelatin (Society of American Bacteriologists, 1957).

Incubation at 0, 15 and 25°C.

Society of American Bacteriologists (1957). *Manual of microbiological methods.* McGraw-Hill Book Co., Inc., New York.

Protein digestion—*Dermatophilus*

M. A. Gordon (1964). J. Bact. 88, 509–522.

The authors used Loeffler's coagulated serum slants.

The inoculum was taken from Brain Heart Infusion agar slants.

The medium was incubated at 36°C (± 1) and readings were made at 48 hours, 5 days, 1 week and 2 weeks or longer.

Protein digestion—*Dermatophilus*

M. A. Gordon (1964). J. Bact. 88, 509–522.

The authors used casein agar plates prepared and read as described by Gordon and Smith (1955) and Gordon and Mihm (1957).

The inoculum was pipetted from a 4 to 5 day broth culture.

The medium was incubated at 36°C (± 1) and readings were made at 48 hours, 5 days, 1 week and 2 weeks.

R. E. Gordon and J. M. Mihm (1957). J. Bact. 73, 15–27.

R. E. Gordon and M. M. Smith (1955). J. Bact. 69, 147–150.

Protein digestion—*Pseudomonas aeruginosa*

K. Morihara (1964). J. Bact. 88, 745–757.

The authors give detailed directions for the study of proteinase and elastase by *Pseudomonas aeruginosa.*

Protein digestion—*Vibrio marinus*

R. R. Colwell and R. Y. Morita (1964). J. Bact. 88, 831–837.

Casein: The authors used the method of J. K. Demeter (1943). *Bakteriologische Untersuchungs Methoden von Milch, Milcherzeugnissen, Molkereihilfsstoffen und Wersandsmaterial,* p. 48. Berlin – modified by the addition to the media of the following salts to produce a synthetic seawater:— sodium chloride, 2.4%; potassium chloride, 0.07%; magnesium chloride (hydrated) 0.53% and magnesium sulfate (hydrated) 0.7%.

A standard inoculum was 1 drop from a pasteur pipette (*c.* 0.05 ml) of a 24 hour artificial seawater broth.

Cultures were incubated at 18°C.

Protein digestion—*Staphylococcus*

S. I. Jacobs, A. T. Willis and G. M. Goodburn (1964). J. Path. Bact. 87, 151–156.

Washed 'chocolate' agar. This medium was used to detect proteolysis. It was prepared in the usual way from whole horse blood (Burroughs Wellcome and Co.) except that the cells were freed from the plasma proteins by washing three times in sterile saline (0.9 per cent NaCl) solution before incorporation into the medium. It was found that heated blood agar prepared in this way was a more sensitive indicator of proteolysis, and gave more consistent results, than the conventional medium, which contains plasma proteins and oxalate. Results were read after 18 hours' incubation at 37°C.

The nutrient agar base was a heart infusion broth (Difco) solidified with 1.5 per cent Davis New Zealand agar. The medium was sterilised at 121°C for 15 min, and reagents were mixed with the sterilized base medium when it had cooled to 50–55°C.

The medium was inoculated from a fresh 18 hr broth culture with a sterile throat swab.

Protein digestion—*Aerococcus catalasicus*

O. G. Clausen (1964). J. gen. Microbiol. 35, 1–8.

Cultures were incubated aerobically at 37°C unless otherwise stated.

The ability to liquefy inspissated ox serum was determined by incubation at 22°C for up to 14 days.

Protein digestion—*Moraxella*

W. J. Ryan (1964). J. gen. Microbiol. 35, 361–372.

Serum liquefaction. Liquefaction of coagulated serum was tested on Loeffler's serum slopes. Readings were taken after 24 and 48 hr incubation.

Casein hydrolysis – Organisms were grown on serum agar plates containing 10% (v/v) skim milk. A positive reaction was indicated by clear zones around the colonies after incubation for 2–3 days.

Fibrinolysis. Ten % (v/v) fresh human plasma was added to melted nutrient agar and the fibrinogen precipitated at 56°C (Christie and Wilson, 1941). Plates were poured, dried and inoculated. Clear zones around colonies after 2–3 days of incubation indicated fibrinolysin production.

R. Christie and H. Wilson (1941). Aust. J. exp. Biol. med. Sci. 19, 329.

Protein digestion—*Bacillus*

G. R. F. Hilson (1965). J. gen. Microbiol. 39, 407–421.

Egg albumen hydrolysis: liquefaction of Löwenstein-Jensen medium by form 2 cultures was confirmed, and compared with the three type strains of *B. licheniformis,* by inoculation on to slopes of this medium made up without malachite green, with incubation up to one week.

Protein digestion—*Alcaligenes*

R. G. Mitchell and S. K. R. Clarke (1965). J. gen. Microbiol. 40, 343–348.

The medium of Hayward and Hodgkiss (1961) was used.

A. C. Hayward and W. Hodgkiss (1961). J. gen. Microbiol. 26, 133.

Protein digestion—marine bacteria

J. R. Merkel (1965). J. Bact. 89, 903–904.

The procedure employs either nutrient agar or 0.5% peptone-2% agar medium prepared with seawater, to which enough algal chromoprotein is added to impart color. Proteolysis is detected by observing the decoloration of the medium in zones surrounding active colonies. This is possible because a correlation exists between the loss of chromophore color and protein degradation.

Partially purified phycocyanin preparations were used in the present study. Purification was achieved by using the method described earlier (Merkel, Braithwaite, and Kritzler, J. Bact. 88, 974, 1964), but batch adsorption and elution were substituted for column techniques. The phycocyanins were used as they were extracted by 0.07 M phosphate buffer, or they were concentrated by ammonium sulfate precipitation and redissolved in 1% NaCl. Phycocyanin stock solutions, which contained 5 to 20 mg of chromoprotein per ml, were sterilized by Millipore filtration, and portions (10 to 15 ml) were added aseptically to approximately 100 ml of melted nutrient agar-seawater or peptone-agar-seawater medium at 48°C. The protein was mixed with the medium, and plates were immediately poured.

Samples of seawater were streaked on the surface of these plates for direct isolation of proteolytic bacteria, or isolated cultures were spot-inoculated on the surface. The plate was incubated at 22°C for 48 hr, but zones of proteolysis can usually be detected as soon as colonies are visible.

Protein digestion—*Streptomyces*

T. F. Fryer and M. E. Sharpe (1965). J. Dairy Res. 32, 27–34.

Casein: The authors used the methods of Gordon and Smith (1955) and Gordon and Mihm (1957).

0.05 ml of an 18–24 hr culture of the organism grown in 5 ml Yeastrel nutrient broth, centrifuged and resuspended in 5 ml distilled water was used as inoculum.

R. E. Gordon and J. M. Mihm (1957). J. Bact. 73, 15.
R. E. Gordon and M. M. Smith (1955). J. Bact. 69, 147.

Protein digestion—*Nocardia*

F. Mariat (1965). Annls Inst. Pasteur, Paris 109, 90–104.

Caractères physiologiques. – a) *Hydrolyse de la caséine:* Plusieurs méthodes ont été proposées pour étudier cette réaction [21, 31, 32]. La plus simple consiste à mélanger dans une boîte de Petri des volumes égaux de gélose à l'eau à 25 p. 1000 et de lait écrémé, maintenus à 50°. Après solidification, le milieu est ensemencé par un inoculum de ± 4 X 4 mm. Lorsque la caséine est hydrolysée, un halo clair se forme plus ou moins rapidement autour de la colonie.

[21] Gordon (R. E.) et Smith (M. M.). *J. Bact.,* 1955, 69, 147.
[31] Mariat (F.). *Ann. Soc. Belge Med. trop.,* 1962, 4, 651.
[32] Mariat (F.). *Arch. Inst. Pasteur Tunis,* 1962, 39, 309.

Protein digestion—*Mycobacterium*

A. Tacquet, F. Tison and B. Devulder (1965). Annls Inst. Pasteur, Paris 108, 797–801.

Peptide decomposition: Used the method of Tacquet, Gillaume and Lefebvre (Zbl. Bakt. Orig., 194, 551–553.)

Protein digestion—*Micrococcus*

I. J. McDonald (1965). Can. J. Microbiol. 11, 693–701.

The author describes a quantitative study of casein hydrolysis by cultures of a *Micrococcus sp.*

Protein digestion—*Pseudomonas aeruginosa*

P. V. Liu (1966). The roles of various fractions of *Pseudomonas aeruginosa* in its pathogenesis. J. infect. Dis. 116, 112–116.

The organisms were grown on tryptone-glucose-extract agar without the addition of glucose and with the cellophane-covered plate. The extract from the plate was precipitated by 50% saturation with ammonium sulfate, resuspended in a minimum amount of saline, and dialyzed in phosphate buffer (pH 7.6) for 2 days. The contents of the cellophane bag were centrifuged to remove the precipitate that formed during the dialysis and concentrated with Carbowax to one-fiftieth the original volume. This concentrate was used in the chromatography. The titration of protease was done as described previously (Liu *et al.*, 1961).

P. V. Liu, Y. Abe and J. L. Bates (1961). J. infect. Dis. 108, 218–228.

SOME REDUCTION REACTIONS

Acetone to isopropyl alcohol—*Clostridium*

A. F. Langlykke, W. H. Peterson and E. B. Fred (1937). J. Bact. 34, 443–453.

To determine whether acetone when added to the culture would be reduced, a number of experiments were carried out. A series of 750 ml Erlenmeyer flasks containing the basic medium [$(NH_4)_2HPO_4$, 0.07%; peptone, 0.5%; asparagine, 0.1%; tap water] were sterilized at 15 pounds pressure for 40 minutes. Sufficient sterile glucose solution was added to bring the glucose content to 3 per cent, and then varying quantities of a sterile acetone solution were added. In each case the total volume was adjusted to 500 ml by addition of sterile water. A two per cent inoculum of a corn mash culture of the organism was then added to each flask except the controls, the mercury seals were applied, and the flasks allowed to incubate at 37°C. When fermentation was complete, as evidenced by the cessation of gassing, the flasks were analyzed for products. The added acetone was almost completely converted to isopropyl alcohol.

Neutral volatile products were determined on a distillate of the culture. For this purpose an aliquot of the culture was made slightly alkaline and about 50 per cent of the liquid was distilled off and collected under carbon dioxide-free water. The distillate was analyzed for butyl and ethyl alcohols by the method of Johnson (1932), acetone was determined by a modification of Goodwin's method (1920), and isopropyl alcohol by the oxidation procedure previously reported [Langlykke *et al.* (1935)].

L. F. Goodwin (1920). J. Am. chem. Soc. 42, 39–45.
M. J. Johnson (1932). Ind. Engng Chem. analyt. Edn 4, 20–22.
A. F. Langlykke, W. H. Peterson and E. McCoy (1935). J. Bact. 29, 333–347.

Ethanol to caproic acid—*Clostridium*

H. A. Barker and S. M. Taha (1942). J. Bact. 43, 347–363.

C_2H_5OH, 1 vol per cent; K_2HPO_4, 0.5%; $MgSO_4 \cdot 7H_2O$, 0.01%; $(NH_4)_2SO_4$, 0.03%; $FeSO_4 \cdot 7H_2O$, 0.002%; yeast autolysate, 0.5 vol per cent (not essential); $CaCO_3$, 10.0%.

After autoclaving 2 vol per cent of 5 per cent Na_2CO_3 and 1 vol per cent of a 1 per cent $Na_2S \cdot 9H_2O$ solution are added and the pH adjusted to 7.0–7.4. Acetic and caproic acid were determined by Duclaux distillation.

Fructose to Mannitol—*Lactobacillus*

W. B. Moore and C. Rainbow (1955). J. gen. Microbiol. 13, 190–197.

Transformation of fructose: Cultures of strains L3, L4, L5 and L6 in CR medium containing 1.8% fructose as sole carbohydrate, were grown at 28°C for 144 hr. The presence of mannitol in the culture fluids of all strains was demonstrated by paper chromatography. This was confirmed by growing strain L3 on 60 ml of CR medium (containing 1.8% fructose instead of glucose) for 144 hr. The method of Peterson and Fred (1920) was used to isolate from the culture filtrate 125 mg of a substance which resembled an authentic specimen of mannitol in melting point (found, 166°C; authentic specimen 167–168°C; mixed melting point 166–167°C) and in chromatographic mobility.

W. H. Peterson and E. B. Fred (1920). J. biol. Chem. 41, 431.

Glycerol to trimethylene glycol

C. H. Werkman and G. F. Gillen (1932). J. Bact. 23, 167–182.

Glycerol, 20 g; K_2HPO_4, 1 g; peptone or NH_4Cl, 3 g; water, 1000 ml; pH 7.0. Sterilised 15 mins at 20 lbs/sq in.

Trimethylene glycol was determined by removing insoluble salts by filtration and concentrating the liquor to one-third its volume by distillation. This concentrate was further reduced to one-fourth of its volume by vacuum distillation and soluble calcium salts were precipitated by oxalic acid. The calcium oxalate was filtered off and the liquor evaporated to a syrup on the water bath. The syrup was dissolved in alcohol (96 per cent) and the salts removed by filtration. The filtrate was then distilled under a 29-inch vacuum. Everything coming over up to 100°C was discarded. The fraction between 110 to 130°C was a clear liquid which was refractionated under vacuum and the portion coming over between 110° and 120°C was collected separately.

This fraction was further purified by distilling at a constant temperature of 210°C at atmospheric pressure. This distillate was weighed and identified by the preparation of the dibenzoate.

Glycerol to trimethylene glycol—coliforms

M. N. Mickelson and C. H. Werkman (1940). J. Bact. 39, 709–715.

A medium of 2 per cent glycerol, 1 per cent phosphate buffer (pH 6.6–6.8) and 0.3 per cent ammonium sulphate in tap water was used. Incubated anaerobically at 30°C. Determination cited in *Enzymologia* 1940.

***o*-bromo-phenol-indo-2:6-dibromophenol reduction**—*Mycobacterium buruli* n.sp.

J. K. Clancey (1964). J. Path. Bact. 88, 175–187.

Dye-reducing ability was estimated by the method of Wrinkle and Patnode (1952). except that the cultures were grown on Löwenstein-Jensen medium and the readings taken at 15 min and 1 hr.

C. K. Wrinkle and R. A. Patnode (1952). Am. Rev. Tuberc. pulm. Dis. 66, 99.

Selenite—*Pseudomonas, Agrobacterium*

A. A. Hendricksen, I. L. Baldwin, and A. J. Riker (1934). J. Bact. 28, 597–618.

Sodium selenite yeast-water glucose agar medium was prepared as follows: glucose, 5.0 g; sodium selenite, 0.1 g; agar, 15.0 g; and 1000 ml of 1 per cent yeast water. Transfers of the various cultures were made to agar slants and incubated at 20°C for four days. The hairy-root organisms did not grow while the crown-gall and radiobacter cultures produced abundant growth with a distinct red color in the streak, because of the presence of free selenium. The sodium selenite apparently was much more toxic to the hairy-root than to the crown-gall or radiobacter organisms.

Selenite—yeasts

G. Falcone and W. J. Nickerson (1963). J. Bact. 85, 754–762.

Quantitative methods for studying selenite reduction are given.

Selenite—*Salmonella*

R. G. L. McCready, J. N. Campbell and J. I. Payne (1966).

Can. J. Microbiol. 12, 703–714.

When *Salmonella heidelberg* is grown in 0.1% w/v Na_2SeO_3 and examined microscopically during growth, two morphological changes can be seen. Red intracellular granules are seen in most of the population within 10 to 12 hours, and organisms containing granules elongate without cell division. The intracellular granules produced by *S. heidelberg* in selenite broth have been identified by X-ray analysis as amorphous red selenium. The intermediate in the conversion of selenite to elemental selenium has been trapped and identified as divalent selenium ion. Growth studies have shown that selenite toxicity is primarily associated with the lag phase of growth, and also that the divalent intermediate is more toxic than the tetravalent precursor.

Sulphates and **Sulphites**—*Desulfovibrio (Sporovibrio)*

A. R. Prévot (1948). Annls Inst. Pasteur, Paris 75, 571–575.

I. Réduction des sulfites par les anaérobies. – La méthode de Wilson, Blair et Maud est couramment employée dans certains laboratoires de contrôle des eaux pour la détection de *W. perfringens* [5]. Nous avons voulu voir si d'autres anaérobies sont capables de donner des réactions positives avec ce test.

Rappelons brièvement la technique: à des géloses profondes désaérées par vingt minutes d'ébullition et ramenées à 60° au bain-marie, on ajoute 2 cm^3 d'une solution à 20 p. 100 de sulfite de sodium et 1 goutte d'alun de fer à 5 p. 100 (ces solutions étant préalablement stérilisées). Si on ensemence avec une suspension faible de *W. perfringens* et si on recoagule dans l'eau froide, après douze et quinze heures d'étuve à 37°, il apparaît d'énormes colonies sphériques d'un noir absolu.

II. Réduction des sulfates par les anaérobies. – Pour détecter qualitativement la réduction des sulfates par les anaérobies, nous avons remplacé, dans les géloses de Wilson, Blair et Maud, le sulfite de sodium par le sulfate de sodium, et ceci en conservant les mêmes proportions: 2 cm^3 de la solution de SO_4Na_2 à 20 p. 100 et 1 goutte d'alun de fer à 5 p. 100. Nous avons alors constaté la rareté extrême de cette réduction chez les souches de notre collection: sur les 102 souches prospectées, appartenant à 31 espèces différentes, 98 donnaient dans ce milieu des colonies blanches, et suivant qu'elles produisaient ou non des gaz et de l'SH_2 à partir des peptones, des dépôts noirs informes à distance ou des halos noirs.

[5] Gernez-Rieux et Buttiaux. *Ann. Inst. Pasteur Lille* (sous presse).

Sulphate—*Desulfovibrio (Sporovibrio)*

J. Senez (1951). Annls Inst. Pasteur, Paris 80, 395–408.

Pour les expériences rapportées dans le présent article, on a employé une solution minérale de base de composition suivante:

NH_4Cl, 0.2 p. 100; $MgSO_4 \cdot 7H_2O$, 0.4 p. 100; Na_2SO_4, 0.1 p. 100;
K_2HPO_4, 0.1 p. 100; $CaCl_2 \cdot 2H_2O$, 0.02 p. 100; NaCl, 4 p. 100.

Afin d'éviter la formation de sulfure de fer et de pouvoir estimer les croissances bactériennes par néphélométrie, on a omis le sel de Möhr qui est classiquement incorporé dans les milieux destinés à la culture des bactéries sulfato-réductrices. Ainsi que l'indiquent Butlin et ses collaborateurs (1949) [4], les besoins en fer de ces germes sont assez faibles pour être satisfaits par les traces de cet élément toujours présentes, comme impuretés, dans les autres constituants.

Les sels sont dissous dans de l'eau distillée, en respectant l'ordre indiqué, et le pH est ajusté à 7.4. Le trouble formé est éliminé par filtration et la solution est stérilisée à l'autoclave (quinze minutes à 120°). Cette première stérilisation peut modifier le pH, qui en ce cas est réajusté, et provoquer un précipité parfois assez abondant. Après filtration, la solution est à nouveau stérilisée, à température moins élevée (quinze minutes à 110°) et demeure cette fois parfaitement limpide.

Lorsque le milieu a été porté à son volume final et ensemencé, sa concentration saline est identique à celle du milieu de Starkey (1948) [7], c'est-à-dire qu'il contient 1.16 millimoles de sulfates par 100 cm^3.

Le lactate et le pyruvate, dissous dans de l'eau distillée, sont stérilisés séparément par filtration sur filtre Seitz. Il en est de même pour l'extrait de levure (Difco), lorsque celui-ci doit être incorporé à concentrations variables. Dans le cas contraire, l'extrait de levure est stérilisé avec la solution de base.

La solution de base, fraîchement préparée et refroidie brusquement au sortir de l'autoclave, est répartie stérilement, à raison de 2.5 cm^3 par tube, dans une série de tubes à essai stériles. Après addition du substrat carboné et, éventuellement, de l'extrait de levure, le volume est complété à 4.8 cm^3 avec de l'eau distillée stérile, chacune des concentrations en carbohydrate ou en extrait de levure étant répétée six fois. Les tubes sont finalement ensemencés avec 0.2 cm^3 d'une culture sur milieu de Starkey au lactate modifié par addition de 0.1 p. 100 d'extrait de levure.

Les tubes sont aussitôt scellés sous vide et leur densité optique est individuellement relevée par comparaison avec un tube étalon non ensemencé, contenant le même milieu; ils sont enfin placés à l'étuve à 32°.

[4] K. R. Butlin, M. E. Adams et M. Thomas. *J. Gen. Microbiol.*, 1949, 3, 46–59.

[7] R. L. Starkey. *J. Amer. Water Works Assoc.*, 1948, 40, 1291–1298.

Sulfates and **sulfites**—anaerobes

A.-R. Prévot (1966). Techniques pour le diagnostic des Bactéries Anaérobes. Éditions de la Tourelle, St. Mandé.

Réduction des sulfates et sulfites en sulfures.

La deuxième recherche se fait par observation dans les géloses sulfatées ou sulfitées + alun de fer ammoniacal, des colonies noires et leur nombre. Il faut distinguer soigneusement les colonies noires des dépôts amorphes de sulfure de fer à distance des colonies blanches. Le second phénomène est dû, non aux sulfato-réductases ou aux sulfitoréductases (endoenzymes qui produisent le sulfure de fer au contact du corps microbien et colorent en noir la colonie) mais aux désulfhydrilases (exoenzymes qui diffusent dans la gélose et séparent – SH des acides aminés soufrés, celui-ci réagissant sur le fer pour donner des dépôts amorphes, loin des colonies qui restent blanches).

L'emploi des géloses sulfitées-alunées a un autre avantage: tout l'hydrogène qui se dégage habituellement des cultures des anaérobies gazogènes, brise la gélose et rend le travail et l'observation difficiles, est utilisé à la réduction des sulfites, donc ne se dégage pas. Il est plus facile de travailler sur ces géloses pour repiquer les colonies dont on obtient rapidement des clones purs.

La technique primitive de Wilson et Blair a de multiples inconvénients. Nous lui avons fait subir des améliorations successives. La dernière en

date est celle de Yalcin qui nous donne maintenant toute satisfaction.
Gélose profonde sulfitée suivant Yalcin.

1) *Composition:*

– B.V.F. pH 8*	1000 cm^3
– Sulfite de soude	7 g
– Citrate de soude	0.5 g
– Alun de fer et d'ammonium	2 g
– Agar-Agar	8 g
– Glucose	2 g

2) *Technique:*

Faire fondre l'Agar-Agar, ajouter dans cet ordre les produits:
1° Sulfite de soude,
2° Citrate de soude,
3° Alun de fer et d'ammonium.
Précipiter 30 minutes à 122° puis ajouter le glucose.
Filtrer sur laine de verre.
Répartir selon la demande.
Stériliser 30 minutes à 112°–114°.

*See **Nitrate reduction**—anaerobes (p. 631).

Tellurite—*Streptobacillus moniliformis*

F. R. Heilman (1941). A Study of *Asterococcus muris (Streptobacillus moniliformis)* II. Cultivation and biochemical activities. J. infect. Dis. 69, 45–51.

When the organisms were grown on serum veal infusion agar containing 0.16% potassium tellurite, there was no blackening of the colonies and thus no reduction of the tellurite.

Tellurite—*Corynebacterium*

P. Hauduroy (1951). Techniques Bactériologiques. Masson et Cie, Éditeurs, Paris.

MILIEU AU TELLURITE

(Pour *C. diphteriae). –* Faire fondre de l'agar-agar à l'extrait de viande ou de l'agar-agar à 0.2 p. 100 de dextrose, par portions de 10 cm^3 dans des tubes ou par quantités plus importantes dans les fioles, et refroidir à 50°C. Pour chaque 10 cm^3 de milieu de culture, ajouter 1 cm^3 de sang de lapin citraté ou défibriné et 1 cm^3 d'une solution stérile de tellurite de potassium à 2 p. 100. Mélanger et couler dans les boîtes de Petri.

Nota. – On prépare également un excellent milieu de culture au tellurite en ajoutant 5 cm^3 de solution de sang telluritė (marque Bacto) à 100 cm^3 d'agar-agar au dextrose-protéose n° 3 (marque Bacto), en chauffant à 80°C, et en refroidissant à 50°C avant de couler en plaques. *(M. T.).*

Tellurite—*Staphylococcus, Micrococcus*

D. A. A. Mossel (1962). J. Bact. 84, 1140–1147.

Tellurite reduction on tellurite-glycine agar. The suspension of the strain used in the earlier tests was also streaked as 4-cm straight lines on freshly dried plates of tellurite-glycine-lithium chloride agar (Zebovitz, Evans, and Niven, 1955). The dehydrated product marketed by Difco Laboratories Inc., Detroit, Mich., was used for the preparation of this agar, and the manufacturers' instructions for the preparation of this medium were strictly followed. In all tests, a coagulase positive and *Micrococcus*-reference strain were used as controls. Plates were incubated for 48 hr at 37°C.

E. Zebovitz, J. B. Evans and C. F. Niven (1955). J. Bact. 70, 686–690.

Tellurite—*Corynebacterium pseudotuberculosis*

C. H. Pierce-Chase, R. M. Fauve and R. Dubos (1964). J. exp. Med. 120, 267–281.

The authors used Mueller's tellurite serum agar.

Tellurite—enterococci

C. G. Rogers and W. B. Sarles (1964). J. Bact. 88, 965–973.

Reduction of potassium tellurite, 1:2,500 in skim milk. Incubated at 37°C.

Tetrathionate—enterobacteria, *Pasteurella, Aeromonas*

L. Le Minor and F. Pichinoty (1963). Annls Inst. Pasteur, Paris 104, 384–393.

Méthode de culture.

On utilise un milieu complexe tamponné à pH 7 dont la composition est la suivante: Na_2HPO_4, 12 H_2O, 3.575 g; KH_2PO_4, 0.98 g; $MgSO_4$, 0.03 g; NH_4Cl, 0.5 g; $FeSO_4$ et $CaCl_2$, traces; extrait de levure Difco, 0.25 g; bacto-peptone Difco, 0.25 g; eau, 1000 ml. Le glucose (2 g par 1000 ml) stérilisé à part est ajouté au moment de l'ensemencement. Le tétrathionate de potassium (1 g par 1000 ml) est stérilisé par filtration. Les solutions de ce sel ne peuvent être stérilisées à chaud, car elles se décomposent très rapidement au voisinage de 100°. L'oxygène inhibant la biosynthèse et l'activité de la TTR, les cultures doivent être obligatoirement réalisées en anaérobiose.

Description de la méthode de diagnostic.

On répartit le milieu de base décrit ci-dessus dans des tubes à essais de 17 mm/170 mm munis d'une fermeture hermétique (pastille de caoutchouc appliquée sur la partie supérieure par une capsule vissée) à raison de 10 ml par tube. Après stérilisation, chaque tube de milieu refroidi est additionné de 0.2 ml d'un mélange à volumes égaux de solutions aqueuses de glucose à 20 p. 100 et de tétrathionate à 10 p. 100. On ensemence avec une anse d'une culture de 18 heures sur gélose inclinée. Les tubes sont mis à incuber dans une étuve à 37° pendant vingt-quatre heures. On titre le thiosulfate formé par l'iode N/10 en présence de quelques gouttes d'empois d'amidon. Cette opération peut être réalisée avec une microburette graduée ou une pipette compte-gouttes de Duclaux. Dans le cas où tout le tétrathionate serait réduit, on devrait théoriquement verser 0.66 ml d'iode. Les nombreux dosages effectués montrent que l'essai doit être considéré comme négatif lorsqu'on verse un volume d'iode inférieur à 0.03 ml et positif au-delà de 0.2 ml. Cependant, dans la plupart des cas, il faut ajouter aux cultures des souches ayant une TTR un volume compris entre 0.5 et 0.7 ml. Les résultats sont donc extrêmement nets et aucune confusion n'est possible. Nous n'avons jamais observé de souches "faiblement positives". puisqu'aucun des essais ne correspond à un volume compris entre 0.03 et 0.2 ml.

Le milieu décrit est totalement dépourvu de pouvoir réducteur à l'égard de l'iode. D'autre part, les bactéries étudiées ne semblent pas donner lieu à une accumulation de composés organiques réducteurs pendant leur croissance. En effet, plusieurs essais effectués avec des cultures dépourvues de tétrathionate ont constamment fourni des volumes d'iode inférieurs à 0.015 ml. Notons en passant que l'addition au milieu d'un agent réducteur minéral (hydrosulfite) ou organique (réductose, thioglycolate, cystéine) est à proscrire. En milieu alcalin, le tétrathionate se décompose spontanément avec formation de thiosulfate. D'une manière générale il faut, par conséquent, veiller à ce que le pH des cultures ne soit jamais supérieur à 7. Pendant la croissance, le pH ne peut que s'abaisser, puisque la réduction de chaque molécule de tétrathionate libère deux protons provenant de l'ionisation de l'acide thiosulfurique:

$$K_2S_4O_6 + 2H = K_2S_2O_3 + H_2S_2O_3$$

La concentration en peptones et en extrait de levure peut être fortement accrue pour la culture des espèces très exigeantes.

Tetrazolium—PPLO

N. L. Somerson and H. E. Morton (1953). J. Bact. 65, 245–251.

Media. The basal medium used throughout was "Bacto–PPLO agar" (Difco). Each 1,000 ml contained the infusion from 50 g dried Bacto-beef heart for infusion, 10 g Bacto-peptone, 5 g NaCl, and 14 g agar. The pH was 7.8 prior to sterilization in the autoclave. Before use, the basal medium was enriched with Bacto–PPLO bovine serum fraction A to a concentration of one per cent of the final volume. For ordinary cultivation, the medium was dispensed into sterile petri dishes in approximately 20 ml amounts. In some instances sterile solutions of tetrazolium salts and sodium lactate were also added to the medium before dispensing.

The tetrazolium salts were made into one per cent aqueous solutions and sterilized by autoclaving at 120°C for 15 minutes. They were stored in the refrigerator in the dark. Enzymatic activity was demonstrated by incorporating the tetrazoles into the agar. Preliminary experiments were performed to determine toxicity of the tetrazolium compounds. A 0.1 per cent final concentration of triphenyltetrazolium chloride inhibited growth of pleuropneumonialike organisms, but most strains were able to grow on a medium containing 0.05 per cent. The medium containing a final concentration of 0.005 per cent triphenyltetrazolium chloride was found to be satisfactory in these experiments. A similar concentration was used with an iodo-derivative* (Atkinson, Melvin, and Fox, 1950) of triphenyl-tetrazolium chloride.

*Trihalotetrazolium chloride, supplied by Dr. S. W. Fox, Iowa State College, Ames, Iowa.

E. Atkinson, S. Melvin and S. W. Fox (1950). Science, N.Y. 111, 385–387.

Tetrazolium

E. M. Barnes (1956). J. gen. Microbiol. 14, 57–68.

Culture medium (subsequently called TG medium). The composition of the medium was (%, w/v): peptone (Evans's), 1; Lab Lemco, 1: NaCl, 0.5; glucose, 1; distilled water; 2:3:5-triphenyltetrazolium chloride to give 375 μg/5 ml medium. In the experiments described below the initial pH value was varied between 6.0 and 7.6.

The medium containing the peptone, Lemco and salt was adjusted to the required pH value and autoclaved at 15 lb per sq in. for 20 minutes in 95 ml lots. After autoclaving, 5 ml of a 20% (w/v) glucose solution (sterilized by autoclaving at 10 lb/sq in. for 10 min) were added together with 0.75 ml of a 1% solution of the tetrazolium salt. (The tetrazolium solution was sterilized by steaming for 30 min). The TG medium was then aseptically pipetted in 5 ml lots into 1 oz screw-capped bottles.

Cultural conditions. The TG medium (5 ml) was inoculated with 0.025 ml of an 18 hr broth culture to give *c.* 1 x 10^7 organisms/ml. The inoculated bottles were incubated for 24 hr at 37°C and the reduced tetrazolium (insoluble red formazan) was extracted as follows.

Extraction of formazan from cultures. A culture (5 ml) was shaken successively with 5, 2 and 2 ml of n-butanol to extract all the formazan. The butanol extracts were withdrawn with a Pasteur pipette into marked centrifuge tubes and the volume made to 10 ml with butanol. The solution was mixed and centrifuged to obtain a clear extract. The colour intensity of the extract was measured in an EEL (Evans Electroselenium Ltd) colorimeter with filter no. 625. The percentage reduction of tetrazolium was calculated by reference to a standard curve on the assumption that it was directly proportional to the amount of formazan produced.

Errors of the method. Care must be taken that the tetrazolium solution is freshly prepared and stored in the dark. In media where considerable reduction has taken place (e.g. with *Streptococcus faecalis*) sufficient formazan to give a visible colour adheres to the organisms during growth and cannot be removed by the butanol extraction. However, even in these cases virtually all the formazan may be recovered.

A medium with an initial pH 6.0 distinguishes between *Streptococcus faecalis* and *Streptococcus faecium.*

Tetrazolium—*Streptococcus*

P. Morelis and L. Colobert (1958). Annls Inst. Pasteur, Paris 95, 568–587.

La réduction du 2.3.4.triphényl-tetrazolium a été éprouvée sur le milieu de Miss Barnes [2]. Cet auteur a établi que le pH le plus favorable pour permettre la discrimination entre les souches réductrices et non réductrices se situait dans une zone étroite, entre 6.0 et 6.2. Comme le potentiel d'oxydo-réduction varie fortement avec des variations minimes du pH, il nous a paru nécessaire de tamponner le milieu par l'addition de phosphate. Le milieu que nous avons finalement utilisé est le suivant: macération de viande de boeuf, 1000 cm^3; peptone Vaillant 5 B, 10 g; chlorure de sodium, 5 g; phosphate monopotassique, 7.260 g; phosphate disodique, 4.776 g; gélose en poudre, 16 g. Le milieu réparti à raison de 96 cm^3 par flacon fut autoclavé vingt minutes à 115°. Dans le milieu, refroidi aux alentours de 50°, on ajoute stérilement par flacon 3.3 cm^3 de solution de glucose à 30 p. 100, puis 1 cm^3 de solution à 1 p. 100 de 2.3.5.triphényl-tétrazolium préalablement stérilisée par un chauffage de trente minutes à vapeur fluente. Dans ces conditions, les souches qui réduisent le tétrazolium en son dérivé le formazan produisent des colonies rouge cerise, alors que les souches faiblement ou non réductrices donnent des colonies rose pâle ou blanches.

[2] Barnes (E. M.). *J. gen. Microbiol.*, 1956, 14, 57.

Tetrazolium—*Streptococcus*

C. W. Langston, J. Gutierrez and C. Bouma (1960). J. Bact. 80, 693–695.

Reduction of 2, 3, 5 triphenyltetrazolium was determined by the method of Barnes (1956).

E. M. Barnes (1956). J. gen. Microbiol. 14, 57–68.

Tetrazolium

Y. S. Halpern and H. E. Umbarger (1961). J. gen. Microbiol. 26, 175–183.

Reduction of triphenyltetrazolium chloride. The capacity of bacterial suspensions, and bacteria disintegrated by sonic oscillation (15 ml of a cell suspension of OD x 10 = 60 at 550 mμ in M/15 phosphate buffer, pH 7.5 was treated for 8 min in a Ryatheon 10 KC magnetostrictive sonic oscillator) to reduce triphenyltetrazolium chloride in the presence of glutamate was determined under both aerobic and anaerobic conditions. For the anaerobic experiments Thunberg tubes were used. The reaction was stopped by the addition of 0.1 ml 4 N-HCl. The formazan formed was extracted with 2 ml isobutanol and the colour intensity measured at 485 mμ in a Coleman Junior spectrophotometer.

Tetrazolium—*Vibrio*

G. H. G. Davis and R. W. A. Park (1962). J. gen. Microbiol. 27, 101–119.

Growth and reduction with 0.2 or 1% (w/v) triphenyltetrazolium chloride (T.T.C.) (Meitert, Meitert and Horodniceanu, 1960).

E. Meitert, T. Meitert and T. Horodniceanu (1960). Archs roum. Path. exp. Microbiol. 19, 149.

Tetrazolium—*Streptococcus*

F. M. Ramadan and M. S. Sabir (1963). Can. J. Microbiol. 9, 443–450.

The authors used the method of E. M. Barnes (1956) J. gen. Microbiol. 14, 57–68.

Tetrazolium

G. S. Wilson and A. A. Miles (1964). Topley and Wilson's *Principles of Bacteriology and Immunity* 5th ed., Arnold.

Reduction of triphenyltetrazolium chloride

This substance is added to tubes of broth in concentrations ranging from 1/50 to 1/50,000. The broth is inoculated with the organism to be tested and incubated. Organisms that grow, reduce the dye to formazan, which settles as an insoluble red deposit. Some organisms, such as the clostridia, the corynebacteria, and Shiga's dysentery bacillus, are very susceptible and fail to grow in concentrations of even 1/5,000 or less (Wundt 1950).

W. Wundt (1950). Dt. med. Wschr. 75, 1471.

Tetrazolium—enterococci

C. G. Rogers and W. B. Sarles (1964). J. Bact. 88, 965–973.

Reduction of tetrazolium chloride (Barnes, 1956) in TYE agar at pH 6.0. Incubated at 37°C.

E. M. Barnes (1956). J. gen. Microbiol. 14, 57–68.

Tetrazolium—*Streptococcus*

R. Whittenbury (1965). J. gen. Microbiol. 38, 279–287.

This test was adapted from that described by Barnes (1956). Liquid medium containing glucose 0.5% (w/v) was adjusted to pH 6.0. Tetrazolium, 0.01% (w/v) final concentration, was added from a stock solution heated at 100°C for 15 min. A heavy inoculum, 3 capillary pipette drops of turbid culture, was added to 4 ml medium in a test tube. The cultures, incubated at 30°C, were examined at 8 hr. Positive cultures were coloured a deep magenta, whilst cultures regarded as negative were either colourless or faintly pink.

The basal liquid medium contained meat extract (Lab-Lemco), 0.5 g; peptone (Evans), 0.5 g; yeast extract (Difco), 0.5 g; Tween 80, 0.05 ml; $MnSO_4 \cdot 4H_2O$, 0.01 g; in 100 ml tapwater, adjusted to pH 6.5 and autoclaved at 121°C for 15 min.

E. M. Barnes (1956). J. gen. Microbiol. 14, 57.

Tetrazolium—*Pseudomonas aeruginosa*

A. H. Wahba and J. H. Darrell (1965). J. gen. Microbiol. 38, 329–342.

The organisms were tested for growth on 1% triphenyltetrazolium chloride in nutrient agar No. 1.

Nutrient agar No. 1 (Oxoid No. 1) contained Lab Lemco beef extract, 1 g; yeast extract (Oxoid L 20), 2 g; peptone (Oxoid L 37), 5 g; sodium chloride, 5 g; agar, 15 g; distilled water to 1 L; pH 7.4.

Tetrazolium—*Staphylococcus aureus*

P. A. Pattee (1966). J. Bact. 92, 787–788.

The agar-layer technique for the plaque assay of staphylococcal phage 83 was conducted in the usual manner with *S. aureus* Ps 83, P and D agar, and P and D soft agar (P. A. Pattee and J. N. Baldwin, J. Bact. 82, 875, 1961). After incubation of the assay plates for 8 hr at 37°C, the plaques were sufficiently developed to be scored. The assay plates containing the fully developed plaques were then flooded with 10 ml of Trypticase Soy Broth (TSB, BBL) containing 0.1% 2, 3, 5, -triphenyltetrazolium chloride (TTC, Nutritional Biochemicals Corp., Cleveland, Ohio). After incubation at 37°C for 20 min, the broth was poured off. Each plaque was now a sharp, clear area against the intense red background produced by the reduction of TTC to the insoluble formazan by the indicator cells. In addition to TSB, distilled water, 0.5% NaCl, 1% glucose, and P and D broth were examined as suspending media for TTC. Although identical results were eventually obtained in all cases, only P and D broth supported the reduction of TTC at a rate similar to that obtained with TSB. Similarly, incubation of the flooded plates at 37°C merely accelerated the rate at which TTC was reduced.

Trimethylene oxide to trimethylene glycol—*Vibrio marinus*

R. R. Colwell and R. Y. Morita (1964). J. Bact. 88, 831–837.

The authors used the method of A. J. Wood and E. A. Baird (1943).

J. Fish. Res. Bd Can. 6, 194–201 – modified by the addition to the media of the following salts to produce a synthetic seawater:– sodium chloride, 2.4%; potassium chloride, 0.07%; magnesium chloride (hydrated), 0.53% and magnesium sulfate (hydrated), 0.7%.

A standard inoculum was 1 drop from a pasteur pipette (*c.* 0.05 ml) of a 24 hour artificial seawater broth.

Cultures were incubated at 18°C.

Trimethylene oxide to trimethylene glycol—*Pseudomonas aeruginosa*

R. R. Colwell (1964). J. gen. Microbiol. 37, 181–194.

Cultures were incubated for 14 days in Wood and Baird's medium (1943) and the presence of trimethylamine was then tested for according to the method given by the authors.

A. J. Wood and E. A. Baird (1943). J. Fish. Res. Bd Can. 6, 194.

Trimethylene oxide to trimethylene glycol—RM bacterium

A. J. Ross, R. R. Rucker and W. H. Ewing (1966). Can. J. Microbiol. 12, 763–770.

Trimethylamine oxide was reduced by the method of Wood and Baird (1943).

A. J. Wood and E. A. Baird (1943). J. Fish. Res. Bd Can. 6, 194–201.

REDUCTION OF AND INHIBITION BY METHYLENE BLUE AND OTHER DYES

Dye reduction tests—*Streptococcus*

J. M. Sherman and W. R. Albus (1918). J. Bact. 3, 153–174.

Tests were made by adding the dye to sterilized whole milk, the advantage of unskimmed milk being that the fat forms a layer over the surface which excludes the air quite effectively and thus reduction is not hindered. The litmus milk was prepared in the ordinary way by adding sufficient litmus solution to the milk to give a rather dark lavender color and then sterilizing. The other dyes were made as follows:

Methylene blue

Medicinal methylene blue	0.5 g
Distilled water	1000 ml

Indigo carmine

Indigo carmine (Kahlbaum's)	1.0 g
Distilled water	1000 ml

Neutral red

Neutral red (Grübler's)	0.1 g
Distilled water	1000 ml

The stain solutions and milk were sterilized separately and then mixed in the proportion of 1 ml of stain to 10 ml of milk.

In making the tests twenty-four hours old cultures of the organisms in milk were used to inoculate from, and the stain culture so prepared was incubated at 37°C. Observations were then made on three points, (1) reduction of stain, (2) time required to reduce, and (3) whether reduction was before or after curding of milk. When no reduction was evident the cultures were allowed to remain six days before final examination was made.

Methylene blue reduction—*Xanthomonas cannae*

M. K. Bryan (1921). J. agric. Res. 21, No. 3, 143–152.

Milk to which methylene blue was added to make it robin's egg blue shows reduction in color on the second day. In 10 days reduction is complete.

Methylene blue reduction—*Escherichia, Aerobacter*

G. A. Lindsey and C. M. Meckler (1932). J. Bact. 23, 115–121.

Added 1 drop of a saturated aqueous solution of methylene blue to 10 ml of the 24 hour cultures of *A. aerogenes* and *E. coli* in lactose-broth, shook the tubes vigorously for one minute and observed for reduction up to one hour.

Methylene blue tolerance—*Streptococcus*

J. M. Sherman, P. Stark and J. C. Mauer (1937). J. Bact. 33, 483–494.

0.1% medicinal methylene blue in sterile skim milk was used to test inhibition by methylene blue.

Methylene blue reduction—*Streptococcus*

P. L. Boisvert (1940). J. Bact. 39, 727–738.

Tubes of sterile milk containing 1.0% methylene blue to make a final dilution of 1/5000 are inoculated and incubated and daily readings taken for 7 days. Control tubes of plain milk and of uninoculated methylene blue milk were used.

Methylene blue tolerance—*Streptococcus*

Y. Abd-el-Malek and T. Gibson (1948). J. Dairy Res. 15, 233–248.

Inhibition by 0.1% methylene blue

Skim milk, sterilized in bulk, and 1% medicinal methylene blue in distilled water, sterilized in screw-capped bottles, were mixed and distrubuted in tubes with aseptic precautions. The tubes were inoculated immediately and were stoppered in order to prevent evaporation.

Methylene blue reduction—general

P. Hauduroy (1951). Techniques Bactériologiques. Masson et Cie, Éditeurs, Paris.

BLEU DE METHYLENE

(Réduction). – Préparer le milieu suivant:

Glucose	5	g
Protéose peptone	10	g
Gélose	2.5	g
Bleu de méthylène	0.004	g
Eau distillée	1 000	ml

On obtient un milieu semi-solide. Répartir en tubes de 15 mm de diamètre à raison de 8 ml par tube. Ensemencer. Examiner aux 1er, 3e, 5e, 7e, 10e, 14e et 21e jours. *(A. T. C. C.)*.

A une culture de 24 heures en bouillon nutritif, ajouter une goutte d'une solution aqueuse de bleu de méthylène à 1.0 p. 100. Placer 24 heures à 37°C. Une décoloration complète est l'indice d'une réaction fortement positive, une coloration verte, l'indice d'une réaction douteuse, l'absence de décoloration, une réaction négative. *(W. I. C.)*.

Methylene blue reduction—*Staphylococcus*

C. Shaw, J. M. Stitt and S. T. Cowan (1951). J. gen. Microbiol. 5, 1010–1023.

One drop 1% (w/v) aqueous methylene blue was added to 5 ml broth culture (1 day at 37°C, 2 days at 30°C or 3 days at 22°C) shaken well to mix, and incubated at 37°C for 2 hr. Complete decolorization below the top half-centimetre was recorded as positive, partial decolorization as a trace reaction.

Methylene blue reduction—microtests

P. H. Clarke and S. T. Cowan (1952). J. gen. Microbiol. 6, 187–197.

Standardized methylene blue (British Drug Houses Ltd.) in concentrations of 0.1 and 0.01% drawn into capillaries with suspension and sealed. Read at 4 and 24 hours at 37°C.

Methylene blue reduction—*Bacillus*

N. R. Smith, R. E. Gordon and F. E. Clark (1952). U.S.D.A. Agr. Monograph No. 16.

Glucose, 5 g; proteose-peptone, 10 g; agar, 2.5 g; methylene blue, 0.004 g; and distilled water, 1000 ml. Tubes 15 mm in diameter containing 8 ml of the semisolid medium were inoculated and examined for the reduction of methylene blue at 1, 3, 5, 7, 14 and 21 days.

Methylene blue reduction—*Pseudomonas (Malleomyces) pseudomallei*

P. de Lajudie and E. R. Brygoo (1953). Annls Inst. Pasteur, Paris 84, 509–515.

On ajoute I goutte d'une solution aqueuse à 1 p. 100 de colorant à 10 ml de culture en bouillon de quarante-huit heures. La lecture se fail après trente minutes, puis douze heures à 37°.

Methylene blue reduction—various

N.C.T.C. Methods 1954.

Incubate cultures at optimal temperature.

Method 1. Grow the organism in broth for 24 hours; add 1 drop of 1% aqueous methylene blue and incubate at 37°C. Read after 1 and 4 hours.
W. W. C. Topley and G. S. Wilson (1929). *The Principles of Bacteriology and Immunity.* London: Arnold.

Method 2. Grow in semisolid medium (glucose 5 g, peptone 10 g, agar 1.2 g, methylene blue 0.004 g, distilled water 1 litre; tubed 8 ml in 6 x 5/8 tube) and read after 1, 3, 5, 7, 10, 14 days.
N. R. Smith, R. E. Gordon and F. E. Clark (1946). *Aerobic mesophilic spore-forming bacteria.* U.S. Dept. Agr. Misc. Pub. No. 559.

Methylene blue reduction—*Streptococcus*

N.C.T.C. Methods 1954.

Add 0.1 ml of a sterile 2% solution of methylene blue to 10 ml of sterile fat free milk. This gives a final concentration of 1/5,000.

Add 0.1 ml of 24 hour broth culture of strain to be tested. Incubate at 37°C. Read at 6, 8, 24 and 48 hours.

R. C. Avery (1929). J. exp. Med. 50, 463.
K. Skadhauge. *Studies on Enterococci,* p. 58.

Methylene blue reduction—*Bacillus*

K. L. Burdon (1956). J. Bact. 71, 25–42.

The speed and completeness in the reduction of methylene blue was observed in a soft agar medium containing: tryptose, 20 g; NaCl, 5 g; and agar, 8 g per L. After autoclaving, the medium was cooled to 55°C and a sufficient amount of a 1 per cent aqueous solution of methylene blue was added to give a distinct, but faint blue color. The completed medium was adjusted to a final pH of 7.2 to 7.4, and distributed aseptically into 13- by 100-mm (or larger) test tubes, to make an agar column several inches deep. Inoculations were made by a single stab with a straight needle, and the cultures were checked for growth and loss of color, in comparison with un-inoculated control tubes, after incubation for 24 to 48 hr.

Methylene blue reduction—*Streptococcus*

M. Moreira-Jacob (1956). J. gen. Microbiol. 14, 268–280.

Methylene blue milk (1/5000) examined after 5 days incubation at 37°C with preliminary readings at 24 and 48 hours.

Methylene blue reduction—*Streptococcus* and *Lactobacillus*

C. W. Langston and C. Bouma (1960). Appl. Microbiol. 8, 212–222.

Skimmed milk (BBL) containing 0.1 per cent methylene blue was inoculated with test cultures and observed for changes from light blue to colorless. The incubation period was for 1 week.

Methylene blue reduction—rhizosphere bacteria

P. G. Brisbane and A. D. Rovira (1961). J. gen. Microbiol. 26, 379–392.

Methylene blue agar. Basal agar + 1% (w/v) glucose and 1 ml/L of 0.5% (w/v) methylene blue was tubed in 10 ml lots and inoculated before setting. Those isolates which reduced more than half the depth of the agar were considered positive.

The basal medium. The yeast-extract peptone nitrate broth of Sperber and Rovira (1959), with 15 g agar/L for solid medium, was used.

J. I. Sperber and A. D. Rovira (1959). J. appl. Bact. 22, 85.

Janus green B reduction—*Streptococcus*

F. M. Ramadan and M. S. Sabir (1963). Can. J. Microbiol. 9, 443–450.

The authors used the method of K. E. Cooper and F. M. Ramadan (1955) J. gen. Microbiol. 12, 180–190.

Methylene blue reduction—*Actinobacillus*

P. W. Wetmore, J. F. Thiel, Y. F. Herman and J. R. Harr (1963). Comparison of selected *Actinobacillus* species with a hemolytic variety of *Actinobacillus* from irradiated swine. J. infect. Dis. 113, 186–194.

The semisolid medium of Burdon (1956) was used.

K. L. Burdon (1956). J. Bact. 71, 25–42.

Methylene blue tolerance—enterococci

C. G. Rogers and W. B. Sarles (1964). J. Bact. 88, 965–973.

Resistance to 0.1% methylene blue was tested in sterile skim milk.

Methylene blue reduction—*Aerococcus catalasicus*

O. G. Clausen (1964). J. gen. Microbiol. 35, 1–8.

Cultures were incubated aerobically at 37°C unless otherwise stated.

Methylene-blue reductase was tested in milk culture to which had been added 0.1% methylene blue. The reduction or lack of reduction was recorded after 20 and 48 hr and finally after 5 days' incubation.

Methylene blue reduction

G. S. Wilson and A. A. Miles (1964). Topley and Wilson's *Principles of Bacteriology and Immunity* – 5th ed. Arnold.

Tested on a 24 hours' broth culture at 37°C. Add 1 drop of 1 per cent aqueous methylene blue, and incubate at 37°C.

Complete decolorization = strong positive
Green coloration = weak positive
No decolorization = negative

Methylene blue reduction and tolerance—*Streptococcus*

R. Whittenbury (1965). J. gen. Microbiol. 38, 279–287.

A heavy inoculum, 3 capillary pipette drops, was added to the milk media which were incubated at 30°C and read at 8 hr.

Methylene blue reduction—*Bacillus anthracis*

R. F. Knisely (1965). J. Bact. 90, 1778–1783.

The reduction of methylene blue was done according to Burdon (1956). A soft agar medium (pH 7.3) containing 1% methylene blue was inoculated by a single stab with a straight needle. The cultures were incubated at 37°C for 24 hr and observed for growth and for reduction of the dye.
K. L. Burdon (1956). J. Bact. 71, 25–42.

RENNIN

Rennin—various

H. J. Conn (1922). J. Bact. 7, 447–448.

Inoculate the culture under investigation into the milk in the usual manner; then incubate for twenty-four hours or such time as is necessary to allow the organism in question to produce vigorous action in the milk with at least 0.5 ml of whey on the surface. At the end of this incubation period obtain fresh milk and place 10 ml of it in a test tube, do not sterilize it. Warm this milk to about 37°C. Then add to the milk a measured quantity, say 0.5 ml, of whey from the incubated culture and place in a 37° incubator. Examine every five minutes for the first half hour and, if not curdled then, at less frequent intervals for a few hours longer. If rennet is present in any abundance, the milk is ordinarily curdled inside of half an hour.

Rennin—*Bacillus*

J. G. Waklin (1928). J. Bact. 16, 355–373.

The clotting time was determined at room temperature using 1 ml of culture mixed with 1 ml of phenol milk (milk 40 ml; 10% phenol 2 ml). Clotting time was recorded up to 1½ hours.

Rennin—various

C. Gorini (1932). J. Path. Bact. 35, 637.

"Take well developed slant agar cultures, pour over them some sterilised milk till the surface is covered, mix with platin needle the culture in the milk and put the culture into the incubator in order to promote the coagulation. The milk must be sterilised at a temperature not exceeding 100°C in order to bring about as little change possible and to maintain its white color.

I have experimented with 12 different strains of *B. typhosus;* all these strains have curdled the milk poured over their slant agar cultures after an incubation from 8 to 23 days at 38°C. The clot is soft, alkaline, quite of a chymasic nature."

Rennin—*Microbacterium*

V. Bolcato (1957). Antonie van Leeuwenhoek 23, 351–356.

Rennet production: Rennet formation was clearly visible using the method of Gorini (1932). J. Path. Bact. 35, 637.

RESPIRATORY TYPE

Respiratory type—*Corynebacterium*

R. F. Brooks and G. J. Hucker (1944). J. Bact. 48, 295–312.

Oxygen relations were determined by the use of shake cultures in tryptone glucose beef extract agar, incubated 3 days at 37°C and observed for location of growth.

Respiratory type—*Bacillus*

N. R. Smith, R. E. Gordon and F. E. Clark (1952). U.S.D.A. Agr. Monograph No. 16.

The medium used (containing 1.0 percent tryptose, 0.5 percent K_2HPO_4, 0.3 percent beef extract, 0.2 percent yeast extract and 1.0 percent glucose) was prepared by adding a sterile 50 percent solution of glucose aseptically to each tube. Before use the medium was steamed to drive off free oxygen, quickly cooled, inoculated, capped with 10-15 mm of sterile melted vaspar (equal parts of vaseline and paraffin) and incubated at 32°C. Growth was recorded at 7 and 14 days, and the colorimetric determination of the pH made at the fourteenth day. The pH of the controls was about 7.6. For certain species this was too alkaline and a near neutral or slightly acid medium was necessary.

For species requiring incubation temperatures of 45° or 50°C the technique had to be modified as follows: Melted vaspar was poured on top of the inoculated medium to a depth of 3 or 4 mm and allowed to harden. Melted 1.5 percent water agar was then added to a depth of 15 or 20 mm, the tubes cooled, and the culture finally incubated at the high temperatures. Results were usually obtained in 2 or 3 days before the agar developed cracks caused by drying. A layer of vaspar may also be placed on top of the agar.

Respiratory type—*Bacillus*

K. L. Burdon (1956). J. Bact. 71, 25–42.

The capacity of the organisms to multiply in the depths of glucose agar was tested in a soft (0.8 per cent agar) 1 per cent glucose tryptose medium, distributed in 18- by 180-mm test tubes to make a column about 4 in. deep in each tube. If not inoculated immediately after autoclaving, the medium was re-melted, then solidified again by rapid cooling, just before use. Cultures were inoculated to the bottom of the agar column by a single stab, and observed for growth during incubation for 72 hours.

Respiratory type—*Microbacterium*

V. Bolcato (1957). Antonie van Leeuwenhoek 23, 351–356.

Relation to free oxygen. A tube of deep saccharose agar* was inoculated while in fluid condition at 45°C with an inoculum not too heavy to permit discrete colonies, rotated to mix the inoculum with the medium and cooled. After incubation at 35°C growth was observed throughout the medium but was heavier upon the surface and in the upper layers.

The organism can grow under anaerobic condition but the best growth occurs in aerated cultures by bubbling air or oxygen in the medium. Therefore, the strain can be considered aerobic to facultative anaerobic.

*peptone, 5 g; beef extract, 3 g; saccharose, 10 g; distilled water, 1000 ml; pH 6.5.

Respiratory type—*Pseudomonas*

M. E. Rhodes (1959). J. gen. Microbiol. 21, 221–263.

Yeast extract broth + 1.0% (w/v) glucose cultures were incubated for 7 days in a McIntosh and Fildes's jar containing a mixture of 95% (v/v) H_2 + 5% (v/v) CO_2.

Respiratory type—*Gaffkya homari; Aerococcus,* tetrad forming cocci, *Pediococcus*

R. H. Deibel and C. F. Niven, Jr. (1960). J. Bact. 79, 175–180.

Growth in relation to oxygen. Determined in agar shake cultures using trypticase soy broth (BBL) plus 1.5% agar.

Respiratory type—*Pediococcus*

H. L. Günther and H. R. White (1961). J. gen. Microbiol. 26, 185–197.

Duplicate broth cultures were incubated aerobically, and anaerobically in an atmosphere of 95% (v/v) hydrogen + 5% (v/v) carbon dioxide. Visual estimation of growth was made after incubation for 24 or 72 hr for slow growing strains.

Maintenance of stock cultures and methods of cultivation. For maintenance of stock cultures, preparation of inocula and in all experimental work, 'Oxoid' tomato juice (TJ) broth or tomato juice (TJ) agar, adjusted to pH 6.6, were used unless otherwise stated. The following were exceptions to this rule: for strain Tc. 1 sodium chloride (5%, w/v) was added to the medium; and for the aerococci glucose Lemco broth (Shattock and Hirsch, 1947) or glucose yeast extract (GY) agar (containing, as %, w/v; peptone, 1.0; Yeastrel, 0.3; glucose, 1.0; NaCl, 0.25; agar, 1.0; at pH 7.4) was used.

Cultures were incubated aerobically at 30°C with specified exceptions.

P. M. F. Shattock and A. Hirsch (1947). J. Path. Bact. 59, 495.

Respiratory type—*Pseudomonas*

M. Véron (1961). Annls Inst. Pasteur, Paris 100, Suppl. 6, 16–42.

Nous utilisons donc des milieux sans nitrate: en pratique, soit la gélose-gélatine sans nitrate, soit la gélose VF (glucosée à 2 p. 100).

La formule de la gélose-gélatine sans nitrate est la suivante: macération de viande (500 g de viande par litre), 1000 ml; gélatine, 35 g; gélose, 6 g; peptone trypsique de caséine, 10 g; KCl, 5 g. Ajuster à pH 7.6. Laisser refroidir à 50°C et ajouter 50 ml de sérum. Précipiter à 115°, filtrer et ajouter: glucose, 10 g. Répartir 15 ml en tubes de 170 × 17 mm et stériliser à 110°C.

Pour utiliser ces géloses profondes, on les régénère dans un bain-marie bouillant pendant 15 à 20 min, puis on les refroidit à 45–50°C, et on les ensemence dans la masse, avec une pipette Pasteur fermée et trempée dans le bouillon de culture, en utilisant le nombre de tubes nécessaires pour obtenir par dilution des colonies isolées. Apres ensemencement, les tubes sont aussitôt refroidis dans l'eau, puis placés à l'étuve à 30°C.

Dans ces conditions, toutes les souches de *Pseudomonas* que nous avons examinées sont aérobies strictes, c'est-à-dire qu'elles ne cultivent qu'à la surface du milieu, et dans les 3 ou 4 mm de hauteur situés juste sous cette surface.

Respiratory type—*Escherichia aurescens*

H. Leclerc (1962). Annls Inst. Pasteur, Paris 102, 726–741.

Type respiratoire. – Il est défini après ensemencement en tube de 8 × 180 mm contenant une colonne de gélose nutritive d'une hauteur de 10 cm, préalablement régénérée.

Respiratory type—*Staphylococcus, Micrococcus*

D. A. A. Mossel (1962). J. Bact. 84, 1140–1147.

Oxygen tolerance: This test was carried out following the procedure recommended by Buttiaux and Gagnon (1959). Tubes (8 mm external diam) containing an agar of the following composition was prepared: Trypticase, 10 g; yeast extract, 2.5 g; glucose, 1 g; cysteine HCl (analytical reagent grade), 0.5 g; agar, 15 g; water, 1,000 ml; pH = 7.2. All tubes were heated in a boiling-water bath for 10 min, then cooled to 47°C ± 2°C and evenly inoculated with a drop (*c.* 0.05 ml) of a slightly turbid suspension containing approximately 10^6 viable cells/ml of the strain under investigation.

After incubation for 48 hr at 30°C, the extent of the zone in which colonies developed was observed. A strain was called obligately aerobic if it had formed colonies in the upper few millimeters of the agar only; when uniform development of colonies throughout the tubes occurred, the strain was considered to be facultatively anaerobic.

R. Buttiaux and P. Gagnon (1959). Annls Inst. Pasteur Lille 10, 121–149.

Respiratory type—lactic acid bacteria

R. Whittenbury (1963). J. gen. Microbiol. 32, 375–384.

The media used was that described under **Fermentation** – lactic acid bacteria (p. 310).

The soft agar medium provided environments ranging from aerobic to anaerobic in the one culture. This was indicated by the site of growth and by the reduction of methylene blue below the top 0.5–1.0 cm of the uninoculated medium. Reducing conditions persisted, at least in the lower half of the medium, throughout the 14-day period of incubation.

Requirement for aerobic conditions.

Although lactic acid bacteria have been described as microaerophilic or anaerobic, a number of these organisms were found to require aerobic conditions when utilizing certain substrates. In such instances, acid-forming growth was restricted to the surface layers of the soft agar culture. Slight, non-acid forming growth, similar to that which appeared in the basal medium, frequently developed throughout the remainder of the culture. On continued incubation, growth increased in density in the surface layers and occasionally developed as secondary disks below the surface as the medium became oxygenated. Incubation of soft agar cultures under an atmosphere of 95% (v/v) H_2 + 5 (v/v) CO_2 confirmed that growth restricted to the surface reflected a requirement for aerobic conditions.

Requirement for anaerobic conditions.

When substrates were fermented only under anaerobic conditions, acid-forming growth was restricted to a sharply defined zone in the anaerobic part of the soft agar culture. The height of the zone, dependent upon substrate and organism, varied from the bottom fifth of the culture to within 1.0–0.5 cm of the surface.

Where cases of anaerobic requirements were observed in soft agar comparative liquid cultures produced variable results. Shallow liquid cultures frequently showed no acid formation whilst deep cultures frequently became acid. When incubated anaerobically, however, acid was always formed.

Requirement for a diminished oxygen concentration.

In the unsupplemented basal medium many organisms produced a thin disk of growth without obvious acid formation about 0.25 cm below the surface. With a fermentable substance added, acid-forming growth was in certain cases confined to the same position.

No strict aerobic, microaerophilic or anaerobic requirement.

In such instances growth and acidity appeared throughout the whole culture and in most cases uniformly and simultaneously.

Respiratory type—*Spirillum*

W. A. Pretorius (1963). J. gen. Microbiol. 32, 403–408.

Oxygen requirements. These were tested by the shake tube method with 15 ml nutrient agar in 6 x 5/8 in. test tubes, and examined after 7 days. Nitrate broth (Difco) tubes were placed in boiling water for 15 min, immediately cooled down, inoculated, and sealed with wax.

Respiratory type—*Micrococcus roseus*

R. C. Eisenberg and J. B. Evans (1963). Can. J. Microbiol. 9, 633–642.

Relation to oxygen was determined with the use of agar shake cultures and cultures in Trypticase soy broth (BBL) incubated under a vaspar seal. The agar was prepared from the Trypticase soy broth by the addition of 1.5% agar.

Respiratory type—*Moraxella*

W. J. Ryan (1964). J. gen. Microbiol. 35, 361–372.

Atmospheric conditions. The optimal atmospheric conditions for growth were determined by incubation of inoculated blood agar plates in air, in a candle-jar for increased CO_2 tension, and in a McIntosh and Fildes jar for anaerobiosis (gas: 95% H_2 + 5% CO_2, by vol).

Respiratory type—*Streptococcus*

R. Whittenbury (1965). J. gen. Microbiol. 38, 279–287.

Soft agar medium, which was used for showing oxygen relationships and fermentative activities, was prepared and inoculated as described previously (Whittenbury, 1963) with the exception that the bromcresol purple content was doubled. Sugars and polyhydroxy alcohols were prepared as Seitz-filtered distilled water solutions and added to give a final concentration in the media of 0.5% (w/v).

R. Whittenbury (1963). J. gen. Microbiol. 32, 375.

Respiratory type—*Pseudomonas*

R. Y. Stanier, N. J. Palleroni and M. Doudoroff (1966). J. gen. Microbiol. 43, 159–271.

Tests for the capacity to grow anaerobically with glucose as a fermentable substrate were conducted in tubes of yeast extract medium supplemented with 1% (w/v) glucose and 0.1% (w/v) Ionagar, and overlayered after inoculation with a plain agar seal.

Respiratory type—*Providencia, Rettgerella*

C. Richard (1966). Annls Inst. Pasteur, Paris 110, 105–114.

Pour des raisons de commodité, les milieux ont été incubés à 37°C. Type respiratoire (ensemencement en gélose profonde VF régénérée). Exigences en facteurs nutritionnels (milieu synthétique glucosé) [10]. Action des inhibiteurs du système respiratoire: milieu de Brown modifié [4].

[4] Edwards (P.), Fife (M.) et Ewing (W.). *Amer. J. med. Technol.*, 1956, 22, 28.

[10] Jacob (F.) and Wollman (E.). *Sexuality and the genetics of bacteria.* Academic Press, New York et Londres, 1961, 62.

Respiratory type—*Brevibacterium*

R. Chatelain and L. Second (1966). Annls Inst. Pasteur, Paris 111, 630–644.

Type respiratoire: en gélose-gélatine profonde sans nitrate [15].

Epreuve de Hugh et Leifson: le métabolisme oxydatif ou fermentatif du glucose est recherché en milieu macération-gélose [25]. L'acidification du milieu est notée après 1, 2, 4, 8 et 16 jours.

La température d'incubation des cultures est de 30°C.

[15] Olivier (H. R.). *Traité de biologie appliquée,* tome II, Maloine édit., Paris, 1963.

[25] Véron (M.) et Chatelain (R.). *Ann. Inst. Pasteur,* 1960, 99, 253.

Respiratory type—anaerobes

A.-R. Prévot (1966). Techniques pour le diagnostic des Bactéries Anaérobes. Éditions de la Tourelle, St. Mandé.

Mais les anaérobies ne réagissent pas tout de la même façon au pouvoir bactéricide ou bactériostatique de l'O_2 et pour bien se rendre compte de cette diversité de comportement, il faut étudier l'aspect de leur croissance en gélose profonde V F. On s'aperçoit alors qu'ils croissent électivement à des niveaux différents de ce milieu très précieux pour leur étude. C'est ce que nous avons appelé leur "type respiratoire", l'un des bons caractères physiologiques de l'espèce et l'un des premiers que l'on observe par la technique française.

Sur le schéma ci-dessous, les 4 types respiratoires anaérobies sont (fig. 1):

Type I: Anaérobies stricts, ne poussant qu'au-dessous de la zone critique.

Type II: Anaérobies microaérophiles, poussant moyennement dans la profondeur, mieux dans la zone critique et envahissant une faible couche de la zone aérobie.

Type III: Anaérobies en zones alternantes, respectant la zone critique, mais envahissant la partie inférieure de la zone aérobie.

Type IV: Anaérobies croissant exclusivement dans la zone critique.

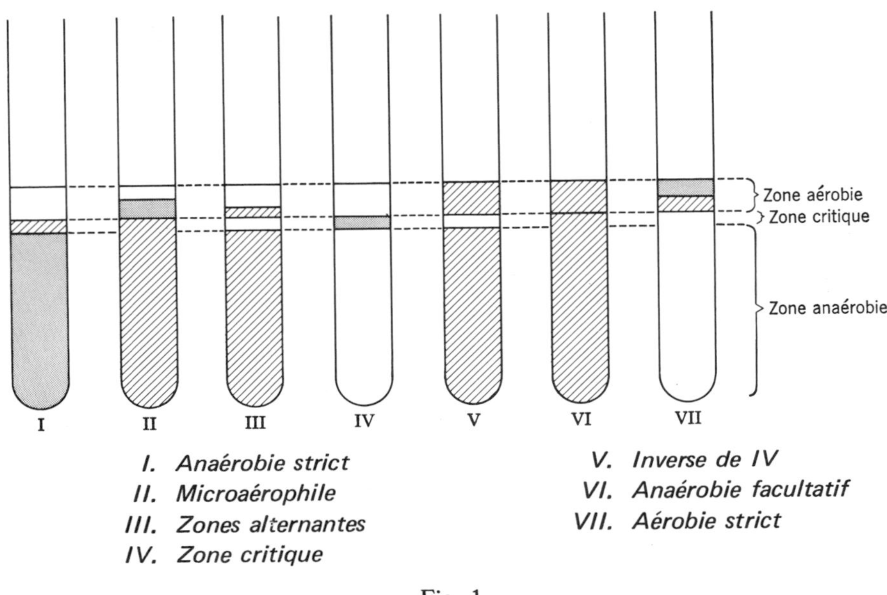

Fig. 1
Types respiratoires des Bactéries

Dans ce schéma les types I à IV représentent les anaérobies stricts, les types V et VI représentent les anaérobies facultatifs et le type VII les aérobies.

Ces niveaux peuvent être mesurés aussi bien en symbole Eh par les potentiomètres électriques qu'en symbole rH par les indicateurs colorés d'oxydoréduction. La première technique est très délicate et ne peut être mise en oeuvre que dans des laboratoires de recherche richement équipés. Elle donne des mesures très exactes exprimées en millivolts. La surface libre de la gélose profonde anaérobie a un redox de – 0.006 mv. La zone aérobie descend jusqu'à – 0.068 mv. La zone critique s'étend entre – 0.068 mv et – 0.120 mv. La zone anaérobie stricte se trouve au-dessous de – 0.252 mv. Ces mesures sont les seules qui soient exactes. Mais les laboratoires ne possédant pas de potentiomètre de redox peuvent néanmoins faire des mesures approximatives grâce aux indicateurs colorés et au symbole rH, dont le rapport avec le symbole Eh est donné par l'équation:

$$rH = \frac{E'o}{30} + 2\ pH \text{ ou } rH = \log \frac{1}{pH_2}$$

On sait que: rH = 0 pour une atmosphère de H_2 pur
rH = 41 pour une atmosphère de O_2 pur

La mesure de rH se fait par l'observation du passage du dérivé oxydé, coloré, de l'indicateur à son leucodérivé (réduit, incolore ou fluorescent).

Un tableau d'équivalence entre les valeurs exprimées en rH et Eh a été dressé; nous en tirons quelques chiffres relatifs aux indicateurs usuels dont les 4 premiers sont utiles pour mesurer les redox de départ et les 3 derniers pour mesurer les redox d'arrivée, le tout étant fonction du pH qui doit être de 7.

		rH	Eh
Indicateurs utilisés pour le rH initial	bleu de méthylène	14	– 0.006 mv
	tétrasulfo-indigotine	12	– 0.068 mv
	bleu de Nil	9.2	– 0.160 mv
	indigo carmin	7.4	– 0.230 mv
Indicateurs utilisés pour le rH terminal	vert-Janus	6	– 0.252 mv
	rouge-neutre	3.1	– 0.344 mv
	safranine	2.5	– 0.389 mv

Les niveaux de la gélose profonde ont pu être mesurés par ces deux méthodes, qui d'ailleurs coïncident avec une approximation suffisante. Ainsi on a:

rH de la surface de la gélose: 14.

rH de la limite inférieure de la zone aérobie: 12.

rH de la zone critique: 12 à 10.

rH de la zone anaérobie strict: 7.4 à 6.

C'est donc un gradient décroissant de redox qui va de la surface de la gélose vers la profondeur et offre à chaque espèce les zones de croissance optimum convenant à son système desmolasique (fig. 2 et 2 bis).

Cette pratique de mesure du redox des cultures anaérobies n'a pas qu'un intérêt purement théorique. Nous avons montré qu'elle a plusieurs applications pratiques.

Les lecteurs intéressés par la question de la respiration des bactéries trouveront un exposé très développé de la question dans le tome VII de la collection "Biocytologia" dirigée par J. A. Thomas: "Aspect du métabolisme respiratoire des Bactéries" A) Respiration aérobie par A. R. Prévot et P. Kaiser; B) Fermentation par P. Kaiser.

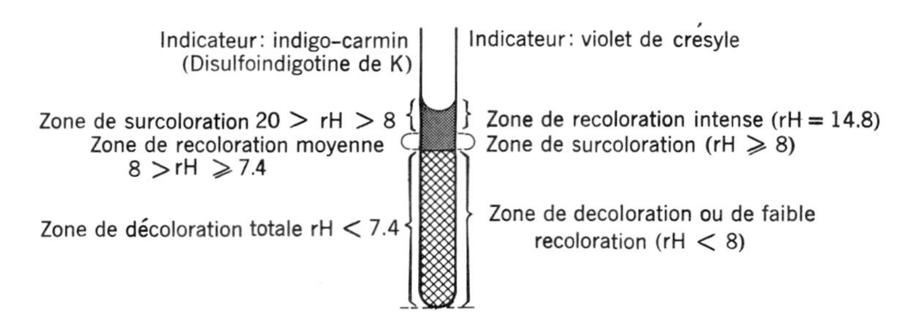

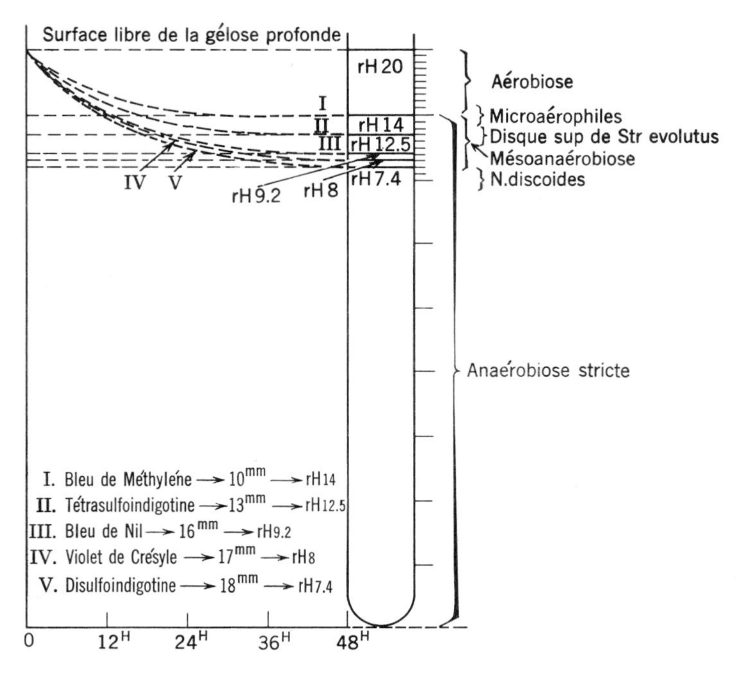

Fig. 2 et 2 *bis*
Niveaux d'oxydoréduction de la gélose profonde

1° Mesure du type oxydoréducteur d'une espèce.

Quand on ensemence un anaérobie dans un milieu qui lui convient, le redox de ce milieu évolue parfois très rapidement (en quelques heures) puis finit par se fixer à une valeur terminale qui est spécifique.

La courbe de variation en fonction du temps a également une forme spécifique. C'est cet ensemble que nous avons appelé "type oxydoréducteur d'une espèce". En étudiant avec 3 indicateurs l'évolution de plusieurs milliers de souches appartenant à plusieurs centaines d'espèces, nous avons vu que pour une espèce donnée, toutes les souches ont une courbe semblable ou voisine. Le type oxydoréducteur d'un anaérobie est donc un des bons caractères de l'espèce. C'est ainsi qu'avec Reinert nous avons délimité 6 groupes oxydoréducteurs présentant 6 types de courbes (fig. 3).

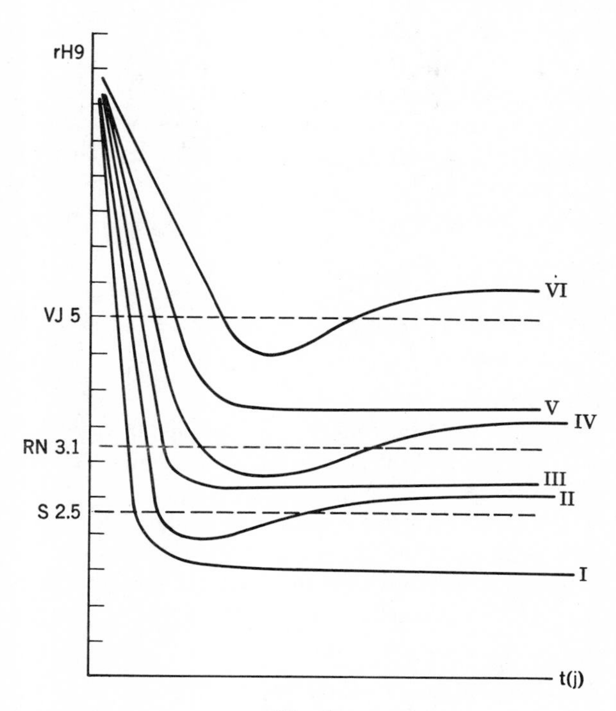

Fig. 3.
Evolution du redox dans les cultures des 6 types oxydoréducteurs anaérobies

SALT (NaCl) TOLERANCE

Salt tolerance—*Pseudomonas*

L. McCulloch (1924). J. agric. Res. 29, 159–177.

In peptone-beef bouillon containing 2 per cent sodium chloride, the growth is practically normal. Growth becomes less in 3 per cent, very scanty in 3½ per cent, and in 4 per cent no growth occurred.

Salt tolerance—*Xanthomonas*

F. R. Jones and L. McCulloch, (1926). J. agric. Res. 33, 493–521.

A medium of peptone-beef bouillon of pH value 6.8, and containing, respectively 2, 3, 4 and 5 per cent of chemically pure sodium chloride was inoculated from 6-day-old beef-bouillon cultures. Examined after four days.

Salt tolerance—*Streptococcus*

J. M. Sherman and P. Stark (1934). J. Dairy Sci. 17, 525–526.

Inhibition of growth by sodium chloride: In poured lactose nutrient agar plate cultures all of the lactis cultures were completely inhibited by 6 per cent NaCl and only three of the 27 cultures were able to grow in the presence of 5.5 per cent of this salt. On the other hand, all of the fecalis cultures grew vigorously in the same medium containing 6.5 per cent NaCl. (In these tests the NaCl was sterilized separately in concentrated solution and added to the melted agar medium before use.)

Salt tolerance—*Lactobacillus*

J. M. Sherman and H. M. Hodge (1940). J. Bact. 40, 11–22.

Lactose, 1%; peptone, 1%; Bacto yeast extract, 1%; NaCl, 2.5%: was used to test salt tolerance.

Salt tolerance—*Streptococcus*

L. A. Rantz (1942). The serological and biological classification of hemolytic and nonhemolytic streptococci from human sources. J. infect. Dis. 71, 60–68.

Growth in the presence of 6.5% sodium chloride.

Six and five tenths g of sodium chloride were added to 100 ml of peptic digest broth and autoclaved. Tubes of this medium were inoculated with streptococci. Visible clouding of the medium after incubation at 37°C for 48 hours was used as the criterion of growth.

Salt tolerance—*Clostridium pasteurianum*

C. H. Spiegelberg (1944). J. Bact. 48, 13–30.

Used tryptone (0.5%), glucose (1%), agar (0.1%) sterilized in 10 ml quantities at 15 lb for 20 minutes as a base. The salt concentration was measured chemically after sterilization. Maximum tolerance was reported as 4%.

Salt tolerance—*Staphylococcus*

G. H. Chapman (1945). J. Bact. 50, 201–203.

7.5% NaCl in nutrient media is tolerated by staphylococci (Koch 1942, Zentbl. Bakt. ParasitKde I, Orig. 149, 122–124).

When 75 g of NaCl are added to 1 liter of bacto phenol red mannitol agar and the sterilized and poured medium is inoculated with material containing staphylococci and then incubated 36 hours at 37°C, nearly all the organisms that grow luxuriantly are staphylococci that coagulate plasma, and almost all of them are surrounded by yellow zones. Nonpathogenic staphylococci, on the contrary, produce small colonies surrounded by red or purple zones.

Salt tolerance—*Streptococcus*

E. J. Hehre and J. M. Neill (1946). J. exp. Med. 83, 147–162.

The medium was as in Test 6*, except for the final 6.5% NaCl .5 ml portions were given the same inoculum as in Test 6 and incubated 10 days at 37°C.

*See **Temperature of growth**—*Streptococcus* (p. 819).

Salt tolerance—*Streptococcus*

K. E. Hite and H. C. Hesseltine (1947). A study of aerobic streptococci isolated from the uterus and vagina. J. infect. Dis. 80, 105–112.

Tolerance for salt was tested in veal infusion broth containing 6.5% sodium chloride.

Salt tolerance—*Streptococcus*

Y. Abd-el-Malek and T. Gibson (1948). J. Dairy Res. 15, 233–248.

The medium contained 0.5% each of lactose, tryptone and meat extract and the chosen w/v percentage of salt. The salt and the other ingredients, both in concentrated solution, were sterilized separately in screw-capped bottles. Immediately before use, measured volumes were combined and tubed aseptically. Tubes were inoculated with a loopful of a 24 hour lactose broth culture and were incubated for 48 hours.

Salt tolerance—*Streptococcus*

P. F. Swartling (1951). J. Dairy Res. 18, 256–267.

4% and 6.5% NaCl was added to a basal medium containing 0.5% lactose, 0.5% tryptone (Difco) and 0.5% beef extract (Lemco) at pH 7.2.

The tubes containing 5 ml were inoculated with a loopful (4 mm) of an 18 hr culture in D.L.B.* of the relevant strain. Control tubes without sodium chloride were also used. Cultures were incubated at 30°C.

*See **Temperature of growth**—*Streptococcus* (p. 820).

Salt tolerance—*Mycobacterium*

R. E. Gordon and M. M. Smith (1953). J. Bact. 66, 41–48.

The cultures were inoculated into three tubes of glycerol broth containing 0, 5 and 7 per cent NaCl, respectively. Care was taken to avoid transfer of a large amount of inoculum. The growth in the three tubes was compared after 14 and 28 days' incubation at 28°C.

Salt tolerance—*Lactobacillus*

M. Briggs (1953). J. gen. Microbiol. 9, 234–248.

Used final 4, 6 and 8% salt in tomato glucose broth – 14 days incubation.

Salt tolerance—*Streptococcus*

G. Andrieu, L. Enjalbert and L. Lapchine (1954). Annls Inst. Pasteur, Paris 87, 617–634.

Le bouillon hypersalé (6.5 p. 100) est conservé en petits flacons bouchés d'une capsule, et non en tube, pour éviter l'évaporation qui perturbe très vite la valeur de l'épreuve.

Salt tolerance—*Lactobacillus*

D. M. Wheater (1955). J. gen. Microbiol. 12, 123–132.

Tomato glucose broth: tomato juice, 10%; neopeptone, 1.5%; glucose, 2.0%; NaCl, 0.5%; Tween 80, 0.1%; Yeastrel, 0.6%; soluble starch, 0.05%; NaCl, 2.0%.

Salt tolerance—*Cytophaga*

B. J. Bachmann (1955). J. gen. Microbiol. 13, 541–551.

NaCl in concentrations from 2 to 3% permits good growth immediately, but the organism can be "trained" to grow in concentrations as low as 1%.

Salt tolerance—*Chromobacterium*

P. H. A. Sneath (1956). J. gen. Microbiol. 15, 70–98.

Nutrient agar and nutrient broth containing a total of 3 and 6.5% (w/v) of sodium chloride were inoculated with a loopful of young broth culture and incubated at 25°C for 7 days.

Salt tolerance—*Microbacterium*

V. Bolcato (1957). Antonie van Leeuwenhoek 23, 351–356.

The culture was inoculated into 4 tubes of saccharose broth containing 0, 1, 2.5 and 5 per cent of NaCl, using a small inoculum. Tubes were incubated for 48 hr at 50°C and compared. The saccharose broth contained peptone, 5 g; beef extract, 3 g; saccharose, 10 g; distilled water, 1000 ml; pH 6.5.

Salt tolerance—*Streptococcus*

P. Morelis and L. Colobert (1958). Annls Inst. Pasteur, Paris 95, 568–587.

L'épreuve de la culture sur milieu contenant 65 p. 1000 de chlorure de sodium a été réalisée en ensemençant II gouttes de culture de 24 heures sur bouillon ordinaire dans un tube contenant 10 cm^3 du bouillon nutritif ordinaire additionné de sel à cette concentration. Toutes les souches étudiées s'y sont développées en vingt-quatre heures en donnant un léger dépôt au fond du tube.

Salt tolerance—*Lactobacillus*

J. Naylor and M. E. Sharpe (1958). J. Dairy Res. 25, 92–103.

Screw cap bottles of T.D.B. (tryptone dextrose broth) containing different concentrations of sodium chloride were inoculated and observed for growth.

Salt tolerance—*Pseudomonas*

M. E. Rhodes (1959). J. gen. Microbiol. 21, 221–263.

Five ml volumes of yeast extract broth containing 6.5, 5.0 and 3.0 (%, w/v) NaCl were inoculated and incubated.

Salt tolerance—*Lactobacillus* and *Streptococcus*

C. W. Langston and C. Bouma (1960). Appl. Microbiol. 8, 212–222.

The organisms were tested for their ability to grow in sodium chloride in medium A. The salt concentration was increased from 5 g to 65 g per L. The medium was dispensed in 9 ml quantities into test tubes and incubated for 2 weeks.

Medium A: Trypticase, 10 g; phytone, 5 g; yeast extract, 5 g; glucose, 5 g; sodium chloride, 5 g; potassium phosphate (dibasic), 1.5 g; "Tween 80" (sorbitan monooleate), 0.5 ml; tomato juice, 200 ml; bromcresol purple, 0.016 g; and distilled water to make 1L: pH 7. Autoclaved at 15 lb pressure for 15 min.

Salt tolerance—*Pediococcus*

H. L. Günther and H. R. White (1961). J. gen. Microbiol. 26, 185–197.

Ability to grow in 4 and 6.5% (w/v) sodium chloride was tested in TJ broth culture.

Maintenance of stock cultures and methods of cultivation. For maintenance of stock cultures, preparation of inocula and in all experimental work, 'Oxoid' tomato juice (TJ) broth or tomato juice (TJ) agar, adjusted to pH 6.6, were used unless otherwise stated. The following were exceptions to this rule: for strain Tc. 1 sodium chloride (5%, w/v) was added to the medium; and for the aerococci glucose Lemco broth (Shattock and Hirsch, 1947) or glucose yeast extract (GY) agar (containing, as %, w/v; peptone, 1.0; Yeastrel, 0.3; glucose, 1.0; NaCl, 0.25; agar, 1.0; at pH 7.4) was used.

Cultures were incubated aerobically at 30°C with specified exceptions.

P. M. F. Shattock and A. Hirsch (1947). J. Path. Bact. 59, 495.

Salt tolerance—*Xanthomonas*

A. C. Hayward and W. Hodgkiss (1961). J. gen. Microbiol. 26, 133–140.

The following medium was dispensed in 10 ml quantities in 1 oz screw-cap bottles, and inoculated with two drops of a light suspension of the test organism in distilled water: peptone (Oxoid), 5.0 g; sucrose (British Drug Houses Ltd., Analar), 5.0 g; K_2HPO_4, 0.5 g; $MgSO_4$, 0.25 g; sodium chloride (Analar), 20.0, 30.0, 40.0 or 50.0 g; distilled water, 1 L; adjusted to pH 7.2. The culture medium was examined for turbidity during incubation for 14 days.

Salt tolerance—*Streptococcus*

W. E. Sandine, P. R. Elliker and H. Hays (1962). Can. J. Microbiol. 8, 161–174.

The sodium chloride concentration of lactic broth* was increased to 4.0 and 6.5% in two different lots of media. Tubes were inoculated with each culture as described in tests on growth temperatures, incubated at 30°C for 48 hours, and observed for growth.

*P. R. Elliker, A. W. Anderson and G. Hannesson (1956). J. Dairy Sci. 39, 1611–1612.

Salt tolerance—*Lactobacillus*

R. E. Smith and J. D. Cunningham (1962). Can. J. Microbiol. 8, 727–735.

Growth was also assessed in APT (Difco) broth containing 6.0, 8.0, 10.0 and 12.0% (w/v) sodium chloride. Tubes were loop-inoculated with 24-hour lactic broth* culture, incubated for 2 weeks and examined for growth, using appropriate controls for comparison.

*P. R. Elliker, A. W. Anderson and G. Hannesson (1956). J. Dairy Sci. 39, 1611–1612.

Salt tolerance—*Vibrio*

G. H. G. Davis and R. W. A. Park (1962). J. gen. Microbiol. 27, 101–119.

Growth within 7 days on nutrient agar containing 5% (w/v) NaCl.

Salt tolerance—*Aeromonas salmonicida*

I. W. Smith (1963). J. gen. Microbiol. 33, 263–274.

Examined in peptone water with 1, 3 and 8% NaCl after 7 days' incubation.

Salt tolerance—*Pseudomonas odorans*

I. Málek, M. Radochová and O. Lysenko (1963). J. gen. Microbiol. 33, 349–355.

Broth culture used – 28°C.

Salt tolerance—various

H. L. Chance (1963). J. Bact. 85, 719–720.

The cells were grown on agar slants for 24 to 48 hr. Sterile water was added to the slants and a suspension of cells made in the water. The suspension was poured into a sterile screw-cap tube and centrifuged. The supernatant liquid was decanted, and 10 ml of a 1% sterile salt solution were added. The screw cap was replaced tightly to minimize evaporation, and the tube was set aside on the shelf at room temperature.

The following organisms survived storage at room temperature for periods of 14 months to 46 months: *Salmonella typhosa, Serratia marcescens, Escherichia coli, Pseudomonas aeruginosa, Streptococcus pyogenes, Aerobacter aerogenes, Bacillus cereus, Proteus vulgaris, Gaffkya tetragena, Staphylococcus aureus, Sarcina lutea* and *Bacillus subtilis.*

Salt tolerance—*Vibrio*

L. Ringen and F. W. Frank (1963). J. Bact. 86, 344–345.

This was determined by adding 3.5% NaCl to Albimi Brucella broth containing 0.15% agar.

Salt tolerance—*Pediococcus*

R. Whittenbury (1963). J. gen. Microbiol. 32, 375–384.

For medium see **Fermentation** – lactic acid bacteria (p. 310), or R. Whittenbury (1963). J. gen. Microbiol. 32, 375–384.

Salt concentrations up to 18% (w/v) were tested. The extent and the nature of growth were affected by increasing salt concentration.

Salt tolerance—*Micrococcus roseus*

R. C. Eisenberg and J. B. Evans (1963). Can. J. Microbiol. 9, 633–642.

Salt tolerance was determined by incubation in stationary flasks containing Trypticase soy broth (BBL – which contains 0.5% NaCl) with additions of NaCl to give the final concentrations of 5.0, 7.5 and 10.0% (w/v).

Salt tolerance—*Staphylococcus*

J. J. Iandolo, Z. J. Ordal and L. D. Witter (1964). Can. J. Microbiol. 10, 808–811.

The growth of *Staphylococcus aureus* MF 31 was studied in Trypticase Soy broth and Trypticase Soy broth made up to 4% and 8% NaCl.

Salt tolerance—*Aerococcus catalasicus*

O. G. Clausen (1964). J. gen. Microbiol. 35, 1–8.

Cultures were incubated aerobically at 37°C unless otherwise stated.

Growth tests were carried out in beef-infusion peptone phosphate broth with 6.5% NaCl; incubation period, 72 hr.

Salt tolerance—*Moraxella*

W. J. Ryan (1964). J. gen. Microbiol. 35, 361–372.

Sodium chloride was incorporated in serum agar in the following concentrations (%, w/v): 5.5, 4.5, 3.5, 2.5, 1.5, 0.5. Plates were inoculated with a drop of a dilute bacterial suspension, sufficient to give semi-confluent growth on control plates, i.e. those containing 0.5% NaCl. Readings were taken after 3–4 days of incubation.

Salt tolerance—*Pseudomonas aeruginosa*

R. R. Colwell (1964). J. gen. Microbiol. 37, 181–194.

The strains were inoculated into YE broth tubes, with final NaCl concentrations of 0, 0.5, 3.0, 5.0 and 10.0%. The tubes were incubated for 14 days after inoculation.

The standard inoculum for all tests was a single drop (1/20 ml) from a sterile disposable pipette (Fisher Scientific Company) of a 24–48 hr broth culture. The stock culture medium (YE) was a modification of that used by Rhodes (1959): Difco yeast extract, 0.3 g; Difco Bacto-Proteose Peptone, 1.0 g; NaCl, 0.5 g; agar, 1.5 g (agar omitted from liquid stock media); distilled water, 1 L; adjusted to pH 7.2–7.4 with NaOH. Tests were carried out at room temperature (25°C) except where otherwise indicated.
M. E. Rhodes (1959). J. gen. Microbiol. 21, 221.

Salt tolerance—enterococci

C. G. Rogers and W. B. Sarles (1964). J. Bact. 88, 965–973.

Ability to grow in tryptose yeast-extract broth containing 6.5% NaCl was tested.

Salt tolerance—*Bacteroides oralis*

W. J. Loesche, S. S. Socransky and R. J. Gibbons (1964). J. Bact. 88, 1329–1337.

Tolerance to 6.7% salt was determined in the basal medium supplemented with 0.5% glucose.

Thioglycollate Medium without Dextrose (BBL), supplemented with 0.2% yeast extract (Difco) and hemin, was used as a basal medium. All cultures were incubated at 35°C in Brewer jars containing an atmosphere of 95% H_2 and 5% CO_2.

Salt tolerance—*Staphylococcus*

I. A. Parfentjev and A. R. Catelli (1964). J. Bact. 88, 1–3.

The authors transplanted 0.1 ml of 24-hr broth cultures of each strain to 10 ml of tryptose phosphate broth to which sodium chloride was added in various concentrations, including a saturated solution. After the first 4 days of incubation, the tubes were examined for visible growth, and transplants were made on blood-agar slants. The slants were incubated for 24 hr at 37°C, and the colonies were used for a check of purity of the culture, antibiotic sensitivity, and coagulase activity. The tubes with the original growth were examined again on the seventh day.

The authors found that *S. aureus* grows at 37°C in tryptose phosphate broth saturated with sodium chloride. No difference was noticed between possibly pathogenic and nonpathogenic strains. Under the conditions of our tests, no changes in the original properties of *S. aureus* strains occurred. In contrast, solutions of sodium chloride in distilled water were injurious to staphylococci and killed most of these organisms in 1 hr. Staphylococci were killed faster at 37°C than at room temperature in a solution of 0.85% sodium chloride in water. Addition of traces of tryptose phosphate broth had a protective effect and prolonged the life of these organisms in physiological saline. All tests were performed at pH 7.2.

Salt tolerance—*Vibrio sputorum*

W. J. Loesche, R. J. Gibbons and S. S. Socransky (1965). J. Bact. 89, 1109–1116.

Thioglycollate medium without added dextrose (BBL) supplemented with 0.2% yeast extract (Difco) and 0.1% KNO_3 was selected as a basal medium because of the characteristic growth of the oral vibrios in the upper third of the tube. To this medium 2, 2.5 and 3.5% NaCl was added.

All cultures were incubated at 35 to 37°C in Brewer jars containing an atmosphere of 95% H_2 and 5% CO_2.

Salt tolerance—*Mycobacterium* and *Bacillus*

G. R. F. Hilson (1965). J. gen. Microbiol. 39, 407–421.

Parallel cultures of *Bacillus* and 'form 2' (FT) strains were made in nutrient broth with and without 4% NaCl (w/v) and incubated for 2 days. In addition, Dubos-Davis fluid medium (Mackie and McCartney's Handbook, 1960) was made up with and without 4% NaCl (w/v) and distributed in 10 ml amounts in ½ oz screw-capped bottles. These were inoculated with the four FT strains and the three *Bacillus licheniformis* strains, and with 6 strains of *Mycobacterium tuberculosis* (the H37 Rv strain and 5 strains recently isolated from sputum). The mycobacterial cultures were incubated for 28 days (when all the control cultures without added NaCl showed good growth) and the remainder for 2 days.

Mackie and McCartney's Handbook of Bacteriology (1960), 10 ed. Ed. by R. Cruickshank, p. 214. Edinburgh and London: Livingstone.

Salt tolerance—*Mycobacterium parafortuitum* n.sp.

M. Tsukamura (1966). J. gen. Microbiol. 42, 7–12.

This was tested in glycerol broth containing 5% (w/v) NaCl.

Salt tolerance—*Clostridium botulinum*

W. P. Segner, C. F. Schmidt and J. K. Boltz (1966). Appl. Microbiol. 14, 49–54.

The paper deals with the effect of salt on the outgrowth of spores. The term 'outgrowth' as used here indicates the first appearance of gas in Vaspar stratified tubes and a change in turbidity of the culture medium.

Spore suspensions. Cultures of the Beluga, 8E, Alaska, and Minneapolis strains were used to prepare type E spore suspensions. Suspensions were prepared and standardized according to the procedures described by Schmidt, Nank, and Lechowich (1961).

Media. The basal medium (TPG) consisted of 5% Trypticase, 0.5% peptone, and 0.4% glucose. The desired weight of sodium chloride, calculated on a weight basis, was dissolved in the medium, and the pH was adjusted to 7.0. The medium was dispensed in 200-ml quantities into 8-oz screw-cap bottles and autoclaved at 121°C for 15 min. Sterile concentrated sodium thioglycolate (20%) was added to the medium just prior to use to give a final concentration of 0.2%. Since the bottles containing the medium were weighed before and after sterilization, the balance of the water lost by evaporation was restored to the nearest gram by the addition of sterile deionized water.

Inoculation and outgrowth techniques. Either 16 by 125 or 16 by 150 mm sterile screw-cap tubes were inoculated with 0.2 ml of a preheated (60°C, 13 min) spore inoculum equivalent to 2 million viable spores. With the use of a five-tube replicate test per variable, the inoculated tubes were poured with about 10 ml of medium, sealed with sterile melted Vaspar, and tempered in either cold tap water or ice water when incubation at low temperatures was desired. Each tube was dried and carefully examined for entrapped air bubbles. Converted to the nearest degree centigrade, tubes were incubated at 46°F (8°C), 50°F (10°C), 60°F (16°C), 70°F (21°C), 85°F (30°C), and 98°F (37°C). During incubation, tubes were examined on a Monday, Wednesday, and Friday schedule.

C. F. Schmidt, W. K. Nank and R. V. Lechowich (1961). Fd Sci. 27, 77–84.

SALT (NaCl) DEPENDENCE

Salt dependence—*Halobacterium salinarium*

V. Mohr and H. Larsen (1963). J. gen. Microbiol. 31, 267–280.

Transformations and lysis in hypotonic and isotonic solutions.

Growth medium

NaCl	25	g
$MgSO_4 \cdot 7H_2O$	5	g
$CaCl_2 \cdot 6H_2O$	0.02	g
Oxoid tryptone	0.25	g*
Yeast autolysate	5.0	g
Tap water	100	ml
pH	7.0	

Yeast autolysate

Steep 500 g of baker's yeast in 500 ml of tap water at 45°C for 24 hours. Heat to boiling; adjust to pH 7.0 with NaOH and filter.
*Autoclaved separately.

SERUM OPACITY REACTION

Serum opacity reaction—*Streptococcus*

H. Gooder (1961). J. gen. Microbiol. 25, 347–352.

Krumwiede (1954) reported that some group A streptococci formed a lipoproteinase which was capable of producing opalescence in horse serum. A survey of a large number of strains of group A streptococci has shown that those strains which carry an easily identifiable M antigen rarely produce the serum opalescence reaction, except for Types 2 and 9.

Serum broth consisting of one part Hartley broth and 3 parts horse serum, sterilised by filtration at pH 7.5 used to demonstrate the serum opacity reaction.

Organisms just grown overnight in Todd-Hewitt broth to a density of *c.* 10^8/ml.

0.01 ml inoculated into the serum broth (3 ml) and incubated 18–24 hours at 37°C and examined for opalescence.

May be due to the lipoproteinase (not proven).

E. Krumwiede (1954). J. exp. Med. 100, 629.

STARCH HYDROLYSIS

Starch hydrolysis

P. W. Allen (1918). J. Bact. 3, 15–17.

The determination of diastase production by bacteria has been replaced by a simple plate method as follows: A starch agar is made by adding 0.2 per cent of water-soluble starch to the regular plain agar. This starch agar can be sterilized in the autoclave along with other media. Some hydrolysis takes place, but not enough to interfere with the test. The agar is poured into Petri dishes and allowed to cool, when a stroke 2 inches long is made with a loopful of an agar slant growth of the organism to be tested. The plates are incubated for two days at 37°C and five days at 20°C, after which they are flooded with a saturated solution of iodine in 50 per cent alcohol.

Starch hydrolysis—*Pseudomonas*

L. McCulloch (1924). J. agric. Res. 29, 159–177.

To ordinary peptone-beef agar 1 per cent of potato starch was added. Streaks were made on this medium and colonies also were grown by the usual poured-plate method. Growth was abundant and the marginal growth was unusually wide, 5 to 20 mm, and thin on both the streaks and colonies. After 10 days the plates were flooded with iodine potassium iodide.

Starch hydrolysis—*Pseudomonas*

C. Elliott (1924). J. agric. Res. 29, 483–490.

Streaks were made on beef peptone agar plates containing 0.2 per cent starch. Tested with a saturated solution of iodine in 50 per cent alcohol. Tested after 2 and 8 days.

Starch hydrolysis—*Xanthomonas beticola*

N. A. Brown (1928). J. agric. Res. 37, 155–168.

Plates were poured with beef-infusion agar, pH 8, containing 0.2 per cent of corn starch. When the agar had hardened, smears of the organism were made across the plates. After seven days there was a fair amount of growth, and the surfaces of some of the plates were flooded with iodine solution.

Tests were made with the iodine again at 14 days.

Starch hydrolysis—mustiness in eggs

M. Levine and D. Q. Anderson (1932). J. Bact. 23, 337–347.

Streaks were made over the surface of starch agar plates to determine diastase production. After forty-eight hours the plates were flooded with an iodine solution. A positive reaction was indicated by a clear zone of hydrolysis around the area of the growth.

Starch hydrolysis—*Streptococcus*

C. E. Safford, J. M. Sherman and H. M. Hodge (1937). J. Bact. 33, 263–274.

Action on starch was determined by plating in proper dilution (not streaking) in starch agar, the plates being flooded with iodine solution after incubation for three days at 37°C.

Starch hydrolysis—*Phytomonas syringae*

M. A. Smith (1944). J. agric. Res. 68, 269–298.

Streak inoculations were made in beef-extract agar containing 0.2 per cent soluble starch. After 5 days' growth the surfaces of the plates were flooded with a saturated solution of iodine in 50 per cent alcohol. Another series of plates was similarly treated after 10 days.

Starch hydrolysis—*Corynebacterium*

R. F. Brooks and G. J. Hucker (1944). J. Bact. 48, 295–312.

Determined by 2-day streak cultures at 37°C on plates of nutrient agar containing 1% soluble starch. The plates were flooded with a saturated solution of iodine in 50% alcohol and observed for colourless area surrounding the streak.

Starch hydrolysis—*Staphylococcus* and *Micrococcus*

Y. Abd-el-Malek and T. Gibson (1948). J. Dairy Res. 15, 249–260.

Plates of nutrient agar + 0.2% soluble starch, inoculated on the surface, were incubated at 30°C for 7 days and were then flooded with iodine.

Starch hydrolysis—*Streptococcus*

Y. Abd-el-Malek and T. Gibson (1948). J. Dairy Res. 15, 233–248.

Poured plates of 0.2% starch agar were incubated for 5 days at 30°C and then flooded with iodine solution.

Starch hydrolysis—*Pseudomonas alboprecipitans*

A. G. Johnson, A. L. Robert and L. Cash (1949). J. agric. Res. 78, 719–732.

Streak inoculations were made on beef-peptone agar containing 0.2 per cent soluble starch. After 10 days a clear zone showed outside the area of growth when the plates were flooded with Lugol's iodine.

Starch hydrolysis—*Azotobacter*

P. Hauduroy (1951). Techniques Bactériologiques. Masson et Cie, Éditeurs, Paris.

Hydrolyse de l'amidon. – Les cultures sont ensemencées sur des plaques du milieu n° 1 (milieu n° 2 pour *Azotobacter indicum*) contenant 1 p. 100 d'amidon de pomme de terre comme seul hydrate de carbone. Après 4 à

5 jours d'incubation, les plaques sont aspergées avec de l'alcool à 95°. Au cas où l'amidon est hydrolysé, une zone claire apparaît autour de la culture. Si l'amidon n'est pas hydrolysé, le milieu devient blanc ou opaque.

Milieu pour *Azotobacter agile* et *Azotobacter chrococcum.*

N° 1:

K_2HPO_4	1.0 g
$MgSO_4$	0.2 g
Acétate de calcium	0.2 g
Citrate de soude	0.2 g
$FeSO_4$	traces
Molybdate de sodium	assez pour obtenir une concentration de 1 pour un million de molybdène dans le milieu,
Gélose	15 g
Eau de robinet	1000 cm^3

La gélose est fondue et on ajoute 5 g de mannite et 5 g de dextrine.

Dans le cas où les *Aztobacter* ne peuvent pas utiliser la mannite ou la dextrine, on ajoute 5 g de glucose et 5 g de saccharose.

Le *p*H final est ajusté à *p*H 7.6.

Milieu pour *Azotobacter indicum.*

N° 2:

KH_2PO_4	1.0 g
$MgSO_4$	0.2 g
$FeSO_4$	0.2 g
Molybdate de soude	de telle sorte qu'on ait une concentration de 1 pour 1 million de molybdène,
Gélose	15 g
Eau de robinet	1 000 cm^3

Faire dissoudre la gélose. Ajouter 10 g de glucose.

Le *p*H final est ajusté à 6.0. *(A.T.C.C.).*

Starch hydrolysis—actinomycetes

P. Hauduroy (1951). Techniques Bactériologiques. Masson et Cie, Éditeurs, Paris.

ACTION DIASTASIQUE

(Recherche). – Des plaques de gélose nutritive contenant 1.0 p. 100 d'amidon (spot) sont ensemencées avec des cultures. Placer à l'étuve à 37°C. 2 à 4 jours. Recouvrir les plaques avec une solution iodoiodurée de Gram. L'activité diastasique est révélée par la présence d'une zone claire autour des colonies. *(W. I. C.).*

Starch hydrolysis—general

P. Hauduroy (1951). Techniques Bactériologiques. Masson et Cie, Éditeurs, Paris.

HYDROLYSE DE L'AMIDON

Préparer des plaques de gélose nutritive contenant 1 p. 100 d'amidon de pomme de terre, ajouté après filtration. Ensemencer en stries. Mettre à l'étuve de 1 à 8 jours, le temps de séjour à l'étuve dépendant de la vitesse de croissance de la culture. Répandre sur la culture de l'alcool à 95 p. 100. Si l'amidon n'a pas été hydrolysé, le milieu devient blanc et opaque. Si l'amidon a été hydrolysé, des zones translucides apparaissent autour ou sous les cultures. *(A. T. C. C.)*.

Starch hydrolysis—*Clostridium, Pseudomonas, Shigella, Vibrio* (micromethods)

P. H. Clarke and S. T. Cowan (1952). J. gen. Microbiol. 6, 187–197.

0.04 ml suspension is pipetted into each of two Durham tubes containing 0.4 ml starch-agar base (0.2% agar New Zealand and 0.05% potato starch) and incubated 4 and 24 hours 37°C.

Starch hydrolysis—*Bacillus*

N. R. Smith, R. E. Gordon and F. E. Clark (1952). U.S.D.A. Agr. Monograph No. 16.

Cultures were streaked once across plates of nutrient agar containing 1 per cent of potato starch. (Filtration of the agar, when necessary, was done before addition of starch). After incubating from 1 to 8 days, depending on the growth rate of the organism, the plates were flooded with 95 per cent alcohol. If the starch was hydrolyzed a translucent zone appeared around or underneath the growth, but if it remained unchanged the medium became white and opaque. The spreading of some cultures over the plate was inhibited by storing the plates 2 or 3 days before inoculation.

Starch hydrolysis—*Pseudomonas*

P. Villecourt, H. Blachère and G. Jacobelli (1952). Annls Inst. Pasteur, Paris 83, 316–322.

Après quinze jours de culture dans de l'eau peptonée renfermant 1 p. 100 d'amidon, I goutte de lugol donnait une coloration bleue si l'amidon n'avait pas été totalement utilisé.

Starch hydrolysis—*Mycobacterium*

R. E. Gordon and M. M. Smith (1953). J. Bact. 66, 41–48.

Duplicate plates of nutrient agar with 1 per cent potato starch were streaked and incubated at 28°C. Hydrolysis of starch was determined by flooding one plate at 5 days and the other at 10 days with 95 per cent alcohol (Kellerman and McBeth, 1912). The amount of hydrolysis was indicated by a clear zone underneath and bordering the growth; unchanged starch became white and opaque.

The potato starch was suspended in approximately 40 ml of cold water before being mixed with the hot, melted agar. After autoclaving, the agar was shaken thoroughly to ensure proper distribution of the starch in the plates.

K. K. Kellerman and I. G. McBeth (1912). Zentbl. Bakt. ParasitKde II, 34, 485–494.

Starch hydrolysis—general

N.C.T.C. Methods 1954.

Plates should be inoculated with a spot inoculum in each quadrant and incubated at the optimal temperature.

Medium: Lemco agar containing 0.2% raw potato starch. The starch solution should not be filtered, nor should the medium be filtered after the addition of the starch.

Method 1. Flood plate with Lugol's iodine solution. Medium turns blue where starch not hydrolysed. Hydrolysis is indicated by a clear colourless zone.

Method 2. Flood plate with 95% alcohol. The undigested starch takes on a milky white appearance. Hydrolysis is indicated by a clear zone.

Society of American Bacteriologists (1947). *Manual of Methods for Pure Culture Study of Bacteria.* Leaflet V, p. 19. Geneva, N.Y. Biotech. Publications.

Starch hydrolysis—*Lactobacillus*

D. M. Wheater (1955). J. gen. Microbiol. 12, 123–132.

Starch tests. The most suitable medium for this test was found to be: Neopeptone 1.5%, sodium chloride 0.5%, Tween 80 0.1%, Yeastrel 0.6%, galactose 0.5%, soluble starch 0.2% and agar 2%. It was noticed that the presence of glucose diminished the hydrolysis of starch; the addition of a carbohydrate, however, greatly stimulated the growth of lactobacilli, and galactose, which was not found to affect starch hydrolysis, was added to the medium. The cultures were streaked on the surface of plates of this medium and incubated in an atmosphere of 90% hydrogen + 10% (v/v) carbon dioxide at 37°C for 3 days. They were tested for hydrolysis of starch by flooding the plates with a solution of iodine in potassium iodide.

Starch hydrolysis—*Bacillus*

K. L. Burdon (1956). J. Bact. 71, 25–42.

Cultures in tryptose broth containing 0.2 per cent soluble starch were examined for hydrolysis of the starch at intervals during incubation for 7 days. Small amounts of the culture removed aseptically to a porcelain spot plate were tested by the addition of a few drops of dilute iodine solution.

Starch hydrolysis—*Nocardia*

R. E. Gordon and J. M. Mihm (1957). J. Bact. 73, 15–27.

The medium for determining hydrolysis of gelatin* was used with the substitution of 10 g of potato starch for gelatin. The starch was also suspended in 40 ml of cold water before its addition to the melted agar. After autoclaving, the agar was carefully mixed and poured into plates. Duplicate plates were streaked and stored at 28°C. After 5 days' incubation, one plate was flooded with 8 to 10 ml of 95 per cent alcohol; the second plate, after 10 days. A clear zone surrounding the growth, in contrast to the opaque, unchanged starch, indicated the extent of hydrolysis (Kellerman and McBeth, 1912).

*See **Gelatin hydrolysis**—*Mocardia* (p. 373).

K. F. Kellerman and I. G. McBeth (1912). Zentbl. Bakt. ParasitKde Abt. II 34, 485–494.

Starch hydrolysis—*Leptotrichia buccalis*

R. D. Hamilton and S. A. Zahler (1957). J. Bact. 73, 386–393.

Tests for starch hydrolysis and utilization were made with soluble starch (Merck) which had been washed with 20 vol of cold distilled water.

Starch hydrolysis—general

Society of American Bacteriologists. *Manual of Microbiological Methods* – McGraw-Hill Book Co. Inc., New York 1957.

A satisfactory method has been proposed by Eckford (1927) for learning the type of action on starch brought about by organisms capable of making good growth in broth. The same method may be adapted to organisms which prefer some other liquid medium by substituting it for broth in Eckford's method. The procedure, however, is not well adapted to those bacteria that fail to grow well in liquid medium. The technic is as follows:

Add 0.2 per cent soluble starch to broth and incubate cultures a week to 10 days. Examine on second, fourth, seventh and tenth days for hydrolysis of starch, production of acid, and reduction of Fehling's solution. For this test a drop is placed in a depression on a porcelain plate and a larger quantity in a serological test tube. The latter is tested for acid production with an indicator of the proper pH range. To the drop on the plate add a drop of dilute iodine solution and read reaction as follows: if blue, no hydrolysis; if reddish brown, partial hydrolysis with production of erythrodextrin; if clear, hydrolysis complete, with production of dextrin or perhaps glucose. The tubes showing complete hydrolysis may be tested for reducing sugar with Fehling's solution.

For bacteria that do not grow well in liquid media, no better method has yet been proposed than the plate technic given in all previous editions of the manual with little modification. This method has its disadvantages but is often useful; it is as follows:

Use beef-extract agar containing 0.2 per cent of soluble starch. Pour it into a petri dish, and after hardening make a streak inoculation on its surface. Incubate at optimum temperature for the organism under investigation. Observations are to be made on the second day for rapidly growing organisms but not until the seventh day for the more slowly growing ones. To make the test, flood the surface of the petri dishes with Lugol's iodine or with a saturated solution of iodine in 50 per cent alcohol. The breadth of the clear zone outside the area of growth indicates the extent of starch destruction. By means of a simultaneous inoculation on another plate containing the same medium with bromcresol purple as an indicator one may at the same time learn whether or not acid is produced as an end product. M. O. Eckford (1927). Am. J. Hyg. 7, 201–221.

Starch hydrolysis

V. B. D. Skerman, *A Guide to the Identification of the Genera of Bacteria.* 2nd ed. The Williams and Wilkins Book Co., Baltimore, 1967.

The hydrolysis of starch is dependent on the growth of the organism on the basal medium in which the starch is incorporated.

Prepare the basal medium both in single and double strength.

Pour a layer (15 ml) of the sterile single strength medium in a petri dish and allow it to set.

Prepare a 10 per cent suspension of starch and bring to a boil over an open flame with continuous stirring. Autoclave at 121°C for 25 minutes.

Dissolve the double strength agar base; mix with an equal quantity of the sterile starch suspension, and pour as a thin layer (5 ml) over the surface of the single strength agar base.

Refrigerate the plates for 2 days before use. This increases the opacity.

Inoculate the medium and incubate at the desired temperature. Hydrolysis may be indicated by a clearing of the medium around or under the colony. This may be checked by flooding the plate with dilute iodine. Unhydrolyzed starch will turn blue.

Starch hydrolysis—*Pseudomonas fluorescens*

M. E. Rhodes (1959). J. gen. Microbiol. 21, 221–263.

Streak cultures were prepared in 1.0, 0.5 and 0.1% (w/v) starch + yeast extract agar, and the plates flooded with Gram's iodine solution after incubation for 7 days.

Starch hydrolysis—*Mycobacterium*

R. E. Gordon and J. M. Mihm (1959). J. gen. Microbiol. 21, 736–748.

Duplicates plates of the following medium were streaked and incubated at 28°C: peptone, 5 g; beef extract, 3 g; agar, 15 g; potato starch, 10 g; distilled water, 1000 ml; pH 7.0. The starch was suspended in 50 ml of cold water prior to its addition to the melted agar. Before the plates were poured, the agar was thoroughly mixed to distribute the starch evenly. One plate was flooded with 95% ethanol at 5 days, and the other at 10 days. After 20-30 min, the unchanged starch became white and opaque, while a clear zone underneath and around the growth measured the hydrolysis of the starch. K. K. Kellerman and I. G. McBeth (1912). Zentr. Bakt. ParasitKde (Abt. 2) 34, 485.

Starch hydrolysis—*Fusobacterium*

A. C. Baird-Parker (1960). J. gen. Microbiol. 22, 458–469.

Medium (%, w/v): proteose peptone no. 3, 1.0; Oxoid yeast extract, 0.1; Lab. Lemco, 0.3; L-cysteine HCl, 0.05; glucose, 0.5; starch, 0.02; agar, 2.0; modified from Eckford's (1927) medium. Plates were incubated for 21 days and the breakdown of starch detected by flooding with Lugol's iodine.

M. O. Eckford (1927). Am. J. Hyg. 7, 201.

Starch hydrolysis—soil bacteria

J. Augier and R. Moreau (1960). Annls Inst. Pasteur, Paris 99, 131–141.

Nous utilisons un extrait de terreau préparé de la façon suivante: on mélange à parties égales (en poids) du terreau de jardinier et de l'eau du robinet; on laisse en contact pendant vingt-quatre heures puis on autoclave à 120-130°C pendant trente minutes. On filtre sur papier, puis à la bougie (L5 si possible), afin d'obtenir un liquide limpide. On le garde après une nouvelle stérilisation à 120°C pendant vingt-trente minutes. A la longue, il peut se former un léger dépôt que l'on élimine. L'extrait de terreau est toujours facile à filtrer et à obtenir très clair. Il n'en est pas de même pour toutes les terres, surtout les terres argileuses. De plus, nous obtenons ainsi un extrait dont la composition semble assez régulière et homogène d'une fois à l'autre. L'extrait que nous utilisons contient en moyenne de 0.3 à 0.4 g p. 1000 de substances organiques dont l'analyse donne:

C = 23 p. 100 H = 3.1 p. 100 N = 2.0 p. 100

Le rapport C/N de l'extrait est donc de 11.5.

Comme nous introduisons 3 ml de cet extrait par litre de milieu, la quantité de substance organique apportée est au maximum de 1.2 mg par litre, ce qui est absolument négligeable par rapport à la quantité d'amidon mise en jeu.

De ces considérations, il suit que la formule du milieu que nous utilisons est la suivante:

Amidon soluble (Prolabo)	1 g
Nitrate de potassium	0.3 g
Carbonate de calcium	1 g
Solution d'oligo-éléments*	1 ml
Extrait de terre	3 ml
Solution standard de Winogradsky	50 ml
Eau distillée Q. S. P.	1 000 ml

Le pH est ajusté à 7.0. On répartit en tubes de 17, à raison de 5 ml par tube. On stérilise à 112°C, pendant vingt minutes.

Mode de lecture des résultats.

Les tubes sont ensemencés chacun de 1 ml d'une suspension-dilution de terre, à raison de 4 ou 5 tubes par dilution (de 10 en 10).

Chaque jour, on prélève stérilement 0.2 à 0.3 ml environ du contenu de chacun des tubes. Cette petite quantité est répartie dans deux tubes à hémolyse: on a donc environ de 0.1 à 0.15 ml de prélèvement par tube à hémolyse. Dans l'un de ces tubes, on ajoute I à II gouttes du mélange à parties égales des réactifs A et B de Griess (Réact. A: acide sulfanilique en solution à 1 p. 100 dans l'acide acétique à 30 p. 100. Solubiliser à chaud. Réact. B: α-naphtylamine 0.3 g que l'on fait bouillir dans 70 ml d'eau distillée. On laisse quelques heures. On décante et l'on ajoute 30 ml.

*La composition de la solution d'oligo-éléments est la suivante (J. Augier):

Molybdate de K	0.5 g	Iodure de K	0.1 g
Borate de Na	0.2 g	Fluorure de Na	0.1 g
Sulfate de Co	0.2 g	Bromure de Na	0.1 g
Sulfate d'Al	0.1 g	Chlorure de Zn	0.05 g
Sulfate de Cu	0.1 g	Eau distillée Q. S. P.	1 000 ml
Sulfate de Cd	0.1 g		

On ajoute ensuite 1 ml de silicate de sodium à 30° C, puis on fait passer un courant de CO_2 dans la solution.

Starch hydrolysis—rhizosphere bacteria

P. G. Brisbane and A. D. Rovira (1961). J. gen. Microbiol. 26, 379–392.

Starch agar. Starch (0.2%, w/v) in basal agar. Hydrolysis was observed by flooding the plates with iodine solution after 7 days at 26°C.

The basal medium. The yeast-extract peptone nitrate broth of Sperber and Rovira (1959), with 15 g agar/L for solid medium, was used.
J. I. Sperber and A. D. Rovira (1959). J. appl. Bact. 22, 85.

Starch hydrolysis—*Xanthomonas*

A. C. Hayward and W. Hodgkiss (1961). J. gen. Microbiol. 26, 133–140.

The following agar medium was dispensed in 45 ml quantities in 2 oz bottles: peptone (Oxoid), 5.0 g; yeast extract (Difco), 3.0 g; agar, 20.0 g; distilled water, 1L; adjusted to pH 7.2. Five ml quantities of 2.0% (w/v) soluble starch (British Drug Houses Ltd., Analar) were added to the molten agar base from which three plates were poured. Three or four organisms were inoculated to each plate, into the centre of 0.5 cm diameter cavities made with a surface sterilized cork borer. After incubation for 6 days starch plates were flooded with Gram's iodine solution. Zones of hydrolysis were recorded.

Starch hydrolysis—*Clostridium*

H. Beerens, M. M. Castel and H. M. C. Put (1962). Annls Inst. Pasteur, Paris 103, 117–121.

Liquéfaction du gel de maïs: Préparer une décoction de 5 g de flocons de maïs dans 100 ml d'eau distillée. Répartir à raison de 12 ml par tube de 160 × 16. Stériliser à l'autoclave à 115° pendant trente minutes. Au cours du refroidissement, une gelée blanche et opaque apparaît; la partie insoluble des flocons forme un sédiment jaunâtre. On ensemence en introduisant au fond du tube 1 ml environ d'une culture de vingt-quatre heures de la souche à étudier. On recouvre d'une couche de paraffine fondue. La liquéfaction du gel se traduit par la formation d'un liquide surnageant clair et de gaz abondant soulevant le disque de paraffine.

Starch hydrolysis—*Vibrio*

G. H. G. Davis and R. W. A. Park (1962). J. gen. Microbiol. 27, 101–119.

Within 5 days, tested upon nutrient agar containing 0.2% (w/v) soluble starch by flooding with iodine solution (Allen, 1918).

P. W. Allen (1918). J. Bact. 3, 15.

Starch hydrolysis—*Streptococcus equinus*

L. K. Dunican and H. W. Seeley (1962). J. Bact. 83, 264–269.

Starch hydrolysis was studied initially by following the disappearance of starch from starch plates. The medium used was of the following percentage composition: tryptone, 1.0; yeast extract, 0.5; K_2HPO_4, 0.2; starch (Lintner's soluble), 0.3; glucose, 0.05; agar, 1.5 (pH 7.0 to 7.2). Streak plates were incubated for 3 days at 37°C, after which the disappearance of starch was noted by flooding the plates with Gram's iodine. A positive result was denoted by the absence of the usual blue-black color of the starch-iodine complex in the zones surrounding the colonies.

Atmospheres of 5% CO_2 and 95% N_2 were achieved by replacing the air in desiccators with this gas mixture. The temperature of incubation in all experiments was 37°C. Utilization of starch for growth was denoted by acid production in tubes with starch alone or starch plus glucose as metabolites. The indicator used was bromcresol purple (32 mg in 1 ml of water and 1 ml of ethanol) at a concentration of 2 ml per liter.

Starch hydrolysis—*Lactobacillus*

R. E. Smith and J. D. Cunningham (1962). Can. J. Microbiol. 8, 727–735.

Streak plates were prepared of Hayward's medium which contained 2.5% soluble starch and 1.8% agar (w/v). Hayward's (1957) basal medium contained Peptone (Oxoid), 0.5% (w/v); yeast extract (Difco), 0.3% (w/v); salts A, 0.5% (v/v); salts B, 0.5% (v/v); Tween 80, 0.1% (v/v); sodium acetate (hydrated), 0.5% (w/v). The constituents were diluted in glass-distilled water and the pH value of the medium adjusted to 6.8–7.0. Solution A contained 10 g KH_2PO_4 and 10 g K_2HPO_4 in 100 ml of distilled water; solution B contained 11.5 g $MgSO_4 \cdot 7H_2O$, 2.4 g $MnSO_4 \cdot 2H_2O$ and 0.68 g $FeSO_4 \cdot 7H_2O$ in 100 ml of distilled water.

After incubation for 2 weeks, starch hydrolysis was determined by flooding the plates with dilute Lugol's iodine.

A. C. Hayward (1957). J. gen. Microbiol. 16, 9–15.

Starch hydrolysis—*Aeromonas*

I. W. Smith (1963). J. gen. Microbiol. 33, 263–274.

Starch hydrolysis was tested on nutrient agar containing 0.4% starch by flooding the plate with Lugol's iodine after incubation for 1 day.

Starch hydrolysis—*Pseudomonas odorans*

I. Málek, M. Radochová and O. Lysenko (1963). J. gen. Microbiol. 33, 349–355.

Detected by Lugol's solution after incubation for 5 days cultivation on meat peptone agar with 0.2% (w/v) starch – 28°C.

Starch hydrolysis—*Actinomyces*

L. K. Georg, G. W. Robertstad and S. A. Brinkman (1964). J. Bact. 88, 477–490.

Starch hydrolysis was determined by growing the strains on slants of basal medium to which had been added 5 g of soluble starch and 15 g of agar per liter. Hydrolysis was determined by flooding the slants with Gram's iodine solution.

The basal medium consisted of: Heart Infusion Broth, 25 g; pancreatic digest of casein, 4 g; and yeast extract, 5 g (per liter of distilled water at pH 7.0).

The inoculum was taken from 3 day old cultures in AM broth. The latter contains KH_2PO_4, 30.0; $(NH_4)_2SO_4$, 2.0; $MgSO_4 \cdot 7H_2O$, 0.4; $CaCl_2$, 0.04; Heart Infusion Broth, 25; dextrose, 5; cysteine hydrochloride, 1; pancreatic digest of casein (Difco Casitone or BBL Trypticase), 4; yeast extract, 5; soluble starch, 1; distilled water, 1 liter. Adjust pH to about 7.2. Dispense 8 ml per tube (18 x 150 ml), cotton-plugged. Autoclave at 15 psi (120°C) for 10 min. The final pH should be 6.8 to 7.0.

The cultures were incubated at 37°C under pyrogallol-carbonate seals and read at 3 and 10 days.

Starch hydrolysis—*Dermatophilus*

M. A. Gordon (1964). J. Bact. 88, 509–522.

The authors used starch agar plates prepared and read as described by Gordon and Smith (1955) and Gordon and Mihm (1957), except that Lugol's iodine was used to test for hydrolysis of starch.

The inoculum was pipetted from a 4 to 5 day broth culture.

The medium was incubated at 36°C (± 1) and readings were made at 48 hours, 5 days, 1 week and 2 weeks.

R. E. Gordon and J. M. Mihm (1957). J. Bact. 73, 15–27.

R. E. Gordon and M. M. Smith (1955). J. Bact. 69, 147–150.

Starch hydrolysis—*Pseudomonas aeruginosa*

K. Morihara (1964). J. Bact. 88, 745–757.

Streak cultures of bouillon-agar plates containing 0.2% soluble starch were treated with iodine solution after 1, 3 and 5 days of incubation. Cultures were incubated at 30°C.

Starch hydrolysis—*Vibrio marinus*

R. R. Colwell and R. Y. Morita (1964). J. Bact. 88, 831–837.

The authors used the method of the Society of American Bacteriologists (1957) *Manual of Microbiological Methods,* McGraw-Hill Book Co., Inc., New York – modified by the addition of the media to the following salts to produce a synthetic seawater:– sodium chloride, 2.4%; potassium chloride, 0.07%; magnesium chloride (hydrated), 0.53% and magnesium sulfate (hydrated), 0.7%.

A standard inoculum was 1 drop from a pasteur pipette (*c.* 0.05 ml) of a 24 hour artificial seawater broth.

Cultures were incubated at 18°C.

Starch hydrolysis—*Bacteroides oralis*

W. J. Loesche, S. S. Socransky and R. J. Gibbons (1964). J. Bact. 88, 1329–1337.

Starch hydrolysis was determined by adding 0.5 ml of a solution of 5% iodine in 10% potassium iodide to the fermentation tube and observing the presence or absence of purplish-black coloration.

Thioglycollate Medium without Dextrose (BBL), supplemented with 0.2% yeast extract (Difco) and hemin, was used as a basal medium. All cultures were incubated at 35°C in Brewer jars containing an atmosphere of 95% H_2 and 5% CO_2.

Starch hydrolysis—*Hyphomicrobium neptunium*

E. Leifson (1964). Antonie van Leeuwenhoek 30, 249–256.

The agar medium consisted of peptone (Casitone, Difco) 0.2%, yeast extract 0.1%, tris buffer (tris(hydroxymethyl) amino methane) 0.05%, soluble starch 0.1%, agar 1.5%, artificial seawater half strength, pH 7.5. Starch hydrolysis was tested with iodine.

Starch hydrolysis—marine bacteria

R. M. Pfister and P. R. Burkholder (1965). J. Bact. 89, 863–872.

Starch hydrolysis was carried out by preparing a 0.2% solution of soluble starch in the basal agar medium. Cultures were streaked on the surface of the agar and incubated for 1 week, at which time a dilute iodine solution was poured over the surface, and the cleared zones were recorded.

The seawater medium contained (per liter): Trypticase (BBL) or N-Z-Case, 2.0 g; Soy-tone (Difco), 2.0 g; yeast extract, 1.0 g; vitamin B_{12}, 1.0 μg; aged seawater; agar (if desired), 16 g. The pH was adjusted to 7.0 prior to autoclaving.

Starch hydrolysis—*Pseudomonas*

G. L. Bullock, S. F. Snieszko and C. E. Dunbar (1965). J. gen. Microbiol. 38, 1–7.

Hydrolysis of starch was determined in nutrient agar containing 0.2% (w/v) soluble starch. Medium flooded with Lugol's iodine solution after incubation. Cultures incubated at 20-22°C for 1 week.

Starch hydrolysis—*Bacillus*

G. R. F. Hilson (1965). J. gen. Microbiol. 39, 407–421.

For starch hydrolysis a 5% (w/v) solution of soluble starch (Analar) in distilled water was sterilized by steaming for 1 hr on three successive days. Five drops of this solution were added with a Pasteur pipette to 10 ml nutrient broth, which was then inoculated. After incubation for 2 days a few drops of the culture were placed on a white tile and Gram's iodine solution added: failure to develop a blue-black coloration indicated complete hydrolysis of the starch. In cases of doubt, the culture was re-incubated for 2 days after the addition of 5 more drops of starch solution and re-tested. Negative controls were provided by the use of uninoculated starch broth incubated similarly, and of cultures inoculated with *Bacillus pumilus* (Knight and Proom, 1950).

B. C. J. G. Knight and H. Proom (1950). J. gen. Microbiol. 4, 508.

Starch hydrolysis—*Alcaligenes*

R. G. Mitchell and S. K. R. Clarke (1965). J. gen. Microbiol. 40, 343–348.

The medium of Hayward and Hodgkiss (1961) was used.

A. C. Hayward and W. Hodgkiss (1961). J. gen. Microbiol. 26, 133.

Starch hydrolysis—*Streptomyces*

T. F. Fryer and M. E. Sharpe (1965). J. Dairy Res. 32, 27–34.

The authors used the methods of Gordon and Smith (1955) and Gordon and Mihm (1957).

0.05 ml of an 18-24 hr culture of the organism grown in 5 ml Yeastrel nutrient broth, centrifuged and resuspended in 5 ml distilled water was used as inoculum.

R. E. Gordon and J. M. Mihm (1957). J. Bact. 73, 15.
R. E. Gordon and M. M. Smith (1955). J. Bact. 69, 147.

Starch hydrolysis—*Thermoactinomyces*

M. J. Kuo and P. A. Hartman (1966). J. Bact. 92, 723–726.

Each sample was plated on a partially dried plate* of Tendler and Burkholder's medium 1a, supplemented with 0.2% soluble starch. After incubation at 55°C for 48 hr on the Trypticase-yeast extract-dung extract-salts medium 1a, colonies were picked and streaked on the same medium containing 1.0 μg/ml of penicillin G. Amylase production was observed after flooding the plates with Lugol's iodine.

M. D. Tendler and P. R. Burkholder (1961). Appl. Microbiol. 9, 394–399.
*J. E. Uridil and P. A. Tetrault (1959). J. Bact. 78, 243–246.

Starch hydrolysis—*Pseudomonas*

R. Y. Stanier, N. J. Palleroni and M. Doudoroff (1966). J. gen. Microbiol. 43, 159–271.

Starch hydrolysis was tested by patching strains on yeast agar supplemented with 0.2% (w/v) soluble starch, and flooding the plates with Lugol's solution after incubation for 48 hr.

STARCH PRODUCTION

Starch production—*Corynebacterium*

E. B. Carrier and C. S. McCleskey (1962). J. Bact. 83, 1029–1036.

Nutrient agar (Difco) was used as the basal medium for the qualitative determination of compounds utilized in the formation of intracellular starch, with the exception that the phosphorylated carbohydrates were tested in filter-sterilized yeast-extract medium (Hehre *et al.*, 1947).

Sugars were added to nutrient agar in 1% concentrations; natural starches, Ramalin (Stein-Hall & Co., New York, N.Y.), maltoheptose and Nägeli amylodextrins (Dexter French, Iowa State University) were added in 0.3% concentration. Sterilization was performed in an autoclave at 121°C for 15 min. Superlose (Stein-Hall & Co., New York, N.Y.), a highly purified amylose fraction of potato starch, and rice amylose were dissolved in cold 0.1 N NaOH and neutralized with 1 N HCl immediately before they were added to media; they partially retrograded when added to nutrient agar. Aqueous solutions of the Schardinger alpha- and beta-dextrins (Corn Products Co., Argo, Ill.) were filter-sterilized and added to sterile nutrient agar to give 1% concentrations. Petri plates were prepared with the various media, inoculated by the streak method, and incubated for 5 days at 37°C. Starch formation was tested by flooding the colonies with Gram's iodine solution.

The intracellular location of the starch-like material was determined by microscopic examination of cells suspended in Gram's iodine and by the treatment of heated and unheated washed cells with saliva.

A quantitative method is also described.

Starch production—*Bacteroides oralis*

W. J. Loesche, S. S. Socransky and R. J. Gibbons (1964). J. Bact. 88, 1329–1337.

Production of iodophilic polysaccharides was tested by the method of Gibbons and Socransky (1962).

R. J. Gibbons and S. S. Socransky (1962). Arch. oral Biol. 7, 73–80.

TEEPOL SENSITIVITY

Teepol sensitivity—enterobacteria

J. E. Jameson and N. W. Emberley (1956). J. gen. Microbiol. 15, 198–204.

*Teepol agar**

1%	Eupeptone No. 2 (Allen and Hanburys)
1%	Evans peptone
1%	Lactose
0.5%	Sodium chloride
0.1%	Teepol 610 (B.D.H.)
0.005%	Bromthymol blue
0.9%	New Zealand (Davis) agar

pH 7.6

When any grade of Teepol, other than Teepol 610 is used in culture media,

the authors recommend that it should first be tested for suitability.
*Minor alterations to the original formula were recommended by the authors (personal communication).

Teepol sensitivity—*Pediococcus*

J. C. Dacre (1958). J. Dairy Res. 25, 409–413.

The cocci grew well in the tomato-extract medium with the addition of 0.05% Teepol but not in the presence of 0.1% Teepol (Jameson & Emberley, 1956).

J. E. Jameson and N. W. Emberley (1956). J. gen. Microbiol. 15, 198.

Teepol sensitivity—*Lactobacillus*

J. Naylor and M. E. Sharpe (1958). J. Dairy Res. 25, 92–103.

Screw cap bottles of tryptic digest broth (T.D.B.) containing different concentrations of Teepol were inoculated and observed for growth. The Teepol was found more reliable than the bile salts used by Wheater (1955) 0.1% Teepol corresponding to 4% bile. The homofermentative lactobacilli were incubated in this medium for 2 days and the heterofermentative lactobacilli for 1 week.

D. M. Wheater (1955). J. gen. Microbiol. 12, 123.

Teepol sensitivity—*Pediococcus*

H. L. Günther and H. R. White (1961). J. gen. Microbiol. 26, 185–197.

Ability to grow in 0.01, 0.05 or 0.1% Teepol was tested in TJ broth culture.

Maintenance of stock cultures and methods of cultivation. For maintenance of stock cultures, preparation of inocula and in all experimental work, 'Oxoid' tomato juice (TJ) broth or tomato juice (TJ) agar, adjusted to pH 6.6, were used unless otherwise stated. The following were exceptions to this rule: for strain Tc. 1 sodium chloride (5%, w/v) was added to the medium; and for the aerococci glucose Lemco broth (Shattock and Hirsch, 1947) or glucose yeast extract (GY) agar (containing, as %, w/v; peptone, 1.0; Yeastrel, 0.3; glucose, 1.0; NaCl, 0.25; agar, 1.0; at pH 7.4) was used.

Cultures were incubated aerobically at 30°C with specified exceptions.

P. M. F. Shattock and A. Hirsch (1947). J. Path. Bact. 59, 495.

Teepol sensitivity—*Lactobacillus*

F. Gasser (1964). Annls Inst. Pasteur, Paris 106, 778–796.

La recherche de la résistance au teepol est effectuée en milieu M. R. S. M.* additionné de 0.6 p. 100 de teepol. Le résultat de la culture est jugé en quarante-huit heures.

*See **Basal medium for growth**—*Lactobacillus* (p. 112).

TEMPERATURE OF GROWTH

Temperature of growth—*Streptococcus*

J. M. Sherman and P. Stark (1934). J. Dairy Sci. 17, 525–526.

As was to be expected from previous knowledge, none of the lactis cultures was able to grow at 45°C and their maximum temperatures were found to range from 41 to 43°C. The fecalis cultures all grew vigorously at 45°C and had maximum temperatures between 48 and 52°C.

Temperature of growth—*Mycobacterium*

R. E. Gordon (1937). J. Bact. 34, 617–630.

Glycerol-phosphate agar slants were inoculated, partially sealed with notched corks, warmed to 47°C in a water bath, then incubated at this temperature for two weeks. Growth or absence of growth was readily determined.

Temperature of growth—*Bacteroides, Lactobacillus*

K. H. Lewis and L. F. Rettger (1940). J. Bact. 40, 287–307.

Incubation of duplicate serum-tomato-semisolid agar stab cultures at 5 degree intervals covering a range of 5–55°C, and an observation period of 28 days was used to determine the temperature growth range.

Temperature of growth—*Lactobacillus*

J. M. Sherman and H. M. Hodge (1940). J. Bact. 40, 11–22.

Litmus milk was used, supplemented with 0.5% peptone and 100 ml tomato juice per liter. pH 6.5. Inoculated with one drop of 24-48 hour vigorous culture, sealed with sterile rubber stoppers. The 15°C tubes were examined after 2 months, whereas the high temperature cultures were observed for acid production after one week.

Temperature of growth—*Streptococcus*

L. A. Rantz (1942). The serological and biological classification of hemolytic and nonhemolytic streptococci from human sources. J. infect. Dis. 71, 60–68.

The ability of the streptococci to grow at 10°C and 45°C was determined by the incubation of tubes of peptic digest broth for 48 hours at these temperatures, after inoculation. Definite clouding of the tube was used as the criterion of growth. It was found important to bring the mediums to the test temperature before inoculation in order to obtain constant results.

Temperature of growth—*Corynebacterium*

R. F. Brooks and G. J. Hucker (1944). J. Bact. 48, 295–312.

Daily observations for 2 weeks of streak cultures on tryptone glucose beef-extract agar slants, incubated at 7, 18, 21, 25, 30, 37 and 45°C were used to determine temperature of growth.

Temperature of growth—*Streptococcus*

E. J. Hehre and J. M. Neill (1946). J. exp. Med. 83, 147–162.

The medium was 1 per cent tryptose peptone, 0.5 per cent bacto-yeast extract, 0.5 per cent NaCl, 0.1 per cent K_2HPO_4, plus 1 per cent glucose. It was chilled to 10°C before inoculation. The inoculum for 5 ml of the broth was 0.1 ml from a young blood broth culture, which was much larger than needed for initiation of growth of these streptococci in this medium when incubated at 37°C. The 10°C cultures were incubated 10 days before considered negative; all of the cultures which grew at all, showed obvious growth within 4 days.

Temperature of growth—*Streptococcus*

K. E. Hite and H. C. Hesseltine (1947). A study of aerobic streptococci isolated from the uterus and vagina. J. infect. Dis. 80, 105–112.

Growth at 45°C and 10°C was tested in 0.1% dextrose-veal infusion broth. In the former instance inoculated tubes were incubated in a water bath adjusted to 45°C readings of turbidity being made at the end of 48 hours. The latter tests were conducted in an incubator regulated at 10°C, and readings were made during 9 days incubation.

Temperature of growth—*Streptococcus*

Y. Abd-el-Malek and T. Gibson (1948). J. Dairy Res. 15, 233–248.

Growth at 8–12°C

Temperatures between these limits were obtained in a basement. They appear to be an adequate substitute for 10°C, a temperature that other workers have used. Tubes of litmus milk (containing glucose and yeastrel when testing certain organisms) were inoculated, chilled in cold water, and then placed at 8–12°C. If no change was detected within 2–3 weeks the result was interpreted as negative.

Temperature of growth—*Streptococcus*

Y. Abd-el-Malek and T. Gibson (1948). J. Dairy Res. 15, 233–248.

Tests were made in litmus milk (free from spores of thermophiles) incubated in a water bath at 45°C. Litmus milk containing glucose (0.25%) and yeastrel (0.25%) was used when testing organisms that show little activity in plain milk.

Temperature of growth—general

P. Hauduroy (1951). Techniques Bactériologiques. Masson et Cie, Éditeurs, Paris.

TEMPERATURE MAXIMA DE CROISSANCE

(Mesure). – Ensemencer les bactéries sur de la gélose nutritive inclinée et maintenir ces tubes de 1 à 3 jours dans un bain d'eau à température constante. La température du bain est élevée ou abaissée de 2° et 2° jusqu'au moment où on atteint la température la plus haute permettant le développement de la culture. *(A. T. C. C.)*.

Temperature of growth—*Staphylococcus*

C. Shaw, J. M. Stitt and S. T. Cowan (1951). J. gen. Microbiol. 5, 1010–1023.

Three nutrient agar slopes, spread with a loopful of broth culture, were incubated at 22, 30 and 37°C. Growth was recorded after 24 and 48 hours.

Temperature of growth—*Streptococcus*

P. F. Swartling (1951). J. Dairy Res. 18, 256–267.

Two loopfuls (4 mm diameter) of a 18-20 hr old culture were transferred to the test media. These were yeast-dextrose litmus milk (0.3% Yeastrel, a commercial yeast autolysate, 1% dextrose, and litmus as indicator in separated milk), and a broth (D.L.B.) containing 1% peptone (Evans), 1% beef extract (Lab. Lemco), 1% dextrose, 0.5% NaCl, tap water, with pH = 7.0. The tubes were incubated in an accurate thermostatically controlled water-bath and growth observed after 6 days (10°C) and 24 hr (40°C and 45°C) respectively.

Temperature of growth—*Bacillus*

N. R. Smith, R. E. Gordon and F. E. Clark (1952). U.S.D.A. Agr. Monograph No. 16.

As most of the strains studied grew well at 28°C, this temperature, unless otherwise noted, was used for the incubation of cultures prepared for all routine tests and observations.

To determine growth at temperatures higher than 37°C, nutrient agar slants were placed immediately after inoculation in a water bath of the desired temperature. After reaching equilibrium they were transferred to a similar water bath kept inside a constant-temperature incubator.

Attention should be called to the fact that, in high temperature incubators, evaporation, radiation from the heater, stratification of the air, and other conditions alter the actual temperature of the medium on which the micro-organism is growing. The differential between the temperature of the medium and that shown by the air thermometer may be considerable, depending on the characteristic of the incubator. The air thermometer usually being higher than the culture, it is recommended that for accurately measuring the temperatures of growth above 37°C the cultures be incubated in water baths inside an incubator.

Temperature of growth—*Mycobacterium*

R. E. Gordon and M. M. Smith (1953). J. Bact. 66, 41–48.

Glycerol agar slants were inoculated and immediately placed in a water bath at the desired temperature. After the temperature of the cultures had reached that of the water bath, they were transferred to another water bath inside a constant temperature incubator, previously adjusted to maintain the proper temperature in the bath. Growth was observed after 7 days' incubation at temperatures of 37°C or above and after 14 days at temperatures below 20°C.

Temperature of growth—*Streptococcus*

N.C.T.C. Methods 1954.

Inoculate a small tube containing 2 ml serum broth with one drop of 20 hour serum broth culture of the test organism. Immerse in water-bath at 45°C, which must be kept constant, (± 1°C). Read at 24 and 48 hours.

For streptococci Abd-el-Malek and Gibson recommend litmus milk in place of serum broth, and add 0.25% glucose and 0.25% yeastrel for organisms that show little activity in plain milk.

Y. Abd-el-Malek and T. Gibson (1948). J. Dairy Res. 15, 233.

Temperature of growth—*Streptococcus*

M. Moreira-Jacob (1956). J. gen. Microbiol. 14, 268–280.

Tubes of glucose 1% Lemco broth and yeast glucose litmus milk were seeded, incubated in a 45°C (± 0.1) water bath and examined for growth after 24 hours.

Temperature of growth—*Bacillus*

K. L. Burdon (1956). J. Bact. 71, 25–42.

Tested with agar slant cultures at 45°C in the same manner as described below for similar trials at 55 to 56°C.

Temperature of growth—*Bacillus*

K. L. Burdon (1956). J. Bact. 71, 25–42.

Tests of value primarily for differentiation of *B. licheniformis* from *B. subtilis*. The ability of the culture to grow at 55 to 56°C was determined by use of tryptose agar slants in tubes not larger than 13 by 100 mm. The slants were warmed in a water bath, adjusted to 55 to 56°C inoculated with a straight needle by a single streak up the centre of the slant and returned at once to the bath. The desired test temperature was controlled by means of a thermometer placed in an uninoculated agar slant tube and was carefully maintained throughout 24 to 48 hours of incubation.

Temperature of growth—*Nocardia*

R. E. Gordon and J. M. Mihm (1957). J. Bact. 73, 15–27.

Subcultures on yeast dextrose or Bennett's agar were made and immediately placed in a water bath at the desired temperature. After the cultures had reached the proper temperature, they were transferred to another water bath inside a constant temperature incubator. The water level and the temperature of the bath were carefully maintained. Growth was recorded after 5 to 7 days at temperatures of 35°C or above and after 3 weeks at 10°C.

Temperature of growth—*Microbacterium*

V. Bolcato (1957). Antonie van Leeuwenhoek 23, 351–356.

Optimum growth in saccharose broth* occurred at 50°C while it was scant at 80°C and did not occur at 10°C after 20 days of incubation. The organisms survived 85°C for 10 minutes and the thermal death point was found at 90°C to be 2 minutes.

*peptone, 5 g; beef extract, 3 g; saccharose, 10 g; distilled water, 1000 ml; pH 6.5.

Temperature of growth—*Acetobacter*

J. de Ley (1958). Antonie van Leeuwenhoek 24, 281–297.

The bacteria were grown on beer-2% agar and incubated for one week at 20°, 25°, 30°, 35° and 40°C.

Temperature of growth—*Pseudomonas*

M. E. Rhodes (1959). J. gen. Microbiol. 21, 221–263.

The ability of isolates to grow at 5°, 12–15°, 22°, 30°, 37° and 42°C (in water baths; ± 0.5°) in 5 ml yeast extract broth was examined.

Temperature of growth—*Mycobacterium*

R. E. Gordon and J. M. Mihm (1959). J. gen. Microbiol. 21, 736–748.

Cultures were made on slants of glycerol agar, heated or cooled to the desired temperature in a water bath then transferred to a water bath at the same temperature inside a constant temperature incubator. The water level and temperature of the water bath were carefully watched. The cultures were examined for growth after 5-7 days at temperatures of 35°C or above, and after 3 weeks at 10°C.

Temperature of growth—*Mycobacterium* (atypical)

A. Beck (1959). J. Path. Bact. 77, 615–624.

Tested by incubating subcultures on Lowenstein medium in water baths set at 20°, 37°, 45° and 52°C.

Temperature of growth—*Streptococcus* and *Lactobacillus*

C. W. Langston and C. Bouma (1960). Appl. Microbiol. 8, 212–222.

Growth at 15° and 45°C

Medium A was used with the following modifications: five grams of agar were added; the tomato juice was filtered and bromcresol purple deleted. The medium was dispensed in 9 ml quantities into test tubes. After sterilization and inoculation (inoculating loop), the tubes were sealed with rubber stoppers to avoid evaporation. A water bath was used to determine growth at 45°C (1 week of incubation) and growth at 15°C (2 weeks of incubation) was determined in a Cenco refrigerating incubator.

Medium A.

Trypticase, 10 g; phytone, 5 g; yeast extract, 5 g; glucose, 5 g; sodium chloride, 5 g; potassium phosphate (dibasic), 1.5 g; "Tween 80" (sorbitan monooleate), 0.5 ml; tomato juice, 200 ml; bromcresol purple, 0.016 g; and distilled water to make 1 L. pH 7. Autoclaved at 15 lb pressure for 15 min.

Temperature of growth—*Xanthomonas*

See p. 716.

Temperature of growth—*Pediococcus*

H. L. Günther and H. R. White (1961). J. gen. Microbiol. 26, 185–197.

To find the optimum growth temperature the amount of growth after incubation for 24 hr at 22°, 30° and 37°C was estimated visually. With slow growing strains the results were read after 72 hr of incubation. To indicate the range of growth temperatures, cultures were incubated at 10°, 40° and 45°C in water baths controlled to within ± 1°C.

Maintenance of stock cultures and methods of cultivation. For maintenance of stock cultures, preparation of inocula and in all experimental work, 'Oxoid' tomato juice (TJ) broth or tomato juice (TJ) agar, adjusted to pH 6.6, were used unless otherwise stated. The following were exceptions to this rule: for strain Tc. 1 sodium chloride (5%, w/v) was added to the medium; and for the aerococci glucose Lemco broth (Shattock and Hirsch, 1947) or glucose yeast extract (GY) agar (containing, as %, w/v; peptone, 1.0; Yeastrel, 0.3; glucose, 1.0; NaCl, 0.25; agar, 1.0; at pH 7.4) was used.

Cultures were incubated aerobically at 30°C with specified exceptions.

P. M. F. Shattock and A. Hirsch (1947). J. Path. Bact. 59, 495.

Temperature of growth—*Lactobacillus*

R. E. Smith and J. D. Cunningham (1962). Can. J. Microbiol. 8, 727–735.

Tubes containing lactic broth were loop-inoculated with 24 hr (30°C) lactic broth culture, and incubated at 15, 45 and 48°C for 2 weeks. Uninoculated control tubes were used for comparison in determining growth, in order to avoid misinterpretation caused by heat turbidity. Acid production in this medium causes the precipitation of casein, which acts as an aid in growth assay.

Temperature of growth—*Streptococcus*

W. E. Sandine, P. R. Elliker and H. Hays (1962). Can. J. Microbiol. 8, 161–174.

Growth at 10°C, 40°C and 45°C

One drop of inoculum from a 24 hour lactic broth* culture was added aseptically to 10 ml of sterile lactic broth. The tubes were incubated in water baths maintained at the desired temperatures and cultures observed for growth after 48 hours, except those at 10°C, which were read after 2 weeks.

*P. R. Elliker, A. W. Anderson and G. Hannesson (1956). J. Dairy Sci. 39, 1611–1612.

Temperature of growth—*Escherichia aurescens*

H. Leclerc (1962). Annls Inst. Pasteur, Paris 102, 726–741.

Température de croissance. – Nous avons déterminé la température optima de culture, la croissance à 44°, enfin le caractère de psychrotrophie selon Mossel et Zwart [35] et Eddy [8]. Une goutte d'une culture en bouillon de la souche étudiée est diluée dans 5 à 6 ml d'eau physiologique. Des quantités identiques de cette suspension (une öse) sont étalées à la surface de 6 tubes de gélose à la tryptose (Difco). Les tubes capuchonnés sont portés quinze jours à + 4°, 10°, 20°, 30°, 37°, quatre jours à 44°. Ils sont examinés chaque jour. La température optima de culture est celle culture est celle qui permet d'obtenir la croissance maxima dans le minimum de temps.

La pigmentation a été observée en même temps que la croissance aux différentes températures d'essai.

[8] Eddy (B. P.). *J. appl. Bact.*, 1960, 23, 216.
[35] Mossel (D. A. A.) and Zwart (H.). *J. appl. Bact.*, 1960, 23, 185.

Temperature of growth—*Aeromonas salmonicida*

I. W. Smith (1963). J. gen. Microbiol. 33, 263–274.

This was examined on solid media and the results recorded after 1 day at 22°C and 37°C and after 7 days at 5°C.

Temperature of growth—*Pseudomonas odorans*

I. Málek, M. Radochová and O. Lysenko (1963). J. gen. Microbiol. 33, 349–355.

Ability of organisms to grow at 42°C and at 5°C was tested in broth cultures incubated in water baths.

Temperature of growth—*Pediococcus*

R. Whittenbury (1963). J. gen. Microbiol. 32, 375–384.

Growth at 30°C, 35°C and 37°C was tested in soft agar in which glucose was autoclaved, in soft agar with a fermentable pentose and in liquid media.

The temperature limit, the ability to grow and the nature of the growth varied with the nature of the medium.

For the composition of the soft agar and the liquid medium see **Fer-**

mentation—lactic acid bacteria (p. 310) or R. Whittenbury (1963). J. gen. Microbiol. 32, 375–384.

Temperature of growth—*Lactobacillus*

F. Gasser (1964). Annls Inst. Pasteur, Paris 106, 778–796.

Les cultures à 15° et à 45°C sont effectuées en milieu liquide (M. R. S. M.)* préalablement porté à la température désirée. Le bain-marie ne doit pas admettre de variations de températures supérieures à ± 0.1°C. Les tubes y sont plongés de façon que le niveau de l'eau du bain-marie dépasse d'au moins 2 cm le niveau du milieu de culture dans le tube d'épreuve. La température du milieu de culture est ainsi homogène et égale à celle de l'eau du bain-marie. La température de 15°C est obtenue en plaçant un bain-marie réglé à cette température dans l'enceinte d'un réfrigérateur à + 4°C. Les tubes sont inoculés avec deux gouttes d'une culture abondante de vingt-quatre heures en milieu M. R. S. M. incubée à 30 ou 37°C, bouchés au coton et capuchonnés au caoutchouc. L'incubation n'est pas prolongée au-delà de six jours. Une première observation après quarante-huit heures donne le plus souvent le résultat définitif.

*See **Basal medium for growth**—*Lactobacillus* (p. 112).

Temperature of growth—*Lactobacillus casei* var. *rhamnosus*

W. Sims (1964). J. Path. Bact. 87, 99–105.

Growth at 15°C and 45°C was tested in MRS broth†.

†J. C. deMan, M. Rogosa and M. E. Sharpe (1960). J. appl. Bact. 23, 130.

Temperature of growth—*Clostridium*

N. A. Sinclair and J. L. Stokes (1964). J. Bact. 87, 562–565.

Determinations of the optimal and maximal growth temperatures were made from 0 to 35°C at intervals of 5°C in tubes of Trypticase Soy Broth which contained 0.2% yeast extract, 0.05% sodium thioglycolate, and 0.075% agar. Although the cultures were exposed to air, the thioglycolate and agar maintained anaerobic conditions. Growth always began about 0.5 in. below the surface of the medium. A detailed study was made of the kinetics of growth of strain 61, in the range of 0 to 30°C at 5-degree intervals. For this purpose, the bacteria were grown in 35-ml glass-stoppered bottles completely filled with Trypticase Soy Broth containing 0.2% yeast extract and 0.05% sodium thioglycolate. Growth was followed turbidimetrically with a Klett-Summerson colorimeter (red filter).

Incubation at 0, 15 and 25°C.

Temperature of growth—*Vibrio marinus*

R. R. Colwell and R. Y. Morita (1964). J. Bact. 88, 831–837.

The ability of the strains to grow in seawater broth at 0, 5, 10, 15, 20, 25, 30, and 37°C was recorded after incubation periods of 1 day to 2 weeks, depending on the incubation temperature employed.

Temperature of growth—*enterococci*

C. G. Rogers and W. B. Sarles (1964). J. Bact. 88, 965–973.

This was determined in tryptose yeast extract broth at 10°C and 45°C.

Temperature of growth—general

E. H. Battley (1964). Antonie van Leeuwenhoek 30, 81–96.

A thermal-gradient block is described which can be used to determine the maximum, optimum and minimum temperatures of growth of microorganisms. Methods are given for the use of this instrument, together with data pertaining to several yeasts.

Temperature of growth—*Aerococcus catalasicus*

O. G. Clausen (1964). J. gen. Microbiol. 35, 1–8.

Cultures were incubated aerobically at 37°C unless otherwise stated.

Blood agar plates were prepared by addition of 5% citrated horse blood to a medium consisting of 1% peptone (Danish, Orthana Bacteriological Brand), 0.3% NaCl, 0.2% Na_2HPO_4, and 1.8% agar (Japanese, quality Kobe I) in aqueous beef-infusion (pH = 7.4). Plates were poured to a constant depth, about 4 mm. Each individual strain was subcultured to three blood agar plates; one was incubated at 22°C, the others at 30°C, and 37°C. Growth was recorded after 24 and 48 hr.

Temperature of growth—*Lactobacillus*

M. Gemmell and W. Hodgkiss (1964). J. gen. Microbiol. 35, 519–526.

Temperature requirements. The maximum temperatures for growth were determined by incubating cultures in tubes of basal medium to which were added glucose (0.5%, w/v) and one Pasteur pipette drop of a 24-hr culture. The tubes were plugged with sterile rubber stoppers and placed in water baths at 15°, 37°, 40°, 43°, 45° and 50°C (all ± 1°C). Incubation was for 2 weeks at 15°C and 4 days at the other temperatures.

The basal medium which was used throughout with minor modifications had the following constituents in 1 L tap water: meat extract (Lab Lemco), 5 g; Evans peptone, 5 g; Difco yeast extract, 5 g; Tween 80, 0.5 ml; $MnSO_4 \cdot 4H_2O$, 0.1 g; potassium citrate, 1 g; pH 6.5.

Temperature of growth—*Pseudomonas aeruginosa*

R. R. Colwell (1964). J. gen. Microbiol. 37, 181–194.

Temperature range of growth. The ability of strains to grow on YE agar slopes and in YE broth at 0°, 5°, 10°, 15°, 25°, 30°, 35° and 40°C was screened by incubation of the inoculated media in incubator rooms temperature-controlled to ± 0.5°C, and the upper and lower limits of growth determined by incubation in a water bath at the desired temperature. As suggested by Haynes & Rhodes (1962), a strain was considered incapable of growth at a given temperature unless it could survive at least three serial transfers at that temperature. In addition, where possible, the medium was pre-incubated at the test temperature for 6–12 hr before inoculation. The incubation period extended from 5 days to 4 weeks, depending on the temperature of incubation.

The standard inoculum for all tests was a single drop (1/20 ml) from a sterile disposable pipette (Fisher Scientific Company) of a 24–48 hr broth culture. The stock culture medium (YE) was a modification of that used by Rhodes (1959): Difco yeast extract, 0.3 g; Difco Bacto-Proteose Peptone, 1.0 g; NaCl, 0.5 g; agar, 1.5 g (agar omitted from liquid stock media); distilled water, 1 L; adjusted to pH 7.2–7.4 with NaOH.

W. C. Haynes and L. J. Rhodes (1962). J. Bact. 84, 1080.
M. E. Rhodes (1959). J. gen. Microbiol. 21, 221.

Temperature of growth—*Pseudomonas*

G. L. Bullock, S. F. Snieszko and C. E. Dunbar (1965). J. gen. Microbiol. 38, 1–7.

Growth at different temperatures. All cultures were tested for ability to grow at 0, 6, 12, 20, 30, 37 and 42°C. The preparation of inocula was a modification of Klinge's (1960) method in that, for each culture tested, a 1/100 dilution, in 0.85% (w/v) sterile saline, was made from a 24 hr turbid broth culture. After shaking, 0.1 ml of the dilution was inoculated into each tube of nutrient broth, which contained 5 ml broth. Inoculated and control tubes were placed in covered pans of water at the indicated temperatures. Glycerol was added to the pan water at 0°C to keep the water from freezing. Temperatures were checked at least twice a day during the 2-week incubation period and did not fluctuate more than 0.5–1.0°C. It took 30–45 min from the time the first tubes of a series were inoculated until all tubes of that series had reached proper temperature. Results were recorded daily for the first week and three times during the second week. Growth was measured by visually comparing turbidity of inoculated tubes with uninoculated controls. Tubes were not removed from the incubator for more than 5 min to record results. Ability to grow at 6, 12, 20 and 30°C was determined once, while ability to grow at 0, 37 and 42°C was determined twice.

K. Klinge (1960). J. appl. Bact. 23, 442.

Temperature of growth—*Streptomyces*

T. F. Fryer and M. E. Sharpe (1965). J. Dairy Res. 32, 27–34.

Inoculated nutrient agar slopes were incubated at 10, 22, 30 and 37°C for 2 weeks.

0.05 ml of an 18–24 hr culture of the organism grown in 5 ml Yeastrel nutrient broth, centrifuged and resuspended in 5 ml distilled water was used as inoculum.

Temperature of growth—*Pseudomonas aeruginosa*

A. H. Wahba and J. H. Darrell (1965). J. gen. Microbiol. 38, 329–342.

Growth on nutrient agar No. 1 at 42°C for three consecutive subcultures (Haynes, 1951, 1962). Slopes were incubated in a covered tin immersed in a 43°C water bath, thus ensuring that the actual temperature inside the culture tubes, as recorded by a thermometer in a blank tube, was in fact 42°C. Control cultures were also incubated at 37°C and 22°C.

Nutrient agar No. 1 (Oxoid No. 1) contained Lab Lemco beef extract, 1 g; yeast extract (Oxoid L 20), 2 g; peptone (Oxoid L 37), 5 g; sodium chloride, 5 g; agar, 15 g; distilled water to 1 L; pH 7.4.

W. C. Haynes (1951). J. gen. Microbiol. 5, 939.
W. C. Haynes (1962). J. Bact. 84, 1080.

Temperature of growth—*Bacillus*

G. R. F. Hilson (1965). J. gen. Microbiol. 39, 407–421.

Cultures were made in nutrient broth and incubated at 45°C, 37°C and room temperature (18–20°C). A thermostatically controlled water bath was used for incubation at 45°C, the cultures being totally immersed in it. The cultures were examined daily up to 3 days, and the amount of growth (turbidity and pellicle formation) recorded.

Temperature of growth—*Alcaligenes*

R. G. Mitchell and S. K. R. Clarke (1965). J. gen. Microbiol. 40, 343–348.

Growth at 42°C was tested on nutrient agar slopes immersed in a water bath.

Temperature of growth—*Pseudomonas*

R. Y. Stanier, N. J. Palleroni and M. Doudoroff (1966). J. gen. Microbiol. 43, 159–271.

Ability to grow at 4, 37, 39 and 41°C was determined in tubes of yeast extract medium, inoculated with a loopful of culture grown for 12–18 hr in the same medium at 30°C. The tubes incubated at 4°C were read after 10 days, the others after 24 hr. Growth was recorded as abundant (+) or slight (±). Water baths were used for all determinations; the cultures were not shaken.

Temperature of growth—*Mycobacterium*

M. Tsukamura (1966). J. gen. Microbiol. 45, 253–273.

Growth at 28, 37, 45 and 52°C was observed on Löwenstein-Jensen medium.

TITRATABLE ACIDITY

Titratable acidity—*Lactobacillus*

G. H. G. Davis (1955). J. gen. Microbiol. 13, 481–498.

Titratable acid production from glucose. Medium: peptone, 0.5% (w/v); yeast extract, 0.3% (w/v); glucose, 0.5% (w/v); Tween 80, 0.1% (v/v); Salts B*, 0.5% (v/v) at pH 7.2, accurately dispensed in 10 ml quantities and sterilized by steaming for 20 min on 3 consecutive days. The inoculum was 2 to 3 drops; tests were incubated for 3 days and titratable acid was recorded in terms of 0.1 N-sodium hydroxide required to neutralize to phenolphthalein. All strains were tested in duplicate and the corrected mean value recorded.
*See A. C. Hayward (1957). J. gen. Microbiol. 16, 9–15.

Titratable acidity—*Microbacterium*
V. Bolcato (1957). Antonie van Leeuwenhoek 23, 351–356.
10 ml of skim-milk were sterilized in 15 by 180 mm tubes, inoculated by loop and incubated for a week at 37°C. The contents of the tubes were washed into Erlenmeyer flasks, titrated with N/10 NaOH and the acidity value corrected for control acidity.

Titratable acidity—*Lactobacillus*
R. E. Smith and J. D. Cunningham (1962). Can. J. Microbiol. 8, 727–735.
Titratable Acidity in Milk. Skim milk was dispensed into 30 ml screw-capped bottles in 18 ml quantities, sterilized at 121°C for 10 minutes, and inoculated with a loopful of a 24 hour Difco APT broth culture. After incubation for 1 week, percentage acidity was determined by titrating 9.0 ml quantities of milk with 0.1 N NaOH to the phenolphthalein end point. Coagulum present was thoroughly disrupted, and titration values were corrected for control acidity.

UREA

Urea—*Escherichia* and *Enterobacter*
S. A. Koser (1924). J. Bact. 9, 59–77.
The composition of the uric acid medium was the same as that previously described (Koser, 1918, J. infect. Dis. 23, 377–379). Since development in this medium depends upon the presence of available nitrogen some care must be exercised to exclude free ammonia in so far as possible. Also the medium should be inoculated lightly from young agar cultures in order to avoid carrying over any superfluous quantity of dead cells, enzymes, etc. In one set of experiments 0.2 per cent glucose was substituted for the 3 per cent glycerol as the source of carbon, since it was thought that perhaps glycerol might not be readily utilized by some of the colon group cultures. However, similar results were secured in both cases. The temperature of incubation and the method of observing development were the same as those described for the citrate medium.

Urea—*Streptobacillus moniliformis*

F. R. Heilman (1941). A study of *Asterococcus muris (Streptobacillus moniliformis).* II. Cultivation and biochemical activities. J. infect. Dis. 69, 45–51.

A solution of urea was sterilized by filtration and added, in amounts sufficient to make a final concentration of 1%, to serum veal infusion broth and to the medium composed of starch, salts and proteose peptone. There was no increase in growth or difference in final pH as determined by various indicators on the mediums containing urea as compared to control mediums not containing urea.

Urea—*Proteus*

C. A. Stuart, E. van Stratum and R. Rustigian (1945). J. Bact. 49, 437–444.

Standard Urease Test

In this laboratory the urea medium is usually made up in lots of 400 ml since this amount is contained in the 5" x 1" filter candle and mantle used to sterilize the medium. The medium is made as follows: To 380 ml of distilled water are added 3.64 g of KH_2PO_4, 3.8 g of Na_2HPO_4 (Sorensen buffers) 8 g of urea (highest purity) 40 mg of yeast extract (Difco) and 20 ml of an 0.02 per cent solution of phenol red. This highly buffered medium has a pH of 6.8. The medium is tubed in approximately 3-ml amounts in tubes 14 mm inside diameter and 125 mm long. Inoculations are made from 18- to 24-hour agar slant cultures with a straight needle, and tests are incubated at 37°C in an air incubator. When possible, reactions are recorded at 8, 12, 24 and 48 hr of incubation. In the previous reports on this medium no particular conditions of the test were given since conditions were kept constant and no unusual reactions were encountered.

Rapid Urease Test

The same medium but with only 1% of the buffer is inoculated with 2-3 loopfuls of an agar slope culture; shaken and incubated at 37°C in a water bath. No time is specified for reading the test but a maximum period of 2 hours would seem to be indicated. Readings were taken at 5 minute intervals. The test failed to distinguish between rapid and slow urease producing species of *Proteus.*

Urea—*Corynebacterium*

M. Welsch, G. Demelenne-Jaminon and J. Thibaut (1946). Annls Inst. Pasteur, Paris 72, 203–215.

Fermentation de l'urée. – Puschel [41] a signalé que *C. diphteriae* ne fermentait pas l'urée (301 souches), tandis que *C. Hofmanni* attaquait cette substance (178 souches). Kleinsorgen et Commichau [24] ont préconisé l'utilisation de cette propriété pour le diagnostic pratique de la diphtérie. Pour vérifier la généralité de ce caractère, nous l'avons recherché pour chacune de nos souches. Les résultats de fermentation anormaux étant, en général, plus fréquents sur milieux solides, nous avons utilisé un milieu liquide préparé, cependant, autant que possible, comme la gélose de Kleinsorgen et Commichau.

Technique. – *Milieu de base:* Sérum de Hiss décrit plus haut.*

Indicateur: rouge de crésol filtré sur bougie Chamberland L III.

Solution aqueuse d'urée à 20 p. 100: Stérilisée par filtration.

Solution de cystine à 1 p. 100: Dissoudre 1 g de soude caustique dans 10 c. c. d'eau distillée, puis y ajouter 1 g de cystine; compléter ensuite à 100 c. c. avec de l'eau distillée. Stériliser à l'autoclave.

Solution aqueuse d'acide chlorhydrique à 1 p. 100: Stérilisée à l'autoclave en tube scellé.

A 95 c. c. de milieu de Hiss, ajouter aseptiquement 10 c. c. de la solution d'urée, 1.25 c. c. de la solution de cystine, 4 c. c. de la solution d'acide et *quod satis* d'indicateur. Répartir en tubes à essai, éprouver la stérilité par séjour de quarante-huit heures à 37°.

La lecture du virage alcalin qui indique la présence d'uréase est faite après vingt-quatre puis quarante-huit heures.

Tous nos *C. diphteriae* sont inactifs sur l'urée.

[24] Kleinsorgen (W.) et Commichau (F.). *Zentralbl. Bakt.*, 1937, 139, 1, 57.

[41] Puschel (J.). *Zentralbl. Bakt.*, 1936, 138, 1, 67.

*See **Fermentation**—*Corynebacterium* (p. 273).

Urea—enterobacteria

W. B. Christensen (1946). J. Bact. 52, 461–466.

Bacto-peptone	1	g
NaCl	5	g
KH_2PO_4	2	g
Phenol red	0.012	g
Glucose	1.0	g
Agar	20.0	g
H_2O distilled	1000	ml

The pH of the medium was adjusted to 6.8–6.9 and the medium tubed in 5 ml quantities and autoclaved at 15 lb pressure for 20 minutes. A 20 per cent solution of urea (Mallinckrodt, analytical reagent) was sterilized by Seitz filtration and added to the tubes, cooled to 50°C, to give a final concentration of urea of 2 per cent.

The tubes were slanted so as to leave a butt of about 1 inch in depth with a slant of about 1.5 inches in length. The media were inoculated heavily over the entire slant. Organisms which hydrolyzed urea produced the characteristic red-violet color on the slant, and, as incubation proceeded, the color extended toward the bottom of the tube. The extent of color penetration into the medium was taken as a measure of urease activity and was recorded as plus and minus signs, as follows: If the color had developed just beneath the surface of the slant but had not penetrated further into the medium, the reaction was designated + – –. If it had penetrated to the junction of the bottom of the slant with the butt, it was designated +. Further penetration into the butt was designated ++, +++, and ++++, the latter symbol being used when the color had reached the bottom of the tube.

Urea—enterobacteria

M. D. Schneider and M. F. Gunderson (1946). J. Bact. 52, 303–306.

Bacteriological laboratories engaged in diagnostic or research problems concerning pathogenic enteric bacilli, namely, the *Salmonella* and *Shigella* groups, invariably find that members of the genus *Proteus* are by far the most misleading. A dependable, easily interpreted, routine differential test medium for screening the latter group of bacilli becomes desirable. Urease production is a characteristic activity of the genus *Proteus* (Bergey *et al.*, 1939). The hydrolysis of urea by urease has been described by Werner (1923) as "an 'alkaline fermentation' during which the 'carbonate of ammonia' was formed". Rustigian and Stuart (1941) recommended a urea medium for the detection of this enzyme activity. A modification of the latter medium has been described by Anderson (1945). Since one of the products of the splitting of urea is the formation of ammonia, Rustigian and Stuart suggested that phenol red indicator be employed to detect this alkaline change colorimetrically. Howell and Sumner (1934) reported that the pH optimum for urease activity upon 2.5 per cent urea is 6.9 with phosphate buffer.

The present report describes a new medium for the detection of urea-splitting organisms. Colorimetric changes are sharp and permit easy interpretation of results.

Methods and Materials

I. The medium is prepared in three parts:

A. The buffered semisolid "deep"

Bacto tryptose (Difco)	10.0 g
Sodium chloride	5.0 g
Bacto agar (Difco)	3.0 g
Dipotassium phosphate	1.5 g
Monopotassium phosphate	1.0 g
Distilled water	1000 ml

Heat to boiling to dissolve the medium completely. The pH should be 6.9.

Add meta-cresol sulfon phthalein
(0.4 per cent alcoholic solution) 10.0 ml

Tube in 4-ml amounts in 10 by 1.2 cm test tubes. Sterilize in the autoclave for 15 minutes at 15 pounds' pressure. Cool and store in the refrigerator.

B. The "urea" overlaying solution

Urea	25.0 g
Sodium chloride	5.0 g
Dipotassium phosphate	1.5 g
Monopotassium phosphate	1.0 g
Distilled water	1000 ml

The pH should be 6.9.

Add meta-cresol sulfon phthalein
(9.4 per cent alcoholic
solution) 10.0 ml

Sterilize by Seitz (or Berkefeld) filtration. Distribute in approximately 10 ml amounts and store in appropriate sterile screw-cap containers in the refrigerator.

C. The "urea-free" overlaying solution

Same formula as in B, except for omission of urea. The solution may be sterilized in the autoclave, since urea is not present.

II. The test procedure:

The inoculation is made from a nutrient or Kligler's iron agar slant. A large inoculum on a straight needle is stabbed into the center of each of two tubes of buffered semisolid "deep". The surface of one is covered with 0.2 ml of urea overlaying solution. The second "deep" is overlaid with 0.2 ml of urea-free solution and serves as the control.

III. Interpretation of reactions after 18 to 24 hours' incubation at 37°C.

A. Colorimetric reaction (yellow to purple) diffusing from the surface and beyond half the length of the "deep", ++, positive.

B. Colorimetric reaction diffusing the surface but *not* beyond half the length of the "deep", +, positive.

C. Colorimetric change at the surface of the "deep", ±, and no change in color, –, negative.

T. G. Anderson (1945). Science, N.Y. 101, 470.

D. H. Bergey *et al.* (1939). Bergey's *Manual of determinative bacteriology.* 5th ed. Williams and Wilkins Co., Baltimore.

S. F. Howell and J. B. Sumner (1934). J. biol. Chem. 104, 619–626.

R. Rustigian and C. A. Stuart (1941). Proc. Soc. exp. Biol. Med. 47, 108–112.

E. A. Werner (1923). *The chemistry of urea,* Longmans, Green & Co., London and New York.

Urea—*Proteus*

S. D. Elek (1948). J. Path. Bact. 60, 183–192.

Three methods were employed for the detection of urease activity. The first of these was the procedure used by Rustigian and Stuart, the second a modification of it; the third was a method designed to give rapid results.

Method I (Rustigian and Stuart). The original medium consisted of 2 per cent urea (Merck), 0.01 per cent yeast extract (Difco) and M/15 primary and secondary phosphate buffers (Sorensen) in distilled water, to yield a pH of 6.8. Because of difficulties of supply, Analar urea and marmite were substituted for the Merck and Difco products originally specified. This medium is sterilized by Seitz filtration and distributed in 5-ml amounts in sterile tubes. The test organism is inoculated heavily into the medium from a 24 hour agar-slope culture and the tube incubated for 24 hours at 37°C. Thereafter two drops of brom-thymol-blue indicator are added. Tubes in which urea has been broken down with formation of ammonia develop a blue color; a negative reaction is indicated by a pale green color.

Method II is a modification of the previous method, nesslerisation replacing the use of an indicator. After 24 hours' incubation 0.5 ml of the culture is transferred to a clean Kahn tube and 0.1 ml of Nessler's reagent is added to it. A positive reaction is indicated by the formation of a brown precipitate, negative tubes remaining colorless. A blank control and a test with a known non-urease-producing organism are included in each batch.

Method III is the rapid test designed for use in the identification of the *Proteus* group. The substrate consists of 2 per cent urea in Clark and Lub's phosphate buffer at pH 7.2 and is prepared as follows. To 50 ml of 0.2 M acid potassium phosphate add 35 ml of 0.2 N NaOH and 4 g of pure urea and make up the volume to 200 ml with ammonia-free distilled water.

Sterilization of this substrate is not required. It should be kept in a bottle with a vaseline-smeared glass stopper and stored in the refrigerator when not in use. In this way it keeps for about a month. Freshly prepared substrate should be tested with a known urease-producer but for the actual test only a negative control and an uninoculated blank need be included. The tubes are not sterilized, but the glassware must be scrupulously clean.

To carry out the test a convenient amount of growth is removed with a cold wire from a 24 hour nutrient agar slope and emulsified in 0.5 ml of the substrate in a Kahn tube. The inoculum should be heavy enough to give a slight but distinct opalescence to the fluid viewed against a dark background. The Kahn tube is now placed in a water-bath at 37°C for exactly 3 hours. At the end of that time it is removed and 0.1 ml of Nessler's reagent is added. Readings are taken 3 minutes after the addition of the Nessler reagent. Both the blank and the negative control should be absolutely colorless. A positive reaction is shown by a yellow color which ranges from a pale but distinct yellow to a dark-brown precipitate. The strength of the reaction must not, however, be taken as indicative of the vigour with which urea is attacked, for organisms from the same culture may yield both "weak" and "strong" reactions on repeated testing, since the concentration of ammonia formed from urea under these conditions does not show a linear increase. In fact quantitative tests at 3-minute intervals give an ammonia-concentration curve with spikes at intervals of about 20 minutes. There is a rapid increase in ammonia, followed by an equally rapid decrease, representing utilisation of ammonia. The increase is always greater than the utilisation and in this way ammonia accumulates. It was found by experiment that 3 hours at 37°C is the minimum time of incubation for reliable results. The strength of the reaction is variable because at the time of nesslerisation the culture may be in the ammonia-production stage (strong reaction) or ammonia-utilisation stage (weak reaction). The incubation should be for a minimum of 3 hours, since at this time most of the cultures seem to give a strong reaction.

In using method III as a screen for isolated non-lactose-fermenting colonies a slight modification was introduced. To allow for the smaller inoculum the volume of the substrate was reduced to 0.3 ml and one drop of Nessler's reagent was used in place of 0.1 ml. Readings were taken 4-5 minutes after nesslerisation. The method devised by Ferguson and Hook was also tried, but it gave disappointing results in my hands.

Comparison with other techniques. Stuart, van Stratum and Rustigian claimed that their rapid method gave positive results in times ranging from 40 seconds to 80 minutes with large inocula of the order of 3 loopfuls from agar slopes. It appeared desirable therefore to determine by titration the minimum amount of urea destruction necessary to produce a positive reaction. It was found that hydrolysis of 0.12 mg of urea was necessary to produce the required color change in the medium of Stuart *et al.*, whereas 0.012 mg was sufficient to give a definite yellow color with nesslerisation in the medium described for method III in this paper.

Stuart *et al.* also stressed that in strongly buffered urea medium only members of the genus *Proteus* give evidence of urea utilisation, whereas in weakly buffered medium coliform and paracolon bacilli may also give the reaction. To obtain maximum sensitivity for their rapid test, which is a modification of the Rustigian and Stuart method described under method I, they had to reduce the buffer to a point just sufficient to maintain the reaction of pH 6.8 in ordinary glassware. Thus the medium used for their rapid test has a molar concentration of 0.0013 M as against the medium described for method III of this paper, which has a concentration of 0.05 M and appears to be about ten times more sensitive and possesses the advantage of a high molar concentration of the buffer. The relative sensitivity of the two tests was further illustrated when 35 *Proteus* strains were tested by both methods in parallel with identical inocula of about 500,000 organisms. All were positive with method III, whereas none was positive with the rapid method of Stuart *et al.* in 3 hours.

Schneider and Gunderson described a complicated method for which they claimed that positive results were usually discernible after 6–8 hours' and were very prominent after 18–24 hours' incubation at 37°C. Their medium contained peptone, which is an alternative source of nitrogen, and they stressed the need for a large inoculum. Another modification in which urea was not the sole source of nitrogen was described by Christensen. Heavy inocula were necessary and the earliest recorded results were obtained after 6 hours' incubation. These two methods appear to offer few advantages; they are relatively slow, require large inocula and contain an alternative source of ammonia which may be the cause of their apparently lesser selectivity.

Clark and Lubs (no reference cited).
W. B. Christensen (1946). J. Bact. 52, 461.
W. W. Ferguson and A. E. Hook (1942–43). J. Lab. clin. Med. 28, 1715.
R. Rustigian and C. A. Stuart (1941). Proc. Soc. exp. Biol. Med. 47, 108.
M. D. Schneider and M. F. Gunderson (1946). J. Bact. 52, 303.
C. A. Stuart, E. van Stratum and R. Rustigian (1945). J. Bact. 49, 437.

Urea—*Proteus*

G. T. Cook (1948). J. Path. Bact. 60, 171–181.

Media used

Medium A contained ingredients described by Rustigian and Stuart with the addition of two per cent agar. Medium B was prepared according to the directions of Christensen, apart from a slightly shorter time of sterilization.

Medium A

Yeast extract (Difco)	0.1 g
Potassium dihydrogen phosphate (KH_2PO_4)	9.1 g
Sodium phosphate (Na_2HPO_4)	9.5 g
Agar	20 g
Distilled water	1000 ml

Steam until the solids are dissolved, filter through lint and add about 3 ml of 0.4 per cent phenol-red solution. Tube in 4.5 ml quantities and autoclave at 10 lb for 10 minutes. A 20 per cent solution of urea is sterilized by Seitz filtration and 0.5 ml added to each tube after cooling to 50°C to make a final medium with a concentration of 2 per cent urea. The pH of the medium is 6.8–7.0.

Medium B

Difco bacto-peptone	1 g
Sodium chloride	5 g

Potassium dihydrogen phosphate (KH_2PO_4)	2 g
Agar	20 g
Distilled water	1000 ml

Steam until the solids are dissolved, filter through lint and add 1 g of glucose and about 3 ml of 0.4 per cent phenol-red solution. Adjust pH to 6.8–7.0 tube in 4.5 ml quantities and autoclave at 10 lbs for 15 minutes. A 20 per cent solution of urea is sterilized by Seitz filtration and 0.5 ml added to each tube after cooling to 50°C to make up a final medium with a concentration of 2 per cent urea.

Note: Both media were tested for sterility by incubation at 37°C for 24 hours.

Methods

Inoculations were made from a 24 hr agar* slope culture over the surface of one tube of medium A and one of medium B. The cultures were incubated at 37°C and the urea tubes were read at 3, 6 and 24 hours after inoculation and, if negative, daily for 7 days. A slight pink color on the surface of the medium was recorded as a trace reaction, diffusion of color into the butt as a moderate reaction and complete diffusion throughout the medium as a strong reaction.

**Proteus* cultures were on Dorset's egg medium slopes.

W. B. Christensen (1946). J. Bact. 52, 461.

R. Rustigian and C. A. Stuart (1941). Proc. Soc. exp. Biol. Med. 47, 108.

Urea—*Staphylococcus*

C. Shaw, J. M. Stitt and S. T. Cowan (1951). J. gen. Microbiol. 5, 1010–1023.

A highly buffered medium in which urea was the only nitrogen source (Stuart, van Stratum and Rustigian, 1945) and a feebly buffered nutrient urea medium (Christensen, 1946) were used; the second was controlled by a medium of the same composition save that urea was omitted.

W. B. Christensen (1946). J. Bact. 52, 461.

C. A. Stuart, E. van Stratum and R. Rustigian (1945). J. Bact. 49, 437.

Urea—*Brucella*

G. Renoux and H. Quatrefages (1951). Annls Inst. Pasteur, Paris 80, 183–188.

Dosage de l'activité uréasique: Nous avons utilisé trois milieux pour cette recherche:

a) Le milieu de Bauer (*in* Hoyer): solution à 5 p. 100 d'urée dans NaH_2PO_4 M/8 ajusté à pH 4 (acide chlorhydrique à 10 p. 100) et contenant 0.0015 p. 100 de rouge de phénol comme indicateur: 1 ml par tube;

b) Le milieu de Fergusson, modifié par Anderson: 1 ml par tube;

c) Le milieu de Christensen.

Nous avons noté, pour les deux premiers milieux, le temps au bout duquel l'indicateur coloré virait, après ensemencement très large (tout ce que peut contenir une anse de 2 mm de diamètre), les tubes de milieu de Bauer étant portés au bain-marie à 37° tandis que les tubes de milieu de Fergusson restaient à la température du laboratoire (20-24°).

En ce qui concerne le milieu de Christensen (rappelons qu'il est nécessaire d'avoir un culot important), nous avons noté après dix-huit heures à 37°, les différences de coloration des diverses zones; ce milieu est ensemencé, au fil droit, en piqûre dans le culot et en strie sur la pente.

Urea—general

P. Hauduroy (1951). Techniques Bactériologiques. Masson et Cie, Éditeurs, Paris.

UREASE

(Production et recherche). – Cultiver les bactéries sur de la gélose ordinaire inclinée. On cherche l'uréase après 3 et 7 jours. Émulsionner la culture dans 2 ml d'eau distillée. Diviser en deux parties égales. Mettre chaque partie dans un tube parfaitement propre. Ajouter une goutte de rouge de phénol à chaque tube, et ajuster la réaction jusqu'à *p*H 7.0 par adjonction de quelques gouttes d'HCl ou de NaOH dilué. Ajouter à l'un des tubes environ 0.02 g d'urée cristallisée, l'autre tube restant comme témoin. S'il y a de l'uréase, le tube où l'on a ajouté l'urée cristallisée devient alcalin en quelques minutes. *(A. T. C. C.)*.

Urea—*Escherichia* and *Proteus*

P. Hauduroy (1951). Techniques Bactériologiques. Masson et Cie, Éditeurs, Paris.

DECOMPOSITION DE L'UREE

(Colibacille) = Genre Colibacille.

(Genre Proteus).

Urée (Merck)	2 p. 100	NaCl	0.5 p. 100
KH_2PO_4	0.1 p. 100	Alcool (95 p. 100)	1 p. 100
K_2HPO_4	0.1 p. 100	*p*H = 7	

Filtrer sur filtre Seitz. Distribuer en tubes par 5 cm^3. Ensemencer avec des cultures de 20 heures sur gélose. Mettre 24 heures à 37°. Ajouter quelques gouttes de bromothymol bleu comme indicateur.

Une couleur bleue intense indique une réaction positive, la décomposition de l'urée et la production d'alcali. *(C. I. E.)*.

Urea—*Escherichia*

P. Hauduroy (1951). Techniques Bactériologiques. Masson et Cie, Éditeurs, Paris.

Décomposition de l'urée. – Préparer le milieu suivant (en accord avec Christensen):

Glucose	1 g	KH_2PO_4	2 g
Bactopeptone	1 g	Gélose	20 g
NaCl	5 g	Eau distillée	1000 cm^3

Chauffer tous les ingrédients jusqu'à l'ébullition. Ajuster à *p*H 6.8.

Répartir par 4 cm^3 dans des tubes à réaction de Wassermann.

Autoclaver. Ajouter une solution à 20 p. 100 d'urée stérilisée sur filtre Seitz à la dose de 0.5 cm^3. Mélanger.

Éprouver une nuit à l'étuve pour la stérilité.

Ensemencer avec une suspension en eau physiologique de germes recueillis sur gélose après 20 heures de culture comme pour le milieu de Simmons.

Urea—*Diplococcus, Clostridium, Bifidobacterium*

M. Huet and N. Aladame (1952). Annls Inst. Pasteur, Paris 82, 766–767.

Technique. – La recherche de l'hydrolyse de l'urée a été faite en milieu de Fergusson, c'est-à-dire un milieu comportant un tampon phosphate, de l'alcool éthylique, de l'urée et du rouge de phénol comme indicateur de pH. Pour plus de commodité, nous nous sommes servis du milieu de Roland et Bourbon, qui contient en plus du tryptophane (destiné à la recherche de l'indol). Comme cela a été bien établi, le tryptophane ne modifie en rien la mise en évidence de l'uréase.

La souche est ensemencée dans un flacon de Jaubert et Gory, qui fournit en vingt-quatre ou quarante-huit heures, selon le germe, une culture jeune, suffisamment abondante pour donner à la centrifugation un culot facilement récupérable. Les corps microbiens ainsi recueillis sont lavés deux fois à l'eau physiologique. Ceci est nécessaire pour éliminer toute trace de bouillon, afin de rester en milieu synthétique et de ramener le pH très bas de certaines cultures au voisinage de la neutralité. Le dernier culot est repris, selon son importance, par 1/2 cm^3 ou 1 cm^3 de milieu de Roland et Bourbon, dans un tube à hémolyse. Les corps microbiens doivent être suffisamment abondants pour que la suspension soit nettement opaque. Lorsque l'urée est hydrolysée, il y a formation de carbonate d'ammonium et l'indicateur vire au rouge violet. La réaction se produit dans un délai variant entre quelques minutes et quelques heures.

Urea—microtests

P. H. Clarke and S. T. Cowan (1952). J. gen. Microbiol. 6, 187–197.

Equal volumes of urea-buffer solution (urea 1%, 0.0125 M buffer pH 6.0, phenol red 0.00025%) and suspension are mixed. Sealed capillary tubes are incubated at 37°C and read for alkali production after 4 and 24 hr.

Tests were made in capillary tubes 10 cm long.

Urea—*Bacillus*

N. R. Smith, R. E. Gordon and F. E. Clark (1952). U.S.D.A. Agr. Monograph No. 16.

The production of urease was demonstrated by growing the cultures on slopes of nutrient agar and testing for urease after 3 and 7 days. The growth was washed off with 2 ml of distilled water into a clean test tube. A drop of phenol red indicator was added and the reaction brought to pH 7.0 by a few drops of very dilute HCl or NaOH. The suspension was divided equally between two tubes and approximately 0.1 g of crystalline urea was added to one, the other serving as the control. If urease was present the suspension with the urea became very alkaline in a few minutes.

Urea—*Brucella*

M. J. Pickett, E. L. Nelson and J. D. Liberman (1953). J. Bact. 66, 210–219.

The double strength urea medium, maintained at 4°C over chloroform and not sterilized before use, was similar to that of Rustigian and Stuart (1941). It contained 4 per cent urea, 4.8 per cent $Na_2HPO_4 \cdot 12H_2O$, 1.8 per cent KH_2PO_4, 20 mg per cent of yeast extract (Difco), and 2 mg per cent phenol red. For the test, 0.5 ml of this medium and 0.5 ml of stock suspension were placed in a 12 by 100 mm tube; this mixture was incubated aerobically at room temperature and was observed for color change at 5 to 10 minute intervals during the following two hours. The test was then read at irregular intervals through a total incubation period of forty-eight hours.

Media were inoculated from a suspension of 5 x 10^9 cell/ml. Cultures were incubated at 35°C under 10% CO_2.

R. Rustigian and C. A. Stuart (1941). Proc. Soc. exp. Biol. Med. 47, 108–112.

Urea—*Clostridium*

H. Tataki and M. Huet (1953). Annls Inst. Pasteur, Paris 84, 890–894.

Nous avons suivi les mêmes modalités que précédemment: cultures jeunes sur milieu VF en fioles de Jaubert et Gory, culots recueillis par centrifugation, lavés deux fois à l'eau physiologique et repris dans ½ cm^3 à 1 cm^3 de milieu de Roland et Bourbon, puis portés à l'étuve à 37°C. L'hydrolyse de l'urée se traduit par un virage au rouge du milieu.

Urea—*Pseudomonas (Malleomyces) pseudomallei*

P. de Lajudie and E. R. Brygoo (1953). Annls Inst. Pasteur, Paris 84, 509–515.

Uréase:– Nous avons utilisé le milieu de Fergusson, modifié par Roland et Mlle. Bourbon (in Dumas).

J. Dumas (1951). *Bactériologie médicale,* Flammarion, Paris.

Urea—*Pseudomonas (Malleomyces) pseudomallei*

L. Chambon and P. de Lajudie (1954). Annls Inst. Pasteur, Paris 86, 759–764.

L'hydrolyse de l'urée et son utilisation comme source de carbone et d'azote, ou d'azote seulement, est un caractère d'identification important. Nous l'avons étudié comparativement sur les milieux classiques de Fergusson-Hook et de Christensen tels qu'ils sont donnés par Dumas (1951), et sur les milieux de Lévy-Bruhl et Cado (1937) dont la composition est la suivante:

Milieu I:

Urée	6
PO_4HK_2	1
ClNa	5
SO_4Mg	0.2
Eau bidistillée	1000 ml

Milieu II:

Urée	3
PO_4HK_2	1
ClNa	5
SO_4Mg	0.2
Glucose	1
Eau bidistillée	1000 ml

Le pH est ajusté à 7.5 par adjonction de soude et la stérilisation assurée par trois chauffages à 100°.

Nous n'avons obtenu aucune culture avec le milieu I de L-B et C, ni avec le milieu de Fergusson-Hook. Par contre le milieu II de L-B et C donne avec les 25 souches étudiées un croît microbien moyen avec formation d'un voile.

La mesure du pH, effectuée après vingt-quatre heures, après cinq jours et après huit jours, montre une acidification initiale, suivie d'une nette alcalinisation du milieu. Celle-ci est d'autant plus marquée que les souches donnent un dépôt muqueux plus abondant.

Toutes ces souches cultivent également sur le milieu synthétique à l'urée de Christensen. Effectuant des lectures au bout de vingt-quatre et au bout de quarante-huit heures, nous avons note un virage au rouge pour 17 d'entre elles, au jaune pour 2 et pas de modification de la teinte du milieu pour 6.

Malleomyces pseudo-mallei est donc capable d'utiliser l'urée comme source d'azote associée à un sucre.

J. Dumas. *Bactériologie Médicale,* Flammarion, édit., 1951.

M. Lévy-Bruhl and Y. Cado. *Ann. Inst. Pasteur,* 1937, 58, 498.

Urea—*Clostridium*

M. Huet and F. de Cadore (1954). Annls Inst. Pasteur, Paris 86, 241–243.

La technique utilisée est la suivante: on prépare, à partir de bouillon VF non glucosé, un milieu solide, gélosé à 20 p. 1000, ajusté à pH 7 et auquel on ajoute par litre 0.8 ml d'une solution à 1.5 p. 100 de bleu de bromothymol dans l'alcool. Ce milieu est réparti en culot dans des tubes de 22, à raison de 20 ml par tube, et stérilisé à l'autoclave. Au moment de l'emploi, on régénère ce milieu par un chauffage de trente minutes au bain-marie bouillant, puis on ajoute à chaque tube, préalablement refroidi aux environs de 50°, 0.5 ml de la solution suivante:

Urée		40 g
PO_4KH_2		4 g
PO_4K_2H		4 g
Eau distillée	Q.S	100 ml

Cette solution, ne pouvant être chauffée, est stérilisée par filtration sur bougie.

Le tube est alors ensemencé à partir d'un matériel polymicrobien quelconque et coulé dans le couvercle d'une boîte de Petri. Quelques mouvements circulaires suffisent à assurer un mélange parfait. Puis le fond de la boîte est placé dans le couvercle, concavité en dessus, emprisonnant une couche de gélose à l'abri de l'air, selon une technique analogue à celle de Boëz. Après vingt-quatre ou quarante-huit heures, temps nécessaire au développement des colonies, on observe les boîtes par transparence sur un fond blanc. Les bactéries non uréolytiques donnent des colonies jaunes ou blanches, tandis que celles libérant une uréase donnent des colonies d'un blanc bleuté entourées d'un halo bleu très visible. Ces colonies ainsi repérées sont repiquées et les souches obtenues purifiées selon les techniques propres aux bactéries anaérobies.

Urea—*Mycobacterium*

A. Tacquet, Ch. Gernez-Rieux and B. Gaudier (1954). Annls Inst. Pasteur, Paris 87, 335–339.

Nous avons eu recours, pour mettre en évidence l'hydrolyse de l'urée, à quatre techniques différentes:

1° *Test d'Elek* [6]. – La solution d'urée est répartie en tubes à hémolyse, sous le volume de 0.3 ml. La suspension de bacilles est incubée à 37° pendant trois heures. Le réactif de Nessler décèle l'uréolyse en donnant une coloration jaune en présence d'ammoniaque. Cette technique interdit l'emploi des cultures obtenues sur milieu de Dubos liquide qui, à lui seul, entraîne la réaction. Il faut donc utiliser uniquement le milieu de Löwenstein, en ayant soin de prélever en surface.

2° *Milieu de Ferguson* [7]. – Avec le bleu de bromothymol comme indicateur. La coloration bleue est le témoin de l'uréolyse. Ce milieu est ensemencé, comme précédemment, à partir d'une culture sur Löwenstein.

3° *Milieu de Stuart* [8]. – Ensemencé comme ci-dessus.

4° *Milieu solide de Christensen* [9]. – Ces deux derniers milieux sont particulièrement commodes car ils permettent, à une température donnée, de suivre l'évolution de la réaction chaque jour.

Ces tests ont été effectués à différentes températures (18°, 30°, 37°, 42°) et le délai d'observation, qui a été habituellement de trois à vingt-quatre heures, a été prolongé jusqu'à quinze jours pour 13 souches.

[6] St. D. Elek. *J. Path. Bact.*, 1948, 60, 183.
[7] W. W. Ferguson et A. E. Hook. *J. Lab. Clin. Med.*, 1943, 28, 1715.
[8] C. A. Stuart, E. van Stratum et R. Rustigian. *J. Bact.*, 1945, 49, 437.
[9] W. B. Christensen. *J. Bact.*, 1946, 52, 461.

Urea—*Proteus, Providencia*

R. Buttiaux, R. Osteux, R. Fresnoy and J. Moriamez (1954). Annls Inst. Pasteur, Paris 87, 375–386.

Une confusion semble s'être établie dans l'esprit des bactériologistes au sujet de l'hydrolyse de l'urée par les *Enterobacteriaceae.* Il semble nécessaire de rappeler les faits suivants:

Dans le sens des travaux initiaux, on considérait que les *Proteus* étaient seuls capables d'hydrolyser l'urée, dans un milieu synthétique dont le type est celui de Ferguson et Hook [11]. Il faut, à notre avis, retenir ce caractère pour la détermination des espèces du genre, bien qu'il ne soit pas rigoureusement spécifique. On peut rapprocher du milieu de Ferguson, celui décrit par Stuart, van Stratum et Rustigian [12].

Sur tous les autres milieux contenant de petites quantités de peptone ou acides aminés, de nombreuses *Enterobacteriaceae,* autres que les *Proteus,* peuvent hydrolyser l'urée. C'est le cas de celui de Christensen [13], de Singer [14], de Roland et Bourbon [15]. Le processus d'utilisation de l'urée n'est plus le même que celui intervenant sur milieu synthétique. Ces milieux modifiés sont utilisables pour la détermination des *Salmonella, Shigella* ou *B. paracoli,* par exemple, mais non pour l'étude rigoureuse des espèces du genre *Proteus.*

[11] W. W. Ferguson et A. E. Hook. *J. Lab. Clin. Med.*, 1943, 28, 1715.
[12] C. A. Stuart, E. van Stratum et R. Rustigian. *J. Bact.*, 1945, 49, 437.
[13] B. W. Christensen. *J. Bact.*, 1946, 52, 461.
[14] J. Singer. *Amer. J. clin. Path.*, 1950, 20, 880.
[15] F. Roland, D. Bourdon et S. Szturm. *Ann. Inst. Pasteur,* 1947, 73, 914.

Urea

N.C.T.C. Methods 1954.

Method 1. Inoculate fairly heavily a tube of Stuart, van Stratum and Rustigian's urea medium. Read daily for 7 days.

Method 2. Inoculate heavily a slope of Christensen's urea medium. Read

after 4 hours and daily for 7 days.
Variables: 30° and 37°C incubation.
Red color = urea hydrolysed.
W. B. Christensen (1946). J. Bact. 52, 461.
C. A. Stuart, E. van Stratum and R. Rustigian (1945). J. Bact. 49, 437.

Urea—*Haemophilus, Alcaligenes*

P. Thibault, S. Szturm-Rubinsten and D. Piéchaud-Bourbon (1955). Annls Inst. Pasteur, Paris 88, 246–250.

Uréase:– L'uréase a été recherchée dans le milieu de Ferguson et Hook.

Urea—enterobacteria

R. Buttiaux, J. Moriamez and J. Papavassiliou (1956). Annls Inst. Pasteur, Paris 90, 133–143.

L'hydrolyse de l'urée sera recherchée sur milieu de Stuart et coll. [12] de préférence aux autres. Nous avons observé [2], en effet, que l'uréase de certains *P. rettgeri* était active dans celui-ci sans l'être dans le milieu de Ferguson. Nous rappelons sa composition:

PO_4HK_2	3.64	g
PO_4Na_2H	3.8	g
Urée	8	g
Extrait de levure	0.04	g
Solution de rouge de phénol à 0.02 p. 100	20	ml
Eau distillée	380	ml

On dissout sans chauffer et sans ajuster le pH (il est spontanément de 6.8). On stérilise par filtration sur bougie Chamberland L_3 ou sur filtre Seitz. On répartit aseptiquement 2 ml environ par tube. Après inoculation large avec une partie de la strie microbienne du milieu de Kligler et incubation à 37°, l'hydrolyse de l'urée se manifeste en quelques heures par un virage au rouge pourpre (pH $\geqslant$ 8). Pour obtenir des résultats comparables à ceux décrits, il est préférable de ne pas utiliser le milieu à l'urée de Christensen.

[2] R. Buttiaux, R. Osteux, R. Frenoy et J. Moriamez. *Ann. Inst. Pasteur,* 1954, 87, 375.
[12] C. A. Stuart, E. Van Stratum et R. Rustigian. *J. Bact.,* 1945, 49, 437.

Urea—*Bacillus*

G. H. Bornside and R. E. Kallio (1956). J. Bact. 71, 627–634.
Modified techniques. Twenty-seven known strains of bacilli from the collection of Dr. N. R. Smith were kindly furnished by Dr. Ruth E. Gordon of Rutgers University. Several species and strains of bacteria from other sources served as biological controls. All organisms were maintained as slants on a basal medium of the following composition (in percentages):

enzymatic casein hydrolyzate, 0.4; poly-peptone (BBL), 0.4; yeast extract (BBL), 0.2; agar, 1.5; adjusted to pH 8. Strains of *B. pasteurii* were cultured on the above medium amended with urea (final concentration, 1.5 per cent). The basal medium, adjusted to pH 7.0 and containing thymol blue, indicated activity of urease if a blue color developed at the site of growth and diffused throughout the medium. Thymol blue provides for a slower indication of urease than is possible with phenol red as an indicator in urea media.

Urea

Society of American Bacteriologists *Manual of Microbiological Methods* – McGraw-Hill Book Co. Inc., New York 1957.

Urea agar:

Urea	10 g
Distilled water	100 ml

Sterilize by filtration. Add ½ ml per tube aseptically to nutrient agar after melting and cooling to 45–50°C. Mix well; cool for slants.

Nutrient agar:

Blood base agar (Difco)	20 g
Peptone	2.5 g
Beef extract	1.5 g
Agar	15.5 g
Distilled water	1000 ml
pH	7.4

Urea—enterobacteria

J. G. Heyl (1957). Antonie van Leeuwenhoek 23, 33–58.

Urea agar base according to Christensen.

Bacto peptone	1.0 g
Sodium chloride	5.0 g
Monopotassium phosphate	2.0 g
Phenol red	0.012 g
Distilled water	1000 ml

Dissolve the ingredients by boiling; adjust pH at 6.9 and autoclave 15 minutes at 120°C. After sterilization add 1 g dextrose and 20 g urea in Seitz-filtered solutions.

Urea—*Mycobacterium*

R. E. Gordon and J. M. Mihm (1959). J. gen. Microbiol. 21, 736–748.

Ten ml of a 15% (w/v) solution of urea, sterilized by filtration, was combined with 75 ml of the following autoclaved broth: KH_2PO_4, 9.1 g; Na_2HPO_4, 9.5 g; yeast extract, 0.1 g; phenol red, 0.01 g; distilled water, 1000 ml; pH 6.8 (Rustigian and Stuart, 1941). The mixture was pipetted aseptically to sterile plugged test tubes in 1.5 ml amounts and heavily in-

oculated with actively growing 2- or 3-day old cultures. Urease was demonatrated by the alkaline reaction of the phenol red after 5, 7, 14, 21 and 28 days at 28°C.
R. Rustigian and C. A. Stuart (1941). Proc. Soc. exp. Biol. Med. 47, 108.

Urea—*Pseudomonas fluorescens*

M. E. Rhodes (1959). J. gen. Microbiol. 21, 221–263.

Christensen's (1946) urea + peptone agar method was used to detect urea-decomposing isolates; this medium was not entirely suitable for pseudomonads because of simultaneous alkali formation from the peptone. Therefore a chemically defined medium with 2.0% (w/v) urea (Seitz-filtered) as sole added nitrogen source was also used; the basal medium contained phenol red and it was essentially that of Dowson (1949) except that the carbon source (galactose) was at 0.1% (w/v), to prevent acid production masking or inhibiting the urease action.
W. B. Christensen (1946). J. Bact. 52, 461.
W. J. Dowson (1949). *Manual of Bacterial Plant Diseases* 1st ed. London: A and C Black.

Urea—*Aeromonas*

J. P. Stevenson (1959). J. gen. Microbiol. 21, 366–370.

Tested on Christensen's (1946) urea agar slopes incubated at 37°C for 5 days. A red coloration of the medium was considered to indicate a positive reaction.
W. B. Christensen (1946). J. Bact. 52, 461.

Urea—*Clostridium*

M. E. Brooks and H. B. G. Epps (1959). J. gen. Microbiol. 21, 144–155.

Cultures were grown for 24 and 48 hr at 37°C in 20 ml amounts of Robertson's meat broth. A few strains from each group were also grown in 100 ml amounts of similar medium and tested daily for 4 days.

The reagent used was a slightly modified version of that of Roland, Bourbon and Szturm (1947) and Huet and Aladame (1952) and was made up as follows: 0.1 g KH_2PO_4; 0.1 g K_2HPO_4; 0.5 g NaCl; 2.0 g urea; 1.0 ml 95% (v/v) ethanol in water; 100 ml distilled water; 5.0 ml Universal Indicator; sufficient 0.1 N-HCl to give an orange colour (about pH 6.0).

A 2.5 ml sample of the culture under test was transferred to a test tube (80 mm x 8.0 mm) and centrifuged. The supernatant fluid was discarded and the sediment washed once in 1.0% (w/v) saline and finally resuspended in 1.0 ml distilled water. One ml of urease reagent was added and after thorough mixing, the test was incubated at 37°C.
M. Huet and N. Aladame (1952). Annls Inst. Pasteur, Paris 82, 766.
F. Roland, D. Bourbon and S. Szturm (1947). Annls Inst. Pasteur, Paris 73, 914.

Urea—*Acinetobacter*

L. Enjalbert, J. Brisou and A. K. Kambou (1959). Annls Inst. Pasteur, Paris 97, 112–115.

Cette uréolyse est constante:

Sur le milieu "urée tamponnée" préconisé par Brisou [6], l'attaque a lieu une heure environ après l'ensemencement, le milieu vire au "bleu roi".

Sur l'urée-indole, le virage au rose franc est plus tardif, environ deux à trois heures après l'ensemencement.

Le milieu de Ferguson, rendu solide par addition de 20 p. 1000 de gélose (couramment utilisé au Laboratoire de Bactériologie du C. H. R. de Purpan), vire au rose franc dans les mêmes délais de deux à trois heures après l'ensemencement.

Le germe ne pousse pas sur le Kristensen.

[6] Brisou (J.). *Etude de quelques "Pseudomonadaceae".* Baillet, édit., Bordeaux.

Urea—*Xanthomonas*

A. C. Hayward and W. Hodgkiss (1961). J. gen. Microbiol. 26, 133–140.

The method of Christensen (1946) was used, with incubation for 14 days.

W. B. Christensen (1946). J. Bact. 52, 461.

Urea—*Pasteurella*

H. H. Mollaret (1961). Annls Inst. Pasteur, Paris 100, 685–690.

Décomposition de l'urée. – Cette action a été démontrée en 1950 par Fauconnier [3], en milieu de Ferguson et en milieu urée-indole (Roland-Bourbon-Szturm) et confirmée en 1955 par Thal et Chen [13] sur milieu de Christensen-urée. Nous avons utilisé ces trois milieux ainsi que le bouillon placenta-urée, préconisé par Sohier [11] pour la mise en évidence de l'uréase chez certains *Corynebacterium,* et dont la formule est la suivante:

Bouillon placenta (Sohrab [12])	100 cm^3
Solution d'urée à 50 p. 100	5 cm^3
Solution de rouge de crésol à 0.4 p. 100	5 cm^3

En milieu urée-indole et en milieu de Ferguson, nous avons observé le virage en moins de quatre heures comme l'a montré Fauconnier, le plus souvent dans l'heure, parfois en quinze minutes, voire même presque immédiatement. Dans ces délais, la température ne modifie pas la rapidité de la réaction. Par contre, avec quelques souches donnant une réaction tardive, c'est toujours à 37° que le virage apparaît en premier. Dans ces cas, le décalage selon la température est habituellement net: par exemple, virage en six heures à 37° et en vingt heures à 18°.

Sur milieu de Christensen le virage apparaît vers la vingtième heure comme l'ont observé Thal et Chen; la température ne modifie pas ce délai.

En bouillon placenta-urée, le virage débute vers la sixième heure à 37° et toujours plus tardivement à 28° et surtout à 18°. Ce dernier milieu nous a donné des résultats plus constamment positifs que les trois autres: ainsi, la souche 35-II que Fauconnier, en 1950, avait trouvée négative en milieu urée-indole et en milieu de Ferguson et qui nous est également apparue négative sur ces deux milieux comme sur milieu de Christensen, s'est révélée positive en cinq jours, en milieu placenta (les témoins non additionnés d'urée n'ayant bien entendu pas viré).

D'autre part, trois souches récemment isolées (230-I, 231-I et 232-I) que nous avons trouvées tardivement positives en milieux urée-indole, de Ferguson et de Christensen (plus de vingt-quatre heures et, pour l'une d'entre elles, plus de quarante-huit heures), se sont montrées par contre, dès l'isolement, positives en bouillon placenta (les repiquages successifs de ces souches sur gélose en ont progressivement réduit pour deux d'entre elles les délais de virage en milieu urée-indole et milieu Ferguson à six heures, à partir du troisième repiquage et à quatre heures, à partir du huitième; ceci explique sans doute que nous ayons trouvé positive en quatre heures, en milieux urée-indole et de Ferguson, la souche 25-II qu'en 1950, Fauconnier avait trouvée négative sur ces mêmes milieux).

L'attaque de l'urée par *P. pseudotuberculosis* est donc un caractère constant chez les 327 souches de ce germe que nous avons examinées;

[3] Fauconnier (J.). *Ann. Inst. Pasteur,* 1950, 79, 104–105.
[11] Sohier (R.). [Communication personnelle.]
[12] Sohrab (H.). *Ann. Inst. Pasteur,* 1947, 73, 916–918.
[13] Thal (E.) et Chen (T. H.). *J. Bact.,* 1955, 69, 103–104.

Urea—*Pseudomonas*

M. Véron (1961). Annls Inst. Pasteur, Paris 100, Suppl. 6, 16–42.

L'hydrolyse de l'urée doit se rechercher à 30°C sur gélose-peptone-glucose-urée de Christensen [13], ou en milieu synthétique-urée contenant, en plus, selon les indications de Rhodes [61], 0.1 p. 100 de glucose; le milieu que nous avons utilisé est le suivant: K_2HPO_4 (450 mg), $(NH_4)_2SO_4$ (75 mg), KCl (50 mg), $SO_4Mg.7H_2O$ (5 mg), urée (2 g), glucose (100 mg), eau distillée (q. s. p. 100 ml). Ce milieu stérilisé par filtration, est ensemencé largement à partir d'une culture sur gélose, et incubé à 30°C.

. . . .

L'étude de l'uréase des *Pseudomonas* nécessite quelques remarques. En effet, cette enzyme inductible n'est pas révélée dans le milieu urée-tryptophane utilisé pour les Entérobactéries, milieu qui devient seulement un peu alcalin en six-huit jours.

[13] Christensen (B. W.). *J. Bact.,* 1946, 52, 461.
[61] Rhodes (M. E.). *J. gen. Microbiol.,* 1959, 21, 221.

Urea—*Escherichia aurescens*

H. Leclerc (1962). Annls Inst. Pasteur, Paris 102, 726–741.

Hydrolyse de l'urée – Elle est recherchée sur le milieu de Christensen [7].
[7] Christensen (W. B.). *J. Bact.*, 1946, 52, 461.

Urea—*Vibrio*

G. H. G. Davis and R. W. A. Park (1962). J. gen. Microbiol. 27, 101–119.

Breakdown by Christensen's (1946) method, in cotton-wool plugged tubes was read over 7 days. Controls without urea were included.
W. B. Christensen (1946). J. Bact. 52, 461.

Urea—enterobacteria

W. H. Ewing (1962). *Enterobacteriaceae. Biochemical Methods for Group Differentiation.* U.S. Department of Health, Education and Welfare, Public Health Service Publication, No. 734 (revised).

Christensen's urea agar (1944, 1946)

Peptone	1	g	
Sodium chloride	5	g	
Glucose	1	g	
Monobasic potassium phosphate	2	g	
Phenol red	0.012	g	(6 ml of 1:500 solution)
Urea	20	g	
Distilled water	100	ml	

Adjust to pH 6.8–6.9. Filter sterilize.

Dissolve 15 g agar in 900 ml distilled water and sterilize at 121°C for 15 minutes. Cool to 50–55°C, then add 100 ml urea concentrate (above). Mix and distribute in sterile tubes. Then the medium is slanted with a deep butt. This medium may be employed in fluid form if desired by omitting the agar.

Inoculation: The medium is inoculated heavily over the entire surface of the slant.

Incubation: 37°C. Examine at 2 hr, 4 hr, and after overnight incubation. Negative tubes should be observed daily for 4 days in order to detect delayed reactions given by members of certain groups other than *Proteus.* Urease positive cultures produce an alkaline reaction in the medium evidenced by a red colour.

Rapid Urease Test.

Alternate method (Stuart, *et al.* 1945, modification of the highly buffered medium of Rustigian and Stuart, 1941).

Yeast extract	0.1	g
Monobasic potassium phosphate	0.091	g
Dibasic sodium phosphate	0.095	g
Urea	20	g

Phenol red	0.01 g
Distilled water	1000 ml

This medium is Seitz filtered and tubed in sterile tubes in 3 ml amounts. The basal medium may be made up in 900 ml of distilled water and sterilized at 121°C, 15 mins. After cooling 100 ml of 20% sterile urea solution is added and the medium tubed in sterile tubes in 3 ml amounts.

Inoculation: Three loopfuls (2 mm loop) from an agar slant culture are inoculated into a tube of medium and the tube is shaken to suspend the bacteria.

Incubation: Tests are incubated in a water bath at 37°C and the results are read after 10 minutes, 60 minutes, and 2 hr.

W. B. Christensen (1944). Personal communication.
W. B. Christensen (1946). J. Bact. 52, 461.
R. Rustigian and C. A. Stuart (1941). Proc. Soc. exp. Biol. Med. 47, 108.
C. A. Stuart, E. van Stratum and R. Rustigian (1945). J. Bact. 49, 437.

Urea—*Staphylococcus* and *Micrococcus*

D. A. A. Mossel (1962). J. Bact. 84, 1140–1147.

Inclined tubes of Christensen's (1946) urea-agar were used for this purpose. Incubation was for 48 hr at 30°C.

W. B. Christensen (1946). J. Bact. 52, 461–466.

Urea—*Haemophilus*

P. N. Edmunds (1962). J. Path. Bact. 83, 411–422.

Urea and phenol red in the same concentrations as in Christensen's (1946) medium were added to the plasma digest broth and the pH value was adjusted to about 6.9 by addition of 0.1 N HCl. Six drops of 48 hr cultures in the blood agar and broth medium were inoculated into the urea medium and the reaction was observed after 2 and 5 days at 37°C.

W. B. Christensen (1946). J. Bact. 52, 461.

Urea—*Actinobacillus* and *Haemophilus*

E. O. King and H. W. Tatum (1962). *Actinobacillus actinomycetemcomitans* and *Haemophilus aphrophilus.* J. infect. Dis. 111, 85–94.

Test performed with Christensen's urea agar.

Urea—*Pseudomonas odorans*

I. Málek, M. Radochová and O. Lysenko (1963). J. gen. Microbiol. 33, 349–355.

The authors used Christensen medium incubated for 7 days at 28°C.

Urea—aerobic actinomyces

F. Mariat (1963). Annls Inst. Pasteur, Paris 105, 795–797.

Milieux. – Divers milieux gélosés ou liquides sont utilisés. Leurs différentes formules permettent d'éliminer les risques d'erreur. Les milieux gélosés sont stérilisés à l'autoclave. Lorsque la formule comporte de l'urée, ce produit; stérilisé par filtration, est ajouté stérilement au milieu refroidi à 50°. Les milieux liquides sont stérilisés par filtration. L'indicateur de pH utilisé dans les diverses formules est le rouge de phénol.

Milieux solides (répartis à raison de 5 ml par tubes de 17 × 170 mm).

A: $Na_2HPO_4 \cdot 12\ H_2O$: 1.00 g. KH_2PO_4: 1.00 g. $MgSO_4 \cdot 7\ H_2O$: 0.5 g. KCl: 0.50 g. Glucose: 10.00 g. Solution d'oligo-éléments: X gouttes. Thiamine, acide nicotinique, pantothénate de Ca, pyridoxine: 1×10^{-7}. Biotine: 1×10^{-9}. Rouge de phénol à 1 p. 100: 2.5 ml. Gélose: 15.00 g. Eau distillée Pyrex: 1000 ml. Urée (ajoutée stérilement): 20.00 g (pH ajusté à 6.8–6.9).

B: même milieu que A, mais sans urée, additionné de $(NH_4)_2\ HPO_4$: 0.10 g p. 1000.

C: même milieu que A, mais glucose: 2.00 g au lieu de 10.00 g et $(NH_4)_2\ HPO_4$: 0.10 g p. 1000 (milieu témoin glucosé à 2 p. 1000 contenant de l'urée et une source d'azote complémentaire).

Milieux liquides (répartis à raison de 1 ml par tubes de 7 × 90 mm).

D: Milieu urée-indole: L-tryptophane: 3.00 g. KH_2PO_4: 1.00 g. K_2HPO_4: 1.00 g. NaCl: 5.00 g. Urée: 20.00 g. Alcool éthylique à 96°: 10 ml. Rouge de phénol à 1 p. 100: 2.5 ml. Eau distillée: 1000 ml.

Cultures. – Les milieux solides sont ensemencés en déposant sur la surface un fragment (± 0.5 × 0.5 cm) d'une culture mère sur gélose glucosée de Sabouraud (ajustée à pH 7); âgée de 10 à 19 jours suivant les souches. Le milieu liquide est ensemencé en dilacérant un semblable inoculum dans le liquide. Les tubes sont incubés à l'étuve à 30°.

L'activité uréasique est appréciée par virage au rouge de l'indicateur coloré après 24, 48, 72 heures, 7, 11 et 16 jours. Cette alcalinisation traduit une libération d'ammoniaque par suite de l'attaque enzymatique de l'urée:

$$H_2N\ CO\ NH_2 + H_2O \rightleftharpoons CO_2 + 2NH_3$$

Urea—*Proteus*

L. M. Bergquist and R. L. Searcy (1963). J. Bact. 85, 954–955.

Recently, the Berthelot color reaction has been applied to measurement of urea nitrogen in urease-treated serum (Fawcett and Scott, 1960. J. clin. Path. 12, 156; Searcy *et al.*, 1961. Am. J. med. Technol. 27, 255; Chaney and Marbach, 1962. Clin. Chem. 8, 130). A blue dye is formed by ammonia in the presence of sodium phenate and sodium nitroprusside. The berthelot color reaction has been used to demonstrate urease activity of 110 *Proteus* and *Escherichia* organisms isolated from hospital patients. All strains were identified biochemically by conventional procedures.

Glass fiber strips (5 by 30 mm) were impregnated with 50 μliters of 15% aqueous urea solution. Strips were dried at room temperature and stored in sterile petri plates. Urea from a strip was eluted in 5.0 ml of ammonia-free distilled water. Extraction of urea from glass fiber was rapid and reached a maximum in 10 min at 37°C. Single-colony isolates of test organisms measuring approximately 1 to 3 mm in diameter were used to inoculate urea eluates. After incubation for 5 min in a constant-temperature aluminum heating block maintained at 37°C, 1 ml of sodium hypochlorite and 1 ml of sodium phenate-sodium nitroprusside reagent (Hyland UN-Test kit, Hyland Laboratories, Los Angeles, Calif.) were added to urea-bacteria mixtures. Berthelot color from ammonia was produced by an additional 5 min of incubation at 37°C. All strains of *Proteus,* except those of *P. inconstans,* generated detectable amounts of ammonia.

Urea—*Actinobacillus*

P. W. Wetmore, J. F. Thiel, Y. F. Herman and J. R. Harr (1963). Comparison of selected *Actinobacillus* species with a hemolytic variety of *Actinobacillus* from irradiated swine. J. infect. Dis. 113, 186–194.

The 'standard medium of Stuart *et al.* (1945) and Christensen's (1946) agar slants' were used.

C. A. Stuart, E. Van Stratum and R. Rustigian (1945). J. Bact. 49, 437–444.

W. B. Christensen (1946). J. Bact. 52, 461–466.

Urea—*Corynebacterium pseudotuberculosis*

C. H. Pierce-Chase, R. M. Fauve and R. Dubos (1964). J. exp. Med. 120, 267–281.

The authors used Bacto urea base concentrate.

Urea—*Mycobacterium buruli* n.sp.

J. K. Clancey (1964). J. Path. Bact. 88, 175–187.

Urease production was assessed by the method of Singer and Cysner (1952).

J. Singer and E. Cysner (1952). Am. Rev. Tuberc. pulm. Dis. 65, 779.

Urea—*Dermatophilus*

M. A. Gordon (1964). J. Bact. 88, 509–522.

For hydrolysis the authors used urea agar slants consisting of 0.5% NaCl, 0.2% KH_2PO_4, 0.1% peptone, 0.1% glucose, and 2% agar; the mixture was adjusted to pH 6.9, and 6 ml of 0.2% phenol red solution were added per liter.

The inoculum was taken from Brain Heart Infusion agar slants.

The medium was incubated at 36°C (± 1) and readings were made at 48 hours, 5 days, 1 week and 2 weeks.

Urea—*Pseudomonas aeruginosa*

K. Morihara (1964). J. Bact. 88, 745–757.

Young cells grown on bouillon-agar were suspended in 0.5% urea solution and kept for several hours. Ammonia produced was detected by use of Nessler's reagent. Cultures were incubated at 30°C.

Urea—*Bacteroides oralis*

W. J. Loesche, S. S. Socransky and R. J. Gibbons (1964). J. Bact. 88, 1329–1337.

The authors used the methods in the *Manual of Microbiological Methods* (1957).

Society of American Bacteriologists *Manual of Microbiological Methods* – McGraw-Hill Book Co. Inc., New York 1957.

Urea—*Aerococcus catalasicus*

O. G. Clausen (1964). J. gen. Microbiol. 35, 1–8.

Cultures were incubated aerobically at 37°C unless otherwise stated.

Christensen (1946) urea medium with indicator was used to demonstrate formation of NH_3 by urease activity. Incubation period: 14 days.

W. B. Christensen (1946). J. Bact. 52, 461.

Urea—*Moraxella*

W. J. Ryan (1964). J. gen. Microbiol. 35, 361–372.

Christensen's medium (Christensen, 1946) was used. A heavy inoculum was placed on the centre of a slope and incubated 24 hr at 37°C.

W. B. Christensen (1946). J. Bact. 52, 461.

Urea—rumen contents

G. A. Jones, R. A. MacLeod and A. C. Blackwood (1964). Can J. Microbiol. 10, 371–378.

The authors fractionated the contents of the rumen and tested each for urease activity.

Estimation of Urease Activity

The urease activity of the various fractions was estimated by the manometric method of Huhtanen and Gall (1955). In this method, the hydrolysis is carried out in conventional Warburg vessels under an atmosphere of CO_2 which is in equilibrium with CO_2 in the aqueous phase. Urea is hydrolyzed by urease to NH_3 and CO_2, 2 moles of NH_3 being released for every mole of CO_2 produced. Owing to the rise in pH of the aqueous phase, CO_2 is absorbed from the atmosphere and the volume of CO_2 absorbed is proportional to the amount of urea hydrolyzed. The main compartment of each Warburg vessel contained 3 ml of rumen fluid preparation. After equilibrium at 37°C, 0.2 ml of 0.3 M urea solution was added to it from the side-arm. Determinations were carried out at least in duplicate and control flasks without added urea were included for each determination.

C. N. Huhtanen and L. S. Gall (1955). J. Bact. 69, 102–103.

Urea—*Malleomyces, Pseudomonas*

Bach-Toan-Vinh (1965). Annls Inst. Pasteur, Paris 109, 460–463.

Recherche d'une Uréase – Milieu de Ferguson et Hook. Milieu de Christensen.

Urea—*Vibrio comma*

J. C. Feeley (1965). J. Bact. 89, 665–670.

Urease activity was determined in Urea Broth (Difco).

Urea—*Mycobacterium* and *Bacillus*

G. R. F. Hilson (1965). J. gen. Microbiol. 39, 407–421.

The rapid test of urease activity of Elek (1948) was used. A well-loaded loopful of the surface growth formed the inoculum.

S. D. Elek (1948). J. Path. Bact. 60, 183.

Urea—*Alcaligenes*

R. G. Mitchell and S. K. R. Clarke (1965). J. gen. Microbiol. 40, 343–348.

Christensen's (1946) method was used; incubation for 21 days.

W. B. Christensen (1946). J. Bact. 52, 461.

Urea—*Pseudomonas*

D. J. Stewart (1965). J. gen. Microbiol. 41, 169–174.

Media frequently used for the detection of urease activity were found to be unsuitable for fluorescent pseudomonads since the presence of free ammonia was found to suppress the formation of urease in growing cultures.

The isolates were inoculated by a straight wire from nutrient agar cultures into 2 ml of basal MU medium (in 100 mm x 10 mm tubes). After incubation for 40 hr at 25°C,* good growth was apparent in all tubes and the pH values had decreased to about 6.5. To each tube was then added aseptically 0.02 ml of 20% (w/v) urea solution, the tubes were shaken and re-incubated at 37°C. Alkalinity developed within a few hours in all tubes; in no case did the incubation period for a pH value of 9.0 (violet) to be reached exceed 6 hr.

The basal MU medium contains, casitone (Difco), 0.02 g; yeast extract (Difco), 0.02 g; glucose, 0.05 g; NaCl, 0.3 g; K_2HPO_4, 0.02 g; distilled water, 98 ml; mixed indicator solution, 2 ml. Adjusted to pH 7.4. Sterilized for 15 min at 120°C.

Glucose was chosen in preference to galactose, which was used by Rhodes (1959), because manometric studies showed that glucose was more rapidly oxidized by the isolates. The buffer salt (K_2HPO_4) was kept at a low concentration to allow a more rapid increase in pH value with base production. The mixed indicator system proposed by Singer (1950) was chosen since its green color at pH 7.4 changes to blue at about pH 8.2 and to a distinct violet at about pH 9.0, thus allowing relative degrees of

base production to be assessed. It consists of three indicator solutions prepared by dissolving 0.20 g of bromothymol blue, cresol red and thymol blue in 6.4 ml, 10.6 ml and 8.6 ml, respectively, of NaOH (0.05 N) and adding 100 ml of distilled water to each solution; the three solutions are then mixed in the proportions 12.5:4:10 by vol.
*The initial incubation for growth of *Ps. aeruginosa* was at 37°C.
M. E. Rhodes (1959). J. gen. Microbiol. 21, 221.
J. Singer (1950). Am. J. clin. Path. 20, 880.

Urea—*Staphylococcus*

M. Kocur, F. Přecechtěl and T. Martinec (1966). J. Path. Bact. 92, 331–336.

Urease production was studied in cultures grown in liquid medium (Christensen, 1946) for 24 hr.
W. B. Christensen (1946). J. Bact. 52, 461.

Urea—*Erwinia*

A. von Graevenitz and A. Strouse (1966). Antonie van Leeuwenhoek 32, 429–430.

The authors used the method of W. H. Ewing, B. R. Davis and P. R. Edwards. Publ. Hlth Lab. 18, 77–83.

Urea—*Providencia, Rettgerella*

C. Richard (1966). Annls Inst. Pasteur, Paris 110, 105–114.

Pour des raisons de commodité, les milieux ont été incubés à 37°C.

Hydrolyse de l'urée: en milieu urée-indole [25] et en milieu de Christensen [2].

[2] Christensen (W.). *J. Bact.,* 1946, 52, 461.
[25] Rolland (F.), Bourbon (D.) et Szturm (S.). *Ann. Inst. Pasteur,* 1957, 73, 914.

Urea—*Brevibacterium*

R. Chatelain and L. Second (1966). Annls Inst. Pasteur, Paris 111, 630–644.

Uréase: en milieu urée-indole [15].

La température d'incubation des cultures est de 30°C.

[15] Olivier (H. R.). *Traité de biologie appliquée,* tome II, Maloine édit., Paris, 1963.

Urea—anaerobes

A.-R. Prévot (1966). Techniques pour le diagnostic des Bactéries Anaérobes. Éditions de la Tourelle, St. Mandé.

Recherche des uréases.

Cette recherche doit se faire en mettant en contact les corps microbiens centrifugés et lavés d'une fiole de Jaubert et Gory de 24 à 48 heures en

présence de réactif de Ferguson (voir plus loin). Il se produit une coloration rouge-violacé en présence d'uréase. Il faut éviter le réactif de Roland et Bourbon car si le germe contient une tryptophane-désaminase et pas d'uréase, le virage se produit et donne une fausse réaction.
Réactif à l'urée.

La détection des uréases se fait en mettant en contact des corps microbiens avec le réactif de Ferguson:

Urée	2	g
PO_4KH_2	0.1	g
PO_4K_2H	0.1	g
NaCl	0.5	g
Alcool à 95°	1	cm^3

à compléter par:

Eau	100	cm^3

Rouge de phénol à 2% quantité suffisante pour obtenir une teinte jaune dont le rougissement indique une activité uréasique.

VOLUTIN

Volutin—*Corynebacterium, Mycobacterium*

P. Hauduroy (1951). Techniques Bactériologiques. Masson et Cie, Éditeurs, Paris.

COLORATIONS DES GRANULATIONS

Méthode de Löffler. – Colorer quelques minutes avec le bleu de Löffler, laver à l'eau et traiter rapidement par la méthode de Gram. Laver. Les granulations sont noir verdâtre et les bacilles verdâtres.

Méthode de Neisser. – Faire agir sur les frottis séchés et fixés pendant quelques secondes le bleu de méthylène suivant la formule de Neisser (colorant de Neisser). Laver. Colorer pendant 3 à 5 secondes dans une solution de vésuvine à 0.2 p. 100. Laver à l'eau. Sécher. Examiner.

Méthode au bleu de Roux. – Colorer une minute avec le bleu de Roux. Laver. Passer rapidement dans la vésuvine à 1 p. 250. Les granulations sont bleu foncé, le corps bacillaire jaune.

Méthode de Fontès. – Permet de mettre en évidence les granulations gramophiles du bacille tuberculeux. Opérer comme suit.

a) Étaler le produit. Fixer par la chaleur.

b) Colorer 2 minutes à chaud avec la fuchsine de Ziehl.

c) Laver à l'eau.

d) Colorer au cristal-violet phéniqué 2 minutes.

e) Sans laver, recouvrir la lame avec la solution de Lugol. Laisser une minute. Renouveler trois fois.

f) Traiter par l'alcool-acétone jusqu'à décoloration complète.

g) Laver à l'eau. Colorer au bleu de méthylène. Laver. Sécher.

Les granulations apparaissent en violet foncé dans le corps rouge du bacille.

Méthode de Müch. – Permet de mettre en évidence les granulations gramophiles du bacille tuberculeux.

a) Colorer 24 à 48 heures avec:

Sol. alcool concentré de violet de méthyle B. V.	10 cm^3
Sol. d'acide phénique à 2 p. 100	100 cm^3

b) Colorer 12 minures avec la solution de Lugol.

c) Sans laver, traiter une minute avec une solution à 5 p. 100 d'acide azotique.

d) Puis 10 secondes avec une solution à 3 p. 100 d'acide chlorhydrique.

e) Laver à l'alcool-acétone jusqu'à élimination des colorants.

f) Recolorer avec de la fuchsine diluée. Lavée. Sécher.

Méthode d'Albert (modifiée par Laybourn). – 1° Recouvrir le frottis pendant 3 à 5 minutes avec le colorant d'Albert.

a) Bleu de toluidine	0.15 g
b) Vert malachite	0.2 g
c) Alcool à 95°	2 cm^3

Ajouter à 100 cm^3 d'eau distillée contenant 1 cm^3 d'acide acétique.

2° Laver et sécher.

3° Colorer pendant une minute avec la solution iodo-iodurée de Lugol.

4° Laver et sécher.

Les granulations sont noir bleuâtre, le protoplasma vert. *(C. C. T. M.).*

COLORATION DE NEISSER

Solution A.		*Solution B.*	
Bleu de méthylène	0.1	Crystal violet	0.3
Alcool	5	Alcool	3.0
Acide acétique glacial	5	Eau distillée	100
Eau distillée	100		

Colorant de contraste. Brun de Bismark à 0.2 p. 100

Fixer la préparation avec un mélange de deux parties de A et une partie de B. Colorer ensuite avec le brun de Bismark. Laver à l'eau. *(W. I. C.).*

Volutin—*Aerobacter*

J. P. Duguid, I. W. Smith and J. F. Wilkinson (1954). J. Path. Bact. 67, 289–300.

After different periods of growth, small blocks of agar were cut from the culture plate and rubbed on a slide to make a smear of the culture without addition of water. The smears were fixed by flaming and stained by Albert's method as modified by Laybourn (1924), with toluidine blue and malachite green followed by iodine. The amount of volutin was assessed according to the proportion of cells containing black granules, the

number of granules per cell and the size of the granules; the amount was recorded on an arbitrarily standardized scale (– for none, ⊥ for a trace, e.g. small granules in a few cells, and +, ++ and +++ for increasingly large amounts).
R. L. Laybourn (1924). J. Am. med. Ass. 83, 121.

Volutin—*Spirillum*
W. A. Pretorius (1963). J. gen. Microbiol. 32, 403–408.
Demonstrated by the method described by Jörgensen (1948, *Microorganisms and Fermentations* Charles Griffen and Co. London).

XANTHINE DECOMPOSITION

Xanthine decomposition—*Mycobacterium*
R. E. Gordon (1966). J. gen. Microbiol. 43, 329–343.
Xanthine (0.4 g) was suspended in 100 ml nutrient agar (peptone, 5 g; beef extract, 3 g; agar, 15 g; distilled water, 1000 ml; pH 7.0); the suspension was then autoclaved, cooled to 47°C, mixed thoroughly, and poured into sterile Petri dishes (20 ml per plate). Care was taken to distribute the insoluble crystals of xanthine evenly throughout the agar. Each culture was streaked once across a plate and checked at 14 and 21 days for the disappearance of the crystals underneath and around the growth.

APPENDIX

Substances which have been employed as substrates in studies on (1) utilisation for growth (2) 'fermentation' tests (3) manometric studies

A. Carbon compounds
  I. Carbohydrates
    a. Monosaccharides
      Trioses: dihydroxyacetone, glyceraldehyde
      Tetroses: L-erythrulose, D-threose
      Pentoses: D-arabinose, L-arabinose, D-lyxose, L-lyxose, D-ribose, D-ribulose, D-xylose, D-xylulose
      Methyl pentoses: D-fucose, L-fucose, D-rhamnose, L-rhamnose
      Hexoses: D-allose, D-fructose, D-galactose, D-glucose, D-mannose, L-sorbose
      Heptoses: α-D-galaheptose, β-D-galaheptose, α-D-glucoheptose, sedoheptulose
    b. Glucosides: aesculin, amygdalin, arbutin, coniferin, α-methyl-D-glucoside, β-methyl-D-glucoside, α-methyl-D-mannoside, salicin

c. Disaccharides: D-cellobiose, lactose, D-maltose, D-melibiose, sucrose, trehalose
d. Trisaccharides: D-melezitose, D-raffinose
e. Polysaccharides
Pentosans: xylan
Hexosans: cellulose, chitin, dextran, dextrin, glycogen, inulin, starch
Other polysaccharides: alginic acid, bacterial polysaccharide, gum arabic, hemicellulose, lignocellulose, mannitan

II. Alcohols
a. Aliphatic
Monohydric: allyl alcohol, methanol, ethanol, n-propanol, iso-propanol, secondary propanol, n-butanol, iso-butanol, secondary butanol, tertiary butanol, n-pentanol, iso-pentanol, n-hexanol, n-octanol
Dihydric: 1,2-ethanediol, di-ethylene glycol, thiodiethylene glycol, triethylene glycol, polyethylene glycol (–200, –300, –400, –1500), 1,3-propanediol, 1,5-propanediol, DL-1,2-propanediol, D(-)1,2-propanediol, 2-amino-2-ethyl-1,3-propanediol, 2-nitro-2-ethyl-1,3-propanediol, dipropylene glycol, 1,4-butanediol, DL,-1,3-butanediol, D(-)2,3-butanediol, *meso*-2,3-butanediol, L(+)2,3-butanediol, 2-butene-1,4-diol, 1,4-butenediol, 1,3-pentanediol, 1,5-pentanediol, 1,6-hexanediol, 2,5-hexanediol, D(+)-3,4-hexanediol, *meso*-3,4-hexanediol, 1,7-heptanediol, geraniol
Trihydric: 1,2,4-butanetriol, 1,2,3-propanetriol (glycerol), 1,2,6-hexanetriol
Tetrahydric: erythritol (anti-1,2,3,4-butanetetrol), *meso*-erythritol, pentaerythritol
Pentahydric: adonitol, isoadonitol, arabitol, D-arabitol, L-arabitol, *meso*-ribitol, *meso*-xylitol
Hexahydric: *meso*-allitol, dulcitol, L-fucitol, *meso*-galactitol, D-glucitol, L-glucitol, D-iditol, L-iditol, lactositol, mannitol, D-mannitol, L-mannitol, L-rhamnitol, sorbitol, D-sorbitol
Polyhydric: D-glycero-D-galactoheptitol, *meso*-glycero-gulo-heptitol, perseitol, polygalitol, primulitol
Derived from cyclic paraffins: cyclopentanol, cyclohexanol, cycloheptanol, cyclooctanol, geraniol, inositol, *meso*-inositol, iso-inositol, phenylethanediol, pinitol, quercitol
b. Aromatic
Monohydric: anisyl alcohol, benzyl alcohol, cinnamyl alcohol,

coniferyl alcohol, *m*-cresol, phenol

Dihydric: aesculetin, catechol, resorcinol, saligenin

c. Alcohols with additional groups

Ether-linked groups: 2(2-butoxy ethoxy) ethanol, 2-ethoxy ethanol, 2(2-ethoxy ethoxy) ethanol, 2(2-methoxy ethoxy) ethanol, ethylene glycol monoethyl ether, diethylene glycol monoethyl ether

Keto-alcohols: D(-)acetoin, acetoin (acetyl methyl carbinol), D(-)ethylpropionyl carbinol, L(+)ethylpropionyl carbinol

III. Aldehydes

acetaldehyde, n-butyraldehyde, formaldehyde, hydroxy-pyruvic aldehyde, propionaldehyde

IV. Carboxylic acids

a. Saturated fatty acids

Monocarboxylic: formic (methanoic), acetic (ethanoic), propionic (propanoic), butyric (butanoic), isobutyric, valeric (pentanoic), isovaleric, caproic (n-hexanoic), capyrlic (n-octanoic = octoic), pelargonic (n-nonanoic), undecylic (hendecanoic), lauric (n-dodecanoic), tridecylic (tridecanoic), palmitic (n-hexadecanoic), 10-methyl-hexadecanoic

Ester: methyl formate

Decarboxylic: adipic, azelaic, eicosandedioic, glutaric, malonic, oxalic, pimelic, sebacic, suberic, succinic

b. Unsaturated fatty acids

Monocarboxylic: erucic, linoleic, linolenic, cis-8-octadecanoic, trans-8-octadecanoic, cis-10-octadecanoic, 8-octadecyanoic, 9-octadecyanoic, 10-octadecyanoic, oleic, petroselaidic, petroselenic

Hydroxymonocarboxylic: ricinoleic

Esters: arachidonic (methyl ester), oleic (ethyl ester), oleic (methyl ester)

Dicarboxylic: citraconic, fumaric, itaconic, maleic, mesaconic

Tricarboxylic: aconitic, cis-aconitic

c. Hydroxy-acids

Monocarboxylic: glyceric, DL-glyceric, glycollic, β-hydroxy-butyric, DL-β-hydroxybutyric, poly-β-hydroxybutyric, D-lactic, DL-lactic, L-lactic

Dicarboxylic: D-malic, L-malic, mucic, oxalacetic, tartaric, D(-)tartaric, L(+)tartaric, *meso*-tartaric

Tricarboxylic: citric isocitric

d. Keto-acids

Monocarboxylic: levulinic, β-oxobutyric, pyruvic

Decarboxylic: 2-keto-gluconic, α-keto-glutaric, oxoglutaric
Tricarboxylic: oxalosuccinic

e. Cyclic acids
   Monocarboxylic: shikimic
   Decarboxylic: phthalic, isophthalic, terephthalic

f. Aromatic acids
   benzoic, *m*-hydroxybenzoic, *o*-hydroxybenzoic, *p*-hydroxybenzoic, benzoylformic, 2-chlorophenoxyacetic, 4(2,4-dichlorophenoxy) butyric, mandelic, D-mandelic, L-mandelic, phenylacetic, β-phenylpropionic, quinic, salicylic

g. Acids with other groups attached
   2,2,dichloropropionic, 2-ethoxy ethyl acetic, ethoxyacetic

h. Mixtures
   valerianic

i. Sugar-derived acids
   arabonic, ascorbic, galactonic, galacturonic, gluconic D-gluconic, glucuronic, α-methyl glucuronic, hydroxy methyl glutaric, pectin, saccharic

V. Fats
   beef tallow, butter fat, coconut oil, corn oil, cottonseed oil, lard, linseed oil, olive oil, tributyrin, tricaprin, tricaproin, tricapyrlin, trilaurin, trimyristin, triolein, tripalmitin, tripropionin, tristearin, Tween 20, Tween 40, Tween 60, Tween 80

VI. Hydrocarbons

a. Paraffins
   Straight chain: methane, n-propane, n-butane, n-pentane, n-hexane, n-heptane, n-octane, n-nonane, n-decane, undecane, dodecane, tridecane, tetradecane, pentadecane, hexadecane, octadecane, eicosane, docosane, acetone, 2-butanone, 2-pentanone, 2-tridecanone, 2-octadecanone
   Branched: dimethyl octane, 3-ethyltetradecane, 2-methyl heptane, 4-methyl heptane, 2-methyl hexane, 2-methyl pentadecane, 3-methyl pentadecane, 4-methyl pentadecane, 5-methyl pentadecane, 6-methyl-pentadecane, 7-methyl pentadecane, 8-methyl pentadecane
   Cyclic: cyclohexane, dimethylcyclohexane
   Mixtures: asphalt, gasoline, kerosine, paraffin oil, paraffin wax, petroleum

b. Olefines
   n-caprylene, hexadecylene, cyclic olefines, pinene

c. Aromatics

benzene, cymene, mesitylene, naphthalene, phenanthene, pseudocumene, pyrogallol (pyrogallic acid), toluene, xylol (xylene)

d. Paraffin with additional groups

propylene oxide

B. Nitrogen compounds

I. Amino acids and derivatives: D-alanine, D-α-alanine, L-alanine, L-α-analine, β-analine, m-aminobenzoic acid, *p*-amino-benzoic acid, DL-α-aminobutyric acid, γ-aminobutyric acid, α-amino-capronic acid, DL-α-aminovaleric acid, δ-aminovaleric acid, anthranilic acid, arginine, DL-arginine, asparagine, L-aspartic acid, betaine, DL-citrulline, creatine, cystine, glutamic acid, L-glutamic acid, glycine, hippurate, L-histidine, kynurenic acid, L-kynurenine, leucine, L-leucine, isoleucine, L-iso-leucine, DL-norleucine, lysine, L-lysine, methionine, DL-ornithine, phenylalanine, L-phenylalanine, proline, L-proline, sarcosine, serine, L-serine, L-threonine, tryptophan, D-trypto-phan, L-tryptophan, tyrosine, L-tyrosine, valine, L-valine.

II. Amines: allylamine, α-amylamine, iso-amylamine, benzylamine, butylamine, iso-butylamine, di-iso-butylamine, cadaverine, ethylamine, di-ethylamine, ethanolamine, galactosamine, glucosamine, histamine, methylamine, trimethylamine, β-methylbutylamine, β-phenylamine, phenylethylamine, propylamine, di-propylamine, iso-propylamine, tri-propyl-amine, putrescine, spermine, tryptamine, tyramine

III. Heterocyclic amines: allantoin

IV. Amides: acetamide, allantoin, benzamide, caproamide, iso-nicotinamide, malonamide, nicotinamide, propionamide, pyrazinamide, salicylamide, succinamide

V. Cyclic amides: γ-butyrolactam, ϵ-caprolactam, pyroglutamic acid, δ-valerolactam

VI. Purines, pyrimidines, and derivatives: cytosine, guanine, thiamine hydrochloride, uracil, xanthine

VII. Sterols: acetyl cholesterone, cholesterol, Δ-4-cholesterone, desoxycortisone acetate, digitonin, testosterone

VIII. Vitamins: nicotinic acid, pantothenic acid, trigonelline

General Index

Index of Tests Applied to Specific Genera